Environmental Governance in a Populist/Authoritarian Era

This volume explores the many and deep connections between the widespread rise of authoritarian leaders and populist politics in recent years, and the domain of environmental politics and governance – how environments are known, valued, and managed; for whose benefit; and with what outcomes.

The volume is explicitly international in scope and comparative in design, emphasizing both the differences and commonalties to be seen among contemporary authoritarian and populist political formations and their relations to environmental governance. Prominent themes include the historical roots of and precedents for environmental governance in authoritarian and populist contexts; the relationships between populism and authoritarianism and extractivism and resource nationalism; environmental politics as an arena for questions of security and citizenship; racialization and environmental politics; the politics of environmental science and knowledge; and progressive political alternatives. In each domain, using rich case studies, contributors analyse what differences it makes when environmental governance takes place in authoritarian and populist political contexts.

This book was originally published as a special issue of *Annals of the American Association of Geographers*.

James McCarthy is a Professor in the Graduate School of Geography at Clark University, Worcester, USA. His work analyses the interactions of political economy and environmental politics. He has published three major edited volumes and over 50 articles and chapters. His current research explores the relationships between climate change, renewable energy, and the future of capitalism.

Environmental Governance in a Populist/Authoritarian Era

Edited by
James McCarthy

Routledge
Taylor & Francis Group

LONDON AND NEW YORK

First published 2020
by Routledge
2 Park Square, Milton Park, Abingdon, Oxon, OX14 4RN

and by Routledge
52 Vanderbilt Avenue, New York, NY 10017

Routledge is an imprint of the Taylor & Francis Group, an informa business

First issued in paperback 2021

British Library Cataloguing-in-Publication Data
A catalogue record for this book is available from the British Library

ISBN13: 978-0-367-34653-9 (hbk)
ISBN13: 978-1-03-208710-8 (pbk)

Typeset in Goudy
by codeMantra

Publisher's Note
The publisher accepts responsibility for any inconsistencies that may have arisen during the conversion of this book from journal articles to book chapters, namely the inclusion of journal terminology.

Disclaimer
Every effort has been made to contact copyright holders for their permission to reprint material in this book. The publishers would be grateful to hear from any copyright holder who is not here acknowledged and will undertake to rectify any errors or omissions in future editions of this book.

Contents

PART VI
Progressive Alternatives

Citation Information

The chapters in this book were originally published in *Annals of the American Association of Geographers*, volume 109, issue 2 (March 2019). When citing this material, please use the original page numbering for each article, as follows:

Chapter 30

Reparation Ecologies: Regimes of Repair in Populist Agroecology
Kirsten Valentine Cadieux, Stephen Carpenter, Alex Liebman, Renata Blumberg, and Bhaskar Upadhyay
Annals of the American Association of Geographers, volume 109, issue 2 (March 2019) pp. 644–660

Chapter 31

Development and Sustainable Ethics in Fanjingshan National Nature Reserve, China
Stuart C. Aitken, Li An, and Shuang Yang
Annals of the American Association of Geographers, volume 109, issue 2 (March 2019) pp. 661–672

Chapter 32

A Manifesto for a Progressive Land-Grant Mission in an Authoritarian Populist Era
Jenny E. Goldstein, Kasia Paprocki, and Tracey Osborne
Annals of the American Association of Geographers, volume 109, issue 2 (March 2019) pp. 673–684

For any permission-related enquiries please visit:
http://www.tandfonline.com/page/help/permissions

Authoritarianism, Populism, and the Environment: Comparative Experiences, Insights, and Perspectives

James McCarthy

Recent years have seen the widespread rise of authoritarian leaders and populist politics around the world, a development of intense political concern. This special issue of the *Annals* explores the many and deep connections between this authoritarian and populist turn and environmental politics and governance, through a range of rich case studies that provide wide geographic, thematic, and theoretical coverage and perspectives. This introduction first summarizes major commonalities among many contemporary authoritarian and populist regimes and reviews debates regarding their relationships to neoliberalism, fascism, and more progressive forms of populism. It then reviews three major connections to environmental politics they all share as common contexts: roots in decades of neoliberal environmental governance, climate change and integrally related issues of energy development and agricultural change, and complex conflations of nation and nature. Next, it introduces the six sections in the special issue: (1) historical and comparative perspectives (two articles); (2) extractivism, populism, and authoritarianism (six articles); (3) the environment and its governance as a political proxy or arena for questions of security and citizenship (seven articles); (4) racialization and environmental politics (five articles); (5) politics of environmental science and knowledge (six articles); and (6) progressive alternatives (five articles). It concludes with the suggestion that environmental issues, movements, and politics can and must be central to resistance against authoritarian and reactionary populist politics and to visions of progressive alternatives to them.

近年来，全球见证了威权领导人和民粹政治的广泛兴起，该趋势并带来了极度的政治忧虑。本刊特辑通过一系列在地理、主题和理论上涵盖广泛范围与视角的丰富案例研究，探讨威权主义与民粹政治转向和环境政治及治理之间众多且深刻的关联。本引文首先摘要诸多当代威权和民粹主义政体的主要共通处，并回顾有关其与新自由主义、法西斯主义以及更为激进的民粹主义形式的关系之辩论。本文接着回顾其所共享的与环境政治的三大连结作为共通脉络：数十年来新自由主义环境治理的根源、气候变迁和能源发展与农业变迁之整体相关议题，以及国族与自然的复杂结合。再者，本文引介本特辑的六大部分：（1）历史与比较性的视角（两篇文章）；（2）资源榨取主义、民粹主义，以及威权主义（六篇文章）；（3）环境及其治理作为政治代理或安全与公民权的问题场域（七篇文章）；（4）种族化与环境政治（五篇文章）；（5）环境科学与知识的政治（六篇文章）；（6）激进的另类方案（五篇文章）。本文于结论中主张，环境议题、运动与政治，能够且必须作为抵抗威权和反动的民粹政治、以及替代该政治的激进另类愿景之核心。关键词：威权主义，环境治理，环境政治，民粹主义。

Los años recientes han sido testigos de la recurrente aparición de líderes autoritarios y política populista alrededor del mundo, un desarrollo de seria preocupación política. Este número especial de *Annals* explora las numerosas y profundas conexiones entre ese giro autoritario y populista, y la política y la gobernanza ambiental, con una gama de ricos estudios de caso que suministran amplia cobertura y perspectivas geográficas, temáticas y teóricas. Esta introducción resume primero las principales características compartidas entre muchos de los regímenes autoritarios y populistas contemporáneos, y reseña los debates que abocan sus relaciones con el liberalismo, el fascismo y las formas más progresistas de populismo. Se hace luego la revisión de las tres principales conexiones con las políticas ambientales, compartidas por todos como contextos comunes: sus raíces en décadas de gobernanza ambiental neoliberal, cambio climático y cuestiones integralmente relacionadas de desarrollo energético y cambio agrícola, y complejas combinaciones de nación y naturaleza. Luego, se presentan las seis secciones de que consta el número especial: (1) perspectivas históricas y comparadas (dos artículos); (2) extractivismo, populismo y autoritarismo (seis artículos); (3) el medio ambiente y su gobernanza como una proxy política o arena para cuestiones de seguridad y ciudadanía (siete artículos); (4) racialización y política ambiental (cinco artículos); (5) políticas sobre ciencia y conocimiento ambiental (seis artículos); y (6) alternativas de progreso (cinco artículos). Se concluye con la sugerencia de que las cuestiones ambientales, movimientos y políticas pueden y deben ser centrales en la resistencia contra la política populista autoritaria y reaccionaria, y a las visiones de alternativas progresistas. *Palabras clave: autoritarismo, gobernanza ambiental, política ambiental, populismo.*

The rise of authoritarian leaders and populist politics around the world and the multiple configurations in which those associated yet distinct political developments manifest have been the subjects of intense concern and analysis over the past several years. The spatial and temporal extent of this tide is terrifying: Authoritarian and populist political configurations have emerged and either taken control of the state or come increasingly close to doing so in a very large and growing number of polities around the world over the past decade, including many of the world's largest, most powerful, and most iconic democratic countries. Although the specific trajectories and genealogies of these political formations are always unique at some level, they also share many general features: nationalism articulated and justified in the name of frighteningly exclusive and often racialized iterations of "the people"; the demonization of alleged enemies internal and external; support for and selection of authoritarian leaders who rise to power by exciting such fears and promising simple, direct, often brutal action to protect and strengthen the nation; and contempt for and direct assaults on democratic norms and institutions. At the same time, though, genuinely progressive movements, leaders, and parties have seen increased support over the same period in many countries. Although we hear largely about alleged polarization, what those superficially opposed movements have in common is a rejection of neoliberal hegemony and the articulation of genuine alternatives. That suggests that this could be a moment of hope and opportunity as well, if the left is able to articulate positive radical alternatives that are broad, inclusive, and sustainable.

So much has been widely discussed. What has received far less analytical attention are the myriad connections between authoritarianism, populism, and environmental politics and governance, the topic of this special issue of the *Annals*, "Environmental Governance in a Populist/Authoritarian Era." An immediate list would include the ways in which populist and authoritarian politics and regimes often arise directly from tensions between rural and urban areas; assert "blood and soil" claims of indissoluble links between the nation and the biological and physical environment; deploy resurgent tropes of territorialized bodies politic, contagion, and disease; exploit national natural resources to buy political support and underwrite their political agenda; attack environmental protections and activists to give extractive capital free reign; eliminate or attack environmental data and science in a "posttruth" era; and are especially dysfunctional political responses to the security threats, fears, and divisions associated with climate change. On the positive side, environmental movements and politics remain both a critical front of resistance to authoritarian and populist politics in many places and one of the chief sources of visions of progressive alternatives to them. These and other actual and potential relationships between authoritarianism, populism, and environmental politics and governance are explored in this special issue's six sections, detailed here: (1) historical and comparative perspectives (two articles); (2) extractivism, populism, and authoritarianism (six articles); (3) the environment and its governance as a political proxy or arena for questions of security and citizenship (seven articles); (4) racialization and environmental politics (five articles); (5) politics of environmental science and knowledge (six articles); and (6) progressive alternatives (five articles). First, though, a slightly more in-depth discussion of the origins and contours of the contemporary turn toward authoritarian and populist politics and their relevance to environmental politics and governance is warranted, to put the articles in a common context.

The Rise of Authoritarianism and Populism

Bolsonaro in Brazil. Battulga in Mongolia. Duterte in the Philippines. Erdoğan in Turkey. Putin in

Russia. Modi in India. Xi in China. Trump in the United States. The list of authoritarian leaders who have recently won or consolidated power over their country's central state, often by deploying or harnessing some variant of populism, is soberingly long and appears to still be growing. In many other countries, perhaps most clearly in Europe, populist and authoritarian parties, leaders, and movements have had growing electoral success and political effect (e.g., Brexit), even if they have not yet been elected to the highest offices. Several things about this trend are noteworthy. First, it spans many usual divides, encompassing countries in every major world region and category. Second, it includes many of the world's largest and most powerful countries. Third, it includes many of the world's largest and most regionally symbolic democracies. Fourth, as that implies, this trend has widespread popular support: Although many elections have had some questionable aspects (e.g., in the United States, gerrymandering and voter suppression preclude truly democratic elections), in many instances it is clear that these leaders and their parties really were chosen by at least very large portions of their electorates.

Authoritarianism and populism can each take many forms, be allied with nominally right or left politics, and articulate with each other in multiple ways (Hall 1980, 1985; Bello 2018; Borras 2018). In the wave of authoritarian and populist politics we are currently experiencing, each national instance, of course, has vitally important specificities and a trajectory that is unique at a sufficient level of resolution. Yet, the political figures and regimes mentioned share a great many common features, as many have noted (Bessner and Sparke 2017; Fraser 2017; Snyder 2017; Albright 2018; Bello 2018; Bigger and Dempsey 2018; Collard et al. 2018; Scoones et al. 2018). They advance militant, often economically protectionist forms of nationalism, insisting on the precedence of national self-interest and sovereignty over shared global interests and institutions. They use bellicose rhetoric and gestures in theatrical efforts to project strength. They promise to take quick and decisive action on highlighted issues, in contrast to liberal democratic administrations portrayed as weak, passive, and indecisive. They make the central populist move of claiming to speak and act in the name of and with the support of "the people," who are typically identified in nativist, xenophobic, and often explicitly racialized

terms. Following closely from that, they often identify internal enemies—ethnic or religious minorities, immigrants, refugees, drug users—as scapegoats and targets for public anger. They use populist rhetorical tropes of resentful antielitism, suspicion of experts and complexity, and celebration of direct action to promise simple, immediate solutions to complex, long-term problems. They present themselves as being, and often truly are, willing and even eager to use violence against opponents internal and external. They engage in direct and indirect assaults on the norms and institutions of democratic societies, including the rule of law, freedom of the press, and opponents' rights of speech and assembly—directly through the centralization and consolidation of power in the executive branch, efforts to test or even actively subvert resistant institutions, and punishment of political critics or opponents and indirectly through the contempt that they exhibit for norms, institutions, and people who oppose them. Moreover, they claim and celebrate a direct connection with "the people" that purportedly bypasses just such potential obstacles. Alongside these many commonalties, they exhibit one last, somewhat ironic common feature: an opportunistic lack of ideological coherence or consistency.

This tide of authoritarian populism has prompted much soul-searching on the left and a few key analytical debates. What is the relationship between the authoritarian populist turn and decades of neoliberalism? Is the turn we are seeing more accurately labeled as fascism or as a clear step in that direction? Finally, is populism inherently conservative or are genuinely progressive populisms possible? A brief sketch of these debates is necessary before considering how each relates to questions of environmental politics and governance.

The politics and political economy of the relationships between neoliberalism and the turn toward authoritarian and populist regimes are clearly complex: Many of these regimes came to power on a platform of reversing major elements of neoliberal globalization, yet they are often continuing to pursue and deepen neoliberal policies in many areas. A number of articles in this special issue examine precisely that tension. Whether such contradictions reflect a coherent underlying strategy or constellation of interests remains unclear in many cases (see Bessner and Sparke 2017; Scoones et al. 2018), although an argument can be made that maximizing capitalists' flexibility and accumulation appears to be a consistent principle through these trajectories, one pursued through different scalar strategies at different moments in time. Most analysts agree, though, that the turn toward authoritarian and populist politics is directly rooted in the failures and successes of neoliberal globalization. Starting as far back as the 1970s but with pronounced acceleration in the 1990s, decades of increasing economic and institutional integration failed to deliver the promised broad-based economic growth, producing instead wrenching economic restructuring, deindustrialization, intensified competition, and accelerating economic inequality that left many workers, sectors, and regions behind. These trends were dramatically intensified during and after the financial crisis beginning in 2008 and the increased volatility and imposition of austerity that followed it. It is entirely understandable that many people felt betrayed and sought leadership that would clearly prioritize their self-interests over some promised-yet-never-realized greater good whose fruits seemed in practice to accrue entirely to the already wealthy. At the same time, however, it became clear how deeply neoliberal ideology's delegitimation of the state as a potentially legitimate or competent owner, manager, or representative of public goods and interests had taken hold: Even as people demanded recognition of their needs and desires, many took for granted that the state could never truly represent "the people" or even their interests and so turned instead to charismatic leaders promising to repudiate elites, including those currently in power. In a widely cited piece, Fraser (2017) diagnosed this conjuncture as representing the failures of what she termed "progressive neoliberalism," which she defined as a Gramscian hegemonic bloc centered on an alliance between certain fractions of capital (notably finance capital but also other technology- and information-centered industries) and cosmopolitan elites, who used a superficial commitment to the politics of recognition and meritocracy to mask neglect of or direct assaults on the interests of the industrial working class and many rural populations, a position further justified by the cultural denigration of the latter groups as backward and reactionary. Fraser argued that perhaps the key feature of the current moment is that protest and resentment against these decades-long trends are now producing electoral effects, through the replacement or

dramatic realignment of major political leaders and parties.

The electoral successes of authoritarian and populist leaders, parties, and movements, most but not all strongly right wing, bring us to another major debate: What, if any, are the inherent politics and trajectories of such formations? In a nutshell, would these current political developments be more precisely or accurately characterized as fascism or steps on a clear road toward fascism or can populism, at least, ever be genuinely progressive? These questions turn out to be tightly linked, inasmuch as both turn on what is at stake in shaping political identities, claims, and agendas in terms of some polity understood as "the people." On the strongly cautionary side, Albright (2018) and Snyder (2017) both drew explicit parallels between the 1930s and the present, particularly between the rise of fascism in Germany and Italy and the trajectories of many contemporary authoritarian leaders who trade in the politics of populism. Albright argued that the three key conditions that paved the way for fascism in the 1930s were economic and political decline and uncertainty; the failures of existing administrations to effectively govern and address key problems; and the collaboration of conservatives who believed that fascist leaders would serve them rather than the other way around. She contended that we see quite similar conditions today in many countries. Snyder, meanwhile, dug deeply into the cultural politics—of identification, fear, scapegoating, demonization, spectacle, and more—through which fascist leaders either actively enrolled people in their movements or at least led them to remain quiescent (see also Arendt [1951] 1973) and drew chillingly precise analogies to specific utterances and actions by President Trump and his administration in particular. For both, the essence of fascism lies in its division of the world into us versus them, with the us articulated in extremely nationalist, xenophobic, and often explicitly racialized terms. It is alleged existential threats to that us that require and justify the extreme political centralization and repression that form fascism's other essential elements. Others offer a somewhat different and more analytically cautious view. Bessner and Sparke (2017) suggested that comparisons between support for Trump and perhaps other contemporary populists and Nazism are perhaps too facile and decontextualized and that they miss something vitally important: The historical experiences of mid-century fascism led to an elite suspicion of public involvement in politics and policy-making, a sentiment that in turn directly shaped the establishment of the elite international institutions, from the Bretton Woods framework onward, which provided the foundations of the neoliberal global order against which many contemporary populist movements are now rebelling. In other words, many people are not wrong in thinking that economic and foreign policy have been shaped by elites rather than the voices and interests of the majority: That was precisely true, by design. To dismiss the resulting resentment, however marred by other political admixtures, as simply and only fascism is both unjust and a missed political opportunity. The convergence with Fraser's (2017) argument is clear.

Turning to populism, many authors contend that its core logic is entirely too close to that of fascism for any version of populism ever to be truly progressive. Claiming to speak for and from "the people" is a move that, ultimately, requires the drawing of a political boundary between those who are included in that group and those who are not. For precisely that reason, Swyngedouw (2010), Rancière (2016), Hofstadter (1960), Müller (2016) and many others reject arguments that there can be truly left or progressive populisms, suggesting instead that in the end, populism is always necessarily antidemocratic, usually constructed and deployed by and for elites despite its superficial opposition to them and all too often enacted along lines of racialized identities. Yet a substantial and growing body of theorists (e.g., Laclau 1977, 2005; Hardt and Negri 2005; Badiou 2016; Grattan 2016; Mouffe 2016; Gerbaudo 2017) have argued that truly progressive, democratic, and inclusive versions of populism are both possible and politically promising. The core of these arguments comes from Laclau (1977, 2005), who emphasized populism as a political activity and process that can symbolically and affectively link disparate groups in a society into a common counterhegemonic struggle. In short, and in direct counterpoint to the preceding critiques, the emphasis is on alliances and inclusion rather than on exclusion, and the organizing principle and goal is the subversion of the dominant order in the name of genuinely greater democracy. This perspective has been most strongly developed in Latin America, where examples of left-leaning populist movements, leaders, and administrations are perhaps most abundant. An argument can also be made that, in a political moment entirely too characterized by nihilism and dystopic visions, populism's powerful affective and emotional elements might be useful or even critical in catalyzing or sustaining political engagement. These questions, along with many others, turn out to be central to the multiple ways in which the rise of authoritarian and populist politics articulate with the environmental politics and governance.

Connections to Environmental Politics and Governance

As Gramsci (1971; see also Ekers, Loftus, and Mann 2009) and Williams (1980) each argued in different registers, hegemony over society cannot be separated from hegemony over nature: They function through the same political formations. Yet the ways in which they do so can be far from transparent. The connections between the widespread rise of authoritarian and populist leaders, administrations, and movements on the one hand and destructive trends in environmental politics and governance on the other are legion. Some are obvious—the Trump administration's withdrawal from the Paris Accord and approval of the Dakota Access Pipeline against the wishes and territorial claims of Indigenous people, the use of revenues from extractive industries to fulfill populist pledges throughout Latin America, the repression and murder of environmental activists—and others are perhaps less so, such as the ways in which emphasizing the credentialed expertise underpinning environmental science might fuel populist resentments in politically counterproductive ways.

Several themes stand out as deeply relevant to this special issue as a whole and to nearly the full breadth of relationships between contemporary authoritarianism and populism and environmental politics and governance. Although some are highlighted in particular articles, all three form inescapable and consequential contexts for all of the cases examined in the issue. Therefore, rather than use them as section headings applying to only some articles, I discuss them here briefly as structuring contexts for the entire issue.

The first is the continued salience of neoliberal capitalism in relationship to the environment to these political developments (see McCarthy and Prudham 2004; Heynen et al. 2007; Bigger and Dempsey 2018). A strong case can be made that deepening rural–urban disparities in the neoliberal era were central to the emergence of the recent populist wave, as many rural areas reacted against the particular burdens increasingly mechanized resource extraction, globalization of primary commodity markets, volatility, austerity, and declining prosperity have imposed on them over the past several decades (for in-depth explorations of this thesis, see Bello 2018; Scoones et al. 2018; see more generally the Emancipatory Rural Politics Initiative at www.iss.nl/erpi). From this perspective, it is striking and telling that in the United States, "four hundred and eighty-nine of the wealthiest counties in the country voted for Clinton; the remaining two thousand six hundred and twenty-three counties, largely made up of small towns, suburbs, and rural areas, voted for Trump" (Remnick [2017]. Equally telling and more hopeful, however, is that many of those rural Trump voters had voted for the socialist Bernie Sanders in the primaries just months earlier; Kojola, this issue.) This argument, which overlaps with Fraser's (2017) presented earlier, is centrally about political contestation over how and for whose benefit particular environments and natural resources have been used and governed. Likewise, many—including a number of authors in this special issue—have argued that many contemporary authoritarian regimes are pursuing and deepening long-standing neoliberal goals with respect to the environment, removing restrictions on capitalist production by withdrawing from constraining international agreements and standards, rolling back domestic environmental protections, and appointing heads of polluting corporations to head the very agencies that are supposed to regulate those corporations (see, e.g., Monbiot 2017; Mansfield 2018). There is a superficial tension in this set of arguments—withdrawals from the European Union (EU), the North American Free Trade Agreement (NAFTA), the World Trade Organization, and other iconic institutions of the era of trade liberalization are interpreted at different points as both a reaction against neoliberalism and as a way to further and deepen neoliberalism; globalization is interpreted at different points as both a way to increase flexibility for capitalists and a way of imposing constraints on them—but that tension is resolvable if we focus on the fact that the consistent goal of capitalists is to maximize their flexibility and accumulation. That goal is best pursued through different scalar strategies at different moments in time: Withdrawing from the EU or from NAFTA does not

mean that the United Kingdom or United States will go back to the union membership levels, labor protections, taxation levels, or social protections of the 1970s or 1980s that preceded those agreements.

The second is climate change, and integrally related issues of energy development and agricultural change (see Zimmerer 2011). Much current work on climate change (a bit too much of it uncritically neo-Malthusian) emphasizes the chances that it will create or exacerbate conflicts and lead to political destabilization—via conflicts over scarce resources, due to climate-induced migration across national borders, or through direct conflict over responsibility for climate change itself. More recently, the potential for conflicts over proposed geoengineering actions has been added to this list (Surprise 2018). Work focused on more explicitly theorizing the potential political trajectories that could follow from climate change (e.g., Mann and Wainwright 2018) considers the possibility that balancing demands for continued economic growth with responses to the security threats associated with climate change could present genuine, existential threats to democratic liberalism and smooth the way toward authoritarian political responses and formations. The continued legitimacy of states might rest on their ability to respond effectively to the threats associated with a changing climate and energy transitions, and authoritarian regimes might promise to take strong action and address the critical issues at which democracy has failed. Indeed, many authors suggest that such trends are already evident (e.g., Beeson 2010; Fritsche et al. 2012; Gilley 2012). Even before such overt junctures, mounting awareness of climate change, even when the latter is consciously denied, might contribute to a generalized sense of insecurity and instability that can find expression in populist and nationalist sentiments (McCarthy et al. 2014). At the very same time, authoritarian discourses, state violence, and state-sanctioned private violence are increasingly evident in efforts to keep fossil fuels flowing, exacerbating the problem.

The third is, broadly, the conflation of nature and nation: the multiple ways in which physical and biological environments and resources become politically understood as inextricably linked to national identities, fortunes, and prospects (Koch and Perrault 2018). Very old and very dangerous links between ideas about the environment and ideas about governance are resurfacing in the authoritarian and populist turn around the globe. Current politics of nativism, masculinism, white supremacy, and the hardening of borders are deeply intertwined with ideas linking racialized, gendered, and national identities to specific environments, territories, and the alleged existential struggle for scarce resources. Likewise, metaphors of the nation as an organism that can be healthy or diseased, contaminated or cleansed, are closely linked to particular imaginaries of national environments. In a more straightforwardly economic register, natural resources within indigenous or otherwise contested territories are being claimed as assets both critical for, and rightly belonging to, the "nation" to be used for purposes of national development. Among the many problems with such frameworks, the intense impulse to recode "nature" as "national"—the national territory, national resources, national self-sufficiency in energy or food, and so on—tends to obscures global and transboundary connections and processes.

Articles and Organization of the Special Issue

The articles in this special issue analyze these and many other topics and dynamics linking authoritarianism and populism to environmental politics and governance. They examine a truly global range of cases of complex socionatures, from a diverse set of theoretical and political positions. No set of organizing categories could do justice to their richness and complexities, and some themes, such as those already presented, run through nearly all of the articles to one degree or another. Still, some quite distinct and more specific themes emerged as well, and they are used to organize and introduce the articles in the issue next.

The first has to do with the need for historical and comparative perspectives. Although geography often excels at producing detailed, intensive case studies, there is also great value in explicitly comparative studies and frameworks that look across larger stretches of space and time. Two articles in the special issue take such an approach. The first article in the section, by Wilson, takes an explicitly comparative historical approach, examining how environmental governance of key sectors functioned under authoritarian regimes in the Soviet Union, Maoist China, and Nazi Germany to see what history can tell us about environmental governance under authoritarian regimes. Several other articles,

although in other sections, also offer much longer term historical perspectives, although typically with respect to only one country or location. The second article in this section, by Middeldorp and Le Billon, provides a comparative perspective across a large number of cases in the present, examining the widespread violent, often deadly, repression of environmental activism and dissent by authoritarian regimes. Although Middeldorp and Le Billon focus in particular on one case in Honduras that also speaks to dynamics around extractives examined in the next section, they emphasize the broader pattern into which that case fits, including the complex ways in which populist and authoritarian politics can interact around questions of environmental governance. Comparative perspectives can be useful, but the article by Arefin in the third section cautions against the temptation of simplistic typologies and the importation and application of Western analytical categories onto states in other regions and political cultures. By contrast, the article by Clarke-Sather in the same section argues that Foucault's characterization of the relationship between liberalization and security does in fact offer sharp insights into the trajectory of agricultural policy under China's authoritarian government.

The six articles in the second section on extractivism, populism, and authoritarianism demonstrate how complex and polyvalent the relations among those categories can be (see also Koch and Perrault 2018). The first article in the section, by Kenney-Lazar, examines rapid economic growth in Laos over the past decade as a case of "neoliberal authoritarianism," arguing that authoritarian rule has been essential to the commodification of rural lands and resources—from mining to industrial tree plantations—that has fueled neoliberal accumulation, yet also fostered populist resistance in the countryside. By contrast, the second article, by Lyall and Valdivia, argues that "petro-populism" in Ecuador has turned on a populist regime gaining and maintaining power precisely by promising to reverse neoliberal policies but in ways that hinge on not only maintaining the flow of oil but actively speculating its price in international markets. The third article, by Myadar and Jackson, likewise examines the interplay of populism and resource nationalism with the legacies of neoliberalism, in their case taking the recent election of a populist strongman as president in Mongolia as an entry point. They argue that populist claims to resource nationalism—or resource sovereignty—in relation to Mongolia's mineral resources, particularly copper and gold, are, in context, an articulation of a critique of neoliberal inequality and structural dispossession, whereas dismissals of such frameworks amount to defenses of neoliberal structures of production and distribution. The fourth article, by Kojola, picks up the thread of the interweaving of mining, authoritarian populism, and resource nationalism. Kojola uses the concept of a moral economy to link political ecology and analyses of populism, arguing that the sense of crisis felt by displaced and marginalized mine workers and their communities is the key to understanding both their attachment to identities that are deeply racialized, gendered, nationalist, and nostalgic but also very place and resource based and their responsiveness to the promise of hope they heard in Trump's rhetoric couched in precisely those terms. Related questions are central to the fifth article, by Graybill, on the relationships among extractive industries, governance, and emotions in Russia. Graybill explicitly links the literatures on affect and emotion with those on authoritarianism and extractive industries and economies, arguing that Russia's authoritarian government actively crafts emotional nationalist narratives in support of extractive development and that the resulting desires and emotions contribute to popular support for extractive industries and activities in the country. Finally, in the sixth article in the section, Graddy-Lovelace takes a longer perspective on the relationships between populism and extractivism by tracing the history of U.S. agricultural policy over the past century, including the fact that extreme populist and nationalist narratives have consistently been used to justify policies whose substance, which supported accumulation and overproduction, contradicted and undermined their professed populist goals.

The seven articles in the third section, on environment as political proxy and arena for security and citizenship, all examine ways in which environmental politics can be ways of advancing or contesting politics around these other fundamental political concerns and categories in modern societies. Such dynamics are, of course, deeply connected to those around populism and nationalism in the previous section and those around racialization in the following section, but at the center of these pieces are cases in which questions of security and citizenship are particularly close to the surface of environmental politics and management. In the first article in this

section, as part of a larger argument about how we should theorize political ecologies of the state, Arefin analyzes how the Egyptian state attempted to blame recent urban floods on terrorism rather than climate change or decaying infrastructure, as the former could be used to justify increased repression, whereas the latter would imply failures of the state to fulfill its core functions. The second article in the section, by Acara, examines water management in Turkey in the context of neoliberal authoritarianism and urbanization. Acara argues that the goal of exploiting and degrading water and other natural resources in the name of urban growth has been pursued in part through the centralization but also obfuscation of decisive aspects of water governance and that such centralization and mystification of control over a vital natural resource has functioned to help normalize authoritarian and arbitrary governance in the country more generally. The third article in the section, by Saguin, examines the management and elite capture of fisheries in Laguna Lake in the Philippines under two authoritarian governments, that of Marcos in the 1970s and 1980s and Duterte in the present. Saguin argues that both leaders fell into the same pattern of politicizing the problem, using populist narratives that emphasized conflict and social justice, but depoliticizing the solution by relying on technocratic management frameworks and techniques that elide fundamental social conflicts and goals. The fourth article, by Kantel, likewise links the management of lakes and fisheries directly to efforts to win national elections and consolidate power. Kantel argues that the Ugandan government recently dissolved community-based, more democratic fisheries management bodies as part of a direct effort to consolidate the ruling elite's increasingly authoritarian hold on power. By using discourses of security and citizenship to cast some, more artisanal, fishers as suspect citizens and potential threats to state security, state officials justified the reallocation of resources and the direction of control and wealth to the country's elite. As in Kenney-Lazar's case, though, Kantel suggests that such strategies might ultimately backfire by fueling opposition. The fifth article, by Clarke-Sather, examines a shift in northwest China from irrigated, subsistence-oriented agriculture to drought-resistant, market-oriented agricultural production, all within the context of what has been termed China's "fragmented authoritarianism," in which many actors within an overall authoritarian state are relatively isolated and set to compete with one another. Clarke-Sather argues that Foucault's concepts of disciplinary power and security apparatuses can help us to understand the particular combination of liberal market mechanisms and authoritarian governance evident in contemporary Chinese agricultural and environmental policy. A key question in that framework, shared with the next article, is what a central state undertaking a broader developmental strategy takes on itself to provide to citizens in the way of environmental entitlements versus what it devolves to or demands of citizens or local governments. The sixth article, by Balls and Fischer, takes up related questions by examining the ways in which electricity provision, development, and democratic politics have been linked in modern India, producing clientelist politics in which promises of cheap or free electricity have been linked to electoral support, perhaps at the expense of more broad-based and inclusive development. Balls and Fischer examine how private solar microgrids—a superficially apolitical and environmentally progressive means of producing and distributing electricity—are disrupting such politics, yet often producing new forms of economic and political exclusion in the process. In the seventh and final article in this section, Chang, Bae, and Park compare the spatial and environmental effects of liberal (1997–2007) versus conservative (2007–2017) South Korean administrations on the landscape near South Korea's border with North Korea. Undertaking the difficult and perhaps too rarely attempted task of empirically documenting and analyzing the environmental outcomes of different governance regimes, they demonstrate that the effects are complex, contingent, and highly variable across space and scale, with notably different dynamics and trajectories in the two areas they analyze.

The five articles in the fourth section, on racialization and environmental politics, take up many of the previous questions about citizenship, security, neoliberalism, and authoritarianism in relationship to environmental politics but with a strong and explicit focus on how racial ideologies and the racialization of particular groups of people figure in those dynamics. The first, by Bledsoe, uses the example of three Afro-Brazilian communities in Brazil's Bay of Aratu to argue that, despite what are commonly perceived as major differences, putatively progressive populist and conservative administrations and political formations

in Brazil over the past two decades have in fact shared a reliance on and commitment to extractivism and racialized violence. The second, by Mullenite, analyzes the politics of infrastructure to unearth how targeted flood control and irrigation measures have been used to help build and maintain an authoritarian and racialized state in Guyana, by selectively directing wealth and protections to some while increasing the tax burdens and vulnerability of others. Taking a long historical perspective, Mullenite argues that these infrastructure measures were used to divide laborers along ethnic lines in the colonial era in ways that undermined anticolonial sentiment and enabled authoritarian rule, whereas in the postcolonial period selective neglect of the same infrastructure was used to marginalize groups who might otherwise have resisted an authoritarian administration. The third article in the section, by Wright, analyzes politics around Trump's promised wall on the Mexico–U.S. border to understand both its visceral appeal to a certain kind of nativist, isolationist imaginary and the rising opposition to the wall on ecological, practical, economic, and political grounds. Wright argues that these contrasting views of the proposed wall represent contrasting understandings of the border: one of the border as the clear, visible, and hardened edge of a discrete, territorialized, and deeply racialized white supremacist national space and identity and the other of the border as a zone of diverse social and natural life, connection, and exchange. The fourth article, by Pulido, Bruno, Faiver-Serna, and Galentine, takes up the theme of the extreme racism of the Trump era and administration and connects it to the wave of environmental deregulation the administration has undertaken. Pulido et al. argue that the highly visible, public, and controversial racism and white nationalism of Trump and many of his supporters—what they term "spectacular racism"—helps to obscure the often more mundane, concrete actions that the administration has taken as part of an enormous wave of environmental deregulation. Both are part of the ongoing unfolding of environmental racism in the United States but in new and complex forms. In the fifth and final article in this section, Sparke and Bessner, building from critiques of Nazi logics of governance, suggest that the Trump administration is not only rolling back environmental regulation but also very selectively reworking neoliberal notions of resilience through a hypernationalist and racially exclusionary framework in which the security of a wealthy elite in an exclusionary homeland is pursued through the market mechanisms of disaster capitalism as other racialized people and places are abandoned to the mounting impacts of climate change. The result, they argue, is an odd, exceptionalist, and dangerous hybrid of the discourses and imaginaries of resilience thinking and "America First."

The six articles in the fifth section explore the politics of environmental science and knowledge in populist and authoritarian contexts: from the difficulties of making scientific knowledge claims in a "posttruth" era dominated by easily and endlessly manipulated digital and social media (MacDonald 2016), to those of asking particular questions in severely repressive and dangerous authoritarian countries, to investigations of the active production of doubt or ignorance regarding environmental quality or change (Proctor and Schiebinger 2008; Oreskes and Conway 2010). The first article in the section, by Dillon et al., follows particularly from the lattermost point. It details the efforts of a group of environmental justice and science and technology studies researchers in the United States and Canada to respond to the Trump administration's active removal of environmental data from federal Web sites and purging or constraining of federal agencies with environmental governance responsibilities. The group, working collectively as the Environmental Data and Governance Initiative, has responded by archiving environmental data, interviewing agency personnel, and monitoring changes to Web pages and environmental policy. In connection with these efforts, Dillon et al. develop and articulate a concept of environmental data justice. Continuing with the theme of the active suppression of environmental data, the second article in the section, by Kopack, examines the difficulties of getting and analyzing data about toxic pollution from the Baikonur Cosmodrome, a legacy of the Soviet space program located in what is now Kazakhstan but still run by the Russian space program. Multiple rocket explosions have contaminated the area with toxic debris, but the Russian government's continuing tight control over both the immediate territory and all directly relevant research and data, as well as active suppression and intimidation of activists by the authoritarian Russian and Kazakh governments, dramatically demonstrates how secrecy can be territorialized in ways that render organizing and dissent

both difficult and dangerous. (On this note, it is important to mention that one prospective contributor to this issue dropped out after deciding that publishing the results of recent research on environmental politics in another severely repressive country would be directly and significantly dangerous to the author and interview subjects alike.) The third article in the section, by Koslov, examines "agnostic adaptation" in response to Hurricane Sandy in New York's Staten Island. Koslov notes that this relatively conservative community has in fact taken significant steps to adapt to future effects of climate change in the aftermath of the storm, yet it has done so with almost no explicit reference to or discussion of climate change because of the charged and polarized politics around that term in the United States: Strategic decisions to not frame adaptive actions as responses to climate change specifically allowed for community agreement and action around them that might not have occurred otherwise. Koslov argues that this dynamic reverses the formulation of the relationship between the postpolitical and action on climate change posited by Swyngedouw (2010): Whereas Swyngedouw argued that focusing on technocratic and practical steps stands in the way of genuine responses to climate change, Koslov suggests that focusing on precisely such steps allowed for meaningful actions, and politics of a sort, that would have been precluded by an insistence on explicit discussions of climate change and climate science. The fourth article in this section, by Bosworth, examines the production of expertise among movements opposing the new Keystone XL and Dakota Access pipelines in the north-central United States. Bosworth argues that the construction and public deployment of environmental expertise by activists in these movements, as a counter to the forms and sources of expertise deployed against them, was itself a progressive form of populist politics that helped to constitute and strengthen the movements themselves. The fifth article, by Forsyth, examines the coproduction of environmental knowledge and narratives, environmental movements, and political power and authority in Thailand. Countering simplistic claims that environmentalism tends on the whole to contribute to the democratization of authoritarian regimes, Forsyth notes that environmental narratives can in fact be deeply conservative and contribute to the reproduction of existing social arrangements rather than substantive democratization and argues that it is essential to analyze specific narratives within the framework of civic epistemologies, which seeks to analyze the broader dimensions a given political order within which particular narratives retain political and epistemic authority. In the sixth and last article in this section, Neimark et al. explore the potential tensions between political ecology as an approach that has often emphasized the social construction and political nature of scientific environmental knowledge claims and political ecology as a field in which many practitioners presumably want to hold on to many central tenets of environmental science in the face of their dismissal or outright denial by many contemporary authoritarian regimes. The keys to resolving this tension, they argue, lie in recognizing the blatant power relations and agendas at work in many contemporary denials of scientific knowledge and distinguishing between the politically motivated production of posttruth and the genuine recognition of sincerely held diverse epistemologies and ontologies.

The sixth and final section of the special issue ends on a positive note: The five articles in the section explore and advocate progressive alternatives to authoritarian and reactionary populist environmental politics and governance. The first, by Andreucci, makes an explicit, grounded argument for the possibility and potential of a genuinely progressive, indeed Gramscian, version of populism in relation to environmental politics in Bolivia. Echoing Bosworth's affirmation of the possibility of a progressive populism, Andreucci builds on Laclau's (1977, 2005) and Fraser's (2017) visions of populism as a potential strategy to enable the construction of a broad counterhegemonic bloc out of disparate particular struggles. Continuing with the theme of left and progressive populisms, the second article in the section, by Knuth, examines the past and potential contributions of populist movements to clean energy transitions in the United States and elsewhere. Knuth argues that left populist movements have already helped to shape clean energy programs in California and in the United States as a whole, in the context of calls for "green jobs" and a "green New Deal" in the wake of the financial crisis. At the same time, Knuth insists that more fully realizing the potential of populist contributions to just transitions will require engagement with the populist politics of grievance and reparation as well and a

strategy that engages with the full breadth of the economy, not just niche sectors. The third article in the section, by Cadieux et al., likewise emphasizes the progressive potential of populism, in their case through comparing major 1930s agrarian populist initiatives in the Midwestern United States to highly diverse and inclusive contemporary urban agriculture movements in the same region. Cadieux et al. use these examples to argue that focusing agroecological social movements on the repair of social and ecological relationships offers opportunities to use their power to counter capitalism and authoritarianism, avoiding many of populism's potentially more reactionary elements. The fourth article in this section, by Aitken, An, and Yang, examines how environmental governance of and around the Fanjingshan National Nature Reserve in China has changed following the election of President Xi, as an authoritarian government has professed a greater commitment to sustainability even as rapid development proceeds apace. Contrasting the trajectories of two development projects, one inside the park and one outside, Aitken et al. express cautious optimism regarding the potential of approaches rooted in increasing local capacities and sustainable ethics to produce real improvements in people's lives even under challenging circumstances. Finally, the fifth article in the section, and the last in the issue, by Goldstein, Paprocki, and Osborne, suggests that inasmuch as the attraction of authoritarian populism in the contemporary United States and beyond often appears to be strongest in areas hit hard by deindustrialization, agrarian dispossession, and climate change, scholars at public land grant universities have distinctive organizational affordances and obligations to respond to those forces. They respond by articulating a manifesto for a progressive mission for land grant institutions.

Looking Ahead

The current conjuncture is grim, but it also contains significant grounds for hope. The articles in this special issue demonstrate widespread rejection of major elements of neoliberal capitalism and deep desires for true alternatives. Although those sentiments have gone, or been taken, in deeply reactionary directions in many instances, they are also suggestive of a window of opportunity for truly broad-based, inclusive, and progressive coalitions and alternatives, along the lines called for by Fraser (2017), Scoones et al. (2018), and many others. Indeed, such movements are having significant success in many places around the world, often although certainly not necessarily through the use of populist frameworks and strategies. Geographers have much to offer efforts to create a truly broad-based, inclusive, historically and geographically aware progressive politics, as we see in the kinds of work highlighted in this special issue. We are adept at analyzing and explaining how any environmental project is always also a social one and vice versa and, more particularly, at understanding how particular sorts of socioenvironmental projects—the liberalization and globalization of agricultural production, for example—relate to broad social tensions and trends. We are especially well equipped, and indeed have an obligation given our disciplinary history, to continue to remind publics of the moral and intellectual bankruptcy and consequences of conflating physical environments and social identities. We can advocate as well as analyze and add our voices and knowledge to the many others attempting to create realistic, grounded, yet ambitious visions of more just, equitable, and sustainable futures (see Braun 2015). In short, environmental issues, movements, and politics can and indeed must be central both to resistance against authoritarian and reactionary populist politics and to visions of progressive alternatives to them. The articles in this issue provide many promising starting points for such visions.

Acknowledgments

My deep and sincere thanks to Jennifer Cassidento and Lea Cutler at the *Annals* for the prodigious amount of excellent work they put into keeping this issue on track; to members of the editorial board for sharing their suggestions, expertise, and labor in relation to the special issue; to Bruce Braun, Nik Heynen, and Karl Zimmerer for generous and incisive comments on this introduction and the issue as a whole; to Ned Resnikoff for compelling me to articulate and justify taken for granted academic frameworks; and to the great many reviewers who reviewed first, second, and in some cases third versions of all of the articles in the issue.

References

Albright, M. 2018. *Fascism: A warning.* New York: HarperCollins.

Arendt, H. [1951] 1973. *The origins of totalitarianism.* New York: Harcourt Brace Jovanovich.

Badiou, A. 2016. Twenty-four notes on the uses of the word "people". In *What is a people?*, ed. A. Badiou, P. Bourdieu, J. Butler, G. Didi-Huberman, S. Khiari, and J. Rancière, 21–31. New York: Columbia University Press.

Beeson, M. 2010. The coming of environmental authoritarianism. *Environmental Politics* 19 (2):276–94.

Bello, W. 2018. Counterrevolution, the countryside and the middle classes: Lessons from five countries. *The Journal of Peasant Studies* 45 (1):21–58.

Bessner, D., and M. Sparke. 2017. Nazism, neoliberalism, and the Trumpist challenge to democracy. *Environment and Planning A* 49 (6):1214–23.

Bigger, P., and J. Dempsey. 2018. The ins and outs of neoliberal natures. *Environment and Planning E: Nature and Space* 1 (1–2):25–43.

Borras, S. M. 2018. Understanding and subverting contemporary right-wing populism: Preliminary notes from a critical agrarian perspective. Conference Paper No. 47 presented at the Emancipatory Rural Politics Initiative International Conference Authoritarian Populism and the Rural World, The Hague, The Netherlands, March 17–18.

Braun, B. 2015. Futures: Imagining socioecological transformation—An introduction. *Annals of the Association of American Geographers* 105 (2):239–43.

Collard, R. C., L. M. Harris, N. Heynen, and L. Mehta. 2018. The antinomies of nature and space. *Environment and Planning E: Nature and Space* 1 (1–2):3–24. doi:2514848618777162.

Ekers, M., A. Loftus, and G. Mann. 2009. Gramsci lives! *Geoforum* 40 (3):287–91.

Fraser, N. 2017. The end of progressive neoliberalism. *Dissent* 64 (2):130–34.

Fritsche, I., J. C. Cohrs, T. Kessler, and J. Bauer. 2012. Global warming is breeding social conflict: The subtle impact of climate change on authoritarian tendencies. *Journal of Environmental Psychology* 32 (1):1–10.

Gerbaudo, P. 2017. *The mask and the flag: Populism, citizenism, and global protest.* New York: Oxford University Press.

Gilley, B. 2012. Authoritarian environmentalism and China's response to climate change. *Environmental Politics* 21 (2):287–307.

Gramsci, A. 1971. *Selections from the prison notebooks of Antonio Gramsci.* London: Lawrence and Wishart.

Grattan, L. 2016. *Populism's power: Radical grassroots democracy in America.* Oxford, UK: Oxford University Press.

Hall, S. 1980. Popular-democratic versus authoritarian populism. In *Marxism and democracy*, ed. A. Hunt, 157–87. London: Laurence and Wishart.

———. 1985. Authoritarian populism: A reply to Jessop et al. *New Left Review* 151:115–24.

Hardt, M., and A. Negri. 2005. *Multitude: War and democracy in the age of empire.* New York: Penguin.

Heynen, N., J. McCarthy, S. Prudham, and P. Robbins. 2007. *Neoliberal environments: False promises and unnatural consequences.* London and New York: Taylor & Francis.

Hofstadter, R. 1960. *The age of reform.* New York: Vintage.

Koch, N., and T. Perreault. 2018. Resource nationalism. *Progress in Human Geography.* doi:10.1177/0309132518781497.

Laclau, E. 1977. *Politics and ideology in Marxist theory: Capitalism, fascism, populism.* London: New Left Books.

———. 2005. *On populist reason.* London: Verso.

MacDonald, G. 2016. Geography in a post-truth world. *American Association of Geographers Newsletter,* December 7. doi:10.14433/2016.0020.

Mann, G., and J. Wainwright. 2018. *Climate leviathan: A political theory of our planetary future.* London: Verso.

Mansfield, B. 2018. From the commons to the body to the planet: Neoliberalism/materiality/socionatures. *Environment and Planning E: Nature and Space* 1 (1–2):58–61.

McCarthy, J., C. Chen, D. López-Carr, and B. L. E. Walker. 2014. Socio-cultural dimensions of climate change: Charting the terrain. *GeoJournal* 79 (6):665–75.

McCarthy, J., and S. Prudham. 2004. Neoliberal nature and the nature of neoliberalism. *Geoforum* 35 (3):275–83.

Monbiot, G. 2017. Freeing up the rich to exploit the poor—That's what Trump and Brexit are about. *The Guardian,* April 3–4.

Mouffe, C. 2016. The populist challenge. Accessed November 26, 2018. https://www.opendemocracy.net/democraciaabierta/chantal-mouffe/populist-challenge.

Müller, J. W. 2016. *What is populism?* Philadelphia: University of Pennsylvania Press.

Oreskes, N., and M. Conway. 2010. *Merchants of doubt.* London: Bloomsbury.

Proctor, R., and L. Schiebinger, ed. 2008. *Agnatology: The making and unmaking of ignorance.* Palo Alto, CA: Stanford University Press.

Rancière, J. 2016. The populism that is not to be found. In *What is a people?,* ed. A. Badiou, 100–106. New York: Columbia University Press.

Remnick, D. 2017. A hundred days of trump. *The New Yorker,* May 1. Accessed July 6, 2019. https://www.newyorker.com/news/john-cassidy/after-a-hundred-days-trump-is-trump-is-trump

Scoones, I., M. Edelman, S. M. Borras, Jr., R. Hall, W. Wolford, and B. White. 2018. Emancipatory rural politics: Confronting authoritarian populism. *The Journal of Peasant Studies* 45 (1):1–20.

Snyder, T. 2017. *On tyranny: Twenty lessons from the twentieth century.* New York: Tim Duggan Books.

Surprise, K. 2018. Preempting the second contradiction: Solar geoengineering as spatiotemporal fix. *Annals of the American Association of Geographers* 108 (5):1228–44. doi:10.1080/24694452.2018.1426435

Swyngedouw, E. 2010. Apocalypse forever? Post-political populism and the spectre of climate change. *Theory Culture Society* 27 (2–3):213–32.

Williams, R. 1980. *Problems in materialism and culture.* London: Verso.

Zimmerer, K. 2011. New geographies of energy: Introduction to the special issue. *Annals of the Association of American Geographers* 101 (4):705–11.

JAMES McCARTHY is a Professor in the Graduate School of Geography at Clark University, Worcester, MA 01610. E-mail: JaMcCarthy@clarku.edu. His research interests include environmental politics, political economy, and political ecology.

Authoritarian Environmental Governance: Insights from the Past Century

Robert Wilson

For over a decade, nature–society geographers have focused on neoliberal and, more recently, postneoliberal environmental governance. Meanwhile, regimes in many nations have become less democratic and other countries, such as the United States, have elected leaders sympathetic to autocrats. Yet despite the spread of authoritarianism, nature–society geographers have as of yet devoted little attention to the subject, which hampers us as we confront this authoritarian moment. This article addresses this oversight but by examining the past rather than the present. Drawing on work by historians in general and environmental historians in particular, I explore authoritarian environmental governance in the Soviet Union, Maoist China, and Nazi Germany, three countries and eras largely overlooked by nature–society geographers. I focus in particular on agricultural collectivization, industrialization and river development, and nature conservation under authoritarian regimes. Understanding past authoritarian environmental governance will enable nature–society geographers to better reckon with the environmental ramifications of a possible new authoritarian era.

近十年来，自然—社会地理学者聚焦新自由主义的环境治理，更晚近则聚焦后新自由主义的环境治理。于此同时，诸多国家的政体已变得更不民主，诸如美国等其他国家，则选出了同情独裁者的领导者。尽管威权主义有所扩散，自然—社会地理学者却仍尚未对该主题投入足够关注，并使我们在面对此一威权时刻时受到束缚。本文应对此一疏忽，但是是通过检视过往、而非当下。我运用普遍的历史研究、特别是环境史研究，探讨苏联、毛时代的中国以及纳粹德国中的环境治理，这三个国家与时代受到自然—社会地理学者大幅忽略。我特别聚焦威权政体下的农业集体化、工业化与河川发展及自然保育。理解过往的威权环境治理，将能让自然—社会地理学者更佳地应付可能的新威权年代中的环境后果。 关键词: 威权主义, 环境治理, 纳粹主义, 社会主义。

Durante más de una década los geógrafos que se especializan en la relación naturaleza–sociedad han concentrado su atención en la gobernanza ambiental neoliberal y, más recientemente, en la gobernanza ambiental posneoliberal. Entretanto, en muchas naciones los regímenes de gobierno se han hecho menos democráticos y otros países, tal como los Estados Unidos, han elegido líderes que simpatizan con los autócratas. Pero a pesar de la propagación del autoritarismo los geógrafos de la naturaleza-sociedad hasta el momento han prestado poca atención a este asunto, lo que nos debilita cuando tenemos que lidiar con el momento autoritario actual. Este artículo aboca este descuido, aunque examinando más el pasado que el presente. Con base en el trabajo de los historiadores en general, y de los historiadores ambientales en particular, exploro la gobernanza ambiental autoritaria en la Unión Soviética, la China maoísta y la Alemania nazi, tres países y eras en gran medida ignorados por los geógrafos de la naturaleza–sociedad. Me centro particularmente en la colectivización agrícola, la industrialización y desarrollo fluvial, y la conservación de la naturaleza bajo regímenes autoritarios. Entender la gobernanza ambiental autoritaria pasada capacitará a los geógrafos de la naturaleza–sociedad para lidiar mejor con las ramificaciones ambientales de una posible nueva era autoritaria. *Palabras clave: autoritarismo, gobernanza ambiental, nazismo, socialismo.*

Liberal democracy is in peril. Russia's democratic reforms, which burgeoned after the collapse of the Soviet Union, have faltered and President Vladimir Putin has enshrined an autocratic government (Gessen 2017). In the Middle East, hope in 2011's Arab Spring has faded as countries such as Egypt returned to authoritarian rule and dictators such as President Bashar al-Assad used the Syrian civil war to further consolidate power. Venezuela, once a leading democracy in South America, has lurched toward dictatorship under President Nicolás Maduro (Naim and Francisco 2016; Aleem 2017).

Meanwhile, in the United States, President Donald Trump has revealed authoritarian leanings in numerous ways, such as by labeling news outlets "fake news" and accusing the mainstream media of being "the enemy of the American people" (Remnick 2017). President Trump has also vilified his political opponents and expressed admiration for strongmen such as Putin, Philippine President Rodrigo Duterte, and North Korean leader Kim Jong-un. Many citizens and scholars see these as dire threats to liberal democracy (Snyder 2017, 2018; Levitsky and Ziblatt 2018; Mounk 2018).

Although such developments have aroused deep concern, especially among vulnerable groups such as immigrants and those who champion freedom of the press, fewer commentators have examined what these developments mean for environmental governance. To be sure, in the United States, environmental activists and scholars have noted how President Trump and his cabinet, particularly former Environmental Protection Agency Chief Scott Pruitt, have tried to undermine environmental regulations and agreements drafted by President Barack Obama and his administration (Frontline 2017; Langston 2018). Hardly any commentators, however, have linked the Trump administration's authoritarian tendencies and contempt for rule of law with its attack on the environmental management state.

Given the long-standing interest among environmental geographers in governance, they would seem well positioned to study and critique how environmental rules, protections, and practices are changing in this populist and authoritarian era. As of yet, few have examined the connection between authoritarianism and the environment. For instance, in two exhaustive recent surveys of political ecology scholarship, there are no entries on authoritarianism and the environment or on similar subjects (Perreault, Bridge, and McCarthy 2015; Bryant 2017). Other nature–society geography surveys, textbooks, and edited collections over the past decade have not devoted space or chapters to authoritarianism and the environment (Castree et al. 2009; Heynen et al. 2009; Peet, Robbins, and Watts 2010; Robbins 2011; Moseley, Perramond, and Hapke 2013; Robbins, Hintz, and Moore 2014). Allied fields such as environmental historical geography have not engaged with the environment and authoritarianism either (Colten 2012; Wynn et al. 2014; Buckley and Youngs 2018).

This special issue focusing on environmental governance in our current age of populism and authoritarianism partially corrects this oversight. To better understand authoritarian environmental governance now, though, we also need to examine the past. In the twentieth century, authoritarianism dominated in a number of countries and eras, perhaps most notably in the communist Soviet Union and Maoist China as well as in Nazi Germany. Although there were many differences among these nations' governments, all were authoritarian regimes with extreme levels of state power, no free elections, and limited or no sanctioned means of public political dissent.

Yet applying the term *environmental governance* to analyze the environmental management systems in these countries requires some translation. Environmental governance is a concept that emerged in the mid-1990s as neoliberalism came to dominate approaches to environmental matters in many countries and sectors of resource management (McCarthy and Prudham 2004; Himley 2008; Bridge and Perreault 2009; Bakker 2010). Using the term *neoliberal environmental governance* in these studies is somewhat redundant because most environmental governance work address neoliberalism to one degree or another. In more recent years, a few scholars have debated whether some places, particularly in Latin America, have entered a postneoliberal era (Bakker 2013; Ruckert, Macdonald, and Proulx 2017). Even these recent studies, though, rely on and modify the framework crafted by scholars examining neoliberal environmental governance. Yet if we understand environmental governance more broadly as a concept referring to forms of politics and social control under which nature is managed, then we can use the concept to investigate society–environmental relations in earlier eras.

In what follows, I examine some of the features of authoritarian environmental governance in three countries: the Soviet Union, Nazi Germany, and Maoist China. I draw on work by environmental historians and other historians who have studied the environmental history of these regimes. My analysis is also an outgrowth of more than a decade of teaching where I have synthesized this scholarship for undergraduate and graduate students, most of whom are entirely unfamiliar with the environmental histories of these nations. To do so, I venture into the allied discipline of history because there is little scholarship about environmental governance, broadly defined, in political ecology or historical geography about these countries during these eras. This speaks more broadly to the lack of research over the past two decades by political ecologists and historical geographers about the Soviet Union, Nazi Germany, or Maoist China. Political ecology emerged as a field beginning in the 1980s, focusing mostly on peasants in what was then called the third world, particularly Africa, South Asia, and Latin America. Historical geographers, too, were largely silent about these authoritarian countries, as evidenced by articles published in the *Journal of Historical Geography* since it was founded in the 1970s.

Certainly there are challenges in examining these three countries and time periods in one article. Doing so risks homogenizing the different histories of these nations by overlooking their numerous differences. Some scholars of authoritarianism, such as Arendt ([1951] 1973) in her classic work on totalitarianism, saw important similarities between fascist Nazi Germany and the Soviet Union. More recent scholarship, however, has questioned the totalitarianism framework, seeing the term in many ways as a Cold War relic (Geyer and Fitzpatrick 2008). Nevertheless, despite the differences among the Soviet Union, Nazi Germany, and Maoist China, it is worth highlighting the commonalities among the environmental governance strategies in these authoritarian regimes. In doing so, I am inspired by Snyder's (2017) recent book *On Tyranny: Lessons from the Twentieth Century*, which explores some of the similar features of authoritarian regimes in the last century, particularly Nazi Germany, the Soviet Union, and Eastern Bloc nations. His book also serves as a cautionary tale about how authoritarian regimes come to power and a warning to Americans in the age of Trump. In a similar manner, this article identifies distinguishing characteristics—insights rather than lessons—of environmental governance in twentieth-century authoritarian regimes. This history helps us better understand the possible social and environmental repercussions when democracy erodes and the environmental policies of authoritarian leaders proceed unchallenged.

Authoritarianism and Collectivization

Although environmental historians have produced histories of environmental management for the Soviet Union, Maoist China, and Nazi Germany,

few have examined the common aspects of environmental governance in authoritarian contexts. An important exception is Josephson, who in a series of monographs, most notably in *Resources under Regimes* (Josephson 2005), identified some of the key features of authoritarian environmental management (see also Josephson and Zeller 2003). They include one officially sanctioned political party, state-directed development, tight control of media, and a group venerated by the ruling party, such as workers in communist countries or Aryans under the Nazi regime (Josephson 2005). As we shall see, however, one-party rule did not preclude the existence of conservation and environmental organizations, although their freedom was often highly circumscribed, their ability to challenge the ruling party limited, and critique came with great risk.

In twentieth-century authoritarian regimes, Josephson recognized a common fondness for large-scale projects such as hydroelectric dams, river rerouting, and the construction of industrial or nuclear complexes. Josephson called these projects "brute-force technologies," which he defined as large-scale systems including both their technological and administrative components (Josephson 2002, 4–8, 255–63). Of course, more democratic societies also undertook such projects—the U.S. 1930s New-Deal–era dams on the Columbia River and in the Tennessee Valley come readily to mind (White 1995; Sneddon 2015)—but state-socialist regimes and the Nazis were particularly enamored with them.

Perhaps the most egregious examples of disastrous twentieth-century environmental governance were collectivization campaigns in the Soviet Union and China during the Mao period. In both cases, the communist governments sought to collectivize agriculture as part of larger efforts to consolidate power, squash opponents, and modernize their countries (Wemheuer 2014a, 2014b). Collectivization entailed transferring formerly privately controlled land to the hands of the state. The first collectivist programs in what would become the Soviet Union began shortly after the Bolsheviks seized power from the Provisional Government in the fall of 1917 (A. Brown 2009; Priestland 2009; Josephson et al. 2013; Wemheuer 2014a, 2014b). Joseph Stalin inaugurated a collectivization program when he launched his first Five-Year Plan in 1928. With this plan, the Soviet Union sought to inaugurate a Great

Break with the past and catapult the nation into a place among the world's modern industrial nations (Josephson et al. 2013; Wemheuer 2014a, 2014b). For the Communist Party and Stalin in particular, collectivizing agriculture was essential to transform the nation and to feed workers in growing industrial cities. Stalin saw peasants working these lands as impediments to his aspirations, however. The communist government expected them to surrender their land, livestock, and farm implements to work on communal fields; live in collective housing; and eat in common dining halls (Applebaum 2017). They sought to transform these peasants into agricultural workers who would labor on vast collective farms, use modern machinery like tractors, and toil under the gaze of Soviet officials in watchtowers. Those who refused were executed or sent to labor concentration camps—later known as the Gulags (Snyder 2010 Applebaum 2017).

Few peasants willingly joined collective farms by 1930, so the Soviets adopted further coercive measures to drive peasants onto state farms and renewed targeting those it deemed hostile to collectivization. Most of these were *kulaks*, supposedly wealthy peasants who resisted official decrees and who communist officials considered enemies of the state (Applebaum 2017). In reality, few of these *kulaks* were wealthy. Many only had a few more livestock or hectares of land than others in their communities. As government quotas to eliminate *kulaks* increased in the early 1930s, though, officials began arbitrarily labeling unwanted peasants as *kulaks*. By the end of 1931, the Soviets had branded 1.8 million citizens as *kulaks*, and 300,000 of them died during expulsion to labor camps (Scheidel 2017). From 1930 to 1933, Soviet-led collectivization became, in the words of Josephson et al. (2013), "violent, brutal, and murderous coercion, a revolution of totality, rapidity, and violence" (97).

Peasants opposed this transfer of private land to the communist state as best they could. Their resistance took many forms. Some simply refused to join the collectives, and others slaughtered their livestock rather than let the Soviets confiscate them. Other peasants, such as in Ukraine, fled to cities or, in some cases, tried to escape the Soviet Union itself. Demonstrators, particularly women, protested the Soviet government. The most risky strategy was to take up arms, kill party activists, and raid grain storage facilities (Snyder 2010; Applebaum 2017). Such

resistance served as a convenient pretext for Stalin to crack down on *kulaks* even further, to arrest them, and deport them to labor camps (Applebaum 2017). By late 1932 and early 1933, hundreds of thousands of peasants were dying, particularly in the Ukraine. The death toll from the famine ultimately reached 5 million in the Soviet Union, with perhaps 3.9 million deaths in Ukraine alone, a calamity known to Ukrainians as the Holodomor (Applebaum 2017; see also Naimark 2017; Scheidel 2017).

Twenty-five years later, the Chinese Communist Party (CCP) under the leadership of Chairman Mao Zedong undertook its own environmental program and collectivization campaign, the Great Leap Forward, with even more devastating consequences. In recent years, journalists and scholars have accessed CCP archives, allowing a much richer understanding of this calamity to emerge (Priestland 2009; Jisheng 2012; Dikötter 2013; Wemheuer 2014a; McNeill and Engelke 2016; Naimark 2017; see also Shapiro 2001; Xun 2013). As with Stalin's collectivization drive in the late 1920s and early 1930s, Chairman Mao's Great Leap Forward sought to increase grain production and harness a more efficient agricultural sector to support industrialization. Collectivization entailed confiscating land that peasants acquired after the Chinese revolution of 1949 and herding peasants into massive communes. Also, in an effort to boost the country's steel production, the communist government forced rural Chinese to build thousands of backyard furnaces to melt farming equipment, cooking implements, and other items to forge steel. Nearly all of this metal was useless, and in the process, peasants lost key tools needed for farming and cooking food. As the Soviets did in Ukraine during the 1930s, Chinese communists continued to confiscate grain to supply cities or to sell it internationally, in this case, to the Soviet Union. Chinese peasants died during the Great Leap Forward on a scale even greater than Ukrainian peasants did during the Holodomor. An accurate tally of people who perished from starvation, disease, or execution might never be known. Using the best available evidence, though, historians now estimate the total dead between 20 and 45 million (Dikötter 2011; Walder 2015; Scheidel 2017).

The authoritarian context of China during the late 1950s and early 1960s both made the famine possible and likely prolonged it. Mao undertook the Great Leap Forward after the Hundred Flowers movement in 1956 in which he encouraged citizens and party officials to critique CCP policies. With the government facing a torrent of criticism, Mao launched an anti-rightist program to punish those who denounced the regime. The party branded tens of thousands as counterrevolutionaries, who then lost their jobs or were deported to labor in rural reeducation camps (Shapiro 2001; Dikötter 2011; 2013). In the wake of this, both senior communist leaders and low-level officials were loath to criticize Mao's directives, fearing that they might meet the same fate (Walder 2015). The climate of fear fostered by the post–Hundred Flowers campaign promoted an atmosphere in which Mao could pursue his deeply misguided Great Leap Forward with few repercussions for him. The campaign continued into 1960 and 1961 even as it became clear the program was an utter failure and millions of Chinese were dying (Dikötter 2011).

Industrialization and River Development

In the decades after the Russian and Chinese revolutions, Lenin, Stalin, and Mao encouraged rapid industrial development of their countries in hopes of matching or even surpassing the United States and countries in Western Europe. In doing so, they tried to achieve in a few years what these capitalist societies had accomplished over decades or a century. This quest for rapid industrialization along with state control and the lack of political opposition would seemingly lay the groundwork for widespread environmental degradation. Indeed, older scholarship on the environmental politics and management of the Soviet Union and postrevolutionary China fiercely critiqued these authoritarian regimes for what they argued were dismal environmental records (Goldman 1972; Pryde 1991; Shapiro 2001). More recent scholarship, particularly about Soviet industrial development in the Arctic and river development in China, does not discount the pollution and environmental degradation in these countries during these times, but it does paint a more nuanced picture of the environmental costs of modernization.

Nations with Arctic regions such as the United States and Canada undertook or fostered exploration, mining, extractive development, and the creation of military bases during the twentieth century to better incorporate these northern areas into their respective nations (Haycox 2002; Wynn 2007; Piper

2010; Desbiens 2013). Arctic development was also crucial for authoritarian states such as the Soviet Union. As it did elsewhere in the country, the central government pursued collectivization programs in the Arctic but over reindeer herding rather than agriculture. Unlike in other circumpolar nations, though, the Soviets depended on thousands of slave laborers to harness the region's resources (Josephson 2014). Although some environmental historians see this modernizing push in the region as an attempt to conquer the Soviet Arctic (Josephson 2014), Bruno (2016) offers a more nuanced view of the Soviet state's efforts to industrialize the area. He argued that parts of nonhuman nature facilitated Soviet attempts to transform the region, whereas others hampered the government's efforts to incorporate the Soviet Arctic into the communist society it sought to build.

Soviet-led modernization and industrial development in the Russian Arctic had similarities to such occurrences in liberal democracies elsewhere in the circumpolar north. River development in China during the Mao era also bore the hallmarks of comparable processes elsewhere. Although the Chinese had managed its major rivers for centuries, water control entered a new phase after the 1949 revolution with the ascent of Mao and the creation of the CCP. As we saw, modernizing China and forging a communist state entailed collectivizing the nation's agriculture. In addition to this, though, the CCP embarked on a water development program closely modeled on Soviet water management, which included construction of massive, multipurpose dams. The Soviet approach, in turn, was influenced by other global developments in water management, particularly the Tennessee Valley Authority in the United States. Engineers who questioned the wisdom of these schemes of the long-term viability of some dams were silenced and labeled "rightists" (Shapiro 2001, 2016; Pietz 2015). Engineers did not face this level of intimidation in liberal democracies. Maoist water management also departed from watershed transformations in North America and Western Europe with its mass mobilization of Chinese laborers to construct water holding ponds and dig canals. During this era, but especially during the Great Leap Forward, millions of Chinese peasants were put to work constructing irrigation works on the North China Plain and elsewhere (Pietz 2015). By reworking China's waterscape, Mao hoped to foster "a 'second creation,' namely, the creation of Communist China" (Pietz 2015, 231).

Green Authoritarianism?

The limited space for dissent in these authoritarian societies would seem to hinder nature protection and render conservation impossible. Certainly, the lack of sanctioned political opposition parties and a free media made criticism and protests against industrial development and large-scale environmental projects difficult. Despite these obstacles, conservationists sometimes found room to pressure these regimes to undertake environmental reform, create protected areas, and conserve wildlife. Also, in some cases, authoritarian states and their leaders supported conservation initiatives and the creation of protected areas.

The case of Nazi Germany is instructive. In the 1932 election, the Nazi Party gained a plurality in the German parliament, and after a fire decimated the Reichstag in early 1933, parliament passed an enabling act granting Chancellor Adolph Hitler sweeping powers. The Nazis outlawed other political parties and imprisoned political opponents in the first German concentration camp, Dachau. From then until the downfall of the regime in 1945, the Nazis ruled Germany. Despite the ascendance of this authoritarian regime and the repression that came with it, conservationists were able to make progress. As Frank Uekötter and other environmental historians of Germany showed, Nazi leaders sympathetic to conservationist aims, most notably Herman Göring, used their clout to help enact environmental decrees (Brüggemeier, Cioc, and Zeller 2005; Uekötter 2006). The most important of these was the Reich Nature Protection Law of 1935. This far-reaching law enabled officials to identify areas worthy of protection, improve the state's administrative capacity to manage the environment, and enact measures to conserve threatened wildlife and habitat (Closmann 2005; Uekötter 2006). After having little success in parliament during the preceding Weimar Republic, German conservationists felt that at least some Nazi leaders appreciated their ideas and concerns (Uekötter 2014).

Despite the seeming success of the new nature protection decree and other measures to conserve forests, the Nazis proved fickle conservationists. As the 1930s progressed, the Nazi regime focused on rearming the country and preparing for war, and this growth in industrial production inevitably affected the environment and conservation measures did little to curtail it (Uekötter 2014). Also, although Hitler was a vegetarian, celebrated alpine landscapes, and owned a beloved German shepherd, Blondi, there is little evidence that he cared much for conservation. The reign of the Third Reich proved the high tide for conservation up to that point in German history, but after the war, conservationists had to reckon with their close association with the Nazis and for being "complicit in a genocidal regime" (Uekötter 2014, 56).

Given its association with political purges, breakneck industrialization, Gulags, and forced labor, the early decades of the Soviet Union would seem unfertile ground for conservation. Even under this dictatorship, though, state-led conservation programs grew during Lenin's short reign, and he expanded a nature preserve system begun under the Czars known as the *Zapovendniki*. Unlike national parks in the United States, which were established both to protect wildlife and sublime landscapes and provide spaces for recreation, the *Zapovendniki* were more akin to natural laboratories and inviolate sanctuaries where scientists could study biological processes. By 1929, more than sixty-one preserves existed covering nearly 4 million ha (Josephson et al. 2013).

Recent work by historian Stephen Brain further challenges the declenionist narrative of much scholarship on the Soviet Union's environmental record. "Dictators like trees," wrote Brain, and perhaps few dictators appreciated the values of trees more than Stalin (Brain 2011, 115). During his time as ruler of the Soviet Union, Stalin endorsed the creation of protected forest reserves, the largest forest reserve system in the world at that time. Although forests in Siberia and the north were opened to wholesale development, Stalin decreed that some forests in Russia remain inviolate reserves. The Stalinist environmentalism, as Brain called it, protected these forests from economic development on a scale unmatched in most other countries. The vision of the communist dictator as a proto-environmentalist seems odd given his push to rapidly modernize the country and his brutal purges of real and imagined political opponents. Yet Stalin was swayed by conservationists who argued that overlogging in some forested areas would negatively affect the country's rivers, which in turn might threaten cities with flooding and harm the hydropower system the regime was developing. This utilitarian-oriented forest protection endured during the Stalin years but in the service of the state's larger goals of industrialization and modernization (Brain 2010, 2011).

Unlike in the case of Nazi Germany, where conservation groups consisted of ordinary citizens interested in environmental protection, Soviet conservation organizations were closely associated with scientists, such as the All-Russian Society for Conservation (VOOP). When advocating for environmental reforms or protecting landscapes, these groups had to carefully couch their language and aims so as not to appear counterrevolutionary (Weiner 1988). Despite the loosening of political restrictions after Nikita Khrushchev came to power and denounced Stalin in 1956, the communist government monitored groups such as VOOP. They certainly were not able to halt or even alter massive development initiatives such as Khrushchev's Virgin Lands campaign, inaugurated in 1954, which opened millions of acres of Kazakhstan steppe to industrial farming. Plowing the grasslands helped foster Dust Bowl–like conditions by the mid-1960s, and by then, the communist government realized that the campaign was a colossal failure (Josephson et al. 2013). Truly independent environmental groups and a more vibrant civil society did not really emerge in the country until Mikhail Gorbachev enacted *glasnost* (openness) reforms during what, as it turned out, were the waning years of the Soviet Union before its collapse in 1991 (Weiner 1999; Josephson et al. 2013).

Conclusion

What insights can nature–society geographers glean from this brief survey of twentieth-century authoritarian environmental governance? There is ample evidence that these authoritarian regimes caused massive environmental destruction and social dislocation, whether through the development of mining and smelting operations, industries, or land collectivization campaigns. Certainly the environmental history of liberal democracies such as the United States, Canada, and Western Europe have also shared the same history of industrialization and modernization with its associated environmental consequences. Indeed, one is struck more by the many similarities between the environmental

consequences of development in communist dictatorships such as the Soviet Union and Maoist China and Western democracies. As Soviet environmental historian Bruno (2016) noted, the "relentless impulse to modernize society and the natural world" resulted in "similar environmental trajectories" (274) that transcended the purported differences between capitalist democracies and authoritarian communist systems (Bonhomme 2013; see also K. Brown 2001, 2013; Johnson et al. 2013).

Perhaps most surprising is that during some periods, these authoritarian states implemented progressive and far-reaching environmental reforms. With the support of conservationists, the Nazis enacted sweeping conservation legislation, made possible, in part, by the suppression of opposition parties and political groups that might have opposed it prior to the 1930s. In the Soviet Union, scientists were able to carve out a "little corner of freedom" within a repressive state and establish nature preserves for scientific research (Weiner 1999). Even Stalin, one of the twentieth century's most brutal authoritarian rulers, supported the creation of a vast forest reserve network. Indeed, the fact that these countries were authoritarian made implementing such reforms more straightforward because they did not have to contend with opposition parties who might have stymied their plans.

What if scientists, conservationists, and ordinary citizens questioned or opposed government plans for industrialization, modernization, and river development, however? Compared to liberal democracies, citizens in the Soviet Union, Maoist China, and Nazi Germany had fewer avenues of recourse or redress if industries polluted their communities or development schemes displaced them. In liberal democracies, citizens could vote for more environmentally progressive elected officials, demonstrate, and freely join conservation and environmental groups. Yet these options were often unavailable in these authoritarian regimes. Even when they permitted conservation groups or environmental scientific societies to exist, members had to exercise caution lest their advocacy be construed as counterrevolutionary or hostile to the regime. Capitalist liberal democracies and authoritarian communist countries undertook or enabled environmentally destructive projects, but the "main distinction lies less in the different economic systems (socialist versus capitalist) and more in the variant political cultures (authoritarian versus democratic)" (Bonhomme 2013, 27). Although these twentieth-century authoritarian regimes did sometimes take measures to curtail pollution or reduce deforestation, citizens had fewer political tools to influence government practices, particularly if doing so ran counter to the regimes' goals.

To confront our own authoritarian moment, we need to better understand this past. "History does not repeat, but it does instruct," wrote Snyder (2017, 9), but history is what has been lacking from much environmental governance scholarship with its resolute focus on contemporary neoliberalism and postneoliberalism. Although most nature–society geographers have vigorously critiqued current neoliberal environmental governance, they have perhaps unwittingly overlooked history, or at least the history of twentieth-century forms of environmental governance under socialism and Nazism. If we have indeed entered a new authoritarian era, then it is all the more necessary to understand the environmental dimensions of tyranny in the not-so-distant past.

Acknowledgments

Many thanks to special issue editor James McCarthy and the anonymous reviewers for their helpful comments on earlier versions of this article.

References

Aleem, Z. 2017. How Venezuela went from a rich democracy to a dictatorship on the brink of collapse. *Vox*, September 19. Accessed October 4, 2018. https://www.vox.com/world/2017/9/19/16189742/venezuela-maduro-dictator-chavez-collapse.

Applebaum, A. 2017. *Red famine: Stalin's war on the Ukraine*. New York: Doubleday.

Arendt, H. [1951] 1973. *The origins of totalitarianism*. New York: Harcourt, Brace, Jovanovich.

Bakker, K. 2010. The limits of "neoliberal natures": Debating green neoliberalism. *Progress in Human Geography* 34 (6):715–35.

———. 2013. Neoliberal versus postneoliberal water: Geographies of privatization and resistance. *Annals of the Association of American Geographers* 103 (2):253–60.

Bonhomme, B. 2013. Writing the environmental history of the world's largest state: Four decades of writing on Russia and the USSR. *Global Environment* 6 (12):12–37.

Brain, S. 2010. Stalin's environmentalism. *The Russian Review* 69 (1):93–118.

———. 2011. *Song of the forest: Russian forestry and Stalinist environmentalism, 1905–1953*. Pittsburgh, PA: University of Pittsburgh Press.

Bridge, G., and T. Perreault. 2009. Environmental governance. In *A companion to environmental geography*, ed. N. Castree, D. Demeritt, D. Liverman, and B. Rhoads, 475–97. Malden, MA: Wiley-Blackwell.

Brown, A. 2009. *The rise and fall of communism*. New York: Ecco.

Brown, K. 2001. Gridded lives: Why Kazakhstan and Montana are nearly the same place. *The American Historical Review* 106 (1):17–48.

———. 2013. *Plutopia: Nuclear families, atomic cities, and the great Soviet and American plutonium disasters*. New York: Oxford University Press.

Brüggemeier, F.-J., M. Cioc, and T. Zeller. 2005. Introduction. In *How green were the Nazis? Nature, environment, and nation in the Third Reich*, ed. F.J. Brüggemeier, M. Cioc, and T. Zeller, 1–17. Athens: Ohio University Press.

Bruno, A. 2016. *The nature of Soviet power: An Arctic environmental history*. New York: Cambridge University Press.

Bryant, R. L., ed. 2017. *The international handbook of political ecology*. Northampton, MA: Edward Elgar.

Buckley, G. L., and Y. Youngs eds. 2018. *The American environment revisited: Environmental historical geographies of the United States*. Lanham, MD: Rowman & Littlefield.

Castree, N., D. Demeritt, D. Liverman, and B. Rhoads, eds. 2009. *A companion to environmental geography*. Malden, MA: Wiley-Blackwell.

Closmann, C. 2005. Legalizing a Volksgemeinschaft: Nazi Germany's Reich Nature Protection Law of 1935. In *How green were the Nazis? Nature, environment, and nation in the Third Reich*, ed. F.-J Brüggemeier, M. Cioc, and T. Zeller, 18–42. Athens: Ohio University Press.

Colten, C. E. 2012. Environmental historical geography: A review. In *Encyclopedia of life support systems (EOLSS)*. Paris, France: UNESCO, EOLSS Publishers. Accessed November 29, 2017. http://www.eolss.net.

Desbiens, C. 2013. *Power from the North: Territory, identity, and the culture of hydroelectricity in Quebec*. Vancouver, BC, Canada: UBC Press.

Dikötter, F. 2011. *Mao's great famine: The history of China's most devastating catastrophe, 1958–1962*. New York: Walker & Company.

———. 2013. *The tragedy of liberation: A history of the Chinese revolution, 1945–1957*. New York: Bloomsbury.

Frontline. 2017. War on the EPA. PBS, October 17. Accessed November 29, 2017. https://www.pbs.org/wgbh/frontline/film/war-on-the-epa/.

Gessen, M. 2017. *The future is history: How totalitarianism reclaimed Russia*. New York: Riverhead Books.

Geyer, M., and S. Fitzpatrick. 2008. *Beyond totalitarianism: Stalinism and Nazism compared*. New York: Cambridge University Press.

Goldman, M. 1972. *The spoils of progress: Environmental pollution in the Soviet Union*. Cambridge, MA: The MIT Press.

Haycox, S. 2002. *Frigid embrace: Politics, economics, and the environment in Alaska.* Corvallis: Oregon State University Press.

Heynen, N., J. McCarthy, S. Prudham, and P. Robbins. 2009. *Neoliberal environments: False promises and unnatural consequences.* London and New York: Routledge.

Himley, M. 2008. Geographies of environmental governance: The nexus of nature and neoliberalism. *Geography Compass* 2 (2):433–51.

Jisheng, Y. 2012. *Tombstone: The great Chinese famine, 1958–1962.* New York: Farrar, Straus and Giroux.

Josephson, P. R. 2002. *Industrialized nature: Brute force technology and the transformation of the natural world.* Washington, DC: Island.

———. 2005. *Resources under regimes: Technology, environment, and the state.* Cambridge, MA: Harvard University Press.

———. 2014. *The conquest of the Russian Arctic.* Cambridge, MA: Harvard University Press.

Josephson, P. R., N. Mnatsakanian, R. Cherp, A. Efremenko, D. Efremenko, and V. Larin. 2013. *An environmental history of Russia.* New York: Cambridge University Press.

Josephson, P. R., and T. Zeller. 2003. The transformation of nature under Hitler and Stalin. In *Science and ideology: A comparative history,* ed. M. Walker, 124–54. London and New York: Routledge.

Langston, N. 2018. DOCUMERICA and the power of environmental history. *Environmental History* 23 (1):106–16.

Levitsky, S., and D. Ziblatt. 2018. *How democracies die.* New York: Crown.

McCarthy, J., and S. Prudham. 2004. Neoliberal nature and the nature of neoliberalism. *Geoforum* 35 (3):275–83.

McNeill, J. R., and P. Engelke. 2016. *The great acceleration: An environmental history of the Anthropocene since 1945.* Cambridge, MA: Belknap.

Moseley, W. G., E. Perramond, and H. M. Hapke. 2013. *An introduction to human-environment geography: Local dynamics and global processes.* Malden, MA: Wiley-Blackwell.

Mounk, Y. 2018. *The people vs. democracy: Why our freedom is in danger and how to save it.* Cambridge, MA: Harvard University Press.

Naim, M., and T. Francisco. 2016. Venezuela's democratic façade has completely crumbled. *Washington Post,* July 1. Accessed October 4, 2018. https://www.washingtonpost.com/opinions/global-opinions/hugo-chavezs-longcon/2016/07/01/26e8b690-3f8c-11e6-80bc-d06711fd2125_story.html?utm_term=.1c584dadaced.

Naimark, N. M. 2017. *Genocide: A world history.* New York: Oxford University Press.

Peet, R., P. Robbins, and M. Watts, eds. 2010. *Global political ecology.* London and New York: Routledge.

Perreault, T., G. Bridge, and J. McCarthy, eds. 2015. *The Routledge handbook of political ecology.* London and New York: Routledge.

Pietz, D. 2015. *The Yellow River: The problem of water in modern China.* Cambridge, MA: Harvard University Press.

Piper, L. 2010. *The industrial transformation of subarctic Canada.* Vancouver, BC, Canada: UBC Press.

Priestland, D. 2009. *The red flag: A history of communism.* New York: Grove.

Pryde, P. 1991. *Environmental management in the Soviet Union.* Cambridge, MA: Cambridge University Press.

Remnick, D. 2017. Donald Trump and the enemies of the American people. *The New Yorker,* February 17. Accessed October 4, 2018. https://www.newyorker.com/news/news-desk/donald-trump-and-the-enemies-of-the-american-people.

Robbins, P. 2011. *Political ecology: A critical introduction.* 2nd ed. Malden, MA: Wiley-Blackwell.

Robbins, P., J. Hintz, and S. A. Moore. 2014. *Environment and society.* 2nd ed. Malden, MA: Wiley-Blackwell.

Ruckert, A., L. Macdonald, and K. R. Proulx. 2017. Post-neoliberalism in Latin America: A conceptual review. *Third World Quarterly* 38 (7):1583–1602.

Scheidel, W. 2017. *The great leveler: Violence and the history of inequality from the stone age to the twenty-first century.* Princeton, NJ: Princeton University Press.

Shapiro, J. 2001. *Mao's war against nature: Politics and the environment in revolutionary China.* New York: Cambridge University Press.

———. 2016. Environmental degradation in China under Mao and today: A comparative reflection. *Global Environment* 9 (2):440–57.

Sneddon, C. 2015. *Concrete revolution: Large dams, Cold War geopolitics, and the U.S. Bureau of Reclamation.* Chicago: University of Chicago Press.

Snyder, T. 2010. *Bloodlands: Europe between Hitler and Stalin.* New York: Basic Books.

———. 2017. *On tyranny: Twenty lessons from the twentieth century.* New York: Tim Duggan Books.

———. 2018. *The road to unfreedom: Russia, Europe, and America.* New York: Tim Duggan Books.

Uekötter, F. 2006. *The green and the brown: A history of conservation in Nazi Germany.* New York: Cambridge University Press.

———. 2014. *The greenest nation? A new history of German environmentalism.* Cambridge, MA: The MIT Press.

Walder, A. 2015. *China under Mao: A revolution derailed.* Cambridge, MA: Harvard University Press.

Weiner, D. R. 1988. *Models of nature: Ecology, conservation, cultural revolution in Soviet Russia.* Bloomington: Indiana University Press.

———. 1999. *A little corner of freedom: Russian nature protection from Stalin to Gorbachёv.* Berkeley: University of California Press.

Wemheuer, F. 2014a. Collectivization and famine. In *The Oxford handbook of the history of communism,* ed. S. A. Smith, 407–23. Oxford, UK: Oxford University Press.

———. 2014b. *Famine politics in Maoist China and the Soviet Union.* New Haven, CT: Yale University Press.

White, R. 1995. *The organic machine: The remaking of the Columbia River.* New York: Hill and Wang.

Wynn, G. 2007. *Canada and Arctic North America: An environmental history.* Santa Barbara, CA: ABC-Clio.

Wynn, G., C. Colten, R. M. Wilson, M. V. Melosi, M. Fiege, and D. K. Davis. 2014. Reflections on the American environment. *Journal of Historical Geography* 43:152–68.

Xun, Z. 2013. *Forgotten voices of Mao's great famine, 1958–1962: An oral history.* New Haven, CT: Yale University Press.

ROBERT WILSON is an Associate Professor in the Department of Geography in the Maxwell School of Citizenship and Public Affairs at Syracuse University, Syracuse, NY 13244-1020. E-mail: rmwilson@maxwell.syr.edu. His research interests include environmental history, historical geography, and histories of environmental management and governance.

Deadly Environmental Governance: Authoritarianism, Eco-populism, and the Repression of Environmental and Land Defenders

Nick Middeldorp ⓘD and Philippe Le Billon

Environmental and resource governance models emphasize the importance of local community and civil society participation to achieve social equity and environmental sustainability goals. Yet authoritarian political formations often undermine such participation through violent repression of dissent. This article seeks to advance understandings of violence against environmental and community activists challenging authoritarian forms of environmental and resource governance through eco-populist struggles. Authoritarianism and populism entertain complex relationships, including authoritarian practices toward and within eco-populist movements. Examining a major agrarian conflict and the killing of a prominent Indigenous leader in Honduras, we point to the frequent occurrence of deadly repression within societies experiencing high levels of inequalities, historical marginalization of Indigenous and peasant communities, a liberalization of foreign and private investments in land-based sectors, and recent reversals in partial democratization processes taking place within a broader context of high homicidal violence and impunity rates. We conclude with a discussion of the implications of deadly repression on environmental and land defenders.

环境与资源治理模型, 强调在地社区与公民社会的参与以达到社会公平与环境可持续性目标的重要性。但威权政治的形成, 却经常通过对异议的暴力压迫, 破坏此般参与。本文企图推进我们对于对抗挑战威权的环境与资源治理形式之环境与社区行动者的暴力之理解, 该暴力是通过生态民粹主义的斗争。威权主义与民粹主义存在着复杂的关系, 包含迈向生态民粹主义运动、并存在于该运动中的威权实践。我们检视洪都拉斯国内一起重大的农业冲突与一位重要的原住民领导者的谋杀事件, 指向在经历高度不平等的社会中经常发生的致命压迫、原住民与农民社区在历史中的边缘化、以土地为基础的部门对外国与私人投资的自由化, 以及晚近在高度杀人暴力与免责率的更广泛脉络下民主化进程的部分倒退。我们于结论中讨论对环境与土地保卫者的致命压迫之意涵。 关键词: 威权主义: 环境保卫者, 洪都拉斯, 民粹主义, 压迫。

Los modelos de gobernanza ambiental y de los recursos enfatizan la importancia de la comunidad local y la participación de la sociedad civil para alcanzar las metas de equidad social y sustentabilidad ambiental. No obstante, las formaciones políticas autoritarias a menudo socaban tal participación por medio de la represión violenta del disentimiento. Este artículo busca avanzar en el entendimiento de la violencia contra activistas ambientales y comunitarios que retan las formas autoritarias de la gobernanza ambiental y de los recursos por medio de luchas ecopopulistas. El autoritarismo y el populismo albergan relaciones complejas, incluso prácticas autoritarias, hacia y dentro de los movimientos eco-populistas. Con el examen de un conflicto agrario mayor y del asesinato de un prominente líder indígena en Honduras, señalamos la ocurrencia frecuente de represión letal en sociedades que experimentan altos niveles de desigualdad, marginación histórica de comunidades indígenas y campesinas, una liberalización de inversiones extranjeras y privadas en los sectores basados en la tierra y recientes reveses en los procesos de democratización parcial que ocurren dentro de un contexto más amplio de altas tasas de violencia homicida e impunidad. Concluimos con una discusión de las implicaciones de la represión letal contra los defensores del medio ambiente y de la tierra. *Palabras clave: autoritarismo, defensores ambientales, Honduras, populismo, represión.*

Environmental and resource governance models emphasize public participation to achieve social equity and environmental sustainability goals (Ribot 2002; Bryan 2011). Yet, authoritarian forms of environmental and resource governance frequently undermine such principles through inconsequential

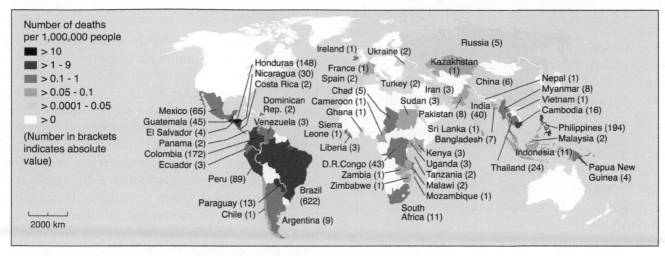

Figure 1. Reported killings of land and environmental defenders worldwide, 2002–2017. *Source:* Global Witness (2017a) data set.

consultation processes, criminalization of dissent, and violent repression (Perreault 2015). In turn, populist forms of emancipatory politics potentially lead to further escalation as they seek to broaden social mobilization beyond directly affected communities to challenge privileged elites and oppressive institutions.[1] At least 1,570 people were killed globally between 2002 and 2017 while seeking to protect their land, community, and the environment through socioenvironmental movements (see Figure 1). Many of them are Indigenous people, thereby pointing at the colonial dimensions of many resource development projects. Beyond these reported cases, many other people have likely lost their lives in more individual and anonymous struggles over lands, resources, and the environment, and countless individuals and communities have experienced harm as a result of the social and environmental impacts of resource projects (Temper et al. 2015; Le Billon and Sommerville 2017).

Studies of repression mostly come from political science, pointing at impunity factors, uncertainty about behavioral norms, and the rise of contentious politics (Earl 2003; Hill and Jones 2014); however, these rarely engage specifically with socioenvironmental conflicts. In contrast, such conflicts are the focus of political ecology studies interpreting repression as one of the violent expressions of uneven power relations, diverging value systems, and the dispossession of agrarian and Indigenous communities (Bury and Kolff 2002; Escobar 2006; Le Billon 2015; Martinez-Alier et al. 2016). Yet few political ecology studies have focused on killings as part of

authoritarian forms of resource governance, which also include the intimidation and criminalization of activists, the securitization of "resource development," and coercive forms of conservation (Peluso and Lund 2011; Roa-García 2017). Finally, studies from anthropology have shown how violence has a profound impact on the lives of targeted activists and community members, but less scholarly attention has been given to the interplay of repression and resistance shaping environmental governance (Rasch 2017).

In this article, we focus on repression in relation to populist forms of socioenvironmental movements and authoritarian forms of environmental and resource governance. Following this introduction, we discuss eco-populism and associated forms of repression. We then examine in more detail the repression of land and environmental defenders in Honduras, based on eight months of fieldwork between May 2013 and March 2018. We conclude with a brief discussion of the effects and implications of violence on community and civil society participation in environmental and resource governance and suggest an agenda for further research.

Eco-Populism, Authoritarianism, and Socioenvironmental Struggles

Eco-populism is defined as socioenvironmental movements scaling up their struggle and inscribing their demands into a "more universal rhetoric and strategy for change" (Griggs and Howarth 2008, 123;

see also Szasz 1994; Leonard 2011). Eco-populism thus broadens social mobilization beyond directly affected communities and often seeks to unite the people against ruling elites and dominant corporations. *Authoritarianism* refers to political formations demanding obedience, punishing dissent, and generally proving to be inflexible and oppressive (Levitsky and Way 2010). Authoritarian forms of environmental and resource governance by states and corporations seek to impose authority over territories, resources, and ecosystems at the expense of local communities' values, uses, and rights.[2]

Authoritarianism and populism, however, should not be considered binary and mutually exclusive categories. Both can take many forms, be asserted to different degrees, and be associated with either right-wing or left-wing values and political regimes (Borras 2018). Most political formations involve hierarchical relations instrumentalizing forms of discrimination, coercion, and restrictions on political freedoms (Levitsky and Way 2010). Most sociopolitical movements tend to homogenize and unify the voices of those they seek or pretend to represent (Laclau 2007). Interactions between populism and authoritarianism can take the form of populist authoritarianism involving the use of popular rhetoric and practices by authoritarian parties or authoritarian populism whereby populist parties drift toward authoritarian discourses and practices. In Bolivia, the populist discourse that had challenged neoliberal forces and brought Evo Morales's Indigenous–popular coalition to power was later used by the Morales government to legitimize the repression of Indigenous movements threatening extractivist accumulation (Andreucci 2018; Marston and Perreault 2017). Interactions can also involve populist responses to authoritarianism, as seen in the case of popular revolutions against dictatorships, as well as authoritarian responses to populism, with, for example, weak democratic regimes responding to populist challenges through increasingly authoritarian behavior. The application of these concepts thus needs to be highly contextualized to recognize some of the contradictions, antagonisms, overlaps, and synergies involved in their relationships (Borras 2018). Here, we briefly nuance the concept of eco-populism in relation to authoritarianism, especially in the context of repressed socioenvironmental struggles.

Eco-Populism

The concept of eco-populism can be interpreted as an emancipatory form of social mobilization seeking to broaden solidarities against a dominant elite-based system governing resources and the environment. Whereas environmental modernization approaches promoted by mainstream development agencies seek to achieve "sustainability" through fine-tuning the status quo, socioecological populism generally seeks to terminate environmentally destructive projects rather than derive benefits from them; to promote eco-centric or alternative local development models; to reaffirm environmental and local, rural, or Indigenous subjectivities; and to pursue a common front among social justice movements challenging systems of domination (Szasz 1994; Leonard 2011; Antal 2017; Condé and Le Billon 2017). If eco-populism is generally associated with left-wing environmental movements struggling against destructive resource use, some eco-populist movements also pursue conservative right-wing values or involve authoritarian practices, both within and outside the movements (McCarthy 2002; Scoones et al. 2018). Populist rural organizations include peasant movements seeking to (re)gain control of lands and community governance, union-based agrarian and miners' movements pursuing better working conditions and control over means of production, and socioenvironmental movements reclaiming notions of indigeneity and traditional livelihoods (Borras 2018). Populist eco-authoritarianism can also take the form of state-imposed obedience to strict environmental behavior and resource use, and follow populist strategies to scapegoat particular resource users.

Political ecology studies have documented both the causes and practices of resistance against land-based, large-scale projects, yet only quite rarely mobilized the concept of eco-populism (Szasz 1994; Luke 1995; Dietz 1999; McCarthy 2002; Robbins and Luginbuhl 2005). By early 2018, the Environmental Justice Atlas had documented about 2,400 cases of environmental justice movements throughout the world, half of them less than a decade old (see EJAtlas.org; Temper et al. 2015). Many of these movements are eco-populist insofar as they intertwine environmental concerns with wider human rights and (differentiated Indigenous) citizen rights and constitute broad networks calling into question dominant models of resource exploitation

(Martinez-Alier et al. 2016). As discussed later, such networks become all the more important in the context of a repressive apparatus denying basic individual rights. Eco-populism, in this regard, also constitutes a safety-in-numbers strategy reducing individual vulnerabilities and consolidating solidarities. Yet, although eco-populism can help broaden coalitions across political divides through local values and interest-based collective identities (Rice and Burke 2018), these same identities can also prove exclusionary and feed in turn authoritarian discourses and practices, including within the movements themselves (see Levitsky and Loxton 2013).

Eco-Populism, Authoritarianism, and Repression

Socioenvironmental conflicts arise in large part due to the inflexible and repressive character of political formations and their denial of effective community or civil society participation in environmental governance. In turn, such authoritarianism is frequently responded to by a reassertion of alternative modes of governance, such as Indigenous or customary laws and institutions. Because it challenges established state and corporate authority, eco-populism is frequently perceived as *insurgency*, a form of rebellion bordering on the belligerent. Eco-populism also frequently relies on public protests as a mode of political engagement, which can bring about violent interactions between parties. Furthermore, whereas many forms of eco-populism mobilize nonviolent forms of struggle, some eco-populist groups violently reject state authority, thereby further constituting a clash of authoritarian practices, as seen with cases of right-wing environmental populism in the western United States, such as the Sagebrush rebellion (McGregor Cawley 1996), the Wise Use Movement (McCarthy 2002), and, most recently, the occupation of Oregon's Malheur National Wildlife Refuge by armed militia members rejecting U.S. federal government control of western lands (Gallagher 2016). In Peru, self-defense peasant organizations (*rondas campesinas*) involved hierarchical, authoritarian, and at times violent internal practices (Gustafsson 2018).

To quell contestation and deter mobilization, governments and corporations use a range of counterinsurgency strategies reflecting their relative impunity and the sophistication of their coercive apparatus and the level of perceived threats to their interests.

Liberal formations generally respond to opposition through inclusion and buy-in strategies, often consisting of public participation processes channeling resistance toward what Blaser (2013, 21) called the "house of reasonable politics," within which only minor differences amenable to compromises are allowed. Outside of the house, authoritarian spaces of criminalization and forceful policing often reign, thereby exposing the authoritarian character of actually illiberal regimes. Authoritarianism is also demonstrated through exclusionary rules and biased judicial systems undermining environmental and community struggles, including burdensome registration processes for civil society organizations, restrictions on foreign nongovernmental organization (NGO) funding, or strict conditions for the expression of dissent (Deonandan and Dougherty 2016), as well as defamation, harassment, spying, infiltration, and disruption through biased investigations, criminal accusations, and long-term detention before trial (Smith 2008; Birss 2017; Vasconsela Rocha and Barbosa 2018).

Deadly forms of state repression are generally understood as the result of impunity for perpetrators associated with the lack of independent and effective judiciary and media reporting; tight and unaccountable networks between political, economic, and military elites; and social habituation to homicides on the part of authorities—including as a result of recent wars and state-tolerated or -encouraged vigilante activity (Cruz 2011; Hill and Jones 2014). Deadly escalation also often results from high uncertainty in the capacity and behavioral norms among protagonists in a context of contentious politics (Leitner et al. 2008), a situation characterizing intermediary political regimes falling between "full" autocracies and democracies (Davenport 2007; Pierskalla 2010). In such contexts, government authorities and corporations are frequently unwilling to follow the praxis of negotiated conflict settlement, and social movements refuse to back down on the premise that sustained contestation will further erode authoritarian power, even if at the cost of deadly repression. The likelihood of killings of environmental and land defenders thus seems higher among middle-income countries with semiauthoritarian regimes (see Figure 2A), a recent history of armed conflicts or high homicide rates (Figure 2B), and frequent conflicts around resource exploitation projects, as seen in

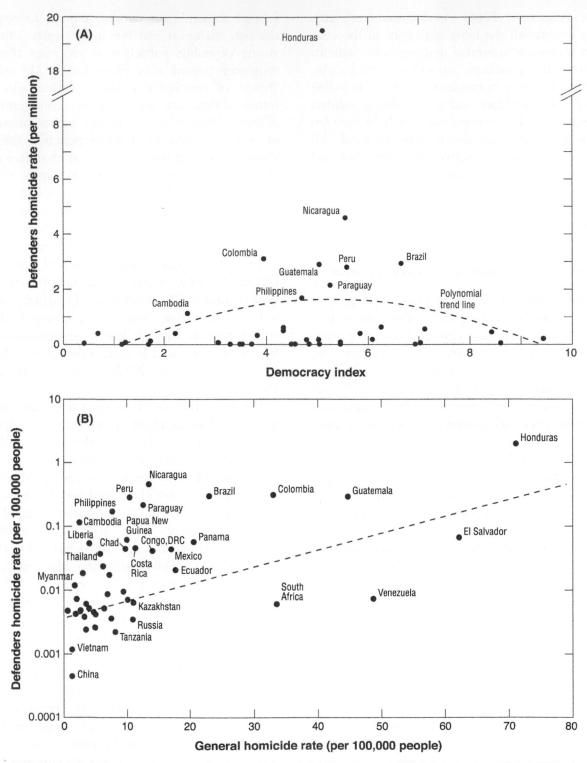

Figure 2. Defenders' homicide rates, democracy levels, and general homicide rates, 2002–2016. *Sources:* Global Witness, Combined Index of Democracy, United Nations Office on Drugs and Crime, and World Bank data sets.

Latin America (see Bebbington and Bury 2013; Himley 2013; Temper et al. 2015; Jeffords and Thompson 2016; Wayland and Kuniholm 2016; McNeish 2018).

Whereas both states and corporations have directly instrumentalized their own security organizations to exert deadly repression, notably in the context of public protests, more insidious forms of repression—including targeted killing—are generally conducted through intermediaries including former military officers, private security contractors, and criminal entrepreneurs. As discussed later, our interviews and observations suggest that activists often fear insidious forms of repression the most, because these do not easily lend themselves to collective forms of protection and reduce accountability likelihood. As expressed by a Honduran antimining activist in a personal interview:

> It is quite difficult for us, facilitators of a process of opposition to mining, to work in the region. We have to take care of our steps, and have a lot of caution, a lot of caution. Live anonymously, not make yourself public, because the threat will always exist. It will exist. Those people who are in the mining companies will contract those [people] who dedicate themselves to assassination, to threats, to force, or to persecute people, in order for those movements that rise up, to dissolve again.

Insidious forms of repression are thus experienced not only by activists and their families but also by broader members of affected communities as fear and suspicion come to permeate everyday life. A local Honduran farmer, pressured by miners to sell his land, and pressured by his community to retain it, expressed deep concerns: "They took a photo of my family from a black car, it was parked at my family's [home]. I feel threatened, I want to leave the country because I am afraid." Protection from the state is rarely viable, not only because state institutions tend to favor resource exploitation and might lack capacity to provide effective protection but also because they are often captured by corrupt ruling elites, corporations, and criminal organizations (Davenport 2007; Cruz 2011; Garay-Salamanca and Salcedo-Albarán 2012; Aguilar-Støen and Bull 2016; Middeldorp et al. 2016). Eco-populist struggles exposing these links and seeking to bring about a new political order constitute a "vital" threat to those interests, thus further motivating deadly repression.

Eco-populism can produce tangible effects on the state (Wapner 1995), with the scaling up of mobilization influencing policies and pushing the state to grant greater decision-making power to communities (Hanna et al. 2016). Eco-populism, and counterreactions to repression, can also affect the balance of power within a government, motivate some civil servants to resign or provide support, help bring a greater degree of accountability, and contribute to delegitimize

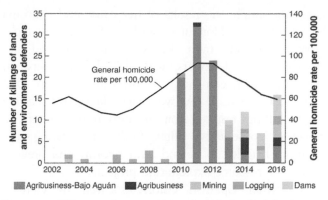

Figure 3. Killings of environmental and land defenders in Honduras, 2002–2017. *Source:* Global Witness (2017a) and United Nations Office on Drugs and Crime data sets, complemented by press and civil society reports.

the authoritarian state and bring about a change in political regime (Haggard and Kaufman 2016). Eco-populism can also exert pressure through increased costs and delayed investments into resource projects and reduce government revenues or motivate a nationalization of resource industries (Franks et al. 2014; Klimbovskaia and Diab 2015). More broadly, the rise of emancipatory mass values in competitive authoritarian regimes generally results in the defeat of the incumbent government and leads to massive postelectoral protests and deadly repression when incumbents stay in power (Zavadskaya and Welzel 2015), especially if electoral fraud is involved. Finally, eco-populist mobilization can antagonize not only the ruling elite but also part of the population (McCright et al. 2014), leading in turn to increased repression by populist countermovements, including progovernment militias, with a further risk of repression.

Eco-Populism and Repression in Honduras

In March 2016, thousands of people took to the streets of Honduras's capital city to protest the killing of Berta Cáceres, a prominent Indigenous social movement leader opposing a hydroelectric dam project (IACHR 2016; GAIPE 2017). "Berta did not die, she multiplied" was a common outcry in Honduras in response to a murder that shook the country and the world, and it became a rallying call for the political opposition during the 2017 electoral campaign, demonstrating the populist appeal of Berta's struggle. Widespread outrage and calls for change in Honduras were not simply the results of her cause and charisma but also because Honduras had been for the past eight years the world's most deadly country for land and environmental defenders, with 128 killings between 2010 and 2017, compared to only nine between 2002 and 2009 (Figure 3). As we suggest later, this massive rise in killings resulted from an authoritarian return to power of conservative elites, further motivating eco-populist struggles to defend land reforms and prevent large-scale resource exploitation projects.

A former Spanish colony located in Central America, Honduras formally transitioned from military dictatorship to civilian rule in the early 1980s, yet remained under the influence of U.S.-backed and -trained Honduran military officers and served as a staging ground for U.S. counterinsurgency wars in the region. Domestically, suspected communists were imprisoned or killed under the National Security Doctrine and Antiterrorist Law (Barahona 2005). The 1980s and 1990s nonetheless showed some signs of democratization and gave rise to Indigenous or agrarian movements (Metz 2010). As U.S. assistance propped up the Honduran government during the 1980s to mid-1990s, the country became affected by youth gangs (*maras*) and the increased presence and influence of drug cartels using Honduras for drugs trans-shipment between Colombia and the United States. Rising homicide rates were confronted with repressive zero-tolerance policies that led to massive incarceration and a militarization of policing (Rivera 2013), as well as a narrowing space for civil and human rights organizations following the 2009 coup d'état against the populist and increasingly left-wing President Manuel Zelaya (Van der Borgh 2016). Having recently joined the Bolivarian Alliance for the Peoples of our America (ALBA), in a clear departure from traditional U.S. alignment, Zelaya was ousted by the military as he planned to organize a referendum on constitutional change that could have allowed him to seek reelection, something prohibited by the Honduran constitution.

The 2009 coup d'état reversed democratic progress and turned Honduras into a deadly country for the opposition and critics, with fifty-eight political killings and the murder of fourteen journalists in 2009 alone (Yoder et al. 2013). Impunity increased and homicide rates climbed from 44.5 per 100,000 in 2005 to 93.2 per 100,000 in 2011 (United Nations Office on Drugs and Crime 2017). The coup also abruptly ended progress made on socioenvironmental justice issues under the Zelaya administration: Agrarian

reforms in favor of peasant cooperatives were annulled in favor of the landed elite and their expanding palm oil plantations, a standing moratorium on new mining projects was canceled, the energy market was privatized, and many new mining and hydroelectric licenses were awarded. The conservative National Party that had returned to power channeled mining revenues toward the repressive apparatus of the state, including a new military police force in 2013 (Middeldorp, Morales, and Van Der Haar 2016). This repressive turn led socioenvironmental movements to broaden their mobilization, joining progressive social movements to denounce the "sale of Honduras."

These conflicts led to two major waves of killings of environmental and land defenders between 2010 and 2017 (Figure 3). The first one, culminating in 2011 and accounting for 68 percent of total killings, involved struggles against agribusinesses, mostly peasants struggling for agrarian reform and the recognition of cooperative farmland in a region known as the Lower Aguán (Global Witness 2017b) and with strong connections with eco-populist resistance against large landowners backing the 2009 coup. The second one, culminating in 2016 and accounting for 32 percent of killings, mostly involved community and environmental activists resisting logging, mining, and dam projects, among them Berta Cáceres. In contrast to the first wave of killings, this one mostly involved dams and mining projects with higher international exposure and resulted in transscalar mobilization. Whereas the first wave followed the general trend of homicide rates, with a common peak in 2010 to 2012, the second wave was marked by a later rise and a peak in 2014 to 2016, confirming a separate dynamic than those of general homicides and a close link to targeted killings following growing resistance in the face of advancing land-based resource projects.

The Bajo Aguán Palm Oil Conflict

Honduras is among the world's most unequal countries in terms of land distribution, with 1 percent of landowners holding a quarter of agricultural land, a similar proportion to the bottom 70 percent of landowners (United States Agency for International Development 2011). The current conflict in the Bajo Aguán—a river valley on the Atlantic coast—opposes peasant cooperatives and large landowners over the ownership and control of land and palm oil. The conflict finds its roots in banana-related land acquisitions by the United Fruit and Standard Fruit companies in the first half of the twentieth century, which were then followed by a series of land abandonment actions by companies, peasant land invasions, progressive land reforms in the 1960s, and a peasant resettlement and rural development program. This program was sponsored by the Organization of American States (OAS)—a regional cooperation body initially founded to combat communism in the Western Hemisphere—to create palm oil cooperatives in the 1970s (Edelman and León 2013). Economic liberalization, structural adjustment, and agrarian counterreforms throughout the late 1980s and 1990s impoverished peasant enterprises and enabled some of Honduras's wealthiest families to take control of large land holdings, notably Miguel Facussé and his corporation Dinant (León 2015). Initially mobilized by the Jesuits, the now landless peasants of the Aguán—most of them organized in the Unified Peasant's Movement of the Aguán (MUCA)—tried to reverse the concentration of land by demanding land redistribution based on the 1962 Law of Agrarian Reform, which saw limited implementation until President Zelaya issued an executive decree in 2008 that clearly threatened the interests of large landowners.

The 2009 coup backed by the landed elite intensified and politicized the peasant movement and its claims, with peasants actively participating in demonstrations against the coup in the capital city. Lethal repression against demonstrations in Tegucigalpa led peasants to argue that "if we are going to die anyway, we might as well do it in the lands" (León 2015, 308), and on 9 December 2009, 600 families occupied two dozen plantations that peasants considered to be rightfully theirs. The movement to retake the land became popular, quickly drawing in 2,400 additional families in the Bajo Aguán alone. The Honduran state and landed elite responded through a joint military–police operation and the increased use of private guards and paramilitaries. Complicated by drug trafficking involvement and personal revenges, the conflict resulted in the killings of at least eighty-nine people between 2010 and 2013, with the vast majority being cooperative peasants and their relatives, including two of MUCA's presidents, a journalist, a human rights lawyer, and a judge but also four security guards and an employee from Dinant, a World Bank–financed palm oil company reportedly associated with some of the killings (Human Rights Watch 2014; Global Witness 2017b). With growing international attention to the conflict, the government awarded a small amount of land to most MUCA members, but several peasant communities split and successfully pursued a legal path instead, with the Supreme Court recognizing the illegal character of land purchases in the 1990s and giving back to peasants control over several plantations in June 2012. Yet within a year their lawyer was killed, the ruling overturned, and cooperatives once again evicted, demonstrating "the ease with which powerful interests in Honduras can bend the law" (León 2015, 310).

By 2013, killing rates had declined, seemingly the result of the "suffocation" of the peasant movement by repression, the end of the main military operation once President Hernandez assumed office, and the passing of many plantations into the hands of smallholders as palm oil corporations retained control of the sector and most of its revenues through their oligopoly over processing plants and palm oil exports. Yet, with fourteen evictions and four killings in 2017, violence continues and the peasant associations still struggle to reclaim their land, end impunity for the killings, and help bring a change in government through the broader struggle of landless peasants.

The Agua Zarca Dam Conflict

The Agua Zarca hydroelectric dam is a project of Desarrollo Energéticos S.A (DESA), a private Honduran company headed by David Castillo, a former officer of the Honduran military's intelligence service, and backed by a powerful Honduran family (Willems and de Jonghe 2016; Lakhani 2017). Construction of the 21.3 MW dam was contracted to Chinese dam-building company Sinohydro and started in 2012 on the Gualcarque River in the department of Intibucá, home to Indigenous Lenca communities. DESA attempted to socialize the project through community meetings, but no legally required consultation process was undertaken. Supported through investments from the Central American Bank for Economic Integration (BCIE), and the development banks FMO (The Netherlands) and Finnfund (Finland), DESA delegated construction to the Guatemalan company COPRECA after the exit of Sinohydro in 2013 following a series of blockades by local communities and grave human rights abuses by Honduran military operating on behalf of the project, including the murder of a local leader.

The Lenca have no recognized autonomous governance structures beyond the village level, but COPINH, cofounded in 1993 by Berta Cáceres, spans different Lenca communities through its *consejo de ancianos* (council of elders) and a network of community radio stations. Although formally constituted as an NGO, COPINH could more accurately be typified as a social movement organization attempting to revitalize Indigenous (Lenca) identity (Metz 2010), reclaim and obtain titles for ancestral lands, and create an autonomous governance structure. It has strongly criticized the government since the 2009 coup d'état, has a fierce antimining and antidam agenda, and has resisted the construction of the Agua Zarca dam from the outset, despite facing harassment, defamation, "preventive" imprisonment of prominent members, death threats, and a history of killings. These notably included local radio stations airing false allegations of involvement by Cáceres in the violent murder of a minor and the killing of a Rio Blanco community leader—Tomas Garcia—by a soldier during a peaceful protest in 2013. In an interview with *Al Jazeera* (Lakhani 2013), Cáceres said:

> The army has an assassination list of eighteen wanted human rights fighters with my name at the top. I want to live, there are many things I still want to do in this world but I have never once considered giving up fighting for our territory, for a life with dignity, because our fight is legitimate. I take lots of care but in the end, in this country where there is total impunity I am vulnerable … when they want to kill me, they will do it.

Investors played down these issues for years: In response to NGO reports on human rights violations and the criminalization of COPINH, and despite the Inter-American Commission on Human Rights's decision to order the Honduran state to support Berta Cáceres with protection measures, the Dutch development bank FMO wrote in 2015 that "for such serious statements to be made, please do provide evidence from an independent and recognized party such as the Judicial System of Honduras for FMO to consider" (Willems and de Jonghe 2016, 9).

Through her resistance, Berta Cáceres became known as one of Honduras's fiercest government critics and Indigenous rights defenders, which led her to win the Goldman Environmental Prize in 2015. Despite her renown, however, Cáceres was assassinated on 2 March 2016 by gunmen entering her home at night. This murder, and that of her colleague Nelson Garcia two weeks later, put the Agua Zarca dam struggle in the international spotlight and led to widespread outrage reminiscent of the killings of Brazilian rubber-tapper Chico Mendes and Nigerian activist Ken Saro-Wiwa. The investors, DESA, and the Honduran government publicly condemned and denied responsibility for the murder, framing it first as a violent robbery, then as a crime of passion, and later as the tragic result of intracommunity tensions. Local journalists were being pressured to stick to these versions: One commented in an interview that "the Ministry of Security was calling me, asking how the story on the crime of passion was advancing." The investors were also quick to legitimize and defend the project, notably by using social media to broadcast videos framing the dam as sustainable, small-scale, and beneficial for local communities; and claiming to count on free, prior, and informed consent by communities (FMO 2016). Yet the murders rallied further support for COPINH: An alliance of international NGOs brought COPINH members, including Cáceres's daughters, to Europe. Speeches, interviews, meetings with parliamentarians, and media-covered protests outside the banks' headquarters were organized, newspaper advertisements demanded the banks' withdrawal, and political lobbying in The Netherlands and Finland started to yield results. Key investors, including FMO, Finnfund, and then BCIE finally divested from the Agua Zarca project in the summer of 2017, paralyzing DESA's project at the time of writing (Lakhani 2017).

Nine men have since been arrested, including one active and two retired military officers with ties to DESA (Malkin 2017). On 2 March 2018, DESA's executive president, David Castillo, was arrested as he was trying to board a plane to the United States. Informants, including a prosecutor, affirmed that the latest arrest would likely not have happened without the investigative work of GAIPE, a civil society–led committee that investigated the murder, and pressure from MACCIH, an OAS-led investigative body seeking to combat corruption in Honduras. Castillo's arrest could also reflect an attempt by the Honduran government to signal some accountability to restart foreign investment into hydroelectric projects; with a lawyer for the "clean energy sector" arguing that "the biggest problem with Berta's death, is that the projects have now halted," and that Castillo was a "sacrificial lamb."

The long-term impacts of Cáceres's murder, beyond the immediate scope of COPINH and the project in question, are twofold. First, as local activists commented, "If even the county's most renowned activist can be killed, nobody is safe [anymore]." Given Cáceres's high profile, no other murder could have been more intimidating. From a calculative perspective of the repressive state, then, the assassination could have been seen by the national elite owning such projects as a worthwhile trade-off: The viability of one project is sacrificed, but hundreds of other potentially conflictive projects can be pursued with less resistance. Second, it appears that the killing of Berta Cáceres was counterproductive to the goal intended by its instigators: Instead of overcoming local protestors, the opposition gained sufficient impetus to force international "divestment" from the project. The murder also consolidated environmental subjectivities among rural communities that are deeply distrustful of mine and dam projects, often describing them as "projects of death." The murder also served as a rallying event for a broad political opposition alliance; increased international pressure to adopt a law on free, prior, and informed consultation; and left many dam projects pending without secure financing. The number of defenders opposing large-scale projects who were killed during 2017 also declined, in contrast to those in the Bajo Aguan conflict, although this might not last given the renewed political polarization that followed the elections that year.

Postelection Populism and Insidious Repression

The contested electoral process in November and December 2017, which left dozens dead in protests against alleged electoral fraud by the ruling National Party (Seligson and Pérez 2017), maintained the conservative party in power and further exposed the authoritarian character of the Honduran regime. In the words of a local NGO worker, "The people became even more outraged because of the electoral fraud, and now, the only way for JOH [Juan Orlando Hernandez] to maintain power, is to repress." Indeed, criminalization—facilitated by recent penal reforms—has targeted people with a "double profile" as political opponent and environmental activist,

often individuals from Indigenous and peasant communities who have a history of defending their territory against state-promoted resource projects. Targeted individuals are facing charges of land or water usurpation, attempted murder, terrorism, and illicit association (in the past only used to combat the youth gangs). OFRANEH, a *Garífuna* (black Indigenous) social movement organization, alone faces close to 500 trials as ancestral *Garífuna* lands are being awarded to hydroelectric, palm oil, and tourist resort projects, despite the Interamerican Court of Human Rights sentencing the State of Honduras in 2015 for not protecting these lands (Corte Interamericana de Derechos Humanos 2015). Evictions for development projects in free trade zones are underway, military incursions are reported in Indigenous territories of economic interest to intimidate local populations, some populations have been forcedly displaced by resource-related violence, and a sixteen-year-old environmental defender was tortured and brutally killed in February 2018.

Conclusion

This article examined three main arguments relating to the deadly interplay of authoritarianism and eco-populism within socioenvironmental conflicts. First, that deadly repression within socioenvironmental conflicts seems more likely when authoritarian formations are challenged by movements with an eco-populist agenda and popular appeal, as they come to not only constitute a threat to individual resource projects but also embody defiance of the established political order. As ruling elites and associated interests seek to reaffirm their sovereign power through killings, violent repression can also consolidate eco-populism by amplifying mobilization through outcry and solidarity, resulting in a deepening of repression and resistance cycles and international calls to end the impunity of the killers and complicit investors. Second, semiauthoritarian formations are more likely to use insidious forms of repression, including targeted killings, in contrast with "full" authoritarian regimes relying more on open forms of repression and with "mature" democracies using subtler coercive means such as circumscribed public participation and antidefamation laws.[3] Such semiauthoritarian formations are often dominated by coalitions of interests associating senior government officials, corporate managers, and violent entrepreneurs, including former and active members of the security apparatus. Third, such a deadly form of environmental and resource governance is particularly prominent in societies experiencing high levels of inequalities, historical marginalization of Indigenous and agrarian communities, a liberalization of foreign and private investment into land-based sectors, and recent reversals in partial democratization processes taking place within a broader context of high homicidal violence and impunity rates.

As suggested by the two conflicts examined here, there is some variation in processes and outcomes between eco-populist movements and authoritarian formations. The assassination of Berta Cáceres demonstrated that the killing of prominent activists can backfire, by reinforcing eco-populist resistance instead of breaking it and generating an international outcry leading to the cancellation of this specific project, as well as affecting sector-wide investments. The assassination resulted in major funding and legitimacy crises, leading to a (short-term) reduction in killings but an increase in criminalization. Globally, the assassination brought greater media attention on environmental defenders killed and contributed to a global initiative for the protection of environmental defenders (United Nations Environmental Program 2018). In contrast, peasant communities engaged in palm oil cultivation and possibly embroiled in the violence of narcotics trafficking received less international media coverage and solidarity networks than communities fighting the Aqua Zarca dam. They also lacked a charismatic and internationally renowned figurehead and could not easily be identified as eco-populist: They are looking for lands to plant palm oil, a struggle unlikely to find support from international environmental organizations. Although the struggle of peasant communities resonated with the populist appeal of land reforms, it has mostly remained a largely domestic or even local matter and sees continued violence including killings. Yet, the impunity of some of the international actors supporting major palm oil corporations is being challenged (Chavkin 2017), with seventeen Honduran farmers filing a case in a U.S. court in March 2018 against the World Bank's International Financial Corporation for providing them with "critical capital and moral cover."

The killing of environmental and land defenders in Honduras was largely the outcome of a resource exploitation agenda supported by powerful domestic and foreign actors, including prominent Honduran families, Western development banks like FMO, and U.S. strategic interests, pitted against a left-wing coalition rooted in social movements and populist leaders. This interpretation, however, should not be overgeneralized, as eco-populist movements have challenged left-wing governments that turned more authoritarian, such as the continuous mobilizations of Nicaraguan peasants against a proposed interoceanic canal (Huete-Perez et al. 2015) and Indigenous opposition to a highway crossing the TIPNIS reserve in Bolivia (Canessa 2012). As such, we join calls for more nuanced interpretations of interactions between populism and authoritarianism, including across the political spectrum (Borras 2018; Scoones et al. 2018).

Much remains to be done, on different levels, to avoid a deepening of deadly environmental and resource governance, including meaningful consultation processes for affected communities and greater efforts to address corruption, root out shadow governance structures violently protecting private interests, and end the impunity of the perpetrators and beneficiaries of violence, including international corporations and financial institutions. In this respect, further research is needed to shed more light on the effects of assassination of land and environmental defenders (for both prominent activists and low-profile community members), the different forms and scales of repression, and the processes through which killings can have rallying effects and bring about deeper forms of accountability and an end to cycles of violence. More research is also needed to better assess and devise protection mechanisms for environmental and land defenders at multiple scales and evaluate whether more effective protection reduces cycles of violence and the risk of eco-populism becoming more authoritarian.

ORCID

Nick Middeldorp ⓘ http://orcid.org/0000-0002-0204-108X

Notes

1. As Borras (2018) suggested, *populism* can be defined as "the deliberate political act of aggregating disparate and even competing and contradictory class and group interests and demands into a relatively homogenized voice, i.e., 'us, the people', against an

'adversarial them' for tactical or strategic political purposes" (3; see also Laclau 2007).

2. On the concept of environmental authoritarianism, see Beeson (2010); on authoritarian resource governance and its relational dimensions between states, companies, and local communities in the case of Laos, see Kenney-Lazar (2018).

3. These insidious forms of repression are complemented by various forms of divide-and-rule and cooptation strategies not discussed in this article (see Schilling-Vacaflor and Eichler 2017; Brock and Dunlap 2018).

References

Aguilar-Støen, M., and B. Bull. 2016. Protestas contra la minería en Guatemala: ¿Qué papel juegan las élites en los conflictos? [Protests against mining in Guatemala: What role do elites play in conflicts?]. *Anuario de Estudios Centroamericanos* 42 (1):15–44.

Andreucci, D. 2018. Populism, hegemony, and the politics of natural resource extraction in Evo Morales's Bolivia. *Antipode* 50 (4):825–45. doi:10.1111/anti.12373.

Antal, A. 2017. Ecopopulism and environmental justice in eastern and south Europe. Paper presented at the Environmental Justice in the Anthropocene Symposium, Colorado State University, Fort Collins, April 24–25.

Barahona, M. 2005. *Honduras en el siglo XX: Una síntesis histórica* [Honduras in the 20th century: A historical synthesis]. Tegucigalpa, Honduras: Editorial Guaymuras.

Bebbington, A., and J. Bury. 2013. *Subterranean struggles: New dynamics of mining, oil, and gas in Latin America.* Austin: University of Texas Press.

Beeson, M. 2010. The coming of environmental authoritarianism. *Environmental Politics* 19 (2):276–94.

Birss, M. 2017. Criminalizing environmental activism. *NACLA Report on the Americas* 49 (3):315–22.

Blaser, M. 2013. Notes towards a political ontology of "environmental" conflicts. In *Contest ecologies. Dialogues in the South on nature and knowledge,* ed. L. Green, 13–27. Cape Town, South Africa: HSRC Press.

Borras, S. M. 2018. Understanding and subverting contemporary right-wing populism: Preliminary notes from a critical agrarian perspective. Paper presented at the Emancipatory Rural Politics Initiative International Conference Authoritarian Populism and the Rural World, The Hague, The Netherlands.

Brock, A., and A. Dunlap. 2018. Normalising corporate counterinsurgency: Engineering consent, managing resistance and greening destruction around the Hambach coal mine and beyond. *Political Geography* 62:33–47.

Bryan, J. 2011. Walking the line: Participatory mapping, indigenous rights, and neoliberalism. *Geoforum* 42 (1):40–50.

Bury, J., and A. Kolff. 2002. Livelihoods, mining and peasant protests in the Peruvian Andes. *Journal of Latin American Geography* 1 (1):1–16.

Canessa, A. 2012. *Conflict, claim and contradiction in the new indigenous state of Bolivia.* Research Network on Interdependent Inequalities in Latin America. Berlin: desiguALdades.net.

Chavkin, S. 2017. Lawsuit: World Bank arm aided firm that hired "death squads." Accessed July 29, 2018. https://www.icij.org/blog/2017/03/lawsuit-world-bank-arm-aided-firm-hired-death-squads/.

Combined Index of Democracy. 2016. Accessed July 29, 2018. https://www.politikwissenschaft.uni-wuerzburg.de/en/lehrbereiche/vergleichende/research/combined-index-of-democracy-cid/.

Condé, M., and P. Le Billon. 2017. Why do some communities resist mining projects while others do not?. *The Extractive Industries and Society* 4 (3):681–97.

Corte Interamericana de Derechos Humanos. 2015. Caso comunidad garífuna de Punta Piedra y sus miembros vs. Honduras [Case of the Garífuna community of Punta Piedra and its members vs. Honduras]. Accessed March 27 2018. http://www.corteidh.or.cr/docs/casos/articulos/seriec_304_esp.pdf

Cruz, J. M. 2011. Criminal violence and democratization in Central America: The survival of the violent state. *Latin American Politics and Society* 53 (4):1–33.

Davenport, C. 2007. State repression and political order. *Annual Review of Political Science* 10 (1):1–23.

Deonandan, K., and M. L. Dougherty. 2016. *Mining in Latin America: Critical approaches to the new extraction.* London and New York: Routledge.

Dietz, T. 1999. Political environmental geography of the tropics. *Development* 42 (2):13–19.

Earl, J. 2003. Tanks, tear gas, and taxes: Toward a theory of movement repression. *Sociological Theory* 21 (1):44–68.

Edelman, M., and A. León. 2013. Cycles of land grabbing in Central America: An argument for history and a case study in the Bajo Aguán, Honduras. *Third World Quarterly* 34 (9):1697–1722.

Escobar, A. 2006. Difference and conflict in the struggle over natural resources: A political ecology framework. *Development* 49 (3):6–13.

FMO. 2016. Agua Zarca Project. Accessed October 27, 2017. https://www.youtube.com/watch?v=hdTRxwk3Vg8.

Franks, D. M., R. Davis, A. J. Bebbington, S. H. Ali, D. Kemp, and M. Scurrah. 2014. Conflict translates environmental and social risk into business costs. *Proceedings of the National Academy of Sciences* 111 (21):7576–81.

GAIPE. 2017. *Represa de violencia. El plan que asesinó a Berta Cáceres* [Repression of violence: The plan that murdered Berta Cáceres]. Accessed March 31, 2018. http://censat.org/es/publicaciones/represa-de-violencia-el-plan-que-asesino-a-berta-caceres.

Gallagher, C. 2016. Placing the militia occupation of the Malheur National Wildlife Refuge in Harney County, Oregon. *ACME: An International Journal for Critical Geographies* 15 (2):293–308.

Garay-Salamanca, L. J., and E. S. Salcedo-Albarán. 2012. Institutional impact of criminal networks in Colombia and Mexico. *Crime, Law and Social Change* 57 (2):177–94.

Global Witness. 2017a. *Defenders of the earth: Global killings of land and environmental defenders in 2016.* London: Global Witness.

———— 2017b. *Honduras: The deadliest place to defend the planet.* London: Global Witness.

———— 2018. Dataset available upon request from Global Witness and communicated to authors. Accessed July 29, 2018. https://www.globalwitness.org/en/.

Griggs, S., and D. Howarth. 2008. Populism, localism and environmental politics: The logic and rhetoric of the Stop Stansted Expansion campaign. *Planning Theory* 7 (2):123–44.

Gustafsson, M. T. 2018. *Private politics and peasant mobilization.* Berlin: Springer.

Haggard, S., and R. R. Kaufman. 2016. *Dictators and democrats: Masses, elites, and regime change.* Princeton, NJ: Princeton University Press.

Hanna, P., E. J. Langdon, and F. Vanclay. 2016. Indigenous rights, performativity and protest. *Land Use Policy* 50:490–506.

Hill, D. W., and Z. M. Jones. 2014. An empirical evaluation of explanations for state repression. *American Political Science Review* 108 (3):661–87.

Himley, M. 2013. Regularizing extraction in Andean Peru: Mining and social mobilization in an age of corporate social responsibility. *Antipode* 45 (2):394–416.

Huete-Perez, J. A., A. Meyer, and P. J. Alvarez. 2015. Rethink the Nicaragua canal. *Science (New York)* 347 (6220):355.

Human Rights Watch. 2014. *There are no investigations here: Impunity for killings and other abuses in Bajo Aguán, Honduras.* New York: Human Rights Watch.

IACHR. 2016. IACHR deplores killing of Nelson Noé García in Honduras. Accessed March 31, 2018. http://www.oas.org/en/iachr/media_center/preleases/2016/039.asp.

Jeffords, C., and A. Thompson. 2016. An empirical analysis of fatal crimes against environmental and land activists. *Economics Bulletin* 36 (2):827–42.

Kenney-Lazar, M. 2018. Governing dispossession: Relational land grabbing in Laos. *Annals of the American Association of Geographers* 108 (3):679–94.

Klimbovskaia, A., and J. Diab. 2015. Populist movements: A driving force behind recent renationalization trends. Policy Brief No. 9, CIGI Graduate Fellows, Waterloo, ON, Canada.

Laclau, E. 2007. *On populist reason.* London: Verso.

Lakhani, N. 2013. Honduras dam project shadowed by violence. *Al Jazeera,* December 24.

————. 2017. Backers of Honduran dam opposed by murdered activist withdraw funding. *The Guardian,* June 4. Accessed October 26, 2017. https://www.theguardian.com/world/2017/jun/04/honduras-dam-activist-berta-caceres.

Le Billon, P. 2015. Environmental conflict. In *The Routledge handbook of political ecology,* ed. T. Perreault, G. Bridge, and J. McCarthy, 598–608. London and New York: Routledge.

Le Billon, P., and M. Sommerville. 2017. Landing capital and assembling "investable land" in the extractive and agricultural sectors. *Geoforum* 82:212–24.

Leitner, H., E. Sheppard, and K. M. Sziarto. 2008. The spatialities of contentious politics. *Transactions of the Institute of British Geographers* 33 (2):157–72.

León, A. 2015. Rebellion under the palm trees: Memory, agrarian reform and labor in the Agúan, Honduras. PhD diss., Graduate Center, City University of New York.

Leonard, L. 2011. Ecopopulism. In *Green culture: An A-to-Z guide,* ed. K. Wehr, 121–22. Thousand Oaks, CA: Sage.

Levitsky, S., and J. Loxton. 2013. Populism and competitive authoritarianism in the Andes. *Democratization* 20 (1):107–36.

Levitsky, S., and L. A. Way. 2010. *Competitive authoritarianism: Hybrid regimes after the cold war.* Cambridge, UK: Cambridge University Press.

Luke, T. 1995. Searching for alternatives: Postmodern populism and ecology. *Telos* 1995 (103):87–110.

Malkin, E. 2017. Who ordered killing of Honduran activist? Evidence of broad plot is found. *New York Times,* October 28. https://www.nytimes.com/2017/10/28/world/americas/honduras-berta-caceres-desa.html

Marston, A., and T. Perreault. 2017. Consent, coercion and cooperativismo: Mining cooperatives and resource regimes in Bolivia. *Environment and Planning A: Economy and Space* 49 (2):252–72.

Martinez-Alier, J., L. Temper, D. Del Bene, and A. Scheidel. 2016. Is there a global environmental justice movement? *The Journal of Peasant Studies* 43 (3):731–55.

McCarthy, J. 2002. First world political ecology: Lessons from the wise use movement. *Environment and Planning A* 34 (7):1281–1302.

McCright, A. M., C. Xiao, and R. E. Dunlap. 2014. Political polarization on support for government spending on environmental protection in the USA, 1974–2012. *Social Science Research* 48:251–60.

McGregor Cawley, R. 1996. *Federal land, Western anger: The Sagebush Rebellion and environmental politics.* Lawrence: University Press of Kansas.

McNeish, J. A. 2018. Resource extraction and conflict in Latin America. *Colombia Internacional* 93:3–16.

Metz, B. E. 2010. Questions of indigeneity and the (re)-emergent Ch'orti' Maya of Honduras. *The Journal of Latin American and Caribbean Anthropology* 15 (2):289–316.

Middeldorp, N., C. Morales, and G. Van Der Haar. 2016. Social mobilisation and violence at the mining frontier: The case of Honduras. *Journal of Extractive Industries and Society* 3 (4):930–38.

Peluso, N. L., and C. Lund. 2011. New frontiers of land control: Introduction. *Journal of Peasant Studies* 38 (4):667–81.

Perreault, T. 2015. Performing participation: Mining, power, and the limits of public consultation in Bolivia. *The Journal of Latin American and Caribbean Anthropology* 20 (3):433–51.

Pierskalla, J. H. 2010. Protest, deterrence, and escalation: The strategic calculus of government repression. *Journal of Conflict Resolution* 54 (1):117–45.

Rasch, E. D. 2017. Citizens, criminalization and violence in natural resource conflicts in Latin America.

European Review of Latin American and Caribbean Studies 103:131–42.

Ribot, J. 2002. *Democratic decentralization of natural resources: Institutionalizing popular participation.* Washington, DC: World Resources Institute.

Rice, J. L., and B. J. Burke. 2018. Building more inclusive solidarities for socio-environmental change: Lessons in resistance from Southern Appalachia. *Antipode* 50 (1):212–32.

Rivera, L. G. 2013. *Territories of violence: State, marginal youth, and public security in Honduras.* New York: Palgrave Macmillan.

Roa-García, M. C. 2017. Environmental democratization and water justice in extractive frontiers of Colombia. *Geoforum* 85:58–71.

Robbins, P., and A. Luginbuhl. 2005. The last enclosure: Resisting privatization of wildlife in the western United States. *Capitalism Nature Socialism* 16 (1):45–61.

Schilling-Vacaflor, A., and J. Eichler. 2017. The shady side of consultation and compensation: "Divide-and-rule" tactics in Bolivia's extraction sector. *Development and Change* 48 (6):1439–63.

Scoones, I., M. Edelman, S. M. Borras, Jr., R. Hall, W. Wolford, and B. White. 2018. Emancipatory rural politics: Confronting authoritarian populism. *The Journal of Peasant Studies* 45 (1):1–20.

Seligson, M. A., and O. J. Pérez. 2017. Hondurans are in the streets because they don't believe their election results. *The Washington Post*, December 19. Accessed March 11, 2018. https://www.washingtonpost.com/news/monkey-cage/wp/2017/12/19/hondurans-are-in-the-streets-because-they-dont-believe-their-election-results/?utm_term=.d01ed7b29d32.

Smith, R. K. 2008. Ecoterrorism: A critical analysis of the vilification of radical environmental activists as terrorists. *Environmental Letters* 38:537–76.

Szasz, A. 1994. *Ecopopulism: Toxic waste and the movement for environmental justice.* Minneapolis: University of Minnesota Press.

Temper, L., D. del Bene, and J. Martinez-Alier. 2015. Mapping the frontiers and front lines of global environmental justice: The EJAtlas. *Journal of Political Ecology* 22 (1):255–78.

United Nations Environment Program. 2018. UN Environment's Policy on Environmental Defenders. Environmental Rights Initiative. Accessed April 17, 2018. https://wedocs.unep.org/bitstream/handle/20.500.11822/22769/Environmental_Defenders_Policy_2018_EN.pdf?sequence=1&isAllowed=y.

United Nations Office on Drugs and Crime. 2017. Statistics. Accessed March 28 2018. https://www.unodc.org/unodc/en/data-and-analysis/statistics/index.html.

United States Agency for International Development. 2011. USAID country profile: Property rights and resource governance. Honduras. Accessed March 28, 2018. http://www.usaidlandtenure.net/sites/default/files/country-profiles/full-reports/USAID_Land_Tenure_Honduras_Profile_0.pdf.

Van der Borgh, G. J. C. 2016. EU support for justice and security sector reform in Honduras and Guatemala. Working paper, Centre for Conflict Studies, Utrecht University. Accessed March 28, 2018. https://dspace.library.uu.nl/handle/1874/345698.

Vasconsela Rocha, P., and R. Barbosa, Jr. 2018. To criminalize is to govern: A theoretical proposal for understanding the criminalization of rural social movements in Brazil. *Colombia Internacional* 93:205–32.

Wapner, P. 1995. Politics beyond the state environmental activism and world civic politics. *World Politics* 47 (3):311–40.

Wayland, J., and M. Kuniholm. 2016. Legacies of conflict and natural resource resistance in Guatemala. *The Extractive Industries and Society* 3 (2):395–403.

Willems, J., and A. de Jonghe. 2016. Protest and violence over the Agua Zarca Dam. Accessed October 27, 2017. https://www.banktrack.org/download/report_agua_zarca_february_2016_pdf/report_agua_zarca_february_2016.pdf.

World Bank. n.d. Data. Accessed July 29, 2018. https://data.worldbank.org.

Yoder, E. M., L. C. Nieto Garcia, M. A. Perla, A. Pérez, N. Cortiñas, C. Scott, F. Houtard, F. J. Aguilar, H. Umaña, and F. Milla. 2013. *The voice of greatest authority is that of the victims.* Accessed October 27, 2017. https://ccrjustice.org/sites/default/files/attach/2015/01/TrueCommission_Report_English_04_13.pdf.

Zavadskaya, M., and C. Welzel. 2015. Subverting autocracy: Emancipative mass values in competitive authoritarian regimes. *Democratization* 22 (6):1105–30.

NICK MIDDELDORP holds a master's degree in International Development Studies from Wageningen University and is currently a doctoral student in geography at the University of British Columbia, Vancouver, BC V6T 1Z2, Canada. E-mail: nick.middeldorp@alumni.ubc.ca. He works on extractive projects, consent processes, violence, environmental justice, and coping strategies of land and environmental defenders.

PHILIPPE LE BILLON is a Professor in the Department of Geography at the University of British Columbia, Vancouver, BC V6T 1Z2, Canada. E-mail: lebillon@geog.ubc.ca. His research interests include conflicts over natural resources, political economy of war, and corruption.

Neoliberalizing Authoritarian Environmental Governance in (Post)Socialist Laos

Miles Kenney-Lazar (iD)

The (post)socialist nation of Laos has pursued neoliberal economic reforms over the past decade that have facilitated the concession of state lands to foreign resource investors for mining, hydropower, and plantation projects. Five percent of the national territory has been ceded and tens of thousands of peasants have been displaced from their customary lands. In this article, I argue that the development of the resource sector has been facilitated by a political–economic regime of neoliberal authoritarianism. Resource extraction is driven by neoliberal economic policies that prize rapid gross domestic product growth, foreign resource investment, and wage-based rural development. This emerging neoliberalism, however, is matched with and dependent on state authoritarianism. The state seeks to assert control over rural lands throughout the country and often peasants are displaced from using these lands when heavy-handed state coercion and repression of peasant resistance are applied. This is particularly apparent in the establishment of industrial tree plantation territories in southern Laos. Efforts by civil society organizations to highlight these injustices and protect rural land rights are often silenced by the state. Fissures in the neoliberalization of authoritarian development are being exposed, however, due to new forms of resistance among the peasantry that threaten its future viability.

过去十年来，（后）社会主义国家老挝追求将国家土地转让给矿业、水力发电和发电厂计画的外国资源投资者之新自由主义经济改革。国土的百分之五被割让，而数以千万计的农夫从惯常生活的土地上被迫迁徙。我于本文中主张，资源部门的发展受到新自由主义威权主义的政治经济体制所推进。资源搾取是由重视国内生产总值的快速增长、外国资源投资，以及基于工资的乡村发展的新自由主义经济政策所驱动。然而此一浮现中的新自由主义，却是与国家威权主义相符合并依赖其生存。国家寻求对全国农村土地进行控制，而农民使用这些土地而经常遭受迫迁时，则面临国家对农民反抗的粗暴胁迫与镇压。此般境况在老挝南部发展产业植林的领土上特别显着。公民社会组织凸显这些不公义和保护农村土地权利的努力，经常被国家噤声。但由于农民所採取的崭新反抗形式，威权主义发展的新自由主义化的内部分歧遭到暴露，并威胁其未来的可行性。*关键词：威权主义，老挝，新自由主义，后社会主义，资源搾取。*

La nación (post)socialista de Laos ha perseguido reformas económicas neoliberales durante la pasada década que han facilitado la concesión de tierras del estado a inversionistas extranjeros en recursos para minería, hidroelectricidad y proyectos de plantaciones. El cinco por ciento del territorio nacional ha sido cedido y decenas de miles de campesinos han sido desplazados de sus tierras habituales. En este artículo, sostengo que el desarrollo del sector de los recursos ha sido facilitado por un régimen político–económico de autoritarismo neoliberal. La extracción de recursos es orientada por políticas económicas neoliberales que valoran el rápido crecimiento del producto nacional bruto, la inversión foránea en recursos y el desarrollo rural basado en salario. Este neoliberalismo emergente, sin embargo, va emparejado con el autoritarismo estatal y depende del mismo. El estado busca reafirmarse en el control de las tierras rurales a través de todo el país y a menudo los campesinos son desplazados del uso de estas tierras cuando se aplica contra la resistencia campesina la mano dura de la coerción y represión del estado. Esto es particularmente aparente en el establecimiento de territorios para plantaciones de árboles industriales en el sur del país. Los esfuerzos de organizaciones de la sociedad civil para destacar estas injusticias y proteger los derechos a la tierra rural son a menudo silenciados por el estado. Sin embargo, fisuras en la neoliberalización del desarrollo autoritario están siendo expuestas, debido a nuevas formas de resistencia entre el campesinado que amenazan viabilidad futura de aquel modelo de desarrollo. *Palabras clave: autoritarismo, extracción de recursos, Laos, neoliberalismo, postsocialismo.*

Over the past three decades, the Lao People's Democratic Republic (hereafter Lao PDR or Laos), a small, (post)socialist nation in mainland Southeast Asia, has increasingly opened its once centralized, command-and-control economy to market forces of regional and global trade, investment, and commodity production. With a gross domestic product (GDP) growth rate of 7 to 8 percent between 2005 and 2015 (World Bank 2017a), Laos has quickly stepped in line with the Asian "miracle" of capitalist economic growth that it had previously shunned in its strictly socialist era (from 1975 to the mid-1980s). Such growth has been led by large-scale investment in the extractive sector, particularly mining, hydropower dam construction, logging, and industrial agricultural and tree plantations. Land has become progressively commodified as the government has leased and conceded land ostensibly owned by the state to investors for resource extraction projects as well as infrastructure development, special economic zones, and urban real estate development. Over 1 million ha of so-called state land has been conceded in such fashion, equivalent to 5 percent of the national territory (Schönweger et al. 2012).

Due to this economic transformation, GDP per capita has increased from $1,617 in 1990 to $6,073 in 2016.[2] At the same time, new economic opportunities have engendered widespread corruption and wealth inequality has rapidly widened (Warr, Rasphone, and Menon 2015). Much of the new capital accumulation has flowed into urban areas, whereas the externalities from resource extraction have accumulated in the countryside. Rural people, especially upland ethnic minorities, have been displaced and resettled, have been dispossessed of their ancestral agricultural and forestry lands, have lost access to valuable forest products and ecosystem services, and have become alienated from culturally important territories and lands (Lawrence 2008; Kenney-Lazar 2012; Delang et al. 2013; Smirnov 2015). They have become increasingly dependent on wage labor as their access to rural means of production declines (Baird 2011; Molina 2011; Kenney-Lazar 2012). Concurrently, deforestation has accelerated rapidly throughout the country (Thomas 2015).

This rapid and dramatic transformation of Laos's countryside, resource landscape, and rural nature–society relations is often framed as an outcome of the country's transition away from socialism toward a market-based economy (Stuart-Fox 1997; Rigg 2005;

Pholsena 2006). Such a narrative is characteristic of scholarship on neoliberalism that prioritizes the general expansion of globalized neoliberal capitalism as the driving force of political–economic transformation in postcolonial and postsocialist contexts such as Laos and China (Bond 2000; Harvey 2005; Goldman 2005; Sharma and Gupta 2006). Such a perspective, applied to the Lao case, fails to see the ways in which (1) this transformation came about as a state project, rather than one imposed by external market forces, facilitated not by the socialist state drawing back but by directly intervening, and (2) how the contemporary Lao economy is mixed and hybrid, whereby the socialist state plays as significant a role in its functioning as does the market. Scholarship that emphasizes neoliberalization as a specific process rather than neoliberalism as a general political–economic system has better captured the ways in which neoliberal economic policies and projects materialize across space in uneven, variegated, incomplete, locally contingent, and even hybrid ways (Peck and Tickell 2002; Ong 2006; Brenner et al. 2010; Springer 2011). I build on this literature to examine the articulation of neoliberal economic policies with coercive political power to argue that the hybrid state–market economy that has developed in Laos hinges on the production and deployment of authoritarian state power. Acting in various undemocratic, top-down, controlling, coercive, and repressive ways, the authoritarian Lao state has forced neoliberal reforms on the Lao economy and population and continues to play a dominant role in ensuring that a model of rapid economic growth based on large-scale resource extraction projects continues unabated, despite contestation.

There are key elements of neoliberalism and authoritarianism that are well suited for rapid capital accumulation and economic growth, especially in resource-based economies. These close links have been well recognized in the literature on neoliberal authoritarianism, such as in the Pinochet (Chile) and Fujimori (Peru) regimes of Latin America (Mauceri 1995; Roberts 1995; Kay 2002); Turkey and Egypt in the Middle East (Oğuz 2009; Roccu 2012); and Laos, Cambodia, and Myanmar in Southeast Asia (Springer 2011; Hirsch and Scurrah 2015; Creak and Barney 2018). Authoritarian power is particularly important for rearranging rural spaces to make way for large-scale resource investment projects, coercively resettling, displacing, and dispossessing rural people of their lands, resources, and territories. Neoliberal reforms allow for the creation of the forms of private property that investors require and the rights to commodify labor and nature to create resource commodities for export.

Yet, the pairing of neoliberalism and authoritarianism can meet its own limits and resistance, thus only facilitating short periods of capital accumulation. In Laos, this has occurred due to three interrelated processes that are characteristic of Polanyi's (1944) countermovement, in which society (and government) reacts to the unrestricted marketization of land and labor. First, large-scale projects do not necessarily match the goals of efficient rapid economic growth and state revenue generation, as many projects are inefficient in their use of land and their profitability, failing or falling far short of their targets (Schönweger and Messerli 2015; Lu and Schönweger 2017). Second, the disruptions that such projects create between rural people and their ancestral lands and resources, especially when they lead to increased hardship and suffering, are generating resistance to them and even threatening authoritarian control. Third, these threats to the model of large-scale resource extractive growth are forcing the Lao state to consider limiting the most authoritarian forms of neoliberal economic development, particularly widespread, coercive land dispossession.

The article's arguments are developed as follows. First, I connect literatures from political ecology, economic geography, and political geography to conceptualize the neoliberalization of authoritarian environmental governance. Next, I trace the history of the neoliberalization of the Lao resource regime, showing how contemporary economic transformations reflect the integration between a neoliberal economic model and an authoritarian state. I then show how authoritarian power both facilitated and is a part of contemporary capital accumulation projects in Laos by reflecting on cases of industrial tree plantations (rubber and eucalyptus) in southern Laos. Thereafter, I reflect on the barriers and resistance to a neoliberal authoritarian model that have emerged in the last few years and what they mean for the future direction of resource capital accumulation and economic change in Laos.

These arguments emerge from field research conducted in Laos between 2013 and 2015. The research consisted of interviews and focus groups with Chinese and Vietnamese industrial plantation investors, government officials at multiple administrative levels, civil society organizations, and villages in the zones of investment. Government and investor documents and maps related to tree plantation projects were also collected.

Neoliberal Authoritarian Environmental Governance

Much has been written about the neoliberalization of nature and neoliberal environmental governance by political ecologists, particularly the expansion of capitalist accumulation and commodification into hitherto untouched realms of nature, the transformation of public and common spaces into private property, the deregulation of the environment, the governance of the environment by market logics, and the various forms of accumulation by dispossession that enable these processes (Boyd et al. 2001; Harvey 2003; McCarthy 2004; Bakker 2005; Heynen et al. 2007; Castree 2008; Smith 2009). The neoliberalization of nature is framed as a key element of its destruction and the creation of environmental injustices, whereby environmental "goods" tend to be controlled and enjoyed by the wealthy, whereas environmental "bads" are borne by the poor, especially communities of color (Holifield 2001, 2004).

The neoliberal natures literature, however, tends to sidestep analyzing the political forms that accompany and are imbricated with such neoliberal transformations, especially the role of the state in its most dominant and authoritarian forms, representative of political ecology's ambivalent approach toward theorizing the state (Robertson 2015). Economic geography scholarship on neoliberalism has recognized the role of the state in facilitating and maintaining neoliberal transformations. Peck and Tickell (2002) wrote about the ways in which neoliberalism "rolls back" certain elements of the state while "rolling out" new forms of regulation that facilitate capital accumulation. "Actually existing neoliberalism" is interrogated to distinguish between how neoliberalism operates in practice and how neoliberalism is framed as an intellectual or ideological project (Brenner and Theodore 2002). Such approaches recognize the different, variegated, specific, and contingent forms of neoliberalism that manifest in distinct places (Brenner et al. 2010) and that neoliberalism is a moving target, a dynamic and ongoing process (Peck and Tickell 2002).

Closer examinations of the economic transformations under way in postsocialist and postcolonial contexts show that neoliberal reforms are only part of the picture and are integrated into existing authoritarian power structures (see Springer 2011; Lim 2014). As Ong (2006) demonstrated, East Asian countries have only selectively adopted and applied neoliberalism to particular sectors, populations, and spaces and thus have remained exceptional. Selective neoliberal interventions are integrated into the governing dynamics of a range of postcolonial, authoritarian, and postsocialist regimes across East and Southeast Asia. In Southeast Asia, particularly Cambodia and Laos, neoliberal economic reforms become wrapped up and embedded within elite state and party patronage networks that ensure the endurance of authoritarian regimes (Cock 2010; Hughes and Un 2011; Barney 2013; Creak and Barney 2018). In postsocialist contexts, the rapid entrance of market relations into previously socialist spaces can link neoliberal economic opportunities with the unchecked power of political elites (Sikor et al. 2009; Stahl 2010). Similarly, the transition from socialist collectivization to capitalist private property is often accompanied by a period of unclear property rights that the state and other elites can exploit to grant land and resources to private investors (Verdery 2003; Sturgeon and Sikor 2004).

An emerging literature on authoritarian neoliberalism sheds light on the importance of linkages between economic transformations and undemocratic political power (Bruff 2014, 2016). In reflecting on the social and democratic resistance to neoliberal transformation, Bruff (2016) wrote that "state-directed coercion insulated from democratic pressures is central to the creation and maintenance of this politico-economic order, defending it against impulses towards greater equality and democratization" (105). Similarly, Hickel (2016) argued that the radical market deregulation of neoliberalism requires the "dismantling or circumvention of the very democratic mechanisms that neoliberal ideology claims in theory to support and protect" (142), in part by enabling corporate elites to capture political institutions at the expense of voters. Such perspectives contribute to Polanyi's (1944, 147) famous statement that "laissez-faire was planned" in that such planning is often coercive and repressive.

In this article, I focus on the ways in which neoliberal reforms are mapped onto already existing authoritarian sociopolitical relations rather than the ways in which they produce new forms of authoritarianism. For that reason, I frame such integration as the neoliberalization of authoritarian governance, rather than Bruff's (2014) "authoritarian neoliberalism," defined as the merging of neoliberal economic rationales and objectives with coercive, top-down, repressive political power. At times they might sit awkwardly in contradiction with one another, inhabit different parts of the state, or manifest in different geographies. At other times, though, the logics of neoliberalism and state authoritarianism operate in concert, indistinguishable from one another. I use Bruff's (2014) argument that authoritarianism should not be understood as only the exercise of brute coercive force but as the ways in which state and institutional power are reconfigured to insulate government and corporate policies and institutional practices from social and political dissent. Neoliberal authoritarianism can be applied more specifically to the ordering of nature and socioecologies to facilitate nature-based accumulation strategies. In the case of resource extraction, authoritarian power is particularly important for removing communities from the sites of extraction, cutting their ties to the natural resources of the area, and repressing any resistance or protest over these actions, including by civil society and media actors.

An Emerging Neoliberal Authoritarian Resource Regime in Laos

The history of the Lao PDR, since its establishment in 1975, has been an often-messy process of merging the political structures and goals of state socialism with a suitable underlying economic system to lead it there. Initially, the government sought to achieve this through the mechanisms of a nonmarket, centralized, command-and-control economy, but when the failures of state socialism became apparent, dragging down the economy and starving the regime of valuable economic resources, they sought to find a different economic mechanism to achieve these goals: market-based regional and global integration (Stuart-Fox 1997). Like China and Vietnam before it, Laos has sought to achieve economic growth through market reforms while keeping its political structure intact, to develop something along the lines of what the Chinese communist party has referred to as "market socialism" (Nonini 2008). The result, which I refer to as (post)socialist, combines the market reforms of neoliberalism with authoritarian political control, generating a hybrid economy, especially in the political–economic governance of land and natural resources (Andriesse 2011; Barney 2013; Yamada 2018).

When the Lao PDR was established in 1975, any form of market activity, whether private trade, business, or investment, was prohibited in an effort to develop a socialist economy. Businesses were nationalized as state-owned enterprises, such as state logging companies, which provided one of the main sources of cash incomes for the new government. The government also required that rural Lao people work in agricultural collectives, combining land and tools to produce rice for themselves and the state. Such collectives failed due to inefficiency and resistance from peasants who preferred their old ways of life (Evans 1990). Economic collapse was only avoided due to the provision of significant amounts of aid by the Soviet Union and Vietnam that helped prop up the Lao regime (Stuart-Fox 1997).

Politically, the early years after the establishment of the Lao PDR were characterized by attempts to secure the stability of the new regime (Creak and Barney 2018). In some areas of the country, this was characterized by ongoing conflicts with rebel groups that had been aligned with the U.S.-backed Royalist regime (Evans 2002; Baird 2018). In areas of the country firmly controlled by the new government, perceived enemies of the state, particularly officials and soldiers associated with the prior regime, were sent to reeducation camps in remote areas of the country, especially the northeastern provinces near Vietnam (Stuart-Fox 1997; Creak 2018). The government set up new forms of surveillance throughout the country by putting people in power at the village level who were friendly to the government and could report on any suspicious activity. Residents of Vientiane were required to attend regular meetings at which they were supposed to criticize their reactionary behavior and thinking, which was surely surveilled by the state (Khamkeo 2006).

As early as the late 1970s, it was apparent that the state socialist economy was faltering (Yamada 2018), exacerbated in the mid-1980s by the dwindling aid provided by the economically collapsing Soviet Union. In concert with Vietnam—Laos's closest political ally that was undergoing a similar economic crisis—the Lao government initiated market-based economic reforms in 1986, termed the New Economic Mechanism. The concept was framed in politically acceptable terms by the revolutionary leader and then Prime Minister Kaisone Phomvihane (Yamada 2018). In a piecemeal fashion over the course of many years, it eased restrictions on foreign investment, trade, and business operations and led to the revival of the economy (Evans 2002).

In the mid-1990s, the first foreign land and resource investments were made, enabled by the 1988 Law on Foreign Investment. New laws were passed in the 1990s, including a rewritten constitution, intended to attract foreign investment by showing that Laos would become a "rule of law" state (Creak 2018), governed by consistent and stable rules rather than by arbitrary decrees from the party leadership. As the government sought to provide legal stability for foreign investors, with equal importance it has projected an image of political stability, often cited as one of its greatest assets (Ministry of Planning and Investment 2017). The government continued to resettle ethnic minority groups from upland to lowland areas as part of a strategy to keep track of villages, areas, and groups associated with rebel activity (Baird and Shoemaker 2007). Any sign of dissent was quashed, such as a peaceful student-led democracy protest in 1999 that led to the arrest of four protest leaders (Inthapannha and Souksavanh 2014; Baird 2018).

In the mid-2000s, the government sought to further facilitate foreign investment in land by developing a policy concept of turning land into capital (TLIC), which can be interpreted as generating revenue from land (Pathammavong et al. 2017). The policy was never issued as an official legal document, but it acted as a form of political support for various types of land commodification projects, such as land titling and the long-term lease and concession of state land to the private sector. Although TLIC-inspired investments aim to generate economic growth by allowing the private sector to profit from the commodification of the country's land, they are based on the deployment of state power over land, using various degrees of coercion and state authority. This is particularly the case for state land concessions, which rely on the leasing of "state" land for periods of up to ninety-nine years. The Lao legal framework gives the government significant powers of management and control over the country's land

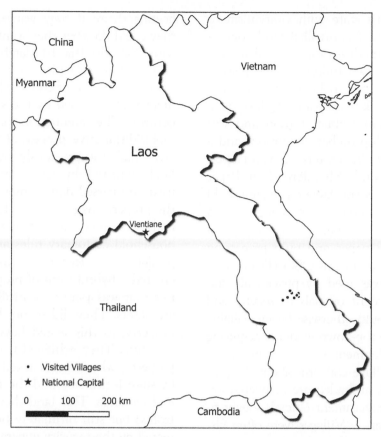

Figure 1. Location of visited villages in Laos. *Source:* Author drawing.

and thus it can be claimed that there are large swaths of state land available for investors to control. In reality, however, most land throughout the country is occupied and used by Lao people (Barney 2009). Thus, as Dwyer (2013) showed, state land must be produced before it can be transferred to investors, and this is done by using authoritarian powers to coercively expropriate such land from the Lao peasantry.

Not surprising, the expropriation of land has led to anger and frustration among the peasantry, even meeting resistance in some areas. Thus, dispossession is a socially disruptive process that the state must manage (Dwyer 2014). Most open forms of resistance have been met with state repression and detention (Baird 2017; Kenney-Lazar, Suhardiman, and Dwyer 2018). The government has also intimidated many who consider resisting and the state does not provide any effective means of addressing grievances or pressing legal cases when companies have practiced social and environmental abuses (Gindroz 2017). Such pressure has also been extended to civil society organizations and their staff who work with communities on these issues. In 2012, a prominent Lao

civil society member, Sombath Somphone, was forcibly disappeared during a routine police traffic stop when driving home from work, due to his prominent role in hosting a civil society forum where land concessions were hotly debated (FORUM-ASIA and AEPF-IOC 2014). Several of the Lao participants who spoke up at the event were investigated and harassed by government officials in their home provinces (Kenney-Lazar 2016). In the years following this event, civil society organizations were afraid to work on these issues and tended to keep a low profile.

The Neoliberal-Authoritarian Production of Industrial Plantations

Authoritarian power has been essential for creating the "state land" essential for attracting foreign investors to develop industrial tree plantations. Nationally, more than 440,000 ha of land have been conceded for industrial crop plantations, including sugar cane, cassava, rubber, and eucalyptus (Schönweger et al. 2012). The combination of multinational capital with state authoritarianism to produce state and corporate controlled land concessions and to export commodities for a global market was on full display with two industrial tree plantation projects studied in southern Laos. The first, the Quasa-Geruco Joint Stock Company (QSG), is a subsidiary of the Vietnam Rubber Group and was granted 8,650 ha to develop rubber plantations and a latex processing facility in eastern Savannakhet, southern Laos. The second, Shandong Sun Paper Industry Company (SP), is the largest privately held Chinese paper and pulp producer and was granted a concession of 7,324 ha to develop eucalyptus and acacia trees as well as a paper and pulp processing facility in the same general area as QSG (Figure 1).

Authoritarian power was used by provincial- and district-level Lao government officials to secure land for QSG and SP that would otherwise be unavailable for lease or purchase. The top-down model of acquiring land for plantation development was apparent in the coercive way in which village concerns about the project impacts, loss of land, and adequate compensation were brushed aside and government officials sought to repress any form of resistance. Villages were often visited by company managers and staff and government officials from the central to district levels who presented the project as a done deal, signed and approved by the central-level government and approved at all lower levels, to cover lands within the village territory that were claimed to belong to the state.

When villagers refused to concede their lands to the project outright, especially if they were not given the chance to negotiate the terms of the project and their compensation, government officials threatened to single out and name uncooperative villagers. At one point, the district governor visited some of the hesitant villages targeted for the QSG and asked for a list of the names of those who refused to accept the project, threatening to bring those villagers to the district office for education, reminiscent of the reeducation camps that officials from the Royalist regime were sent to after the war. In the case of SP, threats were made toward a village that consistently refused to concede land to the project. Ignoring village concerns, the clearance of village land was approved and the company arrived with its bulldozers as well as soldiers and police officers to escort them and prevent any problems with villagers. One villager expressed clearly how such action had repressed their attempts at resistance: "Villagers were afraid of the soldiers beating them, the police officers beating them. If they wanted to arrest the villagers, they could do what they want" (Focus group interview with village leaders, 10 March 2014, Xaylom village).

The state also plays a critical role in creating a hybrid form of property used for plantation development that is owned by the state but under corporate control. The production of concession lands is a rapid, disjunctive process of marketization that these areas had not previously experienced. Most of the lands targeted by QSG and SP were customarily used and passed down among generations. Although the property had yet to be formalized by the state, it had been used for many years under village and household customary rules and systems. Thus, in the production of tree plantations, the state effectively created a hybrid form of property out of such territories that was jointly controlled by state and corporate actors. They did so not by claiming it as private property, as this would have been easy for villagers to refuse. They achieved this by claiming it as state property, which was backed up by the law but also by state legitimacy and the threat of state repression and violence. This land then became corporate controlled but still ultimately belonged to and was protected by the security apparatus of the state.

A Neoliberal Authoritarian Resource Regime in Crisis?

Recent events suggest that there are fissures emerging in the neoliberal authoritarian land regime of Laos that could threaten its perpetuation. Due to village resistance and discontent, widely recognized negative socioenvironmental impacts, limited ability for the government to collect revenues from land investments, and international pressure from bilateral and multilateral development donors, the Lao government has placed several successive moratoriums on the approval of new land concessions (the most recent of which is still in place) and has been reconsidering the role of the TLIC policy (Kenney-Lazar, Dwyer, and Hett 2018). Rather than charting a new path that diverges away from neoliberal authoritarianism, it is likely that the Lao government will seek to curb its most extreme elements to stave off crises of socioenvironmental destruction and state illegitimacy that could generate widespread dissent. Nonetheless, these reactions from society and the state reflect a Polanyian countermovement to the authoritarian marketization of land and labor imposed on the Lao countryside by the socialist state.

Despite the authoritarian nature of the development of land concessions throughout the country, there is an increasing number of cases of resistance by communities that refuse to concede their village territories (see McAllister 2015; Baird 2017; Kenney-Lazar, Suhardiman, and Dwyer 2018). Although most of the villages targeted by QSG and SP lost significant amounts of village land to such projects, some were able to put up an effective front of resistance and refusal, often by working through back channels and political connections with sympathetic government officials to protect some areas of village land,[3] especially lands recognized by the state as property of individual households for agricultural production. This type of resistance has led to a recognition by government officials at all levels that the idea of available or empty state land awaiting investors is largely a myth and that most land targeted for concessions is not easily produced without creating some sort of conflict with villagers who currently occupy, use, and govern such land.

Increasing frustration with land conflicts nationwide has filtered up to the government via the few semidemocratic avenues available. The National Assembly, members of which are elected but in a highly controlled and closed process, has opened a telephone hotline for the public to call in and make complaints about issues that concern them during the legislative sessions, and land issues have been one of the top concerns of callers over the past decade. In facing a crisis of legitimacy, not only over land but also illegal logging and corruption, the Lao People's Revolutionary Party selected a new prime minister in 2016 who was charged with changing the people's image of the government and party by halting illegal logging, clamping down on corruption, and addressing chronic land issues throughout the country (Sayalath and Creak 2017). Although these goals are still incomplete, the new leadership has shown that the government is willing to take them seriously.

Reacting to the government's moratoriums on land concessions, investors have turned to alternative forms of land investment that bypass the state, particularly contract farming and leasing land directly from individuals and communities (Dwyer and Vongvisouk 2017). A recent boom in Chinese banana plantations in northern Laos was based on the model of leasing land from households (Friis and Nielsen 2016). Even SP eventually moved toward a community leasing and contract farming model because they lost political support from the district government, limiting the degree to which the company could mobilize state power to coercively expropriate land from communities. Thus, as the private sector runs up against the limits of neoliberal authoritarianism, they might seek exclusively private sector and neoliberal forms of investment. Yet, such land and agricultural arrangements between plantation companies and farmers carry as many social–environmental risks as large-scale land investments (see Grossman 2000; Dwyer 2013; St. Laurent and Le Billon 2015).

Conclusion

In this article, I have argued that neoliberal reforms have been integrated into an authoritarian regime of environmental governance in ways that show the close connections between neoliberalism and authoritarianism. This idea builds on a longstanding and evolving literature that recognizes the importance of top-down political structures for implementing and maintaining neoliberal economic transformations, including the repression of resistance to such projects. It also contributes to a broader literature on neoliberalism that demonstrates how neoliberal reforms are dependent on and integrated with state power and that neoliberalization is a project developed unevenly across geographical contexts.

In Laos, the government has selectively adopted neoliberal measures to facilitate foreign investment and rapid economic growth in an attempt to reduce poverty and raise the country's general prosperity. Such measures have been integrated into the Lao governance context, however, in ways that have led to a hybrid state–market economy. For example, private property in the resource extractive sector has been established through long-term concessions of state lands and joint-venture projects between state agencies and foreign investors. Additionally, foreign investors often employ authoritarian state power to develop their projects, particularly in the resource sector for coercively separating Lao peasants from their land and resources.

The linkages between neoliberalism and authoritarianism might be reaching their limits as the coercive dispossession of peasant lands has begun to create a wave of frustration and various forms of resistance across the Lao countryside, putting pressure on the government to change its policies to maintain its popular legitimacy. The government is currently considering what reforms to implement to reduce the impacts of foreign investments on rural communities while investors are increasingly avoiding a moratorium and limits on land concessions by arranging land deals directly with rural communities and households. Thus, the beginnings of political change might be under way, due not to neoliberal economic transformations but to reactions to them, particularly their authoritarian mode of implementation.

ORCID

Miles Kenney-Lazar (iD) http://orcid.org/0000-0001-6977-8345

Notes

1. Although Laos, like China and Vietnam, is typically referred to as postsocialist, I place *post* in parentheses to reference that the socialist political regime founded in 1975 is still in place and continues to pursue the goals of socialist development but now by market means.
2. GDP per capita is adjusted for purchasing power parity (PPP) in constant 2011 U.S. dollars (World Bank 2017b).
3. Similar processes have been recognized in Baird and Le Billon (2012) and Kenney-Lazar, Suhardiman, and Dwyer (2018).

References

Andriesse, E. 2011. Laos: A state coordinated frontier economy. Working Paper No. 522, International Institute of Social Studies, The Hague, The Netherlands.

Baird, I. G. 2011. Turning land into capital, turning people into labour: Primitive accumulation and the arrival of large-scale economic land concessions in the Lao People's Democratic Republic. *New Proposals: Journal of Marxism and Interdisciplinary Inquiry* 5 (1):10–26.

———. 2017. Resistance and contingent contestations to large-scale land concessions in southern Laos and northeastern Cambodia. *Land* 6 (1):1–19.

———. 2018. Party, state and the control of information in the Lao People's Democratic Republic: Secrecy, falsification and denial. *Journal of Contemporary Asia*, 48 (5):739–60.

Baird, I. G., and P. Le Billon. 2012. Landscapes of political memories: War legacies and land negotiations in Laos. *Political Geography* 31 (5):290–300.

Baird, I. G., and B. Shoemaker. 2007. Unsettling experiences: Internal resettlement and international aid agencies in Laos. *Development and Change* 38 (5):865–88.

Bakker, K. 2005. Neoliberalizing nature? Market environmentalism in water supply in England and Wales. *Annals of the Association of American Geographers* 95 (3):542–65.

Barney, K. 2009. Laos and the making of a "relational" resource frontier. *Geographical Journal* 175 (2):146–59.

———. 2013. Locating "green neoliberalism," and other forms of environmental governance in Southeast Asia. *CSEAS Newsletter* 66:25–28.

Bond, P. 2000. *Elite transition: From apartheid to neoliberalism in South Africa.* London: Pluto.

Boyd, W., S. Prudham, and R. A. Schurman. 2001. Industrial dynamics and the problem of nature. *Society and Natural Resources* 14 (7):555–70.

Brenner, N., J. Peck, and N. Theodore. 2010. Variegated neoliberalization: Geographies, modalities, pathways. *Global Networks* 10 (2):182–222.

Brenner, N., and N. Theodore. 2002. Cities and the geographies of "actually existing neoliberalism." *Antipode* 34 (3):349–79.

Bruff, I. 2014. The rise of authoritarian neoliberalism. *Rethinking Marxism* 26 (1):113–29.

———. 2016. Neoliberalism and authoritarianism. In *The handbook of neoliberalism*, ed. S. Springer, K. Birch, and J. MacLeavy, 107–17. London and New York: Routledge.

Castree, N. 2008. Neoliberalising nature: The logics of deregulation and reregulation. *Environment and Planning A* 40 (1):131–52.

Cock, A. R. 2010. External actors and the relative autonomy of the ruling elite in post-UNTAC Cambodia. *Journal of Southeast Asian Studies* 41 (2):241–65.

Creak, S. 2018. Abolishing illiteracy and upgrading culture: Adult education and revolutionary hegemony in socialist Laos. *Journal of Contemporary Asia* 48 (5):761–82.

Creak, S., and K. Barney. 2018. Party-state governance and rules in Laos. *Journal of Contemporary Asia* 48 (5):693–716 doi:10.1080/00472336.2018.1494849.

Delang, C. O., M. Toro, and M. Charlet-Phommachanh. 2013. Coffee, mines and dams: Conflicts over land in the Bolaven Plateau, Southern Lao PDR: Coffee, mines and dams. *The Geographical Journal* 179 (2):150–64.

Dwyer, M. B. 2013. Building the politics machine: Tools for "resolving" the global land grab. *Development and Change* 44 (2):309–33.

———. 2014. Micro-geopolitics: Capitalising security in Laos's golden quadrangle. *Geopolitics* 19 (2):377–405.

Dwyer, M. B., and T. Vongvisouk. 2017. The long land grab: Market-assisted enclosure on the China–Lao rubber frontier. *Territory, Politics, Governance* 1–19. doi:10.1080/21622671.2017.1371635.

Evans, G. 1990. *Lao peasants under socialism.* New Haven, CT: Yale University Press.

———. 2002. *A short history of Laos: The land in between.* Crows Nest, Australia: Allen & Unwin.

FORUM-ASIA and AEPF-IOC. 2014. *Universal periodic review second cycle—Lao PDR: Stakeholders' submission.* Vientiane, Laos: Asian Forum for Human Rights and Development (FORUM-ASIA) and International Organising Committee of the Asia-Europe People's Forum (AEPF-IOC).

Friis, C., and J. Ø. Nielsen. 2016. Small-scale land acquisitions, large-scale implications: Exploring the case of Chinese banana investments in northern Laos. *Land Use Policy* 57:117–29.

Gindroz, A.-S. 2017. *Laos, the silent repression: A testimony written after being expelled.* Self-published.

Goldman, M. 2005. *Imperial nature: The World Bank and struggles for social justice in the age of globalization.* New Haven, CT: Yale University Press.

Grossman, L. S. 2000. *The political ecology of bananas: Contract farming, peasants, and agrarian change in the Eastern Caribbean.* Chapel Hill: University of North Carolina Press.

Harvey, D. 2003. *The new imperialism.* New York: Oxford University Press.

———. 2005. *A brief history of neoliberalism.* New York: Oxford University Press.

Heynen, N., J. McCarthy, and P. Robbins, eds. 2007. *Neoliberal environments: False promises and unnatural consequences.* London and New York: Routledge.

Hickel, J. 2016. Neoliberalism and the end of democracy. In *The handbook of neoliberalism,* ed. S. Springer, K. Birch, and J. MacLeavy, 142–52. London and New York: Routledge.

Hirsch, P., and N. Scurrah. 2015. *The political economy of land governance in Lao PDR.* Vientiane, Laos: Mekong Region Land Governance.

Holifield, R. 2001. Defining environmental justice and environmental racism. *Urban Geography* 22 (1):78–90.

———. 2004. Neoliberalism and environmental justice in the US Environmental Protection Agency. *Geoforum* 35:285–98.

Hughes, C., and K. Un. 2011. Cambodia's economic transformation: Historical and theoretical frameworks. In *Cambodia's economic transformation,* ed. C. Hughes and K. Un, 1–26. Copenhagen: NIAS Press.

Inthapannha, S., and O. Souksavanh. 2014. Exiled Lao activist recalls 1999 protest, says democracy still attainable. *Radio Free Asia,* October 29.

Kay, C. 2002. Chile's neoliberal agrarian transformation and the peasantry. *Journal of Agrarian Change* 2 (4):464–501.

Kenney-Lazar, M. 2012. Plantation rubber, land grabbing and social-property transformation in southern Laos. *Journal of Peasant Studies* 39 (3–4):1017–37.

———. 2016. *Resisting with the state: The authoritarian governance of land in Laos.* Worcester, MA: Clark University.

Kenney-Lazar, M., M. B. Dwyer, and C. Hett. 2018. *Turning land into capital: Assessing a decade of policy in practice.* Vientiane, Laos: Land Issues Working Group.

Kenney-Lazar, M., D. Suhardiman, and M. Dwyer. 2018. State spaces of resistance: Industrial tree plantations and the struggle for land in Laos. *Antipode* 50 (5):1290–1310.

Khamkeo, B. 2006. *I little slave: A prison memoir from communist Laos.* Spokane, WA: Eastern Washington University Press.

Lawrence, S. 2008. *Power surge: The impacts of rapid dam development in Laos.* Vientiane, Laos: International Rivers.

Lim, K. F. 2014. 'Socialism with Chinese characteristics' uneven development, variegated neoliberalization and the dialectical differentiation of state spatiality. *Progress in Human Geography* 38 (2):221–47.

Lu, J., and O. Schönweger. 2017. Great expectations: Chinese investment in Laos and the myth of empty land. *Territory, Politics, Governance.* doi:10.1080/21622671.2017.1360195.

Mauceri, P. 1995. State reform, coalitions, and the neoliberal autogolpe in Peru. *Latin American Research Review* 30 (1):7–37.

McAllister, K. 2015. Rubber, rights and resistance: The evolution of local struggles against a Chinese rubber concession in northern Laos. *The Journal of Peasant Studies* 42 (3–4):817–37.

McCarthy, J. 2004. Privatizing conditions of production: Trade agreements as neoliberal environmental governance. *Geoforum* 35 (3):327–41.

Ministry of Planning and Investment. 2017. Why Laos. Accessed December 6, 2017. http://www.investlaos.gov.la/index.php/why-laos.

Molina, R. 2011. *Camps, children, chemicals, contractors & credit: Field observations of labour practices in plantations & other social developments in Savannakhet and Champassak.* Pakse, Laos: Global Association for People and the Environment; Japan International Volunteer Center.

Nonini, D. M. 2008. Is China becoming neoliberal? *Critique of Anthropology* 28 (2):145–76.

Pathammavong, B., M. Kenney-Lazar, and E. V. Sayaraj. 2017. Financing the 450 Year Road: Land expropriation and politics 'all the way down' in Vientiane, Laos. *Development and Change* 48 (6):1417–38.

Peck, J., and A. Tickell. 2002. Neoliberalizing space. *Antipode* 34 (3):380–404.

Oğuz, S. 2009. The response of the Turkish state to the 2008 crisis: A further step towards neoliberal authoritarian statism. Paper presented at the International Initiative for Promoting Political Economy Conference, Ankara, Turkey. Accessed November 30, 2018. http://www.iippe.org/wiki/images/a/ac/Oguz_IIPPE_Ankara.pdf.

Ong, A. 2006. *Neoliberalism as exception: Mutations in citizenship and sovereignty.* Durham, NC: Duke University Press.

Pholsena, V. 2006. *Laos: From buffer state to crossroads?* Chiang Mai, Thailand: Silkworm Books.

Polanyi, K. 1944. *The great transformation: The political and economic origins of our time.* Boston: Beacon Press.

Rigg, J. 2005. *Living with transition in Laos: Market integration in Southeast Asia.* London and New York: Routledge.

Roberts, K. M. 1995. Neoliberalism and the transformation of populism in Latin America: The Peruvian case. *World Politics* 48 (1):82–116.

Robertson, M. 2015. Environmental governance: Political ecology and the state. In *The Routledge handbook of political ecology*, ed. T. Perreault, G. Bridge, and J. McCarthy, 457–66. London and New York: Routledge.

Roccu, R. 2012. Gramsci in Cairo: Neoliberal authoritarianism, passive revolution and failed hegemony in Egypt under Mubarak, 1991–2010. Master's thesis, London School of Economics.

Sayalath, S., and S. Creak. 2017. Regime renewal in Laos: The tenth congress of the Lao People's Revolutionary Party. *Southeast Asian Affairs* 2017:179–200.

Schönweger, O., A. Heinimann, M. Epprecht, J. Lu, and P. Thalongsengchanh. 2012. *Concessions and leases in the Lao PDR: Taking stock of land investments.* Vientiane, Laos: Ministry of Natural Resources and Environment and Centre for Development and Environment, University of Bern, Switzerland.

Schönweger, O., and P. Messerli. 2015. Land acquisition, investment, and development in the Lao coffee sector: Successes and failures. *Critical Asian Studies* 47 (1):94–122.

Sharma, A., and A. Gupta. 2006. Introduction: Rethinking theories of the state in an age of globalization. In *The anthropology of the state: A reader*, ed. A. Sharma and A. Gupta, 1–42. Malden, MA: Blackwell.

Sikor, T., J. Stahl, and S. Dorondel. 2009. Negotiating post-socialist property *and* state: Struggles over forests in Albania and Romania. *Development and Change* 40 (1):171–93.

Smirnov, D. 2015. *Assessment of scope of illegal logging in Laos and associated trans-boundary timber trade.* Vientiane, Laos: Worldwide Fund for Nature.

Smith, N. 2009. Nature as accumulation strategy. *Socialist Register* 43:16–36.

Springer, S. 2011. Articulated neoliberalism: The specificity of patronage, kleptocracy, and violence in Cambodia's neoliberalization. *Environment and Planning* 43:2554–70.

Stahl, J. 2010. The rents of illegal logging: The mechanisms behind the rush on forest resources in southeast Albania. *Conservation and Society* 8 (2):140–50.

St. Laurent, G. P., and P. Le Billon. 2015. Staking claims and shaking hands: Impact and benefit agreements as a technology of government in the mining sector. *The Extractive Industries and Society* 2 (3):590–602.

Stuart-Fox, M. 1997. *A history of Laos.* Cambridge, UK: Cambridge University Press.

Sturgeon, J. C., and T. Sikor. 2004. Post-socialist property in Asia and Europe: Variations on fuzziness. *Conservation & Society* 2 (1):1–17.

Thomas, I. L. 2015. *Drivers of forest change in the greater Mekong subregion Lao PDR country report.* Washington, DC: United States Agency for International Development.

Verdery, K. 2003. *The vanishing hectare: Property and value in postsocialist Transylvania.* Ithaca, NY: Cornell University Press.

Warr, P. G., S. Rasphone, and J. Menon. 2015. Two decades of rising inequality and declining poverty in the Lao People's Democratic Republic. Asian Development Bank Economics Working Paper Series No. 461. Manila: Asian Development Bank.

World Bank. 2017a. GDP growth (annual %). Accessed March 28, 2017. http://data.worldbank.org/indicator/NY.GDP.MKTP.KD.ZG?locations¼LA.

———. 2017b. GDP per capita, PPP (constant 2011 international $). Accessed March 28, 2017. http://data.worldbank.org/indicator/NY.GDP.PCAP.PP.KD?locations¼LA.

Yamada, N. 2018. Legitimation of the Lao People's Revolutionary Party: Socialism, Chintanakan Mai (New thinking) and reform. *Journal of Contemporary Asia* 48 (5):717–38.

MILES KENNEY-LAZAR is an Assistant Professor in the Department of Geography at the National University of Singapore, Singapore 117570. E-mail: geokmr@nus.edu.sg. His research is focused on the political ecology of agro-industrial plantation expansion in Laos and Myanmar, especially peasant contestations of land grabbing.

The Speculative Petro-State: Volatile Oil Prices and Resource Populism in Ecuador

Angus Lyall ⓘ and Gabriela Valdivia ⓘ

In petro-states, the governance of flows of oil and oil money is vital to state legitimacy (e.g., regulations, contracts with companies, social compensation in sites of oil extraction). This article explores how contemporary oil price volatility shapes oil governance and the terms of petro-state legitimacy in Ecuador. In recent years, a technocratic, populist regime, led by President Rafael Correa, promised to return national oil resources to "the people" and inaugurate a "postneoliberal" era of sovereign, oil-driven development. The performance of this promise, through augmented public spending, was contingent on international oil prices. We track the emergence of what we call a speculative petro-state, in which state actors claimed to successfully gamble on volatile markets on behalf of the nation, as an emergent strategy for cultivating popular legitimacy. Such claims took the form of petro-populist discourses and practices. First, the Correa administration characterized new contractual relations with oil companies and capital as evidence of Correa's leadership in complex oil markets, seeking political legitimacy for the state through perceptions of Correa's personal capacity to manage market risk. Second, as prices surged, the Correa administration channeled rents into building spectacular public works or "petro-populist landscapes," as material verification of Correa's petro-leadership in volatile markets. We track how market risk management became one key organizing factor of populist rule in Ecuador and we analyze how this case illuminates relations between populist politics and economic spheres.

在石油国家中, 石油与石油金流的治理, 对国家正当性而言至关重要（例如规范、公司契约、採油现址的社会补偿等）。本文探讨当代的油价波动如何形塑厄瓜多尔的石油治理与石油国家的正当性。近年来, 一个由总统拉斐尔．科雷亚所带领的技术民粹主义政体, 承诺将国家的石油资源归还给"人民", 并开展据独立主权、由石油驱动的发展之"后新自由主义"时代。此一通过增加公共支出的承诺之实践, 则取决于国际油价。我们称之为投机石油国家并追踪其浮现, 其中国家行动者主张代表全国, 成功地地对市场波动进行博弈, 作为浮现中的培植大众正当性之策略。此般宣称, 採取石油—民粹主义的论述与实践形式。首先, 科雷亚政府将与石油公司与资本的崭新契约关系视为科雷亚在复杂的石油市场中的领导力之明证, 并通过对科雷亚管理市场风险的个人能力之看法, 寻求国家的政治正当性。再者, 当价格上涨时, 科雷亚政府将地租引导至发展奇观公共工程、亦或所谓的"石油民粹主义地景", 作为科雷亚在市场波动中的石油领导力之证明。我们追溯市场风险管理如何成为厄瓜多尔民粹主义治理的组织要素, 并分析此一案例如何启发民粹主义政治和经济领域之间的关联性。关键词: 厄瓜多尔, 治理, 油价, 资源民粹主义, 波动。

En los petro-estados, la gobernanza de los flujos de petróleo y del dinero petrolero es vital para la legitimidad del estado (por ej., las regulaciones, los contratos con las compañías, la compensación social en los lugares de extracción del petróleo). Este artículo explora el modo como la volatilidad contemporánea del precio del petróleo configura su gobernanza y los términos de la legitimidad del petro-estado en Ecuador. En años recientes, un régimen populista tecnocrático encabezado por el Presidente Rafael Correa, prometió devolver los recursos petroleros nacionales "al pueblo" e inauguró una era "posneoliberal" de desarrollo soberano basada en el petróleo. La representación de esta promesa, por medio del incremento en el gasto público, era contingente a los precios internacionales del crudo. Le seguimos el paso al surgimiento de lo que nosotros denominamos un petro-estado especulador, en el que actores del estado proclamaron haber jugado exitosamente en los mercados volátiles a nombre de la nación, como una nueva estrategia con la cual cultivar la legitimidad popular. Tales afirmaciones tomaron la forma de discursos y prácticas petro-populistas. Primero, la administración Correa caracterizó las nuevas relaciones contractuales con las compañías y el capital petroleros como evidencia del liderazgo de Correa en los complejos mercados petroleros, buscando legitimidad política para el estado a través de las percepciones de la capacidad personal de Correa para manejar el riesgo del mercado. Segundo, cuando los precios aumentaron, la administración Correa canalizó las rentas para construir obras púbicas espectaculares o "paisajes petro-populistas", a título de verificación material del petro-liderazgo del presidente en los mercados volátiles. Trazamos el modo como el manejo del riesgo de mercado se convirtió en un factor organizativo clave del orden populista en Ecuador y analizamos cómo este caso ilumina las relaciones entre la política populista y las esferas económicas. *Palabras clave: Ecuador, gobernanza, precio del petróleo, populismo de los recursos, volatilidad.*

In 2013, Ecuadorian President Rafael Correa inaugurated the "Millennium City" in the indigenous parish of Playas del Cuyabeno, one of several urbanization projects sponsored by the Ecuadorian state in oil-rich areas of the Amazon. Hailed as the beginning of a "new Amazon," the resettlement was built to compensate indigenous residents directly affected by nearby oil activities. It included sixty-eight furnished, cement homes along paved roads with streetlamps; a school, medical clinic, and police station; and potable water and household electricity, among other services. Playas del Cuyabeno had "the facilities of a proper modern city ... in the middle of the forest" (Petroamazonas 2013). Although its official cost was just over $20 million—an average of $300,000 per household—community leaders and construction workers involved in the project suggest that the real cost might have been twice that figure, due to difficulties of building urban infrastructure in remote, swampy jungle. Correa argued that such spectacular projects validated recent, state-led renegotiations of contracts with private oil companies and substantiated government claims that the state, under Correa's leadership, was a legitimate custodian of national resources.

By 2016, however, Correa appeared regularly on television to explain that an unpredictable "perfect storm," including a major earthquake and, above all, free-falling oil prices, had expunged state capacities to invest in projects like Millennium Cities. In his weekly radio and television show, Correa assured that public services throughout the country could be funded by nonpetroleum revenues. His assurances did not apply to Playas del Cuyabeno or other compensation projects in "strategic" areas of oil extraction. In July 2016, personnel of Petroamazonas, the Ecuadorian oil firm that oversaw state-led oil exploration and extraction, visited Playas del Cuyabeno and informed residents that the firm could no longer fund the city's electricity. That night, the streetlamps went dark and the community's youth suspended their nightly soccer match in the sports coliseum, seemingly with nothing to blame but international oil prices.

We begin with this brief moment in Playas del Cuyabeno to inquire about the emerging impacts of contemporary oil price volatility on petro-state governance. In her widely influential book *The Paradox of Plenty*, Karl (1997) described "petro-states" as countries rich in oil reserves where national economies, politics, and institutions appear to be "petrolized" (64)—that is, where booms and busts in international oil prices trigger profound economic, political, and institutional transformations, shaping the state and the everyday lives of citizens.[1] In 2014, at the height of the Correa regime, the Ecuadorian oil sector accounted for 9.5 percent of overall gross domestic product; by comparison, the growing mining sector accounted for just 2.3 percent (Wacaster 2016). Data on state revenues more clearly reflect the state's reliance on oil. In 2014, the state projected revenues from oil sales and rents of US$17.5 billion—over half of its total projected budget, in comparison with less than US$75 million from mining royalties and rents (Ministerio de finanzas 2014).

In petro-states like Ecuador, the governance of oil, or the negotiation of multiple interests to secure flows of oil (Mitchell 2011) and oil money (Appel et al. 2015; Bridge and Le Billon 2017), is central to political legitimacy (Coronil 1997; Watts 2004). Diverse policies governing oil flows, from contracts with private oil companies to social compensation in sites of oil extraction, are foci of intense political scrutiny, negotiation, and legitimation (Valdivia 2008; Lu, Valdivia, and Silva 2016). Although petro-states negotiate the interests of political factions, oil companies, communities, unions, and environmentalists at distinct "nodes" (Mitchell 2011, 5) of oil governance, these negotiations are largely contingent on state capacities to capture and redistribute oil rents, which, in turn, are tied to the price of oil. We observe that high oil price volatility not only complicates investment in oil activities but also affects negotiations at each node, influencing the terms of exploration and extraction and social compensation practices, as well as oil transport, refining, and storage (see Labban 2010; Zalik 2010).

Oil price volatility has intensified since the turn of the century. In the 1980s and 1990s, the international benchmark price West Texas Intermediate (WTI) generally fluctuated narrowly between $25 and $40 per barrel; however, in 1998, prices fell to $17 and then climbed by fits and starts over the next decade to $157—that is, by 900 percent. Prices plummeted to $48 in 2008 and then hovered between $70 and $120 in the following years. They collapsed again in 2014—hitting $26 by 2016, before recovering to over $70 in 2018. Economists, business scholars, and oil executives scrutinize this volatility, unprecedented over the previous half-century, in terms of possible drivers, including market financialization, geopolitics, vacillating demand in developing countries, OPEC production goals (and realities), and growth in unconventional production, especially shale (e.g., Liu et al. 2016; Ewing and Malik 2017). Unlike previous booms and busts since the 1970s, any single variable is insufficient to account for contemporary volatility.

Yet, in this period, critical scholarship on oil politics has largely overlooked how oil governance changes in relation to price volatility. Instead, it has taken high or low oil prices as contextual matters and directed critical attention to the discourses, policies, and contradictions of petro-state leaders and institutions. In Latin America, for example, amid a crisis in neoliberal rule in the 2000s, charismatic populist leaders such as Hugo Chávez in Venezuela, Rafael Correa in Ecuador, and Evo Morales in Bolivia consolidated power as political outsiders, claiming to represent "the people" and oppose neoliberal austerity amid relatively high commodity prices (De la Torre 2013; Moffitt 2016). They promised to return sovereignty—and resource rents—to the people, by increasing spending on welfare, infrastructure, and human capital, within an inclusive, state-driven development model (Bielschowsky 2009; Purcell et al. 2017). Researchers have highlighted the uneven development that unfolded under these postneoliberal regimes that depended on extractivism (Acosta 2013; Svampa 2015) and the expansion of extractive frontiers (Gudynas 2012; Bebbington 2015). In Ecuador, researchers described Millennium Cities to examine this unevenness in ethnographic detail (Espinosa Andrade 2017; Lyall 2017; Cielo and Carrión Sarzosa 2018). Wilson and Bayón (2018) vividly described empty houses and idle infrastructures, suggesting that these projects were facades of modernity that shrouded

primitive accumulation. Such critiques of postneoliberal oil governance have unveiled legitimizing discourses of national inclusion by mapping ongoing relations of exclusion and dispossession (Silveira et al. 2017).

In this article, we direct our attention to how oil price volatility—and associated risks—reshaped postneoliberal strategies of oil governance in specific ways. Modern states often use diverse discourses on risk to consolidate political legitimacy in relation to uncertain futures (O'Malley 2004; Rose and Miller 2010). For companies, risks associated with market volatility amount to the likelihood of not generating returns on capital; by contrast, for a petro-state, economic risks do not necessarily translate into political risks. In fact, modern states have often used discourses on risk—economic or otherwise—to consolidate political legitimacy (Diprose et al. 2008; Emel and Huber 2008), control populations, or induce individuals to control themselves (Swyngedouw 2005), developing strategies to rule through risk (Muller 2010). Our article traces how, in Ecuador, the Correa administration developed populist strategies to rule through the risks associated with heightened oil price volatility. On the one hand, the regime cultivated an image of Correa as a willful and capable leader, partially through a rhetoric that featured his ability to manage market risks. On the other hand, the government channeled windfall oil rents into building spectacular public works in contested sites of oil extraction, as "petro-populist landscapes" that materially verified Correa's petro-leadership. Although this government augmented funding for many less spectacular, everyday investments into social welfare and human capital programs, as well as infrastructure, we highlight how the volatility of contemporary global oil markets produced conditions for particular representational practices in contested sites. Ultimately, we observe transformations in petro-populist discourses and practices amid unprecedented oil price volatility that mark the emergence of a *speculative petro-state*, a state in which state actors claim to successfully gamble on volatile oil markets on behalf of the people.

We proceed in three stages. First, we describe the postneoliberal reorganization of the oil sector in Ecuador, focusing on the relative riskiness of new policies related to contracts, price hedging, and insurance. Then, we track the formation of new state discourses and practices of populist communication that highlighted market risk management. We draw on speeches, propaganda, official statements, and interviews with state actors conducted between 2014 and 2018, especially actors involved in oil governance, such as the state construction firm Ecuador Estratégico (Strategic Ecuador), the Planning Ministry (SENPLADES), and the Ministry of Finances. Although these sources offer a partial perspective on petro-governance, conditioned by class and political privileges, they also represent key state "nodes" in the translation of flows of oil money into the discourses and practices of a speculative petro-state and they allow us to trace how oil risk management turned into an important organizing factor of populist strategies of rule in Ecuador. We conclude with reflections on the broader relevance of this emerging dynamic for studying relations between populist politics and contemporary economic spheres.

Postneoliberal Oil Governance in Ecuador

Through much of the 2000s and 2010s, postneoliberal oil governance in Latin America was widely referred to as "resource nationalism" (Haslam and Heidrich 2016), as leaders like Correa discursively drew stark distinctions between "sovereign" resource control and "weak" neoliberal policies; however, oil producers like Ecuador, Argentina, Venezuela, and Brazil depended on foreign companies and lenders to extract their marginal, hard-to-access reserves (e.g., Espinasa et al. 2015; Monaldi 2015). With few exceptions, postneoliberal states in the region did not nationalize resources; rather, many renegotiated with foreign oil companies and lenders to increase state revenues from rising oil prices. As long as oil prices rose and remained high, petro-states could augment public spending and project imaginaries of endogenous development, driven ostensibly by sovereign control over resources and powerful, charismatic leaders.

Some governments forced contract renegotiations and replaced contracts in which they had shared a percentage of rents with oil companies for so-called service contracts, which paid companies a fixed price per barrel for extracting oil. That is, the state would pay companies a fixed fee for each barrel of oil, regardless of its market value. A number of petro-states have favored service contracts in recent years, such as Ecuador, Bolivia, and Mexico (Ghandi and Lin 2014), because they benefit the landlord state while prices rise. By the same token, they can be risky. For example, in 2015, as prices collapsed, the

Ecuadorian state paid companies an average fixed fee of $39 per barrel but received as little as $30 per barrel, producing oil at a net loss in the case of several contracts ("Rafael Correa Says" 2015). Thus, under these contracts, price volatility presents opportunity and risk, as price increases lead to enhanced revenues and declines might lead to untenable financial relations with companies and creditors.

In Ecuador, the Correa regime pursued a multipronged reengineering of the oil sector. It forced contract renegotiations between the state and foreign companies in 2010–2011 to increase revenues, generating eighteen service contracts that paid companies between $16 and $41 per barrel for durations of six or ten years ("The Noble Americas Group" 2014). In addition, to increasing revenues, the government rejected investing in common but costly practices to hedge against falling prices, such as private insurance or sovereign wealth and stabilization funds (i.e., savings and investment accounts). Instead, it liquidated Ecuador's Petroleum Stabilization Fund and two similar funds in 2008, transferring $4.5 billion from these funds into the state spending budget. The government also pursued new sources of credit to expand national oil production and refining. In 2010, it converted Ecuador's social security administration into Petroamazonas's largest source of credit. Petroamazonas also pursued a series of high-interest, short-term loans from Chinese state banks, which used lending to secure access to natural resources for the Chinese government (Gallagher and Irwin 2015). Furthermore, Ecuador took a $1 billion loan from a U.S. hedge fund to renovate its main refinery (and sought $13 billion for a new refinery; "The Noble Americas Group" 2014).

Correa's government took on these risks to be able to respond to citizens' demands for the development of social welfare programs, human capital, and infrastructure (Lu, Valdivia, and Silva 2016; Lyall, Colloredo-Mansfeld, and Rousseau 2018) and, in this sense, its risky maneuvers proved highly effective in the short term. Between 2007 and 2013, as the WTI oil price rose from $72 to well over $100 per barrel, Correa witnessed a threefold increase in the state budget (Muñoz Jaramillo 2014). Public spending totaled $129 billion in his first six years in office, compared to just $48 billion in the six previous years (Arévalo Luna 2014).

Increased public spending was featured in television, radio, and billboard propaganda to consolidate the political legitimacy of a postneoliberal state. A decade of political unrest, in which neoliberal governments had witnessed seven different presidents rise and fall, was followed by a decade of political stability under Correa. As we explore in the following section, the Correa regime framed its spending in populist terms, highlighting Correa's personal will and leadership capacities to confront changing, volatile markets on behalf of the nation.

Petro-Populism in Volatile Times

The Citizens' Revolution movement, as an anti-neoliberal, anti-austerity movement, emerged in the early 2000s out of growing cries to reconstitute state capacities for resolving socioeconomic inequalities (Acosta 1998; Acosta and Falconí 2005). Its political party, Alianza PAIS (Country Alliance), selected Correa as its representative for the 2006 elections and militants of the Citizens' Revolution quickly reorganized around the figure of Correa himself, seeking to sustain the legitimacy of the postneoliberal state by espousing Correa's personal will and capacities to micromanage and discipline the state apparatus to benefit all citizens (De la Torre 2013; Becker 2014). This strategy of characterizing Correa as the architect of the state was later reflected in calls within the Citizens' Revolution at the end of Correa's presidency to reverse the "Correa-ization" of the movement and refound it on shared political ideals.

Academics in Latin America have long debated the characteristics, dangers, and emancipatory potential of dominant, populist political figures like Correa (Germani 1971; Laclau and Mouffe 2001; Cadahia, Coronel, and Ramírez 2018). Ramírez Gallegos and Stoessel (2018) categorized this literature into several types, including analyses of populist styles; critiques of populist rhetoric as demagogy; and explorations of the potential of populist figures to cultivate progressive subjectivities. Each of these groupings draws from historical experiences in the region of political strategies that have aimed to unite diverse constituents behind a single leader and against established powers. Some analysts have argued that populism undermines democracy by concentrating power in a single leader (Roberts 2015; De la Torre 2013). Proponents observe how this strategy has constituted and moved "the political" (Laclau and Mouffe 2001), through the slow cultural work of constructing hegemony and wresting the

state from oligarchic control (Cadahia, Coronel, and Ramírez 2018; Coronel and Cadahia 2018). Here, we take no normative stance on populism; rather, we examine contemporary transformations in populism as a particular political performance and communication style—that is, a "political strategy" (Weyland 2013, 20)—that aims to constitute and stabilize political power by representing a personal relationship between a leader and a diverse constituency, united against a common threat.

Correa and his administration consistently used Manichean rhetoric to position him as defending "the people" from the threatening opposition of conservatives, the media, and environmentalist and indigenous critics (De la Torre 2013; Becker 2014). Since the mid-twentieth century, many successful politicians and political regimes in Latin America have used similar rhetoric, drawing dichotomies between the leader—often a political outsider bent on redeeming "the nation"—and internal or external threats, often represented by the oligarchy or foreign interests. In Ecuador, Correa's government presented him as a willful leader against neoliberals, as well as a technocratic expert, equipped with a doctorate in development economics and thus capable of extending the benefits of economic progress to marginalized segments of the population. The government projected this "techno-populism" (De la Torre 2013, 36) into the contentious terrain of oil governance, promising to give not only power but also oil back to the people. In propaganda, including his weekly television show, Correa described rising state oil revenues in terms of his will and capacities as a leader, rather than the whims of changing markets: "We were able to capture the oil bonanza thanks to the sovereign renegotiation of oil contracts. It was not luck!" (Andes 2017). As a savvy defender of national resources, Correa had reverted international power relations through renegotiations, placing "the people" in charge of oil flows and rents.

In addition, the Correa government celebrated 2010 reforms to the Hydrocarbons Law, which secured a 12 percent tax on private oil and mining profits to fund local development in "strategic" sites of resource extraction. Correa decreed that the national oil company also pay this percentage, and a Petroamazonas statement declared, "For the first time in petroleum history, the Amazon is taken into account and is beneficiary of 12 percent of the profits" ("The Neighbors of the Oil Camps" 2017). Correa, it seemed, had personally dealt marginal spaces into the oil boom. Although our research reveals that the government charged companies this 12 percent tax but did not channel those revenues directly into local development (instead budgeting compensation flexibly, according to political necessity), the 12 percent discourse served Correa and his administration to position him as a technocratic defender of resource rents in benefit of the people.

Correa's government sought to consolidate this image of petro-leadership in relation to a series of constituted "threats" to the nation and its resources. First, Correa drew a stark distinction between the postneoliberal present of service contracts and the neoliberal past, when national resources were shared with transnational companies because they were, he argued, controlled by "the bankers, the creditors, the international bureaucracies" ("President Correa Explains" 2014). Second, the government framed conservative economists as a threat, for urging greater caution in relation to changing oil markets. Correa held televised debates with critics whom he named "pseudo-economists," and he described their calls for stabilization funds and other critiques of oil contracts as direct threats to public spending and national well-being or *buen vivir* ("All of the Key Opinions" 2015). Correa ridiculed stabilization funds as "*fonditos*" (little funds) and "lazy money" that did not address the needs of the nation and qualified contract insurance as "clumsy" ("All of the Key Opinions" 2015). In short, Correa and the Correa regime framed the reengineering of the oil sector in terms of his moral and intelligent oil governance on behalf of the people, despite warnings from a seemingly brutish oligarchy and its conservative analysts.

In his weekly communiqués, Correa also attacked Ecuadorian anti-oil environmentalist and indigenous critics as external threats. He cast them outside the nation, as "anarchists" or dangerous, primitive "rock-throwers," framing environmental critiques as threats to national wealth and environmentalists as enemies of the people. Correa increasingly deployed this rhetoric, particularly after he asked the National Assembly to authorize oil extraction in Yasuní National Park in 2013 (Fiske 2017). The following year, the National Assembly, which was controlled by Alianza PAIS, drafted and approved a new penal code that featured harsher punishments for "sabotage" and "terrorism," categories that would be used to persecute environmentalist protesters and indigenous leaders. The government closed down one prominent environmentalist nongovernmental

organization, Pachamama, and threatened to close another called Acción Ecológica (Ecological Action). In a context of increasing socioenvironmental conflict, television propaganda repeated populist refrains such as, "for the people, what belongs to the people." The government's overall message regarding the oil sector was that reengineered oil contracts and financial arrangements had captured rising oil rents that belonged to the people and that conservatives and environmentalists threatened.

Proclaiming the postneoliberal development model the "Ecuadorian Miracle," Correa featured oil contract renegotiations as a key factor of postneoliberalism, which had freed the state "from the hands of financial power" and that engendered "the supremacy of the human being over capital" ("President Correa Explains" 2014, translated by authors). In April 2014, Correa gave a speech at Harvard University, where he highlighted new oil contract relations, particularly contracts that tied state revenues to international oil prices, as evidence of renewed sovereignty and popular power. Again, through such discourses, he endeavored to convert volatile oil market conditions into a source of political legitimacy.

Just three months later, however, oil prices collapsed. In June 2014, the WTI benchmark price began a decline from $105 per barrel. By 2015, the state received as little as $30 per barrel, despite paying oil companies an average of $39 ("Rafael Correa Says" 2015). Thus, the state paid some companies more to produce oil than the oil was worth. By early 2016, the WTI price slumped to $26 and Ecuadorian crude, which sells at a slightly lower price due to quality differences, dropped to under $20. The Ecuadorian economy entered into a recession, national oil companies experienced spiraling debt, and the state accumulated fiscal deficits that it only partially covered by selling off portions of future oil production to the Chinese government.

In the next and final section, we examine how oil price volatility entered not only into the content of petro-populist discourse but also into material practices of populist communication in contested sites of oil extraction, transport, and refining.

Petro-Populist Landscapes

In the 2000s, state-delivered compensation packages to contested sites of oil flows—including cash payments and development projects—became key strategies for reducing local conflict, securing oil flows, and legitimating oil governance more broadly (Brosio and Singh 2015). Within a discursive framework of recovering popular sovereignty over resources, petro-states took over the practice of negotiation and social compensation from private oil companies (Billo 2015). Venezuela, for example, increased compensation spending from $249 million in 2003 to $13.26 billion in 2006 (Franco 2008). In Ecuador, the Planning Ministry (SENPLADES) proposed prioritizing development in contested sites through centralized planning, but Correa and his closest advisers, such as the Coordinating Minister of Strategic Sectors Jorge Glas and the general manager of Petroamazonas Wilson Pástor, rejected this model in favor of one that would more rapidly and visibly mark the gains of Correa's petro-leadership.

Already under pressure to demonstrate that renegotiations of oil contracts could generate increased production, revenues, and development, the government was witnessing large protests from environmentalists and indigenous sectors in 2010 and 2011 that questioned the state's ongoing reliance on resource extraction. With electoral campaigns approaching, one former SENPLADES official recalls, "The electoral pragmatism of 2011 and 2012 was extreme … the president said, 'Planners are great, but I need things urgently'" and "with effect." If the Citizens' Revolution was the promise of state-driven development, achieved largely through Correa's willful and intelligent control of oil market relations, then its government aimed to quickly and conspicuously demonstrate this vision in response to opponents, particularly in contested sites of oil extraction.

In 2011, Correa and his top advisers planned and created the state entity Ecuador Estratégico, staffed mainly by architects and engineers who swiftly built public works projects in remote spaces of resource extraction to secure local consent and materialize spectacular landscapes for a national audience. In 2012, this public company had a staff of eighty-seven and compensation budget of $115 million (Ecuador Estratégico 2016). The following year, the budget doubled. By 2014, 296 functionaries had planned or completed 1,214 infrastructure projects and spent three-quarters of a billion dollars, largely in oil logistics areas. Some interventions to reorder spaces took place in non-Amazonian places, such as neighborhoods surrounding the Esmeraldas refinery, where Ecuador Estratégico built health clinics, a school, and new streets. In general,

Ecuador Estratégico's diverse interventions did not approach the scale of many of the Correa regime's most extensive and emblematic public works projects, such as new highways or regional irrigation canals; rather, they were targeted, visually impactful projects. By the end of 2015, Ecuador Estratégico had intervened in nearly 1,000 communities with infrastructure projects, including 226 electrification projects and fifty-four modern schools (Ministerio Coordinador de Sectores Estratégicos 2016).

Government propaganda featured these projects on billboards, television, and state newspapers, as evidence of how proper leadership amid changing oil markets expands citizenship rights. For example, in 2015, Correa inaugurated a "Millennium Educational Unit" or modern school in Dayuma, a site in the Amazon that had witnessed a deadly clash between the military and indigenous groups over oil extraction when Correa had recently taken office. The school of 430 students cost $5.6 million, a staggering sum for a single school in a country where most schools have just one or two teachers. On that occasion, Correa repeated to the cameras that due to the "renegotiation of the oil contracts, 12 percent of the profits from the extraction of oil stays" in local communities (Andes 2015).

Again, the dominant political strategy aimed to further consolidate this imaginary of Correa's petro-leadership in relation to specific threats to the nation and its resources. In the televised inauguration of the first Millennium City, for example, Correa described it as his "personal revenge" and the "revenge of an entire nation against the conspirators … the rock-throwers"; that is, against antioil activists. Oil, he argued, is "a blessing, not a curse … here and now, we are demonstrating it" (Presidencia 2013). Although he only witnessed the construction of five Millennium Cities, these settlements supplied images that played repeatedly on state-run television to contradict conservatives and environmentalists alike and unite constituents against the traditional right and the anti-oil left.

With the collapse in oil prices, Ecuador Estratégico's spending budget plummeted from $216 million to $37 million (Ecuador Estratégico 2016). As we mentioned in the opening of this article, the lights went out in Playas del Cuyabeno and other strategic sites. By 2017, Ecuador Estratégico was reassigned almost exclusively to building ordinary public housing. In 2018, the new government declared that the institution would be fully absorbed by a public housing initiative.

In retrospect, oil price volatility had spatial and temporal impacts on the governance of oil flows in contested sites. First, volatility facilitated the appearance of an entirely new institution and budget, geared toward representing new state–citizen relations and securing oil flows in geographically variegated, strategic landscapes. A former vice-minister explains in an interview that it is not only high oil prices that enable the creation of new institutions like Ecuador Estratégico but also the fast rate of increasing oil revenues. In moments of stable state revenues, ministries compete over available funds, generally for existing national programs; however, in moments in which revenues rapidly increase, in-flows of revenues challenge institutional spending capacities, facilitating the creation of new, flexible, or strategic institutions with overtly political aims. Second, price volatility enabled accelerated forms of governance, such as the rapid construction of targeted, conspicuous public works, often with no plans or budgets for long-term maintenance. Ecuador Estratégico's interventions were designed for immediate visual impact and were often not integrated into local planning or operating budgets. In turn, price volatility also ensured that Ecuador Estratégico would not be able to maintain interventions over the long term.

Following the 2014 price collapse, Correa's government reframed the notion of oil market risk, characterizing Ecuador's recession as the result of a "perfect storm"—an unpredictable, destructive force that had befallen the nation and its resources. Correa's credibility declined, however (CEDATOS 2015). In 2017, Correa's former vice-president, Lenín Moreno, narrowly won the presidency over a conservative banker and consolidated popular support by investigating corruption scandals under Correa, particularly in the oil sector (Lyall 2018). Moreno has since constituted a new petro-populist grid of intelligibility in relation to oil market risk, positioning himself as defender of national resources against risky policies. He has frequently referred to the fundamental uncertainty of economic dynamics (even citing the uncertainty principle in quantum mechanics to make his case), and he has reframed Correa as a corrupt and overconfident gambler, who exposed the nation to volatile prices and risky loan agreements. Thus, oil risk has continued to function as an organizing principle of populist performances. Today, Moreno draws popular support by criticizing his immoral, risk-taking predecessor and the "white elephants" he built ("Moreno Questions Investment in Schools" 2016).

Conclusion

In this article, we have examined the entanglements of oil price volatility and oil governance in Ecuador, tracking how volatility entered into populist claims to political legitimacy. Following years of neoliberal austerity and political instability, Rafael Correa and the Citizens' Revolution promised to return the nation and its resources to the people by inaugurating a postneoliberal era of sovereign, oil-driven development. A collection of state entities and institutions were enabled by sharply increasing oil rents to construct and project a narrative about the personal will and capacity of Correa to manage changing, volatile oil markets on behalf of the nation. Despite criticisms from the traditional right and the environmentalist and indigenous left, they reinforced this narrative through the production of spectacular public work projects in once-contested sites of oil extraction, transport, and refining.

These transformations in populist discourse and practices in relation to price volatility speak not only to the Ecuadorian experience but also to the emergent problem among contemporary petro-states of how to project authority over national resources in increasingly unpredictable conditions. For petro-states, the management and control of flows of oil and oil money has long been vital to state legitimacy, as diverse regimes have used oil rents to foster support (Coronil 1997; Karl 1997; Bridge and Le Billon 2017). Our analysis of the relationship between price volatility and populist regimes suggests that oil price risk might become a threat to political stability, but it can also generate new idioms and material practices of populist legitimacy. Governments might be more successful at leveraging risk in their favor during price upswings, as they position charismatic leaders as savvy and successful managers of market uncertainties, but, as we observe in the case of Moreno, volatility could also be used to seek legitimacy during downswings, as politicians denounce risk-taking predecessors and position themselves as prudent managers. In either context, the Ecuadorian case demonstrates how market risk management can function as an organizing principle of contemporary populist petro-rule.

This article also sheds light on broad questions of contemporary populism in Latin America. During the postneoliberal moment, much attention was paid to political iconoclasts that hailed the supremacy of human beings over capital. How market dynamics interpolated and enabled postneoliberal discourses and institutions has gone understudied. Although the Correa government deployed multiple legitimizing strategies, in this article we examined how petro-populist communication enrolled oil price volatility into new discourses and practices of oil governance and, in turn, new claims to political legitimacy organized around risk.

Stuart Hall (1986) was influenced by Laclau and Mouffe's opposition to economic reductionism but nonetheless questioned them for going too far, for reducing politics to a "totally open discursive field" (56) in which populists constitute political subjectivities. Similarly, we have demonstrated how economics shape the conditions of possibility for populism's representational field in Ecuador. This suggests the need to examine the economic conditions that not only produce booms and busts into which populist leaders emerge but also shape the very terms of populist political legitimation. Persistently turbulent oil markets and other manifestations of market volatility under contemporary global capitalism might induce further openings for charismatic leaders, with new claims to understand, mediate, or conquer risk on behalf of the people.

Acknowledgments

We thank Christian Lentz for his helpful comments on an early version of this article. Jeremy Rayner and Benjamin Rubin also provided insightful feedback on later versions. Elizabeth Havice provided suggestions that were invaluable for getting the article into its final stage. We are very grateful to James McCarthy and two anonymous reviewers for their patience, encouragement, and comments. Any shortcomings or omissions in the final product are our own responsibility.

ORCID

Angus Lyall ⓘ http://orcid.org/0000-0002-9199-5415
Gabriela Valdivia ⓘ https://orcid.org/0000-0001-9442-8005

Note

1. Although petro-states like Ecuador might share similar development dependencies and characteristics with mining countries (e.g., Bolivia), petro-dollars and oil wealth powerfully shape governance institutions in petro-states (Karl 1997). Petro-states, of course, might follow different paths. In Ecuador, the recent rise of the mining sector might lead to different modalities of dependence on extractivist economies, although this is still a nascent development (van Teijlingen et al. 2017; Vela-Almeida, Kolinjivadi, and Kosoy 2018). By the same token, mining-dependent countries also have the potential to become petro-states (Perreault and Valdivia 2010; Anthias 2017).

References

Acosta, A. 1998. *El estado como solución* [The state as the solution]. Quito, Ecuador: Friedrich Ebert/ILDIS.

———. 2013. Extractivism and neoextractivism: Two sides of the same curse. In *Beyond development*, ed. M. Lang and D. Mokrani, 61–86. Quito, Ecuador: Fundación Rosa Luxemburgo.

Acosta, A., and F. Falconí. 2005. *Asedios a lo imposible: Propuestas económicas en construcción* [Sieges of the impossible: Economic proposals in construction]. Quito: FLACSO-Ecuador.

Andes. 2015. Presidente Correa destaca uso de recursos petroleros en la Amazonía en inauguración de UEM en Dayuma [President Correa emphasizes the use of oil resources in the Amazon in the innauguration of the UEM in Dayuma]. *Andes*, May 7. Accessed April 20, 2018. http://www.andes.info.ec/es/noticias/sociedad/1/38957/presidente-correa-destaca-usorecursos-petroleros-amazonia-inauguracion-uem-dayuma

———. 2017. Presidente Correa destaca que el verdadero cambio fue posible cuando se modificaron las relaciones de poder [President Correa emphasizes that true change was possible when power relations were modified]. *Andes*, March 24. Accessed April 19, 2018. http://www.andes.info.ec/es/noticias/actualidad/1/56085/presidente-correadestaca-verdadero-cambio-fue-posible-cuando-modificaron-relaciones-poder

Anthias, P. 2017. Ch'ixi landscapes: Indigeneity and capitalism in the Bolivian Chaco. *Geoforum* 82:268–75. doi:10.1016/j.geoforum.2016.09.013.

Appel, H., A. Mason, and M. Watts, eds. 2015. *Subterranean estates: Life worlds of oil and gas*. Ithaca, NY: Cornell University Press.

Arévalo Luna, G. 2014. Ecuador: Economía y política de la revolución ciudadana, evaluación preliminar [Ecuador: Economy and policies of the Citizens' Revolution, primary evaluation]. *Revista Apuntes del CENES* 33 (58):109–34.

Bebbington, A. 2015. Political ecologies of resource extraction: *Agendas pendientes*. *European Review of Latin American and Caribbean Studies* 100:85–98. doi:10.18352/erlacs.

Becker, M. 2014. Rafael Correa and social movements in Ecuador. In *Latin America's radical left: Challenges and complexities of political power in the twenty first century*, ed. S. Ellner, 127–47. New York: Rowman and Littlefield.

Bielschowsky, R. 2009. Sesenta años de la CEPAL: Estructuralismo y neoestructuralismo [Seventy years of ECLAC: Structuralism and neo-structuralism]. *Revista Cepal* 97:173–94.

Billo, E. 2015. Sovereignty and subterranean resources: An institutional ethnography of Repsol's corporate social responsibility programs in Ecuador. *Geoforum* 59 (1):268–77. doi:10.1016/j.geoforum.2014.11.021.

Bridge, G., and P. Le Billon. 2017. *Oil*. New York: Wiley.

Brosio, G., and R. J. Singh. 2015. *Raising and sharing revenues from natural resources*. Washington, DC: World Bank.

Cadahia, L., V. Coronel, and F. Ramírez. 2018. *A contracorriente: Materiales para una teoría renovada del populismo* [Against the current: Material for a renewed theory of populism]. La Paz: Vicepresidencia de Bolivia (Vice-Presidency of Bolivia).

CEDATOS. 2015. *Aprobación Presidente* [President approval]. Accessed April 19, 2018. http://www.cedatos.com.ec/detalles_noticia.php?Id=231.

Cielo, C., and N. Carrión Sarzosa. 2018. Transformed territories of gendered care work in Ecuador's petroleum circuit. *Conservation and Society* 16 (1):8–20. doi:10.4103/cs.cs_16_77.

Coronel, V., and L. Cadahia. 2018. Populismo republicano: Más allá de Estado versus pueblo [Republican populism: Beyond the state versus the people]. *Nueva Sociedad* 273:72–82.

Coronil, F. 1997. *The magical state: Nature, money, and modernity in Venezuela*. Chicago: University of Chicago Press.

De la Torre, C. 2013. In the name of the people: Democratization, popular organizations, and populism in Venezuela, Bolivia, and Ecuador. *European Review of Latin American and Caribbean Studies* 95 (1):27–48.

Diprose, R., N. Stephenson, C. Mills, K. Race, and G. Hawkins. 2008. Governing the future: The paradigm of prudence in political technologies of risk management. *Security Dialogue* 39 (2–3):267–88. doi:10.1177/0967010608088778.

Ecuador Estratégico (Strategic Ecuador). 2016. *Plan estratégico: 2015–2017* [Strategic plan: 2015–2017]. Accessed February 5, 2017. http://www.ecuadorestrategicoep.gob.ec/images/leytransparencia/Enero2016/Plan_Estrategico2015-2017.pdf.

El grupo Noble Americas presto USD 1000 millones a Petroecuador [The Noble Americas Group lent 1000 million USD to Petroecuador]. 2014. *El Comercio*, September 27. Accessed April 20, 2018. http://www.elcomercio.com/actualidad/noble-americas-prestomilmillones.html.

Emel, J., and M. Huber. 2008. A risky business: Mining, rent and the neoliberalization of risk. *Geoforum* 39 (3):1393–1407. doi:10.1016/j.geoforum.2008.01.010.

Espinasa, R., E. Marchan, and C. Sucre. 2015. Financing the new silk road: Asian investment in Latin America's energy and mineral sector. Technical Note 834, 118, Inter-American Development Bank, Washington, DC.

Espinosa Andrade, A. 2017. Space and architecture of extractivism in the Ecuadorian Amazon region.

Cultural Studies 31 (2–3):307–30. doi:10.1080/09502386.2017.1303430.

Ewing, B., and F. Malik. 2017. Modelling asymmetric volatility in oil prices under structural breaks. *Energy Economics* 63:227–33. doi:10.1016/j.eneco.2017.03.001.

Fiske, A. 2017. Natural resources by numbers: The promise of "El uno por mil" in Ecuador's Yasuní-ITT oil operations. *Environment and Society* 8 (1):125–43. doi:10.3167/ares.2017.080106.

Franco, R. 2008. Venezuela: Energy, the tool of choice. In *Energy and development in South America: Conflict and cooperation*, ed. C. Arnson, C. Fuentes, and F. Rojas Aravena, 35–40. Santiago, Chile: FLACSO.

Gallagher, K. P., and A. Irwin. 2015. China's economic statecraft in Latin America: Evidence from China's policy banks. *Pacific Affairs* 88 (1):99–121.

Germani, G. 1971. *Política y sociedad en una epoca de transición* [Politics and society in an epoch of transition]. Buenos Aires: Paidós.

Ghandi, A., and C. Lin. 2014. Oil and gas service contracts around the world: A review. *Energy Strategy Reviews* 3:63–71. doi:10.1016/j.esr.2014.03.001.

Gudynas, E. 2012. Estado compensador y nuevos extractivismos: Las ambivalencias del progresismo sudamericano [Compensation state and new extractivisms: The ambivalences of South American progressivism]. *Nueva Sociedad* 237:128–46.

Hall, S. 1986. On postmodernism and articulation: An interview with Stuart Hall. *Journal of Communication Inquiry* 10 (2):45–60. doi:10.1177/019685998601000204.

Haslam, P., and P. Heidrich, eds. 2016. *The political economy of natural resources and development: From neoliberalism to resource nationalism*. London and New York: Routledge.

Karl, T. 1997. *The paradox of plenty: Oil booms and petrostates*. Berkeley: University of California Press.

Labban, M. 2010. Oil in parallax: Scarcity, markets, and the financialization of accumulation. *Geoforum* 41 (4):541–52. doi:10.1016/j.geoforum.2009.12.002.

Laclau, E., and C. Mouffe. 2001. *Hegemony and socialist strategy: Towards a radical democratic politics*. New York: Verso.

Liu, L., Y. Wang, C. Wu, and W. Wu. 2016. Disentangling the determinants of real oil prices. *Energy Economics* 56:363–73. doi:10.1016/j.eneco.2016.04.003.

Los vecinos de campos petroleros accedieron a servicios [The neighbors of the oil campsaccessed services]. 2017. *El Telégrafo*, May 20. Accessed February 30, 2018. http://www.eltelegrafo.com.ec/noticias/politiko2017/49/losvecinos-de-campospetroleros-accedieron-a-servicios.

Lu, F., G. Valdivia, and N. Silva. 2016. *Oil, revolution, and indigenous citizenship in Ecuadorian Amazonia*. New York: Springer.

Lyall, A. 2017. Voluntary resettlement in land grab contexts: Examining consent on the Ecuadorian oil frontier. *Urban Geography* 38 (7):958–73. doi:10.1080/02723638.2016.1235933.

———. 2018. A moral economy of oil: Corruption narratives and oil elites in Ecuador. *Culture, Theory, and*

Critique 9 (4):380–99. 1–20. doi:10.1080/14735784.20
18.1507752.

Lyall, A., R. Colloredo-Mansfeld, and M. Rousseau. 2018.
Development, citizenship, and everyday appropriations of buen vivir: Ecuadorian engagement with the
changing rhetoric of improvement. *Bulletin of Latin
American Research* 37 (4):403–16. doi:10.1111/
blar.12742.

Ministerio coordinador de sectores estratégicos
(Coordinating Ministry of Strategic Sectors). 2016.
Rendición de Cuentas: 2015 [Accountability: 2015].
Quito, Ecuador: Ministerio coordinador de sectores
estratégicos.

Ministerio de Finanzas (Ministry of Finances). 2014.
Presupuesto general del estado: 2014 [General budget of
the state: 2014]. Quito, Ecuador: Ministerio de
Finanzas. Accessed August 1, 2018. http://www.finanzas.
gob.ec/wp-conent/uploads/downloads/2014/05/Justificativo-
de-Ingresos-y-Gastos_2014.pdf.

Mitchell, T. 2011. *Carbon democracy: Political power in the
age of oil.* New York: Verso.

Moffitt, B. 2016. *The global rise of populism: Performance,
political style, and representation.* Palo Alto, CA:
Stanford University Press.

Moreno cuestiona inversion en escuelas [Moreno questions investment in schools]. 2016. *Expreso*, October
30. Accessed April 19, 2018. http://www.expreso.ec/
actualidad/moreno-cuestiona-inversion-en-escuelas-
MJ812439.

Monaldi, F. 2015. Latin America's oil and gas: After the
boom, a new liberalization cycle? *ReVista: Harvard
Review of Latin America: Energy, Oil, Gas and Beyond*
15 (1):2–7.

Muller, B. 2010. *Security, risk and the biometric state:
Governing borders and bodies.* London and New York:
Routledge.

Muñoz Jaramillo, F., ed. 2014. *Balance crítico del gobierno
de Rafael Correa* [Critical balance of the government
of Rafael Correa]. Quito, Ecuador: Arcoiris
Producción Gráfica.

O'Malley, P. 2004. The government of risks. In *The
Blackwell companion to law and society*, ed. A. Sarat,
292–308. New York: Wiley.

Perreault, T., and G. Valdivia. 2010. Hydrocarbons, popular protest and national imaginaries: Ecuador and
Bolivia in comparative context. *Geoforum* 41
(5):689–99. doi:10.1016/j.geoforum.2010.04.004.

Petroamazonas. 2013. Comunidades del Milenio:
Pañacocha—Playas del Cuyabeno [Communities of
the millennium: Pañacocha—Playas del Cuyabeno].
Accessed April 19, 2018. http://www.petroamazonas.
gob.ec/wp-content/uploads/downloads/2015/08/comu-
nidades-del-milenio6.pdf.

Presidencia (Presidency). 2013. Comunidad del Milenio
Playas de Cuyabeno [Community of the Millennium
Playas de Cuyabeno]. Accessed January 5, 2014. http://
www.presidencia.gob.ec/wp-content/uploads/downloads/
2013/10/2013-10-01-ComunidadMilenioCuyabeno.pdf.

Presidente Correa explica en Harvard como libero al
Ecuador del poder financiero [President Correa
explains in Harvard how he liberated Ecuador from
financial power]. 2014. *El Ciudadano*, April 9.

Accessed April 20, 2018. http://www.elciudadano.gob.
ec/presidente-correa-explica-en-harvard-como-liberoal-
ecuador-del-poder-financiero/.

Purcell, T., N. Fernandez, and E. Martinez. 2017. Rents,
knowledge and neo structuralism: Transforming the
productive matrix in Ecuador. *Third World Quarterly*
38 (4):918–38. doi:10.1080/01436597.2016.1166942.

Rafael Correa dice que se viven momentos duros por vol-
can y caída de precio del petroleo [Rafael Correa says
that difficult moments are experienced due to the
volcano and the fall of the oil price]. 2015. *El
Universo*, August 25. Accessed April 15, 2018. http://
www.eluniverso.com/noticias/2015/08/25/nota/5085482/
presidente-dice-que-se-viven-momentos-durosvolcan-
caida-precio.

Ramírez Gallegos, F., and S. Stoessel. 2018. El incómodo
lugar de las instituciones en la "populismología" latin-
oamericana [The uncomfortable place of the institutions in the Latin American "populismology"].
Estudios Políticos, Universidad de Antioquia 52:106–27.

Roberts, K. 2015. Populism, political mobilizations, and
crises of political representation. In *The promise and
perils of populism: Global perspectives*, ed. C. de la
Torre, 140–58. Lexington: University of Kentucky
Press.

Rose, N., and P. Miller. 2010. Political power beyond the
state: Problematics of government. *The British Journal
of Sociology* 61 (Suppl. 1):271–303. doi:10.1111/
j.1468-4446.2009.01247.x.

Silveira, M., M. Moreano, N. Romero, D. Murillo, G.
Ruales, and N. Torres. 2017. Geografías de sacrificio
y geografías de esperanza: Tensiones territoriales en el
Ecuador plurinacional [Geographies of sacrifice and
geographies of hope: Territorial tensions in plurinational Ecuador]. *Journal of Latin American Geography*
16 (1):69–92. doi:10.1353/lag.2017.0016.

Svampa, M. 2015. Commodities consensus:
Neoextractivism and enclosure of the commons in
Latin America. *South Atlantic Quarterly* 114
(1):65–82. doi:10.1215/00382876-2831290.

Swyngedouw, E. 2005. Governance innovation and the
citizen: The Janus face of governance beyond-the-
state. *Urban Studies* 42 (11):1991–2006. doi:10.1080/
00420980500279869.

Todas las opiniones clave del debate entre Correa y econ-
omistas críticos [All of the key opinions from the
debate between Correa and critical economists].
2015. *El Telégrafo*, October 29. Accessed May 6,
2018. http://www.eltelegrafo.com.ec/noticias/politica/
2/todas-las-opiniones-claves-del-debate-entre-correa-
yeconomistas-criticos-resumen.

Valdivia, G. 2008. Governing relations between people
and things: Citizenship, territory, and the political
economy of petroleum in Ecuador. *Political Geography*
27 (4):456–77. doi:10.1016/j.polgeo.2008.03.007.

van Teijlingen, K., E. Leifsen, C. Fernández-Salvador, and
L. Sánchez-Vasquez. 2017. *La Amazonía minada:
Minería a gran escala y conflictos en el sur de Ecuador*
[The mined Amazon: Large-scale mining and conflicts
in the south of Ecuador]. Quito, Ecuador:
Universidad de San Francisco de Quito Press.

Vela-Almeida, D., V. Kolinjivadi, and N. Kosoy. 2018. The building of mining discourses and the politics of scale in Ecuador. *World Development* 103:188–98. doi:10.1016/j.worlddev.2017.10.025.

Wacaster, S. 2016. *2014 Minerals yearbook, U.S. Department of the Interior, U.S. Geological Survey: ECUADOR.* Washington, DC: U.S. Geological Survey. Accessed August 1, 2018. http://www.minerals.usgs.gov/minerals/pubs/country/2014/myb3-2014-ec.pdf.

Watts, M. 2004. Antinomies of community: Some thoughts on geography, resources and empire. *Transactions of the Institute of British Geographers* 29 (2):195–216. doi:10.1111/j.0020-2754.2004.00125.x.

Weyland, K. 2013. The threat from the populist left. *Journal of Democracy* 24 (3):18–32. doi:10.1353/jod.2013.0045.

Wilson, J., and M. Bayón. 2018. Potemkin revolution: Utopian jungle cities of 21st century socialism. *Antipode* 50 (1):233–54. doi:10.1111/anti.12345.

Zalik, A. 2010. Oil "futures": Shell's scenarios and the social constitution of the global oil market. *Geoforum* 41 (4):553–64. doi:10.1016/j.geoforum.2009.11.008.

ANGUS LYALL is a PhD Candidate in the Department of Geography at the University of North Carolina–Chapel Hill, Chapel Hill, NC 27514. E-mail: angusl@live.unc.edu. His current research examines the intersection of resource governance and urban expansion in Latin America, exploring how embodied postcolonial histories shape urban aspirations in rural spaces.

GABRIELA VALDIVIA is an Associate Professor in the Department of Geography at the University of North Carolina–Chapel Hill, Chapel Hill, NC 27514. E-mail: valdivia@email.unc.edu. Her research focuses on the political ecology of natural resource governance in Latin America; how states, firms, and civil society appropriate and transform resources to meet their interests; and how capturing and putting resources to work transforms cultural and ecological communities.

Contradictions of Populism and Resource Extraction: Examining the Intersection of Resource Nationalism and Accumulation by Dispossession in Mongolia

Orhon Myadar and Sara Jackson

We examine contradictions of populism and resource extraction in Mongolia in the context of the recent presidential election of Khaltmaa Battulga, who is often portrayed as dangerously populist. We consider Battulga's victory as an echo of Mongolian voters' sense of dispossession and discontent driven by gross wealth disparity and precarious livelihoods. Rather than treating these concerns as mere tools of the populist political agenda, we view them as moments of resistance to the asymmetry between accumulation and dispossession in Mongolia, a central outcome of twenty-five years of the neoliberal regime. We situate our analysis of Mongolia's resource politics through an examination of the world's second largest undeveloped copper–gold mine, Oyu Tolgoi. The mine offers a window into the country's turbulent resource politics that has concentrated wealth among a powerful few while nearly one third of Mongolians remain trapped in vicious poverty. Relying on fieldwork conducted over several years, the article argues that public grievances against the asymmetry of accumulation and dispossession are routinely discounted by discursive tools within the populist paradigm. "Resource nationalism," in particular, is used by those who promote neoliberalism and the open market as a pejorative label to silence public grievances.

我们检视经常被描绘成高度危险的民粹主义者——哈勒特马．巴特图勒嘎（Khaltmaa Battulga）在蒙古的晚近总统选举脉络下，民粹主义与资源搾取之间的冲突。我们将巴特图勒嘎的胜利，视为蒙古选民对严重的财富不均和不稳定的生计所导致的剥夺感和不满之反应。我们并非将这些考量视为仅只是民粹政治议程的工具，而是将其视为对蒙古的积累与剥夺之间的不对称之反抗时刻，该现象是长达二十五年的新自由主义体制所造成的主要结果。我们通过检视世界第二大尚未开发的铜金矿奥尤陶勒盖（Oyu Tolgoi）来定位蒙古的资源政治。该矿场提供窥探蒙古财富聚焦少数有权力者、但近乎三分之一的蒙古人仍深陷极度贫穷的动盪资源政治的一扇窗。本文根据若干年的田野工作，主张公众对于不对等的积累与剥夺之不满，一再受到民粹主义范式的论述工具所贬抑。特别是"资源国族主义"被提倡新自由主义者与市场开放者用来作为使公众不满消音的贬抑标籤。*关键词：掠夺式积累，蒙古，奥尤陶勒盖，民粹主义，资源国族主义。*

Examinamos las contradicciones del populismo y la extracción de recursos en Mongolia en el contexto de la reciente elección presidencial de Khaltmaa Battulga, quien a menudo es retratado como peligrosamente populista. Consideramos la victoria de Battulga como un eco del sentido de desposesión y descontento de los votantes de Mongolia, motivado por la desigualdad bruta de la riqueza y los precarios niveles de vida. Más que tratar estas preocupaciones como simples herramientas de la agenda política populista, las consideramos como momentos de resistencia a la asimetría entre acumulación y desposesión en Mongolia, un resultado principal de veinticinco años de régimen neoliberal. Situamos nuestro análisis de la política de los recursos de Mongolia por medio de un examen de la segunda más grande mina subdesarrollada de cobre-oro del mundo, Oyu Tolgoi. La mina es una ventana hacia la turbulenta política de los recursos del país que ha concentrado la riqueza entre unos pocos poderosos mientras que cerca de un tercio de los mongoles siguen atrapados en una pobreza descomunal. Dependiendo del trabajo de campo conducido durante varios años, el artículo arguye que las quejas públicas contra la asimetría de la acumulación y la desposesión son rutinariamente descartadas con herramientas discursivas dentro del paradigma populista. EL "nacionalismo de los recursos," en particular, es usado por quienes promueven el neoliberalismo y el libre mercado como etiqueta peyorativa para silenciar los reclamos públicos. *Palabras clave: acumulación por desposesión, Mongolia, nacionalismo de los recursos, Oyu Tolgoi, populismo.*

Once hailed as the world's fastest growing economy, Mongolia's economy has rapidly declined since 2013 as Chinese demand for key export minerals has slowed (Hoyle 2016; Bloomberg 2017). Amidst the ongoing crisis, voters elected Khaltmaa Battulga as the country's fifth president since Mongolia transitioned from the Soviet-backed one-party regime in the 1990s. Battulga's victory put the small northeast Asian nation of 3 million into the political spotlight as his victory unsettled and awed spectators and pundits.

By many measures, Battulga is an unorthodox candidate. He is popularly known as Genco Battulga, a nod to his business empire named after Genco Abbandando, the Corleone family's consigliere in the *Godfather* series. Like the fictional character, Battulga is both alluring and dubious. Enkhtsetseg Davga of the Open Society Forum stated that Battulga appealed to voters because he spoke the "people's language." As a wrestler, Battulga is relatable to many voters because wrestling is one of Mongolia's most popular national sports, symbolizing strength, sportsmanship, and integrity (E. Davga, personal communication, 9 March 2018). His personal envoy Gansukh Amarjargal shared a similar sentiment. Unlike his opponents and all former presidents, Battulga did not go to university, nor is he a political elite who shaped Mongolia's post-Soviet political landscape. He spent most of the post-Soviet years building his business empire, becoming one of Mongolia's richest men. His wealth, not political prowess, propelled him into politics in 2004 as a member of Parliament. He later served as Minister of Roads, Transportation, Construction and Urban Development. His political tenure, however, was marred by allegations of corruption, leading to the arrests of many of his associates.

Given his shady background, Battulga's 2017 presidential campaign against Miyeegombyn Enkhbold, the ruling party candidate and a seasoned politician,[1] seemed a long shot. Battulga stunned the political establishment with a victory running the campaign slogan "Mongolia Will Triumph."[2] Both within and outside Mongolia, Battulga's win was seen as another populist stab at establishment politics, pitting the "people" against the "power bloc" (Hart 2013).[3] Battulga's appeal, however, cannot be reduced simply to populism, nor can he be easily dismissed as an opportunistic leader manipulating the mindless and alienated masses. To do so would underestimate the complex forces that played into his popular support and ultimate victory against the odds.

This article considers Battulga's victory in the context of Mongolian voters' sense of dispossession and discontent, driven by gross wealth disparity, precarious livelihoods, and uncertain future conditions. Rather than treat these concerns as mere tools of the populist political agenda, we consider them salient grievances that emerged in response to the neoliberal regime of the last twenty-five years. Echoing the public outcry, Battulga highlighted poverty and unemployment as the most critical national security issues in his inaugural address (Office of the President of Mongolia 2017).

We situate our analysis within resource politics that propelled Mongolia's spectacular economic growth and drove its equally remarkable downturn. Harvey's (2004) work on "new" imperialism and accumulation by dispossession informs our conceptualization of Mongolia's resource politics and booms and busts. He suggested that neoliberal economic policies concentrate wealth among a powerful few while dispossessing the masses of their wealth, land, and natural resources. Dispossession illuminates not only the popular struggle against the hegemonic structure of Mongolia's political establishment but also voters' response to global capitalism and neoliberal policies.

According to our field research[4] and the work of other scholars (Rossabi 2005, 2009; Empson and Webb 2014), neoliberal policies have allowed the accumulation of wealth and power into the hands of a few—both domestic and foreign—whereas the majority have not benefited from Mongolia's mining riches but have had to bear more negative externalities. Public grievances against accumulation and dispossession have been discounted as populist cries. To silence public grievances, opponents frame these policies as resource nationalism to discredit them by those who promote neoliberalism and open markets.

Surprisingly, the intersection between resource nationalism and accumulation by dispossession is largely absent from the literature on resource politics. Although there is discussion of resource nationalism as it relates to the theorization of sovereignty (Emel et al. 2011) and postneoliberal order in the spaces of resource curse and resurgent leftist governments in various countries around the world (Perreault 2006; Kohl and Farthing 2012; Bebbington and Bury 2013; Ciccantell and Patten 2016), there is little discussion of how resource nationalism is used to push and perpetuate neoliberal politics while silencing objections to structural dispossession. We hope to fill this gap.

Global Capitalism, Neoliberalism, and Minegolia

In the early 1990s, the collapse of the Soviet Union allowed the "triumphant" return of the neoliberal agenda on a global scale (Harvey 2006), including unprecedented influence in Mongolia. This experience articulates the struggles faced by many developing countries situated along the global core–periphery divide (Holden, Nadeau, and Jacobson 2011; Veuthey and Gerber 2012). In dismal condition after the Soviet Union collapsed, Mongolia, along with other former Soviet satellites, experienced soaring unemployment, astronomical inflation, and civil unrest. It soon became dependent on aid through neoliberal experiments of the Washington Consensus (Myadar and Rae 2014).

Much-needed aid came with the adoption of "shock therapy." Among conditions attached to donor aid, privatization of state properties has been a signature policy restructuring Mongolia's economy, allowing private appropriation of public goods and property (Rossabi 2005; Myadar and Rae 2014). Harvey (2004) called this the cutting edge of accumulation by dispossession as resources and assets were transferred from state to private hands.

The nexus of neoliberal policy and resource politics is well documented. Numerous scholars have examined neoliberalism's consequences on the global political economy in general (Chomsky 1999; McCarthy and Prudham 2004; Harvey 2005; Heynen et al. 2007) and resource politics specifically (Sawyer 2004; Himley 2008; Fletcher 2010). As McCarthy

and Prudham (2004) argued, "Connections between neoliberalism, environmental change, and environmental politics are all deeply if not inextricably interwoven" (277). These dialectics of neoliberalism and nature frame Mongolia's rapidly evolving geographies in direct response to neoliberal initiatives and the resource-based economy.

Under the auspices of global capitalist institutions, Mongolia restructured its economy to become a new resource frontier for and through foreign capital. Valuable natural resources are the material and symbolic manifestations of Mongolia's current resource-based economy. As the country opened its market, numerous mining companies clamored to exploit Mongolia's resources. The nickname Minegolia allegorically captures the country's volatile resource politics and fragile dependence on mining (Bulag 2009; Jackson 2015b).

Oyu Tolgoi (OT) is the most famous by-product of Minegolia. The OT agreement is Mongolia's signature deal and reflects contradictions of global capitalism and local interests. The joint venture between Australian Rio Tinto, Canadian Ivanhoe Mines, and the Mongolian government owns the mining rights to the largest and highest grade copper and gold deposits in Mongolia's South Gobi Desert. Mongolia's turbulent mining politics and antipathy toward many foreign mining companies centers on this $6.6 billion megamine. Ivanhoe's former chairman Robert Friedland's reported promise that investors would reap colossal profits from Mongolia's resources inflamed growing discontent toward foreign investment (Combellick-Bidney 2012).[5] Following large-scale protests, during which Friedland's effigy was burned, Parliament imposed a 68 percent windfall tax on copper and gold ores in 2006. Ivanhoe's stock plummeted and foreign interest in mining in Mongolia halted. Parliament continued to push for a 51 percent Mongolian share, but Ivanhoe and Rio Tinto refused. More protests broke out in Ulaanbaatar following the July 2008 Parliamentary elections, leading to five deaths. "Although small in number, the riots and five deaths owed ... to growing public perceptions that the country's mineral wealth was being managed and sold off to foreigners" (Bulag 2009, 130).

In the 2008 Parliamentary elections, neither party won a majority, requiring a coalition government. The coalition government navigated competing demands from the public and foreign partners and investors.[6] As many Mongolians pushed for greater control over strategic resources, investors accused the government of economic nationalism and communist-era practices (Bulag 2009). Despite this friction, the government of Mongolia signed the OT Investment Agreement in October 2009, resulting in Ivanhoe's 66 percent ownership and the Mongolian government's 34 percent share.

Contestations over OT focus on how foreign entities benefit. Various studies note that low-income countries typically are not equally situated at negotiation tables with capital- and technology-rich multinational firms (Sach and Warner 2001; Kohl and Farthing 2012). These countries are often left with contracts not necessarily in the best interests of their people and environments.

The OT agreement was deemed a "new era" for Mongolia by many government officials, investors, and international financial circles. Consequently, Mongolia experienced foreign direct investment influxes from 2010 to 2012 with double-digit economic growth, peaking at 17 percent in 2011 (World Bank 2013). Then the economy cooled in 2013, dropping to 7.8 percent growth in 2014 and 1 percent in 2016 as commodity prices slumped (Asian Development Bank 2017). Public perceptions of dispossession simultaneously increased. Despite public contestations against OT, international and domestic business interests labeled these efforts resource nationalist, effectively silencing the opposition.

In his campaign, Battulga capitalized on the sentiments that Mongolia received a bad deal and promised greater government control of strategic mines. Battulga was also seen as a wild card by many, however, compared to Enkhbold, who signified stability and continuation of policies such as a contested International Monetary Fund bailout ("Mongolians prepare" 2017). Battulga's promises to push for a greater share of resources alarmed foreign investors, who resumed alarms of populism in general and resource nationalism specifically.

Since Battulga assumed office, several prominent politicians have been charged with misuse of power and corruption in negotiating with Rio Tinto on the OT agreement. In April 2018, two former prime ministers, Bayar Sanj and Saikhanbileg Chimed, were arrested and detained in connection with an ongoing investigation of their handling of two agreements related to the OT. Bayar served as the prime minister when the OT agreement was signed in 2009[7] and Saikhanbileg served as the prime minister when the expansion agreement was signed in 2015 (Hornby 2018; "Mongolia anti-graft agency" 2018).[8] Although this investigation could be seen as Battulga's populist pandering to his base, it is also an indication of how resource politics in general and the OT case in particular remains a critically important political and public issue.

Contradictions of Populism and Resource Extraction

As a political label, populism is uniquely oxymoronic. Although centered on a "people" driven idea, populism's deployment is less romantic. Invoked by all political camps, populism generally references a movement that is radical, fringe, and often dangerous. Unlike leftist or rightist or even centrist ideologies, populism is slippery, easily morphing into various political and ideological contexts in tension with the establishment (Laclau 2005). Populists often emerge to fight existing circumstances when the ruling block has an unfair advantage.

A leader with a dominant personality is considered central in a populist movement (Roberts 2006). Hart (2013, 194, 302) argued that populism glorifies an authoritarian and anti-intellectual leader as the protector of the "manipulated mindless masses." Looking at Battulga's election with this understanding of populism, we have two basic assumptions: (1) his voters were embittered against the ruling bloc and (2) he manipulated disgruntled voters to do his bidding. This fails, however, to appreciate contradictory forces shaping Mongolia's political landscape and divergent concerns crucial to Battulga's voter base. We also risk essentializing his voters as a mindless mass who succumbed to an opportunistic politician. Trivializing Battulga's voter base not only takes away voters' agency but also silences their concerns.

We argue that Battulga's victory was an expression of broad popular discontent, driven by Mongolia's neoliberal experiment that began in the 1990s: an economic recession, a growing wealth gap, decreasing standards of living for many, and dissipating social safety nets. Although varied in aims, the factors that created Battulga's base are tied to Mongolia's resource-driven economy (Montsame 2017). Battulga attracted voters who felt that outsiders were taking unfair advantage of Mongolia. His

"Mongolia Will Triumph" slogan spoke to this sentiment. "His nationalist-tinged campaign struck a chord with young voters hoping for a greater slice of Mongolia's natural resource wealth" (Bloomberg 2017b, 1). Similarly, Soni (2017) suggested that Battulga convinced younger voters that resource nationalism would restimulate the economy. He also appealed to those who believe that Mongolia's strategic resources should be majority state owned. On this platform, Battulga won Umnugovi province, where OT operates, in the second round against the investment-friendly Enkhbold.[9]

Resource Nationalism

Contestations over strategic resources have led to labeling Battulga's—and his opponent Ganbaatar's—campaigns as populist ones, propelled by resource nationalism, but this is too simple a label. To understand what drives popular mobilizations against the status quo, we argue that two mutually reinforcing themes produce and shape these contradictions: resource nationalism and accumulation by dispossession.

Resource nationalism is the assertion of greater national control over extractive industries (Stevens 2008; Kohl and Farthing 2012), less pejoratively discussed as resource sovereignty. For the sake of brevity, we focus on how the concept of resource nationalism is used to characterize so-called reckless behavior by national governments to foment populist resentment; for example, by Putin's Russia, Chavez's Venezuela, and Morales's Bolivia (Bremmer and Johnson 2009).

The critique of resource nationalism is typically made by and for the interests of economists, financiers, creditors, and investors who warn of an impending danger. *The Economist* suggested that Mongolian state control over mining "may kill the goose before it has laid any golden eggs" ("Before the gold rush" 2013, 29). The article featured an image of shadowy figures strangling a goose laying eggs, illustrating the nefarious character of resource nationalists. This discursive and symbolic articulation captures the precise sentiment invoked by resource nationalism. Paradoxically, Childs (2016) noted that its negative meaning is only affiliated with developing countries, whereas it is considered an acceptable and rational practice in Norway, Canada, and the United States.

In Mongolia, any mineral deposit expected to contribute 5 percent or more to gross domestic product (GDP) is called a *strategic deposit*. This name suggests its importance to national security. This is how many people view OT, which is expected to produce up to 30 percent GDP by 2021, making it one of Mongolia's most visible and important resources.

The OT agreement epitomizes Mongolia's tensions regarding resource politics, featuring prominently in public discontent locally and nationally. Although agreement provisions were considered balanced and fair by financial markets and industry standards, public opinion is often to the contrary (Shapiro 2009). In national surveys between 2014 and 2017, upwards of 85 percent of respondents believed that strategic deposits such as OT should be at least 51 percent state owned, with 36 and 40 percent believing that they should be fully state owned (Sant Maral 2014, 2015, 2016, 2017).

The government has attempted to revisit the agreement several times. In September 2012, some newly elected members of parliament sent an open letter to Rio Tinto to renegotiate a 50 percent stake (Hook 2012). In response to pressure to renegotiate, Rio Tinto withheld $4.2 billion in financing for the underground operation. Pro-mining interests lamented that "any attempts to step away from the most neoliberal provisions embedded in the mining regime [are perceived] as a burst of 'populism'" (Hatcher 2014, 133).

In 2015, the Mongolian government was muscled into compliance, and Mongolia was "back in business" ("Rio Tinto and Mongolia" 2015). Rio Tinto's actions are not isolated or unique. Many countries around the world are facing millions of dollars in claims made by multinational mining companies via international arbitration (Clarke and Cummins 2012).

The government's agreement to 34 percent is also a focal point for many Mongolians. Although there are no immediate signs to nationalize OT, the new government and public continue to push back. One of the contradictions that OT embodies is the identity of a national resource that is not controlled by the state. The state profits from OT but not as much as corporations and shareholders. OT has made promises that it cannot fulfill and the current economic downturn has done little to improve its image as the answer to Mongolia's economic insecurities. Reactions to OT demonstrate how the discourse against resource nationalism has been used repeatedly to rhetorically delegitimize and silence public grievances of mining practices and discredit politicians who rally on their behalf. Dispossession drives these grievances and serves as a salient counterdiscourse to the hegemonic apparatus of the global financial elite and their domestic partners.

Accumulation by Dispossession

Accumulation by dispossession and global resource politics have been discussed widely (Sneddon 2007; Benjaminsen and Bryceson 2012; Veuthey and Gerber 2012) as mining companies and local communities have battled over land and resource rights (Olivera and Lewis 2004; Bebbington 2011; Hall 2013). Typically, local communities' livelihoods depend on the land and mining companies need land to acquire natural resources. Conflicts over the same territory have long pitted mining companies against local communities (Girvan 1976; Perreault 2006; Holden and Jacobson 2008).

Local communities near OT contend that they bear disproportionate negative impacts while wealth is funneled to local elites and abroad. Their pleas have extensive emotional resonance given Mongolia's complicated relationships with mining nationwide and cultural reverence of ancestral lands. Local communities in Mongolia's South Gobi region, including Khanbogd soum,[10] where OT is located, have raised several key issues (McGrath et al. 2012; Jackson 2015a, 2015b, 2018). Among the most critical is water, given its scarcity and the colossal demand for water by mega mines in the region. Controversy has shrouded the pipelines constructed to redirect water both to and from the mine (Meesters and Behagel 2017; Jackson 2018).

Access to pasture land is another major concern. Land rights in Mongolia have been customarily negotiated and pasture land is not privately owned (Myadar 2009, 2011). In the Gobi's harsh terrain, herders rely on seasonal migrations to locate scarce water and pasture resources. For most herder families, livestock are their livelihood. De facto privatization of water and land resources for mining and transportation infrastructure dramatically limits herders' resource access. Interviews conducted by Jackson with more than 100 herders from 2011 to 2015 in Khanbogd and other nearby soums surrounding OT reveal that many have abandoned traditional grazing lands or have sought alternative livelihoods

because of infrastructure development from OT and other mines. In 2015, no respondent could name any herders under forty years old. Growing environmental and social issues related to resources compounded their sense of displacement.

OT has also increased regional traffic, with many industrial-sized trucks transporting raw materials to and supplies from China. These trucks grind dirt on unpaved roads into a fine powder, producing great dust plumes. As a result, people and livestock suffer from severe and frequent respiratory problems (Jackson 2015a). Although many roads used by OT are now paved, many dirt roads still crisscross the region connecting different mines and markets.[11]

Generally, mining does not provide appreciable employment as companies prefer to hire workers from outside the soum or foreigners (World Bank 2009). These concerns articulate an inherent contradiction of market principles. When unequal partners enter into a market, the powerful tend to benefit more, often at the disadvantaged's expense (McCarthy 2004). OT illustrates this asymmetry, particularly because the World Bank is one of its major financiers.

These contestations over space, resources, and ultimately identity are similar to experiences of communities around the world that have been enveloped by global capitalism through resource extraction. The clash between herders and mining companies in Mongolia exemplifies the fundamental contradiction that Harvey (2014) argued emerges as a direct consequence of global capitalism.

Conclusion: Resource Miracle or Resource Curse?

Resource politics in the Global South swing between two poles—economic miracle or resource curse. Too often peripheral countries with rich natural resources are plagued by underdevelopment, corruption, political instability, and even brutal internal strife (McCarthy 2004; Watts 2004; Frankel 2012). Mining companies obtain wealth as local communities bear lasting costs including environmental destruction followed by social and economic devastation. Harvey (2005) placed his hope in social movements that challenge neoliberal politics and accumulation by dispossession—the ideal form being a "democratic collective management structure" (50) where the commons are collectively managed rather than privately owned.

Mongolia's case offers a cautionary tale as it negotiates a future largely defined by a resource-based economy. Bearing scars symbolically and physically from twenty years of tumultuous mining politics, Mongolian voters have placed their hope in an unorthodox candidate accused of being dangerously populist. Battulga is a peculiar candidate who not only emerged from a shady background but galvanized massive popular support against the ruling party's candidate. He also alienated those who accused him of running on a populist platform driven by ethnic and resource nationalism. In protest, some 18,000 voters cast blank ballots (*tsagaan songolt*), claiming that neither candidate was worthy.[12] Rather than defining the last elections as about Battulga or his political platform, however, geographers should look beyond the political label and see Mongolian voters' choice as a moment of resistance to structural dispossession. We contend that Battulga's election was a response to the neoliberal regime.

At various scales, the Mongolian public expresses resistance against unequal distribution of costs and benefits of mining. On one hand, local community concerns about mining affect fuel counternarratives to unrealized promises of resource wealth. On the other hand, the broad public perception that Mongolia was treated unfairly in major mining deals, specifically OT, shapes discontent. Public resentment is directed toward the disparity between the very few whom mining benefits and the third of the population trapped in poverty. Any resistance to this lived dispossession of the public has been labeled resource nationalist—effectively silencing real concerns as resource nationalism inflames popular resentment and, as Battulga's election demonstrates, providing conditions for massive public mobilizations.

Acknowledgments

All findings and conclusions expressed in this article are those of the authors. We thank Ronald Davidson, Rebecca Watters, and Brent White for reading various versions of the article and providing much appreciated suggestions and edits. We also thank the three anonymous reviewers and James McCarthy for their helpful feedback.

Funding

We thank the following funding sources for making various phases of the research possible: the Canadian Social Sciences and Humanities Research Council, York University; a Field Research Fellowship from the American Center for Mongolian Studies; the Henry Luce Foundation; the U.S.–Mongolia Field Research Fellowship Program (sponsored by the American Center for Mongolian Studies, the Council of American Overseas Research Centers and the U.S. State Department Educational and Cultural Affairs Bureau); Metropolitan State University of Denver; Department of Social Anthropology, University of Cambridge; and School of Geography and Development, University of Arizona.

Notes

1. Enkhbold served as prime minister of Mongolia and is the current speaker of the Parliament. His party won a landslide in 2016 parliamentary elections.
2. Battulga's "Mongolia First" slogan appealed to anti-Chinese sentiment that runs deep in Mongolia, alluding to the fact that Enkhbold was half-Chinese with questionable national allegiance (Soni 2017).
3. The election was one of the most contentious elections in Mongolia's recent history. In the first ever run-off election, Battulga won with 50.6 percent.
4. Orhon Myadar has been studying Mongolian politics since 2005 and Sara Jackson has been studying Mongolian resource politics since 2009.
5. At an investor conference in 2005, Friedland reportedly compared OT to making a $5 T-shirt and selling it for $100.
6. Radchenko and Jargalsaikhan (2017) argued that Mongolia's relatively peaceful transitions have been due to fractured micropolitics among political parties, preventing one person or party from emerging as the dominant force. This fractured context arguably facilitated Battulga's emergence.
7. Bayartsogt Sangajav, who served in the Bayar cabinet as finance security, was also arrested and detained.
8. At the time of writing this article, all three were released on bail and barred from leaving the country.
9. In the first round, Mongolian People's Republic Party's Ganbaatar Sainkhuu received the most votes in Umnugovi. Formerly an independent candidate, Ganbaatar was recruited and nominated at the last minute when MPRP's only viable candidate, Nambar Enkhbayar, could not run because of his prior conviction for corruption. Ganbaatar is an ardent resource nationalist and a vocal critic of OT. Changing his party affiliation did not play in his

favor—nor did a video that implicated him in accepting illicit campaign donations from a member of the Unification Church of Korea, which is considered a cult in Mongolia.

10. A soum is the Mongolian equivalent of a county.
11. Although OT is not the only mining company in the area and should not be solely blamed, it remains the biggest mining operation in the area.
12. If 10 percent of voters submit *tsagaan songolt*, the presidential election is invalidated.

References

Asian Development Bank. 2017. Asian development outlook 2017: Sustaining development through public-private partnership. Accessed January 15, 2018. https://www.adb.org/countries/mongolia/economy.

Bebbington, A., ed. 2011. *Social conflict, economic development and the extractive industry: Evidence from South America.* Vol. 9. London and New York: Routledge.

Bebbington, A., and J. Bury, eds. 2013. *Subterranean struggles: New dynamics of mining, oil, and gas in Latin America.* Vol. 8. Austin: University of Texas Press.

Before the gold rush. 2013. *The Economist*, February 16. Accessed August 15, 2015. https://www.economist.com/news/asia/21571874-mongolias-road-riches-paved-shareholders-tiffs-gold-rush.

Benjaminsen, T. A., and I. Bryceson. 2012. Conservation, green/blue grabbing and accumulation by dispossession in Tanzania. *Journal of Peasant Studies* 39 (2):335–55.

Bloomberg. 2017a. How the world's fastest-growing economy went bust. Accessed September 13, 2017. https://www.bloombergquint.com/markets/2017/02/12/how-the-world-s-fastest-growing-economy-went-bust.

———. 2017b. Nationalist candidate Battulga wins Mongolian election in runoff. Accessed September 13, 2017. https://www.bloomberg.com/news/articles/2017-07-07/nationalist-candidate-battulga-wins-mongolian-election-in-runoff.

Bremmer, I., and R. Johnston. 2009. The rise and fall of resource nationalism. *Survival* 51 (2):149–58.

Bulag, U. E. 2009. Mongolia in 2008: From Mongolia to Minegolia. *Asian Survey* 49 (1):129–34.

Childs, J. 2016. Geography and resource nationalism: A critical review and reframing. *The Extractive Industries and Society* 3 (2):539–46.

Chomsky, N. 1999. *Profit over people: Neoliberalism and global order.* New York: Seven Stories Press.

Ciccantell, P. S., and D. Patten. 2016. *The new extractivism, raw materialism and twenty-first century mining in Latin America.* London and New York: Routledge.

Clarke, M., and T. Cummins. 2012. Resource nationalism: A gathering storm? *International Energy Law Review* 6:220–25.

Click, R. W., and R. J. Weiner. 2010. Resource nationalism meets the market: Political risk and the value of petroleum reserves. *Journal of International Business Studies* 41 (5):783–803.

Combellick-Bidney, S. 2012. Mongolia's mining controversies and the politics of place. In *Change in democratic Mongolia*, ed. J. Dierkes, 271–96. Boston: Brill.

Emel, J., M. T. Huber, and M. H. Makene. 2011. Extracting sovereignty: Capital, territory, and gold mining in Tanzania. *Political Geography* 30 (2):70–79.

Empson, R., and T. Webb. 2014. Whose land is it anyway? *Inner Asia* 16 (2):231–51.

Fletcher, R. 2010. Neoliberal environmentality: Towards a poststructuralist political ecology of the conservation debate. *Conservation and Society* 8 (3):171–81.

Frankel, J. A. 2012. The natural resource curse: A survey of diagnoses and some prescriptions. HKS Faculty Research Working Paper Series RWP12-014, John F. Kennedy School of Government, Harvard University.

Girvan, N. 1976. *Corporate imperialism.* New York: Monthly Review Press.

Hall, R. 2013. Diamond mining in Canada's Northwest Territories: A colonial continuity. *Antipode* 45 (2):376–93.

Hart, G. 2013. Gramsci, geography, and the languages of populism. In *Gramsci: Space, nature, politics*, ed. M. Ekers, G. Hart, S. Kipfer, and A. Loftus, 301–20. New York: Wiley.

Harvey, D. 2004. The "new imperialism": Accumulation by dispossession. *Actuel Marx* 35 (1):71–90.

———. 2005. *Spaces of neoliberalization: Towards a theory of uneven geographical development.* Vol. 8. Stuttgart, Germany: Franz Steiner Verlag.

———. 2006. *Spaces of global capitalism.* New York: Verso.

———. 2014. *Seventeen contradictions and the end of capitalism.* Oxford, UK: Oxford University Press.

Hatcher, P. 2014. Fighting back: Resource nationalism and the reclaiming of political spaces. In *Regimes of risk*, 125–43. Basingstoke, UK: Palgrave Macmillan.

Heynen, N., J. McCarthy, S. Prudham, and P. Robbins, eds. 2007. *Neoliberal environments: False promises and unnatural consequences.* London and New York: Routledge.

Himley, M. 2008. Geographies of environmental governance: The nexus of nature and neoliberalism. *Geography Compass* 2 (2):433–51.

Holden, W. N., and R. D. Jacobson. 2008. Civil society opposition to nonferrous metals mining in Guatemala. *Voluntas: International Journal of Voluntary and Nonprofit Organizations* 19 (4):325–50.

Holden, W., K. Nadeau, and R. D. Jacobson. 2011. Exemplifying accumulation by dispossession: Mining and indigenous peoples in the Philippines. *Geografiska Annaler: Series B, Human Geography* 93 (2):141–61.

Hook, L. 2012. Mongolia seeks to renegotiate Oyu Tolgoi deal. *Financial Times*, October 15. Accessed August 8, 2015. https://www.ft.com/content/f74417d0-16dc-11e2-8989-00144feabdc0.

Hornby, L. 2018. Mongolia arrests 2 former PMs linked to mining probe. *Financial Times*, April 11. Accessed April 11, 2018. https://www.ft.com/content/b46ae610-3d5e-11e8-b7e0-52972418fec4.

Hoyle, R. 2016. Mongolia: Land of lost opportunity. *The Wall Street Journal*, March 21. Accessed March 18, 2018. https://www.wsj.com/articles/mongolia-land-of-lost-opportunity-1458518881.

Jackson, S. L. 2015a. Dusty roads and disconnections: Perceptions of dust from unpaved mining roads in Mongolia's South Gobi province. *Geoforum* 66:94–105.

———. 2015b. Imagining the mineral nation: Contested nation-building in Mongolia. *Nationalities Papers* 43 (3):437–56.

———. 2018. Abstracting water to extract minerals in Mongolia's South Gobi Province. *Water Alternatives* 11 (2):336–56.

Jackson, S. L., and D. Dear. 2016. Resource extraction and national anxieties: China's economic presence in Mongolia. *Eurasian Geography and Economics* 57 (3):343–73.

Kohl, B., and L. Farthing. 2012. Material constraints to popular imaginaries: The extractive economy and resource nationalism in Bolivia. *Political Geography* 31 (4):225–35.

Laclau, E. 2005. Populism: What's in a name? In *Populism and the mirror of democracy*, ed. F. Panizza, 32–49. New York: Verso.

McCarthy, J. 2004. Privatizing conditions of production: Trade agreements as neoliberal environmental governance. *Geoforum* 35 (3):327–41.

McCarthy, J., and S. Prudham. 2004. Neoliberal nature and the nature of neoliberalism. *Geoforum* 35 (3):275–83.

McGrath, F., V. Martsynkevych, D. Hoffman, R. Richter, S. Dugersuren, and A. Yaylymova. 2012. *Spirited away—Mongolia's mining boom and the people that development left behind.* Prague: CEE Bankwatch Network, Regine Richter, Urgewald, OT Watch and Bank Information Center.

Meesters, M. E., and J. H. Behagel. 2017. The social license to operate: Ambiguities and the neutralization of harm in Mongolia. *Resources Policy* 53:274–82.

Mongolia anti-graft agency arrests two prime ministers amid mine probe. 2018. Accessed April 11, 2018. https://www.reuters.com/article/mongoliaoyutolgoi/mongolia-anti-graft-agency-arrests-two-primeministers-amid-mine-probe-idUSL3N1RO3OP.

Mongolians prepare to elect a new president. 2017. *The Economist*, July 22. Accessed September 13, 2017. https://www.economist.com/news/asia/21723885-disillusionment-politics-rife-mongolians-prepare-elect-new-president.

Montsame. 2017. Монгол улсын ерөө нхийлөө гчийн сонгууль: тулааны урлагийн тамирчин Х. Баттулгын ялалт [Mongolia's presidential election: The victory of Kh. Battulga—The athlete of art of fight]. Accessed September 13, 2017. http://www.montsame.mn/read/60433.

Myadar, O. 2009. Nomads in fenced land: Land reform in post-socialist Mongolia. *APLPJ* 11:161–203.

———. 2011. Imaginary nomads: Deconstructing the representation of Mongolia as a land of nomads. *Inner Asia* 13 (2):335–62.

Myadar, O., and J. D. Rae. 2014. Territorializing national identity in post-socialist Mongolia: Purity, authenticity, and Chinggis Khaan. *Eurasian Geography and Economics* 55 (5):560–77.

Office of the President of Mongolia. 2017. Inauguration address by Khaltmaagiin Battulga, the President of Mongolia, at the ceremony of presidential swearing into office. Accessed March 9, 2018. http://eng.president.mn/newsCenter/viewEvent.php?cid=44&news-Event=Khaltmaagiin%20Battulga%20Takes%20Oat-h%20of%20Office%20as%20the%20President%20of%20Mongolia#.

Olivera, O., and T. Lewis. 2004. *Cochabamba! Water war in Bolivia.* Boston: South End Press.

Organization of Security and Cooperation in Europe. 2017. Limited Election Observation Mission. Presidential Election, Second Round. Accessed September 13, 2017. https://www.osce.org/odihr/elections/mongolia/328381?download=true.

Perreault, T. 2006. From the Guerra del Agua to the Guerra del Gas: Resource governance, neoliberalism and popular protest in Bolivia. *Antipode* 38 (1):150–72.

Radchenko, S., and M. Jargalsaikhan. 2017. Mongolia in the 2016–17 electoral cycle: The blessings of patronage. *Asian Survey* 57 (6):1032–57.

Rio Tinto and Mongolia sign multibillion dollar deal on mine expansion. 2015. *The Guardian*, May 19. Accessed September 19, 2017. https://www.theguardian.com/global/2015/may/19/rio-tinto-and-mongolia-signmultibillion-dollar-deal-on-mine-expansion

Roberts, K. M. 2006. Populism, political conflict, and grass-roots organization in Latin America. *Comparative Politics* 38 (2):127–48.

Rossabi, M. 2005. *Modern Mongolia: From khans to commissars to capitalists.* Berkeley: University of California Press.

———. 2009. Mongolia: Transmogrification of a communist party. *Pacific Affairs* 82 (2):231–50.

Sachs, J. D., and A. M. Warner. 2001. The curse of natural resources. *European Economic Review* 45 (4–6):827–38.

Sant Maral. 2014. Politbarometer. Accessed January 19, 2018. http://www.santmaral.mn/sites/default/files/SMPBE14%20Apr.pdf.

———. 2015. Politbarometer. Accessed January 19, 2018. http://www.santmaral.mn/sites/default/files/SMPBE15%20Apr.pdf.

———. 2016. Politbarometer. Accessed January 19, 2018. http://www.santmaral.mn/sites/default/files/SMPBE16.Mar%20(updated).pdf.

———. 2017. Politbarometer. Accessed January 19, 2018. http://www.santmaral.mn/sites/default/files/SMPBE17%20Mar_Extended%20version.pdf.

Sawyer, S. 2004. *Crude chronicles: Indigenous politics, multinational oil, and neoliberalism in Ecuador.* Durham, NC: Duke University Press.

Shapiro, R. D. 2009. *Analysis of concerns about the draft agreement to undertake the Oyu Tolgoi agreement.* Ulaanbaatar, Mongolia: World Growth Mongolia.

Sneddon, C. 2007. Nature's materiality and the circuitous paths of accumulation: Dispossession of freshwater fisheries in Cambodia. *Antipode* 39 (1):167–93.

Soni, S. K. 2017. Mongolia's new president is Mongolia first and China last. Accessed January 19, 2018. http://www.eastasiaforum.org/2017/08/11/

Stevens, P. 2008. National oil companies and international oil companies in the Middle East: Under the shadow of government and the resource nationalism cycle. *Journal of World Energy Law & Business* 1 (1):5–30.

Veuthey, S., and J. F. Gerber. 2012. Accumulation by dispossession in coastal Ecuador: Shrimp farming, local resistance and the gender structure of mobilizations. *Global Environmental Change* 22 (3):611–22.

Watts, M. 2004. Resource curse? Governmentality, oil and power in the Niger Delta, Nigeria. *Geopolitics* 9 (1):50–80.

World Bank. 2009. The potential social impacts of mining development in Southern Mongolia. Accessed June 23, 2015. http://siteresources.worldbank.org/MONGOLIAEXTN/Resources/Southern_Mongolia_Social_Impacts.pdf.

———. 2010. Mongolia ground water assessment of the Southern Gobi region. Accessed June 23, 2015. http://documents.worldbank.org/curated/en/2010/04/16330609/mongolia-groundwater-assessment-southern-gobi-region.

———. 2013. Mongolia economic update. Accessed June 23, 2015. http://www.worldbank.org/content/dam/Worldbank/document/EAP/Mongolia/MQU_April_2013_en.pdf.

ORHON MYADAR is an Assistant Professor in the School of Geography and Development at the University of Arizona, Tucson, AZ 85721. E-mail: orhon@email.arizona.edu. Her research interests include border, mobility, displacement, and resource politics.

SARA JACKSON is a Geography Lecturer in the Department of Earth and Atmospheric Sciences at Metropolitan State University of Denver, Denver, CO 80217. E-mail: sjacks62@msudenver.edu. Her research interests include political and cultural geographies of resource extraction in Mongolia and sustainable infrastructure in Denver, Colorado.

Bringing Back the Mines and a Way of Life: Populism and the Politics of Extraction

Erik Kojola 🆔

Conflicts over resource extraction are key political issues in the contemporary United States and are a mobilizing issue for right-wing populism exemplified by President Donald Trump's claims of ending the "war on coal." Through a political ecology framework attentive to culture, discourse, and history, I examine how mining is symbolic of broader cultural, geographic, and class divides. Mining is mobilized in extractive populism through rhetoric of giving power back to the people, insiders versus outsiders, resource nationalism, and cutting burdensome government regulations. I study the emblematic case of the rural Iron Range mining region in northern Minnesota and the recent rightward swing in this historically Democratic stronghold, which I argue is intertwined with the micropolitics of struggles over proposed copper mines. Through ethnographic observations and interviews with local community and political leaders, workers, and residents, I find that support for mining among white, working-class, and rural residents is made meaningful through nostalgia for preserving mining as a way of life and anger at outsiders disrupting their livelihoods and extractive moral economy. These discourses resonated with the populist, nationalist, and racist rhetoric of Trump's campaign. I argue that place-based and class identities and social imaginaries linked to mining are an important dynamic in emergent authoritarian populism and for understanding what motivates reactionary political movements and why populist politicians use mining to construct authenticity.

有关资源搾取的冲突，是美国当代的关键政治议题，并且是右翼民粹主义的动员议题，并由唐纳德. 特朗普总统宣称终结"煤炭战争"为代表例子．我通过关注文化、论述与历史的政治生态架构，检视矿业如何作为更为广泛的文化、地理与阶级划分之象徵。矿业在搾取式民粹主义中，通过还权于民、内部人与外部人的对立、资源国族主义，以及减少繁重的政府规范之修辞进行动员。我研究明尼苏达北部乡村的铁矿山脉矿业区的象徵性案例，以及此一历史上为民主党的版图在晚近的政治右转，我主张其与所提出的铜矿之微政治斗争相互交织。我通过民族志观察，以及对地方社群和政治领导者、工人与居民的访谈，发现白人工人阶级与农村居民对矿业的支持，通过保存矿业作为生活方式的怀旧，以及对外地人破坏其生计和对搾取式道德经济的愤怒来製造意义。这些论述与特朗普所宣传的民粹主义、国族主义和种族主义修辞相互呼应。我主张，根据地方及阶级的身份认同，以及连结至矿业的社会想像，是对于浮现中的威权民粹主义、以及理解什麼驱动反动式政治运动和为何民粹主义政客能够运用矿业来建构本真性的重要动态。 关键词: 阶级，论述，政治生态学，民粹主义，资源搾取。

Los conflictos asociados con la extracción de recursos son asuntos políticos claves en los Estados Unidos contemporáneos y constituyen un tema movilizador para el populismo de derecha, cuyo ejemplo es el clamor del Presidente Donald Trump de terminar "la guerra del carbón". A través de un marco de ecología política atento a la cultura, el discurso y la historia, examino cómo la minería es simbólica de las más amplias divisorias culturales, geográficas y de clase. La minería es movilizada en el populismo extractivo por medio de retóricas como devolver el poder al pueblo, nacionales contra extranjeros, nacionalismo de los recursos y la supresión de molestas regulaciones gubernamentales. Estudio el caso rural emblemático de la región minera de la Iron Range en el norte de Minnesota y el reciente giro a la derecha de este bastión históricamente Demócrata, lo que a mi parecer está entrelazado con la micropolítica de las luchas sobre propuestas relacionadas con minas de cobre. Por medio de observaciones etnográficas y entrevistas en la comunidad local y con líderes políticos, obreros y residentes, encuentro que el apoyo a la minería entre los blancos, la clase trabajadora y los residentes rurales se hace significativo a través de la nostalgia por preservar la minería como un medio de vida, y por la ira contra los foráneos que perturban su sustento y la economía moral extractiva. Estos discursos resonaron en la campaña populista, nacionalista y de retórica racista de Trump. Yo argumento que las identidades basadas en lugar y clase, lo mismo que los imaginarios relacionados con la minería, son una dinámica importante en el emergente populismo autoritario y sirve para entender lo que motiva los movimientos políticos reaccionarios y por qué los políticos populistas usan la minería para construir autenticidad. *Palabras clave: clase, discurso, ecología política, extracción de recursos, populismo.*

Since the mid-2000s there has been renewed interest in populism with the rise of populist right-wing politicians in Western Europe and the United States (Solty 2013; Moffitt 2016; Bonikowski 2017). Scholars and activists have sought to understand the political–economic conditions, political strategies, and discourses that have contributed to contemporary right-wing populism, particularly the election of U.S. President Donald Trump (Azari and Hetherington 2016; Oliver and Rahn 2016). I argue that resource extraction is an important mobilizing issue for populism in the United States because of how mining is connected to place-based and class identities, nationalism, and masculinity (Scott 2010; Emel, Huber, and Makene 2011; Himley 2014; Li 2015). Trump's declarations that he will put coal miners back to work as part of his campaign to "Make American Great Again" are indicative of how mining is deployed in populist themes of anti-elitism, speaking for the people, and us versus them (Oliver and Rahn 2016) and used to advance conservative and nationalist political projects.

In this article, I explore the discursive, cultural, and affective dynamics of resource extraction that are mobilized by conservative politicians in a form of extractive populism that has contributed to rightward swings in rural, industrial, and predominantly white regions. Through a political ecology approach (Peet and Watts 1996; Nesbitt and Weiner 2001; Moore 2005; Peluso 2012; Li 2015), I conceptualize conflicts over resource extraction as political, economic, and cultural struggles over rights, identity, and livelihoods. I contribute to growing political ecology scholarship that uses theories of collective memory and affect (Legg 2004; Sultana 2011; Lundgren and Nilsson 2018; Perreault 2018; Threadgold et al. 2018) by emphasizing the role of people's racialized and gendered emotions and memories of place in reproducing the hegemony of extractive capitalism (Ekers, Loftus, and Mann 2009; Marston and Perreault 2017) rather than motivating resistance.

I also expand on analysis of contemporary right-wing populism in the United States (Cramer 2016; Oliver and Rahn 2016) that has largely neglected

the role of place-based identities and socionatures. I argue that explanations for white working-class conservatism and antienvironmentalism need to move beyond general national-level accounts and explanations based on single dimensions to develop nuanced analysis of the ways in which power and ideology operate through the interconnections of nature, class, race, and gender (Spruyt, Keppens, and Van Droogenbroeck 2016). Research needs to consider the particular socionatures and histories that shape why reactionary populists like Trump gain support in particular places and how rural and working-class communities understand support for extractive populism as common sense and a way to defend moral economies (Jenkins 2016; McQuarrie 2017; Scoones et al. 2018).

I use the upper Midwest, specifically the northern Minnesota Iron Range—a rural mining region—as an emblematic site to examine how and why a historically Democratic region swung rightward in ways connected to the cultural and class politics of mining. In the 2016 election, Trump won in towns across the region that had not supported a Republican presidential candidate since the 1930s. At the same time, there is a growing conflict over developing new copper mines that has split the Democratic coalition over the tensions between job creation and environmental conservation. Through ethnographic fieldwork in rural mining towns; indepth interviews with residents, workers, and local leaders; and discourse analysis of social media and newspaper coverage, I explore how place-based and class identities, collective memories, and moral economies shape the social imaginaries of proposed mining. How are these imaginaries and discourses constituted by right-wing populism in ways that affect political shifts in northern Minnesota?

I find that mine supporters frame copper mining through themes of insiders versus outsiders, excessive government regulation, renewing heritage, and nationalism. Copper mining is made meaningful through desires to bring back a prosperous past based on a moral economy in which mining provided stable jobs for white men and to defend the right of local communities to create livelihoods from extracting natural resources. Opposition to mining is then interpreted by Iron Rangers as elite urban environmentalists and government bureaucrats interfering in their livelihoods and disparaging their way of life while privileging urban people of color and immigrants. Trump's populist message appealed to people's sense of powerlessness and anger while drawing on nostalgia to provide hope for reasserting the moral and economic worth of rural places (Lewin 2017) and renewing masculinity through industrial labor.

Building on Huber's (2013) concept of energy populism in which expanding cheap fossil fuels is seen as standing up for the people, I point toward a broader extractive populism that promotes extractive capitalism and antienvironmentalism as protecting the people, renewing the "good life," and promoting national security. Extractive populism also relies on a discourse of what Perreault (2018) called "masculine nationalism," in which heroic male miners can provide material resources to secure the nation against foreign enemies while restoring heteronormative middle-class families.

Populism and Resource Extraction

Populism is a complex, contested, and fraught term (Laclau 1977), and I build on approaches that theorize populism as a political style, strategy, and ideology, rather than a coherent set of political philosophies or policies. Jansen (2011) conceptualized populism as a mode of political practice that is used as a means to gain support. These approaches focus on the rhetoric, emotional appeals, embodied practices, and narratives used by politicians across the political spectrum (Jansen 2011; Moffitt 2016; Agnew and Shin 2017). Populism involves claims to stand up for and give voice to "the people" while critiquing the elite (Laclau 1977; Badiou 2016). Populist rhetoric often constructs an outside threat and a sense of crisis and loss that can then be solved by empowering the people and a return to an idealized past (Agnew and Shin 2017; Balthazar 2017; Kenny 2017).

Populism can be deployed in conservative and progressive movements. Claims to the people can deploy inclusive or exclusionary visions based on classed, racialized, and gendered notions of ideal citizens (Badiou 2016). I focus on what has been called right-wing (Green et al. 2016), nationalist

(Gusterson 2017), nostalgic (Balthazar 2017), or authoritarian populism (Hall 1985) to highlight the ways in which populism has been mobilized in contemporary conservative politics in the United States to blame economic and social woes on government bureaucrats, immigrants, racial others, and environmental regulations while claiming to stand up for working-class and rural people (Oliver and Rahn 2016; Koch 2017; Scoones et al. 2018). Right-wing populists ofen claim that traditional political parties and urban elites ignore the working class and favor urban communities, people of color, and immigrants (Inglehart and Norris 2016; Bobo 2017; Lamont, Park, and Ayala-Hurtado 2017).

Integrating theories of populism with political ecology's attention to regional-level political–economic, discursive, and cultural dynamics of natural resources (Walker 2003; McKinnon and Hiner 2016) is productive for understanding the role of place-based and class identities and environmental imaginaries in populist movements. Populism is linked to natural resource extraction in the United States because miners and mining regions are symbolic of the people and the idealized heartland (Taggart 2000). This is based on a colonial and racial imaginary of white settlers and pioneers making a living out of an empty land (Ekers 2009; Perreault 2018). Rural spaces are romantically imagined as idyllic and white landscapes (Ageyman and Spooner 1997; Holloway 2007). Mining is often made meaningful through narratives of masculine labor and nostalgia for when white men could depend on the stability of industrial jobs to sustain vibrant communities (Scott 2010; Rolston 2014). Environmental protections are presented as hurting working-class people and threatening the masculinity of male workers in rural extractive industries (Foster 1993; Loomis 2015). Environmental regulations are also framed as outside elites and bureaucrats interfering in local communities' rights to maintain livelihoods (Harvey 1996; McCarthy 2002).

Natural resources are deeply connected with nationalist imaginaries and the material embodiment of the nation (Perreault 2013; Forchtner and Kølvraa 2015). Managing natural resources is a major state-making project (Whitehead, Jones, and Jones 2007), and notions of citizenship, belonging, and identity are navigated through land and nature (Coronil 1997; Valdivia 2008; Perreault and Valdivia 2010). Resource nationalism is a discourse that claims to protect the nation by developing domestic supplies of natural resources (Forchtner and Kølvraa 2015). Mining development is often justified as empowering the nation to compete with foreign countries and provide the materials necessary for security (Kohl and Farthing 2012; Bridge 2014; Arsel, Hogenboom, and Pellegrini 2016; Rosales 2017; Arbatli 2018). Rhetoric about resource independence generates political legitimacy for expanding extraction to protect the people in the face of material scarcity and uncertain geopolitics (McCarthy 2002; Scott 2010; Phadke 2011). Perreault (2013) argued that resource nationalism must be understood within place-based histories, historical memories, and social imaginaries of nature and the nation. Subnational identities tied to regions and cities also motivate mobilization to demand local communities' control over resources (Tidwell and Tidwell 2018).

The Minnesota Iron Range

This article provides empirical research on an understudied sociopolitical context, the northern Minnesota Iron Range, and how and why this historically leftist region has undergone a rightward shift. Much of the scholarly and media attention to mining politics and working-class conservatism in the United States is on Appalachia, Western states, and the Pacific Northwest (Walker 2003; Prudham 2005; Lewin 2017), whereas research on working-class politics has focused on postindustrial landscapes of the Rust Belt (Cramer 2016). Yet these places have unique characteristics and do not represent all rural, industrial, and resource extraction regions.

The Iron Range is a sociocultural region and geographic area that spans over 100 miles from Grand Rapids, Minnesota, to Ely, Minnesota. The Iron Range is the largest iron ore–producing area in the United States, with mines operating since the 1880s. The region is emblematic of extractive regions with geographic isolation, strong community identity, a predominantly white population, and an economically dominant mining industry. Currently, the region is struggling with unemployment and population loss largely due to shifts in the global steel industry and increased mechanization (Manuel 2015).

The region has a unique sociopolitical history of strong unionism and populist, and at times radical, politics as well as being in a politically progressive state with relatively strong environmental protections (Manuel 2015). In Minnesota, the Democrats are the Democratic Farmer-Labor (DFL) Party, which reflects the populist tradition of an alliance among farmers, workers, and urban progressives. Even as other extractive, industrial, and rural regions like Appalachia began shifting to the right in the 1970s and 1980s (Edsall and Edsall 1992; Davis 2017), the postwar liberal alliance between labor and the Democratic Party has been maintained, although tenuously, for longer in the Iron Range (Manuel 2015). Unions like the United Steelworkers continue to have some strength, but industrial decline and institutional and cultural shifts are weakening the liberal coalition.

Political Shifts in the 2016 U.S. Presidential Election

The 2016 U.S. presidential election marked a major political shift, as Trump won across the Iron Range, where a Republican presidential candidate had not won since Herbert Hoover in 1928 (Kraker 2016). Statewide, Hillary Clinton narrowly beat Trump (46.4 percent to 44.9 percent), but Trump carried much of northern Minnesota, including winning some Iron Range precincts by 15 percent. The 2016 election was a reversal, as Trump narrowly won precincts that Barack Obama won in 2012 with over 60 percent of the vote.[1]

Conflicts over Proposed Copper Mining

Since the early 2000s, multinational mining companies have been exploring copper deposits in northeastern Minnesota—what is believed to be one of the world's largest reserves—and several companies have proposed projects. Yet, copper mining is embroiled in public controversy with contested regulatory processes and mobilization from supporters and opponents (Kojola 2018). The proposed mines bring the prospect of job creation and economic growth. On the other hand, Minnesota does not have any copper mines, and this type of mining is more hazardous than the existing iron ore mines. Copper mining would extract small amounts of metal from ores that contain sulfides that create hazardous sulfuric acid when exposed to air and water. The proposed mines are also located in more environmentally sensitive areas, including the Boundary Waters Canoe Area Wilderness (BWCAW), which is one of the most visited wilderness areas in the

United States. Conflicts over copper mining have challenged the long-standing Democratic coalition between Iron Range labor and Twin Cities progressives.

Populism and the Cultural Politics of Copper Mining on the Iron Range

I examine the cultural meanings, affects, and discourses that animate contention over copper mining and support for right-wing populism in the Iron Range through ethnographic observations of daily life and public events and in-depth interviews. I spent the summer of 2017 living in Ely, Minnesota, and took several shorter trips to the Iron Range from 2015 to 2017 to conduct interviews, attend government hearings, and visit operating iron ore mines. I attended pro-mining rallies and marches and other public events, like summer festivals, where pro-mining groups had a presence. I also conducted twenty-nine in-depth interviews with community leaders, elected officials, retired miners, local residents, industry representatives, and union staff. Analysis of Web sites and social media of mining advocacy groups and newspaper coverage also provided insights into public debates and rhetorical strategies. I collected state and local news articles from 2004, when the first mine was proposed, through 2017, when the Trump administration made decisions to advance the projects.

Through analyzing interviews, textual data, and field notes, I find four emergent themes in the discourses around copper mining: outsiders versus insiders, excessive government regulation, renewal of a way of life, and resource nationalism. In the following sections I elaborate on these themes and provide emblematic examples.

Outsiders versus Insiders

Pro-mining activists and leaders use us versus them rhetoric to frame mining opponents as outsiders—wealthy urban liberals—who want to preserve their wilderness playground and dictate how local communities—insiders—use the land. A retired white male miner and community leader told me,

> But that's part of the people that are moving here, they are like my age, retired people, that have had good jobs, so they don't need a job. So they're fine and dandy with nothing happening here. They don't see any need for the industry or anything like that.

Opponents are interpreted as upper-class outsiders who want to stop development and preserve the environment for their enjoyment, which is disrupting the local extractive moral economy that provides residents with livelihoods through using natural resources. These tensions are part of broader rural–urban divides and reactions against perceived judgments of rural people as backward and competing visions of rural places as landscapes of consumption or production (Lichter and Brown 2011).

Anger at outsiders is linked to a sense that the government and Democrats are focused on the needs of urban communities, which are seen as either educated and elite or poor minorities and immigrants. Mining supporters often expressed anger at the Democratic Party for drifting from its working-class base and becoming a party of the educated urban elite and immigrants. Defending an implicitly white rural "us" then motivated voting Republican and turning against Democrats who opposed mining, which symbolizes a way of life and a set of values. In an op-ed for a regional newspaper, a man who identified as an Iron Ranger, third-generation union member, and lifelong Democrat wrote,

> Not only do the Twin Cities Democrats not care about the Iron Range economy, they actually despise and look down upon us. The time for talking with these people is over. It's time to flip the Range to red and let them know we won't tolerate their anti-mining nonsense. The Iron Range is not going to be their playground. (Chezick 2017)

This writer framed the debate over mining as a rural versus urban issue and constructed a sense of rural people being disparaged. Voting Republican was presented as a way for the Iron Range to defend itself and reinvigorate masculine jobs in resource extraction and the economic and moral worth of rural regions that depend on extraction—not outdoor recreation tourism.

Trump's appeals to stand up for rural America and everyday working people resonated with Iron Rangers, who felt that establishment Democrats were ignoring their contributions and livelihoods. A city council member of a small Iron Range town and former miner told me that he was a lifelong union guy and Democrat but voted for Trump, who he saw as standing up for the "working man." Democrats and environmentalists had gone too far in opposing "everything," whereas Trump was addressing the important "meat and potatoes" issues.

Support for Republicans and Trump is also contested, however, even by some who support the mines. Another male union member from the Iron Range responded to the op-ed quoted earlier and argued that people should not vote solely on the copper mining issue and should be wary of Republicans who support policies that weaken unions and give tax cuts to the wealthy (Pliml 2017).

Anger with the Democrats and urbanites was also intertwined with concerns about racial others and immigrants. Urban versus rural divides are often racialized (Scoones et al. 2018) and rhetoric about the people is coded as white and rural (Holloway 2007). I found that some Iron Rangers expressed a sense that Democrats were more concerned about helping poor urban people of color than white, rural, and working-class people. For example, an Iron Range state DFL representative whom I interviewed made a passing critique of a newly elected Latina state representative from Minneapolis who replaced a longtime white DFL stalwart. This revealed a resentment that the DFL was shifting toward urban people of color.

Excessive Government Regulation

Mine supporters often described federal government actions delaying mine development as bureaucratic red tape and government overreach, which constrains local livelihoods and rights to make decisions about resource use. Within this logic, expansion of resource extraction and elimination of environmental protections becomes a populist cause of giving power back to the people and allowing the wise use of natural resources. One local pro-mining activist, a middle-aged white man, told me that one of the biggest problems is that the permitting process takes too long and there are "too many fingers in the pie." He remarked how this frustration led him and others to support Trump.

Mine supporters say they are against environmental pollution but claim that locals, not outside regulators or environmentalists, are the best stewards of the land who will not let mining pollute their backyard. A white male retiree and community leader remarked,

> Nobody wants to pollute our air or our water. That's why we live here. We've been pretty good stewards of our land over the years. ... We probably have the cleanest water, not only in the state of Minnesota, but probably in the whole country.

Iron ore mining is relatively less hazardous than other forms of mining such as coal, gold, and uranium mining, therefore, Iron Range communities have experienced less pollution than places like Appalachia or Montana. The proposed projects are copper mines, though, which brings new risks from acid mine drainage and development near wilderness areas.

Renewal of a Mining Way of Life

Support for right-wing populism and expansion of resource extraction is also mobilized by emotional appeals to nostalgia and place-based identities. One white male pro-mining activist told me that his vision for the region was to "go back to what we used to be." This is indicative of how resource extraction is imbued with social meaning through romantic nostalgia and imaginaries of the good life that are based on patriarchal and white supremacist ideologies (Bouzarovski and Bassin 2011; Smith and Tidwell 2016; Ulrich-Schad and Duncan 2018). Experiences of work in rural areas are shaped by race, class, and gender and represented through narratives of domesticity and gendered divisions of labor (Randall 2005; Scott 2007; O'Shaughnessy and Krogman 2011). A middle-aged white male Iron Range resident recounted the past as a more prosperous time:

> When I was growing up everybody's parents, or dad at least, worked there. Women didn't work really. Everybody had a stay-at-home mom. But at that point in time, those were good union jobs. So, everybody was making good money. It was a middle-class town. We didn't have any extreme poverty, and we didn't have any extreme wealth.

This memory constructs an idealized imaginary of the past as the good life, which Huber (2013) argued is an exclusionary vision that privileges a particular type of white and rural culture connected to masculine extractive labor. As Ekers (2009) argued, the loss of family breadwinning jobs for men in extractive industries can generate a crisis of masculinity, which I find contributes to the appeal of nostalgia in extractive populism.

Iron Range residents expressed a longing for a time when their communities were vibrant. A retired white male miner and community leader told me,

> Back when I was growing up, we had like nine grocery stores in Ely. We had six women's clothing stores, five men's clothing stores. We had five car dealerships.

The extractive moral economy meant that mining companies provided jobs for local men and revenue for local businesses and schools. Downtown shopping districts and schools are now a symbol of the region's struggles. Copper mines are then presented as a fix that can breathe life back into "dying mining towns."

Right-wing populist rhetoric, such as Trump's "Make America Great Again" slogan, provides a sense of security in the face of economic dislocation through an appeal to nostalgia (Ulrich-Schad and Duncan 2018). One white male pro-mining activist told me that he thought Trump would help the working class:

> He [Trump] was for the American people. He was for the American workers. ... He wanted to bring business back to this country. Those are the things that your normal everyday working guy wanted to hear instead of watching their jobs go overseas because of excessive taxes and restrictions.

Taxes and regulations are blamed for job loss and deregulation and development offer the hope of putting men back to work and protecting the U.S. economy from overseas competition.

Resource Nationalism

Support for mining is also driven by nationalist concerns about resource independence that are shaped by racialized and xenophobic fears of "unfriendly" foreign countries. A retired miner told me that the copper mines were needed to compete with China:

> Your grandchildren will be working for the Chinese and they will mine it. Let's look at the world as a whole, when I was in high school they said, "One fifth of the world lived in China." Do you think you're going to be able to hold them people back forever? In my lifetime they've started getting enough to eat and they're going to progress. ... The Chinese will not put up with welfare and all this other bullshit. If you don't make the muster in the morning son you ain't eating no rice.

His racist and xenophobic discourse presents China as a threat to the United States that is driven to expand and is not constrained by human rights and environmental protections. Thus, copper mines need to be constructed in Minnesota, so that the United States can compete and control its own supplies.

China serves as a scapegoat for economic struggles and job loss, rather than the multinational mining companies and corporate-friendly trade policies.

Conclusion

This article traces the cultural, affective, and discursive dynamics of conflicts over proposed copper mining in northern Minnesota in relation to the upsurge of right-wing populism. Many residents and leaders understand federal government actions and environmentalism as an affront to their sense of a moral economy based on local communities' rights to make decisions about resource use and the ability to sustain livelihoods by using nature. Right-wing extractive populism acknowledged people's sense of marginality and anger while providing a target—environmentalists and environmental regulations—and a vision of hope—a return to the heyday of mining and white masculine industrial labor. Copper mining is framed as a way to reassert the Iron Range as central to the nation's security and prosperity. Thus, the dominant framings of copper mining resonated with right-wing populist discourses of us versus them, giving power back to the people, and nationalism (Hochschild 2016; Oliver and Rahn 2016) that are intertwined with whiteness, xenophobia, and masculinity.

Drawing on political ecology perspectives (McCarthy 2002; Robbins 2002; Kosek 2006), I argue that class, race, and gender dynamics of place-based identities and moral economies tied to mining are a key part of the micropolitics of right-wing populism. Yet, these dynamics are often overlooked in the literature on populism in the Global North that focuses on a narrow concept of racism or xenophobia and national-level explanations (Inglehart and Norris 2016; Oliver and Rahn 2016; Spruyt, Keppens, and Van Droogenbroeck 2016; Bonikowski 2017). I add to emerging analysis of how right-wing populist movements mobilize place-based and class identities, emotions, and collective memories (Green et al. 2016; Hochschild 2016; Balthazar 2017; Gusterson 2017; McQuarrie 2017; Ulrich-Schad and Duncan 2018) by emphasizing the role of environmental imaginaries and the social meanings of land, labor, and natural resources. Right-wing politicians' support for extractive populism and anti-environmentalism is not due simply to industry's political–economic power but also to the cultural and affective power of mining and how it provides legitimacy and mobilizes white and rural people. Whereas much of the political ecology literature on collective memory and place has focused on resistance (Legg 2004; Sultana 2011; Lundgren and Nilsson 2018; Perreault 2018; Threadgold et al. 2018), I emphasize how these dynamics can reaffirm the hegemony of extractive capitalism. I show how defending rural livelihoods and sense of place can motivate support for nationalist, racist, and capitalist political projects, which demonstrates the contradictions in moral economies of resource extraction.

Extractive populism also resonates in rural regions because of political and economic conditions (Scoones et al. 2018). Populist messages resonate when there is a sense of crisis (Moffitt 2016), which is common in resource-dependent communities that face economic depressions and a sense of insecurity created by boom-and-bust cycles. In the Iron Range, populist politicians spoke to people's hardships through a narrative of hope for returning to a prosperous past. The current rightward swing also emerges from the particular political conjuncture and reaction to shifts in the Democratic party toward urban, upper class, and financial constituencies, the "New Democratic Party," that was personified in Hillary Clinton (Frank 2016; McQuarrie 2017). Copper mine supporters framed Democrats who opposed mining as turning away from their labor base and becoming the party of urban liberal environmentalists. The 2016 presidential vote was also a broader reaction against the status quo, as progressive populist candidate Bernie Sanders defeated Clinton in the Minnesota primary, and I met several people who supported Sanders and then voted for Trump.

Although rural, white, working-class communities are not to blame for the election of Trump—wealthy and urban white voters were a large part of his success (Gusterson 2017; Walley 2017)—their support for right-wing populism has ideological and material implications. This has bolstered politicians who push privatization and deregulation to expand environmentally hazardous extractive development and advance racist, sexist, and anti-immigrant projects. Although right-wing populists claim that they will help rural and poor communities, their pro-industry policies accelerate the processes of neoliberal capitalism and automation that produce economic struggles in extractive regions and will reduce social and environmental protections. The mobilization of nostalgia, anger, and connections to place has reproduced commonsense beliefs that the Iron Range cannot exist without mining and has bolstered an extractive populist alliance between conservative politicians, mining companies, and rural, working-class, white residents.

ORCID

Erik Kojola ⓘ http://orcid.org/0000-0001-6229-1817

Note

1. Data from Office of Minnesota Secretary of State, 2016 General Election Results by Precinct for precincts 6A and 6B, which include Hibbing and Virginia, Minnesota. http://electionresults.sos.state.mn.us/Results/FedStatebyLEGDistrict/100?districtid=366

References

Ageyman, J., and R. Spooner. 1997. Ethnicity and the rural environment. In *Contested countryside cultures: Otherness, marginalisation and rurality*, ed. P. Cloke and J. Little, 197–217. London and New York: Routledge.

Agnew, J., and M. Shin. 2017. Spatializing populism: Taking politics to the people in Italy. *Annals of the American Association of Geographers* 107 (4):915–33.

Arbatli, E. 2018. Resource nationalism revisited: A new conceptualization in light of changing actors and strategies in the oil industry. *Energy Research & Social Science* 40 (June):101–18.

Arsel, M., B. Hogenboom, and L. Pellegrini. 2016. The extractive imperative in Latin America. *Extractive Industries and Society* 3 (4):880–87.

Azari, J., and M. J. Hetherington. 2016. Back to the future? What the politics of the late nineteenth century can tell us about the 2016 election. *The Annals of the American Academy of Political and Social Science* 667 (1):92–109.

Badiou, A., ed. 2016. *What is a people*. New York: Columbia University Press.

Balthazar, A. C. 2017. Made in Britain: Brexit, teacups, and the materiality of the nation. *American Ethnologist* 44 (2):220–24.

Bobo, L. D. 2017. Racism in Trump's America: Reflections on culture, sociology, and the 2016 U.S. presidential election. *The British Journal of Sociology* 68 (November):S85–S104.

Bonikowski, B. 2017. Three lessons of contemporary populism in Europe and the United States. *The Brown Journal of World Affairs* 23 (1):9–24.

Bouzarovski, S., and M. Bassin. 2011. Energy and identity: Imagining Russia as a hydrocarbon superpower.

Annals of the Association of American Geographers 101 (4):783–94.

Bridge, G. 2014. Resource geographies II: The resource–state nexus. *Progress in Human Geography* 38 (1):118–30.

Chezick, B. T. 2017. Let's help the anti-mining dems lose. *Mesabi Daily News*, August 12. Accessed November 28, 2017. https://www.virginiamn.com/opinion/letters_to_editor/let-s-help-the-anti-mining-dems-lose/article_a7c07eb4-7fb5-11e7-b0dc-0f113ed38264.html.

Coronil, F. 1997. *The magical state: Nature, money, and modernity in Venezuela.* Chicago: University of Chicago Press.

Cramer, K. J. 2016. *The politics of resentment: Rural consciousness in Wisconsin and the rise of Scott Walker.* Chicago: The University of Chicago Press.

Davis, M. 2017. The great god Trump and the white working class. *The Catalyst.* Accessed September 21, 2018. https://catalyst-journal.com/vol1/no1/great-god-trump-davis.

Edsall, T. B., and M. D. Edsall. 1992. *Chain reaction: The impact of race, rights, and taxes on American politics.* New York: Norton.

Ekers, M. 2009. The political ecology of hegemony in depression-era British Columbia, Canada: Masculinities, work and the production of the forestscape. *Geoforum* 40 (3):303–15.

Ekers, M., A. Loftus, and G. Mann. 2009. Gramsci lives! *Geoforum* 40 (3):287–91.

Emel, J., M. T. Huber, and M. H. Makene. 2011. Extracting sovereignty: Capital, territory, and gold mining in Tanzania. *Political Geography* 30 (2):70–79.

Forchtner, B., and C. Kølvraa. 2015. The nature of nationalism: Populist radical right parties on countryside and climate. *Nature and Culture* 10 (2):199–224.

Foster, J. B. 1993. The limits of environmentalism without class: Lessons from the ancient forest struggle of the Pacific Northwest. *Capitalism, Nature, Socialism* 4 (1):11–40.

Frank, T. 2016. *Listen, liberal, or, what ever happened to the party of the people?* New York: Picador.

Green, S., C. Gregory, M. Reeves, J. K. Cowan, O. Demetriou, I. Koch, M. Carrithers, et al. 2016. Brexit referendum: First reactions from anthropology. *Social Anthropology* 24 (4):478–502.

Gusterson, H. 2017. From Brexit to Trump: Anthropology and the rise of nationalist populism. *American Ethnologist* 44 (2):209–14.

Hall, S. 1985. Authoritarian populism: A reply. *New Left Review* 1 (151):115–24.

Harvey, D. 1996. *Justice, nature and the geography of difference.* Cambridge, MA: Blackwell.

Himley, M. 2014. Mining history: Mobilizing the past in struggles over mineral extraction in Peru. *Geographical Review* 104 (2):174–91.

Hochschild, A. R. 2016. The ecstatic edge of politics: Sociology and Donald Trump. *Contemporary Sociology: A Journal of Reviews* 45 (6):683–89.

Holloway, S. L. 2007. Burning issues: Whiteness, rurality and the politics of difference. *Geoforum* 38 (1):7–20.

Huber, M. T. 2013. *Lifeblood: Oil, freedom, and the forces of capital.* Minneapolis: University of Minnesota Press.

Inglehart, R., and P. Norris. 2016. *Trump, Brexit, and the rise of populism: Economic have-nots and cultural backlash.* Rochester, NY: Social Science Research Network.

Jansen, R. S. 2011. Populist mobilization: A new theoretical approach to populism. *Sociological Theory* 29 (2):75–96.

Jenkins, J. 2016. Contested terrain of extractive development in the American West: Using a regional political ecology framework to understand scalar governance, biocentric values, and anthropocentric values. *Journal of Political Ecology* 23 (1):182–96.

Kenny, M. 2017. Back to the populist future? Understanding nostalgia in contemporary ideological discourse. *Journal of Political Ideologies* 22 (3):256–73.

Koch, I. 2017. What's in a vote? Brexit beyond culture wars. *American Ethnologist* 44 (2):225–30.

Kohl, B., and L. Farthing. 2012. Material constraints to popular imaginaries: The extractive economy and resource nationalism in Bolivia. *Political Geography* 31 (4):225–35.

Kojola, E. 2018. Indigeneity, gender and class in decision-making about risks from resource extraction. *Environmental Sociology.* Advance online publication. doi: 10.1080/23251042.2018.1426090.

Kosek, J. 2006. *Understories: The political life of forests in Northern New Mexico.* Durham, NC: Duke University Press.

Kraker, D. 2016. Iron range voters turn to Trump to boost region's struggling economy. *MPR News*, November 17. Accessed January 6, 2017. http://www.mprnews.org/story/2016/11/18/iron-range-voters-turn-to-trump-to-boost-regions-struggling-economy.

Laclau, E. 1977. *Politics and ideology in Marxist theory: Capitalism, fascism, populism.* London: New Left Books.

Lamont, M., B. Y. Park, and E. Ayala-Hurtado. 2017. Trump's electoral speeches and his appeal to the American white working class. *The British Journal of Sociology* 68 (November):S153–S180.

Legg, S. 2004. Memory and nostalgia. *Cultural Geographies; London* 11 (1):99–107.

Lewin, P. G. 2017. "Coal is not just a job, it's a way of life": The cultural politics of coal production in Central Appalachia. *Social Problems.* Advance online publication. doi:10.1093/socpro/spx030.

Li, F. 2015. *Unearthing conflict: Corporate mining, activism, and expertise in Peru.* Durham, NC: Duke University Press.

Lichter, D. T., and D. L. Brown. 2011. Rural America in an urban society: Changing spatial and social boundaries. *Annual Review of Sociology* 37 (1):565–92.

Loomis, E. 2015. *Empire of timber: Labor unions and the Pacific Northwest forests.* New York: Cambridge University Press.

Lundgren, A. S., and B. Nilsson. 2018. Civil outrage: Emotion, space and identity in legitimisations of rural protest. *Emotion, Space and Society* 26 (February):16–22.

Manuel, J. T. 2015. *Taconite dreams: The struggle to sustain mining on Minnesota's iron range, 1915–2000.* Minneapolis: University of Minnesota Press.

Marston, A., and T. Perreault. 2017. Consent, coercion and *cooperativismo*: Mining cooperatives and resource regimes in Bolivia. *Environment and Planning A: Economy and Space* 49 (2):252–72.

McCarthy, J. 2002. First world political ecology: Lessons from the wise use movement. *Environment and Planning A* 34 (7):1281–1302.

McKinnon, I., and C. C. Hiner. 2016. Does the region still have relevance? (Re)considering "regional" political ecology. *Journal of Political Ecology* 23 (1):115–22.

McQuarrie, M. 2017. The revolt of the Rust Belt: Place and politics in the age of anger. *The British Journal of Sociology* 68:S120–S152.

Moffitt, B. 2016. *The global rise of populism: Performance, political style, and representation.* Stanford, CA: Stanford University Press.

Moore, D. S. 2005. *Suffering for territory: Race, place, and power in Zimbabwe.* Durham, NC: Duke University Press.

Nesbitt, J. T., and D. Weiner. 2001. Conflicting environmental imaginaries and the politics of nature in Central Appalachia. *Geoforum* 32 (3):333–49.

Oliver, J. E., and W. M. Rahn. 2016. Rise of the Trumpenvolk: Populism in the 2016 election. *The Annals of the American Academy of Political and Social Science* 667 (1):189–206.

O'Shaughnessy, S., and N. T. Krogman. 2011. Gender as contradiction: From dichotomies to diversity in natural resource extraction. *Journal of Rural Studies* 27 (2):134–43.

Peet, R., and M. Watts. 1996. *Liberation ecologies: Environment, development, social movements.* London and New York: Routledge.

Peluso, N. L. 2012. What's nature got to do with it? A situated historical perspective on socio-natural commodities. *Development and Change* 43 (1):79–104.

Perreault, T. 2013. Nature and nation: Hydrocarbons, governance, and the territorial logics of "resource nationalism" in Bolivia. In *Subterranean struggles: New dynamics of mining, oil and gas in Latin America,* ed. A. Bebbington and J. Bury, 67–89. Austin: University of Texas Press.

———. 2018. Mining, meaning and memory in the Andes. *The Geographical Journal* 184 (3):229–41.

Perreault, T., and G. Valdivia. 2010. Hydrocarbons, popular protest and national imaginaries: Ecuador and Bolivia in comparative context. *Geoforum* 41 (5):689–99.

Phadke, R. 2011. Resisting and reconciling big wind: Middle landscape politics in the New American West. *Antipode* 43 (3):754–76.

Pliml, G. 2017. Despite concerns, staying blue. *Mesabi Daily News,* August 19. Accessed March 29, 2018. https://www.virginiamn.com/opinion/letters/despite-concerns-staying-blue/article_93b28e0a-853f-11e7-be14-c7dce4708c08.html.

Prudham, S. W. 2005. *Knock on wood: Nature as commodity in Douglas-fir country.* London and New York: Routledge.

Randall, R. 2005. Wilderness wives: Domestic economy and women's participation in nature. In *This elusive land: Women and the Canadian environment,* ed. M. Hessing, R. Raglon, and C. Sandilands, 35–56. Vancouver: University of British Columbia Press.

Robbins, P. 2002. Obstacles to a first world political ecology? Looking near without looking up. *Environment and Planning A* 34 (8):1509–13.

Rolston, J. S. 2014. *Mining coal and undermining gender: Rhythms of work and family in the American West.* New Brunswick, NJ: Rutgers University Press.

Rosales, A. 2017. Contentious nationalization and the embrace of the developmental ideals: Resource nationalism in the 1970s in Ecuador. *The Extractive Industries and Society* 4 (1):102–10.

Scoones, I., M. Edelman, S. M. Borras, Jr., R. Hall, W. Wolford, and B. White. 2018. Emancipatory rural politics: Confronting authoritarian populism. *The Journal of Peasant Studies* 45 (1):1–20.

Scott, R. R. 2007. Dependent masculinity and political culture in pro-mountaintop removal discourse: Or, how I learned to stop worrying and love the dragline. *Feminist Studies* 33 (3):484–509.

———. 2010. *Removing mountains: Extracting nature and identity in the Appalachian coalfields.* Minneapolis: University of Minnesota Press.

Smith, J. M., and A. S. D. Tidwell. 2016. The everyday lives of energy transitions: Contested sociotechnical imaginaries in the American West. *Social Studies of Science* 46 (3):327–50.

Solty, I. 2013. The crisis interregnum: From the new right-wing populism to the Occupy movement. *Studies in Political Economy* 91 (1):85–112.

Spruyt, B., G. Keppens, and F. Van Droogenbroeck. 2016. Who supports populism and what attracts people to it? *Political Research Quarterly* 69 (2):335–46.

Sultana, F. 2011. Suffering for water, suffering from water: Emotional geographies of resource access, control and conflict. *Geoforum* 42 (2):163–72.

Taggart, P. A. 2000. *Populism.* Buckingham, UK: Open University Press.

Threadgold, S., D. Farrugia, H. Askland, M. Askew, J. Hanley, M. Sherval, and J. Coffey. 2018. Affect, risk and local politics of knowledge: Changing land use in Narrabri, NSW. *Environmental Sociology* 4 (4):393–404.

Tidwell, J. H., and A. S. D. Tidwell. 2018. Energy ideals, visions, narratives, and rhetoric: Examining sociotechnical imaginaries theory and methodology in energy research. *Energy Research & Social Science* 39 (May):103–07.

Ulrich-Schad, J. D., and C. M. Duncan. 2018. People and places left behind: Work, culture and politics in the rural United States. *The Journal of Peasant Studies* 45 (1):59–79.

Valdivia, G. 2008. Governing relations between people and things: Citizenship, territory, and the political economy of petroleum in Ecuador. *Political Geography* 27 (4):456–77.

Walker, P. A. 2003. Reconsidering "regional" political ecologies: Toward a political ecology of the rural American West. *Progress in Human Geography* 27 (1):7–24.

Walley, C. J. 2017. Trump's election and the "white working class": What we missed. *American Ethnologist* 44 (2):231–36.

Whitehead, M., R. Jones, and M. Jones. 2007. *The nature of the state: Excavating the political ecologies of the modern state.* Oxford, UK: Oxford University Press.

ERIK KOJOLA is an Assistant Professor in the Department of Sociology and Anthropology at Texas Christian University, Fort Worth, TX 76129. His research interests are in labor–environment relationships, the politics of resource extraction, and political ecology.

Emotional Environments of Energy Extraction in Russia

Jessica K. Graybill

The association among extraction, emotion, and governance has received little attention in human and environmental geography and social science more generally. Extraction in authoritarian environments has long been studied regarding energy and extractive environments, with special attention to political ecological struggles and conflicts over rights and access to lands and critical resources. Yet little attention has been paid to how affect and emotions could shape the socioeconomic political and ecological aspects of extraction under authoritarian rule. Geographers have argued that emotions matter in making sense of place, politics, and environmental transformation for at least a decade, and it is now time to craft theoretical insights that examine emotional geographies of extraction, their politics, and their governance in ways that move toward deeper understanding of how affect and emotion are produced and mobilized in different kinds of extractive environments and under varying socioeconomic and political conditions. Thus, this article explores affect and emotion related to extractive environments in the Russian Federation with the aim of pushing emotional geographies—and extractive studies—forward in new, synergistic ways. Using discourse and content analysis, I explore the roles of affect and emotion in the production and reproduction of narratives about extraction in one post-Soviet and hybrid authoritarian-democratic regime. I argue that a tacitly endorsed, pervasive state emotional geography about extraction in an era of increasingly authoritarian rule acts to (re)create desire for continued pursuit of extraction by this energy superpower.

搾取、情感和治理之间的关联，在人文与环境地理学和更为广泛的社会科学中鲜少受到关注。有关威权环境中的搾取之研究，长期关乎能源与搾取环境，并特别关照取得土地和关键资源的权益与管道之政治生态斗争与冲突。但情感与情绪如何能够在威权治理下形塑搾取的社会经济政治与生态面向，却鲜少受到关注。地理学者主张情绪在理解地方、政治与环境变迁中具有作用，至少已有十年的历史，当下则是打造理论洞见来检视搾取的情绪地理及其政治和治理之时刻，以更深刻地理解情感和情绪如何在不同种类的搾取环境与各种社会经济和政治情境下进行生产并动员。本文因而探讨与俄罗斯联邦中与搾取环境有关的情绪及情感，旨在以崭新且协力的方式，推进情绪地理学和搾取研究。我运用论述与内容分析，探讨情绪与情感在一个后苏联的涵合威权民主政体中的搾取叙事生产与再生产中所扮演的角色。我主张，在逐渐朝向威权治理的年代中，有关搾取的默许且无所不在的国家情绪地理，用来（再）创造此一能源霸权对持续追求搾取之慾望。 *关键词：威权主义，搾取，涵合体制，俄罗斯，国家情感。*

La asociación entre extracción, emoción y gobernanza ha recibido poca atención en geografía humana y ambiental y, más generalmente, en la ciencia social. La extracción en entornos autoritarios ha sido estudiada por mucho tiempo en lo que concierne a entornos extractivos relacionados con energía, con especial atención a las luchas ecológicas políticas y conflictos sobre derechos y acceso a las tierras y recursos críticos. Con todo, escasa atención se ha prestado a cómo el afecto y las emociones podrían ayudar a configurar los aspectos socioeconómicos, políticos y ecológicos de la extracción bajo orden autoritario. Los geógrafos han sostenido por lo menos durante una década que las emociones importan en la construcción del sentido del lugar, en políticas y transformación ambiental, y ya es tiempo de confeccionar perspectivas teóricas que examinen las geografías emotivas de la extracción, sus políticas y su gobernanza, de tal suerte que nos lleven a un entendimiento más profundo sobre cómo el afecto y la emoción son producidos y movilizados en diferentes tipos de entornos extractivos y bajo varias condiciones socioeconómicas y políticas. Entonces, este artículo explora el afecto y la emoción en relación con entornos extractivos en la Federación Rusa con la mira de impulsar las geografías emotivas—y los estudios extractivos—en nuevas maneras sinérgicas. Usando análisis de discurso y contenido, exploro los roles del afecto y la emoción en la producción y reproducción de narrativas acerca de la extracción en un régimen híbrido autoritario–democrático postsoviético. Yo sostengo que una geografía emotiva omnipresente acerca de la extracción, tácitamente patrocinada, en una era de orden crecientemente autoritario, actúa para (re)crear deseo por la continuada actividad de extracción por esta superpotencia energético. *Palabras clave: autoritarismo, emoción de estado, extracción, régimen híbrido, Russia.*

The critical roles that energy and resource development play in shaping local communities and ecologies, national development trajectories, and global economies has been well researched in multiple sites globally. Much research about energy and resource development (hereafter, extraction) focuses on conflict, social or ecological injustice, and questions of human access or rights to land, critical (i.e., water, subsistence food) resources, and sacred places amidst reduced access and pollution due to extractive activities that often benefit national or global actors instead of local actors (Klare 2001; Le Billon 2001, 2008; Watts 2004; Collier and Hoeffler 2005; Perreault and Valdivia 2010; Coombes, Johnson, and Howitt 2011; Ballvé 2012; Vélez-Torres 2013). Indeed, of recent interest among geographers is the role of national governments in making carbon economies (Bridge 2011), in politicizing resources (Watts 2010; Huber 2018), and in (re)nationalizing resources (Bridge 2014; Childs 2016; Koch and Perreault 2018). These geographic scholars and many others have commonly used political geographic or political ecological thought to formulate understandings of how environments are degraded, people become marginalized, environmental conflict is managed, and identities and social movements are created in places strongly ruled by a national government. Studies show that even in authoritarian regimes, with their urge to control governmental and public discourses about energy and resource development decisions and trajectories, discourses of struggle and conflict cannot be eliminated from the narratives that develop among the populations living in locales of resource extraction and refinement. For example, struggle against the construction of environmentally harmful hydrocarbon infrastructure (e.g., roads, pipelines) on Sakhalin Island in Russia was well documented by scholars (Wilson 2003) and in local to international media due to environmental activism in the mid-2000s (Bradshaw 2006; Graybill 2009). Often, these struggles include an ecological–ethnic discourse when extraction impinges on indigenous rights and access to land and critical resources, as indicated in

the Puno region in Peru related to large-scale mining activities (McDonnell 2015).

Despite this rich body of knowledge about how people might operate—with struggle, using subversive tactics of survival, against degradation—in extractive regimes or about resources as more than just natural, only recently have geographers focused on affect and emotions about resources, energy, or extraction as a way to make sense of their impact on everyday lives, individual subjectivities, and political choices. Regarding access to water resources in Bangladesh, Sultana (2011) found that as environments change due to resource development, how people feel about and thus use the environment might also change, leading to new responses to place, environment, and natural resources. In a study of subjectivity and emotion among fishermen, Nightingale (2013) noted the Scottish fisherman's subjectivity, or "how he comes to understand his role in his occupation and that particular place" as creating a "strong link between his attachments to the sea, the emotional relations produced, and his political work in fisheries management" (2369). Finally, in an exploration of emotions about rapid ecological change due to oil and gas development on Sakhalin Island in Russia, Graybill (2013) wrote that emotions cannot be regarded as separate from economic development or as private individual matters, because they form part of the "foundation of the economic and public spheres on Sakhalin Island, simultaneously provoking resource overharvest, which will affect the long-term economic base of the region, and public outcry against certain kinds of new extractive resource regimes developing in the post-Soviet era" (50).

The intractability and intangibility (Davidson and Bondi 2004) of emotions infuse all aspects of our everyday lives, informing how we feel, think, and act in our physical, material, and social environments. Davidson, Bondi, and Smith (2005) explained emotional geography as a way to "understand emotion—experientially and conceptually—in terms of its socio-*spatial* mediation and articulation" (3, italics in original). In this way, emotions are relational and fluid, embodied, and produced in context among people and in place (Ahmed 2004; Davidson and Bondi 2004; Davidson, Bondi, and Smith 2005; Smith et al. 2009). Within this literature, the terms *affect* and *emotion* can be ascribed different meanings. Emotion is often associated with specific states (e.g., pride, fear, joy) that can be attributed to individual people, whereas affect can be understood as "pre- or extra-discursive, non-individualized and mobilized conceptually" (Bondi 2005, 437). In this article, a focus on affect and emotion deviates from more conventional geographic and political ecological discourses about extraction, instead asking how geographers might engage with somatic (embodied) power, not just with the cognitive or rationalizing powers of the human mind.

With this ever-growing body of literature on emotional geographies generally and about affect and emotion related to resources (represented earlier are water, fish, hydrocarbons) specifically, it should no longer be questionable whether affect and emotions are important to geographies of extraction. Rather, pushing this intersectional field forward requires asking about the kinds of affect and emotion that operate in different energy and resource extraction contexts, what roles they play in communities, and how or whether an understanding of emotional "being-in-common" (Pratt 2012, 177) can be heard in local-global contexts. Neuman et al. (2007) attempted to address what they called the "affect effect," or "the ways in which citizens, through a mixture of impulse and calculation, reckon what is politically significant to them" (9). This article explores emotional responses to extraction in an increasingly authoritarian setting, the Russian Federation, to understand how emotional narratives about extractive development are expressed by those who live in an increasingly authoritarian nation state. How are narratives about extraction expressed by individuals in this society? What kinds of emotions and affect are associated with extractive development in Russia? For all of the focus on human relationships to the environment and resource development to date, geographers have only taken limited steps to tackle such issues and in different kinds of governments. In so doing, I explore the synergy between geographies of extraction and emotion to understand how certain foci of economic development—such as the extraction of energy resources—are accepted and even supported by citizens of nations that are strongly ruled by individual personalities.

Alongside examining scholarly literature relevant to placing affect and emotions about extraction and in authoritarian settings, this case study draws on interviews and conversations I gathered from 2003 to 2017 during multiple field seasons involving semi-structured and open-ended interviews about oil and gas extraction in sites of active or onboarding extraction and from conversations about resource extraction in other places in Russia where discussion has turned seriously to twenty-first-century energy development. As perceptions, discourses, and responses (including emotional) to extraction have been my major research focus during this time period, I have developed a database of more than 100 individual interviews with Russian citizens who reside in places experiencing varying stages of oil and gas exploration and extraction on Sakhalin Island and the Kamchatka and Kola peninsulas; in places of hydrocarbon infrastructure development (i.e., offshore rig foundations) near Vladivostok; and in urban centers where conversations about energy and extraction have been pursued, including Moscow, Murmansk, St. Petersburg, Samara, and Saratov. All conversations occurred in Russian and transcribed interviews note the ways in which things were said (e.g., with bodily gestures or with inflected voices), which is useful for registering emotions. Using NVivo 12 (2018) software, which allows analysis of materials in English and Russian, I conducted a content analysis of this database, explicitly noting and tagging emotions in the transcribed interviews and conversations. Once certain words and phrases related to affect and emotion were located in the database, I reexamined the surrounding sentences to contextualize emotions about hydrocarbons and energy infrastructure in Russia to understand how people might express affect about extractive industrial activity in the twenty-first century. Participants include workers in the oil and gas industry, academic experts, professionals from multiple fields (e.g., teaching, bookkeeping, secretarial work) living in locations of extraction or infrastructure development, and other educated working-class citizens who have knowledge of Russia's focus on extractive (especially oil, gas, and mineral) resources.

The Sensitivities of Authoritarian Energy

Creation of the extractive landscape of the Russian Federation has roots in at least one century of authoritarian legacy—an inheritance that has contributed to ongoing dependence on an extractive relationship with nature, especially of forest, fish, minerals, and hydrocarbon resources, which shape how individuals and society respond to this extractive relationship with nature. The Union of Soviet Socialist Republics, formed in 1917, promoted Soviet Communism across Eurasian space, including plans for the development of large-scale urbanization and industrialization that would colonize Eurasian space over five-year increments of time, methodically and mechanically grinding their way through *chernozem* (black earth), Arctic permafrost, central Asian steppe, and the Siberian plains to implant concrete and steel infrastructure and wholly new urban settlements and manufacturing facilities on the landscape eastward, southward, and northward from the western core. The vertical hierarchy of exceptionally similarly designed urban places was controlled by and connected to Moscow in the overall network by the Trans-Siberian Railway (Medvedkov 1990; Dienes 2013; Dixon and Graybill 2016). This node-and-network system provided a transcontinental web across which raw materials, goods, laborers, and bureaucrats were mobilized, expanding resource extraction across Eurasia and solidifying a modern authoritarian-ruled empire based on industrial growth and resource extraction.

Solidification of authoritarian Soviet rule over society and nature depended a great deal on the control of society's actions and, in turn, the control of affect and emotion. Shalin (1996), a sociologist specializing in emotions in Russian culture, considered that "Soviet civilization represented a concerted effort to harness emotions to an ideological cause, to reshape them in a systematic fashion consistent with the political agenda articulated by the Communist Party of the Soviet Union" (23). Furthermore, Shalin (1996) compared building communism to a "Herculean emotional task" in which the duty of *Homo Sovieticus*, the New Soviet Man, was to be "happy" (26). In describing the duty of being happy, Shalin (1996) drew the conclusion that every society "mobilize[s] affect to accomplish their goals [but] some do this in a more ruthless way than others" (46). Describing how individuals navigate affect and emotion, he described how people who "came of age under the communist regime developed a chameleon-like quality that enabled them to conceal their true feelings, suppress politically incorrect thoughts,

and engage in behavioral gambits dramatizing authorized identities" (Shalin 2004, 416).

The preceding reflections on Soviet communist culture—particularly the concept of duty assigned to particular emotions—ask us to consider that expression of affect and emotion in authoritarian societies might be different than that in more democratic regimes. For example, the use of mass hysteria to bolster policy and the prevention of public dissent (e.g., via demonstrations) are well-known strategies in authoritarian regimes. Indeed, many scholars examining the authoritarian East consider rules about feelings to be culturally scripted (Hwang 2006; Kim and Cohen 2010). Writing about Vietnam, Gillespie (2018) found that such scripts guide individual reflection on and response to emotions. Regarding nationalism prior to the Beijing Olympic games, Yang (2005) reflected on the social norms that are encoded into society and that act as "emotional schema" that "establish a mental aura, where the emotional state of the individual members is resonated with a larger collectivity" (842). Finally, Liao (2013) defined culturally scripted, ideological emotions as "state emotion" in China, which reflects the domestic normative environment and the workings of which "cannot be properly understood without simultaneously focusing on its normative elements ingrained in the national identity and the political culture where it has evolved" (155).

Russia today is not a purely authoritarian regime, however. Collapse of the Soviet state in 1991 led to the breakdown of the Soviet command economy and thus the downfall of state-bolstered heavy and military industries and, accordingly, Soviet-era resource extraction (Gustafson 2012). Upheaval of the command system for extraction, production, mobilization, and use of materials and goods within the Soviet bloc occurred alongside other socioeconomic and political transformations, namely, the introduction of capitalism and democracy as potentially viable economic and political structures operating within Russia. What some scholars call rampant kleptocracy (Dawisha 2014) and "wild east" capitalism (Sergeev 1998) resulted during the Yeltsin era, as Russia turned away from communist ideals of strong central leadership, regional self-sufficiency, and a command economic system toward prodemocratic political processes alongside economic reform. In the early post-Soviet period, however, instability and corruption caused Yeltsin's popularity to decline as political and social pressure to resolve internal issues increased (Pipes 2004; Gorbachev 2017). The oligarchs—many of them former Soviet *apparatchiki*—wielded the same controlling power they had before in politics, only now in the name of capitalism (Hoffman 2002). The fortunes made (and lost) during the Yeltsin era created a realm of political and economic instability and creative maneuvering for nearly two decades until the installation, by Yeltsin, of Vladimir Putin as president.

Becoming Great Again, Emotionally

The Putin era began with hope for the return of Soviet-era stability, with the public understanding Putin to be an ex-KGB strongman who would reign in the oligarchs and the increasingly independent regions (Hoffman 2002). Despite—or perhaps because of—the awkward and unruly opening up of Russia's extractive resources to the global market in the 1990s, extractive energy development in Russia today is driven by global market forces and transnational investment. Yet, it also depends on highly controlled, authoritarian-like decision-making structures of power for the production and mobilization of resources once they are extracted. A return to a highly state-controlled and -negotiated landscape for energy production, especially for oil and gas production, after the mid-2000s has been accomplished as Putin has attempted to (re)build a strong nation state in what has been called (re)nationalization of the extractive industries (Ellman 2006; Tynkkynen 2016).

According to The U.S. Energy Information Administration (International Energy Agency 2017), Russia was the world's largest producer of crude oil and second largest producer of natural gas in 2016, when hydrocarbon activities comprised ~36 percent of the federal budget's revenues and 70 percent and 90 percent of Russian oil and gas exports, respectively, went to European countries. To accomplish these and other economic goals, Putin's Russia has acquired a hybrid form in which authoritarianism and democracy are both used to accomplish political governance over society and the economy (Grishin 2015; Levitsky and Way 2016). Democratic elections in a hybrid regime operate to stabilize autocratic rule while also promoting democratic institutions and protecting civil liberties. Elections, then, are simultaneously regime sustaining and regime subverting (Grishin 2015) and function as expressions of

(dis)satisfaction with government policies and as lines of communication between ruling elites and citizens. Shevtsova (2016) suggested that the partial reversion to authoritarian tactics in the Putin era indicate the enduring quality of "personalized power" (40) in an era characterized by support for the president but tempered by undercurrents of discontent due to foundering economic conditions, higher costs of material goods, and mercurial curtails on personal freedoms. In an exploration of emotions in and about democracies, Shalin (2004) asked us to consider that democracy is "an embodied process that binds affectively as well as rhetorically" (407). If this is true, then we should investigate affect and rhetoric that is created in autocratic or hybrid authoritarian-democratic regimes.

The notion of becoming great—again—is a discourse propagated at a national level in post-Soviet Russia. Indeed, examining political documents from the decades of the 2000s, Bouzarovski and Bassin (2011) wrote that "Russia's energy assertiveness … has acquired an explicitly formulated political connotation" (787). They saw that energy and energy infrastructure play a large role in the age of Putinism, a political movement concerned with the restoration of Russia as a great power (*derzhava*) via reconstruction of a strong domestic economy and respected global status, in the nationalistic push to make Russia a great force again (Migranyan 2004; Gudkov 2011; Lacqueur 2015). Extractive resources—especially oil, gas, and coal—and their flow in, through, and out of Russian economy and territory are key to Putinist doctrine. Fish et al. (2017) described Putinism as "conservative, populist, and personalistic" and "closely intertwined with Russia's extractive, rent-driven economy, where conservatism is related to maintaining the current socioeconomic and political status quo, where populism includes resistance to decadent liberalism" and personalism includes the "hollowing out of parties and institutions for the (re)advent of one-man rule" (61–62).

With increased economic and political stability since 1991 and despite sanctions and devaluation of the ruble in 2014, Russians have gained a renewed sense of pride in their country and hope for its future accomplishments (Levada Center 2017). Pride is accompanied by continued pursuit of development trajectories from the Soviet past, including continuing to reap the benefits of extractive development of raw resources, especially in the northern and eastern resource peripheries (Bradshaw and Prendergrast 2005). This is noted in the energy plans put forward by the Russian government until 2030 (Ministry of Energy of the Russian Federation 2009) and expressed in everyday media. For example, the content of a recent Rosneft advertisement is the following:

In the opening scene, showing the purchase of fuel at a Rosneft gas station, the narrator says, "The road connects us all. With Rosneft, you are choosing the guarantee of quality and control of your fuel." With images of a mother and daughter driving down a scenic road with no civilization in sight, the narrator speaks of "wide open and endless spaces that some call sadness and others call a smile," because the daughter isn't smiling until mom buys her a snack at a Rosneft gas station at the end of that scene. The narrator continues by remarking that "they bring us surprises and we try to do that for others" while a man looks at a Rosneft advertisement with the words "family team" hovering over a stylized image of three polar bears. The man impulsively buys a stuffed polar bear from the Rosneft gas station and drives it home, safely belted in the passenger seat, to the little girl from the previous frame. As the image shifts from a hug between father and daughter, to stars in the sky, to a final rearrangement of all the stars into the shape of the Russian Federation, the narrator ends by saying, "But any road, no matter what kind, is the road home." The advertisement ends with the narrator's more formal voice, reminding us that "Rosneft is the guarantee of quality and control. There is assurance in every kilometer." (Rosneft 2017, author's translation)

This thirty-second advertisement is only one among many television advertisements, largely from Rosneft and Gazprom—Russia's largest and state-owned oil and gas companies, respectively—that provide a constant stream of media reminders of the force of Russia's energy resources and their extraction and that subtly hint at Russia's power in making their modern, civilized society possible via oil and gas streams. The emotional schema connoted in this media clip is echoed in some interviews. One urbanite in the Volga region shrugged, noting in a rather unattached, analytical way, "Oil and gas is the most important part of our economy. It is our future and we should do everything we can to develop it." This quote indicates how extractive development is understood as a fundamental part of the economy, to be pursued as long and as much as possible. Similar, yet stronger sentiment was expressed in Vladivostok, where a scholar noted emphatically, while leaning

into a conversation and banging fists on the table, "*Vpered* [forward]! Russian energy first! It is our duty to exploit it so we can keep developing."

Excitement about the opening up of the Arctic exists within that region as well as in other places within Russia. In Vladivostok, for example, a geographer excitedly pointed upward, indicating north, and shouted, "Let the Arctic melt! Then, then we can go and develop it all, and all for Russia. Then we won't be behind anymore!" A port laborer, also from Vladivostok, commented brusquely, "We are waiting for the ice to melt. Then we can build our ships here and they will command the north. You'll see. It will be great, work for everyone again." Similarly, a university student hoping to work as a bookkeeper within the extractive industry noted, when discussing the difference between natural and anthropogenic climate change, that "any climate change is good for Russia! It means greater development in the north, which will assist us in achieving our energy development goals, and then we will be a truly global leader. I want to work in that world." Finally, an urbanite from Saratov working in the car manufacturing industry said, despairingly, "Well, of course climate change will get rid of the polar bears: their entire ecology is disappearing. Yes, it's sad. My grandchildren won't even know what a polar bear is!" A few minutes later in the same conversation, this environmental tone disappeared, replaced by a resigned yet proud questioning statement: "But, what is to be done? If it's going to change, why shouldn't Russia benefit from the change? Better Russia than another country!"

These quotes speak to the national goal of becoming an energy superpower (*energeticheskaia sverkhderzhava*; see Rutland 2008), which seems not to have changed since 2000, as Putin has seemingly put order back into the country's operations—albeit with an authoritarian hand. Approval for government operations has mostly increased during his tenure in power, especially after 2014 and the annexation of Crimea (Levada Center 2017). Thus, it seems even more imperative to heed the call by Bouzarevski and Bassin (2011) to examine more closely the mechanisms used by the state to create identity as it relates to energy infrastructure, as little attention has been given to "the manner in which energy systems shape, and are shaped by, narratives and practices of identity building at different scales" (784).

As acknowledged earlier, state identity building in authoritarian settings is rife with emotions (Shalin

1996, 2004; Liao 2013; Gillespie 2018) and deserves greater attention and elucidation in any examination of the intersections of energy, emotions, and style of leadership. National pride in energy development is expressed as state emotion by a longtime oil engineer on Sakhalin who noted emphatically, while waving his arms energetically and with great pride, that Russia has been a forerunner in developing hydrocarbon extraction techniques: "It was *Russia* who first developed the technique of hydrofracking. Russia! Now it is used worldwide." The palpable sense of accomplishment at Russia's claim to developing an extraction technique now used worldwide was unforgettable. Similarly echoing the themes of pride and accomplishment, an oilman (*neftyanik*) who has worked in many regions of the former Soviet Union recognized only power in the hydrocarbon development of the last century, noting that it all began with foreigners who came to Tsarist Russia to seek their fortunes. With grumbling pride (because Baku is now located in Azerbaijan and not in Russian-governed territory), the oilman asked rhetorically, "And where did the Nobel family come to invest? They came to Baku of course—to *our* Empire. Because *we* have the best oilfields."

In the twenty-first century, the pride of exploration and extraction is expressed about Russia's northern periphery. Whereas some Russian Arctic locales have been developed under the Soviet rubric of urban–industrial (e.g., Norilsk) or military–industrial (e.g., Murmansk region) complexes, much of this region is relatively pristine and home to multiple indigenous groups, largely nomadic and seminomadic reindeer herders. Indeed, eight of the world's ten largest Arctic cities are in Russia and it is from these locations and from newer fly-in, fly-out (FIFO) locations that twenty-first-century resource extraction occurs. Existing Arctic cities act as "hubs" of information, technology, monetary, and sociopolitical networking for national and transnational oil, gas, and mining operations, thereby providing regional control over resource development in remote northern regions. Headquarters of these companies might be in larger cities (Moscow, St. Petersburg, overseas), but regional operational bases provide hope and anticipation of future development for the residents of these regions and for migrant laborers from across Russia and other countries, especially the near abroad (former Soviet countries) working in urban settlements and FIFO camps

associated with extractive development. In short, the Arctic is buzzing with the anticipation of extractive development, which is well outlined in Russia's strategy for Arctic development. There is no doubt of the coming utility of Arctic resources in this policy: "use of the Arctic zone of the Russian Federation as a strategic resource base of the Russian Federation provides the solution to problems of social and economic development of the country" (Russian Federation Policy for the Arctic to 2020 2009).

Shalin (1996) suggested that a Soviet emotional culture "continues to persist after the coercive institutions supporting it have broken down" and that this has "greatly complicated the transition to civic culture and democracy in Russia" (23). Many participants in this study reveal a nationalistic affect regarding the extraction of energy resources, about the renewed strength of this industry, and Russia's associated economic vitality. An oilman working offshore near Sakhalin Island noted passionately, "When I am near a rig, I can smell our power and I have hope—no, I know!—that everything will be alright." The use of the term "our" (*nash*) in Russian indicates this participant's strong sense of nationalism. This kind of emotional culture, about industry and economy, could be attributed to a new post-Soviet state emotion, where individual sentiments and emotions begin to take shape—or are shaped from—collective culture, media messaging, and authoritative mandates to continue developing extractive resources as a mainstay of the national economy. As Russia's early post-Soviet socioeconomic transformation resulted in uneven development across Russian territory and by type of industry, there is even more hope in the power of extractive resources as the key to (re)fueling more even national growth. Among citizens, it seems that the conditions of "uneven development, increasing disparities and territorial fragmentation" (Bradshaw and Prendergrast 2005, 87) that resulted from the dissolution of the Soviet Union can only be soothed by the balm of extractive resource development.

Despite the decrease in price per barrel of oil in 2014, the potential of extraction is still considered as a cornerstone of the regional and national economies. Partially, this is because of a century of investment in development based on oil, gas, and mineral resources and the lingering notion that heavy industry and extraction remain a viable path forward for the national state (Alekperov 2012; Gustafson

2012). Another persistent notion is the idea of authoritarian control over land and resources. President Putin is characterized by some as a leader seeking total state power (Fish et al. 2017), which Gustafson (2012) found to be "the precondition for stability and economic growth, and control of natural resources is the precondition of state power" (249). Indeed, a quietly patriotic bookkeeper in Murmansk, remembering the wild, unstable capitalism of the 1990s, urgently whispered, "Putin must (dol'zhen) lead us in our national development of resources. Without our strong leader, there will only be chaos and the foreign companies will ruin us." Finally, the culture around the development of oil and gas is deeply ingrained in Russian mythology about the source of power and wealth of the former Soviet states (Rogers 2015). An analyst of the Soviet and Russian oil industry wrote that "it is oil, then as now, that makes the weather in the Russian economy" (Gustafson 2012, 185). Indeed, younger generations on Sakhalin Island consider themselves to be the zolotie deti (golden children) of that region due to the economic and social benefits that hydrocarbon extraction will bring for at least their lifetimes (Graybill 2009).

Chinks in the (re)developing armor of the post-Soviet emotional schema, however, exist. Despite the presence of what seems to be a persistent state emotion about extraction in national, media, and citizen narratives, some participants in this research note that "energy development the way we do it destroys the environment." This sentiment was voiced repeatedly by residents across Sakhalin Island, who are surrounded by visual reminders of oil spills around Soviet-era onshore oil derricks (near Okha), by watching the emplacement of new roads and pipelines that provide service to newer offshore development (near Nogliki), or by lamenting the loss of favorite ocean beaches due to the construction of Russia's first operating liquefied natural gas terminal (in Prigorodnoye). One source of discontent with extraction in the 1990s was the realization that many places in Russia had become heavily polluted due to the focus on extractive and heavy industrial development during the Soviet period. On the heels of the 1986 Chernobyl nuclear disaster, which occurred during glasnost, scholars and journalists in Russia and abroad increasingly wrote about systemic pollution in urban, industrial, and extractive sites that was invasive and threatened ecosystem

and human well-being (Pryde 1991; Brown 2013). Exposé-like reports of environmental catastrophe circulated in the 1990s and international environmental nongovernmental organizations moved into the newly opened territories of the former Soviet Union with the intent of aiding people and environments experiencing environmental stress (Henry 2013). In sites of hydrocarbon extraction, the loss of environmental quality due to poor quality control became well understood only when transnational corporations—with higher environmental standards—began to operate in Russia (personal communication, World Wildlife Federation representative, Kamchatka, December 10, 2009). Alongside embarrassment at the conditions of the environment was humiliation at the state of social conditions as well, leading to protest against hydrocarbon extraction in sites where multinational corporations had some influence (e.g., Sakhalin Island; see Rutledge 2004; Bradshaw 2006).

Within regions of extraction, discontent exists related to the loss of innocence about extractive development, the loss of pristine places, and the loss of nature as a beloved place of respite and spiritual refueling. On the Arctic coast near the onshore infrastructure developed for the now-delayed Shtokman offshore gas field, a librarian noted, "Yes, we are now connected to all of Russia by a road. But what does this road bring? Nothing but uncivilized tourists who trample our tundra and take our berries. Who needs this? Not us. But what is to be done? Nothing." Such losses and the emotions connected to them lead some to think that greater sacrifice of relatively pristine remote areas—such as those found in Siberia or in the Arctic—to resource development is the price to pay to gain stature as a leader in world energy production. Others bemoan the sacrifice of the environment to extraction and have protested or fought against it on local to international scales. One schoolteacher from Kamchatka, commenting on the gas pipelines that snake across the urban and rural landscapes, was disgusted at their disrepair and leakiness:

> It is shameful that we sell our resources to other places and we are not even fully gasified in the place where the resources come from. What do we get from this? A diseased ecosystem and diseased humans ... we've sacrificed our entire lives for energy development, and for whom? Not for ourselves.

Despair associated with the loss of good environmental conditions varies, but individuals for and against hydrocarbon development note environmental loss. For those living in resource peripheries—the perceived *terra nullius* of urbanites and more centrally located Russians—the pain of loss is experienced due to proximity to extraction sites. Fear also exists of the invisible toxins released from extraction that have seeped into the soils and groundwater supply. For those living farther away, the notion that some places can be sacrificed for the overall betterment and development of the nation and its people is stronger. Although the notion of sacrifice zones is not new (Lerner 2012; Hernández 2015), it is important to recall it here as a way to theorize what some interviews reveal: that the idea of environmental sacrifice zones exists in and about extractive landscapes and is consistently put forward as a way of emotionally coping with or rationalizing environmentally harmful hydrocarbon development. As one professional working for a statistics bureau in St. Petersburg noted, "What do we have to develop, if not energy resources?"

Discussion

For the last century, Russia has pursued a relentless pace of extraction-based urban and industrial development. Affect and emotions about socioeconomic and environmental transformations of the country include national pride alongside desire for improved urban and industrial conditions and the willingness to accept authoritarian rule in exchange for a stable development trajectory and to accept environmental degradation from extraction as part of "progress." Other responses, however, include disgust at the environmental and social costs of extractive development. Sometimes, these contradictory responses have been expressed by single participants, making it clear that further exploration of somatic and cognitive processes associated with affect and emotion are necessary to gain more insight into the effects of affect or how feeling and thinking about extractive environments might lead to action, reaction, or even paralysis regarding extractive environments. In the Russian context, it is no wonder that emotions about extraction—pride, loss, disgust, fear, hope (on hold), anticipation—indicate positive and negative valences toward the mostly authoritarian governance over extraction and its impact on people and environment. On the one hand, participants of all types and in all places suggest that an extractive relationship with nature is perfectly "natural" (Graybill 2017), especially for crafting a strong nation-state. On the other hand, the unnatural—polluted, uninhabitable, wasted—landscapes of extraction resulting from Soviet and now post-Soviet development schemes strike fear into many who hope for a healthier, brighter future in places for living, working, and recreation. This research indicates that participants in this research have multiple emotions about extraction that are often strongly expressed in individual statements or that are created by a passionate engagement in a livelihood, whether it is one that is related to industrial extraction (e.g., in the oil and gas sector) or one that is tied to a more subsistence-based existence where the importance of at least a relatively pristine nature is important for a certain level of survival. Often, single individuals provided contradictory emotional responses to questions and in conversations. Rather than understanding this as ambivalence, a more accurate interpretation might see this as the struggle between individual and state emotions. For example, a quiet bookkeeper might see the only option for Russia's economic advancement as one related to extraction and might patriotically desire that as a path forward while also feeling that such development strategies could be ecologically harmful.

In the case of the Russian Federation, the importance of the command economic political system in guiding extractive development must be evoked, because it provides a rubric for understanding the rerise of authoritarian decision making and for understanding how people think about current and future energy use and development. The Soviet communist mode of development pursued rapid urban, military, and industrial production at the cost of environmental and human well-being, including the devastating consequences of plutonium development for nuclear power (Brown 2013) and the flooding of villages for construction of hydroelectric dams to power industry and cities (Bater 1983), to cite just two examples. To fulfill Soviet five-year plans, authoritarian decision making was employed by the ruling communist leader and multiple leaders and regional planners in Moscow, who then implemented plans via commands given to bureaucrats and laborers in Russia's regions. The act of waiting for commands from the federal center for development trajectories still exists as part of the social fabric and cultural infrastructure in Russia today (personal communication, Sakhalin urban planner, Yuzhno-Sakhalinsk, April 21, 2010). Connecting this pattern to the emotional culture of Russia, it is possible to note that "chronic conditions of voicelessness that mark Russian history [create] ironic detachment and emotional littering which help to mask the gap and to protect one's inner core" (Shalin 1996, 46). The reliance on leaders to determine a development trajectory—such as the decision to remain dependent on hydrocarbons in the national economy (see Ministry of Energy 2009)—then is not surprising in a Murmansk university lecturer's matter-of-fact comment: "Our leaders know what is best for the country. They wouldn't focus on extractive resources otherwise. We just need to wait for their guidance and be patient."

When considering the role of extraction despite the desire for environmental stewardship in Russia, we should remember the words of political scientist Karen Dawisha, who wrote that "[i]nstead of seeing Russia as a democracy in the process of failing we need to see it as an authoritarian system in the process of succeeding" (Dawisha 2014, 7). Currently, the most powerful investors and political proponents of development are precisely those actors who are most invested—economically or politically—in extraction. Energy exports amounted to 77 percent of Russia's total exports in 2016 (International Energy Agency 2017), a whopping percentage of which is either owned or operated by oligarchs or well-connected regional bureaucrats who are connected in the vertical hierarchy of the Putin regime (Alekperov 2012; Gustafson 2012). In short, the political will and economic incentives for socioeconomic development along nonextractive trajectories, perhaps those less harmful to the environment and more freeing for society, do not currently exist in Russia on a large scale. A prime example of this is the collapse of two ministries—the Ministry of Natural Resources (responsible for resource exploration) and the Ministry of the Environment (responsible for environmental conservation and management)—into the Ministry of Natural Resources and the Environment in 2008. Political maneuvering such as this indicates increasing verticalization of power and authoritarianism in Russia as a two-pronged modus operandi that promotes hydrocarbon extraction over other economic activities and over environmental protection.

Returning to the relationship between energy and identity and the lack of scholarship about their interconnections leaves many questions unexamined about "the mechanisms through which state-level actors create particular visions of national identity with the aid of, and in relation to, energy" (Bouzarovski and Bassin 2011, 784). This is crucial in examining the rerise of the *derzhava* movement, in which the extraction of hydrocarbon resources is propagandized as part of the Russian identity at the federal scale, thereby soaking the Russian soul, soil, and seas in extracted, transported, and utilized hydrocarbons. My exploration of affect and emotion among Russian citizens also addresses this gap and provides another mode of exploring why or how society accepts the energy and extraction foci as acceptable and even desirable for twenty-first-century economic development.

It is necessary to return to two questions asked within the literature on emotional geographies. First, Bondi (2005) asked how emotional geographies might engage with everyday emotional life without relegating emotion to individualized subjective experience. This question relates to the politics of emotional geographies, where "failure to trouble individualistic understandings of emotion suggests an uncritical relation to wider social trends, while detachment from everyday emotional life suggests an unwillingness to situate knowledge claims" (438). This article explores emotions in one cultural and political setting in an attempt to move beyond the description of individual emotions and to question if a wider societal "emotional schema" about extraction might exist in hybrid authoritarian-democratic Russia. Although every survey or database has limitations, this study suggests that there is a state emotion about extraction in Russia and that the affective signature of this state emotion affects how society thinks and (re)acts to extractive development. This research is not conclusive; instead, it has aimed to provide more avenues for investigation of affect and emotions in geographic studies of people, places, and the phenomena that transform them.

Second, Bondi (2005) asked how emotional geographies might connect and engage with expressions of emotion, wondering how "personal, articulated accounts of emotion [should] be understood" (438). This study suggests that scholarly representation of emotions can provide insight into the effects of affect or how emotions about things lead to everyday reactions to them. No study could possibly account for all of the emotions experienced by any participant in any study, and reliance on participants to provide content—informational and emotional—will always provide incomplete knowledge. When researchers are passionate, diligent, and committed to their research projects and subjects, however, what is often shared in interviews and conversations is earnest and honest. We can only trust in this relationship and in the interpretations of affect and emotions provided by researchers even in the study of locales where emotions have been harnessed in particular ways or where state emotion aims to mobilize certain kinds of citizen affect. Even in this examination of emotions about the environment and energy development in the post-Soviet context, where a state emotion might be present, there is a range of expression that provides a glimpse into the emotional foundations of thoughts that might provoke human actions and shape social, political, and ecological landscapes. In this study of a hybrid Russia, combining political research that emphatically claims, for example, that "Putin did not inherit his status as an autocrat: he created it" (Fish et al. 2017, 68) with the knowledge that intentions of crafting Russia as an energy superpower require coordinated and spatially expansive commitment to extractive development, it would seem that greater understanding of how affect and emotions play out in social and political life is important to understanding the construction of narratives and practices of identity and belonging.

Conclusion

In this article, I have focused on how emotional geographies relate to discourses about extractive development within one sociopolitical context, the Russian Federation. This exploration of the nature of emotions, emotional schema, and the importance of state emotions in one authoritarian setting indicates that the context of energy production influences affect and emotions about resources and their development. Only recently have geographers focused on affect and emotions to investigate the impact of resources or extraction on everyday lives, individual subjectivities, and societal political choices. Further exploration of this intersectional field might include asking what kinds of affect and emotion operate in different energy and resource

extraction contexts, what roles they play in communities, and how or whether they are bolstered by certain sociopolitical environments. Of particular interest in future studies might be the duality of individual and state emotions and how that dynamic plays out in hybrid democratic-authoritarian and authoritarian regimes. The findings of this study—that emotional narratives about extractive development are created by an authoritarian government and are reproduced by society with strong emotions and affect—suggest the need for continued engagement with a wide range of emotional geographies related to extractive development. For all of the focus on human relationships to the environment and resource development to date, geographers have only taken limited steps to tackle such issues. This contribution emphasizes the need for greater attention to emotional geographies of energy and extraction and their relationship to political regimes and discourses, especially where political choices and policy are related to resource governance and energy (in)justice.

References

Ahmed, S. 2004. *The cultural politics of emotion.* New York: Routledge.
Alekperov, V. 2012. *Oil of Russia: Past, present and future.* Minneapolis, MN: East View Press.
Ballvé, T. 2012. Everyday state formation: Territory, decentralization, and the Narco Landgrab in Colombia. *Environment and Planning D: Society and Space* 30 (4):603–22.
Bater, J. H. 1983. *The Soviet city: Ideal and reality. Explorations in urban analysis.* Beverly Hills, CA: Sage.
Bondi, L. 2005. Making connections and thinking through emotions: Between geography and psychotherapy. *Transactions of the Institute of British Geographers* 30 (4):433–48.
Bouzarovski, S., and M. Bassin. 2011. Energy and identity: Imagining Russia as a hydrocarbon superpower. *Annals of the Association of American Geographers* 101 (4):783–94.
Bradshaw, M. J. 2006. Battle for Sakhalin. *The World Today* 62 (11):18–19. Accessed June 23, 2018. http://www.chathamhouse.org/publications/twt/archive/view/166915.
Bradshaw, M. J., and J. Prendergrast. 2005. The Russian heartland revisited: An assessment of Russia's transformation. *Eurasian Geography and Economics* 46 (2):83–122.
Bridge, G. 2011. Resource geographies I: Making carbon economies, old and new. *Progress in Human Geography* 35 (6):820–34.

———. 2014. Resource geographies II: The resource–state nexus. *Progress in Human Geography* 38 (1):118–30.

Brown, K. 2013. *Plutopia: Nuclear families, atomic cities, and the great Soviet and American plutonium disasters.* Oxford, UK: Oxford University Press.

Childs, J. 2016. Geography and resource nationalism: A critical review and reframing. *The Extractive Industries and Society* 3 (2):539–46.

Collier, P., and A. Hoeffler. 2005. Resource rents, governance, and conflict. *Journal of Conflict Resolution* 49 (4):625–33.

Coombes, B., J. T. Johnson, and R. Howitt. 2011. Mere resource conflicts? The complexities in indigenous land and environmental claims. *Progress in Human Geography* 36 (6):810–21.

Davidson, J., and L. Bondi. 2004. Spatialising affect, affecting space: Introducing emotional geographies. *Gender, Place and Culture* 11 (3):373–74.

Davidson, J., L. Bondi, and M. Smith. 2005. *Emotional geographies.* Hampshire, UK: Ashgate.

Dawisha, K. 2014. *Putin's kleptocracy: Who owns Russia?* New York: Simon and Schuster.

Dienes, L. 2013. Reflections on a geographic dichotomy: Archipelago Russia. *Eurasian Geography and Economics* 43 (6):443–58.

Dixon, M., and J. K. Graybill. 2016. Uncertainty in the urban form: Post-Soviet cities today. In *Questioning post-Soviet*, ed. E. C. Holland and M. Derrick, 19–38. Washington, DC: Wilson Center.

Ellman, M., ed. 2006. *Russia's oil and natural gas: Bonanza or curse?* London: Anthem.

Fish, M. S., V. Kara-Murza, L. Aron, L. Shevtsova, V. Inozemtsev, G. Robertson, and S. Greene. 2017. What is Putinism? *Journal of Democracy* 28 (4):61–75.

Gillespie, J. 2018. The role of emotion in land regulation: An empirical study of online advocacy in authoritarian Asia. *Law & Society Review* 52 (1):106–39.

Gorbachev, M. 2017. *The new Russia.* Malden, MA: Polity.

Graybill, J. K. 2009. Places and identities on Sakhalin Island: Situating the emerging movements for "Sustainable Sakhalin." In *Environmental justice of the former Soviet Union*, ed. J. Agyeman and E. Ogneva-Himmelberger, 71–96. Boston: MIT Press.

———. 2013. Mapping an emotional topography of an ecological homeland: The case of Sakhalin Island, Russia. *Emotion, Space and Society* 8:39–50.

———. 2017. Oil and water don't mix: Impacts of oil and gas development on Sakhalin's water resources. In *The politics of fresh water: Access, conflict and identity*, ed. C. M. Ashcraft and T. Mayer, 98–113. London and New York: Routledge.

Grishin, N. 2015. The meaning of elections in the Russian Federation. *Perspectives on European Politics and Society* 16 (2):194–207.

Gudkov, L. 2011. The nature of Putinism. *Russian Politics and Law* 49 (2):7–33.

Gustafson, T. 2012. *Wheel of fortune: The battle for oil and power in Russia.* Cambridge, MA: Belknap.

Henry, L. A. 2013. *Red to green: Environmental activism in post-Soviet Russia.* Ithaca, NY: Cornell University Press.

Hernández, D. 2015. Sacrifice along the energy continuum: A call for energy justice. *Environmental Justice* 8 (4):151–56.

Hoffman, D. E. 2002. *The oligarchs: Wealth and power in the new Russia.* New York: Public Affairs.

Huber, M. 2018. Resource geography II: What makes resources political? *Progress in Human Geography.* Advance online publication. Accessed April 17, 2018. https://doi.org/10.1177/0309132518768604.

Hwang, K. 2006. Moral face and social face: Contingent self-esteem in Confucian society. *International Journal of Psychology* 41 (4):276–81.

International Energy Agency. 2017. *Russia 2016: Energy policies beyond IEA countries.* Paris: Organization for Economic Cooperation and Development/International Energy Agency. Accessed December 20, 2017. https://www.iea.org/publications/freepublications/publication/Russia_2014.pdf.

Kim, Y., and D. Cohen. 2010. Information, perspective, and judgments about the self in face and dignity cultures. *Personality and Social Psychology Bulletin* 36 (4):537–50.

Klare, M. T. 2001. The new geography of conflict. *Foreign Affairs* 80 (3):49–61.

Koch, N., and T. Perreault. 2018. Resource nationalism. *Progress in Human Geography.* Advance online publication. Accessed June 18, 2018. https://doi.org/10.1177/0309132518781497.

Lacqueur, W. 2015. *Putinism: Russia and its future with the West.* New York: St. Martin's.

Le Billon, P. 2001. The political ecology of war: Natural resources and armed conflicts. *Political Geography* 20 (5):561–84.

———, P. 2008. Diamond wars? Conflict diamonds and geographies of resource wars. *Annals of the Association of American Geographers* 98:345–72.

Lerner, S. 2012. *Sacrifice zones: The front lines of toxic chemical exposure in the United States.* Cambridge, MA: MIT Press.

Levada Center. 2017. Approval ratings of government institutions. Accessed January 20, 2018. https://www.levada.ru/en/2017/09/26/approval-ratings-of-government-institutions/.

Levitsky, S., and L. A. Way. 2016. *Competitive authoritarianism: Hybrid regimes after the Cold War.* Cambridge, UK: Cambridge University Press.

Liao, N. 2013. Dualistic identity, memory-encoded norms, and state emotion: A social constructivist account of Chinese foreign relations. *East Asia* 30 (2):139–60.

McDonnell, E. 2015. The co-constitution of neoliberalism, extractive industries, and indigeneity: Anti-mining protests in Puno, Peru. *The Extractive Industries and Society* 2 (1):112–23.

Medvedkov, O. 1990. *Soviet urbanization.* London and New York: Routledge.

Migranyan, A. 2004. What is "Putinism"? *Russia in Global Affairs* 2:28–45.

Ministry of Energy of the Russian Federation. 2009. Energy strategy of Russia for the period up to 2030. Accessed January 10, 2018. http://www.energystrategy.ru/projects/docs/ES-2030_(Eng).pdf.

Neuman, W. R., G. E. Marcus, A. N. Crigler, and M. MacKuen, eds. 2007. *The affect effect: Dynamics of emotion in political thinking and behavior.* Chicago: University of Chicago Press.

Nightingale, A. 2013. Fishing for nature: The politics of subjectivity and emotion in Scottish inshore fisheries management. *Environment and Planning A* 45 (10):2362–78.

NVivo (version 12). 2018. Burlington, MA: QSR International Pty Ltd.

Perreault, T., and G. Valdivia. 2010. Hydrocarbons, popular protest and national imaginaries: Ecuador and Bolivia in comparative context. *Geoforum* 41 (5):689–99.

Pipes, R. 2004. Flight from freedom: What Russians think and want. *Foreign Affairs* 83 (3):9–15.

Pratt, K. 2012. Rethinking community: Conservation, practice, and emotion. *Emotion, Space and Society* 5 (3):177–85.

Pryde, P. 1991. *Environmental management in the Soviet union.* Cambridge, UK: Cambridge University Press.

Rogers, D. 2015. *The depths of Russia: Oil, power, and culture after socialism.* Ithaca, NY: Cornell University.

Rosneft. 2017. рекиама азС роСнефть. уверенноСть в каждом кииометре [Advertisement "Rosneft: Assurance in every kilometer"]. Accessed June 22, 2018. https://www.youtube.com/watch?v=RPs-fSZx79A.

Rutland, P. 2008. Russia as an energy superpower. *New Political Economy* 13 (2):203–10.

Rutledge, I. 2004. *The Sakhalin II PSA—A production "non-sharing" agreement: Analysis of revenue distribution.* Prague, Czech Republic: CEE Bankwatch Network.

Russian Federation Policy for the Arctic to 2020. 2009. Accessed October 15, 2018. http://www.arctis-search.com/Russian+Federation+Policy+for+the+Arctic+to+2020

Sergeev, V. M. 1998. *The wild east: Crime and lawlessness in post-communist Russia.* Armonk, NY: M. E. Sharpe.

Shalin, D. 1996. Soviet civilization and its emotional discontents. *International Journal of Sociology and Social Policy* 16 (9–10):21–52.

———. 2004. Liberalism, affect control, and emotionally intelligent democracy. *Journal of Human Rights* 3 (4):407–28.

Shevtsova, L. 2016. Forward to the past in Russia. In *Authoritarianism goes global*, ed. L. Diamond, M. F. Plattner, and C. Walker, 40–56. Baltimore, MD: Johns Hopkins University Press.

Smith, M., J. Davidson, L. Cameron, and L. Bondi, eds. 2009. *Emotion, place and culture.* Farnham, UK: Ashgate.

Sultana, F. 2011. Suffering for water, suffering from water: Emotional geographies of resource access, control and conflict. *Geoforum* 42 (2):163–72.

Tynkkynen, V. 2016. Energy as power—Gazprom, gas infrastructure, and geo-governmentality in Putin's Russia. *Slavic Review* 75 (2):374–95.

Vélez-Torres, I. 2013. Governmental extractivism in Colombia: Legislation, securitization and the local settings of mining control. *Political Geography* 38:68–78.

Watts, M. J. 2004. Antinomies of community: Some thoughts on geography, resources, and empire. *Transactions of the Institute of British Geographers* 29 (2):195–216.

———. 2010. Resource curse? Governmentality, oil and power in the Niger Delta, Nigeria. *Geopolitics* 9 (1):50–80.

Wilson, E. 2003. Freedom and loss in a human landscape: Multinational oil exploitation and survival of reindeer herding in north-eastern Sakhalin, the Russian far east. *Sibirica* 3 (1):21–48.

Yang, G. 2005. Emotional events and the transformation of collective action: The Chinese student movement. In *Emotions and social movements*, ed. H. Flam and D. King, 79–98. London: Routledge.

JESSICA K. GRAYBILL is an Associate Professor in the Department of Geography at Colgate University, Hamilton, NY 13346. E-mail: jgraybill@colgate.edu. Her research interests include resilience studies and socioecological transformations in postsocialist, urban, and remote spaces and grappling with how environments, livelihoods, and possible futures are cocreated by multiple actors.

U.S. Farm Policy as Fraught Populism: Tracing the Scalar Tensions of Nationalist Agricultural Governance

Garrett Graddy-Lovelace

The scalar tensions of nationalism manifest acutely in agriculture—particularly in the contemporary United States. This is paradoxical because farm policy calls for and enacts nativist governance that undermines the conditions of farming: from labor to water, topsoil, and pollinators, to export markets. At the heart of these scalar contradictions is the fraught, shifting terrain of agrarian populism. The intertwined origin of the U.S. Farm Bill, the American Farm Bureau Federation, and Cooperative Agricultural Extension shows how early twentieth-century fraught agrarian populism drove farm policy but how it also carried a pivotal consensus of recognition about the ecological and economic dangers of overproduction. Drawing on archival research at the U.S. Department of Agriculture's (USDA) National Agricultural Library Special Collections, discourse and policy analysis of U.S. Farm Bills, and qualitative research with farmer organizations, this article traces how racialized xenophobia accentuates the hypocrisy of U.S. agriculture's extreme dependency on migrant labor, as heightened borders also reveal their ecological farce in the face of intrinsically transnational climate change, soil erosion, and water constraints. The America First trade agenda decries imports while sidelining the crisis of commodity crop glut and the spatial fix of subsidizing exports as surplus disposal. Yet, even amidst the scalar contradictions of nativist agricultural governance and the fraught farm populism driving it, there existed a kernel of agrarian populism grounded in a collective honest recognition of the ecological, economic, rural, and social crises of overproduction—and that organized against it. This kernel catalyzed the origin of both the Farm Bill and the Farm Bureau but has been subsumed in and through both since.

国族主义的尺度争议，尖锐地展现在农业上——特别是在当代的美国。其矛盾之处在于，农田政策召唤并执行有损农业环境的本土治理：从劳动、水、表土、受花粉器到出口市场。位于这些尺度冲突核心的，便是令人担忧且不断转变的农业民粹主义领域。美国农田法案、美国农业事务联合会，以及农业合作推广的纠缠起源，显示二十世纪早期令人担忧的农业民粹主义如何驱动农田政策，但同时传达承认生产过剩的生态与经济危险之关键共识。本文运用美国农业部（USDA）国家农业图书馆特别收藏的档案研究，针对美国农田法案的论述与政策分析，以及对农民组织的质性研究，追溯种族化的仇外心理如何处于美国农业对移工的极度依赖之虚伪核心，而强化的边境亦揭露其面对本质上是跨国的气候变迁、土壤侵蚀与水资源限制时的生态闹剧。"美国优先"的贸易议程，在责难进口的同时，却旁观商品作物的过度供应危机，以及补贴出口作为处置生产过剩的空间修补。但即便在本土农业治理的尺度冲突和驱动该治理的令人担忧的农田民粹主义中，仍存在根植于对过度生产的生态、经济、农村与社会危机的诚实集体认识——以及组织进行反对的农业民粹主义核心。此一核心催生了农田法案与农业事务联合会，却也从此被纳入其中。 关键词: 农业危机, 农田法案, 农业会, 生产过剩, 美国农业史.

Las tensiones escalares del nacionalismo—en particular, en los Estados Unidos contemporáneos—se manifiestan agudamente en la agricultura. Tal situación es paradójica por cuanto la política del campo requiere y promulga una gobernanza vernácula que socava las condiciones de la agricultura: desde el trabajo, al agua, el mantillo del suelo y los polinizadores, hasta los mercados de exportación. En la médula de estas contradicciones escalares está el tenso y cambiante terreno del populismo agrario. El entrelazado origen del proyecto de la Ley Agrícola de los EE.UU., la Agencia de la Federación Agraria Americana y la Cooperativa de la Extensión Agrícola, muestra el modo como el inquieto populismo agrario de principios del siglo XX manejó la política agrícola, así como también adelantó un consenso crucial de reconocimiento de los peligros ecológicos y económicos de la superproducción. Con base en investigación de archivos en las Colecciones Especiales de la Biblioteca Agrícola Nacional del Departamento de Agricultura (USDA), en el discurso y análisis político de los proyectos de Leyes Agrarias de los EE.UU., e investigación cualitativa con las organizaciones de agricultores, este artículo rastrea el modo como la xenofobia racializada acentúa la hipocresía acerca de la dependencia extrema de la agricultura americana en el trabajo migratorio, en la medida en que fronteras de sensibilidad exacerbada revelan también su farsa ecológica frente a un cambio climático intrínsecamente transnacional, la erosión del suelo y los limitantes hídricos. La agenda comercial de América Primero condena las importaciones mientras deja de lado la crisis de superabundancia en las mercaderías de cosechas y el amaño espacial de subsidiar las exportaciones para disponer de los excedentes. Con todo, incluso en medio de las contradicciones escalares de la gobernanza agrícola vernácula y el tenso populismo agrario que la orienta, se dio una simiente de populismo agrario fundamentado en el reconocimiento colectivo honesto de las crisis ecológica, económica, rural y social generadas por la superproducción—el que organizaron como respuesta al problema. Esta simiente catalizó el origen tanto del proyecto de Ley Agrícola como de la Agencia Agraria, aunque ha sido subsumida por las dos desde entonces. *Palabras clave: Agencia Agraria, crisis agraria, excedentes, historia agrícola de los EE.UU., proyecto de Ley Agrícola.*

"Our continent was tamed by farmers. So true," President Trump spoke at the American Farm Bureau Federation's ninety-ninth annual meeting in Nashville, Tennessee, in January 2018: "Our armies have been fed by farmers and made of farmers." After four years of farm income decline in the United States, debt, bankruptcy, and foreclosures have risen (Weber 2017), alongside farmer suicide rates (McIntosh et al. 2016). As of August 2018, the U.S. Department of Agriculture (USDA) admitted a decline in net farm income to its lowest level since 2002 (adjusting for inflation) and forecasted that median farm income earned by farm households would decline to $1,691 in 2018 (USDA ERS 2018). Low farmgate prices, labor shortages, and climate change–related natural disasters wreak havoc on U.S. agriculture. Yet, Trump's Farm Bureau remarks lambasted environmental regulation, vilified migration, and overstated the tax reform's benefits for farmers. He lauded Immigration and Customs Enforcement policing, the national anthem, the Second Amendment, and Andrew Jackson—all to applause.

The nationalism of contemporary U.S. policy and political rhetoric leverages the nation as a discrete, bordered political entity; concurrently, it asserts and wields the nation as the dominant and dominating scale of reference, allegiance, and identity. As such, nationalism carries a number of internal contradictions regarding who and what are included within the scale of reference of the nation—and how, where, when, and why exclusions occur. The engine of capitalism has long run on the generativity of contradictions and crisis (Mansfield 2004; Harvey

2005). Building on Gough's (2004) argument that shifting assertions of scale are "underpinned by a number of fundamental contradictions of capitalist reproduction and the state" (185), this article traces how surges in nationalism—deployments of the nation-state as the dominating scale of reference—accentuate these contradictions.

Environmental governance, as a dynamic and evolving process, employs various scales of reference to evade the ecological crises of capitalism. Under neoliberalism, it becomes entangled in the scalar contradictions of environmental regulations (Heynen and Perkins 2005; Carr and Affolderbach 2014). Yet, much of what environmental governance aims to safeguard falls under what could be called agricultural governance, such as land and freshwater use. Accordingly, understanding how environmental governance unfolds—or folds—during a surge of aggressive nationalism requires tracing the scalar contradictions of nationalist agricultural governance.

Scalar contradictions of nationalism become acute in agriculture, where exclusionary assertions of a national scale of reference undermine the very conditions of farming. That the scalar contradictions of nationalism become glaring in agriculture is itself a paradox, given the nativist tendencies of agricultural governance in the United States. This article focuses on the role of agrarian populism in this paradox and how the tensions comprising the contested, shifting terrain of farm populism drive the contested, shifting terrain of convoluted farm policy. U.S. agricultural policy currently answers to and amplifies a farm populism advocating for bordered, ethnonationalist exceptionalism—although to its own detriment. Such political deployments of farm populism are not new, although they differ from previous iterations. This article contextualizes contemporary farm populism and policy within its historical antecedents. It focuses on one unique, although foundational, historical time period: from World War I through the start of World War II, the founding years of both the American Farm Bureau Federation (AFBF) and the U.S. Farm Bill. The intertwined origin of the AFBF and the Farm Bill exemplifies how the contested terrain of agrarian populism relates to the tensions of nationalism, as demonstrated by a content and discourse analysis of the U.S. Department of Agriculture National Agricultural Library Special Collections archives regarding farmer organizations in and between the world wars as well as of contemporary AFBF and state Farm Bureau policy agendas.[1] This historical analysis sheds light on how current fraught agrarian populism causes and emerges from acute scalar contradictions of nationalist agricultural governance.

Current Fraught Agrarian Populism

The term *populism* remains slippery, with its increasingly divergent array of political valences. It animates all ends of political spectrums, either promoting or derailing democracy, decentralizing or centralizing power. The very vagueness of the term contributes to its potency, from authoritatianism (Scoones et al. 2018) to emancipatory potential (Laclau 2005). Amidst this wide, bumpy terrain stands agrarian populism, with its turn-of-the-twentieth century origins farmer uprisings against agro-capitalism—a democratic counter (Gilbert 2015) to liberalism, finance capitalism, and communism (Goodwyn 1978). I describe the term and realm of agrarian populism as *fraught*, both in the sense of being overloaded by the weight of projections and associations—many ominous—and in the sense of causing or being affected by tensions and anxiety.

At the heart of these tensions and anxiety I argue is agrarian inviability—the inability to secure a dignified livelihood and life from agriculture on a community level. The article traces this inability to the overlooked role of surplus, which has long served as a mechanism of agricultural exploitation and accumulation, a driver of scalar fixes and thus contradictions. Amidst the layers of coloniality, classism, racism, and gender hierarchies at play in U.S. farm

policy, there has been a long-standing, although now muffled, conscientious recognition of the crises of commodity crop surpluses. This recognition served as a catalyst for a kernel of what could be called grounded agrarian populism at work in the Farm Bill: a grassroots, agrarian justice–oriented populism, grounded in community viability in land-based life.

That the contradictions of nationalism manifest in agriculture so acutely is itself a paradox, as the rural electorate led the nationalist push leading up to and during the 2016 presidential election (Scala and Johnson 2017). Many factors account for this situation, from enflamed racism and sexism, to the searing frustrations of rural "landscapes of despair" (Monnat and Brown 2017). This article focuses on the agricultural roots of the countryside's economic despair: the overlooked crisis of overproduction and secular price decline amidst rising input costs. Although rural does not necessarily equate to agricultural, the very fact that so much of rural economics has become nonagricultural points to the financial crisis of small and midsize farms and thus the economic limits of such farming. The intertwined early twentieth-century origin of both the AFBF and the Farm Bill hinged on escaping the disasters of commodity crop glut. A century later, the dangers of surplus have been hidden through neoliberal scalar fixes—as has the subsequent agrarian crisis itself. The domestic scalar fixes of "getting big or getting out" of agriculture work alongside the geopolitical scalar fixes of exporting surplus "away." In this one-eyed trade vision, a myopic celebration of exports obscures analysis of the downward price pressure of imports. Farmland and agribusiness consolidation result from this unchecked crisis of surplus even as they exacerbate it (Ray et al. 2003). Attending to the scalar contradictions that hide the problem of overproduction helps elucidate the further scalar contradictions that emerge from it. As such, it helps shed light on the fraught populism driving nativist U.S. agricultural governance and its deleterious social, economic, and political–ecological impacts.

The omnibus U.S. Farm Bill stands as a powerful yet nebulous nexus of issues, actors, alliances, and tensions. Highly complex and contentious, it has long driven land use governance, both explicitly through Conservation Title provisions and implicitly through monocultural production incentives. Under the Trump administration, USDA Secretary Perdue, and contemporary congressional agricultural committees, the farm policy goal of environmental deregulation has gained robust traction, evidenced in the July 2018 House Bill markup. A major source of this deregulatory urgency has been the AFBF, the century-old farmer lobby and insurance organization. Representing 6 million people across fifty states, plus Puerto Rico, the AFBF wields noted political authority under Democratic as well as Republican parties, although particularly under the latter—all in the name of the "The Unified National Voice of Agriculture" (AFBF 2018a, 2018c, 2018d).

The AFBF maintains nonprofit, tax-exempt status but commands a multi-billion-dollar revenue-generating enterprise of insurance companies and for-profit farmer cooperatives and a stock portfolio that includes the major agribusiness companies Archer Daniels Midland, ConAgra, Monsanto (now Bayer), Phillip-Morris, and Dow-Dupont. Yet, over two thirds of its 6 million members necessarily include nonfarmer insurance purchasers, as the United States has fewer than 2 million farmers total. A long line of critics has assailed the elitism of AFBF (Berger 1971): Writing in the 1950s, McConnell (1953) chronicled how its lobbying prowess resulted in "on the one hand, a vast increase in the strength and influence of the Farm Bureau and, on the other hand, a great financial boon to the type of farmers who were the natural clientele of the Farm Bureau" (77). The organization did not originate explicitly to divide classes. In addition to complex, county-level mobilizations, called "grassroots" by some historians (Berlage 2016), the AFBF began with multiclass consensus around an original kernel of what could be called grounded agrarian populism (in the sense that it was grassroots, grounded in land-based life and land justice): political resistance to economic pressures to overproduce. Nevertheless, "the conflict between large and small farmers is in no slight degree the product of the Farm Bureau's rise to power" (McConnell 1953, 162)—and continues to be so. Yet, the AFBF is not monolithic, although it operates at the federal level as if it were. More research is needed to document and understand the internal class dynamics, tensions, cooptations, and contradictions within the AFBF itself—particularly because such demographic information remains publicly unavailable.

Questions of representation become all the more problematic amidst defiantly reactionary social exclusions. The 2017 AFBF Policy Book upholds marriage as the union between one man and one woman and opposes "granting special privileges to those that participate in alternative lifestyles." It demands English-only "Star-Spangled Banner" and Pledge of Allegiance and decries desecration of the U.S. flag or "purging of United States history by the removal of symbols that represent historic events and/or persons from our nation's past" (AFBF 2017a). The 2017 Policy Book updates AFBF's long history of racialized ethnonationalism, opposing "any program which tends to separate, isolate, segregate or divide the people of our country under the guise of emphasizing ethnic diversity." The militarized, racialized, patriarchal overtones unfold into a prosperity theology of individualized accumulation: "We believe in the American capitalistic, private, competitive enterprise system in which property is privately owned, privately managed and operated for profit and individual satisfaction. Any erosion of that right weakens all other rights guaranteed to individuals by the Constitution," the Policy Book asserts, articulating the contours of contemporary U.S. nationalism. "America's unparalleled progress is based on freedom and dignity of the individual, sustained by our founding principles rooted in Judeo/Christian values, commandments and the sanctity of life" (AFBF 2017a). Here, the original coloniality of U.S. agricultural policy (Graddy-Lovelace 2017) tears into the twenty-first century amidst layers of convoluted populism.

A cursory look at current AFBF policy agendas divulges social biases such as racism and xenophobia as well as political–economic tendencies toward labor exploitation and aversion to antitrust protections. Like the Farm Bill, it does not contest—and arguably enables—monocultural production and agribusiness concentration. Yet, a longer durée analysis of both discloses a more complicated story. Both the Farm Bill and the major farm lobby organization driving it were both originally grounded in a shared kernel of grounded agrarian populism: a consensus-by-necessity on the honest recognition of—and resistance to—the ecological and economic destructiveness of overproduction.

The fraught terrain and divergent meanings of agrarian populism have changed over the course of the twentieth and now twenty-first centuries, shifting spatially and temporally as the scalar contradictions of nativist agricultural governance experienced cycles of being heightened, staved off with various scalar fixes, and heightened again. Nevertheless, on some fronts, the contradictions of nationalism have persisted: The coloniality of indigenous erasure and white supremacy marked early twentieth-century AFBF policies as they do today. The 2017 AFBF Policy Book still opposes tribal sovereignty over native-held lands. On the other hand, amidst the Dust Bowl disaster, early AFBF advocacy for the 1938 Soil Conservation & Domestic Allotment Act acknowledged and foregrounded the ecological disaster of soil erosion. After nitrogen fertilizer obscured and deferred the impacts of topsoil erosion on soil fertility, AFBF policy sidelined conservation goals and has since come to defend chemical inputs and decry environmental regulations, impact statements, and even research. Meanwhile, amplifying and calling for increased policing of national borders accentuates the ecological farce of "building walls" in the face of climate change, water constraints, and pollinator loss—all sidelined by AFBF. It also accentuates their hypocrisy in terms of labor: America First policies posit an allegiance to the nation-state scale of reference alongside an extreme dependency on migrant farm work.

Accordingly, the vast ecological, economic, social, and geopolitical ramifications of the Farm Bill, and the AFBF's marked influence therein, demand geographic analysis. Critical geographic and historic research is needed to trace how a tangled early twentieth-century set of meanings for agrarian populism relates to a twenty-first-century set of divergent connotations—which range from AFBF lobbyists to transnational agrarian mobilizations for food sovereignty and agroecology (Desmarais 2007). The former now prefer the identities of producers and agribusiness to the old-fashioned adjective *agrarian*; the latter, although associated with peasant groups across the world, also exist in the United States in such organizations as the Rural Coalition/*Coalición Rural*, the National Family Farm Coalition, and the Federation of Southern Cooperatives/Land Assistance Fund, National Farmers Union (NFU), National Young Farmers Coalition, and others. This article focuses on the organization that has consistently wielded the most political power and influence over the past century—the AFBF—but whose proximity to agrarian populism remains the most tenuous.

Fraught Populist Origins of the Farm Bill and Farm Bureau

World War I catalyzed both the U.S. Farm Bill and the Farm AFBF. Agricultural production served internal nation-building and geopolitical nation-flexing but required local, county-by-county incentivization. With European farmers in the trenches, the U.S. government aggressively encouraged farmers to expand wheat production for export to Europe. A few years of unusually high rainfall in the otherwise arid Midwest—along with advances in machinery—led to a boom in farmland expansion. Growers tilled up the deep-rooted native prairie grass to plant wheat in what would come to be known as "The Great Plow-Up." As prices increased, farmers expanded their fields and new farmers moved westward to join the boom. By the late 1910s, however, European wheat production had rebounded, causing the world market to glut and prices to fall. Prices collapsed by 1920. For a decade, farmers responded by further expanding production, desperate to recoup investments. The U.S. stock market crashed, the Great Depression began, and the real aridity of the southern Midwest returned—but in extreme form. A devastating drought swept the region with prices still nearing nothing, hurling farmers into destitution and desperate migration.

The "farm problem"—the treadmill of surplus, plummeting prices, debt, further overproduction, topsoil loss, further expansion, and more overproduction—was a governance problem on two fronts: It resulted from geopolitical militarized nation-building, and it resulted in demands for domestic government intervention. Regarding the former, AFBF President Edward O'Neal (1939) lauded the "aggressive leadership of extension service in developing county farm bureaus" (3) before and during the war. The government made matching grants, via the 1914 Smith-Lever Act, to support county agents in augmenting commodity crop yields. These county agents served as AFBF representatives, thereby blurring the lines between public and private sectors. The Federal Office of Extension issued a 1917 Farm Bureau Organization Plan that described—and proscribed—AFBF recruitment strategies in fifty-three-page detail. In the "Food Will Win the War" campaign, a federated network of county farm bureaus became crux to the "national defense program to promote food production" (O'Neal 1939, 6). In 1919, the county-level farm bureaus merged into a national federation to secure federal clout. Impact, however, continued to move in both directions, with the USDA exerting influence on rural communities through cooperative extension (Ball 1936).

Although the county-level bureaus were "regarded as instrumentalities by means of which the USDA is brought into friendly, familiar, and collaborative contracts with [white, male, landed] farm operators," Wubnig (1935) noted, AFBF "attitude reflects the mentality of the substantially well-to-do, somewhat 'conservative' farmers who comprise the great bulk of its membership" (10–11). AFBF membership expanded sixfold between 1933 and 1945.

The sheer number of members, however, proves that they were not all elite. The organization as a whole, though, arose as a counterweight to the previous generation's "interludes of agrarian fury" (McConnell 1953, 5): Although the radical anticapitalist agrarian populist movement of the late nineteenth century had dissipated somewhat due to political losses, and then during the throes and farmgate prices of World War I, "the memory of it was still vivid in the minds of members of grain exchanges, heads of farm equipment trusts, and directors of banks. It is even likely that some of these glimpsed the possibility of enlisting organized agriculture or, rather, re-organized agriculture on the side of capitalism" (McConnell 1953, 20). USDA Secretary Houston famously called on farmers to join or form farm bureaus to "stop Bolshevism," and AFBF President Howard asserted, "I stand as a rock against radicalism" (McConnell 1953, 48). As the product of prosperity rather than plight, the AFBF included nonelite farmers but under the broader guise of lifting them into higher echelons of class and political influence. It strove not to include the masses; hence its widespread requirements of high dues. Even in the throes of the Depression, farmers largely opted for government engagement rather than revolution: "In a large sense, the conduct of the Farm Bureau through the great depression can be regarded as a substantial return on the investment in concern for agricultural welfare which business groups had made in the proceeding period" (McConnell 1953, 56).

Nevertheless, the complexity of farm and home bureaus—their procapitalist populism—merits attention. The Farm Bureau Community Handbooks strategically included women and children, thereby expanding the protagonists of the family farm. This also worked to emphasize, reinforce, and define the role of family, home, and family farm following colonial settler gender roles and hierarchies. People joined for political power at the federal level but

also to take advantage of AFBF's scientific, educational, and social projects of professionalization, moralism, and functional specialization. Club culture arose (Berlage 2016) around rural connection and empowerment. Scientization of domestic and farm economies and ecologies countered the inferiority felt in relation to industrialization, urbanization, and international markets—an inferiority exemplified in the AFBF speech "Shall American Farmers Become Peasants?" (Dodd 1938). The AFBF had become a "chamber of agriculture," mirroring city chambers of commerce (O'Neal 1936, 6). It fit farming into nationalistic discourses of modernization and progress to legitimize the agricultural economies and ecologies as technical. Farmers were able to interface with government agencies as "experts" in their own fields (Porter 2000) and as a united front against encroaching powers of railroads, cities, and industry (Kile 1948).

County-level farm bureaus had emerged as mechanisms of government pressure to produce but then became means by which farmers countered the ills of surplus. It was after World War I when the AFBF congealed, as other farmer organizations reckoned with the implications of (over)production. Amidst war:

> the government urged as a patriotic duty that farmers increase their production in every possible way. The result was that a large area of marginal and submarginal land was brought into production. Due to the fact that agriculture cannot readily adjust supply to demand, these marginal acres have been producing surpluses that have demoralized the prices of farm products and brought great economic distress. It has been recognized that inasmuch as the government was largely responsible in bringing into production these marginal acres, there is a direct responsibility to assist in bringing about a readjustment. (National Cooperative Council 1936, 5)

The collapse of agricultural prices in 1920, even as nonagricultural prices and wages remained rigid, aggravated disparity between farm income and costs. Facing plummeting prices, farmers organized out of necessity. A wide array of farmer groups emerged and diverged over how to survive the crisis.

A USDA internal memo surveyed what became known as the "Farm Bloc," from the National Grange to the United Farmers League ("the extreme left wing of the farmers' movement, sought to organize small farmers against rich farmers, land bankers, and New Dealers"; USDA n.d., 20), to the Farmers

Holiday Association ("a full-blown representative of traditional agrarian radicalism"; USDA n.d., 20). The latter called on farmers to stage general strikes in the face of Depression farm foreclosures. As "the only organization which speaks for the 'Class interests' of American farmers," NFU "carries its radicalism a good long way ... [and] maintains further that most of the 'orthodox and respectable' farmers organizations are in effect 'company unions'" (USDA n.d., 15–16). This final chide at the AFBF articulated an allegation and tension that has persisted for a century. Despite internal oppositions, the dire situation forced dialogue among this wide array of farmer organizations. The Farm Bloc ultimately converged on the central demand of "equality for agriculture" through *parity*, a term that originally referred to ratio prices matching pre–World War I farmer viability but then came to refer to equity between agricultural and industrial sector purchasing power. To secure parity, farmer advocate George Peek declared that he would "strip to the waist to fight for remedial legislation which will provide for the disposal of the surplus" (as cited in Porter 2000, 385). Although idiosyncratic as a technical term, parity persisted as a rallying cry, anchoring farmer mobilizations for farm justice even through the early 1980s farm crisis (Naylor 2017).

The AFBF emerged as part of broader phenomenon of necessitated horizontal collaboration among farmers across the national scale of reference—and as unique in its intimacy with the federal government. A 1929 USDA press release chronicling the "Government's Policy towards the Cooperative Movement" surveyed the tenuous situation: "Agriculture has inherent difficulties which cannot be overcome by the individual producer. It is a far-flung industry characterized by small producing units" (Hyde 1929, 1). The inherent difficulties of capitalist agriculture hinged on intrinsic competition, which would only be solved by organization, even as, left to its own allegedly self-regulating devices, it results in consolidation and lack of competition: "We cannot merge six million farms into one gigantic producing corporation." As surpluses mounted, farm prices dropped further: "Circumstances were forcing agriculture toward a unified front" (Gregory 1935, 154)—with each other and with government. With statutory minimum wages and maximum hours achieved by and for labor groups and government aid and price controls for industry, agriculture deserved *parity*: the "most sacred text in the Bureau scriptures" (Saloutos 1947, 315).

Having set up Washington, DC, headquarters and lobbying hard to secure bipartisan leverage, the AFBF led the fight for legislative farm relief with the 1933 Agricultural Adjustment Act (AAA). As the "magna charta of American agriculture" (Gregory 1935, 156), the AAA instituted supply management techniques such as acreage controls and domestic allotments. The 1933 AFBF annual conference address basked in their legislative success: "We rightfully feel that we not only should claim the victory, but ... that we have a primary responsibility and great opportunity in its successful administration" (AFBF 1933). Two years later, President Roosevelt spoke to the 20,000 men and women gathered at the annual AFBF meeting in Chicago, marveling at their size, political power, and political success: "There was more than mere braggadocio in this statement that 'the Farm Bureau and the A.A.A. are inseparable'" (McConnell 1953, 75).

When the Supreme Court overturned the AAA due to an unconstitutional funding mechanism, within days hundreds of farm organization leaders convened on Washington, where they completed recommendations in two days. USDA Secretary Henry Wallace lauded them over national radio: "The most important thing that has been accomplished is the demonstration that farmers do not need to sit helpless while the ruthless forces of unrestrained individualism grind them down" (Wallace 1937, 2). By the 1938 AAA, AFBF had become a fixture in both the seat of federal government and in small-town county seats across the nation. AAA Administrator Evans (1939, speech; author's notes) noted, "I believe that when a farmer joins a farm organization he is joining an insurance society to see that his interests are properly represented in any case where collective action is effective." Indeed, by 1939, AFBF literally expanded to the insurance sector, ultimately selling life, fire, farm, automobile, and health plans, solidifying its centrality in underserved rural realms. As a political powerhouse and expanding insurance empire—with corresponding growth in membership and congressional influence, the AFBF would shape agricultural politics in the decades to come, eventually becoming the primary voice of agriculture on the national political stage.

A century after their intertwined origin, the relationship of the AFBF to the Farm Bill continues to constitute the contested terrain of farm populism, as the widely divergent and convoluted notions of agrarian populism make their way into convoluted farm policy that now drives nationalism. Driving the tensions of this fraught terrain are a set of scalar contradictions: exclusionary assertions of a national scale of reference that undermine the very conditions of farming.

Ecological Scalar Contradictions

The 1930s brought—arguably forced—a pivotal moment of ecologically oriented farm policy: Suddenly, the nation-state scale of reference invoked by U.S. farm policy and populism hinged on topsoil. "Conservation of soil is the last line of defense against national suicide," USDA Secretary of Agriculture Wallace warned (Wallace 1936, 10), as ominous dust settled on President Franklin D. Roosevelt's White House desk. The Dust Bowl brought stark reckoning for bringing 50 million new acres under wartime production and rendered it impossible to "ignore the crime of perpetuating an exploitative agriculture which already is bringing to the front a physical crisis in land—land, which is the heritage of all people" (Wilson 1936, 3). Foregrounding soil allowed the Farm Bloc to rescue the 1933 AAA: "Cooperative conservation is better than competitive destruction" (Tolley 1936, 13). "Under the present Triple A program, restraint upon production of surplus commodities is a by-product of soil conservation" (Tolley 1936, 14). In a 1936 radio program, Wallace admonished "shipping our soil fertility at bargain prices to foreign countries ... merely to satisfy certain special interests which profit by volume" (Wallace 1936a, 7). From the 1938 Soil Conservation & Domestic Allotment Act onward, the Conservation Title has dutifully followed the primary Commodities Title in all subsequent Farm Bills. It has remained a nominal governance priority, although its programs are voluntary and its funding was negligible until the working lands programs in the 1980s, and more recently is minimal and erratic.[2]

By the late twentieth century, soil conservation ranked low in AFBF priorities and thus Farm Bill policies. Nitrogen fertilizer staved off the "accelerating biophysical contradictions" (Weis 2010) of nutrient-extractive crop production. This has allowed a deeper settler-coloniality, original to the AFBF and the Farm Bill, to resume, long after the Dust (Bowl) settled. The ecological short-

sightedness of vastly extractive agricultural production can be understood as a scalar contradiction of nationalism: The biophysical processes so crucial to agriculture—topsoil fertility, fresh water, and pollinators, among others—are themselves undermined. Here, the scale of reference of one's homeland excludes the well-being of the land itself; patriotic allegiance to one's "country" leaves the actual countryside degraded. Land is valued as property but not as the processes that comprise and regenerate the (ecological) services, benefits, and resources that make such property valuable. In the process of rendering land territorial gain, land is rematerialized from its composite ecologies and reduced to a means of production and accumulation.

No issue galvanized AFBF members and Trump supporters more effectively than the alleged overreach of the "Waters of the U.S." (WOTUS) Rule, which defines the jurisdiction of the Clean Water Act over waters that have a "significant nexus" with navigable waters. AFBF erroneously claimed that this meant puddles and ditches, catalyzing a successful "Ditch the Rule" action alert and lobbying campaign (AFBF 2015). In reality, the rule posits a case-by-case determination of water pollution and does not exempt nonpoint source pollution and emissions. Meanwhile, agricultural input runoff contributes to unprecedented levels of hypoxia, eutrophication, and dead zones in U.S. riversheds, deltas, and bays—with decades and centuries of legacy pollutants from upstream agriculture. AFBF has also opposed organic research funding, the Endangered Species Act, mandatory environmental impact statements, and expansions to the conservation reserve working lands programs, despite record farmer demand for them. Arguably, soil conservation does not necessarily contradict with environmental deregulation, as the former can merely allow for more accumulation. In the aggregate, however, the ecological scalar contradictions of the Farm Bill become glaring in current Conservation Title cuts, advocated heartily by AFBF lobbying.

The ecological aspects of the nationalist scalar contradictions reach new heights with climate change. Despite leading 1990s advocacy for cap-and-trade plans, the AFBF now downplays global warming, inviting prominent climate critics to their conventions. At the 2010 annual AFBF meeting, then-President Bob Stallman warned of cap-and-trade proponents: "The days of their elitist power grabs are over" (Abbott

2010). Without denying anthropogenic climate change, the AFBF posits technological, market-based fixes, from genomic editing to agrofuel crops. In an interview, their executive policy director told me that it would be "unfair" to hold U.S. farmers responsible for a global issue. In a slippery deployment of nationalism, the AFBF policy agenda states: "In the absence of an international agreement to which all nations are committed, we do not believe the United States should saddle the U.S. economy with costs and regulations that will not result in a meaningful impact on the climate" (AFBF 2017a); meanwhile, they assert explicit support for "the U.S. coal industry and coal-fired electrical generating plants to help achieve energy independence." As climate change weather events, from droughts to storms to fires, aggravate agricultural vulnerability, questions of climate risk management proliferate.

Scalar Contradictions of Labor, Borders, and Trade

The ecological oversights of nationalist farm policy work alongside even more glaring scalar contradictions of labor. The Farm Bill does not cover farmworker policies directly but, like the AFBF, it emerged from the racialized throes of early twentieth-century farm labor politics. For a century, the AFBF has lobbied effectively to secure an ample reserve of low-wage, nonunion farm labor unhindered by labor protection regulations. It began by working, in the 1910s, alongside other farm organizations, such as the Associated Farmers of California, and citrus and sugar farm organizations that embarked on a "strike breaking and 'union busting' campaign of extraordinary rigor and virulence" (U.S. Department of Agriculture 1920, 22). Characterized by racism and white supremacy, citrus and sugar leagues met regularly "off the record" to ensure that white planters and farm operators would keep wages for black workers extremely low, and they persuaded New Deal agencies to "purge" their rural relief roles of able-bodied black men to keep them desperate for work and thus brutally exploited in "big gangs which toil under the supervision of overseers" (U.S. Department of Agriculture 1920, 30). These tactics continued post-Reconstruction Southern "Black Codes and Labor Control" legislation that sought to keep the valuable labor of freed African Americans "available to the agricultural interests" (Royce 1993, 63). These

schemes festered on in the Farm Bureau's attack on the Farm Security Administration (FSA) for encouraging production among low- and medium-income farmers and "subsidized people who had been failures as farmers" (Saloutos 1947, 331). This classism was racialized and racist: The FSA program bravely aimed to support black farmers, tenant farmers, and sharecroppers, and the AFBF successfully shuttered it (McConnell 1953). All the while, the AAA excluded tenant—and black—farmers from its allegedly populist beneficence.

As farm labor organized in the 1960s, the AFBF counterorganized, denouncing minimum wage and maximum hour demands as "class legislation" (Saloutos 1947, 325). County farm bureaus' positions remained "of course, that of the farm employer seeking a large supply of low wage labor" (Wubnig 1935, 12). The AFBF fought for and gained agricultural exemptions from the National Labor Relations Act. By the 1930s, nearly half a million tenant and sharecropper families had been displaced—"tractored out"—in the South alone. The Southern Tenant Farmers Union (STFU) gained improbable traction resisting exploitative tenant and labor policies (Grubbs 2000), even as key leaders suffered arrest and imprisonment without charges or legal records: "All indications point to an open and shut case of 'Peonage' (holding a man in slavery)" (Mitchell 1940, 12). STFU members countering voter suppression suffered mob attacks by local, prominent landowners as well as eviction. Labor battles expanded in the 1940s, as World War II labor shortages and food demands hamstrung farm production. The Bracero Program leveraged the scalar fix—and contradiction—of ethnonationalism; long-standing labor exploitations gained new ground with racialized hierarchies of citizenship. In their online history, the AFBF (2017b) vilifies Cesar Chavez as instigating "radical solutions to their concerns over perceived farm labor problems," remembers the grape boycott as troublesome for farmers, and positions AFBF as the victim: United Farm Workers "chose Farm Bureau as a direct target of their ... pickets and protests" (AFBF 2018c). In a scalar contradiction of racist, classist nationalism, alleged farm populism belies elite accumulation as it goes to extreme lengths to secure a surplus of exploited labor.

Currently, anti-immigrant farm politics stoke a white ethnonationalism that belies the reality of the estimated 2.5 to 3.0 million, largely immigrant, men, women, and children who toil in U.S. farms, ranches, and animal feeding operations. According to the 2014 Department of Labor's National Agricultural Workers Survey, nearly half lack formal work authorization, although other experts estimate the number at closer to 70 percent. Anti-immigrant policies increase deportation, raids, and border policing, leading to agricultural labor shortages; geographers, however, track the ways in which this legal liminality serves the interests of an agricultural sector dependent on subjugated labor (Clark 2017). Farm operators, needing workers, have turned increasingly to the H-2A Visa Program (which replaced the Bracero Program in 1964) for temporary, foreign agricultural workers. The Department of Labor certified 200,000 H-2A visas in 2017, doubling 2013 rates. At their 2018 annual meeting, AFBF President Zippy Duvall foregrounded the farm labor crisis: "Everywhere I go, no matter which region or state, farmers tell me this is the No.1 problem they face—not enough ag workers to get their crops out of the field. … We hear from livestock producers and dairy farmers that lose all their workers whenever ICE comes looking for one bad guy" (Duvall 2018, 9). AFBF now lobbies for the H-2C Visa Agricultural Guestworker Act to replace H-2A visas to override the latter's minimum wages, basic worker protections from toxic pesticides, and obligations for housing and transportation of workers to and from fields: "increasing immigration enforcement without also reforming our worker visa program will cost America $60 billion in agricultural production" (AFBF 2018d). H-2C visas exemplify the scalar contradictions of nationalist agriculture: "Agriculture needs a program that functions as efficiently as the current free market movement of migrant farm workers." They force workers into short-term, low-wage contracts in farming, forestry, aquaculture, and meat processing facilities, and prohibit family members from joining them. "There are certain farm jobs, like tending livestock and pruning or picking fresh produce, which require a human touch" (AFBF 2018d). Deploying searing double moves, AFBF admits the skills of farmworkers—and even their humanness (to stave off suggestions of mechanization)—even as they advocate inhumane work conditions and policing: "Farmers must be able to keep their experienced workers—their trustworthy, right-hand men and women who have worked with them for years and can get the work of the farm done. Our proposal

offers a tough but fair solution for these workers and their employers. Enforcement is an important part of the solution, but not the whole solution" (AFBF 2018a, 2018c, 2018d). A scalar contradiction erupts: The people doing the work of farming, who increasingly live in and make up U.S. rural communities, become dehumanized, in a searing scalar double move that includes to exploit even as it excludes to evict. The AFBF supports rampant raids for appropriation-by-criminalization even as the hypocrisy of deportations leaves harvests rotting in the field.

The AFBF remains avowedly opposed to labor organizations, supporting laws that would "mandate specific penalties for unions, union members and public employees who engage in illegal strikes, and prohibit the use of amnesty in such situations" (AFBF 2017a,). They call for exemption of foodservice-sector employees from minimum wage, and they support retention of agricultural exemption from the overtime requirements of Fair Labor Standards Act and oppose earned sick leave. They oppose worker protection standards regarding field-entrance pesticide signage and demand that the U.S. Occupational Safety and Health Administration "repeal its farm labor housing regulations, since such housing is not a workplace" (AFBF 2017a). Following a century of conflating unions to monopolies, the 2017 AFBF Policy Book calls for legislation to amend antitrust laws "to further limit the anti-trust immunity of labor unions" (AFBF 2017a). Trump's 2017 Executive Order on Promoting Agriculture and Rural Prosperity in America systematically divorces the people who own and administer the farms from those who physically upkeep their daily operations, the latter of whom remain racialized, dehumanized, and severed from the rural communities they live and work in—and keep afloat.

The racism of this scalar contradiction seeps beyond xenophobic labor hypocrisies. Anti-indigenous settler coloniality stands firm with opposition to the very term indigenous "sovereign nations," the federal designation of reservations as sovereign states, or "any effort of any federally recognized Native American Tribe to extend their reservation status or sovereignty to non-tribal lands" (AFBF 2017a). In yet another scalar contradiction of white supremacy, nativist agricultural governance aims to subjugate Native agriculture and self-governance.

Ethnonationalism also unfolds in trade, the other major AFBF policy agenda. U.S. agricultural policy posits a scalar contradiction of world trade, wherein U.S. farmers outcompete their global counterparts with dumped exports, feed them with aid, but remain immune to the reciprocal risks of imports. In this one-eyed vision of trade, imports remain invisible, and food security discourses wield geopolitical prowess of future market dependence. The original AAA emerged as a Polanyian attempt to protect farmers from post–World War I swings in global commodity prices. "Prior to the AAA, nearly half the farmers' income was subject to the vicissitudes of the world market. The AAA has largely freed American farmers from this price dependence on world markets" (Ezekial 1934, 1). The glut of surpluses needed outlets, however, thereby reviving a scalar tension that has only become starker (Goldstein 1989). The AFBF lobbied effectively to establish the Undersecretary for Trade and Foreign Agricultural Affairs. Yet, by 2018 Farm Bill markups and neo-mercantilist nationalist renegotiations of the North American Free Trade Agreement, the scalar contradictions of nationalist agricultural policy crashed into public view, instigating global backlash, retaliatory tariffs against U.S. farm products—and even AFBF concern.

Hidden Agrarian Crisis of Surplus

If the scalar contradictions of nationalism manifest so acutely in agriculture, what accounts for widespread farmer support for nativist policy? Analysts have documented the xenophobia and racism of resurgent rural nationalism, but a historical analysis of agricultural policy unearths an important cause of the social and economic vulnerabilities that, when left to fester, become vulnerable to enflamed othering: that secret old crisis of surplus.

In 1933, a North Carolina Cotton Growers Cooperative Association farmer testified before Congress: "We taught the American farmer how to operate in high gear but have not taught him how to get back into low gear in production, or even in intermediate." As landlords defaulted on mortgages, tenants "have been reduced to mere serfs," and farm laborers "cannot provide even the bare necessities of life for themselves and families" (Blalock 1933, 2–3). The USDA National Archives retains this history, chronicling the hazards of even "the possibility of an enormous crop, with the threat of resulting price

wreckage," as warned by AAA administrator Tolley (1936, 6):

> In the depression years, farmers were pushed—or if you please, coerced—by sheer competition into ruthless exploitation of soil fertility. Destruction of the remaining [g]ood lands of the country appeared to be the goal of rugged individualism. Did this "riot" of free competition help the farmer? No, it did not. Production ran wild and surpluses piled higher and higher. But the farmer got no pay from producing the excess, and farm income fell lower and lower. (11)

The entwined origin of both the AFBF and the Farm Bill hinged on the then-glaring problem of overproduction. Why, then, a century later do the Farm Bill and AFBF evade the word *overproduction* altogether and in fact call for increased yield?

Over the course of the twentieth century, farmland consolidated as farms grew larger, midsize farms declined, and those still farming turned increasingly to off-farm household income (Cochrane 1993). The late 1970s ushered in a wave of foreclosures. Amidst farmer suicide hotlines, the American Agricultural Movement gained ground, eventually attracting 2.5 million participants. This grassroots network of growers aligned with rural banks and local food processors and distributors to call national attention to the struggles of rural economies and communities. It culminated in the late 1970s "tractorcades," wherein tens of thousands of farmers and allies traveled to Washington, DC, a few on 900 tractors, to gather publicly for their cause: agricultural policies that would support viable and equitable prices for family farms. Such "parity," it was argued, would ensure that the ratio of production cost to farmgate price afforded a livelihood. Nevertheless, U.S. agricultural policy moved in the opposite direction, and by the 1996 Freedom to Farm Bill, any vestige of supply management had been eliminated in the name of liberalization (Graddy-Lovelace and Diamond 2017). It permitted, and "even encourage[d], a free fall in domestic farm prices while simultaneously promoting rapid liberal trade measures to open new markets for US products" (Ray et al. 2003, 1).

The obscuring of the crisis of surplus works alongside the obscuring of the agrarian crisis itself (Gardner 2009). Each year, more than half of the roughly 2 million farm households in the United States report a financial loss from their farming operations (Prager, Tulman, and Durst 2018); for the past five years, most farm households have had negative farm income (USDA Economic Research Service [ERS] 2018). Farming household income has increased—but this is a subtle indicator of losses amidst a veneer of success. For a generation, it has been off-farm income and the capital gains from land appreciation that keep farmers (who stay afloat) afloat—not farming (Lobao and Meyer 2001). Farm production has shifted to million-dollar farms (those with gross cash farm income of $1 million or more), which account for half of all U.S. farm production, up from a third in 1991 (ERS 2016). From the 1980s onward, average farm size remained steady, but this "seeming stability" hides major, continuing shifts of consolidation (McFadden and Hoppe 2017). The midpoint acreage of this average increases each year, with midsize farmers declining precipitously (Lyson and Welsh 2005). In the price–cost squeeze, particularly acute in livestock and dairy sectors, farms seek lower per unit costs by expanding the size of their operation, prompting even the Congressional Research Service to ask, "As large farms produce an increasing share of U.S. agricultural production, some critics have questioned whether current farm policy is reinforcing or accelerating this process" (Shields 2009, 2). In attempts to help, federal supports have arguably enabled consolidation. Such demographic and structural shifts in agriculture toward increased disparity have changed the distribution of income support, which increasingly goes to higher income farm households. Consolidation, unfettered, begets more consolidation.[3]

Land consolidation parallels agro-industry consolidation and the industrialization of farming itself (Fitzgerald 2010), with stark consequences "across an array of indicators measuring socioeconomic conditions, community social fabric, and environmental conditions. Few positive effects of industrialized farming were found across studies" (Lobao and Stofferahn 2008, 219). The ideal of the family farm has turned into the myth of the family farm (Vogeler 1981; Dixon and Hapke 2003), with agribusiness benefiting off of this myth (Appleby 1982; Clapp and Fuchs 2009).

The AFBF sits uncomfortably atop this myth. As we walked out of the penthouse corner office of the Washington, DC, headquarters, the chief communications director asked whether I had seen the recent *Wall Street Journal* article "Supersize Farms Are Gobbling Up American Agriculture." Leaving the plush carpet and elegant reclaimed barnwood interior, he admitted that much of his public relations work is countering the allegation of corporate agriculture: "Ninety-seven percent of our members are family farmers," he argued, as if by rote. Indeed, the AFBF Web site reiterates the statistic that 88 percent of all farms are small, family farms, against a backdrop of smiling white nuclear families in amber-colored grain fields. Rebuffing assertions that it disproportionately serves rich farmers, AFBF has nevertheless supported policies that do. Even when the AFBF lobbied for protected prices in the 1930s, it denounced the FSA's efforts to help small farmers as merely "experimentation in collective farming and other socialistic policies" (Robertson 1942). Currently, AFBF explicitly opposes income means testing, payment limitations, and "targeting of benefits being applied to farm program payment eligibility" in commodity supports, crop insurance, and conservation title supports (AFBF 2018b). It led the fight to permanently repeal estate taxes. At federal and state levels, the AFBF garners robust income: State-level farm bureaus are themselves nonprofits but with multi-million-dollar assets: The Iowa Farm Bureau had $1 billion of assets as of 2010 (Rodriguez 2018).

The question of what constitutes grounded agrarian populism begs the question of what constitutes the agricultural sector at large. The broad political category of farmers gets wielded as a potent oversimplification, in Farm Bill lobbying efforts, by the AFBF, in journalism, and even in scholarship. This vast generalization subsumes the stark class antagonisms within the category of those classified as farmers, and even family farmers. As such, it does a disservice to the diversity of farmers in the United States—and thus to agricultural policy. The AFBF claims to represent the identity of farmers but ends up representing elite farmers' interests and undermining political needs of small, midsize, urban, and diverse farming communities and farmworkers. This fraught representation persists in part because many farmers feel otherwise excluded and belittled (Berry 2017). More geographic research is needed to address the class heterogeneity, tensions, alliances, and cooptations within the broad category of U.S. farmers and especially AFBF members.

The AFBF began in the ashes of a radical agrarian populism to counter the financialized means of wealth concentration; it carried on a kernel of a vision for agrarian equity in the form of parity, but political and economic dominance quickly subsumed it. The AFBF supported the Commodity Exchange Bill to regulate commodity future exchanges and prevent speculation: "Why should a farm crop be placed practically on the level of a horse race?" O'Neal (1936, 6) demanded. Lobbying to "regulate grain gambling" on the Boards of Trade, its 1939 annual meeting launched an antitrust resolution: "Opportunity has been too much curtailed in America by the insidious growth of special privilege, which has been used by speculators and by monopolistic industry" (O'Neal 1939, 1). Other than sustained opposition to railroad monopolies, however, the AFBF has swung away from such antitrust origins (Ogg 1936). My interviews with the AFBF policy director, communications director, and policy directors at five state farm bureaus belied a party line wary of "fear mongering by certain specialty groups" regarding price fixing by horizontally consolidated seed industries and the vertically consolidated meat sector: "There's enough checks and balances out there," I was told.

This scalar contradiction of nationalism blurs the lines between dominating and feeling dominated by federal government—between reviling federal official bureaucrats and thrilling at presidential presence and influence. The tension is not new. Working with and as the USDA agricultural cooperative extension service, the AFBF adopted a quasi-governmental role (Bliss 1920) amidst and through a tangled set of public–private decentralized networks. A 1921 memorandum of agreement finally extricated the AFBF from Federal Extension Service. Continuing the scalar contradictions of antigovernment nationalism, the AFBF 2017 Policy Book begs Congress to "Allow farmers to take maximum advantage of market opportunities at home and abroad without government interference" (AFBF 2017a), even as AFBF President Duvall joins the White House's Advisory Committee for Trade Policy and Negotiations. The fact that farm policy needs appropriations, negotiating, haggling, and lobbying at every step has solidified AFBF influence: "not the least element of the structure of power traced here is the unusual confidence in the art of lobbying possessed by the Washington office of the Farm Bureau" (McConnell 1953, 179)—an ongoing art.

Although reacting to the failures of neoliberalism, the current, authoritative strand of nationalism aggravates neoliberal evisceration of national public safety nets and services for farming and rural communities and it enables enrichment and consolidation of transnational private sector agro-industries. Executive proposals in 2018 eliminate funding for a rural single-family housing direct loan program, small-town wastewater treatment facilities, and rural business and cooperative programs, among others. After dissolving the cabinet post of Undersecretary of Rural Development, Trump launched an interagency Rural Prosperity Task Force; it met once in 2017, to decry governmental regulatory overreach. Meanwhile, USDA agricultural economists calculate that only the top 10 percent of farm households will accrue between 50 and 70 percent of the tax cuts made under the 2017 Tax Cuts and Jobs Act (Bawa and Williamson 2017).

The USDA's ten-year forecast predicts continuation of low commodity crop prices, below costs of production. As farmgate prices dip further below their reference prices, the AFBF has responded with "Winning the Game" marketing workshops for enterprising farmer producers to brand and sell their products more strategically. I asked a Western state farm bureau policy director if there were internal farm bureau debates about the risks of imports to domestic farmgate prices. Such worries are subsumed, he explained, within the need to export: "We are too good at doing what we do. Making cheaply. Surplus. Need to get it out of this country … we do a brilliant job of making and growing"—hence the need to dump it abroad. In this scalar contradiction of America First nationalism, the nation-state chronically encourages overproduction. A range of fixes emerge to convert the crises of surplus into avenues for further accumulation: turning gluts of commodity crops into "flex" agrofuel and feed crops and surplus disposals via food trade and aid.

At the heart of nationalism's scalar contradictions regarding coloniality, ecology, labor, and geopolitics of trade is the systemic erasure of the farm problem of glut. The self-defeating aspect of overproduction is staved off with growth and consolidation, with the winners left to tell the tale. The results of overproduction have been absorbed by grain dealers, processors, and Concentrated Animal Feeding Operations, who wield enormous power and influence in policy, often in conjunction with AFBF. Despite the productivist Malthusian rhetoric, a myopic focus on quantity ignores the deleterious impact of glut on prices, and thus on rural economies, ecologies, communities, and politics.

Conclusion

The original context and kernel of grounded agrarian populism—rooted in grassroots movements for dignity, agrarian justice, and land-based life—are worth excavating from the layers of settler coloniality and agrocapitalism at work in nationalist farm policy and lobbying: "The Farm Act is agriculture's charter of economic equality. It offers promise of economic security to every man and woman who will join with their neighbors in straightening out the mess we have gotten into with blind, heedless and unplanned individual production," the AFBF (1933) wrote: "They want to cut down their production. They realize their enormous, accumulated surpluses are destroying them" (2). A century later, the AFBF and Farm Bill have lost this collective realization and have come to champion individualism.

Geographers have chronicled the far-reaching impacts of overproduction (Guthman 2011) and how they relate to the productivist myopia of dominant food security paradigms and policies (Lang and Barling 2012; Sage 2013), the treadmill of inputs and debt and corresponding cycles of human and ecological exploitations (Marsden 1998), and the resulting rural problems (D. Woods 2014). Amidst the scalar fixes of dumping surpluses abroad (Winders 2012), surplus grain enriches industrial animal feeding operations. Yet, the AFBF aggressively opposes and seeks to prevent discussion of supply management: "Every educational means available should be used to educate farmers and ranchers on the principles of a market-oriented agriculture" (AFBF 2017a). My interviews with state-level farm bureau policy leaders discussed the AFBF Policy Book as "Farm Bureau bible," but long-standing methodological questions remain regarding how the diversity of farmer needs makes its way to federal consensus: "Even more significant is the fact that the success is not that of 'agriculture' as an entity but of one segment of those who speak in its name. And here the measure, indeed the means, of this success is failure and defeat for those who have been excluded" (McConnell 1953, 2). Generations later, the AFBF's disproportionate influence in Farm Bill policies continues to crowd out other grassroots farmer organizations that are upholding, adapting, and expanding agrarian populism grounded in agrarian justice for land-based life. Importantly, such organizations are increasingly returning to the fundamental crises of surplus: the National Family Farm Coalition (2018) centered its 2018 Farm Bill policy proposal around strategic reserves as a foundation for renewed supply management (National Family Farm Coalition 2018), NFU proposes an inventory management system, and member groups defend existing production controls for viable dairy, sugar beet, and cranberry prices.

Environmental governance shifts and strives to counter the harmful impacts of industrial agriculture, even as the latter pushes back and erodes environmental regulations; accordingly, the question of environmental governance necessitates analysis of agricultural governance, particularly in authoritarian populist eras. U.S. farm policy emerged as an attempt to stave off the crisis of surplus but has been subsumed by the scalar contradictions of ethnonationalism and its origins in coloniality. In the contemporary nationalist United States, these contradictions become acute. This article surveyed the entwined origins of the AFBF and Farm Bill, which epitomize the fraught agrarian populism that has come to drive ethnonationalist policy—to its own detriment. More research is needed to trace the kernel of agrarian populism grounded in honest concern for the economic, ecological, social, and political problems of overproduction. This catalyzed the intertwined origin of the Farm Bill and the AFBF: public consciousness of and populist resistance to the crisis of surplus. As the AFBF changed swiftly thereafter and the Farm Bill changed drastically by the late twentieth century, this kernel stays buried in U.S. farm policy and lobbying. As the long-obscured agrarian crisis erupts anew, however, this dormant kernel might take root and grow.

Acknowledgments

I am grateful for the contributions of Kathy Ozer (rest in peace), Brad Wilson, Adam Diamond, Veronica Limeberry, Hoppy Henton, National Family Farm Coalition, Rural Coalition, Paul Lovelace, and the USDA NAL librarians. All errors are mine.

Notes

1. Using NVivo coding software, my research assistant Veronica Limeberry and I analyzed eighteen state-level Farm Bureau policy agendas, twenty-two federal AFBF policy agenda items and meeting transcripts, four 2017 USDA Farm Bill Field Hearing video

transcripts, USDA Farm Bill social media campaigns, and 2017 Farm Bill Congressional hearings. Additionally, in the fall and winter of 2017, I conducted seven key informant interviews (by phone and in person) with executive policy directors of the AFBF and four states, as well as with the AFBF communications director and membership executive.

2. The AAA soil conservation programs invoked the national security of strong federal government, but they were multiscalar efforts that had a strong regional dimension (Gilbert 2015).

3. Between 1991 and 2015, commodity program payments to farms with at least $1 million in gross cash farm income (GCFI) jumped from 11 percent to 34 percent, whereas payments to small operations (with less than $350,000 in GCFI) fell from 61.3 percent to 30.2 percent (McFadden and Hoppe 2017, 24). Working land conservation program payments such as the Environmental Quality Incentives Program shifted from smaller operations in 2006 to 2015 toward midsize and now larger operations.

References

Abbott, C. 2010. Largest U.S. farm group rallies against climate bill. Reuters. January 10. Accessed February 12, 2019. https://www.reuters.com/article/us-farmers-usa/largest-u-s-farm-group-rallies-against-climate-bill-idUSTRE6091WT20100110

American Farm Bureau Federation (AFBF). 1938. Resolutions adopted at the twentieth annual meeting, New Orleans, LA. 15 December 1938. Box IX E 1 a/ III B4 1 e, USDA NAL Archives.

———. 2015. Final "Waters of the U.S." Rule: No, No, No! No clarity, No certainty, No limits on agency power. Accessed June 1, 2018. https://www.fb.org/tmp/uploads/Final_Rule_No_No_No-Detailed_Version-Copy.pdf.

———. 2017a. Farm bureau policies for 2017: Resolutions on national issues adopted by the voting delegates of the member state farm bureaus to the 98th annual meeting of the American Farm Bureau Federation. American Farm Bureau Federation. Accessed February 12, 2019. https://www.idahofb.org/uploads/2017%20AFBF%20Policy%20Book.pdf

———. 2017b. History. Accessed December 1, 2018. https://www.fb.org/about/history.

———. 2017c. Oppose means testing on crop insurance. Accessed April 1, 2017. https://www.fb.org/files/2018FarmBill/Oppose_Means_Testing_on_Crop_Insurance.pdf

———. 2018a. Climate change. Accessed November 15, 2018. https://www.fb.org/issues/regulatory-reform/climate-change.

———. 2018b. Farm Bill. Page 4. Accessed February 12, 2019. https://www.fb.org/files/Farm_Bill_Mar18.pdf.

———. 2018c. Home page. Accessed December 1, 2018. https://www.fb.org/.

———. 2018d. Solutions for ag labor reform. Accessed November 15, 2018. https://www.fb.org/issues/immigration-reform/agriculture-labor-reform/solutions-for-ag-labor-reform.

Appleby, J. 1982. Commercial farming and the "agrarian myth" in the early republic. *The Journal of American History* 68 (4):833. doi: 10.2307/1900771.

Ball, C. 1936. History of the U.S. Department of Agriculture and the development its objectives. BOX IX e1: Farm Groups, 1935–1939, USDA NAL Archives.

Bawa, S., and J. Williamson. 2017. *Tax reform and farm households.* Philadelphia, PA: USDA Economic Research Services.

Berger, S. R. 1971. *Dollar harvest: The story of the farm bureau.* Lexington, MA: Heath Lexington.

Berlage, N. 2016. *Farmers helping farmers: The rise of the farm and home bureaus, 1914–1935.* Baton Rouge: Louisiana State University Press.

Berry, W. 2017. *The art of loading brush: New agrarian writings.* Berkeley, CA: Counterpoint.

Blalock, B. 1933. Testimony before Senate Committee on Ag & forestry. 28 March. BOX IX E 1, Farm Groups, 1917–1934, USDA NAL Archives.

Bliss, R. K. 1920. Cooperative extension work to the new Farm Bureau movement. Association of Public Land Grant Proceedings. 20 October. BOX IX E 1, Farm Groups, 1917–1934, USDA NAL Archives.

Carr, C., and J. Affolderbach. 2014. Rescaling sustainability? Local opportunities and scalar contradictions. *Local Environment* 19 (6):567–71. doi: 10.1080/13549839.2014.894281.

Clapp, J., and D. Fuchs, eds. 2009. *Corporate power in global agrifood governance.* Cambridge, MA: MIT Press.

Clark, G. 2017. From Panama Canal to post-Fordism: Producing temporary labor migrants within and beyond agriculture in the U.S. (1904–2013). *Antipode* 49 (4):997–1014. doi: 10.1111/anti.12218.

Cochrane, W. 1993. *The development of American agriculture: A historical analysis.* 2nd ed. Minneapolis: University of Minnesota Press.

Desmarais, A. 2007. *La via campesina: Globalization and the power of peasants.* Halifax, AB, Canada: Fernwood.

Dixon, D., and H. Hapke. 2003. Cultivating discourse: The social construction of agricultural legislation. *Annals of the Association of American Geographers* 93 (1):142–64. doi: 10.1111/1467-8306.93110.

Dodd, N. E. 1938. Shall American farmers become peasants? AFBF annual meeting, Chicago, IL, December 5.

Duvall, Z. 2018. Annual address: Enduring mission and impact. 99th AFBF annual meeting, Nashville, TN, January 7.

Economic Research Service. 2016. *America's diverse family farms: 2016 edition.* Washington, DC: U.S. Department of Agriculture. Accessed December 1, 2018. https://www.ers.usda.gov/webdocs/publications/81408/eib-164.pdf?v=42709.

Ezekial, M. 1934. The interest of agriculture in reciprocal trade agreements talk before Land Grant College Association. 21 November. BOX IX E 1; Farm Groups, 1917–1934, USDA NAL Special Collections.

Fitzgerald, D. 2010. *Every farm a factory: The industrial ideal in American agriculture*. New Haven, CT: Yale University Press. doi: 10.1086/ahr/109.5.1594.

Gardner, B. 2009. *American agriculture in the twentieth century: How it flourished and what it cost*. Cambridge, MA: Harvard University Press.

Gilbert, J. 2015. *Planning democracy: Agrarian intellectuals and the intended new deal*. New Haven, CT: Yale University Press.

Goldstein, J. 1989. The impact of ideas on trade policy: The origins of U.S. agricultural and manufacturing policies. *International Organization* 43 (1):31–71. doi: 10.1017/S0020818300004550.

Goodwyn, L. 1978. *The populist moment: A short history of the agrarian revolt in America*. Oxford, UK: Oxford University Press.

Gough, J. 2004. Changing scale as changing class relations: Variety and contradictions in the politics of scale. *Political Geography* 23 (2):185–211. doi: 10.1016/j.polgeo.2003.11.005.

Graddy-Lovelace, G. 2017. The coloniality of U.S. agricultural policy: Articulating agrarian (in)justice. *Journal of Peasant Studies* 43 (3):1–22.

Graddy-Lovelace, G., and A. Diamond. 2017. From supply management to agricultural subsidies—And back again? U.S. Farm Bill & agrarian (in)viability. *Journal of Rural Studies* 50:70–83. doi: 10.1016/j.jrurstud.2016.12.007.

Gregory, C. 1935. The American Farm Bureau Federation and the AAA. *Annals of the American Academy of Political Social Science* 179 (1):152–57. doi: 10.1177/000271623517900120.

Grubbs, D. H. 2000. *Cry from the cotton: The Southern Tenant Farmers' Union and the New Deal*. Fayetteville: University of Arkansas Press.

Guthman, J. 2011. Excess consumption of over-production? US farm policy, global warming, and the bizarre attribution of obesity. In *Global political ecology*, ed. R. Peet, P. Robbins, and M. Watts, 51–66. London and New York: Routledge.

Harvey, D. 2005. *A brief history of neoliberalism*. New York: Oxford University Press.

Heynen, N., and H. A. Perkins. 2005. Scalar dialectics in green: Urban private property and the contradictions of the neoliberalization of nature. *Capitalism Nature Socialism* 16 (1):99–113. doi: 10.1080/1045575052000335393.

Hyde, A. 1929. *The government's policy toward the cooperative movement*. Washington, DC: USDA Office of Press Services.

Kile, O. M. 1948. *The Farm Bureau through three decades*. Baltimore, MD: Waverly. Accessed December 1, 2018. http://hdl.handle.net/2027/uc1.$b91270.

Laclau, E. 2005. *On populist reason*. New York: Verso.

Lang, T., and D. Barling. 2012. Food security and food sustainability: Reformulating the debate. *The Geographical Journal* 178 (4):313–26. doi: 10.1111/j.1475-4959.2012.00480.x.

Lobao, L., and K. Meyer. 2001. The great agricultural transition: Crisis, change, and social consequences of twentieth century U.S. farming. *Annual Review of Sociology* 27 (1):103–24. https://doi.org/10.1146/annurev.soc.27.1.103. doi: 10.1146/annurev.soc.27.1.103.

Lobao, L., and C. W. Stofferahn. 2008. The community effects of industrialized farming: Social science research and challenges to corporate farming laws. *Agriculture and Human Values* 25 (2):219–40. doi: 10.1007/s10460-007-9107-8.

Lyson, T. A., and R. Welsh. 2005. Agricultural industrialization, anticorporate farming laws, and rural community welfare. *Environment and Planning A* 37 (8):1479–91. doi: 10.1068/a37142.

Mansfield, B. 2004. Rules of privatization: Contradictions in neoliberal regulation in North Pacific fisheries. *Annals of the American Association of Geographers* 94 (3):565–84. doi: 10.1111/j.1467-8306.2004.00414.x.

Marsden, T. 1998. Agriculture beyond the treadmill? Issues for policy, theory and research practice. *Progress in Human Geography* 22 (2):265–75. doi: 10.1191/030913298669229669.

McConnell, G. 1953. *The decline of agrarian democracy*. Berkeley: University of California Press. doi: 10.1086/ahr/59.3.659.

McFadden, J., and R. Hoppe. 2017. The evolving distribution of payments from commodity, conservation, and federal crop insurance programs. Accessed November 15, 2017. https://www.ers.usda.gov/webdocs/publications/85834/eib-184.pdf?v=43068.

McIntosh, W. L. K. W., E. Spies, D. M. Stone, C. N. Lokey, A.-R. T. Trudeau, and B. Bartholow. 2016. Suicide rates by occupational group—17 states, 2012. *Morbidity and Mortality Weekly Report* 65 (25):641–45.

Mitchell, H. L. 1940. Secretary of SFTU. National Union 1909–1939. National Council of Farmer Cooperatives. File IX E1g; Southern Tenants Farmers Union (Southern Sharecroppers Union) 1935–1940, USDA NAL Archive Special Collections.

Monnat, S., and D. Brown. 2017. More than a rural revolt: Landscapes of despair and the 2016 presidential election. *Journal of Rural Studies* 55:227–36.

National Cooperative Council. 1936. An agricultural policy for the United States: Preliminary report. Special Legislative Committee. 12 February. William Byrd Press, Inc. BOX IX e1; Farm Groups, 1935–1939, USDA NAL.

National Family Farm Coalition. 2018. 2018 Farm bill. Accessed January 30, 2019 http://nffc.net/index.php/2018farmbill/.

Naylor, G. 2017. Agricultural parity for land de-commodification. In *Land justice: Re-imagining land, food, and commons*, ed. J. M. Williams and E. Holt-Gimenez, xviii–xxii. Oakland, CA: Food First.

O'Neal, E. 1936. Man on the land. Speech, AFBF annual meeting, Chicago.

———. 1939. Letter from Edw. O'Neal AFBF Pres to all state farm bureaus. 29 June. BOX IX E 1a; Farm Groups, 1918–1936, USDA NAL Archive Special Collections.

———. 1940. 76th Congress, 3rd Session, House of Rep, Committee on Appropriations, Hearings on Ag Dept of Appropriation Bill for 1941. Notes on Payments, IX E1a, 5, USDA NAL Special Collections.

Ogg. 1936. 76th 1st Session, House of Rep, Committee on Appropriations, Hearings on Ag Dept Appropriation Bill for 1940. 3 March 1939. Notes on Payments, IX E1a, 4, USDA NAL Archives.

Porter, K. K. 2000. Embracing the pluralist perspective: The Iowa Farm Bureau Federation and the McNary-Haugen movement. *Agricultural History* 74 (2):381–92.

Prager, D., S. Tulman, and R. Durst. 2018. How do tax benefits and asset appreciation affect the returns to farming for U.S. farm households? In *Proceedings of the Allied Social Sciences Association annual meeting*, 1–19. Economic Research Report Number 254. Washington, DC: U.S. Department of Agriculture.

Ray, D. E., D. G. De la Torre Ugarte, and K. J. Tiller. 2003. *Rethinking US agricultural policy: Changing course to secure farmer livelihoods worldwide*. Knoxville, TN: Agriculture Policy Analysis Center. Accessed January 1, 2019. https://www.iatp.org/sites/default/files/Rethinking_US_Agricultural_Policy_Changing_Cou.pdf.

Robertson, N. 1942. Wealthy farmers to campaign for continuation of subsidies. Washington Daily News (Series 1 - Box 1.2/40. Folder: I.2.1XEI Farm Groups 1936–1945, 1 of 3).

Rodriguez, S. 2018. Morning Ag. *Politico*, February 6. Accessed November 1, 2018. https://www.politico.com/newsletters/morning-agriculture/2018/02/06/drought-worries-across-ag-regions-094229.

Royce, E. 1993. *The origins of Southern sharecropping*. Philadelphia: Temple University Press.

Sage, C. 2013. The interconnected challenges for food security from a food regimes perspective: Energy, climate, and malconsumption. *Journal of Rural Studies* 29:71–80. doi: 10.1016/j.jrurstud.2012.02.005.

Saloutos, T. 1947. The American Farm Bureau Federation and farm policy: 1933–1945. *Southwestern Social Science Quarterly* 28:313–33.

Scala, D., and K. Johnson. 2017. Political polarization along the rural–urban continuum? The geography of the presidential vote, 2000–2016. *Annals of the American Academy of Political and Social Sciences* 672 (1):162–84. doi: 10.1177/0002716217712696.

Scoones, I., M. Edelman, S. M. Borras, R. Hall, W. Wolford, and B. White. 2018. Emancipatory rural politics: Confronting authoritarian populism. *Journal of Peasant Studies* 45 (1):1–22. doi: 10.1080/03066150.2017.1339693.

Shields, D. 2009. The farm price–cost squeeze and U.S. farm policy. Accessed November 1, 2018. http://www.farmpolicy.com/wp-content/uploads/2009/08/crs-report-farm-price-cost-squeeze-and-us-farm-policy.pdf.

Tolley, H. R. 1936. Soil conservation and agricultural adjustment. Address before AFBF, Chicago, IL, December 10.

U.S. Department of Agriculture (USDA). 1920. Notes on Farm Bloc. Internal USDA document. BOX IX e1; Farm Groups, 1935–1939, USDA NAL Archive Special Collections.

U.S. Department of Agriculture, Economic Research Service. 2018. 2018 Farm household income forecast. Accessed January 1, 2019. https://www.ers.usda.gov/topics/farm-economy/farm-household-well-being/farm-household-income-forecast/.

Vogeler, I. 1981. *The myth of the family farm: Agribusiness dominance of U.S. agriculture*. Boulder, CO: Westview.

Wallace, H. A. 1936. Agricultural security. Address to AFBF, California. 9 December. BOX IX E 1a; Farm Groups, 1918–1936, USDA NAL Archive Special Collections.

———. 1937. Statement in opening conference of farm organization leaders. BOX IX e1; Farm Groups, 1935–1939, USDA NAL.

Weber, B. 2017. Testimony on examining the farm economy: Perspectives on rural America. U.S. Senate Committee on Agriculture, Nutrition & Forestry. Accessed November 1, 2018. https://www.agriculture.senate.gov/hearings/examining-the-farm-economy-per-spectives-on-rural-america.

Weis, T. 2010. Accelerating biophysical contradictions of industrial capitalist agriculture. *Journal of Agrarian Change* 10 (3):315–41. doi: 10.1111/j.1471-0366.2010.00273.x.

Wilson, M. L. 1936. The challenges to agriculture. Address by assistant secretary of agriculture, Iowa Farm Bureau Federation, Des Moines, IA, January 16.

Winders, W. 2012. *The politics of food supply: U.S. agricultural policy in the world economy*. New Haven, CT: Yale University Press.

Woods, D. 2014. The many faces of populism: Diverse but not disparate. In *Many faces of populism: Current perspectives*, ed. D. Woods and B. Wejnert, 1–25. Bingley, UK: Emerald Group.

Wubnig, A. 1935. A brief history of the Outstanding Farmers' Organizations. Prepared for the Labor Relations Division. BOX IX e1: Farm Groups, 1935–1939, USDA NAL Archive Special Collections.

GARRETT GRADDY-LOVELACE is Associate Professor in the Global Environmental Politics program at American University's School of International Service, 4400 Massachusetts Avenue NW, Washington, DC 20016. E-mail: graddy@american.edu. Drawing on geography, political ecology, and decolonial theory, she researches and teaches agricultural policy, agricultural biodiversity, agrarian politics, and food and farm justice.

The State, Sewers, and Security: How Does the Egyptian State Reframe Environmental Disasters as Terrorist Threats?

Mohammed Rafi Arefin (ID)

On 25 October 2015, Alexandria, Egypt, experienced heavy rainstorms, which overwhelmed the city's sewer and drainage systems. The storm flooded the city and caused the death of seven residents. How the causes of the 2015 Alexandria floods would be narrated and explained became a subject of national contestation; the Egyptian state claimed that the floods were caused not by failures of urban environmental governance or climate change as others suggested but by an act of terrorism. In this article, I draw on geopolitical ecology and situated urban political ecology to examine the production and contestation of the Egyptian state's narratives of the floods. In doing so, I argue against using the term *authoritarian*, which predetermines the state's function through an inherited set of characteristics. Rather, I start with the how and why of state power to carefully examine the complex relationship between the state and urban environments. This situated approach reveals how failures of urban environmental governance are being reframed by repressive regimes to further justify their rule.

2015年十月二十五日，埃及亚历山大市遭逢暴风雨，淹没了该城市的下水道与排水系统。暴风雨淹没了城市，并导致七名居民死亡。这场2015年亚历山大市水灾的导因如何被叙述与解释，则成了全国争夺的对象；埃及政府宣称这些洪水并非像若干人所指称的城市环境治理失败或气候变迁所导致，而是由恐怖主义行动所引发。我于本文中运用地缘政治生态学和脉络化的城市政治生态学，检视埃及政府对于水灾叙事的生产与争夺。我通过这麼做，主张不应使用"威权"这个概念，因为该概念预设了国家通过内在的特徵组合进行运作。反之，我从国家权力"如何"以及"为何"之问题开始，详细检视国家与城市环境之间的复杂关系。此一情境化的方法，揭露了城市环境治理的失败如何被压迫的政体重新框架，以进一步合理化其统治。关键词：基础建设，下水道设施，安全，恐怖主义，城市环境治理。

El 25 de octubre de 2015, Alejandría, Egipto, experimentó intensos temporales, que sobrepasaron la capacidad de los sistemas de alcantarillado y drenaje de la ciudad. La tormenta inundó la ciudad y causó la muerte de siete residentes. El modo como las causas de las inundaciones de Alejandría en 2015 serían narradas y explicadas se convirtió en materia de preocupación nacional; el estado egipcio sostuvo que las inundaciones fueron causadas no por fallas de la gobernanza ambiental urbana o por cambio climático, como otros sugirieron, sino por un acto terrorista. En este artículo, me baso en ecología geopolítica y en ecología política urbana situada para examinar la producción y contestación de las narrativas sobre las inundaciones por el estado egipcio. Al hacerlo, arguyo contra el uso del término autoritario, que predetermina la función del estado a través de un conjunto heredado de características. Mejor, empiezo con el cómo y el porqué del poder estatal para examinar cuidadosamente la compleja relación entre el estado y los entornos urbanos. Este enfoque situado revela el modo como las fallas de la gobernanza ambiental urbana están siendo re-enmarcadas por regímenes represivos para justificar aún más su gobierno. *Palabras clave: gobernanza ambiental urbana, infraestructura, saneamiento, seguridad, terrorismo.*

Quickly gathering our notes, my research assistant and I headed for the door; we had just secured an appointment with an Egyptian official responsible for Cairo's sanitation system whom we had been trying to interview for months. As we were leaving, the person who helped us secure the interview warned, "He has been directed to not answer the kinds of questions you are asking." Nonetheless, we rushed to interview the official in his ornate office in downtown Cairo. He did, in fact, evade most of my questions, and he repeatedly returned to a statement to close entire avenues of inquiry: "Terrorists are targeting Egypt's infrastructure—even the sewer system."

This response did not immediately catch my attention. Blaming terrorism for the country's troubles had become an increasingly common government response following the presidential election of former general Abd al-Fatah al-Sisi. Elected in 2014, Sisi ran on a platform of economic stability, national security, and infrastructural development in which terrorism became the regime's enemy par excellence. Blaming terrorists for targeting Egypt's urban sanitation infrastructure, however, revealed how the state's narrative of terrorism and security extended into its repressive approach to urban environmental governance.

Six months earlier on 25 October 2015, the sewerage and drainage system of Alexandria, Egypt's second largest city, failed during a heavy rainstorm, killing seven people. Photographs of Alexandria's residents wading through sewage-flooded streets quickly circulated in news and social media. As these images circulated, the cause of the flooding became the subject of political contestation. A few days after the floods, the Ministry of Interior announced that the flooding was caused by terrorists who had attacked Alexandria's sewerage and drainage systems to encourage antigovernment sentiments. Although the findings of the ministry's investigation were contested and discredited by opposition groups, journalists, engineers, and public officials, the discourse circulated in official statements months later.

In this article, I ask how the failure of Alexandria's sewerage and drainage system was reframed and mobilized by Sisi's repressive regime. Drawing on archival research, interviews, and news and social media sources, I demonstrate how the government sought to entrench its power by offering terrorism and the attendant response of security as the master discourses through which other explanations of the 2015 Alexandria floods were explained and obfuscated. After reviewing how the state's narrative emerged, I outline alternative explanations for the floods from opposition groups, historical engineering documents, and contemporary engineers. This close examination draws on work in urban political ecology to argue that, in the current era, failures of urban environmental governance are being reframed by repressive regimes to further

justify their rule. Turning to the analytic tools of geopolitical and situated political ecology, I resist using the term *authoritarian*; instead, I argue that events such as the Alexandria floods are complicated cases that exceed a priori typologies of the state and call for more specific theorizations of the relationship between, in this case, the state, sewers, and security.

Beyond Authoritarianism: Toward a Situated Urban Political Ecology

Urban political ecology rejects a split between the city and nature by framing cities as sites of metabolism (Swyngedouw and Heynen 2003; Heynen, Kaika, and Swyngedouw 2006). This intervention brought abject and forgotten urban infrastructures and services like sewerage and drainage systems into view for critical urban scholars (Melosi 1993, 2008; Gandy 1994, 1999, 2006; McFarlane 2008; Moore 2008, 2009, 2012; Nagle 2013; Fredericks 2013, 2014). Such studies have argued that urban infrastructures, especially those designed to manage waste, are a central but often overlooked dimension of urban politics (McFarlane and Silver 2016).

Melosi (2008) suggested that the politics of sanitation are often overlooked because sanitation is taken for granted in modern cities. Urban disasters—such as hurricanes, floods, earthquakes, and fires—that disrupt the functioning of the sanitary city, however, expose, even if only for a brief moment, urban life's dependency on such infrastructures. More recently, scholars have explored urban disasters like flooding through a range of approaches including risk (Ribot 2014), citizenship (Ranganathan 2014; Anand 2017), denaturalizing disasters (Smith 2006; Squires and Hartman 2006), assemblage (Ranganathan 2015), site ontology (Meehan and Rice 2011), and slow violence (Nixon 2013). These approaches share a concern with emphasizing the political economic dimensions of disasters in cities that are too often framed as natural and therefore both inevitable and apolitical. This scholarship has been successful in highlighting how disasters are differentially experienced.

Yet the current conjuncture, which is characterized by a global rise of populist and authoritarian regimes, has created a new set of challenges to these frameworks. Deteriorating urban infrastructures have come to symbolize the failures of the Keynesian welfare state in the Global North and the ruins of the postcolonial promise of development in the Global South. These quickly deteriorating infrastructures are providing fodder for the rise of populist and authoritarian regimes that use these failures to call for the entrenchment of increasingly repressive states. In this era, repressive states are no longer ridding themselves of responsibility by claiming that urban disasters and infrastructural failures are natural, inevitable, and unstoppable. Instead, states reposition disasters and infrastructural failures as highly political threats necessitating an increase in state presence, control, and security. From the water crisis in Flint, Michigan, to the cholera outbreaks in Sanaa, Yemen, state violence increasingly operates in and through the management of urban environmental disasters. Although not entirely new, the present configuration of states, violence, and urban disasters enlivens debates about the role of the state in political ecology with renewed urgency.

What began as a call to cross subdisciplinary lines between political ecology and political geography, political ecologies of the state is now a thriving field of inquiry that examines how both the state and nature are complex and contradictory entities that produce unexpected state–nature relations (Robbins 2003, 2008; Harris 2017). In this intersection, Bigger and Neimark (2017) proposed the term *geopolitical ecologies* to combine "the strengths of political ecology with those of geopolitics in order to account for, and gain a deeper understanding of, the role of large geopolitical institutions, like the US military, in environmental change" (14). In this article I expand the purview of this framework to include geopolitical discourses and practices such as the global threat of terrorism (Graham 2004). To properly situate this intervention in the Egyptian context, I bring the insights of geopolitical ecologies of the state together with two other bodies of literature: situated urban political ecology and political economic approaches to the Middle East.

Although the Egyptian state's response to the Alexandria floods could be framed as another action of a resilient authoritarian Arab state, I argue against the use of this commonplace yet dangerously simplistic state typology. Moving beyond the term authoritarian, I work in the framework of situated urban political ecology, which argues against "the application of Northern theories uncritically to Southern contexts to highlight that urban political ecology

tends to overlook the situated understandings of the environment, knowledge and power that form the core of other political ecological understandings ..." (Lawhon, Ernstson, and Silver 2014, 498). Instead of importing Eurocentric geopolitical typologies of the world, urban political ecology can offer situated political ecologies of the state that detail how the state frames environmental governance failures and how the uneven material consequences of state narratives are contested by historical analyses and everyday politics (see Sowers [2012] and Myers [2016] for examples of such work).

Following the call of situated urban political ecology, I turn to the work of Hanieh (2013, 2016) to move beyond the term authoritarian. Hanieh argued that many studies of the Middle East begin with the methodological dichotomy of the state and civil society. This dichotomy undergirds decades of scholarship that frames the Arab world as a site of exceptional authoritarian resilience that has resisted international waves of democratization that began in the 1980s. In this Eurocentric scholarship, Arab states are categorized as authoritarian monarchies or authoritarian republics sustained by forces such as religion, statecraft, natural resource endowment, or militaries. The state–civil society dichotomy and the resulting state typologies become problematic, not only in their inability to explain contemporary politics in the region but also in their normative understandings of political-economic change. What differentiates authoritarian states from democratic states, in this framework, is not only the weakness of political and civil rights but also the weakness of capitalism. Authoritarian states, unlike their democratic foils, prevent the free efficient functioning of markets. The issue then ultimately with this line of reasoning, according to Hanieh (2013), is that "the agency of freedom is neatly located in the realm of the market" (4).

Although the term authoritarian might be attractive in its power to link right-leaning, fascist, and populist trends across the Global North and Global South, political ecology must be wary of the histories of inherited geopolitical terms and the normative assumptions they carry. This caution must be taken not only because the term authoritarian carries assumptions (e.g., locating agency in free markets) that do not align with basic tenets of political ecology but also because such typologies obscure contemporary and historical state–nature relations in the

Middle East. Often depicted in neo-Malthusian terms of water scarcity, population growth, and urban sprawl, state–nature relations in the Middle East are a complex and often contradictory accumulation of historical imperialist and colonial management coupled with contemporary waves of centralization and decentralization (Mikhail 2011; Sowers 2012; Davis 2016). Instead of beginning with an inherited form of the state and endowing this form with determinant explanatory power, I follow Hanieh's (2013) call to start from the how and why of state power.

Taking seriously Hanieh's critique in relation to the project of situated urban political ecology, I use the term *repressive* to describe the Egyptian state. Here the term points to the state as a producer and eraser of environmental knowledge and narratives.[1] The following case demonstrates how the state represses dissent and alternative narratives of environmental disaster by resorting to the master discourse of terrorism. This characterization of the state is inspired by what Robbins (2008) called the knowing state, which in times of crisis is "especially driven to produce, erase, and struggle over environmental narratives" (212–13).

In what follows, I draw from the scholarship just outlined to demonstrate how the Egyptian government transformed a failure of Alexandria's combined drainage and sewerage system into a terrorist threat posing an imminent danger to cities across Egypt. I then trace narratives that contest the state's response to the 2015 Alexandria floods. I argue that this situated approach, which displaces the term authoritarian and contests the state's master discourse of terrorism, is necessary to disrupt the entrenchment of the repressive regime and place further attention not on terrorism but on infrastructural neglect and climate change in coastal cities.

Drowning in Neglect

On 25 October 2015, Alexandria received almost ten inches of rain, exceeding the yearly annual average by about two inches. As water levels across the city rose, the combined drainage and sewerage systems were overwhelmed and wastewater flooded the streets. The flooding was quickly compounded by the city's aging electrical infrastructure; as water came into contact with exposed electrical wiring, large stagnant puddles were electrified. To further exacerbate the situation, uncollected household waste, which littered the street before the storm, festered in stagnant pools of wastewater. Overwhelming Alexandria's urban infrastructure, the heavy rainfall caused the death of seven people: four from electrocution, one from drowning, and two with the specific causes unreported.

After the storm, images of Alexandria's flooded streets circulated on news and social media. On platforms such as Twitter and Facebook, the hashtag "Alexandria is Drowning" in Arabic indexed photographs of cars underwater, flooded buildings, and people wading through sewage and garbage-filled stormwater.[2] Nationally broadcast news shows featured phone interviews with Alexandria's residents complaining about power outages and being trapped in their homes. These frustrations soon coalesced around a specific image. A photo began to circulate on social media that showed a police officer dressed in a white uniform with his hands behind his back, head tilted slightly upward, standing idly next to a sewage-flooded street in Alexandria. Above the street is a massive billboard with the president's face superimposed on top of an Egyptian flag. Focusing on the longtime neglect of Alexandria's drainage and sewerage system, this photo came to symbolize public frustrations with the state in both preparing for and responding to the Alexandria floods.[3] This image implicated both local authorities (signified by the officer) and the larger Egyptian state (with Sisi's image) in neglecting Alexandria's preparedness for the storms. In response, each level of government began to produce different narratives of the floods.

This is where a situated approach to urban environmental crisis becomes important. Beginning from an authoritarian state category simplifies the actions of the state, which are often contradictory. As Bell (2000) outlined, the state's repression of public participation in environmental governance creates the conditions for new modes of environmental action and advocacy. This means that the state's repressive tactics are fragmented and always shifting. Bell traced how the Egyptian state's repressive management of urban environments has evolved using discourses of beautification in the postcolonial era, modernization during times of economic liberalization, and environmentalism in the era of sustainability (Bell 2009). Today, repressive environmental governance has come to incorporate discourses of terrorism and security. To describe and analyze these shifts, I start with an approach that begins with Alexandria's municipal politics.

Calling the floods an environmental disaster, the governor of Alexandria, Hany El-Messiry, submitted his resignation. A state-owned news outlet reported that the governor resigned due to public criticism that he had not done enough to prepare for the storm (Ali 2015). Although the governor did play a role in coordinating the city's urban infrastructures and services, he had little control over the specific management of the sewerage and drainage systems. Beginning in 2004, a large-scale decentralization of urban services throughout Egypt meant that another entity was responsible for the system's maintenance. Through this decentralization of services, independently managed state-funded water and wastewater holding companies were created in Egypt's major governorates. These companies were allocated operating budgets by the government but were expected to make the majority of their operation costs through service fees collected from residents. Through interviews with officials working in these holding companies, I found that this arrangement began to falter in the wake of the 2011 revolution. Economic instability heavily affected the holding companies' budgets and capacity for maintaining water and wastewater systems.

After his resignation, El-Messiry continued to substantiate the narrative that the flood was caused by environmental forces and exacerbated by government neglect. El-Messiry stated that he spoke with the head of the holding company on 5 October, twenty days before the massive rainstorms, to discuss the deteriorated condition of the drainage and sewerage system in response to a smaller scale flood that occurred the day before. In addition, before his resignation, the governor reiterated that little to no development of the sewerage and drainage infrastructure had been done in fifteen years. The system was in dire need of an estimated $10 million upgrade ("Egypt's PM Appropriate" 2015).

The former governor's comments exposed a network of neglect and improper management that plagues urban environmental governance in Egypt, spanning from local officials to budget and policy decisions made at higher levels of the government. This narrative of infrastructural negligence started to circulate widely as made evident by the 26 October front-page headline of a major privately owned national newspaper that read "Drowning in Neglect" ("We Are Drowning" 2015). After the storm, President Sisi tried to quell mounting criticism by ordering his prime minster to allocate the necessary funding to upgrade the system. Furthermore, Sisi's government promised to compensate the families of the people who died in the floods.

Shifting the Narrative: Combined Sewer Overflow or Terrorist Attack?

Amid continued critical media coverage, another event—five days after the Alexandria floods—threatened to shake the public's confidence in Sisi's promises of stability, security, and development. On 31 October 2015, Metrojet Flight 9268, en route from the Sinai resort town of Sharm Al-Sheikh to Saint Petersburg, crashed in the Sinai Peninsula, killing all 224 people on board. Shortly after the plane crash, Ansar Bait al-Maqdis, an Islamic State–aligned group in the Sinai, claimed responsibility for smuggling a bomb onto the flight. Occurring within a week of each other, the Alexandria floods and the bombing of a civilian aircraft demonstrated Sisi's inability to deliver on his promises of stability, security, and development.

In a speech given the day after the crash, Sisi attempted to shore up his legitimacy and relocate blame away from his regime. This speech would signal the beginning of the state's attempt to shift the narrative of the Alexandria floods from a failure of urban environmental governance to terrorism. After discussing a range of military actions that were successful against "terrorist entities," Sisi assured a room full of military and police officers that "we are safeguarding the state, not merely the government. We are protecting Egypt—not anybody else—from an evil which God wanted us to be aware of." In the same speech, he criticized the media for its "unfair" coverage of the flooding in Alexandria, stating that it was dealt with swiftly and he should not be blamed. Sisi ended the speech with a message to the media: "Don't let people lose hope. The media and the state should not disagree. Take care, let us not disagree" (al-Sisi 2015). Addressing both the Alexandria floods and the downing of Metrojet Flight 9268 in the same speech, Sisi provided a detailed explanation of how he was both increasing security by fighting terrorism and investing in infrastructure. Prior to Sisi's speech, the causes of these two events seemed distinct in the state narrative. Soon after, though, the Ministry of Interior would make direct connections between the floods and terrorism.

The Ministry of Interior, responsible for Egypt's domestic law enforcement, released a statement through its official Facebook page on 6 November stating that it arrested three members of the now-outlawed Muslim Brotherhood allegedly involved in causing the Alexandria floods. Along with photos of the three men, the statement charged them with being part of a larger seventeen-man terrorist cell that was planning a series of attacks in Alexandria. The statement detailed the findings of the ministry's investigation that unearthed photos and plans of an operation attempting to:

> block the drains and the pipes of the sewerage system by placing a cement mixture in the system to prevent the drainage of water and to burn and destroy electrical switches and trash cans to start crises in order to encourage a state of popular resentment against the current government. (Ministry of Interior 2015)[4]

Later, the Ministry of Interior released a video that depicted the three men confessing to joining terrorist groups and blocking key sewer and drain pipes ahead of the storm to, as one of the accused confessed, "confuse the government and show it as a failed regime." In addition, the video showed the men reenacting how they blocked the system.[5] These videos and statements established the state's narrative of terrorism as the cause of the flooding, displacing the former governor's narrative and public accounts of neglect. Terrorism became the master discourse through which the floods were now to be understood; this narrative shifted the urgency of improving environmental governance or infrastructural development to a focus on terrorism's attendant state response, increased security.

It is important to note that although the state's narrative endured, it did not go uncontested. For example, journalist Haitham al-Sheikh released a public statement claiming that the suspects named in the Ministry of Interior's statements were arrested weeks before the storm on other charges.[6] In addition, the Arab African Center for Freedom and Human Rights released a statement detailing the conditions under which each of the seventeen men were arrested. Many of those arrested were involved in domestic disputes or other unrelated charges. The center claimed that the members of the alleged terrorist cell were forcibly disappeared by the Ministry of Interior before the floods and then accused of sabotaging Alexandria's infrastructures while they were in detention.[7]

Even though the Ministry of Interior's narrative of terrorism was discredited in detail by journalists and opposition groups, the discourse continued to circulate in official statements. This was evident in my interview with the government official months later, which opens this article. Through presidential addresses and the Ministry of Interior, the state produced a master narrative of terrorism through which alternative narratives of the Alexandria floods were obfuscated. Terrorism became the narrative through which infrastructural neglect, mismanagement, and climate change were to be explained. This state narrative did not just obfuscate the causes of an urban environmental disaster; it also had devastating material consequences. Forced disappearances and recorded confessions have most likely condemned those named in the ministry's statements to state prisons.

After outlining how the state's narrative of terrorism was produced even in the face of contesting narratives from journalists and other groups, I now turn my attention to other sources that undermine the state's narrative: colonial-era planning documents and the reports of Egyptian engineers. Rather than position these as the authentic narratives of the events, I am tracing an alternative explanation for why the system failed and was made available for use as a tactic in the consolidation of Sisi's regime.

Contesting the Master Discourse of Terrorism: Colonial Planning, Infrastructural Neglect, and Climate Change

Opposition groups, journalists, and state officials contested terrorism as the cause of the floods by pointing to wrongful arrests, failures of urban environmental governance, and infrastructural neglect. To further understand how Alexandria's sewerage and drainage system broke down and became available as a site through which the state could further entrench securitization, I contextualize the breakdown of the system in two historical junctures: the colonial assessment and planning of the system in the late nineteenth century (during which Egypt was under British control as a veiled protectorate) and the 2004 reorganization of the sewerage and drainage administration.

Alexandria's first drainage system was built in the 1870s and funded by local merchants. Initially covering only a small part of the city, the system was designed and built for drainage but was soon used also for sewage. Although condemned for sewage use in 1895, the dual use continued out of need (Carkeet James 1917). This continued use did not go unnoticed by the British-controlled government. In the 1917 edition of the book *Drainage Problems of the East*, a British colonial sanitary engineer included a foreword to his chapter on Alexandria that stated:

> The following notes will show that there is no city in the East, with any pretensions to being a city, that cannot hold up the finger of scorn to Alexandria in the matter of sanitation … in spite of its increasing importance, it has spent practically nothing in improving its sanitation for the last thirty years, while, refusing advice and warning, it has persisted in neglecting the first rudiments of sanitation. (Carkeet James 1917, 239)

Called to consult on a scheme to build two separate but connected drainage and sewerage systems in Alexandria by the British-backed government, Carkeet James (1917) made two recommendations. The first was that rainfall was not an important consideration in the design of the system, as "the rainfall of Alexandria is very slight, and is mostly confined to December and January, the average annual fall for the last thirty years (1869–1899) being only 224.8 millimeters or 8.47 inches" (319). The second recommendation concerned Alexandria's proximity to sea level. James argued that although the city was nearly at mean sea level, the range of the tide was minimal and did not pose a risk to the system.

I outline the details of this colonial-era document to highlight the fact that key features that contributed to the 2015 floods were structured into the system almost 100 years prior. The combined sewer system (a system now notorious for stormwater flooding in coastal cities) was condemned in 1895 and the misguided calculations of colonial engineers would later be the weak points of the contemporary system: rising tides and rainfall. This colonial-era planning document describes a slower, more

mundane causal explanation for the floods rather than a spectacular single act of terror.

This document also lends credence to the contemporary consequences of climate change that threaten the already neglected system. In 2013, the World Bank published a report that ranked Alexandria in the top five major cities around the world that are at increased risk of flooding and becoming uninhabitable due to climate change (Hallegatte et al. 2013). As rainy seasons have become more intense and coastal sea levels have also begun to rise, Alexandria's weather has come to pose a significant threat to an already taxed sewerage and drainage system.

In addition, there was a more recent decision in the system's management that exacerbated the faults of an aging urban infrastructure unequipped for climate change. In 2004, the incoming president of the newly created Water and Wastewater Holding Company of Alexandria made the decision to join the previously connected but distinct sewerage system with the drainage system. This decision was reportedly made without any scientific study and was only briefly announced as a measure to mitigate the unlikely risk of a tsunami (Mahmood 2015). Joining the drainage and sewerage system made the city susceptible to the degradation of both systems and presented the possibility of a combined sewer overflow in the event of high tides or heavy rain.

The decline of the system that made the 2015 flash floods so devastating began in 2004 and in effect disabled an already neglected, aging system (Bhattacharya et al. 2017). After the floods, a series of articles written by engineers and journalists sought to explain these mundane technical details to the public; they argued that the government's accusations of terrorism distracted from the actual culprit—corruption in municipal politics (al-Sharnubi 2015). A damning article written by an engineer, titled "In Terms of Engineering, Why Did Alexandria Flood?" called the 2004 decision to join the distinct sewerage and drainage systems a crime against Alexandria's residents (Mahmood 2015).

The 1917 account from British engineer Carkeet James, as well as the series of articles by engineers and journalists dismissing the 2004 reorganization of Alexandria's sewerage and drainage systems, offers a historically informed narrative that contests the spectacular explanation of a terrorist attack. These explanations show a story of the floods as a slower kind of violence not carried out by a terrorist cell but as an accretion of colonial miscalculation and a failure of postcolonial development that was made worse through decentralizations of vital urban governance and services (Anand 2015).

Conclusion

In this article, I have traced how the Egyptian state narrated the causes of the 2015 Alexandria floods as an act of terrorism to further entrench its repressive rule. By claiming that terrorism was the cause of the floods, the state obfuscated alternative environmental narratives that located blame with the state. In following the production of the state's master discourse, I have also detailed alternative narratives that sought to displace the discourse of terrorism. These alternative narratives show a state that has been neglectful of infrastructural development and unprepared for the shifting storm patterns brought on by climate change. This examination shows how Alexandria's urban infrastructures became key sites in which the state sought to quell criticism and expand security efforts in the name of fighting terrorism. Using the case of Alexandria, I argue that in the current political moment in Egypt and across the globe, the discourse of terrorism and its attendant statist response of security have become important forces in urban environmental governance and infrastructural development.

In outlining the case of the 2015 Alexandria floods, I use insights from geopolitical ecologies of the state and situated urban political ecology. Following the call of geopolitical ecologies, I take seriously the discourse of terrorism in structuring the state response to environmental crisis. I choose not to inherit from geopolitics the problematic state typology of authoritarian, which obfuscates the complex relation between different levels of government, urban infrastructure, and security. In the case of the Alexandria floods, an interpretation of an authoritarian state, which begins from the problematic state—civil society dichotomy, would not account for the contradictory narratives about sewers and security between, for example, the former governor of Alexandria and the Ministry of Interior.

Instead, I follow a situated urban political ecology approach that begins with Southern theories, municipal politics, popular news and social media, and historical sources. In doing so, I use the term repressive to understand the Egyptian state's approach to environmental governance. By this, I mean that the state represses other environmental narratives to become the sole arbiter of the causes and necessary responses to the floods. I suggest that this term offers a more specific way to draw international connections between issues of environmental governance in the current era. Struggles can be linked not through a priori state categories but through specific place-based struggles against repressive environmental state narratives that obfuscate regional and international linkages for fighting the impacts of government neglect and climate change on coastal cities.

Acknowledgments

I thank two anonymous reviewers and James McCarthy for helping to refine this article. I must also thank Yara Sultan for her invaluable research assistance and Danya Al-Saleh, Rachel Boothby, Travis J. De Wolfe, and Rebecca Summer for their thoughtful feedback on drafts of this article.

Funding

This research was supported by a National Science Foundation Graduate Research Fellowship; the Center for Arabic Studies Abroad (CASA); and the Center for Culture, History, and Environment.

ORCID

Mohammed Rafi Arefin ⓘD http://orcid.org/0000-0001-9227-9587

Notes

1. Although many theories of the state assert that states are repressive, I am using the term repressive here as shorthand to describe the state's actions in producing and erasing environmental narratives, especially in times of disasters and crisis.
2. To view a sample of these tweets see https://twitter.com/search?q=غرق_ب-ه-ري_كن_أس&src=typd.
3. See this image that circulated widely at https://twitter.com/DrMahmoudRefaat/status/658368752677998592.
4. The online statement can be viewed at https://www.facebook.com/pg/MoiEgy/photos/?tab=album&album_id =981095145267482.
5. The Ministry of Interior's video can be viewed at https://www.youtube.com/watch?v=57ozD0R6uZ4&feature=youtu.be.

6. This statement can be viewed through the public post at https://www.facebook.com/haytham.elshikh.18/posts/10153708691802505.
7. The Arab African Center for Freedom and Human Rights statement was made available at https://www.facebook.com/Araby.african.center/posts/551062858384359.

References

Ali, M. 2015. Breaking. Egypt's PM accepts resignation of Alexandria's governor El-Mesery. *Ahram Online*, October 25. Accessed December 2, 2017. http://english.ahram.org.eg/NewsContent/1/64/161860/Egypt/Politics-/BREAKING-Egypts-PM-accepts-resignation-of-Alexandr.aspx

al-Sharnubi, N. 2015. The reasons for the flooding of Alexandria and al-Buhira. *Bawbat al-Ahram*, November 10. Accessed April 29, 2018. http://www.ahram.org.eg/NewsPrint/452845.aspx

al-Sisi, A. A. 2015. Presidential address. November 1. [Author's trans.] Accessed December 2, 2017. https://www.youtube.com/watch?v=betbukM75z8

Anand, N. 2015. Accretion. *Cultural Anthropology*, September 24. Accessed December 2, 2017. https://culanth.org/fieldsights/715-accretion.

———. 2017. *Hydraulic city: Water and the infrastructures of citizenship in Mumbai*. Durham, NC: Duke University Press.

Bell, J. 2000. Egyptian environmental activists' uphill battle. *Middle East Report* 216:24–25.

———. 2009. Land disputes, the informal city, and environmental discourse in Cairo. In *Cairo contested: Governance, urban space, and global modernity*, ed. D. Singerman, 349–71. Cairo, Egypt: The American University in Cairo Press.

Bhattacharya, B., C. Zevenbergen, R. A. Wahaab, W. A. I. Elbarki, T. Busker, and C. N. A. Salinas Rodriguez. 2017. Characterisation of flooding in Alexandria in October 2015 and suggested mitigating measures. *EGU General Assembly Conference Abstracts* 19:14230.

Bigger, P., and B. Neimark. 2017. Weaponizing nature: The geopolitical ecology of the U.S. Navy's biofuel program. *Political Geography* 60:13–22.

Carkeet James, C. 1917. *Drainage problems of the East*. Bombay: The Times of India.

Davis, D. K. 2016. *The arid lands: History, power, knowledge*. Cambridge, MA: MIT Press.

Egypt's PM appropriate LE75mn to repair rain drainage system in flooded Alexandria. 2015. *Ahram Online*, October 25. Accessed December 2, 2017. http://english.ahram.org.eg/NewsContent/1/64/161846/Egypt/Politics-/Egypts-PM-appropriates-LEmn-to-repair-rain-drainag.aspx

Fredericks, R. 2013. Disorderly Dakar: The cultural politics of household waste in Senegal's capital city. *The Journal of Modern African Studies* 51 (3):435–58.

———. 2014. Vital infrastructures of trash in Dakar. *Comparative Studies of South Asia, Africa, and the Middle East* 34 (3):532–48.

Gandy, M. 1994. *Recycling and the politics of urban waste*. London and New York: Routledge.

———. 1999. The Paris sewers and the rationalization of urban space. *Transactions of the Institute of British Geographers* 24 (1):23–44.

———. 2006. The bacteriological city and its discontents. *Historical Geography* 34:14–25.

Graham, S., ed. 2004. *Cities, war, and terrorism: Towards an urban geopolitics*. Malden, MA: Wiley-Blackwell.

Hallegatte, S., C. Green, R. J. Nicholls, and J. Corfee-Morlot. 2013. Future flood losses in major coastal cities. *Nature Climate Change* 3 (9):802–06.

Hanieh, A. 2013. *Lineage of revolt*. Chicago: Haymarket.

———. 2016. Beyond authoritarianism: Rethinking Egypt's "Long Revolution." *Development and Change* 47 (5):1171–79.

Harris, L. 2017. Political ecologies of the state: Recent interventions and questions going forward. *Political Geography* 58:90–92.

Heynen, N., M. Kaika, and E. Swyngedouw, eds. 2006. *In the nature of cities: Urban political ecology and the politics of urban metabolism*. London and New York: Routledge.

Lawhon, M., H. Ernstson, and J. Silver. 2014. Provincializing urban political ecology: Towards a situated UPE through African urbanism. *Antipode* 46 (2):497–516.

Mahmood, M. A. 2015. In terms of engineering, why did Alexandria flood? *Masr al-Arabiyya*, November 24. Accessed December 2, 2017. http://www.masralarabia.com/807191/ساحة-الحرية-هندسيا-لماذا-غرقت-الإسكندرية؟

McFarlane, C. 2008. Sanitation in Mumbai's informal settlements: State, "slum," and infrastructure. *Environment and Planning A* 40 (1):88–107.

McFarlane, C., and J. Silver. 2016. The political city: "Seeing sanitation" and making the urban political in Cape Town. *Antipode* 49 (1):125–48.

Meehan, K., and J. L. Rice. 2011. Social natures. In *A companion to social geography*, ed. V. J. Del Casino Jr., M. Thomas, P. Cloke, and R. Panelli, 55–71. Malden, MA: Wiley-Blackwell.

Melosi, M. V. 1993. The place of the city in environmental history. *Environmental History Review* 17 (1):1–23.

———. 2008. *The sanitary city: Environmental services in urban America from colonial times to the present*. Pittsburgh, PA: University of Pittsburgh Press.

Mikhail, A. 2011. *Nature and empire in Ottoman Egypt: An environmental history*. Cambridge, UK: Cambridge University Press.

Ministry of Interior. 2015. Ministry of Interior Official Facebook Page. November 6. https://www.facebook.com/pg/MoiEgy/photos/?tab=album&album_id=981095145267482

Moore, S. A. 2008. The politics of garbage in Oaxaca, Mexico. *Society & Natural Resources* 21 (7):597–610.

———. 2009. The excess of modernity: Garbage politics in Oaxaca, Mexico. *The Professional Geographer* 61 (4):426–37.

———. 2012. Garbage matters: Concepts in new geographies of waste. *Progress in Human Geography* 36 (6):780–99.

Myers, G. 2016. *Urban environments in Africa: A critical analysis of environmental politics*. Chicago: University of Chicago Press.

Nagle, R. 2013. *Picking up: On the streets and behind the trucks with the sanitation workers of New York City.* New York: Farrar, Straus and Giroux.

Nixon, R. 2013. *Slow violence and the environmentalism of the poor.* Cambridge, MA: Harvard University Press.

Ranganathan, M. 2014. Paying for pipes, claiming citizenship: Political agency and water reforms at the urban periphery. *International Journal of Urban and Regional Research* 38 (2):590–608.

———. 2015. Storm drains as assemblages: The political ecology of flood risk in post-colonial Bangalore: Stormwater drains as assemblages. *Antipode* 47 (5):1300–20.

Ribot, J. 2014. Cause and response: Vulnerability and climate in the Anthropocene. *The Journal of Peasant Studies* 41 (5):667–705.

Robbins, P. 2003. Political ecology in political geography. *Political Geography* 22 (6):641–45.

———. 2008. The state in political ecology: A postcard to political geography from the field. In *The Sage handbook of political geography,* ed. K. R. Cox, M. Low, and J. Robinson, 205–18. Los Angeles: Sage.

Smith, N. 2006. There's no such thing as a natural disaster. *Perspectives from the Social Sciences,* June 11. http://understandingkatrina.ssrc.org/Smith/.

Sowers, J. 2012. *Environmental politics in Egypt: Activists, experts and the state.* London and New York: Routledge.

Squires, G., and C. Hartman, eds. 2006. *There is no such thing as a natural disaster: Race, class, and Hurricane Katrina.* London and New York: Routledge.

Swyngedouw, E., and N. Heynen. 2003. Urban political ecology, justice and the politics of scale. *Antipode* 35 (5):898–918.

We are drowning in neglect … the government is failing. 2015. *Al-Youm Al-Sabaa,* October 25. Accessed December 2 2017. https://www.youm7.com/story/2015/10/25/2407410/-الإهمال-فى-بنغرق-إحنا-السابع-اليوم حكومة-فاشلة

MOHAMMED RAFI AREFIN is an Assistant Professor/Faculty Fellow supported by the Provost's Postdoctoral Fellowship in the Gallatin School of Individualized Study at New York University, New York, NY 10003. E-mail: mra349@nyu.edu. His research interests include the politics of waste and sanitation, urban geography, and political economy.

Sequestering a River: The Political Ecology of the "Dead" Ergene River and Neoliberal Urbanization in Today's Turkey

Eda Acara (iD)

This article explores the neoliberal authoritarian transformation of Turkey's water sector since 2000 by examining the policies surrounding the Ergene River, a dead river that runs through Turkey's European province of Thrace. Within the context of the accelerating neoliberalization of water resources in Thrace for the benefit of the Istanbul region at the expense of severe environmental pollution, the result was a combination of authoritarian policy prerogatives and priorities that rest on organized irresponsibility and a politics of nongovernance regarding environmental protection. As this article demonstrates, contemporary authoritarian neoliberalism in Turkey has created gray zones of authority involving many public authorities with varying and sometimes overlapping mandates, within which blatant breaches of the law became akin to the metropolitan municipal governance of distant water resources.

本文通过检视围绕着额尔古纳河这条流经土耳其位于欧洲的色雷斯州的死河之政策, 探讨土耳其水资源部门自2000年以来的新自由主义威权转变。在加速色雷斯的水资源新自由主义化以嘉惠伊斯坦堡区域、并以严重的环境污染为代价的脉络中, 该结果是以组织化的不负责任为基础的威权政策特权与优先权和有关环境保护的非治理政治之组合。如同本文所证实, 土耳其当代的威权新自由主义, 已创造出涉及诸多公共职权的灰色权力地带, 并有着各种且有时相互重叠的命令, 其中公然违反法律, 近乎成为大都会市政府有关远距水资源的治理。 关键词: 额尔古纳河, 大都会市政体制, 新自由主义城市化, 河流污染, 色雷斯区域, 土耳其。

Este artículo explora la transformación autoritaria neoliberal del sector del agua de Turquía a partir del 2000 examinando las políticas relacionadas con el Río Ergene, una corriente muerta que fluye a través de la provincia europea de Tracia, en Turquía. En el contexto de una acelerada neoliberalización de los recursos hídricos de Tracia para beneficio de la región de Estambul, a expensas de severa contaminación ambiental, el resultado fue una combinación de prerrogativas políticas autoritarias y prioridades que descansan sobre la irresponsabilidad organizada y una política de desgobierno en lo que concierne a la protección ambiental. Como se demuestra en este artículo, el neoliberalismo autoritario contemporáneo en Turquía ha creado zonas grises de autoridad que involucran a muchas autoridades públicas, con mandatos variados y a veces traslapados, dentro de los cuales las descaradas burlas a la ley se asemejan a la gobernanza municipal metropolitana aplicada a los distantes recursos del agua. *Palabras clave: municipalidad metropolitana, polución fluvial, regímenes, región de Tracia, Río Ergene, Turquía, urbanización neoliberal.*

Accumulation through the privatization and commodification of water resources has been a major concern of the geographical literature on neoliberalization (Swyngedouw 1999; Heynen, Kaika, and Swyngedouw 2006; Castree 2008; Bakker 2010; Linton 2010; Harris 2013; Harris and Işlar 2014). Several of these works challenge and criticize these policies of privatization, commercialization, or liberalization of water resources for enabling the global dissemination of public–private partnerships or control by multinational companies over the developing world's water resources (Bakker 2010). This global strategy to manage water privately started in the early twentieth century with attacks on what Bakker (2010) referred to as the Municipal Hydraulic Paradigm, condemning municipal mismanagement of public water to legitimize a privatization agenda.

Despite the policy-related core positionality of municipalities in the privatization, commercialization, or liberalization of natural resources and the increasing local effects of state rescaling that directly influence municipal interventions in natural resources, particularly water resources, there is little discussion of the effects in the current literature. Even when they are mentioned, it is within a context of misconduct and catastrophe.[1] Thus, more research is needed to understand how municipal intervention in water sources, whether in sewage elimination, irrigation, or energy production, affects the hydrological cycles of rivers (Swyngedouw 1999; Kaika 2005).

Turkey is an important case of authoritarian rule with increasingly predatory use and abuse of natural resources for urban growth. There has been a growing literature on the Adalet ve Kalkınma Partisi's (Justice and Development Party [AKP hereafter]) authoritarian rule in Turkey since 2002, predominantly between the Gezi park protests in 2013 and the militarization of politics after a state of emergency was declared following an attempted coup in 2016, focusing on the authoritarianization of neoliberalism in Turkey (Öniş 2015; Özbudun 2015; Esen and Gumuscu 2016; Somer 2016; Akbulut, Adaman, and Arsel 2017; Kuyucu 2017; Akçay 2018; Ongur 2018). According to Yalman and Bedirhanoğlu (2010), the AKP has consistently reproduced neoliberal authoritarianism since the liberalization of capital accounts in 1989. The contemporary repressive period is different, however, due to the AKP's reintegration of Islam with Turkish neoliberalism (Yalman and Bedirhanoğlu 2010). Other studies have explored this difference to explain the successful hegemony of the AKP's neoliberal authoritarianism. First, the AKP has expanded selective welfare that specifically targets ethnic minorities in metropolitan cities (Yörük 2012; Akçay 2018). Second, it has overseen an increase in debt-based consumption, predominantly in the housing market (Kuyucu 2017; Türem 2017; Akçay 2018). Considering that neoliberalism is "a program to establish unlimited dominance of capital on all societies and on the world system" (Boratav 2016, 1), in Turkey's case this program has been implemented by bolstering the urban economy.

This article examines how sequestering of water as a by-product of neoliberal urbanization has become an increasingly integral part of the authoritarian regime in contemporary Turkey. The fact that urbanization processes are shaped by economic crisis and the accumulation of capital under neoliberalism is well documented. (Brenner 2004, 2009; Harvey 2014; Bayırbağ and Penpecioğlu 2015). The fundamental aim here is documenting and analyzing the processes through which rivers are urbanized at different institutional levels and how policies of abandoning large, polluted bodies of water have continued as a strategy to advance one-party economic and political control of these large cities.

This article first explores the neoliberal and authoritarian transformation of the water sector in Turkey after 2000 by examining state policies regarding the pollution of the Ergene River, a "dead river" that runs through Turkey's European province of Thrace, bordering on Bulgaria and Greece. Two intersecting macropolitical processes converged in this period: On the one hand, it was a time when European accession processes accelerated, widely affecting Turkey's economic, ecological, and political development, with the AKP assuming preeminent power in state institutions. The result was a combination of policy prerogatives and priorities that can be described as "organized irresponsibility" (Prudham 2004). The chaos on the ground in relation to Thrace's municipal interaction with this acutely polluted and officially "dead" river illustrate the degree to which the abandonment of natural resources to environmental destruction are symptomatic of neoliberal urbanization in Turkey and how such policies have furthered the AKP's authoritarian control over urban populations and their ecology. The data sources for the article consist of an analysis of national and regional plans and interviews with experts from various municipalities and water-related institutions in Turkey.

Regional Water Use at the Expense of Istanbul

Production of new water flows includes geographical projects where the terrain is redrawn or transformed (Swyngedouw 1999; Castree and Braun 2001; Watts and Peet 2004). This generates capital accumulation within which rivers flow to the center of accumulation (Swyngedouw 1999). As they flow, however, they create their own social and political relations in addition to power positions and struggles over the governance of newly formed urban terrains and water management. Thus, urbanization and modernization are key to the production of water flows (Swyngedouw 1999; Kaika 2005; Heynen, Kaika, and Swyngedouw 2006) and affect how the state "sees"; that is, how it calculates and builds scientific knowledge and management strategies for water (Scott 1998). The management of regional water use in the Thrace riverscape is best understood through its dependence on the Istanbul metropolitan region's water use and governance.

In this section, I illustrate how the Turkish state sees Thrace's regional water flows in relation to the Istanbul metropolitan region's needs by providing the historical, political, and economic background regarding the expansion of Istanbul city toward Thrace. This is also the background within which the politics of nongovernance were realized in matters of river pollution. I argue that the decentralization of Istanbul's industry, which accelerated in the 1990s, positioned and reproduced Thrace as a region of resource extraction for Istanbul's urbanization needs, which rest on seeing the Ergene River as a water source for the decentralized industry. Thus, "infrastructural power," which Mann (1993) defined as the "institutional capacity of a central state to penetrate its territories and logistically implement decisions" (59), was built on the production of Thrace as a resource region based on its water capacities. In the contemporary context, however, this infrastructural power has extended into the production of drinking water for Istanbul from regional water sources other than the Ergene River in addition to using Thracian waterscapes for urban rents, earned through large housing projects.

Constituting a drainage area of $10,730 \, km^2$, the Ergene River is the region's main surface water source for irrigation, industrial production, and household use. The primary sources of contamination are household and industrial waste, which are directly discharged without any treatment (Çevre ve Orman Bakanlığı [Ministry of Environment and Forestry] 2003, 2006, 2010; Trakya Üniversitesi [Trakya University] 2007). The proportion of contamination from household waste is greater than that of industrial discharge, although industrial waste discharge has longer lasting effects than household wastewater (Muluk et al. 2013). Thrace has six organized industrial zones, one European Free Trade zone, and eighteen small industrial sites,

accommodating approximately 2,500 industrial facilities (Trakya Kalkınma Ajansı [Development Agency of Thrace] 2010). These figures only show legal, registered industrial ownership, however, and it is reported that 70 percent of industrial activity in Thrace occurs outside of these legally established organized zones (Inci 2010). The Ergene River basin constitutes 54 percent of Thrace, where industrial and agricultural production coexist, and pollution at multiple sites along the river affects locations far from the original contamination source. With the increasing effects of climate change, decreasing forest coverage due to Istanbul's construction boom, and growing demands for water to be transferred from Thrace, the Ergene River basin is now considered a water-scarce region (Kantarcı 2004).

Until the 1980 coup d'état, a majority of Turkey's industrial activities, primarily state enterprises, were located in Anatolian cities like Ankara, Antalya, Adana, and Izmir (Quataert 1994; Şenses and Taymaz 2003). With the emergence of economic liberalization and the adoption of structural adjustment programs, however, followed by Turkey's membership negotiations with the European Union during the 1990s, Istanbul became central for Middle Eastern and Eastern European countries for the service, finance, and trade sectors (Keyder 1999). The restructuring of Istanbul's industry in Thrace accelerated dramatically in the early 1990s (Mortan et al. 2003). As early as the 1960s, industry started to be decentralized to make urban centers industry free (Sönmez 1996; Erbil 2017).[2] This spatial policy, aiming to decentralize Istanbul's industry, accelerated during the 1990s due to regional investment incentives, specifically for export-oriented trade (Atiyas and Bakis 2015). In parallel, rising land prices inside Istanbul, cheap labor, and available natural resources in Thrace were the most important reasons why large capital investment migrated to Thrace and Marmara (Sönmez 1996). As a result, Tekirdağ, a county in Thrace, became a significant site for industrial relocation alongside Kocaeli, which is located on the southeastern edge of Istanbul's Anatolian side (Trakya Kalkınma Ajansı [Development Agency of Thrace] 2010). According to a survey conducted in Çorlu, one of Tekirdağ's main industrial towns, half of its companies were started after 1990 (Sazak 2000), and total investment incentives were highest between 1990 and 1999 (Güler and Turan 2013).

During the AKP era, the most significant intervention in the region's hydrosocial cycles has been in the production of drinking water for Istanbul. There are currently fifteen drinking water sources to meet the drinking water needs of Istanbul, with three in Thrace and three in Kocaeli and Sakarya on Istanbul's southeastern periphery. With a high concentration of industry, Kocaeli and Sakarya, much like Çorlu and Çerkezköy in Thrace, have suffered pollution and deterioration in public health.[3] These water sources account for 57 percent of the total water transfer from surrounding regions to Istanbul, with 14 percent coming from various sources in Thrace. Istanbul's total water dependency on these regions rose from 47 percent to 57 percent between 2009 and 2016.[4]

These water transfers to Istanbul have been criticized by both professional and nongovernmental organizations for their adverse effects on the communities from which the transfers occur (İlhan et al. 2014). For example, in 2014, excessive water transfers from Pabuçdere and Kazandere reservoirs caused fisheries in Vize and Kırklareli counties in Thrace to suffer income losses.[5] Seven small reservoir dams built on Istranca streams in İğneada, which is a protected river basin under the Ramsar Convention on Wetlands of International Importance especially as Waterfowl Habitat, have provoked anger and protest since 2010.

Recent interventions in the hydrological cycle have also involved the production of water geographies as a source of urban rent on the outskirts of the Istanbul metropolitan area (Akbulut and Bartu Candan 2014). The most significant example concerns urban development around drinking water basins on the city's peripheries, particularly a so-called megaproject, Kanal Istanbul, which involves the construction of an artificial water channel in the middle of these drinking water basins (Paker 2017). Meanwhile, tourism has also been prioritized in regional economic targets, specifically aiming to develop Thrace's seaside (Trakya Kalkınma Ajansı [Development Agency of Thrace] 2010). Finally, construction of a third Bosporus bridge, an airport in the north of Istanbul, and another bridge to link Thrace to Çanakkale have caused dramatic increases in housing and land prices, specifically along Thrace's coast.[6]

The changes in how the Turkish state sees and acts regarding water since 2000 have only been possible through the reorganization of water-related institutions by the central government and a rescaling of water-related organizations at the local level (municipalities are part of the equation) to make them follow central government orders. The next section explains this new networked authoritarianism (MacKinnon 2011; Moore 2014), which largely rests on politics of nongovernance and organized irresponsibility. These two elements form the basis of state coercion strategies in regional policymaking.

What a State Doesn't See Is What a State Sees: Toward a Networked Authoritarianism

Institutional changes in the early twenty-first century have aggravated and redefined existing politics of nongovernance in the realm of water quality governance. The politics of nongovernance have been used as a central and local government strategy to contain popular unrest concerning water-related policies. This concept of nongovernance is inspired by Prudham's (2004) work on organized irresponsibility as an organizing principle of neoliberalism across different regulations, institutions, and among various other stakeholders within the privatized water sector. In examining the water pollution tragedy in the Walkerton, Ontario, water poisoning incident,[7] for example, Prudham (2004) argued that neoliberalism uses accidents and environmental catastrophes to seed organized irresponsibility into its regulatory systems, and thus to create chronic (not accidental) crisis through which nature is commodified (Prudham 2004; Harris 2013; Harris, Goldin, and Sneddon 2013; Ercan and Oğuz 2015). I slightly modify the concept here as an integral part of what I term the politics of nongovernance regarding pollution of the Ergene River, which has a thirty-year history that the Turkish state has long acknowledged. I argue that, in the 1990s, the politics of nongovernance was shaped by the hegemony of the irrigation sector because of the state seeing water as a raw material for rural development. During the 2000s, institutional transformation, led by the programs of international financial institutions and Turkey's European Union negotiations reterritorialized the state regarding water-related sectors, which resulted in the Turkish state seeing water as an investment opportunity for "green development" (Kalkınma

Bakanlığı [Ministry of Development] 2013, 137). This was overseen by the central government within a networked system using authoritarian strategies.

Protection of the environment has long been seen as hindering economic growth and socioeconomic development in Turkey. There have been frequent warnings that decentralizing Istanbul's industry would be detrimental to environmental and human health since an early report by the Organization for Economic Co-Operation and Development (OECD 1968). Indeed, the history of the water sector's five-year national development plans until the Fourth Five-Year Plan (1979–1983) predicted the problems, further underlined by one of my expert participants, who said that "during the second and third five year plans, water was envisioned to be abundant ... drinking water was the first priority need ... within the context of irrigation, meeting the water demands were seen to be important."[8] This emphasis strengthened the budgets and planning power of irrigation-related institutions over others, such as the General Directorate of State Hydraulic Works (Devlet Su İşleri [DSI]; Kibaroğlu, Baskan, and Alp 2009). The World Bank's irrigation-targeted financial incentives during the 1990s were useful in closing budget deficits in DSI, which shrank considerably until the 2000s (Akıllı 2012). Later studies of Ergene River pollution, however, including regional and national plans after 1990, warned that the wastewater used in irrigation threatened regional and national health.[910]

The DSI's financial reorganization created uneven geographies of power across water-related institutions, such as that between the former Directorate of Rural Services (Köy Hizmetleri Müdürlüğü) and the DSI. According to the same expert mentioned earlier, who worked on irrigation projects for more than twenty years, "The irrigation investments almost hindered other environmental investments ... so much so that even inside the State Planning Organization, we [irrigation department] became estranged from the environment department." This expert experience indicates the agricultural growth ideology in rural areas despite other environmental concerns. This act of prioritizing a growth-focused ideology and related institutional power arithmetic already signaled a politics of nongovernance in the 1990s.

The authority crisis is a pandemic in the politics of nongovernance. Monitoring water pollution is still regulated by law and controlled overall by state agencies, such as the Ministry of Environment and Urban Planning, the Ministry of Forestry and Water Affairs,[11] the Ministry of Health, the Ministry of Internal Affairs and district governments, in addition to, for example, local environmental task forces of the National Gendarmerie (Yasamis 2007). This crowded list of institutions is often presented as a barrier to monitoring water quality and use in national plans (Kibaroğlu and Baskan 2011; Orhan and Scheumann 2011; Kalkınma Bakanlığı [Ministry of Development] 2014). Given the serious authority crisis of several central state institutions, it is difficult to determine which institutions are responsible for the death of a river. The restructuring of 2011 further transformed the chain of power in environmental and urban affairs and the investment limits of institutions like State Hydraulic Affairs, which cascaded more excessive credit borrowing and outsourcing of operating costs onto private firms, municipalities, or both (T. Cinar 2009; Akıllı 2012; Harris and Işlar 2014; Gülen n.d.).

In the case of the Ergene River, the construction of municipal treatment plants was outsourced to construction firms. In the Tenth Five-Year National Development Plan (2014–2018), municipalities of mid- and small-sized cities were encouraged to form consortiums to increase their attractiveness for companies to build the plants (Trakya Kalkınma Ajansı [Development Agency of Thrace] 2010; Kalkınma Bakanlığı [Ministry of Development] 2013). Construction in Thrace is still underway, but based on comparative case studies on Antalya and Izmit (T. Çınar 2009; Harris and Işlar 2014), it seems that water tariffs might rise in the future. The fact that seven wastewater treatment plants were contracted to the same company, namely, Türkerler Holding,[12] also signals monopolization.

There is an authority crisis mixed with arbitrariness over which governmental institution prepares the Ergene River's basin management plans. Although preparation of the basin management plans is a legal duty of the Ministry of Environment and Urban Planning (Çevre ve Şehircilik Bakanlığı, MoEUP), the Ergene Basin Management Plan was arbitrarily prepared by the Ministry of Forestry and Water Affairs (Orman ve Su İşleri Bakanlığı, MoFWA) because of a former prime minister's demands (see also Yılmaz 2012).[13] For one expert participant of this research, this institutional arbitrariness was an act of censorship to conceal the severity of pollution in the region.[14]

Censorship is a widespread institutional symptom of the politics of nongovernance within the DSI. During my in-depth interviews, many of the experts wanted to go off-record or refused to speak to me. I was also unable to get water data or any other environmental measurement data from the DSI because of information leakages in 2012 from the DSI in numerous court cases involving hydroelectric power plants.[15] Similar censorship pressures were present in MoEUP and MoFWA regarding Ergene River pollution, so much so that it was handled as if it were "a terror problem."[16] This definition was also significant as MoFWA gave the Ministry's Ergene River Basin management and pollution control plan a militaristic name: "Operation Dawn" (Şafak harekati; "DSI Works" 2014).

The politics of nongovernance is a recognized strategy of the (neo)liberal state. In the case of Thrace, nongovernance before 2000 was deliberately crafted as governmental rationality so that economic development, involving both industrial and agricultural production, would not be slowed. The authority crisis and policy emphasis over irrigation projects were two state tactics of nongovernance regarding environmental protection in the 1990s. In the 2000s, environmental concerns were made visible by the state as they were rationalized and commodified. In the next section, I focus on the relationship between administrative domination and the politics of nongovernance within the context of metropolitan and district municipalities.

Organized Irresponsibility in Networked Authoritarian Neoliberalism

Cultivating neoliberal hegemony across an urban scale is a global phenomenon (Brenner 2004; Brenner, Peck, and Theodore 2010; Davies 2014). The AKP's power specifically rests on harnessing the powers of the local through an urban neoliberal economic and political system that includes the local natural resources, land, people, and state. The literature on AKP-led neoliberalism underlines its strong cultivation of urban entrepreneurialism, which saw the urban poor and their land as a fundamental part of urban growth. Neoliberal confiscation of land for mass housing has been instrumental in transforming the urban core and expanding the peripheries of metropolitan areas (Türkün 2011; Yüksel 2011;

Kuyucu 2014). Containment of the urban poor was made possible by "the new welfare state," with more systematic social welfare policies (Yörük 2012).[17] Administrative domination not only targets the urban poor and their land but is also a core activity of the local state, including but not limited to policing, performance management, and micromanagement of public services (Davies 2014).

So far in the urban policy literature on Turkey, the administrative domination of the central government has been often documented in the urban regeneration policy (A. K. Çınar and Penpecioğlu 2018; Kuyucu 2018). A very recent study in this regard further underlines that the administrative domination might result in transforming long-lasting voting patterns by constructing a hegemony built on construction and infrastructure (A. K. Çınar and Penpecioğlu 2018). Thus, it is important to explore the policy mechanisms that shape the conflictual relationship between the central and local governments. In the case of Ergene River, a politics of nongovernance is evidently one of these policy mechanisms.

From his research on urban development projects in Istanbul, Kuyucu (2014) found that legal ambiguities regarding squatter settlements have been key to real estate market creation. It is possible to extend this argument throughout the AKP government era into the growing regionalization of several metropolitan municipalities with greater jurisdictional powers (see also Bartu Candan and Kolluoğlu 2008; Demirtaş-Milz 2013; Alkan 2015). The regional conflict between Thracian municipalities (Kırklareli, Tekirdağ, and Edirne) and Istanbul Metropolitan Municipality shows that preparation of a regional development plan works as a tactical urban expansion technology to control, appropriate, and calculate land—in other words, to reterritorialize the power of administrative domination in the subregional extremities of the Istanbul city region. The fact that the main opposition party, Republican People's Party, represents Thracian municipalities might be another reason for the administrative domination. In a different context, Kuyucu (2018) underlined a similar administrative domination of the central government that ultimately shapes the urban regeneration policies in large metropolitan cities. These examples illustrate a reallocation of local power in a highly top-down network from central to local governments while articulating arbitrary decision making and policy appointment in the process.

The first basin-wide environmental plan for the Ergene River was prepared in 2004 by the University of Thrace, commissioned by Thrace Development Association (Trakya Kalkınma Birliği, TRAKAB). TRAKAB was run by the presidents of three municipalities and members of the provincial administration. Several experts who conducted research about pollution at a regional level played active roles in the plan. Yet, despite being legally binding, it was canceled due to the institutional and legal changes in 2010 that allocated regional planning privileges to Istanbul Metropolitan Municipality.[18] The municipality and its company, Istanbul Metropolitan Company (IMP), produced a new plan that, in its early days of preparation, had more than 200 professors working in various Istanbul universities. They prepared a highly controversial Istanbul city plan, which created a public outcry against the location of a new airport and a bridge across an ecologically conserved forest area. After the company's press release supporting the protests, the planning mission and the company were dissolved.[19] IMP's plan[20] was eventually legally canceled in 2012 as the court declared that the plan stood against "the fundamentals of planning, urbanism principles and public good" (Dağlar 2014) due to incorrect scaling of land use.

The planning conflict between the Istanbul Metropolitan Municipality and the rest of Thrace's regional communities, including its municipalities, shows that metropolitan municipalities become a technology of central government to territorialize its administrative domination over the region. The fact that MoEUP granted metropolitan planning powers to the Metropolitan Municipality's company despite agreeing on a previous protocol with a different planning group further illustrates centralized decentralization occurring at the regional scale.

Conclusion

Institutional authoritarianism within the water sector, which articulates contemporary urban contradictions in Turkey, rests on a politics of nongovernance with at least two uses. The first is that the politics of nongovernance is a strategy to maintain state power across shifting strategic priorities in the water sector. This has transformed the responsibilities and duty definitions of state institutions within Turkey's water sector and caused opposition within and outside of state institutions. Examples of this argument are the DSI's decreasing influence over irrigation policy and changes in ministerial design of water governance that centralize decision making with the prime minister and higher levels of the ministries regarding limits to water investment. This institutional authoritarianism no doubt has been strengthened by the current state of emergency and a culture of statutory decree, which are beyond the scope of this article.

Second, the politics of nongovernance is a policy tactic to normalize authoritarian and arbitrary decision-making processes in central government, with a major role taken by municipalities as the implementers of such decisions. The creation and continued implementation of the regional plan prepared by the Istanbul Metropolitan Municipality despite a court order creates legal ambiguity. This leaves the defense of the region's specific land and extraction of natural resources, the most important of which is water, to separate court cases opened by civil society and universities. They obviously lack the capacities and resources of the central and metropolitan municipal state, however. This behavior by the state suggests that its politics of nongovernance is an institutional and governance tactic of metropolitan and central municipal rule to establish administrative domination.

Acknowledgments

The author thanks Audrey Kobayashi, Ariel Salzmann at Queen's University, Elif Ekin Akşit Vural and Hasan Vural at Ankara University, and the anonymous reviewers for reading and commenting on this article. The empirical study in this article is part of the author's dissertation. It is dedicated to the memory of the author's father, Altan Acara (1930–2014).

Funding

Funding for this research was provided by the Social Sciences and Humanities Research Council of Canada Doctoral Fellowship (Award No. 752-2012-1485), Queen's University Dean's Travel Fund, and American Association of Geographers Thesis Dissertation Award.

ORCID

Eda Acara ⓘD http://orcid.org/0000-0002-2918-2624

Notes

1. The Walkerton and Flint catastrophes are among such examples.
2. For an earlier account of industrial policies in Turkey, see Sönmez (1996).
3. For increasing cancer due to air pollution in Dilovası of Kocaeli, see Hamzaoglu et al. (2014).
4. For more on the estimates, see Istanbul Büyükşehir Belediyesi (Istanbul Metropolitan Municipality 2009, 2015).
5. For a parliamentary question regarding the matter, see Türkiye Büyük Millet Meclisi (Grand National Assembly of Turkey 2007).
6. Thrace is among the regions with rapidly increasing housing prices since 2016 (Şat Sezgin and Aşarkaya 2017) in addition to having some of the most rapidly increasing housing prices compared to rents (TCMB [Central Bank of Turkey] 2017).
7. The drinking water was infected by *Escherichia coli*, affecting 2,300 residents of Walkerton (Prudham 2004).
8. She was a retired expert who, at the time of the interview, was employed by the Ministry of Development to transfer her knowledge to incoming personnel.
9. For reports about the extreme pollution in the Ergene River during the 1990s, see Artüz (1990). Reports after 2000 include TBMM Araştırma Komisyonu (Research Commission of the Grand National Assembly 2003); Çevre ve Orman Bakanlığı (Ministry of Environment and Forestry 2003, 2006, 2010); and Olgun and Çobanoğlu (2012). See the study by Ekmekyapar, Karabulut, and Meriç Pagano (2011) in Tekirdağ, Thrace, about contamination of drinking water wells by industrial zones; see Yorulmaz et al. (2012) for a cancer-scan research study.
10. Water from the Ergene River is still used in some parts of the region (Kılıç 2011). From my limited observations, the upper waters of the river remain an irrigation source in Uzunköprü.
11. Çevre Yönetimi Müdürlükleri (The General Directorate for Environmental Management), Çevre Etki Değerlendirme ve Planlama Genel Müdürlüğü (General Directorate for Environmental Impact Assessment and Planning), and Çevre ve Orman Il Genel Müdürlükleri (the Provincial Directorates of Environment and Forestry) or private consultant firms designated by MoEUP are among the local water pollution monitoring systems under the mandate of MoEUP and the MoFWA ("Regulation on monitoring water pollution" 2004; "Regulation on environmental inspection" 2008).
12. See Türkerler (2017).
13. In the Tenth National Development Plan (2014–2018), management of the Ergene River basin was specifically cited to be an exemplary project of opportunity for environmental investment (Kalkınma Bakanlığı [Ministry of Development] 2013).
14. This was an off-record interview with an expert who, at the time of the interview, was working at a government institution.
15. This was also off-record information. Some officers in the ministries and DSI responsible for leaking documents were later reappointed to distant branches of the institution.
16. This was an expression used by one of the experts at the water-related government institution.
17. The AKP's social welfare policies, based on family and social solidarity networks, continues to be a major issue of inquiry. Critical work in this area evaluates these social policies as political technologies to contain growing urban informality and dispossession. For more, see Atasoy (2009), Yılmaz (2013, 2015), Doğan and Yılmaz (2011), and Demirtaş-Milz (2013). See Bayırbağ (2013) for the rise of the charity state at the hands of the municipalities.
18. For outcries, see Gökçen (2010) and İtez (2016).
19. Three of the former and two of the current municipality employees emphasized that there was a direct order from the prime minister.
20. The plan is known as the Revisioned Environmental Plan of the Ergene River Basin in Thracian Sub-Region, presented at a scale of 1/100,000. This plan continues to be a polarizing issue that causes conflict between various planning parties. For a recent example of such conflict, see Türkmen (2017).

References

Akbulut, B., F. Adaman, and M. Arsel, eds. 2017. *Neoliberal Turkey and its discontents: Economic policy and the environment in the justice and development party era.* London and New York: I. B. Tauris.

Akbulut, B., and A. Bartu Candan. 2014. İki Ağacın Ötesinde: Istanbul'a Politik Ekoloji Çerçevesinden Bakmak [Looking beyond two trees: Exploring Istanbul by political ecology]. In *Yeni İstanbul çalışmaları*, ed. A. Bartu Candan and C. Ozbay, 282–303. Istanbul, Turkey: Metis Yayınları.

Akçay, Ü. 2018. *Neoliberal populism in Turkey and its crisis.* Berlin: Institute for International Political Economy. Accessed April 26, 2018. http://www.ipe-berlin.org/fileadmin/downloads/working_paper/IPE_WP_100.pdf.

Akıllı, H. 2012. Türkiye'de su yönetiminin değişen yüzü: Devlet su işleri genel müdürlüğü [The changing face of water governance in Turkey: General directorate of the state water affairs]. *Memleket Siyaset Yönetim* 7 (18):55–85. Accessed April 26, 2018. http://www.msydergi.com/uploads/dergi/47.pdf.

Alkan, A. 2015. Türkiye'nin Yeni Metropoliten Rejimi: Otoriteryen Kentleşmenin "Yerel Yönetimlerde Yeniden Yapılanma" Formu [The new metropolitan regime in Turkey: The restructured shape of authoritarian urbanization in the local governments]. *Ayrıntı Dergi* 10. Accessed April 26, 2018. http://ayrintidergi.com.tr/turkiyenin-yeni-metropoliten-rejimi-otoriteryen-kentlesmenin-yerel-yonetimlerde-yeniden-yapilanma-formu/.

Artüz, I. 1990. *Tekirdağ yöresinde çevre sorunlarının boyutları* [Environmental problems of Tekirdağ]. Istanbul, Turkey: TÜSES Vakfı Çevre Sorunları Çalışma Grubu.

Atasoy, Y. 2009. *Islam's marriage with neoliberalism, state transformation in Turkey.* New York: Palgrave Macmillan.

Atiyas, I., and O. Bakis. 2015. Structural change and industrial policy in Turkey. *Emerging Markets Finance and Trade* 51 (6):1209–29. doi:10.1080/1540496X.2015.1080523.

Bakker, K. 2010. *Privatizing water, governance failure and the world's urban water crisis.* Ithaca, NY: Cornell University Press.

Bartu Candan, A., and B. Kolluoğlu. 2008. Emerging spaces of neoliberalism: A gated town and a public housing project in Istanbul. *New Perspectives on Turkey* 39:5–46. doi:10.1017/S0896634600005057.

Bayırbağ, M. K. 2013. Continuity and change in public policy: Redistribution, exclusion and state rescaling in Turkey. *International Journal of Urban and Regional Research* 37 (4):1123–46. doi:10.1111/1468-2427.12000.

Bayırbağ, M. K., and M. Penpecioğlu. 2015. Urban crisis: Limits to governance of alienation. *Urban Studies* 54 (9): 2056–71. doi:10.1177/0042098015617079.

Boratav, K. 2016. The Turkish bourgeoisie under neoliberalism. *Research and Policy on Turkey* 1 (1):1–10. doi:10.1080/23760818.2015.1099778.

Brenner, N. 2004. *New state spaces, urban governance and rescaling of statehood.* Oxford, UK: Oxford University Press.

———. 2009. A thousand leaves: Notes on the geographies of uneven spatial development. In *Leviathan undone? Towards a political economy of scale*, ed. R. Mahon and R. Keil, 27–51. Vancouver, Canada: UBC Press.

Brenner, N., J. Peck, and N. Theodore. 2010. Variegated neoliberalization: Geographies, modalities, pathways. *Global Networks* 10 (2):182–222. doi:10.1111/j.1471-0374.2009.00277.x.

Castree, N. 2008. Neoliberalising nature: The logics of deregulation and reregulation. *Environment and Planning A* 40 (1):131–52. doi:10.1068/a3999.

Castree, N., and B. Braun, eds. 2001. *Social nature: Theory, practice and politics.* Oxford, UK: Blackwell.

Çevre denetimi yönetmeliği [Regulation on environmental inspection]. 2008. *Official Gazette*, November 21. Accessed January 31, 2018. http://www.resmigazete.gov.tr/eskiler/2008/11/20081121-4.htm.

Çevre ve Orman Bakanlığı [Ministry of Environment and Forestry]. 2003. *Ergene Havzası Çevre Düzeni Planı* [Environmental plan of Ergene Basin]. Ankara, Turkey: Çevre ve Orman Bakanlığı.

———. 2006. *Ergene Havzası Çevre Yönetimi Master Planı (2006-2008)* [The environmental management master plan of the Ergene Basin]. Ankara, Turkey: Çevre ve Orman Bakanlığı.

———. 2010. *Meriç-Ergene Havzası Endüstriyel Atıksu Yönetimi Ana Plan Çalışması Final Raporu* [The final report of the industrial wastewater management plan in the Meric-Ergene Basin]. Ankara, Turkey: Çevre ve Orman Bakanlığı.

Çınar, A. K., and M. Penpecioğlu. 2018. Kentsel dönüşüm ve yerel siyasetin değişen dinamikleri: 2009 ve 2014 İzmir yerel seçim sonuçları bağlamında bir araştırma [Urban transformation and changing dynamics of local politics: Analyzing the results of the 2009 and 2014 local elections in Izmir]. *Planlama* 28 (1):56–75. doi:10.14744/planlama.2017.92485.

Çınar, T. 2009. Privatisation of urban water and sewerage services in Turkey: Some trends. *Development in Practice* 19 (3):350–64. doi:10.1080/09614520902808076.

Dağlar, A. 2014. Trakya'da bir çevre zaferi daha [Another environmental victory in Thrace]. *Hurriyet*, September 6. Accessed September 19, 2018. http://www.hurriyet.com.tr/gundem/trakya-da-bir-cevre-zaferi-daha-2657402.

Davies, J. 2014. Rethinking urban power and the local state: Hegemony, domination and resistance in neoliberal cities. *Urban Studies* 51 (15):3215–32. doi:10.1177/0042098013505158.

Demirtaş-Milz, N. 2013. The regime of informality in neoliberal times in Turkey: The case of the Kadifekale urban transformation project. *International Journal of Urban and Regional Research* 37 (2):689–714. doi:10.1111/1468-2427.12005.

Doğan, A. E., and B. Yılmaz. 2011. Ethnicity, social tensions and production of space in forced migration neighbourhoods of Mersin: Comparing the case of the Demirtaş neighbourhood with newly established ones. *Journal of Balkan and Near Eastern Studies* 13 (4):475–94. doi:10.1080/19448953.2011.623871.

DSI Çalışıyor Ergene güzelleşiyor [DSI works, Ergene gets prettier]. 2014. Ankara, Turkey: Devlet Su Işleri. Accessed June 23, 2018. http://www.dsi.gov.tr/haberler/2014/04/16/dsicalisiyorergeneguzellesiyor

Ekmekyapar, F., A. Karabulut, and S. Meriç Pagano. 2011. Çorlu-Çerkezköy Çevresinde Yeraltı Suyu Seviyelerinin ve Su Kalitesinin Değerlendirilmesi [Evaluation of water quality and quantity of the underground water around Çorlu and Çerkezköy]. In *Kıyı bölgelerinde çevre kirliliği ve kontrolü sempozyumu*, ed. S. Meriç Pagano, E. Güneş, F. Ekmekyapar, F. Uysal, G. Yıldız Töre, Ş. Ordu, T. Tunçal, et al., 703–11. Tekirdağ, Turkey: Cem Davetiyeleri Matbaacılık.

Erbil, T. 2017. Planning dilemmas in deindustrialization process in Istanbul. *ITUA/Z Journal of Faculty of Architecture* 14 (2):43–56. doi:10.5505/itujfa.2017.28291.

Ercan, F., and Ş. Oğuz. 2015. From Gezi resistance to Soma massacre: Capital accumulation and class struggle in Turkey. *Socialist Register* 51:114–35. Accessed June 24, 2018. https://socialistregister.com/index.php/srv/article/view/22097.

Esen, B., and S. Gumuscu. 2016. Rising competitive authoritarianism in Turkey. *Third World Quarterly* 37 (9):1581–1606. doi:10.1080/01436597.2015.1135732.

Gökçen, E. 2010. TRAKAB gerçeğini, ilk taslağın fikir babasından dinledik [We listened to the truth about TRAKAB from firsthand]. *Gazete Trakya*, March 12. Accessed June 16, 2018. http://www.gazetetrakya.com/HaberTRAKAB_gercegini_ilk_taslagin_fikir_babasindan_dinledik-289288.gazetetrakya.

Gülen, F. n.d. *Özelleştirme, Yap-İşlet-Devret, Kamu Özel Sektör İşbirliği* [Privatization, BOT, public–private partnership]. Ankara, Turkey: YAYED. Accessed April 25, 2018. http://www.yayed.org/id81-incelemeler/ozellestirme-yap-islet-devret-kamu-ozel-sektor-isbirligi.php.

Güler, M., and M. Turan. 2013. Trakya bölgesi'nin kentleşmesinde sanayileşme ve demografi ilişkisi [The relationship between industrialization and demography in the context of urbanisation in the Thrace region]. *Çağdaş Yerel Yönetimler* 22 (2):17–43. Accessed April 25, 2018. http://www.todaie.edu.tr/resimler/ekler/bbc8b1044350714_ek.pdf?dergi=Cagdas%20Yerel%20Yonetimler%20Dergisi.

Hamzaoglu, O., M. Yavuz, G. Turker, and H. Savli. 2014. Air pollution and heavy metal concentration in colostrum and meconium in two different districts of an industrial city: A preliminary report. *International Medical Journal* 21 (1):77–82. doi:10.1155/2012/490647.

Harris, L. M. 2013. Framing the debate on water marketization. In *Contemporary water governance in the Global South: Scarcity, marketization and participation*, ed. L. M. Harris, J. A. Goldin, and C. Sneddon, 111–18. London and New York: Routledge.

Harris, L. M., J. A. Goldin, and C. Sneddon. 2013. Placing hegemony: Water governance concepts and their discontents. In *Contemporary water governance in the Global South: Scarcity, marketization and participation*, ed. L. M. Harris, J. A. Goldin, and C. Sneddon, 251–59. London and New York: Routledge.

Harris, L. M., and M. Işlar. 2014. Neoliberalism, nature and changing modalities of environmental governance in contemporary Turkey. In *Global economic crisis and the politics of diversity*, ed. Y. Atasoy, 52–78. New York: Palgrave.

Harvey, D. 2014. *Seventeen contradictions and the end of capitalism*. Oxford, UK: Oxford University Press.

Heynen, N. C., M. Kaika, III, and E. Swyngedouw. 2006. *In the nature of cities*. London and New York: Routledge.

İlhan, A., D. Yıldız, F. Z. Tokaç, M. L. Kurnaz, and M. Türkeş. 2014. *İstanbul'un su krizi ve kolektif çözüm önerileri* [Water crisis in Turkey and collective solutions]. İstanbul, Turkey: Sosyal Değişim Derneği. Accessed June 24, 2018. https://www.suhakki.org/wp-content/uploads/2015/03/istanbuldasukrizi-arastirma.pdf.

Inci, O., ed. 2010. *Trakya (Istanbulun isgaline) direniyor* [Thrace region resists against (the invasion of) Istanbul]. İstanbul, Turkey: Cumhuriyet Kitaplari.

İstanbul Büyükşehir Belediyesi [Istanbul Metropolitan Municipality]. 2009. *İstanbul Büyükşehir Belediyesi 2009 Faaliyet Raporu* [Istanbul Metropolitan Municipality annual report in 2009]. Istanbul, Turkey: İstanbul Büyükşehir Belediyesi. Accessed June 24, 2018. http://www.ibb.gov.tr/tr-TR/BilgiHizmetleri/Yayinlar/Documents/FaaliyetRaporu/2009/1.pdf.

———. 2015. İSKİ 2015 Faaliyet Raporu [Istanbul Water and Sewerage Administration's annual report of 2015]. Istanbul, Turkey: İstanbul Büyükşehir Belediyesi. Accessed June 16, 2016. http://www.ibb.gov.tr/tr-TR/BilgiHizmetleri/Yayinlar/FaaliyetRaporlari/Documents/2015/ibb_faaliyetraporu2015.pdf.

İtez, O. 2016. İçinde yaşayanların çok fark etmediği ama dışardan bakanların olağanüstü bulduğu bir yer Edirne [Edirne, as place of wonder for outsiders]. Accessed October 2017. http://www.arkitera.com/soylesi/839/namik-kemal-doleneken-soylesi.

Kaika, M. 2005. *City of flows*. London and New York: Routledge.

Kalkınma Bakanlığı [Ministry of Development]. 2013. *Onuncu Kalkınma Planı (2014–2018)* [Tenth development plan (2014–2018)]. Ankara, Turkey: Kalkınma Bakanlığı.

———. 2014. *Su Kaynakları Yönetimi ve Güvenliği Özel İhtisas Komisyonu Raporu* [Special Commission report on water resource management and security]. Ankara, Turkey: Kalkınma Bakanlığı.

Kantarcı, D. 2004. Ergene Nehri Havzasının Özellikleri, Su Üretimi ve Sulu Tarımın Geliştirilmesi için Çareler [Features of the Ergene River basin, solutions with regard to production of water and developing irrigated agriculture]. Paper presented at Trakya Su Platformu, Çorlu, Tekirdağ, Turkey, December.

Keyder, C. 1999. The setting. In *Istanbul, between the global and the local*, ed. C. Keyder, 3–31. New York: Rowman & Littlefield.

Kibaroğlu, A., and A. Baskan. 2011. Turkey's water policy framework. In *Turkey's water policy, national frameworks and international cooperation*, ed. A. Kibaroğlu, A. Kramer, and W. Scheumann, 3–27. London: Springer.

Kibaroğlu, A., A. Baskan, and S. Alp. 2009. Neoliberal transition in hydropower and irrigation water management in Turkey: Main actors and opposition groups. In *Water policy entrepreneurs: A research companion to water transitions around the globe*, ed. D. Huitema and S. Meijerink, 287–304. Cheltenham, UK: Edward Elgar.

Kılıç, A. 2011. Ergene Ölüyor Öldürüyor Radikal [Ergene dies and kills]. Accessed April 25, 2018. http://www.radikal.com.tr/turkiye/ergene_oluyor_olduruyor-1059916.

Kuyucu, T. 2014. Law, property and ambiguity: The uses and abuses of legal ambiguity in remaking Istanbul's informal settlements. *International Journal of Urban and Regional Research* 38 (2):609–27. doi:10.1111/1468-2427.12026.

———. 2017. Two crises, two trajectories: The impact of the 2001 and 2008 economic crises on urban governance in Turkey. In *Neoliberal Turkey and its discontents: Economic policy and the environment in the justice and development party era*, ed. B. Akbulut, F. Adaman, and M. Arsel, 44–74. London: I. B. Tauris.

————. 2018. Politics of urban regeneration in Turkey: Possibilities and limits of municipal regeneration initiatives in a highly centralized country. *Urban Geography* 1–25. OnlineFirst. doi:10.1080/02723638.2018.1440125.

Linton, J. 2010. *What is water? The history of a modern abstraction.* Vancouver, Canada: UBC Press.

MacKinnon, R. 2011. China's "networked authoritarianism. *Journal of Democracy* 22 (2):32–46. doi:10.1353/jod.2011.0033.

Mann, M. 1993. *The sources of social power. Vol. II: The rise of classes and nation-states, 1760–1914.* Cambridge, UK: Cambridge University Press.

Moore, S. M. 2014. Modernisaton, authoritarianism and the environment: The politics of China's South-North water transfer project. *Environmental Politics* 23 (6):947–64. doi:10.1080/09644016.2014.943544.

Mortan, K., N. H. Ozgen, M. Ozkan, and B. Tekin. 2003. *Lüleburgaz icin Kent Stratejisi* [Urban strategy of Luleburgaz]. Lüleburgaz, Turkey: Lüleburgaz Belediye Başkanlığı..

Muluk, Ç. B., B. Kurt, A. Turak, A. Türker, M. A. Çalışkan, O. Balkız, S. Gümrükçü, G. Sarıgül, and U. Zeydanlı. 2013. *Türkiye'de Suyun Durumu ve Su yönetiminde yeni yaklaşımlar: Çevresel Perspektif* [An environmental perspective: Waters of Turkey and new perspectives on water management]. Istanbul, Turkey: Iş Dünyası ve Sürdürülebilir Kalkınma Derneği- Doğa Koruma Merkezi.. Accessed June 21, 2018. http://www.dkm.org.tr/Dosyalar/YayinDosya_RnF27jIq.pdf.

Olgun, E., and N. Çobanoğlu. 2012. Türkiye su politikalarının biyoetik değerlendirilmesi: Ergene nehri örneği [Evaluation of Turkish water policies according to bioethical approach: Example of Ergene River]. *Ankara Üniversitesi Sosyal Bilimler Dergisi* 3 (2):139–56. doi:10.1501/sbeder_0000000049

Ongur, H. O. 2018. Plus ça change … re-articulating authoritarianism in the new Turkey. *Critical Sociology* 44 (1):45–59. doi:10.1177/0896920516630799.

Öniş, Z. 2015. Monopolising the centre: The AKP and the uncertain path of Turkish democracy. *The International Spectator* 50 (2):22–41. doi:10.1080/03932729.2015.1015335.

Organization for Economic Co-Operation and Development (OECD). 1968. *Economic and social development plan for Eastern Thrace (Turkey).* Paris: OECD.

Orhan, G., and W. Scheumann. 2011. Turkey's policy for combating water pollution. In *Turkey's water policy: National frameworks and international cooperation*, ed. A. Kibaroglu, A. Kramer, and W. Scheumann, 117–39. Berlin: Springer.

Özbudun, E. 2015. Turkey's judiciary and the drift toward competitive authoritarianism. *The International Spectator* 50 (2):42–55. doi:10.1080/03932729.2015.1020651.

Paker, H. 2017. The "politics of serving" and neoliberal developmentalism: The megaprojects of the AKP as tools of hegemony building. In *Neoliberal Turkey and its discontents: Economic policy and the environment in the justice and development party era*, ed. B. Akbulut, F. Adaman, and M. Arsel, 103–19. London: I. B. Tauris.

Prudham, S. 2004. Poisoning the well: Neoliberalism and the contamination of municipal water in Walkerton, Ontario. *Geoforum* 35 (3):343–59. doi:10.1016/j.geoforum.2003.08.010.

Quataert, D. 1994. *Manufacturing in the Ottoman Empire and Turkey, 1500–1950.* Schoharie, NY: State University of New York Press.

Şat Sezgin, A. G., and A. Aşarkaya. 2017. İnşaat Sektörü [Construction sector]. Iş Bankası, Istanbul, Turkey. Accessed June 20, 2018. https://ekonomi.isbank.com.tr/UserFiles/pdf/sr201702_insaatsektoru.pdf.

Sazak, Ş. 2000. Istanbul Sanayinin Desentralizasyonu ve Bunun Trakya Bölgesine Etkisinin Çorlu-Büyükkarıştıran Sanayi Alanında Değerlendirilmesi [Decentralization of Istanbul's industry and evaluation of the effects of this decentralization on the industrial areas of Çorlu and Büyükkarıştıran in Thrace Region]. In *Trakya'da Sanayileşme ve Çevre Sempozyumu III*, 45–58. Edirne, Turkey: TMMOB Makine Mühendisleri Odası.

Scott, J. 1998. *Seeing like a state: How certain schemes to improve the human condition have failed.* New Haven, CT: Yale University Press.

Şenses, F., and E. Taymaz. 2003. Unutulan Bir Toplumsal Araç: Sanayileşme Ne Oluyor? Ne Olmalı? [A forgotten social vehicle: What happens to industrialization?]. In *Iktisadi kalkınma, kriz ve istikrar*, ed. A. K. Kose, F. Şenses, and E. Yeldan, 429–63. Istanbul, Turkey: Iletişim Yayınları.

Somer, M. 2016. Understanding Turkey's democratic breakdown: Old vs. new and indigenous vs. global authoritarianism. *Southeast European and Black Sea Studies* 16 (4):481–503. doi:10.1080/14683857.2016.1246548.

Sönmez, M. 1996. *İstanbul'un iki yüzü* [Two faces of Istanbul]. Ankara, Turkey: Arkadaş Yayınları.

Su kirliliği kontrol yönetmeliği [Regulation on monitoring water pollution]. 2004. *Official Gazette*, December 31. Accessed January 25, 2018. http://mevzuat.basbakanlik.gov.tr/Metin.Aspx?MevzuatKod¹/47.5.7221&source XmlSearch¹/4&MevzuatIliski ¹/40.

Swyngedouw, E. 1999. Modernity and hybridity: Nature, regeneracionismo, and the production of the Spanish waterscape, 1890–1930. *Annals of the Association of American Geographers* 89 (3):443–65. doi:10.1111/0004-5608.00157.

TBMM [Grand National Assembly of Turkey]. 2007. *Tansel Barış'ın 7/128 Esas no'lu Yazılı Soru Onergesi* [Parliamentary Question by Tansel Barış, by Indictment No. 7/128]. Ankara, Turkey: TBMM. Accessed June 12, 2016. https://www.tbmm.gov.tr/tutanaklar/TUTANAK/TBMM/d23/c003/b010/tbmm230030100111.pdf.

TBMM Araştırma Komisyonu [Research Commission of the Grand National Assembly]. 2003. *TBMM 22.Dönem Araştırma Komisyonu Ergene Hazasına Ait Sorunlar ve Çözüm Onerileri* [Problems of Ergene River and their solutions by Twenty Second Research Commission of the Grand National Assembly]. Ankara, Turkey: TBMM.

TCMB [Central Bank of Turkey]. 2017. *Konut Fiyat Endeksi, Yıl Sonu Değerlendirme Raporu* [Annual assessment report on housing price index]. Istanbul, Turkey: TCMB. Accessed April 21, 2018. http://www.tcmb.gov.tr/wps/wcm/connect/988442bd-18b3-4760-8492-df559ea391e5/Y%C4%B1l+Sonu+De%C4%9Ferlendirme+Raporu-2017.pdf?MOD=AJPERES&CACHEID=ROOTWORKSPACE-988442bd-18b3-4760-8492-df559ea391e5-m8KBbpj.

Trakya Kalkınma Ajansı [Development Agency of Thrace]. 2010. *T21 Trakya Bölge Planı* [T21 regional plan of Thrace]. Tekirdağ, Turkey: Trakya Kalkınma Ajansı. Accessed April 25, 2017. http://www.trakyaka.org.tr/default.asp?page=content&page_id =144.

Trakya Universitesi [Trakya University]. 2007. *Trakya Universitesi Ergene havzası çevre düzeni planı cilt 2* [Trakya University's environmenal plan of the Ergene basin, volume 2]. Edirne, Turkey: Trakya Üniversitesi Rektörlüğü Yayınları.

Türem, U. 2017. The state of property: From the empire to the neoliberal republic. In *Neoliberal Turkey and its discontents: Economic policy and the environment in the justice and development party era*, ed. B. Akbulut, F. Adaman, and M. Arsel, 18–43. London: I. B. Tauris.

Türkerler. 2017. Ergene Atıksu Arıtma Tesisleri [Ergene wastewater treatment facilities]. Accessed June 25, 2018. http://www.turkerler.com/alt-yapi/ergene-atiksu-aritma-tesisleri/.

Türkmen, Ç. 2017. Trakya'yı bir prostat profesörüne teslim etmedik' [We haven't left Thrace region to a prostate professor]. *Görünüm*, November 14. Accessed November 15, 2017. http://www.gorunumgazetesi.com.tr/haber/46877/trakyay-bir-prostat-profesrne-teslim-etmedik.html.

Türkün, A. 2011. Urban regeneration and hegemonic power relationships. *International Planning Studies* 16 (1):61–72. doi:10.1080/13563475.2011.552473.

Watts, M., and R. Peet. 2004. *Liberation ecologies: Environment, development, social movements.* London and New York: Routledge.

Yalman, G., and P. Bedirhanoğlu. 2010. State, class and the discourse: Reflections on the neoliberal transformation in Turkey. In *Economic transitions to neoliberalism in middle-income countries: Policy dilemmas, economic crises, forms of resistance*, ed. A. Saad-Filho and G. Yalman. 107–28. New York and London: Routledge.

Yasamis, F. D. 2007. Assessment of compliance performance of environmental regulations of industries in Tuzla (Istanbul, Turkey). *Environmental Management* 39 (4):575–86. doi:10.1007/s00267-003-0129-8.

Yılmaz, T. 2012. Veysel bey ergene'nin hali ne böyle? [Mr. Veysel, what is the matter with Ergene?]. *Hürriyet*, July 8. Accessed January 31, 2015. http://www.hurriyet.com.tr/gundem/20938382.asp.

———. 2013. AKP ve devlet hayırseverliği [AKP and the state philanthropy]. *Toplum Ve Bilim* 128:32–70.

———. 2015. AKP ve yoksulluk: Paternalizm, muhtaçlaştırma ve yeni tahakküm stratejileri [AKP and poverty: Paternalism, deprivation and the new strategies of domination]. *Ayrıntı Dergisi* 10:58–66. Accessed April 25, 2018. http://ayrintidergi.com.tr/akp-ve-yoksulluk-paternalizm-muhtaclastirma-ve-yeni-tahakkum-stratejileri/.

Yörük, E. 2012. Welfare provision as political containment: Politics of social assistance and the Kurdish conflict in Turkey. *Politics & Society* 40 (4):517–47. doi:10.1177/0032329212461130.

Yorulmaz, F., E. Berberoğlu, E. Seçgin Sayhan, M. Eskiocak, G. Varol Saraçoğlu, and C. B. Demirkan. 2012. Endüstri Yoğun Bölgede Yaşayanlarda Ya da Birinci Derecede Yakınlarında Kanser Bildirenlerin Çevresel Risk Etmenlerine Göre Değerlendirilmesi: Çorlu Örneği [Evaluation of individuals or individuals with first degree family members, living in the industry intensive sites: The case of Çorlu]. Paper presented at 15. Ulusal Halk Sağlığı Kongresi, Uludağ Universitesi, Bursa, Turkey, October.

Yüksel, A. S. 2011. Rescaled localities and redefined class relations: Neoliberal experience in south-east Turkey. *Journal of Balkan and Near Eastern Studies* 13 (4):433–55. doi:10.1080/19448953.2011.621788.

EDA ACARA is an Assistant Professor in the Department of Geography at Bakircay University, İzmir, 35665. Turkey. E-mail: edacara@gmail.com; eda.acara@queensu.ca. She is also an affiliated Researcher and Lecturer in the Urban Policy Planning and Local Governments program at Middle East Technical University. She carried out much of the research for this article while completing her PhD in the Department of Geography and Planning at Queen's University. Her research interests include the effects of neoliberalization processes across the rural–urban nexus.

"Return the Lake to the People": Populist Political Rhetoric and the Fate of a Resource Frontier in the Philippines

Kristian Saguin

In this article, I examine the shifting political ecologies of governance of Laguna Lake, Philippines, in the context of historical and contemporary populist political rhetoric. Rodrigo Duterte, who was elected president in 2016 through a platform of change, brought national attention again to the lake by promising to give it back to the people marginalized by decades-long elite capture. This populist rhetoric is the latest in attempts to manage an urban resource frontier with conflicting demands and uses. By narrating a history of governance of Laguna Lake, I trace parallels between current and past strategies of addressing resource conflicts: from Ferdinand Marcos's authoritarian rule in the 1970s and 1980s and the pluralist modes that followed to Duterte's law-and-order vision of development. By comparing the populist narratives of Marcos and Duterte, I demonstrate that populist rhetoric in authoritarian forms entails the contradictory processes of politicization of the problem and depoliticization of solutions. Authoritarian populist narratives transform the framing of environmental problems through antagonistic politics even as solutions are constrained within existing depoliticized technologies of government that limit the spaces of contestations.

我于本文中检视菲律宾在历史与当代民粹主义政治修辞脉络中, 治理拉古纳湖的转变中的政治生态。罗德里戈. 杜特尔特于2016年通过改变的平台获选为总统, 并通过承诺将数十年来由菁英所佔领的湖归还给受到边缘化的人民, 重新将全国焦点聚焦于该湖。此一民粹主义修辞, 是管理充满冲突的需求与使用的城市资源前沿的最新尝试。我通过叙述治理拉古纳湖的历史, 追溯当代与过往应对资源冲突的策略之间的相似性: 从1970年代与1980年代费迪南. 马科斯的威权统治, 到随着杜特尔特的"法律与秩序"之发展愿景兴起的民粹主义模式。通过比较马科斯与杜特尔特的民粹叙事, 我证实威权形式中的民粹修辞, 引发了该问题的政治化与解决方案的去政治化之矛盾过程。威权民粹叙事通过对抗政治, 改变了环境问题的框架方式, 即便解决方案被限缩于限制争夺空间的既有去政治化治理技术之中。关键词: 威权, 杜特尔特, 拉古纳湖, 马科斯, 民粹主义。

En este artículo examino las cambiantes políticas ecológicas de gobernanza del Lago Laguna, Filipinas, en el contexto retórico de la política populista histórica y contemporánea. Rodrigo Duterte, quien fue elegido presidente en 2016 con base en una plataforma de cambio, atrajo de nuevo la atención nacional hacia el lago prometiendo devolverlo a la gente marginada luego de la captura de la élite, prolongada durante décadas. Esta retórica populista es el más reciente intento de manejar una frontera de recursos urbanos dentro de demandas y usos en conflicto. Narrando una historia de gobernanza del Lago Laguna, hago paralelos entre las estrategias actuales y pasadas para considerar conflictos por recursos: del férreo autoritarismo de Ferdinando Marcos en los años 1970 y 1980 y los modos pluralistas que siguieron, hasta llegar a la visión del desarrollo dentro de la ley y el orden de Duterte. Comparando las narrativas populistas de Marcos y Duterte, demuestro que la retórica populista en formas autoritarias implica los procesos contradictorios de politización del problema y despolitización de las soluciones. Las narrativas populistas autoritarias transforman el marco de los problemas ambientales por medio de políticas antagónicas aun cuando las soluciones están constreñidas dentro de las actuales tecnologías despolitizadas de gobierno que limitan los espacios de las controversias. Palabras clave: el autoritario, Duterte, Laguna Lake, Marcos, populismo.

What is the place of populism and narratives of the people in authoritarian and pluralist modes of governing contentious resources?

How does populist political rhetoric attempt to justify authoritarian and technocratic rule in resource conflict? In this article, I take the problem of

Laguna Lake, the largest in the Philippines and a significant urban resource frontier, to examine how populist narratives have unfolded in more than four decades of state intervention in resource conflicts. Comparing the authoritarian populist political rhetoric of the former president Ferdinand Marcos and the current president Rodrigo Duterte, I find a contradiction: Populist narratives about resource conflicts entail a dual moment of politicization of the problem and depoliticization of solutions that appear to challenge yet maintain existing modes of governance. While politicizing antagonistic relations between the people and its enemy, authoritarian populist narratives also limit the possibilities of political engagement through a recourse to technocratic, depoliticized solutions.

Populism in its authoritarian form is not new in the spaces of governance in the Philippines and in the postcolonial Global South. Its reemergence amid a supposed postpolitical condition of neoliberal, pluralist, consensus-based environmental politics, however, necessitates empirical work sensitive to the multiple historical and contemporary forms of populism. I contribute to this task by investigating the politics of resource governance of Laguna Lake.

Decades of resource conflict in Laguna Lake between small-scale fisherfolk and large-scale fish pen owners have captured the national imagination as an example of elite-driven social justice problem that the state had routinely failed to address. Through the fish pen technology, the state introduced aquaculture in the lake to improve incomes of small-scale fisherfolk and increase fish productivity. Elite capture by middle-class entrepreneurs and fishing corporations from outside the lake, however, led to the unchecked proliferation of large-scale fish pen structures that displaced fisherfolk from their traditional fishing grounds and resulted in recurring conflict. In many ways, Laguna Lake's aquaculture problem has symbolized many of the country's developmental ills and has subsequently attracted exercises in authoritarian interventions.

Seeking to depart radically from the status quo, both Marcos and Duterte deployed populist rhetoric by coupling narratives of conflict and social justice with authoritarian techniques of management. Marcos, the late strongman who was president for twenty years, sought to solve the lake's aquaculture problem by employing a rhetoric that was pro-poor and anti-elite. Yet, his interventions were restricted to the authoritarian developmentalist modes of efficient planning, anticonflict compromise, and the rule of law, which legitimized capitalist aquaculture in the lake.

The promise of change and the specter of a return to authoritarian rule accompanied Duterte's election in 2016. Despite lacking a coherent vision for the environment, Laguna Lake became a testing ground for his law-and-order developmentalism not very different from Marcos. Duterte, in his first State of the Nation address, instructed state agencies to review permits granted to Laguna Lake aquaculture, punctuating a procedural administrative order with the populist language of protecting the country, giving the lake back to the marginalized, and promising fisherfolk "priority in its entitlements" (Duterte 2016). Yet, his government's subsequent actions and plans mirror past strategies that constrict possible socioecological futures for fisherfolk.

Drawing from an analysis of secondary sources and other published materials on Laguna Lake during the Marcos and Duterte presidencies, I probe the contradictions of authoritarian populist political rhetoric where politicization of the problem through antagonism is paired with depoliticized modes of state governance. I begin with a brief discussion of debates linking populism and the political and then describe Laguna Lake resource controversies and the response of Marcos's authoritarian developmentalist regime. Next, I discuss the reemergence of populist narratives in Duterte's revival of the lake's antagonistic relations. Last, I reflect on both moments to illustrate the place of populist political rhetoric in environmental governance and its implications for the future of resource production and relations in the lake.

Populist Politics in Resource Governance

Political ecologists have documented in rich historical–geographical detail the multiple evolving forms that environmental governance has taken, in the process posing questions about what constitutes the state, the political, and the people (Bridge 2014; Robertson 2015). In Latin America, for example, political–ecological work on neo-extractivism and hydrocarbon governance have contributed to these debates by tracing the links between narratives of the people, resources, and the nation (Perreault and Valdivia 2010; Kohl and Farthing 2012). Revisiting the state, the political, and the people becomes even more relevant with the recent global resurgence of populist politics in an era where environmental issues have taken on a postpolitical character.

Swyngedouw (2009) perhaps provided the most explicit theoretical engagement of populism and contemporary environmental politics in geography, arguing how populist tactics are "the symptomatic expression of a post-political condition" (611). Postpolitics is characterized by the replacement of the political—conflict, dissent, and contestations—with the politics of consensus and technocratic management, wherein contentious problems are reduced to matters of policy technologies and administration (Wilson and Swyngedouw 2014). Symptomatic of a liberal democratic order, postpolitics narrows down possibilities and downplays conflict. Good governance discourses via compromise, consensus, inclusion, and participation among various stakeholders have been widely deployed in place of the political. Because much postpolitical literature deploys a highly specific definition of the truly political, the analytical implications and usefulness of such restriction has also been contested by some critics (McCarthy 2013; Beveridge and Koch 2017).

For Swyngedouw (2009), populist imaginaries contribute to construct the postpolitical condition characterized by depoliticized governance and the evacuation of the political in resolving environmental problems. If the political is necessarily antagonistic (Mouffe 2005), his diagnosis, however, misses a crucial element of tactics to deploy populist narratives: They entail moments of politicization as much as depoliticization. Populism relies on antagonistic politics in identifying a problem, which might pave the path for the particular, often depoliticized solutions familiar to scholars of the postpolitical. In challenging the status quo, populist politics builds on a necessary conflict between the people and its other (Panizza 2005), which is assigned blame for (environmental) problems. In authoritarian forms of populism, this is accompanied by a narrowing of the possibilities of how such matters might be addressed. Identifying the enemy and taking action against the enemy lends strength to the power of populist narratives. This dual moment of politicization and depoliticization serves multiple purposes, including the hegemonic legitimation and construction of a political project, a regime, or the state, such as in authoritarian populism (Scoones et al. 2018). Authoritarian populism challenges the dominant pluralist modes of governance by politicizing the

problem, yet it mirrors the latter's strategy of depoliticizing the spaces of engagement.

Defining populism and identifying its place in politics and democracy continue to be the subject of debate in political theory and beyond (Laclau 2005; Ranciere 2016). I find Panizza's (2005) nonessentialist approach to the term useful for this article, though, because it views populism as a "mode of identification" rather than an inherent characteristic possessed by individuals or social movements. Antagonism is the mode of identification central to populist strategies. It constructs both the people to whom the populist speaks (Ranciere 2016) and the other or the political frontier that needs to be defeated for the people (Laclau 2005; Panizza 2005). Both the people and the enemy are empty signifiers, in that they may take various discursive forms. Populist narratives often pose the need to undermine existing social order and disrupt status quo by vanquishing the enemy that oppresses the people and by giving control back to the underdogs. They invoke a "promise of plenitude" once these antagonisms have been resolved (Panizza 2005).

In its authoritarian guise, populism negates politics by creating a shared vision of people and leader working together toward one solution (Panizza 2005), which parallels the postpolitical mode of anticonflict, consensus-based, and inclusive pluralist governance in nonauthoritarian contexts. Populism, however, simultaneously politicizes by elevating antagonistic relations. This duality is crucial in the narrative power and the eventual outcomes of populist politics in different historical contexts. In the Global South with authoritarian postcolonial histories, we need to ask, for example, how a resurgent populism intersects with, undermines, or supports existing structures transformed by adoption of postpolitical modes of governance. The interplay of politicization and depoliticization therefore deserves further place-based empirical investigation to understand how populist narratives shape the contemporary landscape of environmental governance.

Marcos's Populist Political Rhetoric: Authoritarian Developmentalism and Managing Resource Controversies

Laguna Lake has presented the state with a host of problems rooted in its complex historical production as an urban resource frontier. Since 1966, the lake has served as a site for pioneering, foreign-loan-funded projects aimed at harnessing it as a resource—from improving fish production through aquaculture to serving as sink for urban stormwaters and a source of domestic water. The lake has been a site of several governance innovations—state regulation, community-based projects, and hybrid private–public partnerships, among others—to manage such multiple conflicting demands (Saguin 2016). These often came with a developmental and urban justification, reflecting both its location and connections with nearby Metro Manila and its place in the history of twentieth-century high modern, grand plans of controlling nature (Saguin 2017). The modern visions of harnessing the lake as a resource reached a peak under Marcos's infrastructure-driven authoritarian developmentalism in the 1970s and 1980s.

The history of Laguna Lake fisheries presents a classic political ecological case of the state attempting to manage the contradictions of resource frontier production. Aquaculture became an important governance problem for the state because of the rapid, unregulated proliferation of fish pen structures and the social unrest that followed. Initially introduced to improve incomes of poorer fisherfolk communities, it quickly became an elite venture, as knowledge and financial capital of Manila-based entrepreneurs and fishing corporations enabled them to take over fish production in the lake.

Owing to the high profitability and cheap appropriation of the lake's ecological surpluses in the 1970s, several urban middle-class groups (politicians, celebrities, foreigners) also invested in aquaculture operations, contributing to the fish pen rush that by the early 1980s had occupied close to a third of the 90,000-ha lake space. This rush has been met with resentment by lake fisherfolk, whose fishing grounds have been severely reduced. Fisherfolk developed antagonistic relations that have escalated to violent encounters with armed fish pen guards employed to watch over what have now become highly valuable fish. Fisherfolk have continually framed the problem as a social justice issue, claiming prime legitimacy of use and advocating their right to continue to freely make a living off the lake (Saguin 2016).

What came to be known as the fish pen controversy took off in the national imagination after the deaths of fisherfolk and fish pen workers in 1982 and 1983. The state had long attempted to regulate fish pens through demolitions of unlicensed structures outside zoning belts, but institutional confusion between the lake management body and local government units and the strong political clout and connections of fish pen owners and associations have made these piecemeal efforts routinely ineffective.

Plans for Laguna Lake's production as an urban resource frontier preceded Marcos's imposition of martial law in 1972, which paved the way for centralized, authoritarian rule for more than a decade. The lake, however, became an arena for his visions of development, playing a role in his desire to create a "New Society" out of the destruction of the old political and social order. With the support of technocrats, business cronies, and the military, he ruled through constitutional authoritarianism (Noble 1986; Teehankee 2017) and continually deployed elements of populist discourse to legitimate his regime. Appealing to the poor and the people's sense of being marginalized by the ruling elites, he sought to correct structures that caused persistent inequality and hindered development by targeting two enemy groups: the communists and the rural landlord elites (Anderson 1988). The strong antipathy toward elites might have little to do with a genuine concern for the plight of the poor and more with his desire to rule over the established oligarchs and consolidate local political and economic elite power at the national level (Anderson 1988; Teehankee 2017; Bello 2018).

Through his technocrats, he crafted an economic strategy that relied on securing foreign borrowing to fund large-scale, grand infrastructure projects, including several Laguna Lake development projects such as a hydraulic control structure, flood control structures, a fish pen development project, and a cooperative development project. More than his predecessors, Marcos espoused a form of populist nationalism (Webb 2017) that promised change for the nonelites long marginalized and indignified by those in power. This intersected with a developmental regime characterized by technocratic solutions to social problems, effectively cultivating an image of a strongman with political will and efficient planning tools empathetic to the plight of the poor. It must be pointed out, however, that technocracy under the Marcos authoritarian regime was peopled by elites with multiple, sometimes competing, economic and political interests (Tadem 2013).

The fate of the lake during the martial law years articulated Marcos's mixing of populist nationalism and technocratic developmentalism in managing environmental problems. When the fish pen controversy exploded, Marcos intervened through a series of Letters of Instructions, ordering various agencies to regulate fish pen sprawl. He simultaneously secured foreign loans to fund a national cooperative development project designed to enable displaced fisherfolk access to the fish pen technology. Marcos sought "to democratize the benefits derived from the lake by providing the marginal users of the same opportunities to own, manage and operate fishpens" (Marcos 1984) to ensure "the participation of lakeshore dwellers ... in the development of the lake" (Marcos 1983b). To the elites, he was quoted as saying that they would be asked to share the pens with the small fishermen: "This is social justice in action" (Ng 1983, 16).

Although ringing with the high politics of long-overdue redistributive justice, this strategy effectively flattened the highly unequal relations between fish pen owners and fisherfolk in the lake. Rather than a radical reordering of lake social relations and spaces of politics, the selective demolitions and cooperative projects ended up legitimizing elite fish pen presence in the lake as long as they remained within the zoning belts, while building fisherfolk capacity to compete side by side these privately owned pens.

For Marcos and his propagandists, a peaceful coexistence and a "happy compromise" could be reached between the two antagonistic parties through proper planning and management without undermining the ability of fish pens to make money off the lake (Samonte 1983). Then–First Lady Imelda Marcos, after meeting with both sides, concluded that "with better planning and technology, both pen operators and fishermen could make a living off the lake without conflict" ("FL Vows" 1983, 1). What began as a political project that highlighted the antagonism as the root cause of lake problems ended up as a compromise between the two parties, mediated by technologies of government to avoid conflict: characteristics of depoliticized governance. Quelling rural unrest and communist influence over the lake provided immediate justification for the Marcos regime to contain the conflict. As part of his authoritarian developmentalist model, the need for compromise and efficient planning was a reaction to "the increasing pressure of urban growth and development" that needed to be "responsive to the demands of the

various beneficial users thereof" (Marcos 1983a). This marked the beginning of a governance shift from the sectorally specific fisheries regulation to a multiuse lake resource management.

Through the dual strategy of proper zoning and demolition of unruly and illegal fish pens while redistributing freed-up space for fisherfolk to engage in aquaculture, Marcos was able to strike a solution that was both populist and anti-elite but without undermining the status quo of capitalist-driven aquaculture. Addressing the fish pen controversy became politicized through a mode of identification that antagonized the elites and the people while bringing in notions of social justice, both of which aligned with his vision of the New Society. Fish pen governance became immediately depoliticized, however, through a recourse to the technocratic solutions of adhering to strict zoning of use arranged according to neat, geometric belts based on a computed carrying capacity (*rationalization*) complemented by the simultaneous redistribution of demolished fish pen sites and building of financial capacity of fisherfolk (*democratization*). The two anodyne terms of rationalization and democratization—limiting the fish pens to legal limits while attempting to give fisherfolk a chance at the fish pen technology—would recur throughout the next three decades of lake management.

Subsequent administrations would adopt similar technocratic governance mechanisms to lake problems, downplaying the antagonism between pens and fisherfolk. Through a discourse of good governance, the post-Marcos administrations would use the lake as an experimental site for innovations—increasingly turning to neoliberal modes of addressing the environmental problems of the lake—to maintain the multiple resource use of the lake. Fisherfolk interests needed to complement a broad set of other user needs: the narrative shifting to that of integrated management of various stakeholders at the scale of the watershed. The role of a strong state armed with a discourse of returning the lake to the dispossessed fisherfolk would emerge again in the wake of Duterte's election as president.

Duterte's Populist Political Rhetoric: Returning the Lake to the People and the Promise of Plenitude

Duterte shares with Marcos an affinity for populist rhetoric and authoritarian tendencies. The thirty-year interim between Marcos's forced exile that ended his dictatorship and Duterte's rise to the presidency has been characterized by successive governments that promoted the virtues of liberal democracy, employed a discourse of good governance, embraced neoliberal economic policies, and saw elite reshuffling of power at the national level (Thompson 2016; Teehankee 2017; Bello 2018). Duterte's emphasis on change, his illiberal language, and his steadfast promise of ridding the country of the drug menace—constructing drug users and drug lords as the primary enemy—contributed to his broad appeal across all classes discontented with the failures of the status quo.

Duterte has publicly admired Marcos's strongman qualities and has mirrored a few of his policies, including an infrastructure construction push and subscribing to his brand of authoritarian developmentalism. Even if he lacks a coherent strategy, like Marcos before him, he has emphasized the importance of law and order as a prerequisite to addressing poverty and delivering development (Quimpo 2017). His brutal "war on drugs" has been his administration's most radical break from previous governments, even as he has not steered too far from his predecessors in many other aspects, such as his continuation of neoliberal economic policies.

Bringing attention back to the forty-year-old Laguna Lake fish pen controversy provided Duterte with an opportunity to display this strongman vision of development via order and his politics of change. In a little over a year, he and his technocrats have mobilized a populist narrative that targeted elites, where the country's social and environmental ills could be solved through political will coupled with proper management and deployment of police or military force, if necessary.

Duterte would mention seeing from an airplane crowded structures sticking out prominently in the lake's landscape, observing the nature of the problem from afar: "Whenever I look down there … those triangles … you could not see (spaces) … and the fishermen are left with nothing" (Nilles 2016). The sight of geometric fish pen structures crowding out the space for small fisherfolk seems to encapsulate government inaction, unbridled influence of elite power, and neglect of social justice. Part of the appeal of turning attention to the lake's fish pen problem was that it was visual and affective; images of fish pen structures being dismantled showed a

state in action. After months of uncertainty as to whether to remove all pens (zero-fish-pen policy), state agencies used scientific studies on the lake's carrying capacity to rationalize or determine the total size of structures to be demolished (a quarter of the 13,000-ha structures; Geronimo 2017).

Laguna Lake was important to the administration as it was to be a "showcase of social justice" (Department of Environment and Natural Resources 2016). The success of fish pen demolitions became even more crucial to show the change his administration had promised, following the fizzling of a mining operations crackdown that saw the suspension of more than 100 mining operations before being revoked after staunch opposition from the sector and from within the government (Bello 2018). The controversies surrounding mining showed competing interests among technocrats within the administration, which have not played out to the same degree in Laguna Lake.

After Duterte's statements, technocrats would repeat a similar populist discourse on Laguna Lake: With the new administration focused on law and order, it was time for the elites to return the lake to the people. The goal was to bring the lake back to its glory days of bounty (the "promise of plenitude") after decades of exploitation. After the first round of demolitions, the environment secretary was quoted: "(This is) how (the lake) should be—for the people and not for big corporations who don't even give back to the people" (Pazzibugan 2017). The elite constructed as the enemy in this case were fish pen operators, who "have benefited and profited from the lake region since the '70s" (Pazzibugan 2017). It was time for the ordinary people to benefit from the lake.

The rhetoric of antagonism between a simplified elite and people was supported by Duterte's embodiment of the collective desire to take back power from the few. It also justifies the means necessary for this to happen, including using, as one lawmaker put it, "military, navy, all forces of government at the fingertips of the President" (Geronimo 2016). Duterte's threats against fish pen operators resembled similar statements he made against those he identified as enemies of the Filipino people (drug users, vocal critics of human rights violations, opposition politicians), with threats often taking on a personalistic tone: "If you don't want to make it smaller, I will destroy it … it will be my direct order and I will assume full responsibility for them" (Ranada

2016). This reflects a key populist mode of identification that makes the political personal (Panizza 2005). The effect of such rhetoric is to bring dignity to the marginalized and wresting power away from the elites to the benefit of the leader (Thompson 2016). It also shows the continuing process of state construction and legitimization through these narratives and practices.

Fish pen owners have responded by reiterating their contribution to the fish supply and food security in Metro Manila, their economic links with fisherfolk livelihoods, and their minimal ecological impacts on the lake (Cinco 2017). Yet, as with past government threats, they promised to cooperate in regulating fish pens and emphasized their place in any efforts to plan and govern the lake. Fisherfolk, meanwhile, have long lamented the ineffectiveness of attempts to remove illegal fish pens from the lake and have recognized the role of elite power in their continued plight (Saguin 2016). Marcos's failure to resolve the fish pen controversy had made Laguna Lake fisherfolk carefully optimistic when Duterte brought up his fish pen plan. After Duterte's initial statement, the largest national fisherfolk alliance mentioned how the fisherfolk had long waited for "a leader who will address the welfare of small fishers" (Nilles 2016) and expressed support for the demolition drive. Several months later, they commented that is one of the administration's prominent achievements but that it was not nearly enough (Cinco and Celis 2017).

Several fisherfolk expressed worry that a zero-fish-pen policy would create income losses for lake dwellers who have forged livelihood connections with the deeply embedded fish pen economy. Forty years of fish pen presence in the lake has enabled thousands of fishers and lake dwellers to engage in fish-pen-related work such as drag seining for harvest, fish trading and transporting, and seed production. As fishing-based lake dwellers formed strong economic linkages with fish pens in a time of subsiding violence and conflicts, they have developed a more ambiguous opinion about continued fish pen presence in the lake. The simple antagonism between the fisherfolk and fish pens betrays these multiple livelihoods and relations that also condition how particular fisherfolk view the fish pen dismantling issue (Saguin 2018).

Despite the politicization of the lake issue through antagonistic politics, the governance strategy has remained firmly within the technocratic management language and practice of rationalization and democratization reminiscent of Marcos-era strategies. This has taken the form of plans to organize fisherfolk into cooperatives to increase fish pen ownership and provide them with greater space in fish pen allocation (Geronimo 2017). The solution, methods, and spaces of engagement have been predetermined by the state to follow its vision of law-and-order development. It is notable how fisherfolk remained passive and absent apart from being rendered as recipient of state development plans or as one in the many stakeholders in the lake. Similar to the good governance discourses, solutions to the lake's problem have been constrained by a narrow optic of how fisherfolk could participate in shaping their futures. The unambiguous antagonism between fisherfolk and fish pens is a powerful but reductive narrative about resource conflicts in the lake that enabled ready solutions to be implemented. The history of lake governance to address resource conflicts therefore has exhibited a mixture of both depoliticized authoritarian and liberal democratic strategies.

Perhaps the biggest irony of the "return the lake to the people" narrative is the proposal to bring the lake back to its "original, pristine condition" and convert it to an ecotourism zone (Gamil 2017). Dismantling unsightly fish pen structures therefore becomes a precondition for transforming the lake "into a vibrant economic zone showcasing ecotourism" (Duterte 2016), an activity that undermines the productive use of the lake for fisheries. Business interests have expressed support in formulating a new master plan for the lake (Mercurio 2016), which fisherfolk groups have opposed and read as attempts to privatize the lake (Alcober 2017).

Conclusion

In this article, I showed how populist political rhetoric serves to both politicize environmental problems and depoliticize efforts to address them. Laguna Lake under Marcos and Duterte demonstrates this contradiction in resource governance. Unlike leaders who remained firmly within the reformist liberal democracy tradition that extolled the value of pluralist good governance, both promised a change in status quo by eliminating the old configuration of elite power and giving control back to fisherfolk. The lake—a visible social justice problem—served as a showcase for their vision. For Marcos, it was the New Society and authoritarian developmentalism in action that was at stake. For Duterte, it was the change he promised to deliver through political will and law and order. Rather than matching the radical vision promised by their populist narratives, however, the actions they have mobilized and solutions proposed have remained largely within the confines of the existing repertoire of governance strategies. In the lake, this has taken the form of the technocratic language of rationalization and democratization not fundamentally different from the status quo of post-political modes of governance. The politicized antagonistic construction of political identities has been matched by depoliticized methods that flatten conflicts.

Plans for ecotourism similarly undermine the productive use of the lake for fisherfolk. This political ecological tension between a productive and a pristine future signals how crises might serve a crucial role in the regulation of capitalist relations in a resource frontier. In Laguna Lake, populist narratives have enabled depoliticized modes of governance to maintain and legitimize the presence of fish pens. Yet the very same narratives open the possibility for the restructuring of agrarian relations once entrenched institutional frameworks and regimes of accumulation are rendered obsolete.

Authoritarian populist rhetoric transforms the framing of environmental problems even as solutions are constrained within technologies of government that do not open up spaces of contestations. Populist rhetoric draws strength from its ability to navigate this contradiction. Calls for emancipatory politics therefore need to recognize and confront the vision of radical democracy that underpins the power of such rhetoric.

Acknowledgments

I thank Jake Soriano and Jose Javier for providing comments and assistance in the preparation of this article.

References

Alcober, N. 2017. Fisherfolk want businessmen out of Laguna Lake rehabilitation. *Manila Times*, January 5. Accessed March 4, 2017. http://www.manilatimes.net/fisherfolk-want-businessmen-laguna-lake-rehabilitation/305362/.

Anderson, B. 1988. Cacique democracy in the Philippines: Origins and dreams. *New Left Review* 169:3–31.

Bello, W. 2018. Counterrevolution, the countryside and the middle classes: Lessons from five countries. *The Journal of Peasant Studies* 45 (1):21–58.

Beveridge, R., and P. Koch. 2017. The post-political trap? Reflections on politics, agency and the city. *Urban Studies* 54 (1):31–43.

Bridge, G. 2014. Resource geographies II: The resource–state nexus. *Progress in Human Geography* 38 (1):118–30.

Cinco, M. 2017. Operators: "Zero-pen" policy endangers food security. *Philippine Daily Inquirer*, January 28. Accessed February 15, 2017. http://newsinfo.inquirer.net/866092/operators-zero-pen-policy-endangers-food-security#ixzz4Z5wGIZ3U.

Cinco, M., and N. Celis. 2017. Still no room for small fishers as pens remain in Laguna de Bay. *Philippine Daily Inquirer*, July 20. Accessed November 7, 2017. http://newsinfo.inquirer.net/915396/still-no-room-for-small-fishers-as-pens-remain-in-laguna-de-bay#ixzz4t8KNEZ32.

Department of Environment and Natural Resources. 2016. Lopez to sit down with Laguna Lake fishpen operators before permit moratorium. Accessed November 7, 2017. http://www.denr.gov.ph/news-and-features/latest-news/2846-lopez-to-sit-down-with-laguna-lake-fishpen-operators-before-permit-moratorium.html.

Duterte, R. R. 2016. State of the Nation address of Rodrigo Roa Duterte. Accessed October 27, 2017. http://www.officialgazette.gov.ph/2016/07/26/the-2016-state-of-the-nation-address/.

FL vows just lake decision. 1983. *Bulletin Today*, March 29: 1.

Gamil, J. T. 2017. Last call on for self-imposed demolition of Laguna Lake fish pens. *Philippine Daily Inquirer*, January 14. Accessed November 7, 2017. http://newsinfo.inquirer.net/862203/last-call-on-for-self-imposed-demolition-of-laguna-lake-fish-pens#ixzz4t8EnAWBB.

Geronimo, J. Y. 2016. DENR: Massive dismantling of Laguna lake fish pens in 2017. *Rappler*, October 1. Accessed November 7, 2017. https://www.rappler.com/nation/147864-denr-dismantling-laguna-lake-fish-pens-2017.

———. 2017. Atienza to DENR: Laguna Lake still cluttered with fish pens. *Rappler*, August 15. Accessed November 7, 2017. https://www.rappler.com/nation/178828-atienza-laguna-lake-still-cluttered-fish-pens.

Kohl, B., and L. Farthing. 2012. Material constraints to popular imaginaries: The extractive economy and resource nationalism in Bolivia. *Political Geography* 31 (4):225–35.

Laclau, E. 2005. Populism: What's in a name? In *Populism and the mirror of democracy*, ed. F. Panizza, 33–49. London: Verso.

Marcos, F. E. 1983a. Executive order No. 927. Manila, Philippines. Accessed September 26, 2017. http://www.officialgazette.gov.ph/1983/12/16/executive-order-no-927-s-1983/

———. 1983b. Letter of instructions No. 1325. Manila, Philippines. Accessed September 26, 2017. https://www.gov.ph/documents/20147/344512/19830521-LOI-1325-FM.pdf/c97c15fb-ba56-1c98-7b60-14850e3e10b6?version=1.0

———. 1984. Letter of instructions No. 1399. Manila, Philippines. Accessed September 26, 2017. https://www.gov.ph/documents/20147/343618/19840411-LOI-1399-FM.pdf/9950dacc-441f-ebac-e580-1afd1867a9fa?version=1.0

Mercurio, R. 2016. PCCI supports ban on fishing permits for Laguna Lake. *The Philippine Star*, December 31. Accessed November 7, 2017. http://www.philstar.com:8080/business/2016/12/31/1658203/pcci-supports-ban-fishing-permits-laguna-lake.

McCarthy, J. 2013. We have never been post-political. *Capitalism Nature Socialism* 24 (1):19–25.

Mouffe, C. 2005. *On the political*. London and New York: Routledge.

Ng, W. 1983. KKK may run fishpens. *Bulletin Today*, March 30:1, 16.

Nilles, G. 2016. Rody to destroy triangle fish pens in Laguna Lake. *The Philippine Star*, August 5. Accessed February 15, 2017. http://www.philstar.com/headlines/2016/08/05/1610275/rody-destroy-triangle-fish-pens-laguna-lake.

Noble, L. G. 1986. Politics in the Marcos era. In *Crisis in the Philippines: The Marcos era and beyond*, ed. J. Bresnan, 70–113. Princeton, NJ: Princeton University Press.

Panizza, F. 2005. Introduction: Populism and the mirror of democracy. In *Populism and the mirror of democracy*, ed. F. Panizza, 1–31. London: Verso.

Pazzibugan, D. 2017. Gov't starts dismantling fish pens in Laguna de Bay. *Philippine Daily Inquirer*, January 27. Accessed February 15, 2017. http://newsinfo.inquirer.net/865905/govt-starts-dismantling-fish-pens-in-laguna-de-bay#ixzz4t8FA7yhV.

Perreault, T., and G. Valdivia. 2010. Hydrocarbons, popular protest and national imaginaries: Ecuador and Bolivia in comparative context. *Geoforum* 41 (5):689–99.

Quimpo, N. G. 2017. Duterte's "War on Drugs": The securitization of illegal drugs and the return of national boss rule. In *A Duterte reader: Critical essays on Rodrigo Duterte's early presidency*, ed. N. Curato, 145–66. Quezon City, Philippines: Bughaw.

Ranada, P. 2016. Duterte: Mid-December demolition of excessive Laguna Lake fish pens. *Rappler*, November 24. Accessed February 15, 2017. https://www.rappler.com/nation/153456-duterte-laguna-lake-fish-pens-demolition.

Ranciere, J. 2016. The populism that is not be found. In *What is a people?*, ed. A. Badiou, 100–106. New York: Columbia University Press.

Robertson, M. 2015. Political ecology and the state. In *Routledge handbook of political ecology*, ed. T. Perrault, G. Bridge, and J. McCarthy, 457–66. London and New York: Routledge.

Saguin, K. 2016. Blue revolution in a commodity frontier: Ecologies of aquaculture and agrarian change in Laguna Lake, Philippines. *Journal of Agrarian Change* 16 (4):571–93.

———. 2017. Producing an urban hazardscape beyond the city. *Environment and Planning A* 49 (9):1968–85.

———. 2018. Mapping access to urban value chains of aquaculture. *Aquaculture* 493:424–35.

Samonte, S. 1983. Lake law author air views. *Bulletin Today*, April 13:13.

Scoones, I., M. Edelman, S. M. J. Borras, R. Hall, W. Wolford, and B. White. 2018. Emancipatory rural politics: Confronting authoritarian populism. *The Journal of Peasant Studies* 45 (1):1–58.

Swyngedouw, E. 2009. The antinomies of the postpolitical city: In search of a democratic politics of environmental production. *International Journal of Urban and Regional Research* 33 (3):601–20.

Tadem, T. S. E. 2013. Philippine technocracy and the politics of economic decision making during the martial law period (1972–1986). *Social Science Diliman* 9 (2):1–25.

Teehankee, J. C. 2017. Was Duterte's rise inevitable? In *A Duterte reader: Critical essays on Rodrigo Duterte's early presidency*, ed. N. Curato, 37–56. Quezon City, Philippines: Bughaw.

Thompson, M. R. 2016. Bloodied democracy: Duterte and the death of liberal reformism in the Philippines. *Journal of Current Southeast Asian Affairs* 35 (3):39–68.

Webb, A. 2017. Hide the looking glass: Duterte and the legacy of American imperialism. In *A Duterte reader: Critical essays on Rodrigo Duterte's early presidency*, ed. N. Curato, 127–44. Quezon City, 1101 Philippines: Bughaw.

Wilson, J., and E. Swyngedouw. 2014. Seeds of dystopia: Post-politics and the return of the political. In *The post-political and its discontents: Spaces of depoliticization, spectres of radical politics*, ed. J. Wilson and E. Swyngedouw, 1–22. Edinburgh, UK: Edinburgh University Press.

KRISTIAN SAGUIN is an Assistant Professor in the Department of Geography at the University of the Philippines, Diliman, Quezon City, 1101 Philippines. E-mail: kcsaguin@up.edu.ph. His research interests include the political ecologies of urbanization and environmental change in Metro Manila and Laguna Lake.

Fishing for Power: Incursions of the Ugandan Authoritarian State

Anne J. Kantel

A few months before Uganda's 2016 presidential elections, the government issued an executive order dissolving community-based Beach Management Units, the local and democratic governance bodies responsible for managing fishing activities. The official narrative cited rampant corruption and the exploitation of Uganda's valuable fishing resources as justification for the suspension. A popular counternarrative, however—told in carrying whispers at fishing landing sites around Lake Victoria—painted the order as President Museveni's attempt to secure votes during a tough presidential campaign. Drawing on seven months of ethnographic research on fisheries management in Uganda and critical studies on the nexus of coloniality, securitization, and common pool resource theories, this article analyzes sociopolitical narratives around fisheries governance a year before and after the presidential elections in Uganda. The author illustrates how recent policy changes in the country's fisheries governance sector are underlined by a powerful narrative of peace and security and argues that the political intervention can be interpreted as efforts by the national government to secure the ruling elite's increasingly authoritarian hold on state power.

乌干达2016年总统选举的数月前，政府批准一行政命令，解散以社区为基础的海滩管理单位，该单位是负责管理渔业活动的在地且民主治理单位。官方叙事引用猖獗的贪污和剥削乌干达宝贵的渔业资源，作为中止该单位的藉口。但在维多利亚湖周边的渔船码头暗中流行的反叙事，则将该命令描绘为总统穆赛韦尼在艰困的总统竞选活动中巩固得票的企图。本文运用在乌干达为期七个月对渔业管理的民族志研究，以及对殖民性、安全化和共享资源理论轴线的批判研究，分析乌干达总统大选一年前与一年后围绕着渔业治理的社会政治叙事。本文作者描绘出该国渔业管理部门的晚近政策变迁，如何由强而有力的和平与安全叙事所强化，并主张政治介入可诠释为国家政府为确保治理菁英把持逐渐威权化的国家权力之努力。关键词： 威权主义，殖民性，渔业管理，安全化，乌干达。

Pocos meses antes de las elecciones presidenciales de Uganda en 2016, el gobierno expidió una orden ejecutiva por medio de la cual se disolvían las Unidades Administrativas de Playa, de base comunitaria, que eran los cuerpos de gobernanza democrática local encargados del manejo de las actividades pesqueras. La narrativa oficial citó la corrupción rampante y la explotación de los valiosos recursos pesqueros de Uganda como justificación para la suspensión. Sin embargo, una contranarrativa popular—referida en murmullos propagados en los sitios de atracadero alrededor del Lago Victoria—retrataron la orden como el intento del Presidente Museveni por asegurar votos durante una difícil campaña presidencial. Con base en siete meses de investigación etnográfica sobre el manejo de la pesca en Uganda y en estudios críticos sobe los nexos de la colonialidad, la titularización y teorías de los recursos comunes, este artículo analiza las narrativas sociopolíticas en torno a la gobernanza de la pesca un año antes y otro después de las elecciones presidenciales de Uganda. El autor ilustra el modo como los cambios de las políticas recientes en el sector de la gobernanza de la pesca del país están subrayados por una poderosa narrativa de paz y seguridad, y arguye que la intervención política puede interpretarse como los esfuerzos del gobierno nacional para asegurar el control cada vez más autoritario del poder del estado por la élite dominante. *Palabras clave: autoritarismo, colonialidad, gobernanza de la pesca, titularización, Uganda.*

In November 2015, Uganda's reigning president Yoweri Museveni issued an executive order to dissolve all community-based Beach Management Units (BMUs), the local democratic governance bodies responsible for managing the country's decentralized (both artisanal and commercial) fishing activities. The official narrative cited corruption charges against members of various local BMUs and the exploitation of Uganda's valuable fishing resources. A close reading of the government's sociopolitical narratives around fisheries governance from 2015 to 2017, however, permits a different explanation for the state's intervention. Drawing on seven months of ethnographic research on fisheries management in Uganda and critical studies on the nexus of coloniality, securitization, and common pool resource theories, I suggest that recent policy changes in Uganda's fisheries governance sector can be interpreted as efforts by the national government to secure the ruling elite's increasingly authoritarian hold on state power under the guise of a discourse of peace and security.

In his book *Citizen and Subject*, Mamdani (1996) argued that African states' tendency toward authoritarianism finds its origin in the institutional legacy of colonialism. Others have emphasized autocratic opportunism, neo-patrimonialism, and incongruence of colonial administrative structures with preexisting cultural diversity (Schneider 2006; Goh 2015; Boone 2017). Without disputing such contributions, this article speaks to a stream of scholarship that analyzes the embodiment of colonial power dynamics in current-day practices and discourses (Quijano and Michael 2000; Maldonado-Torres 2007). In Uganda, these practices of coloniality serve as part of both the political elite's self-image and the targeted population's "public transcript" (Scott 1992). I argue that the official state narrative of peace and security frames Uganda's artisanal fisherfolk population as one that needs to be disciplined and managed, unlike more legitimate and productive commercialized fishing. Such a narrative of differentiation, which depicts an essential group within the population as undisciplined, illegitimate, and criminal, offers a blueprint for the organization of the Ugandan society into citizens and noncitizens and employs similar language to that used by colonizers in the early twentieth century.

In addition to literature on coloniality, this article draws on and seeks to speak to existing research on fisheries practices in Uganda, which—although limited—offers some notable contributions to scholarship studying the effects of different natural resource governance designs on human–environment systems. This includes Barratt and Allison's work on the questionable premises and (in)effectiveness of participatory common pool resource management approaches that emerged in the late 1990s (Barratt and Allison 2014; Barratt, Seeley, and Allison 2015), as well as Nunan's research analyzing

heterogeneity of fisheries communities and potential implications on effective governance in Uganda (Nunan 2006; Nunan et al. 2012). Their research is complemented by a growing body of literature analyzing fisheries governance in the wider Lake Victoria basin as well as scholarship on the effects of decentralization and privatization in other natural resource sectors (Ribot 2004; Colchester 2006; Cavanagh and Benjaminsen 2014; Nel 2015). An important finding of this literature is that successful (i.e., sustainable) management of common pool resources requires an in-depth understanding of relevant stakeholders' capacities, social and political contexts, and incentives. This article seeks to build on such work by illustrating how state elites' desires to secure power influence the design of government interventions.

To analyze the sociopolitical discourse around Uganda's fisheries governance system, I draw on a variety of empirical data: ethnographic interviews, participant observation, and articles from three leading English-language Ugandan newspapers. The resulting picture is one in which the government and its supporters employ a narrative of peace and security to justify increasingly authoritarian means of state power in the country's fisheries sector.

History and Political System in Uganda: From British Protectorate to a Hybrid Regime

Uganda's colonial history plays an important part in contemporary discursive structures and embedded social, political, and economic relations. Although some have argued that Uganda did not suffer the same degree of colonial oppression as neighboring countries, its experience as a British protectorate is nevertheless entrenched in the country's collective memory and contemporary political institutions. As I illustrate later, this legacy is visible in the ruling elite's justifications of its recent interventions in the country's fisheries sector. In other words, the ruling party carries forth practices of coloniality. Coloniality, as an analytical perspective, allows us to see two dynamic relations of power rooted in colonial experiences. The first is visible through the continued importance of a capitalist mode of production in discourses around economic growth and development; the second manifests itself through a "racialized" classification of the population (Quijano and Michael 2000, 534; Maldonado-Torres 2007).

Regarding the former, capitalist discourses of economic growth and development in Uganda help depict the country's commercialized fisheries as the only sustainable and, therefore, legitimate practice on Lake Victoria. Similar to scholarship outlining the often negative effects of both environmental conservation and privatization policies on customary forest or land rights (Peluso 1994; Colchester 2006; Nel 2015), I show how the perpetuation of a peace and security discourse robs artisanal fisherfolk of their access to the lake's fish resources while benefiting the large-scale commercialized fisheries. Various scholars have pointed toward similar instances of so-called green or blue grabbing as neglected and silencing side effects of environmental conservation projects that go hand in hand with a commercialization of community resources (Benjaminsen and Bryceson 2012; Cavanagh and Benjaminsen 2014; Lyons and Westoby 2014; Carmody and Taylor 2016).

Regarding the latter, racialized frameworks of meaning making are called on that justify authoritarian means to marginalize and silence members of society perceived to threaten the state. In this sense, coloniality speaks to a different but important literature in environmental politics, which emphasizes processes of securitization. Coming out of the literature of international relations, securitization scholars have successfully argued that an issue becomes a security threat not because it constitutes an objective existential threat but because a group of political actors actively construct it as such to legitimize the use of extraordinary means to ensure the group's survival (Buzan, Waever, and de Wilde 1998; McDonald 2008). Although much has been written on securitization of water resources (Lankford et al. 2013), the aspect of fisheries—as a natural resource that could be securitized—remains underexplored. Coloniality adds to this perspective by highlighting the role of power relations based on racialized language and capitalist hierarchies that find their roots in the historical experiences of both the colonized and the colonizers.

Uganda's current political system is often categorized as competitive authoritarianism, a hybrid of democratic and authoritarian elements in everyday governance practices. It can be defined as a civilian regime in which formal democratic institutions are regarded as the primary means of gaining and maintaining power. Elements such as fraud, violation of civil and political liberties, and abuse of state power, however, skew the sociopolitical playing field in a way that precludes genuinely competitive elections or viable legal channels to contest the ruling party's power (Levitsky and Way 2010; Tripp 2010; Kagoro 2016). Some have found that states categorized as hybrid regimes could be regarded as stable systems rather than transitioning democracies (O'Donnell 1996; Carother 2002). Although this approach has general validity, I argue that Uganda is moving ever more toward authoritarianism in its regime-defining dynamics of power. I seek to build on important work in political geography that has illustrated how government decentralization programs in many African states have weakened local participation and democracy and contributed to a de facto recentralization of state power across capital cities (Ribot 2004; Ribot, Agrawal, and Larson 2006; Ece, Murombedzi, and Ribot 2017).

Over the past ten years, Uganda's political, economic, and legal ratings as well as its rankings on freedom of the press have been in steady, rapid decline (Musisi 2016; Freedom House 2017). President Yoweri Museveni, who won his fifth consecutive term in 2016, has been in power for thirty-two years, with questions about the duration of his presidency and succession pushing their way into the public discourse. They first arose in the early 2000s in the context of presidential term limits, a threat that Museveni neutralized by proposing a quid pro quo: removing term limits in exchange for a multiparty system (Tripp 2010). Museveni and his National Resistance Movement (NRM) Party, on taking power in 1986, had established Uganda as a "no-party" state, with political candidates to be elected on merit rather than sectarian support (Government of Uganda 1995). In 2005, the government abandoned the movement system in favor of a multiparty state—officially by public referendum, unofficially in exchange for the abolition of presidential term limits.

Yet, the competitiveness of that system is increasingly questioned. Ogenga Latigo, a former leader of the opposition, criticized the system in 2015, claiming that multipartyism has failed because there is no level playing field in the country (Masaba 2015). Similarly, Kagoro (2016) argued that the Ugandan regime is a tripartite one, in which the government, NRM Party, and the military function essentially as one and the same. This has important consequences for the democratic process, particularly during elections. During the

2001 general elections, now-retired senior military officer Henry Tumukunde allegedly "reminded voters that, irrespective of whom they voted for, President Museveni would still rule because they had the guns" (Tangri and Mwenda 2010, 44).[1]

Moreover, although many opposition parties have their own flaws (corruption charges, internal struggles, unconvincing platforms), the NRM Party finds in Uganda's bureaucracy an in-house election campaign service that reaches even the most remote parts of the country, an unmatchable advantage. In an op-ed published two months after the 2016 elections, Ugandan international lawyer Sam Mayanja summarized the political situation in his country:

> Uganda's third experiment in multiparty political elections has come and gone. ... The tragedy among the political elites began to play itself out even before the election fever took hold. Whereas the political elite knew that Uganda is in a multiparty political dispensation, only the National Resistance Movement (NRM) got the public to know that that party had a single candidate to front for the presidential elections 2016. (Mayanja 2016, 26)

More recently, Museveni's attempts to promote his son and wife to powerful positions within the government and military sparked concerns that the president seeks to institutionalize a dynasty (Izama 2015). In September 2017, the military stormed the parliament building in response to violence between members of the ruling party and the opposition following introduction of a bill that would scrap the constitutional limit on presidential age ("Uganda Introduces Bill to Remove Presidential Age Limit" 2017). A few weeks later, Museveni warned opposition forces to stop their campaign against the regime: "I want to warn all those who are threatening people, *akabwa akasiru kayigga enjovu* [a foolish dog hunts an elephant] ... to think that you can threaten NRM, and you use violence, yet NRM is the master of violence but our violence is disciplined and purposeful" (Kaaya 2017). On 27 December 2017, Museveni signed the age limit bill into law, which allows him—if he chooses—to run for a sixth term in 2021 ("Uganda Enacts Law Ending Presidential Age Limits" 2018).

Fisheries, Narratives, Peace, and Security: A Discourse Analysis

This article focuses on fisheries to illustrate how Uganda's president and the ruling NRM Party use a

discourse of peace and security to intensify their increasingly authoritarian hold on power. The country's fisheries practices, formerly based on purely artisanal fisheries, changed significantly in the late 1980s, when the commercialized fisheries on Lake Victoria experienced momentous growth by targeting a fast-spreading and invasive Nile perch species (introduced in the 1960s and made famous by the 2004 documentary *Darwin's Nightmare* [Sauper 2004]). Increasingly decentralized governance structures, together with liberalization and privatization designed to attract foreign investors, transformed the artisanal fisheries sector into a global market specialized in the production of tilapia and Nile perch (Bruton 1990). Policies favoring commercialized over artisanal fisheries and a growing population go hand in hand with an overexploitation of fish (and other natural) resources, which poses an increasing political and ecological risk to the livelihoods of Ugandans living along the shores of Lake Victoria (Abila 2000; Allison 2003; Njiru et al. 2008).

In the following, I offer an analysis of the sociopolitical discourse around fisheries governance in Uganda from January 2015 to April 2017, spanning a period of approximately one year before and one year after the country's presidential elections in February 2016. Discourses around Uganda's fisheries, I posit, offer important insights into the government's tools and motives regarding regime security. Hajer and Versteeg (2005) noted that discourse analysis "illuminates a particular discursive structure, that might not be immediately obvious to the people that contribute to the debate" (175–76). I follow discourse analysis scholarship grounded in political ecology by exploring power dynamics that illuminate the state's discourse of peace and security as a meaning-making framework for recent changes in the country's fisheries governance (Neumann 2009; Robbins 2009; Doshi and Ranganathan 2017). A political ecology approach to discourse analysis allows us to not only identify sociopolitical narratives but also situate them within the complex web of human–environment interactions that are at the center of all social, political, and economic systems. It is a contextualized analysis that seeks to unravel power asymmetries that constitute and reproduce practices governing access to and management of Uganda's fish resources and therefore allows us to integrate the critical perspectives of coloniality and securitization in the analysis.

Figure 1. Map of Uganda.

I draw on a variety of empirical data collected during seven months of field research in the Ugandan part of Lake Victoria. I conducted 120 in-depth interviews with state representatives and local fishers from three different landing sites along the eastern shore of Lake Victoria (see Figure 1), which vary in size (two relatively large, one very small), principal catch (tilapia, Nile perch, or mukene [silverfish]), and proximity to larger cities (one within city limits, one approximately twenty minutes from the closest trading center, and one in a very remote setting). Interview data are complemented by more than 100 hours of participant observation and in-depth analysis of relevant articles published by the three main English-language newspapers (*Daily Monitor*, *New Vision*, and *The Observer*). Of more than 1,200 articles about natural resource management published between January 2015 and April 2017, I coded 313 as relevant for the analysis of governance tendencies in Uganda's fisheries sector.

Due to the previously mentioned political and ecological risks to livelihoods faced by a growing population along the shores of Lake Victoria in Uganda, access to and management of natural resources, including fisheries, played a central role

during the 2016 presidential campaigns (Lumu 2016). All eight political candidates promised to improve services in water, sanitation, and electricity; address management issues in the forest, fish, and mining sectors; and solve ongoing land conflicts. Several candidates visited fishing landing sites along Lake Victoria, indicating the community's importance across the political spectrum. None were as active as the NRM elite, though, in targeting the fishing community and its governance structures.

In November 2015, Museveni issued an executive order dissolving all community-based BMUs in the country. Established in 2003 by the Ministry of Agriculture, Animal Industries, and Fisheries (MAAIF), BMUs were the lowest governance bodies responsible for management of fishing activities in Uganda's major lakes.[2] The order also temporarily interrupted the activities of all rule enforcement personnel in the sector, such as police patrols and MAAIF's district-level fisheries staff (see Figure 2).

The presidential decree, which surprised even top ministry officials, was framed as necessary to protect fisheries and fishers from rampant corruption and mismanagement by fisheries officers, police, and BMU members: "As the President of Uganda, I am going to dissolve the fish management units. However, bring elders so that we can discuss possibilities of creating a new body led by elders, who know the lake. You [the people] should lead the way and leave these thieves in fisheries" (Museveni, quoted in Lumu 2015a). While scrambling to come up with guidelines to reform the now rule- and authority-less sector, senior officials of MAAIF tried to follow Museveni's confusing signals to create such a new governance body and put in place temporary two-person fishing landing site committees along lake and river shores. The difference was obvious: Whereas the original BMUs allowed for community-based decision making, the new committees were appointed from above by NRM-affiliated district fisheries officers.

The official justification for suspending the BMUs, fisheries officers, and lake patrols was to protect the country's resources ("The fish and the lakes are big treasures that you have"; Museveni, quoted in Kazibwe 2015) and ensure people's peace and security by protecting them from corrupt, immoral local authorities (Lumu 2015a). A government representative explained Museveni's decision: "The

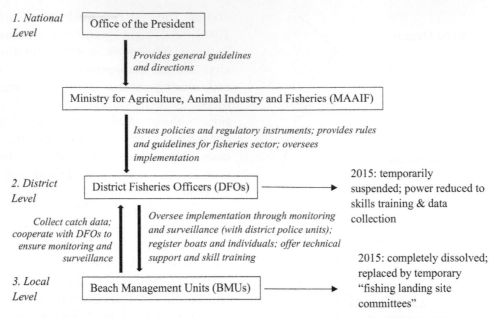

Figure 2. Uganda's executive fisheries governance system prior to November 2015 (simplified). MAAIF = Ministry for Agriculture, Animal Industry and Fisheries; DFOs = District Fisheries Officer; BMU = Beach Management Unit.

President directed that all enforcement officers be removed from the lake because they have been harassing people. … The Beach Management Units [have] also been suspended because they … failed to protect the lake and resorted to conniving with other enforcement operatives to harass fishermen" (Hajusa 2016, 20).

This narrative of peace and security, emphasizing the government's role in protecting its citizens and stabilizing the country, penetrated the entire presidential NRM campaign. "Ugandans," a presidential advisor found in 2015, "do appreciate that under President Museveni, the country is now peaceful and for this, they would rather vote for him other than give their vote to the other persons whose names appeared in the polls" (Byaruhanga 2015, 16). Similarly, one of Museveni's longtime followers and NRM founders in Buganda stated: "During his [Museveni's] time the people have enjoyed 29 years of political stability and tremendous economic development" (Kaggwa 2015, 40). Museveni in power and Uganda's peace and security were depicted as the same.

In turn, potential threats to the regime's power and stability were framed as threats to the country's peace and security, displaying the government's coercive power and authority. In the months leading up to the election, opposition leaders were arrested and jailed regularly, amid credible allegations about torture and harassment by the police

and other government actors (Human Rights Watch 2017). Members of local fishing communities told of violent abuse and arrests by police and military. The government made clear to citizens that development and security would only be achieved through the NRM Party and Museveni's reelection. Voting for the opposition, the NRM campaign emphasized repeatedly, would mean fewer services and less growth, development, and stability (Hajusu and Odeke 2015).

Blame for underdevelopment and small-scale conflict, meanwhile, was shifted to culturally unacceptable behavior by specific groups. This was done through so-called othering moves that created clear demarcations between members of legitimized in-groups and those of delegitimized out-groups. For example, in early February 2016—a few days before election day—Museveni charged that "people in [landing site] Kalangala have betrayed our country. … In western Uganda, the *balaalo* [cattle keepers] do not eat calves but the people here eat young fish, which depletes fish stocks in the lake" (Lubulwa 2016).

Several months after winning reelection, Museveni made another surprising announcement (see Figure 3 for timeline of events). Supported by the Association of Fishers and Lake Users of Uganda (AFALU), a civil society organization representing the commercial large-scale Nile perch fisheries on Lake Victoria, Museveni announced that he had

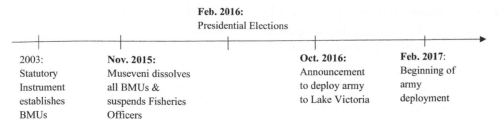

Figure 3. Timeline of events. BMU = Beach Management Unit.

deployed a unit of the Ugandan army to "stamp out bad fishing" and threatened, "All individuals who do bad fishing will see what is going to happen to them" (Sekanjako, Okino, and Apunyo 2017, 3).

A representative of MAAIF summarized the events following the initial BMU suspension as a period of reform, in which the ministry tried to come up with new BMU guidelines that addressed some of the major shortcomings of the original design. He added:

> We took it to the President and still he said no. ... He wanted to deploy the military. He feels that the BMU and the DFOs [district fisheries officers] are corrupt, that is the argument. ... So the President said, let me send my soldiers. They are disciplined. (Government official, interview 2017)

Evidence from the field supports this statement; corruption in the sector is rampant. Overlapping institutions and authorities provide ample opportunities for individuals in powerful positions to abuse the system and ask for bribes in exchange for leniency on rule violations. The situation enables a narrative in which the government is using its power to protect citizens from corruption, crime, and instability.

There is another aspect present in that statement: The president always planned to employ the army on Lake Victoria despite the existence of alternatives. This counternarrative, fed by Museveni's own remarks, tells a story of power grabbing by Uganda's elite: "The strategy for ending [illegal fishing] has been shaped in my head and this unit [the military] will help me implement it. It seems the strategy for Beach Management Units was wrong. The BMUs operate on the principle of elections. Criminals elect their fellow criminals" (Kwesiga 2017a, 3). Within just fifteen months, Uganda's fisheries governance design went from formalized community-based management to a top-down military structure.

Reactions by the population to the government's policy intervention in the fisheries sector were mixed. Official accounts published by the state-owned newspaper painted a positive picture, stating that the fishing community "commended" Museveni for suspending enforcement operatives to curtail corruption (Hajusa 2016). Similarly, a member of AFALU told me:

> The fishermen are actually happy with the [military] operation. They provide fuel, boats, engines, and money for the operation. ... Now the operation is covering Lake Victoria because its biggest export is Nile perch. (Interview 2017)

On the ground, however, former BMU members, small-scale subsistence fishermen, and opposition members criticized the changes as political maneuvering to catch votes and deepen the government's influence. "Since the campaigns began," a fisher stated, "the government [in the form of district fisheries officers] has not come. That's why everyone is using illegal fishing" (interview 2017). Initially, telling people that they can fish and trade without rule enforcement was greeted enthusiastically by the poorest members of the community. When the government used the suspension of the BMUs to impose its own landing site committees and rumors spread that, in the future, members of local fisheries governance institutions would have to be vetted and appointed by MAAIF, though, approval fell significantly. "This way it is going to be politicized," another fisher explained. "The government tries to make everyone part of the same movement: the NRM" (interview 2017).

Village elders now tell stories about past times, when Museveni and his followers fought as rebels in the bush. "To overthrow the government, they used to get arms from other countries through the lake," said one fisher (interview 2017). The current regime, the narrative goes, is trying to position its people along Lake Victoria to monitor weapon smuggling. "There is a lot of politics behind fisheries," a fishing landing site elder explained. He continued:

> Have you heard about the rebels in Kasese? It is believed that the people in Kasese are connected to the rebel group in Central African Republic and the DR [Democratic Republic of the] Congo. And because fish is supposed to support the rebel groups in Kasese and the DR Congo very much, the government has now come in to fight illegalities [i.e., illegal fishing methods]. ... The government is now only pretending to fight illegalities, but in reality, they are fighting the rebels. (Interview 2017)

These are stories people tell only in private—but they do so all along the shores of Lake Victoria.

Several recent developments support this counternarrative, or "hidden transcript," in which people start resisting the perceived power play of the incumbent regime (Scott 1992). For example, in his 2016 State of the Nation address, Museveni announced his plan to register all fishermen operating on Uganda's lake to stop illegal fishing. His government, he said, would invest in radar technology that can watch the lake's surface up to the international borders with Tanzania, Kenya, and the Democratic Republic of the Congo. "No illegal boats will operate on those lakes; we shall encourage the acquisition of ships and steamers that will transport people to and from the islands so that the illegal fishermen will not hide behind the cover of being canoe transporters" (Sekanjako et al. 2016, 7).

In June 2017, *New Vision*, often seen as the regime's official outlet, launched a new series called "Crime on Lake Victoria," with reporters writing about lawlessness, murder, illegal fishing, and armed smuggling on the lake. Having "reached alarming levels," the reporters stated, these episodes warrant the deployment of Uganda's military (Kwesiga and Etukuri 2017). Moreover, Uganda is involved in a long-standing dispute with Kenya over ownership of Migingo Island, located at the Uganda–Kenya border in Lake Victoria. Originally framed as a territorial dispute over valuable fishing grounds, in 2016 the two countries agreed to a joint police force to "look into allegations of human rights violations [and] address cross-border terrorism" and other issues (Etukuri 2016). In early 2017, the Ugandan government announced it would replace its police unit on the island with armed military forces (Kato 2017). A community elder told me:

> It is a very small conflict but because of security issues, they have a big interest. ... Because the guy in power used these islands during his bush war very much. (Interview 2017)

Coloniality, Securitization, and the Road toward Authoritarianism

The preceding analysis illustrates both the regime's narrative of peace and security, justifying increasingly authoritarian policy interventions in the fisheries sector, and the emerging local counternarrative of a regime power play. Figure 4 summarizes the findings, emphasizing three categories of analysis: (1) power and authority, (2) stability and protection, and (3) othering moves.

Whereas the first two categories describe the framing of the two narratives, a closer look at the category of othering moves reveals dynamics of power within the current political system by underscoring how Museveni and the supporting elite are using fisheries policy to protect their grip on state power. They do this by employing a discourse that carries forth a logic of differentiation informed by coloniality. The previous discourse analysis begs a power-laden question: Who is the "fishing community" the president and his supporters seek to protect or discipline? Put differently, who is constituted as an other? Revealingly, in his 2016 presidential address, Museveni stated:

> By 2005 we were exporting 26,615 tonnes, valued at $143,6 million ... per annum. Once it was realised that there was money in fishing, all parasites descended on our lakes and started fishing out all the young fish. ... It is estimated that Uganda loses close to sh300 billion every year in illegal fishing activities on Lake Victoria alone. ... These are enemies of our future and our prosperity and must be treated as such. (Sekanjako et al. 2016, 7)

Several groups could be the target of the labels "parasites" and "enemies of our future," with the intent being a delegitimization of claims to access or rights to the lake's natural resources. It could refer to nonnationals who seek to benefit from the country's wealth and therefore threaten security. This is certainly a narrative that the NRM government promotes, albeit disproportionately emphasizing nonnationals from neighboring Rwanda and the

Narrative Category	Peace and Security	Elite-Power-Grabbing
Category 1: Power & Authority		
Key identifiers	References to military/government/policies…	
Characteristic quote	"The President hailed the UPDF [Uganda People's Defence Force] for the continued peace in the country. He noted that the force has succeeded in its mission due to professionalism and discipline." (Sekanjako, Okino, and Apunyo 2017)	"The whole [fisheries governance] chain is going to be NRM; they are all yellow [the NRM party-color]." (Fisherman, Interview, 2017)
Category 2: Stability & Protection		
Key identifiers	References to crime/illegalities/civil war/violence/corruption…	
Characteristic quote	"His Excellency [Museveni] ordered a stop to the BMU and the policy because they would torture the fishers." (Government official, Interview, 2017)	"There are some prominent illegal fishers here, who were ex-soldiers from the Kony-rebels. The government fears these guys thinking they could go back to the bush and fight the government again." (Village elder, Interview, 2017)
Category 3: "Othering moves"		
Key identifiers	Clear demarcation of different in- and out-groups	
Characteristic quote	"For a long time, fishermen have viewed the government as the enemy, as someone who is interfering. They see the lake as a God-given resource that they can exploit." (Government official, Interview, 2017)	"The military operation exists but only at the big landing sites on the islands with Nile perch and in the upper lake. Since the upper side of the lake has many islands with many big boats. And these people are connected to the government, they pay for the soldiers's food and the fuel. (…) But on this side of the lake, we cannot afford this, so it will take some time for the army to come." (Fisherman, Interview, 2017)

Figure 4. Two narratives: peace and security vs. elite power grabbing. UPDF = Uganda People's Defence Force; NRM = National Resistance Movement; BMU = Beach Management Unit.

Democratic Republic of the Congo (Mugerwa and Bagala 2016; Muhereza 2016). According to a 2017 newspaper report, the officer of the Marine Unit on Lake Albert (shared with the Democratic Republic of the Congo) perceives Uganda's security as threatened by an influx of Congolese fishers: "We even doubt their intention. Why do they have to come in huge numbers?" (Okethwengu 2017, 13).

Interestingly, foreign investors in the fisheries sector are not part of this delegitimizing narrative. Rather, they are referenced as victims of criminality and in need of state protection: "The illegal activities," Museveni decried, "have led to the depletion of fish in the lakes, which has immensely affected the economy in terms of closure of fish factories, loss of jobs and investors" (Kwesiga 2015). Illegal fishing has occasioned enormous loss in the country, Museveni noted, adding that out of the twenty-one fish factories that were operational ten years ago, only eight are afloat today (Kwesiga 2017a). All eight, however, are owned

by non-Ugandans who are well connected for government lobbying. Represented by the Uganda Fish Processors and Exporters Association (UFPEA), the factory owners work with AFALU, an organization representing the commercialized Nile perch fishers, to put in place state regulations for "sustainable fisheries" and to "inform the government about illegalities in the Nile perch fisheries" (civil society member, interview 2017). These organizations have access to Uganda's political elite, who fear a loss in revenue and foreign investment if the fish stock continues to decline.[3]

The loser here is the noncommercial artisanal fishery. Those who catch mukene, tilapia, mudfish, and the occasional Nile perch for subsistence are pushed out of the trade even though Ugandan law classifies fisheries as an open-access resource, allowing, in theory, for people to "descend" on the lake to fish. Artisanal methods (i.e., the use of small, paddle-run boats and small-mesh nets) have been constructed as not only unsustainable but illegal.

Citing environmental and safety regulations, the government has banned both traditional canoe-style boats and modern (imported) small-mesh nets used by locals to catch smaller sized, often immature, fish (Kwesiga 2017b).

Although there are well-founded arguments that the traditional canoe-style boats are a safety risk and the imported nets pose a threat to Lake Victoria's ecosystem and contribute heavily to the decline of Lake Victoria's fish stock, the reasons for their continued use by small-scale fishers constitute a gaping silence in the state's official discourse. Impoverishment of the lake region, as well as an influx of new fishers seeking employment but lacking the knowledge of veteran fisherfolk, are just two of the reasons why subsistence fishers use unsustainable, illegal practices. Instead of addressing these political challenges, though, artisanal fishers are framed as disloyal, "calf-eating," and lazy criminals, exploiting valuable national resources for their own gain and to the disadvantage of all others. They are the "parasites" and "enemies of our future" to whom Museveni's speech refers. Although the regime constantly emphasizes that it seeks to include "stakeholders such as owners of fish factories and fish suppliers to industries, ... all other people must leave the lake" (Kwesiga 2017a, 3). The regime and its allies have successfully constituted a discourse connecting resource wealth and security that differentiates between legitimate and illegitimate lake users. By conceptualizing Ugandan subsistence fishers as needing to be disciplined and managed, and by using a narrative of differentiation that depicts an essential group within the population as undisciplined, criminal, and therefore illegitimate, the official state line offers a blueprint for the organization of Ugandan society into citizens and noncitizens that employs a racialized language very similar to that used by early twentieth-century colonizers.

Benefiting from the official state discourse are the ruling political NRM elite, wealthy fishers, and the non-Ugandan owners of the country's fish factories. In the name of economic growth, development, and—to a certain extent—environmental protection, the centralization of wealth and power through the exploitation of natural resources and the dispossession of large parts of the population through criminalization reveals the underlying influence of a capitalist logic of production. Whereas the commercialized fishery is depicted as the only legitimate user

of the lake's resources and—significantly—offers funding for the functioning of the state, impoverished subsistence fishers are constructed to be illegitimate because they are framed as a threat to the lake's sustainability and the country's stability. This perceived threat to peace and security also allows the ruling elite to justify the use of increasingly authoritarian tools of governance.

Conclusion

This article illustrates how the Ugandan state's ruling elites have justified the abolishment of Uganda's local governance bodies and subsequent state action through an official discourse of peace and security. Drawing on ethnographic research on fisheries management in Uganda and critical studies on the nexus of coloniality, securitization, and common pool resource theories, I suggest that recent policy changes in Uganda's fisheries governance sector can be interpreted as efforts by the government to secure an increasingly authoritarian hold on state power. Its purpose is to secure Uganda's borders and increase surveillance of the state's "potentially" rebellious fisherfolk population. This is done by creating a narrative that essentially depicts part of the population as undisciplined, criminal, and therefore illegitimate. The article illustrates how perspectives of coloniality allow us to see and understand these dynamics of power in light of continuing hierarchies based in racism and capitalism.

Whether the strategy of the regime to hold and build its power will pay off in the long term remains to be seen. Oppositional voices are growing and require an increasingly intense use of state force to be silenced. The fisheries sector is an interesting example to trace and understand these more general authoritarian tendencies within the Ugandan state. Although—for the moment—top-down government control in the fisheries sector might help control potential oppositional voices and secure the wealth of the country's political and economic elites, this strategy might backfire in the long term. As one fishing community representative warned:

> If they have a policy that destroys all the small boats [i.e., the artisanal fisheries canoes], the country is going to face an uprise. They [the artisanal fishers] might not have the voice to do much, but there will be a social uprise. Because people can't eat and feed their children. Because those big boats are purely capital

investments. So there will be an outcry. (Civil society member, interview 2017)

Acknowledgments

I thank the Office of the Provost and the School of International Service at American University for their support of my empirical field-work, the Ugandan Council for Science and Technology for granting me access to research their country's fisheries sector, and the researchers at UNU WIDER, Helsinki, for their valuable input during the early stages of this work. I also thank Ken Conca, Malini Ranganathan, Tarek Tutunji, and the SIS PhD Colloquium community for their helpful comments on previous drafts and Megan Snow for geographic information systems support. I am also greatly indebted to my three reviewers, who provided invaluable and construct-ive feedback for this article. Finally, I am eternally grateful to all of my interview partners in Uganda, who welcomed me and allowed me access to their world. Thank you.

Notes

1. Tumukunde was forced to resign from his military posts in 2005 after he openly opposed the lifting of the presidential term limit (Kasasira 2015).
2. Due to political gridlock that has put a hold on a newly drafted Fisheries Bill since 2005, the current acting Ugandan law governing the fisheries sector is the outdated Fisheries Act from 1964. All other current governance instruments in the sector are based on this Act and are issued by the state's executive in the form of statutory instruments. The 2004 Fisheries Policy released by the Ministry for Agriculture, Animals and Fisheries provides useful guidelines but is not law.
3. In 2016, Uganda exported a value of US$121 million in fish products, contributing 5 percent to the overall export revenue of the country ("Uganda Exports by Category" 2017). The fish export value as been on a steady decline since 2005 (Nakaweesi 2016).

References

Abila, R. 2000. *The development of the Lake Victoria fishery: A boon or bane for food security?* Kisumu, Kenya: Kenya Marine & Fisheries Research Institute.

Allison, E. H. 2003. Linking national fisheries policy to livelihoods on the shores of Llake Kyoga, Uganda. Ladder Working Paper No. 9, Department for International Development, London.

Barratt, C., and E. H. Allison. 2014. Vulnerable people, vulnerable resources? *Development Studies Research* 1 (1):16–27. doi:10.1080/21665095.2014.904079.

Barratt, C., J. Seeley, and E. H. Allison. 2015. Lacking the means or the motivation? *The European Journal of Development Research* 27 (2):257–72. doi:10.1057/ejdr.2014.33.

Benjaminsen, T. A., and I. Bryceson. 2012. Conservation, green/blue grabbing and accumulation by dispossession in Tanzania. *Journal of Peasant Studies* 39 (2):335–55. doi:10.1080/03066150.2012.667405.

Boone, C. 2017. Sons of the soil conflict in Africa: Institutional determinants of ethnic conflict over land. *World Development* 96:276–93. doi:10.1016/j.worlddev.2017.03.012.

Bruton, M. N. 1990. The conservation of the fishes of Lake Victoria: Africa: An ecological perspective. *Environmental Biology of Fishes* 27 (3):161–75.

Buzan, B., O. Waever, and J. de Wilde. 1998. *Security: A new framework for analysis.* Boulder, CO: Lynne Rienner Publishers.

Byaruhanga, M. 2015. Why President Museveni is still a darling of Ugandans. *New Vision*, January 15:16.

Carmody, P., and D. Taylor. 2016. Globalization, land grabbing, and the present-day colonial state in Uganda. *Journal of Environment & Development* 25 (1):100–126. doi:10.1177/1070496515622017.

Carother, T. 2002. The end of the transition paradigm. *Journal of Democracy* 13 (1):5–21.

Cavanagh, C., and T. A. Benjaminsen. 2014. Virtual nature, violent accumulation: The "spectacular failure" of carbon offsetting at a Ugandan national park. *Geoforum* 56:55–65. doi:10.1016/j.geoforum.2014.06.013.

Colchester, M. 2006. Forest peoples, customary use and state forests. Paper presented at the Conference of the International Association for the Study of Common Property, Bali, Indonesia, June 19–23.

Doshi, S., and M. Ranganathan. 2017. Contesting the unethical city: Land dispossession and corruption narratives in urban India. *Annals of the American Association of Geographers* 107 (1):183–99. doi:10.1080/24694452.2017.1419414.

Ece, M., J. Murombedzi, and J. Ribot. 2017. Disempowering democracy: Local representation in community and carbon forestry in Africa. *Conservation & Society* 15 (4):357–70. doi:10.4103/cs.cs_16_103.

Etukuri, C. 2016. Uganda, Kenya agree on Migingo Island. *New Vision*, August 24. Accessed November 30, 2017. https://www.newvision.co.ug/new_vision/news/1433591/uganda-kenya-agree-migingo-island.

Freedom House. 2017. Freedom in the world 2017: Uganda. Accessed October 18, 2017. https://freedomhouse.org/report/freedom-world/2017/uganda.

Goh, D. P. S. 2015. Colonialism, neopatrimonialism, and hybrid state formation in Malaysia and the Philippines. In *Patrimonial capitalism and empire*, ed. M. Charrad and J. Adams, 165–90. Bingley, UK: Emerald Group.

Government of Uganda. 1995. Constitution of the Republic of Uganda. Accessed July 31, 2018. https://ulii.org/node/23824.

Hajer, M., and W. Versteeg. 2005. A decade of discourse analysis of environmental politics. *Journal of*

Environmental Policy and Planning 7 (3):175–84. doi:10.1080/15239080500339646.

Hajusa, E. 2016. Museveni applauded on fisheries officers ban. *New Vision*, January 18:20.

Hajusu, E., and F. Odeke. 2015. Don't waste time voting for the opposition, says Museveni. *New Vision*, January 20:3.

Human Rights Watch. 2017. World report—Uganda. Accessed November 30, 2017. https://www.hrw.org/world-report/2017/country-chapters/uganda.

Izama, A. 2015. Family therapy: Dynasty and change in Uganda. *African Arguments*, June 29. Accessed December 15, 2017. http://africanarguments.org/2015/06/29/family-therapy-dynasty-and-change-in-uganda-by-angelo-izama/.

Kaaya, S. K. 2017. Uganda: NRM is the master of violence—Museveni. *The Observer*, October 18. Accessed November 22, 2017. http://allafrica.com/stories/201710180045.html.

Kaggwa, K. 2015. The secret behind Museveni's power. *New Vision*, January 26:40.

Kagoro, J. 2016. Competitive authoritarianism in Uganda: The not so hidden hand of the military. *Zeitschrift für Vergleichende Politikwissenschaft* 10 (Suppl. 1):155–72. doi:10.1007/s12286-015-0261-x.

Kasasira, R. 2015. Brig Tumukunde promoted, then retired. *Daily Monitor*, September 1. Accessed November 24, 2017. http://www.monitor.co.ug/News/National/Brig-Tumukunde-promoted–then-retired/688334-2854028-k7slfyz/index.html.

Kato, J. 2017. Uganda: Special forces command to take charge of Migingo Island. *The Monitor*, February 14. Accessed November 30, 2017. http://allafrica.com/stories/201702150041.html.

Kazibwe, K. 2015. Museveni dissolves beach management units. *Chimp Reports*, November 11. Accessed September 26, 2017. http://www.chimpreports.com/museveni-dissolves-beach-management-units/.

Kwesiga, P. 2015. Museveni suspends fisheries officers. *New Vision*, November 16:3.

———. 2017a. Illegal fishing: Museveni wants offenders jailed. *New Vision*, March 15:3.

———. 2017b. Trade Ministry bans fishing gear imports. *New Vision*, March 31:10.

Kwesiga, P., and C. Etukuri. 2017. Crime on Lake Victoria: Part I. *New Vision*, June 15:18–19.

Lankford, B., K. Bakker, M. Zeitoun, and D. Conway, eds. 2013. *Water security: Principles, perspectives and practices.* London and New York: Routledge.

Levitsky, S., and L. A. Way. 2010. *Competitive authoritarianism: Hybrid regimes after the cold war.* Cambridge, UK: Cambridge University Press.

Lubulwa, H. 2016. President to give Kalangala residents free fishing gear. *Daily Monitor*, February 8. Accessed February 9, 2017. http://www.monitor.co.ug/Elections/President-to-give-Kalangalaresidents-free-fishing-gear/2787154-3066674-kslsurz/index.html.

Lumu, D. 2015. Museveni disbands fish policing units. *New Vision*, November 11:3.

———. 2016. Key issues that shaped 2016 presidential campaigns. *New Vision*, February 17:17.

Lyons, K., and P. Westoby. 2014. Carbon colonialism and the new land grab. *Journal of Rural Studies* 36:13–21. doi:10.1016/j.jrurstud.2014.06.002.

Maldonado-Torres, N. 2007. On the coloniality of being. *Cultural Studies* 21 (2–3):240–70. doi:10.1080/09502380601162548.

Mamdani, M. 1996. *Citizen and subject: Contemporary Africa and the legacy of late colonialism.* Princeton, NJ: Princeton University Press.

Masaba, J. 2015. Taking stock of multiparty politics. *New Vision*, January 26:79.

Mayanja, S. 2016. Political elite failing multiparty democratic process in Uganda. *New Vision*, April 11:26.

McDonald, M. 2008. Securitization and the construction of security. *European Journal of International Relations* 14 (4):563–87. doi:10.1177/1354066108097553.

Mugerwa, R., and A. Bagala. 2016. Uganda threatens war on DR Congo over Lake Albert attacks. *The East African*, May 26. Accessed July 31, 2018. http://allafrica.com/stories/201605241238.html.

Muhereza, R. 2016. 34 Rwandans deported. *Daily Monitor*, April 25. Accessed November 30, 2017. http://www.monitor.co.ug/News/National/34-Rwandans-deported-/688334-3174490-b0lc61z/index.html.

Musisi, F. 2016. Press freedom: Uganda on the decline. *Daily Monitor*, May 3. Accessed October 18, 2017. http://www.monitor.co.ug/News/National/Press-freedom–Uganda-on-the-decline/688334-3186650-3fkci5/index.html.

Nakaweesi, D. 2016. Why Uganda is not exporting enough products. *Daily Monitor*, December 14. Accessed December 15, 2017. http://www.monitor.co.ug/Business/Prosper/Why-Uganda-is-not-exporting-enough-products/688616-3485226-q94fohz/index.html.

Nel, A. 2015. The neoliberalisation of forestry governance, market environmentalism and re-territorialisation in Uganda. *Third World Quarterly* 36 (12):2294–2315. doi:10.1080/01436597.2015.1086262.

Neumann, R. P. 2009. Political ecology: Theorizing scale. *Progress in Human Geography* 33 (3):398–406. doi:10.1177/0309132508096353.

Njiru, M., J. Kazungu, C. C. Ngugi, J. Gichuki, and L. Muhoozi. 2008. An overview of the current status of Lake Victoria fishery. *Lakes & Reservoirs: Research & Management* 13 (1):1–12. doi:10.1111/j.1440-1770.2007.00358.x.

Nunan, F. 2006. Empowerment and institutions: Managing fisheries in Uganda. *World Development* 34 (7):1316–32. doi:10.1016/j.worlddev.2005.11.016.

Nunan, F., J. Luomba, C. Lwenya, E. Yongo, K. Odongkara, and B. Ntambi. 2012. Finding space for participation: Fisherfolk mobility and co-management of Lake Victoria fisheries. *Environmental Management* 50 (2):204–16. doi:10.1007/s00267-012-9881-y.

O'Donnell, G. 1996. Illusions about consolidation. *Journal of Democracy* 7 (2):34–51.

Okethwengu, B. 2017. Congolese arrested fishing in L. Albert. *New Vision*, April 21:13.

Peluso, N. L. 1994. *Rich forests, poor people: Resource control and resistance in Java.* Berkeley: University of California Press.

Quijano, A., and E. Michael. 2000. Coloniality of power, eurocentrism, and Latin America. *Neplantla* 1 (3):533–80.

Ribot, J. C. 2004. *Waiting for democracy: The politics of choice in natural resource decentralization.* Washington, DC: World Resource Institute.

Ribot, J. C., A. Agrawal, and A. M. Larson. 2006. Recentralizing while decentralizing: How national governments re-appropriate forest resources. *World Development* 34 (11):1864–86. doi:10.1016/j.worlddev.2005.11.020.

Robbins, P. 2009. The practical politics of knowing. *Economic Geography* 76 (2):126–44. doi:10.1111/j.1944-8287.2000.tb00137.x.

Sauper, H. 2004. *Darwin's nightmare.* Accessed October 29, 2018. http://www.darwinsnightmare.com.

Schneider, L. 2006. Colonial legacies and postcolonial authoritarianism in Tanzania. *African Studies Review* 49 (1):93–118. doi:10.1353/arw.2006.0091.

Scott, J. C. 1992. *Domination and the arts of resistance: Hidden transcripts.* New Haven, CT: Yale University Press.

Sekanjako, H., M. Karugaba, M. Mulondo, and M. Walubiri. 2016. Govt to register all fishermen. *New Vision*, June 1:7.

Sekanjako, H., P. Okino, and H. Apunyo. 2017. UPDF to stamp out illegal fishing, says Museveni. *New Vision*, February 7:3.

Tangri, R., and A. M. Mwenda. 2010. President Museveni and the politics of presidential tenure in Uganda. *Journal of Contemporary African Studies* 28 (1):31–49. doi:10.1080/02589000903542574.

Tripp, A. M. 2010. *Museveni's Uganda: Paradoxes of power in a hybrid regime.* Boulder, CO: Lynne Rienner.

Uganda exports by category. 2017. *Trading Economics.* Accessed December 15, 2017. https://tradingeconomics.com/uganda/exports-by-category.

Uganda introduces bill to remove presidential age limit. 2017. *Al Jazeera English*, September 27. Accessed November 22, 2017. http://www.aljazeera.com/news/2017/09/uganda-introduces-bill-remove-presidential-agelimit-170927172204813.html.

Uganda enacts law ending presidential age limits. 2018. *Al Jazeera English*, January 2. Accessed July 31, 2018. https://www.aljazeera.com/news/2018/01/uganda-enacts-lawpresidential-age-limits-180102182656189.html.

ANNE J. KANTEL is a Doctoral Candidate at the School of International Service at American University, Washington, DC 20016. E-mail: anne.kantel@american.edu. Identifying as a scholar of political ecology, her work engages with broader questions of justice, power, and equality in natural resource governance in sub-Saharan Africa.

From the Heavens to the Markets: Governing Agricultural Drought under Chinese Fragmented Authoritarianism

Afton Clarke-Sather (iD)

The applicability of liberal governance models in authoritarian contexts has been widely debated. This study contributes to this debate by using a Foucauldian analytic of apparatuses of security to understand how Chinese state actors used the liberalization of agricultural markets to indirectly mitigate against the risk of droughts. Apparatuses of security, which rely on indirect governance through market and biopolitical mechanisms, were proposed by Foucault to explain the emergence of governmentality in liberal states. The Chinese state has been described as a system of fragmented authoritarianism, wherein state actors are isolated from and compete with one another. Food policy in the early People's Republic of China extended this isolation to the point of encouraging local autarky at the county level. Around the year 2000, local officials in northwest China began promoting the replacement of subsistence agriculture with drought-resistant commercial agriculture as a means of mitigating against drought. This shifted the source of risk of food shortfalls from an environmental risk of drought to a state-mediated market risk and displaced the long-standing model of local autarky providing food security. By illustrating the dynamics of how Chinese state actors govern nature through market mechanisms, this study contributes to theorizations of how liberalization can function to govern nature in authoritarian contexts.

自由主义的治理模式在威权脉络中的应广受辩论。本研究运用傅柯对于安全配置（apparatuses）的分析，理解中国政府行动者如何运用农业市场的自由化，间接减轻乾旱的风险，从而对上述辩论做出贡献。傅柯所提出的安全配置，倚赖通过市场与生命政治机制的间接治理，用来解释治理术在自由主义国家中的浮现。中国政府向来被描绘为破碎的威权主义系统，其中各个国家行动者被各自孤立并相互竞争。中国人民共和国早期的粮食政策，将此般孤立延伸至鼓励农村层级的在地自力更生之程度。西元2000之际，中国西北部的地方官员，开始推动以抗旱的商业作物来取代粮食作物，作为缓解乾旱的工具。这麽做，将粮食短缺的风险来源，从乾旱的环境风险转换成国家媒介的市场风险，并取代了在地自力更生以提供粮食安全的长期模式。通过描绘中国政府行动者如何通过市场机制治理自然之动态，本研究对于理论化自由化如何能够用在威权脉络中用来治理自然做出贡献。 *关键词: 安全配置, 中国, 乾旱, 治理术, 水—粮食安全。*

La aplicabilidad de modelos de gobernanza liberal en contextos autoritarios ha sido ampliamente debatida. Este estudio contribuye en este debate usando una analítica foucaultiana de aparatos de seguridad para entender cómo actores del estado chino usaron la liberalización de los mercados agrícolas para mitigar indirectamente contra el riesgo de sequías. Los aparatos de seguridad, que dependen de la gobernanza indirecta gracias a mecanismos biopolíticos y de mercado, fueron propuestos por Foucault para explicar la aparición de gobernabilidad en estados liberales. El estado chino ha sido descrito como un sistema de autoritarismo fragmentado, en donde los actores del estado están aislados entre sí y compiten unos con otros. Al comienzo de la República Popular de China, la política alimentaria extendió ese aislamiento hasta el punto de estimular la autarquía local a nivel de condado. Alrededor del año 2000, oficiales locales en la China del noroeste empezaron a promover el remplazo de la agricultura de subsistencia con una agricultura comercial resistente a las sequías, como un medio de mitigación contra la sequía. Esto cambió la fuente del riesgo del déficit de alimentos desde un riesgo de sequía ambiental a un riesgo de mercado mediado por el estado, y desplazó el modelo de vieja data de la autarquía local como fuente de seguridad alimentaria. Ilustrando la dinámica de cómo los actores del estado chino gobiernan la naturaleza a través de mecanismos de mercado, este estudio contribuye a la teorización del modo como la liberalización puede funcionar para gobernar la naturaleza dentro de contextos autoritarios. *Palabras clave: aparatos de seguridad, China, gobernabilidad, seguridad de agua-alimentos, sequía.*

In north China, peasants have long said that they "depend on the heavens to eat" (*kaotian chi fan*), a sentiment that belies the risk of agricultural drought that the region has long faced. Yet, during my 2010 fieldwork on agricultural modernization and water, a joke had emerged among peasants in rural northwest China, who said that today they "depend on the markets to eat" (*kao shichang chifan*). This small shift is profoundly important, because it indicates a shift in the perception of food shortage from being a malady arising from natural drought to being a malady with origins that lie in the vagaries of the markets. Yet this shift in peasants' perceptions is also deeply tied to a shift in how the Chinese state has governed the risk of agricultural drought and has much to tell us about how liberalization of environmental governance might coexist with continuing authoritarian forms of governance. I employ Foucault's analytic of apparatuses of security to explain the use of markets to govern drought through agricultural trade in contemporary China. Foucault used apparatuses of security to explain the origins of governmentality and its connections to environmental governance in early liberal states, which he contrasted with sovereign and disciplinary forms of power. I argue that although the mechanisms through which state entities relate to one another, namely, statistical reporting and inspections, remain based on disciplinary expression of power, the way in which state entities relate to nature has come to be governed through apparatuses of security.

I explore these changing relationships between state actors and nature through an examination of state grain policy since the Mao era and ground this analysis in the particular agricultural production of Anding District, in Gansu province, where I have conducted ethnographic and survey-based fieldwork on multiple occasions between 2010 and 2017. This article is based primarily on historical sources documenting the changes that have shifted agricultural governance and drought in this region. Anding District sits high in China's Loess Plateau. It has a semiarid environment (380 mm rainfall annually) and has long been particularly vulnerable to drought. Informants often joked that nine of ten years in Anding are drought, and they are not far off. From 1956 to 1985 only eight years did not experience any seasonal drought (defined as rainfall at least 25 percent below average during one of three growing phases; Dingxi Gazetteer Editing Committee 1990). Drought is a preoccupation of both the local populace and the local state. This article analyzes how the local state has governed drought as follows. First, I examine Foucault's governmental forms of power and place them in conversation with disciplinary and sovereign forms of power. I then examine grain policy and the structure of the rural state during the Maoist period, with particular concern for drought management in Anding District. Finally, I examine how interregional trade was framed as a drought mitigation strategy in the Reform and Opening Period, again focusing on Anding District, before offering some conclusions.

Authoritarian States, Liberal Governmentality, and Apparatuses of Security

Authoritarianism and liberalism are often presented in binary terms. Yet most states contain a mixture of liberal and illiberal elements of rule (Koch 2014), and characterizing states as liberal or illiberal is more often a question of the particular constellations of liberal and illiberal elements than an existential condition. These liberal or illiberal elements can be conceptualized as specific practices of rule, with states employing differing practices of political power with different populations and in addressing different issues. It is productive to think through these differences by contrasting Foucault's analytics of sovereign, disciplinary, and governmental forms of power. To Foucault, governmental forms of power came to supplement but not supplant sovereign and disciplinary forms of power.

Foucault (2007) defined governmentality as "the ensemble formed by institutions, procedures, analyses, and reflections, calculations, and tactics that allow the exercise of ... power that has the population as its target, political economy as its major form of knowledge, and apparatuses of security as its essential technical instrument" (108). Apparatuses of security have several hallmarks that have been examined elsewhere (see Elden 2007); here I emphasize two. First, apparatuses of security emphasize the circulation of objects, ideas, and people. Whereas discipline operates in a centripetal way, through protection and enclosure, apparatuses of security operate centrifugally by enrolling new elements "allowing the development of ever-wider circuits" of power (Foucault 2007, 45). Second, apparatuses of security do not attempt to regulate nature but instead work with and through nature. In contrast to sovereign forms of power, apparatuses of security view nature as "not something on which, above which, or against which the sovereign must impose just laws … nature is such that the sovereign must deploy reflected procedures of government within this nature, with the help of it, and with regard to it (Foucault 2007, 75). In describing apparatuses of security, Foucault examined seventeenth-century French grain policy and the shift from state-controlled policy to prevent famine to the laissez-faire policies of a group of political economists known as the Physiocrats. Foucault identified the Physiocrats' policies as both requiring the use of circulation in place of territorial circumscription and being "natural" insofar as localized grain shortages (scarcity dearness) prevented broader famines (scarcity scourge). These policies sat in contrast to disciplinary and sovereign forms of power, which sought to centralize power, circumscribe space, and strictly regulate nature (see Elden 2007). Foucault's apparatuses of security are particularly well suited to understanding environment governance and have been drawn on by political ecologists of water interested in the indirect management of water and the uses of markets for water management (Alatout 2013; Clarke-Sather 2017). This article contributes to these studies by illustrating how apparatuses of security allow us to think through the liberal–illiberal binary in environmental governance and allows us to see how illiberal states might deploy liberal forms of rule.

Although China is widely viewed as an authoritarian state, Lieberthal and Oksenberg (1988) proposed that it is better understood as a fragmented authoritarian state that, although authoritarian in character, contains multiple and often competing bureaucracies, leading to a decentralized decision-making process. Throughout this article I refer to these bureaucracies as *state actors*, those constituent elements of the state that act with the imprimatur of state support. These state actors are organized both vertically into *tiao* or functional bureaucracies that stretch from Beijing to the county level (e.g., water resource agencies) and horizontally into *kuai* or territorial administrations such as counties and prefectures. Tensions exist between different functional bureaucracies, different levels of territorial administrations, and functional bureaucracies and territorial administrations. One particular realm in which the fragmented authoritarian framework has been applied is environmental governance. The Three Gorges Dam was one of the case studies examined by Lieberthal and Oksenberg (1988), and the framework has seen broad application in the areas of both water (Mertha 2008; Nickum 2010; Magee 2013) and energy and climate (Marks 2010). The case of water management provides an example of the often competing priorities of different functional bureaucracies. For example, both the Ministry of Water Resources and the Ministry of Environmental Protection claim jurisdiction over water quality monitoring and regulation, resulting in both organizations undertaking duplicative monitoring and planning efforts (Nickum 2010). More recent literature on environmental governance in China has conceptualized these groups as the "water machine" of interrelated state interests that at times coordinate and at others compete (Crow-Miller, Webber, and Rogers 2017) and assemblages that work with this water machine (Webber and Han 2017). In Anding District, authority over connections between food and water is managed by a constellation of state actors that resembles Webber and Han's (2017) assemblage. Funding for state projects to manage these connections originates in the central state, particularly the State Leading Council on Poverty Alleviation. Ground-level implementation of projects relating to water–food connections are carried out by prefecture-, county- and township-level governments (the kuai) or by functional bureaucracies, particularly the district-level agricultural and water bureaus (the *tiao*). The arrangement of this constellation of state actors has shifted from the Maoist period to the present day.

The Cellular State, Grain, and Drought during the Maoist Period

The management of food, famine, and drought by rural state actors in China can be placed in conversation with Foucault's analysis of the Physiocrats' approach to famine as a prototypical apparatus of security (Foucault 2007). Prior to the rise of the Physiocrats, the prevention of famine was accomplished through disciplinary approaches used to maintain food price stability: price controls, strict control of movement and export of grain, and careful accounting of where grain was stored. Control of food during the Maoist era (1949–1979) relied on similar techniques, featuring quotas on production, grain taxes in rural areas used to subsidize urban populations, control of storage and distribution, limitations on trade in food outside of the state, and prohibition on nongrain food crops to preserve land for planting grain. The central task of the state in rural areas at this time, the extraction of grain, was carried out in a way that is quite similar to Foucault's description of disciplinary apparatuses for grain management as "a series of controls on prices, storing, export, and cultivation" (Foucault 2007, 32).

During the Maoist period, the Chinese state reorganized the economy of the countryside based on a locally autarkic model that operated territorially through mechanisms of discipline. Beginning in the mid-1950s, the central state emphasized the role of the commune as the building block of the rural economy and aimed for each commune to produce as much of its own materials internally as possible (Shue 1988; Yang 1998; Naughton 2007). The result was a pattern of local units at the commune (contemporary township) level that were isolated from one another both in material flows and in communications and only reported vertically to the center, a pattern that Shue (1988) called the cellular state. Although the cellular political structure of a local elite mediating between the central state and the rural populace predated the People's Republic of China (PRC; Shue 1988), before the Maoist period this cellular structure did not extend to the economic sphere (with the exception of key state monopolies such as salt). PRC policies during the Maoist period cut off existing trade networks between places, favoring instead increased self-dependence within each territorial unit and requiring approval from a higher level to conduct trade between counties (Shue 1988). The result was a reduction of specialization in rural production. As all regions sought to produce grain, specialized crops, such as cotton, were reduced (Yang 1998).

This cellular model of the state exercised territorial power from the center in a disciplinary manner, creating "empty, closed space(s)" (Foucault 2007, 17) that are organized according to the principles of hierarchy, separation, and visibility. In the Maoist era, rural spaces were demarcated into an administrative hierarchy, with each territorial unit occupying a specific place in the hierarchy of the Chinese state. Communication and trade between spaces at the same level (whether commune, township, or county) was sharply limited, with the expectation that all communication and material flows would travel vertically. Finally, the state juridically oversaw the distribution of most materials and goods, including food, housing, education, and medical care.

Expressions of power between state units were accomplished through two disciplinary techniques: statistical reporting and inspection tours. Spaces were classified as territorial units and statistically measured. Reporting requirements were created to make the local state visible and legible to central state actors, specifically in reporting of harvests and availability of grain. This legibility was created in such a way that local state actors would know that their actions were being monitored and would self-regulate accordingly (Oi 1989), a hallmark of Foucault's disciplinary forms of power. Specific grain production targets were set for each work unit of the local state for production, and statistical reporting revealed whether they had been accomplished or not. These statistical reports were supplemented by a second form of disciplinary power: inspection tours by state leaders. Inspection tours function both as a disciplinary form of state power insofar as they create a specific type of visibility of through firsthand observation by higher level state actors and as a performative act of sovereignty with deep roots in imperial China (Chang 2007). Oi (1989) showed that inspection teams were most likely to check on those units that had become exceptional, whether they consistently failed to meet grain production quotas or showed exemplary production. Those local cadres who could keep their reports to the central state within broadly acceptable limits were unlikely to elicit spot inspections.

The management of grain must also be understood as a means of managing natural disasters, particularly floods and droughts. For more than 2,000 years the Chinese state has managed some form of famine relief program (L. Li 1982). A portion of state grain requisitions through grain taxes was distributed during times of famine, approximately 70 percent of which were tied to droughts or floods (Shuie 2004). These grains were moved around the country, primarily from south China to the north, although most famine relief grain was stored in state-backed local granaries. During the Maoist period, the PRC largely adopted the system of famine relief from the imperial period (L. Li 1982). As a strategy for exerting state power over drought and its primary impact, famine relief from the imperial and PRC periods bears the marks of sovereign forms of power: Control of relief grain was concentrated in the central state, which has the power to distribute relief according to the will of the sovereign, and aid was sent to demarcated territories based on statistical reporting of shortages. Like Foucault's descriptions of seventeenth-century French grain policy, the imperial and PRC system managed grain, and indirectly drought and famine, through control of the material elements of grain that was regulated centrally by the state. The granting of relief often also involved the ritual of imperial inspection tours, which, in addition to providing a means of surveillance, allowed for the spectacle of the sovereign providing relief (Chang 2007).

In the case of Anding, the Maoist-era disciplinary mechanisms managing nature can be seen in the food policies to manage drought. In the 1970s the Anding District water bureau attempted to solve the problem of drought by constructing a series of eight reservoirs to provide irrigation water. Importantly, these reservoirs were all constructed within the county boundaries, and the water that was collected from these reservoirs was directed toward irrigation only within the borders of the county (Clarke-Sather 2012), fitting the model of local autarky. This irrigation water was, in turn, used to support the cultivation of wheat, which was the primary staple crop emphasized in annual statistical reports. Irrigation as a solution to the problem of drought represents disciplinary forms of power insofar as nature was something that the sovereign acted against, with dams and diversions widely seen as expressions of modern state power (Swyngedouw

1999). Power over nature was centralized in state actors who attempted to prevent famine through the regulation of nature. This irrigation system was prototypical of widespread irrigation development across north China during the Maoist period, which aimed to "free agricultural productivity from the constraints of nature" (Crow-Miller, Webber, and Rogers 2017, 237). Based more on a politics of mobilization (Perry 2011) than on sound engineering, these dams, which were completed around 1980, were, with one exception, unusable due to siltation · by the early 1990s. One dam remains in use at one third of its original capacity and irrigates a small region of the valley.[1]

Coincident with the failure of the irrigation system in the early 1990s, Anding District experienced the worst drought in sixty years, and this drought revealed the vestiges of another institution of the Maoist period, the granting of state relief grain and visits by state leaders. In 1995, during the worst of the drought, Chinese Premier Li Peng visited Anding to show the concern of the state for those suffering from drought and provide relief to the stricken (Zhang and Chen 1995; Cook 2004). Such trips by state leaders during times of natural disaster have a history dating back millennia and are used by the central state to perform sovereignty during times of natural disaster (Chang 2007). Li Peng's visit to Anding, however, also marked a turning point. Although the reform and opening process had begun in coastal areas fifteen years earlier (1979), liberalization of trade was slow to arrive in western China, which remained relatively isolated. The drought of 1995, however, prompted a series of changes in how the local state approached managing the food–water nexus in times of drought.

Apparatuses of Security and Reform-Era Grain and Drought Management

Many have argued that a central impetus for the Reform and Opening policies that changed China's economy beginning in 1979 was a crisis in food (Yang 1998). In the late 1970s, Chinese grain imports grew rapidly, from 5.69 million tons in 1977 to 10.69 million tons in 1979, and many of the reforms of the early Reform and Opening Period were born of a crisis of food production centered in the rural areas (X. Li, Wang, and Jia 2011). These began with technical changes to how grain was procured and the introduction of the household responsibility system, which disassembled the communes by contracting plots of land to families to farm, and culminated in the 2006 elimination of the millennia-old agricultural production tax (which was initially a means by which the state appropriated grain; Kennedy 2007). Each reform represented a move away from a disciplinary approach to food security and toward relying on apparatuses of security based on the circulation of grain to provide sufficient food supply. The logics involved were similar to those presented by Foucault's Physiocrats, namely, that allowing markets to increase prices would encourage work teams, and later farmers, to plant more grain. Each of these reforms has paralleled the logic that Foucault used to describe the Physiocrats, promoting trade between places and the movement of grain to allow for "natural" processes to take their course. Yet, as Foucault illustrated, the creation of new forms of rule based on circulation have also altered the structure of the state. The Reform and Opening Period allowed new forms of trade between different localities for essential produce and, as such, the cellular structure of isolated state units began to break down, replaced by organizations that were "urged to spread and sprawl, free-form and web like, as they follow the 'natural' networks of commercial exchange between city and countryside" (Shue 1988, 131).

A key element of this agricultural liberalization in the Reform and Opening Period has been to encourage different locales to specialize in locally unique crops, with the expectation that trade between different places will then lead toward rural development. This specialization in agriculture led to a decrease in the percentage of cultivated land planted to staple grains from 80.3 percent to 68.3 percent between 1978 and 2008, with specialty crops expanding correspondingly (Alperman 2011). One place where this specialization has occurred is in Anding, where agriculture has increasingly specialized in producing potatoes as a means of drought mitigation.

Shortly after Li Peng's visit during the drought of 1995, the county government in Anding District began a concerted attempt to switch to growing potatoes as a way to mitigate against drought. Historically subsistence agriculture in Eastern Gansu balanced crops that matured in different seasons of the year to protect against the possibility of drought in any one season (Clarke-Sather 2015). Potatoes were part of this mixture. Summer grains (primarily wheat) required early-season rainfall and often withered in droughts during May and June. Potatoes, in contrast, required late summer rainfall during July, August, and September (Shang 2007; Yan 2008), the three months during which 60 percent of the annual precipitation in Anding District falls (Wei, Li, and Liang 2005). Following the drought of 1995, local government leaders promoted commercial potato agriculture as a means to address the problem of drought in Anding District, discursively framing the switch to potatoes as "going with nature and the seasons" (shunying tianshi; Yan 2008). The first efforts to create potatoes as a way of governing drought emerged in 1996 when the prefecture party committee proposed the "Potato Project" (yangyu gongcheng, literally potato engineering project), which included five goals: increasing the planted area dedicated to potatoes, improving seeds, improving yields, increasing the portion of potatoes used for industrial uses, and increasing the portion of potatoes exported to other areas (Yan 2008).

None of these goals emerged from organic processes but instead emerged through a series of state interventions. Potato breeding facilities were established by the prefecture and provincial governments, which created a new variety, Xindaping, which was specialized to the area and commands a price premium of approximately 10 percent. Agricultural extension was introduced by the county agricultural bureau to encourage farmers to plant new varieties of potatoes. An industrial center for potato food products, primarily potato starch, was established in the town of Chankou. Integration with national markets was the result of specific policies of the county- and prefecture governments as well. Storage and marketing facilities were built by both county- and township-level governments, and the county subsidizes trains that deliver potatoes to markets in eastern China. These measures were successful in transforming Anding's agriculture; by 2005, less than ten years after the potato project was begun, potatoes had surpassed wheat as the most widespread crop in Anding (Clarke-Sather 2012). Although promoting potato agriculture was a way of governing through the markets, the markets themselves did not emerge by "nature" but instead were created through a series of state interventions.

The promotion of potato agriculture in Anding fits the paradigm of a Foucauldian apparatus of security to govern peasants' relationships with drought. Rather than trying to regulate food supply directly within a circumscribed territory as previous efforts at irrigation to grow wheat had done, potato marketing aimed to change the constellation of social and natural forces that related to the food–water nexus. Potatoes were selected as a commercial crop resistant to drought whose demand aligned with seasonal precipitation patterns. The connection between food supply and events of water shortage was now facilitated by markets (which were constructed by state actors) to promote the circulation of goods, leading peasants to say that the risk of food shortage now lay in the markets, rather than the heavens. State discourses that promoted potatoes framed this as a case of allowing nature to take its own course, another facet of apparatuses of security. The term "going with nature and the seasons" was used to describe the process of working with nature, rather than against it, that informed potato promotion efforts. In official discourses around potato promotion in Anding, state actors have called for following the Three Obeys: "going with nature and the seasons," "going with the market," and "going with the epoch" (Wang 2012). Whereas the first two seem somewhat clear, the last is defined by Wang (2012) to mean "going with the laws of science." This idea of going with nature and naturalizing the markets is the hallmark of apparatuses of security, which emphasize working with, rather than against, nature. This application of apparatuses of security to the problem of drought stands in stark contrast to governing through the provision of irrigation water. Potato agriculture does not attempt to shift or discipline how agricultural water behaves, instead taking agricultural water as it comes and shifting the constellation of social forces that surround it.

Conclusion

From the Maoist period to the reform era, state actors in Anding shifted from combating drought through irrigated grain production to ameliorating drought through crop specialization, climate adaptation, and market mechanisms. We see in this shift a move from governing nature through mechanisms of discipline toward governing through apparatuses of security. State actors' direct regulation of nature by providing irrigation water has been replaced by government regulation of drought by means of climatic adaptation, "going with nature and the seasons," and enrolling agriculture into ever-expanding circuits of material circulation. These adaptations would point toward Chinese state actors exercising what are apparently liberal practices of rule. Yet, the operation of the Chinese state itself remains based primarily on disciplinary exercises of power.

One way of understanding this shift is with reference to the "hollowing out" of the local state that occurred during the 1990s and 2000s, when the county and township levels of government became less powerful and in the process were forced to become more creative in finding how to provide services (Smith 2010). Yet despite reduced revenue and a reduced role, local governments are still expected to provide services, still face evaluation based on their ability to meet statistical targets, and still face the possibility of inspection tours. The relationships between levels of the state continue to operate through the disciplinary practices of legibility, namely, statistical reporting and inspection. Township leaders continue to chafe under the reports filled with metrics that they must send to their superiors (Smith 2010), and now these statistics are more likely to feature growth in trade and investment than grain production. The concurrence of state units relating to one another through disciplinary mechanisms, simultaneously governing nature through apparatuses of security, illustrates Foucault's point that governmental forms of power have come to supplement rather than supplant disciplinary and sovereign forms of power and that diverse practices of rule coexist and reinforce one another. This case also builds on Koch's (2014) view that authoritarianism must be viewed through the practices that states employ rather than a liberal–illiberal binary by showing that selective liberalization in environmental governance is indeed constitutive of authoritarian rule. Liberalization of grain trade and drought management in some ways ensured the durability of the PRC's unique authoritarian state. Moreover, this case illustrates that authoritarianism is not synonymous with centralization. Whereas authoritarianism tends to be associated with centralization of state power the deployment of apparatuses of security in Chinese policy toward drought shows how authoritarianism could be accomplished by practices of rule that are themselves diffusing of power.

Funding

This research was financially supported by: the U.S. Department of Education's Fulbright Hays program (award P022A090013), the National Science Foundation (award 0927391), the Political Geography Specialty Group of the American Association of Geographers, the University of Colorado's Beverly Sears program, and the University of Delaware Global Scholars Program.

ORCID

Afton Clarke-Sather ⓘD http://orcid.org/0000-0002-5428-9415

Note

1. Anding District is also the designated beneficiary of the Tao River Transfer, an interbasin transfer project. As of the 2017 growing season (three years after opening), however, farmers were not reliably receiving irrigation water.

References

Alatout, S. 2013. Water scarcity in late modernity. In *Contemporary water governance in the Global South: Scarcity, marketization and participation*, ed. L. Harris, C. Sneddon, and J. Goldin, 101–08. London and New York: Routledge.

Alperman, B. 2011. Introduction. In *Politics and markets in rural China*, 1–13. London and New York: Routledge.

Chang, M. G. 2007. *A court on horseback*. Cambridge, MA: Harvard University Asia Center.

Clarke-Sather, A. 2012. State Development and the Rescaling of Agricultural Hydrosocial Governance in Semi-Arid Northwest China. *Water Alternatives* 5 (1):98–118.

———. 2015. Hydrosocial governance and agricultural development in semi-arid northwest China. In *Negotiating water governance: Why the politics of scale matter*, ed. E. S. Norman, C. Cook, and A. Cohen, 247–62. London: Ashgate.

———. 2017. State power and domestic water provision in semi-arid northwest China: Towards an aleatory political ecology. *Political Geography* 58:93–103.

Clarke-Sather, A., X. Tang, Y. Xiong, and J. Qu. 2017. State development and the rescaling of agricultural hydrosocial governance in semi-arid northwest China. *Water Alternatives* 10 (1):111–18.

Cook, S. 2004. *Rainwater harvesting in Gansu Province, China: Development and modernity in a state-sponsored rural water supply project*. New Haven, CT: Yale University.

Crow-Miller, B., M. Webber, and S. Rogers. 2017. The technopolitics of big infrastructure and the Chinese water machine. *Water Alternatives* 10 (2):233–49.

Dingxi Gazetteer Editing Committee. 1990. *Dingxi xianzhi* [Dingxi gazetteer]. Lanzhou, China: Gansu People's Press.

Elden, S. 2007. Governmentality, calculation, territory. *Environment and Planning D: Society and Space* 25 (3):562–80.

Foucault, M. 2007. *Security, territory, population*, ed. M. Senellart. New York: Palgrave Macmillan.

Kennedy, J. J. 2007. From the tax-for-fee reform to the abolition of agricultural taxes: The impact on township governments in North-west China. *The China Quarterly* 189:43–59.

Koch, N. 2014. Bordering on the modern: Power, practice and exclusion in Astana. *Transactions of the Institute of British Geographers* 39 (3):432–43.

Li, L. 1982. Introduction: Food, famine, and the Chinese state. *The Journal of Asian Studies* 41 (4):687–707.

Li, X., S. Wang, and Y. Jia. 2011. Grain market and policy in China. In *Politics and markets in rural China*, ed. B. Alperman, 89–105. London and New York: Routledge.

Lieberthal, K., and M. Oksenberg. 1988. *Policy making in China*. Princeton, NJ: Princeton University Press.

Magee, D. 2013. The politics of water in rural China: A review of English-language scholarship. *Journal of Peasant Studies* 40 (6):1189–1208.

Marks, D. 2010. China's climate change policy process: Improved but still weak and fragmented. *Journal of Contemporary China* 19 (67):971–86.

Mertha, A. 2008. *China's water warriors: Citizen action and policy change*. Ithaca, NY: Cornell University Press.

Naughton, B. 2007. *The Chinese economy: Transitions and growth*. Cambridge, MA: MIT Press.

Nickum, J. E. 2010. Water policy reform in China's fragmented hydraulic state: Focus on self-funded/managed irrigation and drainage districts. *Water Alternatives* 3 (3):537–51.

Oi, J. C. 1989. *State and peasant in contemporary China: The political economy of village government*. Berkeley: University of California Press.

Perry, E. J. 2011. From mass campaigns to managed campaigns: "Constructing a new socialist countryside." In *Mao's invisible hand the political foundations of adaptive governance in China*, ed. E. J. Perry and S. Heilmann, 30–61. Cambridge, MA: Harvard University Asia Center.

Shang. 2007. Dingxi potatoes go out. *People's Daily*, April 3. Accessed February 29, 2011. http://www.cpad.gov.cn/data/2007/0403/article_333569.htm.

Shue, V. 1988. *The reach of the state: Sketches of the Chinese body politic*. Palo Alto, CA: Stanford University Press.

Shuie, C. H. 2004. Local granaries and central government disaster relief: Moral hazard and intergovernmental finance in eighteenth- and nineteenth-century China. *The Journal of Economic History* 64 (1):100–124.

Smith, G. 2010. The hollow state: Rural governance in China. *The China Quarterly* 203:601–18.

Swyngedouw, E. 1999. Modernity and hybridity: Nature, regeneracionismo, and the production of the Spanish waterscape, 1890–1930. *Annals of the Association of American Geographers* 89 (3):443–65.

Wang, S. 2012. "Three west" poverty alleviation at 30: Dingxi, 30 years of great changes. Accessed July 8, 2012. http://news.xinhuanet.com/politics/2012-06/21/c_112270124_3.htm.

Webber, M., and X. Han. 2017. Corporations, governments, and socioenvironmental policy in China: China's water machine as assemblage. *Annals of the American Association of Geographers* 107 (6):1444–60.

Wei, H., J. L. Li, and T. G. Liang. 2005. Study on the estimation of precipitation resources for rainwater harvesting agriculture in semi-arid land of China. *Agricultural Water Management* 71 (1):33–45.

Yan, Q. 2008. *Tudou de weixiao* [Potato smiles]. Lanzhou, China: Duzhe Press.

Yang, D. L. 1998. *Calamity and reform in China: State, rural society, and institutional change since the great leap famine.* Stanford, CA: Stanford University Press.

Zhang, S., and J. Chen. 1995. Lipeng zai Gansu kaocha gongzuo shi qiangdiao bixu ba jiaqiang nongyo fang zai guomin jingji shauwei [Li Peng emphasizes the importance of putting agriculture first in the national economy while inspecting works in Gansu]. *People's Daily*, July 27, 1. Accessed November 5, 2018. http://rmrb.zhouenlai.info/人民日报/1995/07/1995-07-27.htm#1036506.

AFTON CLARKE-SATHER is an Associate Professor in the Program in Geography at the University of Minnesota, Duluth, Duluth, MN 55812. E-mail afton@d.umn.edu. His research interests include water governance in rural areas of the United States and China.

Electricity-Centered Clientelism and the Contradictions of Private Solar Microgrids in India

Jonathan N. Balls (iD) and Harry W. Fischer (iD)

Most discussions about solar microgrids focus on sustainable energy and development goals and the technical aspects of electricity generation, storage, transmission, and distribution. Very few explicitly examine the ways in which their introduction upsets and reshapes entrenched practices of electoral politics and citizen claim-making around electricity access and development. In India, as in many parts of the world, electricity represents the most visible symbol of economic development and social well-being. Democratic politics in many developing countries are linked to demands for access to electricity. The meshing of electricity, development, and democratic politics in postindependence India has produced a politics of clientelism in which parties have sought to gain voter support with promises of cheap or free electricity. Although this electricity-centered clientelism has expanded supply, it has simultaneously contributed to skewed spatial access, unreliable supply, and high debt burdens for state-owned electricity distribution companies. This article examines histories of clientelism and the contradictions emerging from the introduction of private solar microgrids in rural areas of the northern Indian state of Uttar Pradesh. It shows that although solar microgrids avoid electricity-centered clientelism, significant numbers of poor rural households in their supply areas are both excluded by their user-pays approach and unable to demand fair access through political representatives. The study calls for alternative governance and support programs at local levels that ensure that private solar microgrids can deliver reliable electricity to poor households.

太阳能微电网的相关讨论，多半聚焦可持续能源与发展目标，以及电力生产、储存、传送与分配的技术层面。鲜少有讨论明确检视引进太阳能微电网，如何扰乱并重塑围绕着电力取得与发展的选举政治的长期运作和民众要求。如同在世界上诸多地方一般，在印度，电力是经济发展和社会福祉最显而易见的象徵。在诸多开发中国家，民主政治被连结至取得电力的要求。独立后的印度中，电力、发展和民主政治的结合，生产出侍从主义政治，其中政党通过承诺廉价或免费的电力来寻求选民的支持。尽管此一以电力为核心的侍从主义扩张了供给，但却同时导致偏斜的空间取得管道、不可靠的供给、以及国有电力分配公司的高额债务负担。本文检视侍从主义的历史，以及在北印度的北方邦偏远地区引入私有太阳能微电网所引发的矛盾。本文显示，尽管太阳能微电网避免了以电力为核心的侍从主义，但在其所供电的区域中，为数众多的贫穷偏远家户却被排除在使用者付费的方案之外，同时无法通过代议政治来要求公平的取得管道。本研究呼吁确保私人太阳能微电网能够将电力确实传送至贫困家户的另类治理和地方层级的支援计画。 *关键词：侍从主义，电力，印度，民粹主义，太阳能微电网。*

La mayoría de las discusiones sobre microrredes solares se enfocan en energía sustentable y metas de desarrollo, y sobre los aspectos técnicos de la generación, almacenamiento, trasmisión y distribución de electricidad. Solo en contados casos se examinan explícitamente los modos como su introducción molesta y reconfigura prácticas arraigadas de la política electoral y la elaboración de reclamos ciudadanos en torno al acceso y desarrollo de la electricidad. Como ocurre en muchas partes del mundo, en la India la electricidad representa el símbolo más visible del desarrollo económico y el bienestar social. En muchos países en desarrollo las políticas democráticas van de la mano con el clamor por acceso a electricidad. Después de la independencia, la estrecha relación de la electricidad, el desarrollo y la política democrática ha dado lugar en la India a una política de clientelismo en la que los partidos buscan ganar el apoyo del voto con promesas de electricidad barata o gratis. Aunque este clientelismo centrado en electricidad ha expandido la oferta, simultáneamente ha contribuido al acceso espacial sesgado, suministro inseguro y al peso del alto endeudamiento de las compañías de distribución de electricidad de propiedad estatal. Este artículo examina historias de clientelismo y las contradicciones que surgen de la introducción de microrredes solares privadas en las áreas rurales del norte del estado indio de Uttar Pradesh. El artículo muestra cómo, aunque las microrredes solares contrarrestan el clientelismo centrado en la electricidad, un número significativo de hogares rurales pobres en sus áreas de suministro quedan excluidos por su estilo de suscripción a la vez que se imposibilitan para demandar un acceso justo a través de representantes políticos. El estudio propende por una gobernanza alternativa y programas de apoyo a nivel local para asegurar que las microrredes solares puedan proporcionar electricidad confiable a los hogares pobres. *Palabras clave: clientelismo, electricidad, india, microrredes solares, populismo.*

Over the past decade, policymakers have promoted solar microgrids as a key strategy to expand rural electricity provision in alignment with broader sustainability and development goals. The scale of projected development is staggering: The International Energy Agency (IEA) estimates that more than a third of unelectrified households in the Global South will be electrified with microgrids (IEA 2017). In India, policymakers have heralded private solar microgrids as the decentralized solution for overcoming the challenge of providing a universal and reliable supply of electricity in rural and remote parts of the country (Ministry of New and Renewable Energy [MNRE] 2016).

Although much contemporary discussion on solar microgrids centers on the technical aspects of their expansion and distribution for achieving sustainable development goals, electricity and electricity infrastructure systems are inherently political (Zimmerer 2011). Electricity is one of the most visible and symbolically important forms of development in India. It is loaded with material significance: It provides the ability to do a range of simple, yet eminently modern things that can alter people's economic opportunities, their connections to the broader world, or even their fundamental relationship to day and night (Gupta 2015). Electricity has immense political significance: It is one of the most fundamental signifiers of inclusion within the national project of development and one of the chief ways that state power, both literally and figuratively, is experienced in everyday lives. How electricity is produced, transported, and consumed is therefore core to the popular imagination and sense of citizenship and an integral feature of India's democratic politics (Min 2011; Kale 2014; Gupta 2015).

Electricity has thus been a central object of individual and group claim-making by citizens of postindependence India. Political parties and politicians have routinely made promises to expand the electricity grid and provide access to electricity to mobilize voter support (Kale 2014). Consequently, a politics of patronage that can be called *electricity-centered clientelism* has developed as normal political practice in much of India. Although this electricity-centered clientelism has undoubtedly expanded supply to a broader cross section of Indian society, it has simultaneously contributed to highly skewed patterns of spatial and social access, unreliable supply, and high debt burdens for state-owned electricity distribution companies (discoms; Dubash, Kale, and Bharvirkar 2018). Over the past decade, in response to the limitations of the grid, and in alignment with a broad policy emphasis on renewables, policymakers have looked to private investment in decentralized, market-based, and renewable energy systems as a cost-effective solution to delivering sustainable rural electricity supply (MNRE 2016).

This article examines histories of clientelism and the contradictions emerging from the introduction of private solar microgrids in rural areas of Uttar Pradesh (UP) in India. It argues that although solar microgrids appear to be a benign, apolitical, market-based, and technical solution for supplying electricity to poorly served rural areas, significant numbers of poor households in villages are excluded due to the relatively high cost of electricity and the user-pays systems followed by private providers. Although private solar microgrid systems have managed to avoid being captured by politicians for electricity-centered clientelism, they are certainly not isolated from local politics, and they risk foreclosing possibilities for poor households to make demands of their local political representatives for access to electricity and inclusive development.

In the following sections, we provide a brief background of electricity-centered clientelism in India and the distinctive form it has taken in UP. We then describe the entry of private solar microgrid companies in UP and how these systems of electricity supply operate. The concluding section discusses how alternative local governance approaches could play a more substantive democratic role in ensuring solar microgrids can deliver reliable electricity, while not falling prey to electricity-centered clientelism or defaulting on their public responsibility to provide access to the poor.

Democracy, Development, and Electricity-Centered Clientelism

Modernization and development have always been materially and symbolically associated with states establishing large-scale road, electricity, clean water, sanitation, and housing infrastructure. Indeed, the development of infrastructure for the delivery of what are deemed to be essential services has been the object of state formation around the world (Meehan 2014). In many postcolonial countries, democratic politics has centered on mobilizing support by invoking visions of poverty alleviation and development to key electoral groups (Corbridge and Harriss 2000; Witsoe 2013). Political parties have sought to win elections by guaranteeing their voter base access to modern infrastructure and services in exchange for their votes. These dimensions of political practice often operate through populist appeals, where charismatic politicians build "personal" ties with their constituents by combining popular narratives of progress with the delivery of material benefits, often distributed on the basis of a shared social affiliation (Kitschelt and Wilkinson 2007; Scoones et al. 2018).

The evolution of clientelism in India stems from the structural conditions of democracy at the time of its independence from British colonial rule in 1947. With a new constitution that guaranteed democratic rule based on universal franchise, India's major political parties faced the challenge of mobilizing the support of voters across the country who were differentiated by caste, class, education levels, geography, language, religion, and deprivation. To win elections, political parties evolved large networks of local interlocutors that could invoke narratives of development and channel tangible material benefits to secure the votes of different social groups (Mitra 1992).

Political parties spoke of national development and poverty alleviation primarily in terms of providing access to electricity, roads, clean water, and sanitation (Corbridge and Harriss 2000). Electricity was perhaps the most important of these in establishing patron–client relationships between parties and voters (Min 2015). It was seen as the key to modernization and economic progress, but its development was limited to a few large cities. In the first thirty years following independence, nearly one fifth of all national planned investment was for power generation and distribution (Kale 2014).

Power-sector governance was set as a state-level matter at independence. Between the 1950s and 1970s, the generation and distribution of electricity was gradually nationalized, with each state setting up its own state electricity board (SEB; Kale 2014). It was during this period that electrification and access to cheap or free electricity supply became a key basis of democratic electoral politics and governance across India. An electricity-centered clientelism emerged in varying forms and intensity in most states. Common features of electricity-centered clientelism ranged from carefully targeted subsidies and interference in the flow of electricity, particularly for farmers, to sanctioning of electricity theft by key voter constituencies, to free electricity connections for local political brokers and favored individuals (Kale 2014; Martin 2018).

By the 1980s, despite the overall expansion of electricity supply across India, electricity-centered clientelism proved expensive. Ruling parties in states pressured SEBs to maintain low electricity tariffs and preferential supply for their favored voter constituencies. The underfunded SEBs were thus crippled by large debts and were unable to supply reliable electricity (Chatterjee 2018). The problems of indebted SEBs assumed crisis proportions in the 1990s, leading the central government to reform the sector. Following its agenda for liberalizing India's economy, the central government's reforms aimed to promote private investment and competition. State governments were required to restructure their SEBs, were required to allow private investment in power generation, and were encouraged to privatize discoms. During the 1990s and 2000s, most states broke up their SEBs, separating electricity generation, transmission, and distribution functions. Yet, although significant private investment took place in power generation, nearly all states resisted privatizing discoms. Most remained state-owned and subject to the vagaries of electricity-centered clientelism.

Overall, the history of democratic politics, electricity development, and electricity-centered clientelism of the past seven decades has resulted in an impasse with unreliable supply and highly differentiated access, particularly for rural households. India's 2011 census showed 67 percent of households using electricity. A recent study found that more than half of electrified rural households in UP, Bihar, Madhya Pradesh, Jharkhand, Odisha, and West Bengal receive fewer than twelve hours of supply per day (Jain, Urpelainen, and Stevens 2016).

Although clientelist politics have expanded aggregate access to electricity, they have simultaneously produced a situation akin to what Witsoe (2013) described as "democracy against development." This is where the distribution of political patronage becomes the orienting logic of state administrative practice, giving certain electoral groups preferential access to key state resources while undermining broad and inclusive development gains. Where democracy is seen to work against development, policymakers and development actors often look to avoid state institutions, preferring instead to focus on private investment and ostensibly nonpoliticized channels to pursue policy objectives (Chhotray 2007). This was seen with India's electricity reforms in the 1990s and 2000s and is apparent in support for private solar microgrids.

It is in this context, and in tandem with the global push for renewable energy, that policymakers have advocated state support for private investment in solar microgrid systems. Microgrids are small electricity distribution networks, either self-contained or connected to the grid, that have battery storage. Solar microgrids installed in India typically provide households with six to twelve hours of electricity daily, for lighting and to charge mobile phones. Larger systems deliver metered electricity sufficient to power a range of devices, such as fans and televisions. The draft National Policy on renewable energy (MNRE 2016) outlines a target of 10,000 microgrids for supplying electricity in rural areas. The central government and several state governments offer financial incentives to private companies to set up these decentralized systems. Numerous new businesses and social enterprises have responded, drawing on state and development sector financing. Thousands of solar microgrids operate nationally, with households paying upward of 100 rupees ($1.50) per month to providers. This is not an insignificant sum for households that live close to the margin of subsistence, and, per unit, electricity from solar microgrids is typically two to four times more expensive than grid-based electricity.

Method

Our study is based on qualitative research conducted by the first author in UP between 2014 and 2018 in four periods and over ten months of fieldwork. Semistructured and open interviews were conducted with eight managers from four microgrid providers and with twenty-six individuals involved in UP's power sector, including current and retired bureaucrats associated with the Uttar Pradesh Power Corporation Limited (UPPCL) and Uttar Pradesh Electricity Regulatory Commission (UPERC), consumer representatives, and journalists. Interview information was validated by cross-referencing and coded for analysis. Repeat field visits were made during each fieldwork period to nine microgrid systems in the central districts of Unnao, Sitapur, Barabanki, and Gonda. Secondary material was analyzed, including newspaper reports, policy papers, and legislation.

Electricity-Centered Clientelism in Uttar Pradesh

UP is India's most populous state, with more than 200 million people. Its population is largely rural, with 78 percent of people living outside cities. It is also one of India's poorest states, consistently ranking near the bottom of national tables on education, health, and per capita income. Due to its size, UP has a strong influence on national politics. Of India's fourteen prime ministers, nine have been elected from UP constituencies. UP's electricity grid is part of India's Northern Grid. Coal-based thermal power makes up 74 percent of UP's installed power capacity, renewables comprise 20 percent, and gas and nuclear power make up the remainder.

After independence, the Congress Party was the dominant political party in UP. In the precolonial period, the party had already evolved a large network of landed local political actors and brokers to mobilize support in local elections. In the years following independence, these networks consolidated control over key state institutions and the bureaucracy to channel material benefits to rural constituencies in return for voter support (Brass 1965).

The state government constituted the Uttar Pradesh State Electricity Board (UPSEB) in 1959 and gave it responsibility for electricity generation, transmission, and distribution. In the following decades, substantial investments were made in transmission and generation infrastructure. Electrification was a monumental task, as electricity infrastructure was largely limited to urban centers and resources were scarce. In the 1960s and 1970s, various central and state-level programs extended the electricity grid to farmers. This was done to power irrigation pumps as part of India's green revolution, initiated to increase agricultural productivity. The ruling Congress Party, seeking the support of powerful farmer groups, agreed to provide unmetered and cheap electricity to farmers (Kale 2014). A political compact was born that has been protected by successive governments.

UP's political landscape was transformed in the 1980s and 1990s, with the Bharatiya Janata Party (BJP), Bahujan Samaj Party (BSP), and Samajwadi Party (SP) all becoming strong forces and the latter two parties successfully consolidating lower caste support (Jaffrelot 2003). On gaining power, the BSP and SP used their control over the UPSEB to provide electricity development to their own constituencies. The BSP significantly accelerated village electrification, especially in the rural constituencies of BSP legislators (Min 2011). At the same time, successive governments kept electricity tariff increases below inflation for domestic consumers. Between 1991 and 1999, domestic consumers increased from 28 to 43 percent of overall load, but revenue from these customers only grew from 16 to 17 percent of total revenue (UPPCL 2000).

By the late 1980s, the UPSEB was reporting large annual losses. Neither the UP government nor the UPSEB was able to access capital for much-needed investment in transmission and generation infrastructure. When an acute state-level fiscal crisis hit UP at the end of the 1990s, the then-BJP-led government briefly embraced wholesale reform. These reforms, implemented together with the World Bank, sought to incentivize private investment and reduce political interference. The UPSEB was disaggregated, with generation, transmission, and distribution entrusted to separate public companies. An independent regulator, the UPERC, was tasked with setting electricity tariffs. Five public discoms were constituted serving different geographical areas under the umbrella of the newly formed UPPCL with the intention that they would be privatized. Private investments in UP-based coal generation plants were made. Distribution-side reforms stalled, however. The BJP-led government, having pushed reforms

despite large-scale union-led opposition, intervened in 2002 to stop electricity tariffs from being raised, fearing losses in upcoming state elections. This forced the newly formed UPPCL to take on new debts. To date, a single politically appointed chairman oversees the UPPCL and all five discoms, which remain firmly entrenched within existing structures of political control.

Since these reforms, political parties have continued to exert influence over the UPERC in setting electricity tariffs. Tariffs for domestic and agricultural consumers have been kept below the cost of supply. In 2006, the then-SP government introduced subsidized power for loom weavers in the run-up to state elections. Loom weavers are politically important in the east of UP, which itself is geographically important for the SP's electoral strategy. In contrast, tariffs for industrial consumers are high, reflecting industry's lesser electoral weight. Industry consumers have responded by leaving UP or by securing private power generation sources, further diminishing revenue at the UPPCL.

These macropolitical dynamics are underpinned by local-level clientelist networks that are crucial for mobilizing votes. Because politicians have control over the transfer of bureaucratic employees, they are able to exert significant influence on the day-to-day running of discoms, especially by ordering them to privilege certain districts. So-called VIP districts receive twenty-four hours of daily supply, whereas other districts receive much less. The UPPCL exempts hundreds of local areas from power cuts in response to demands from local politicians (Parashar 2012). One prominent explanation for India's now notorious Northern Grid collapse of 2012, remembered as "the world's largest power cut," was that UP was drawing too much power from outside of the state to meet the demands of local politicians (Denyer and Lakshmi 2012). Weeks after the BJP won a landslide victory in state elections in 2017, the newly formed BJP government announced that key Hindu religious centers would receive twenty-four hours of electricity, reaffirming the importance of identity politics in the state despite the party's stated commitment to development for all. In the run-up to state and national elections, political parties typically order increased hours of supply throughout the state (Min 2015).

Politicians also act to protect power theft by party affiliates. During the last SP government (2012–2017), losses from illegal connections, unpaid bills, and transmission losses in the constituencies of the chief minister and family members regularly accounted for more than 50 percent of electricity distributed in those districts, compared to an average of 30 percent elsewhere (Shah 2015). As Min and Golden (2014) showed, members of the UP Legislative Assembly are more likely to win reelection when electricity theft has increased during their time in office. The documentary film *Katiyabaaz* strikingly documents how citizens and local politicians in Kanpur mobilized to have the local discom's director transferred from that post following a crackdown on illegal connections (Kakkar and Mustafa 2014).

Thus, electricity expansion in UP has been closely tied to the evolution of clientelist politics in the state. On the one hand, the postcolonial period has seen the consolidation of strong multiparty electoral competition and growing political enfranchisement of less powerful groups. Yet these politics have at the same time undermined the state's ability to expand reliable electricity. Only 37 percent of UP's households were electrified in 2011. By 2014 the accumulated debts of the state's discoms had reached Rs. 53,000 crore (US$8.2 billion). To compensate for these losses, discoms implement long hours of power cuts in loss-making rural areas. Despite central government efforts to bail out UP's discoms and force reform on several occasions, at a state level electricity-centered clientelism has continued.

The Apolitical Countermovement: Solar Microgrids

With access to electricity highly uneven and UP required to meet centrally mandated solar generation targets, policymakers in the state have looked to decentralized, privately run energy arrangements to provide electricity to rural areas. The SP government, in power between 2012 and 2017, introduced a state microgrid policy, aiming to simplify bureaucratic processes for providers. The policy outlines the right for providers to sell electricity to the UPPCL, provides a guarantee that providers can sell their microgrids to the UPPCL should they wish to exit the market, and makes available a capital subsidy for developers. Through a process of competitive tendering, the SP government also provided financing to private providers to set up several dozen microgrids in pilot villages. In our interviews,

officials of the UPPCL, UPERC, and the Chief Minister's Office indicated their belief that solar microgrids would be essential for UP to succeed in providing reliable electricity supply to rural areas and to eventually ease the burden public discoms face supplying rural areas.

Our fieldwork focused on three microgrid providers: Boond, NatureTech, and Mera Gaon Power (MGP), as well as microgrids run by local entrepreneurs, funded and supported by the The Energy and Resources Institute (TERI). NatureTech, Boond, and TERI microgrids number between twelve and sixty each, whereas MGP operates more than 1,000 microgrids. NatureTech received grant funding from the UP government to set up the first of its microgrids, whereas bilateral and domestic grant money was invested in Boond, MGP, and TERI's microgrids. Most are basic direct current (DC) systems, typically with a capacity of between one and four kilowatts, connecting several dozen households each (Figure 1). They provide enough electricity for customers to power light-emitting diode (LED) lights and to charge mobile phones. A small number are larger alternating current (AC) systems, which power higher metered consumption. Providers require customers to pay weekly or have "pay-as-you-go" top-up mechanisms. Customers typically pay between 100 and 200 rupees (US$1.50 to $3.00) per month. This amounts to more than a half-day's wage for the poorest households, at the current rates for the government's employment generation scheme.

Microgrids are relatively simple technical arrangements, which can be constructed in several days. Running microgrids on a day-to-day basis is more challenging, as they must be physically maintained, money collected, and illegal connections policed. Our visits to microgrids showed businesses adopting various strategies to avoid social and political interference in their daily operation. First, all put into place formal operational protocols, including strict record-keeping, rules for the payment of fees, and standards for disconnection in the event of nonpayment. Local employees were appointed to collect fees and maintain systems, and in the case of one provider, employees were moved around to make sure that they were not in charge of collecting money in villages where they had family connections. Second, in more recent microgrids, electronic control systems are installed. Customers pay cash to an employee or an entrepreneur and their meter is

Figure 1. Parallel wires; microgrid wiring attached to pylons several feet under grid wiring.

then electronically topped up. These control systems also cut lines when unsanctioned connections are made. Overall, those who pay receive a relatively reliable basic service.

The founders of Boond, NatureTech, and MGP all came from development-sector or corporate backgrounds. In our discussions, they argued that their success depended on operating separately from

Figure 2. Villagers in heated discussion with microgrid entrepreneur about the fees they must pay for their electricity supply.

bureaucratic institutions and political interference. We were told that their businesses avoided getting involved with politicians and officials, largely because of the perceived inefficiency and corruption of government. One interviewee told us how they had previously moved their microgrids from villages where they had experienced problems with local politicians interfering with operations. Engaging with politicians and officials is necessary at times, though. One founder recounted how they had spent months building relationships with key officials and politicians to receive state financing to set up their microgrids. Another described how they at times faced local political interference and had called on higher level political contacts to resolve these situations.

Although microgrid providers claimed to be apolitical, we could see during fieldwork that they were enmeshed in local politics. All need to recruit employees with the right caste and social position within communities to run microgrids and collect money. Solar microgrid modules, controls, and top-up systems are placed under the control of a carefully

selected village household. In the case of Boond and NatureTech, that household has the responsibility for monitoring the system and collecting money and is provided with a financial incentive to undertake this role. Discussions with local employees highlight how personal family and caste-based connections have been essential for successful operations.

Further, despite the attempts of microgrid providers to avoid political interference, customers in villages still feel a strong right to mobilize when they believe that their rights are not being fairly met. This was exemplified by one field visit in January 2015. During the winter, UP experiences dense daytime mists resulting in reduced solar generation. On arriving in the village with an entrepreneur running a microgrid, villagers informed us that the microgrid was generating insufficient power and that they would not pay their monthly fees. Over two hours of heated discussion, it further transpired that money from the elected village council had been invested (Figure 2). People in the village argued that the microgrid was therefore rightly theirs and that the fees they were paying were unjust. The entrepreneur explained at length his costs, and in the end a compromise of lower fees for that month was reached.

This case reflects a tension common in many microgrids: Although public or grant funds had been necessary for entrepreneurs to invest, day-to-day decisions ultimately remain under the control of private actors, who are likely to be motivated first by market logic. When compared against the failure of electricity-centered clientelism, microgrids offer a compelling alternative for those who can pay. Yet there are clear drawbacks. Households are left in the hands of little-regulated, local monopoly suppliers. There is no formal political or administrative oversight over which households in a village get a connection, the terms by which connections (and disconnections) are managed, how prices are set, and how money is collected.

Although potentially unsettling existing electricity-centered clientelism, solar microgrids, as currently seen, are only likely to deliver limited progress toward greater electricity access. In villages we visited, between 30 and 70 percent of households were connected. Many households cannot afford microgrid electricity. Households that can pay receive only basic electricity from systems with a limited capacity. Placing electricity supply in the hands of private entrepreneurs has thus improved access for some but with significant limitations. There is a risk that a broad turn to private solar microgrids, as currently seen, will undermine existing avenues for poor households to demand electricity access from the state as an essential developmental resource, even as ineffective as these avenues currently are. If solar microgrids do proliferate and villagers who can pay turn to them for reliable supply, discoms that already have few incentives to supply rural areas will see little reason to step in to provide access for the remaining households who cannot pay.

Conclusion

Private solar microgirds now appear as a key strategy to expand energy provision in alignment with broader sustainability goals. Our first case documents how electricity infrastructure and supply in UP has developed through dynamics of electricity-centered clientelism, resulting in uneven spatial access and crippling the ability of the state to deliver broad and equitable electricity development. Our second case shows that, although private solar microgrids have often been framed as a market based, apolitical alternative, they remain limited in their capacity to ensure reliable and inclusive electricity access. Although they have made progress in expanding provision in some areas, access remains contingent on ability to pay, which many people cannot do. Citizens do mobilize around their interests to confront private actors in charge of these systems, yet there are no clear and tangible mechanisms of accountability to ensure effective delivery.

To be clear, we do believe that private microgrids have significant potential in UP and elsewhere. Their development cannot be treated purely as a technical matter, however, and the onus for ensuring access cannot rest on private actors alone. The core policy question is what kinds of interventions and reforms can lead to more equitable access, while also enabling transitions to more sustainable energy sources. Our analysis suggests two key issues that deserve particular attention.

First and foremost, the pursuit of electricity development demands explicit attention to issues of distributional justice, with a focus on the arrangements that can bring about and sustain more equitable access (Hall, Hards, and Bulkeley 2013). As we have shown, neither the logic of clientelist-based politics nor private microgrid expansion is naturally oriented toward inclusive development; both fall short of these goals. Solar microgrids can nevertheless play an important role in expanding access where the grid is not reliable. To be effective, these efforts need to be driven not just by the financial interests of private entrepreneurs but anchored within a broader state-directed strategy for renewable energy expansion. This could include both public and private investment as well as subsidies targeting poor households.

Second, although microgrids are perceived as a means to insulate energy from clientelist politics, we argue that what is needed is not less politics but more substantive democracy. Substantive democracy would require ensuring channels for citizens to meaningfully engage in key decision-making processes as well as creating credible mechanisms of public oversight. Framing investments in renewables in these terms would entail asking whether interventions are providing opportunities for individuals to hold those in charge to account for the delivery of electricity.

Policy could devise mechanisms to ensure greater control of private microgrids by the public, with means of enforcement through regulatory authorities at higher scales. This has already been done for a range of other development and public service functions (Fox 2015). Local governance institutions, such as village councils, could play a more substantive role in providing opportunities for democratic participation and oversight of key aspects of microgrid management. Although such measures risk enabling new kinds of electricity-centered clientelism, evidence shows that decentralized governance approaches, if implemented with strong financial and administrative support from the state, can lead to more accountable, responsive, and effective delivery of basic public services (Faguet 2014). So far experiments with alternative governance approaches to microgrid management in India have been limited.

Most fundamentally, the political nature of energy development suggests that simply devising technical solutions, new provision systems, or regulatory reforms is not, by itself, enough. Deeply entrenched forms of political practice suggest that existing forms of clientelism are likely to persist. As investment in renewables grows, the question remains how to channel existing forms of political practice into a more inclusive vision of development. Seeing electricity provision as embedded within an evolving set of political relationships suggests that there might be no final solution to this challenge. It underscores the need for ongoing attention to the ways that different arrangements for energy provision support particular forms of political practice and the extent to which these politics are, in turn, able to bring about more inclusive and sustainable development.

Acknowledgments

We thank Haripriya Rangan for her critical feedback on drafts of this article and Alessandro Antonello, Amanda Gilbertson, Amy Piedalue, Pawan Singh, and Shikha Lakhanpal for their insightful suggestions. We are grateful for the time willingly and generously given by interviewees working in the electricity and microgrid sector in India.

ORCID

Jonathan N. Balls 🆔 http://orcid.org/0000-0002-1855-228X

Harry W. Fischer 🆔 http://orcid.org/0000-0001-7967-1154

References

Brass, P. 1965. *Factional politics in an Indian state: The Congress Party in Uttar Pradesh*. Berkeley: University of California Press.

Chatterjee, L. 2018. The politics of electricity reform: Evidence from West Bengal, India. *World Development* 104:128–39. doi: 10.1016/j.worlddev.2017.11.003.

Chhotray, V. 2007. The "anti-politics machine" in India: Depoliticisation through local institution building for participatory watershed development. *The Journal of Development Studies* 43 (6):1037–56. doi: 10.1080/00220380701466526.

Corbridge, S., and J. Harriss. 2000. *Reinventing India: Liberalization, Hindu nationalism and popular democracy*. Cambridge, UK: Polity.

Denyer, S., and R. Lakshmi. 2012. Huge blackout fuels doubts about India's economic ambitions. *Washington Post*, August 1. Accessed January 16, 2019. https://www.washingtonpost.com/world/asia_pacific/huge-black-out-fuels-doubts-about-indias-economic-ambitions/2012/08/01/gJQAtjeYOX_story.html?noredirect=on&utm_term=.01a3fae5c7d0

Dubash, N., S. Kale, and R. Bharvirkar. 2018. *Mapping power: The political economy of electricity in India's states*. New Delhi: Oxford University Press.

Faguet, J. P. 2014. Decentralization and governance. *World Development* 53:2–13. doi: 10.1016/j.worlddev.2013.01.002.

Fox, J. 2015. Social accountability: What does the evidence really say? *World Development* 72:346–61. doi: 10.1016/j.worlddev.2015.03.011.

Gupta, A. 2015. An anthropology of electricity from the Global South. *Cultural Anthropology* 30 (4):555–68. doi: 10.14506/ca30.4.04.

Hall, S., S. Hards, and H. Bulkeley. 2013. New approaches to energy: Equity, justice and vulnerability. Introduction to the special issue. *Local Environment* 18 (4):413–21. doi: 10.1080/13549839.2012.759337.

International Energy Agency (IEA). 2017. Energy access outlook 2017: From poverty to prosperity. Paris: IEA.

Jaffrelot, C. 2003. *India's silent revolution: The rise of the lower castes in North India*. London: Hurst & Company.

Jain, A., J. Urpelainen, and L. Stevens. 2016. *Energy access in India*. Rugby, UK: Practical Action Publishing.

Kakkar, D., and F. Mustafa, directors. 2014. Katiyabaaz [DVD]. India: Globalistan Films, ITVS.

Kale, S. 2014. *Electrifying India: Regional political economies of development*. Stanford, CA: Stanford University Press.

Kitschelt, H., and S. I. Wilkinson. 2007. *Patrons, clients and policies: Patterns of democratic accountability and political competition.* Cambridge, UK: Cambridge University Press.

Martin, N. 2018. Corruption and factionalism in contemporary Punjab: An ethnographic account from rural Malwa. *Modern Asian Studies* 52 (3):942–70. doi: 10.1017/S0026749X1700004X.

Meehan, K. M. 2014. Tool power: Water infrastructure as wellsprings of state power. *Geoforum* 57:215–24. http://dx.doi.org/10.1016/j.geoforum.2013.08.005

Min, B. 2011. *Electrifying the poor: Distributing power in India.* Ann Arbor: University of Michigan.

———. 2015. *Power and the vote.* Cambridge, UK: Cambridge University Press.

Min, B., and M. Golden. 2014. Electoral cycles in electricity losses in India. *Energy Policy* 65:619–25. doi: 10.1016/j.enpol.2013.09.060.

Ministry of New and Renewable Energy (MNRE). 2016. *Draft national policy on RE based mini/micro grids.* New Delhi: Ministry of New and Renewable Energy.

Mitra, B. 1992. *Power, protest and participation: Local elites and the politics of development in India.* London and New York: Routledge.

Parashar, B. K. 2012. UP politicians grab 18% power for their turfs. *Hindustan Times,* September 17. Accessed January 16, 2019. https://www.hindustantimes.com/india/up-politicians-grab-18-power-for-their-turfs/story-CUfqKtqx7pKIm0yKnZoUHK.html

Scoones, I., M. Edelman, S. M. Borras, Jr., R. Hall, W. Wolford, and B. White. 2018. Emancipatory rural politics: Confronting authoritarian populism. *The Journal of Peasant Studies* 45 (1):1–20. doi: 10.1080/03066150.2017.1339693.

Shah, P. 2015. In VIP districts, only 50% pay for power. *The Times of India,* May 30. Accessed January 16, 2019. https://timesofindia.indiatimes.com/city/lucknow/In-VIP-districts-only-50-pay-for-power/articleshow/48674757.cms

Uttar Pradesh Power Corporation Limited. 2000. *Petition before the Uttar Pradesh Electricity Regulatory Commission.* Lucknow, India: Uttar Pradesh Power Corporation Limited.

Witsoe, J. 2013. *Democracy against development: Lower-caste politics and political modernity in postcolonial India.* Chicago: University of Chicago Press.

Zimmerer, K. 2011. New geographies of energy: Introduction to the special issue. *Annals of the Association of American Geographers* 101 (4):705–11. doi: 10.1080/00045608.2011.575318.

JONATHAN N. BALLS is a Postdoctoral Scholar at the School of Geography, University of Melbourne, and New Generation Network (NGN) Fellow at the Australia India Institute, Australia. E-mail: jonathan.balls@unimelb.edu.au. His research focuses on energy geographies, bottom of the pyramid capitalism, and frugal innovation.

HARRY W. FISCHER is an Associate Senior Lecturer at the Department of Urban and Rural Development, Swedish University of Agricultural Sciences, Uppsala, Sweden, and New Generation Network (NGN) Fellow at the Australia India Institute, Australia. E-mail: harry.fischer@slu.se. His research looks at democratic decentralization, environmental governance, and rural development in India.

Dreams and Migration in South Korea's Border Region: Landscape Change and Environmental Impacts

Heejun Chang, ⓘ Sunhak Bae, and Kyunghyun Park

The border region of South Korea has undergone dramatic social and environmental changes since the late 1990s with shifts in governmental regimes. Under the proliberal government (1997–2007) that enhanced economic ties between North and South Korea, the border region was open for introducing new people and industries. With a new conservative governmental regime in the past decade (2007–2017), social and environmental challenges emerged in the border region. Such challenges were not uniformly present throughout different areas, however. We examined the spatial transformation of the border region using sociodemography, economy, landscape fragmentation, and water quality data with a focus on two gateway regions (Paju and Goseong) as representative cases. Although these two regions are similar in size and served as central nodes of flow between the two Koreas, they experienced different trajectories under disparate national and regional policies. In Paju, a closer region to Seoul, the capital of South Korea, the landscape became more fragmented as a result of urban expansion, but different subcenters were formed to accommodate the growing population and industries that were less dependent on external shocks, contributing to the economic and environmental resilience of the region. In contrast, with continuous declining aging population, Goseong's landscape became less fragmented with one remaining main urban center, but its economy, society, and environment became fragile after the closure of the Kumgangsan tour. These different patterns of regional resilience can be fully understood by considering various social, environmental, and institutional factors acting on multiple scales that helped shape the region's stability.

大众环境主义若再生产社会秩序结构的话, 则可能产生有限的民主成果。本文通过检视公民认识论的分析架构, 补充政治生态学和科学与技术研究的环境叙事的当前使用, 寻求推进我们对于环境民主化的理解。公民认识论是先于国家及其他行动者企图维持不容挑战的政治秩序的存在面向。它们以还原的方式显示叙事形成的结构, 以及不同行动者的知识与政治主体如何共同生产, 因而扩充当前的分析。本文将此一分析运用至泰国的大众环境主义, 并特别关照1968年至今的社区森林与伐木。本文结合访谈与历史新闻报导的内容分析, 展现多样的行动者——包括国家、环境保育菁英, 以及农民运动倡议者——如何遵循适当的社区文化与行为的未受挑战之常规, 组织有关森林的政治倡议与生态宣称。近年来, 这些行动维护有关森林与社会的叙事, 同时反对对社区和森林而言可说更具培力远见的另类方案。本文主张, 揭露公民认识论, 能够较根据既有叙事涉入环境政治、抑或是单独分析叙事的限制而言, 对更深刻的环境民主化形式做出贡献。关键词: 威权主义, 环境主义, 政治生态学, 科学与技术研究, 泰国。

La región limítrofe de Corea del Sur ha experimentado dramáticos cambios sociales y ambientales desde finales de los 1990, con los cambios de los regímenes gubernamentales. Bajo el gobierno pro-liberal (1997–2007), que mejoró los lazos económicos entre las dos Coreas, la región fronteriza estuvo abierta a la llegada de nueva gente e industrias. Con un nuevo régimen de gobierno conservador en la década pasada (2007–2017), en la región limítrofe surgieron retos sociales y ambientales, aunque tales retos no se presentaron uniformemente distribuidos en diferentes áreas. Examinamos la transformación espacial de la región fronteriza utilizando datos de socio-demografía, economía, fragmentación del paisaje y calidad del agua, concentrándonos en dos regiones de acceso (Paju y Goseong), como casos representativos. Aunque las dos regiones son similares en tamaño y sirvieron como nodos centrales de flujo entre las dos Coreas, experimentaron diferentes trayectorias bajo políticas nacionales y regionales discrepantes. En Paju, una región más cercana a Seúl, la capital de Corea del Sur, el paisaje experimentó mayor fragmentación como resultado de la expansión urbana, pero se formaron diferentes subcentros para acomodar la creciente población y las industria que tenían menos dependencia de conmociones externas, contribuyendo a la resiliencia económica y ambiental de la región. En contraste, con la continuada declinación de la población vieja, el paisaje de Goseong se hizo menos fragmentado, con un centro urbano que se mantuvo, aunque su economía, sociedad y medio ambiente se hicieron frágiles después de la clausura del tour de Kumgangsan. Tan diferentes patrones de resiliencia regional pueden entenderse cabalmente tomando en cuenta varios factores sociales, ambientales e institucionales, que actúan a múltiples escalas para ayudar a configurar la estabilidad de la región. *Palabras clave: calidad del agua, escala, fragmentación del paisaje, región fronteriza, resiliencia.*

Borderland studies have gained popularity in recent years as the interaction between border countries has increased or declined in many parts of the world since the 1990s when global and regional geopolitical and economic landscapes changed (e.g., collapse of the Soviet Union, formation of the European Union, new trade agreements, tightening border security after 11 September 2001). Study topics have been diverse, ranging from sociohistorical perspectives on borders (Tagliacozzo 2016) to biodiversity conservation (Apostolopoulou et al. 2014) to residents' lives in a demilitarized zone (Medzini 2016) to cross-border cooperation (Pérez-Nieto 2016; Scott and Ley 2016). Borderland studies have moved from a top-down approach, which emphasizes the role of state authority to form landscape and views borders as objects, to a bottom-up approach, which investigates the transformation of borderlands as a dynamic process (borders in motion) with local places as the main agents of change (Newman 2006; Konrad 2015). The top-down approach, emphasizing power relationships, views borderlands as somewhat static and deterministic, whereas the bottom-up approach focuses on the bordering process that affects the daily lives of people residing in borderlands. Although borderlands scholars are now conceptualizing borderland as transitional space and acknowledge the differences that exist in borderlands between different countries and even within one country's borderlands, few focused on how different geopolitical and socioeconomic conditions helped shape the transition of borderlands within one country.

As the only remaining country in the world that was divided into two during the first phase of the Cold War, South Korea's border region has been relatively unknown to the outside world because of limited information. With the heavy presence of both Korean and U.S. soldiers, the border region has been considered a stagnant or restricted place for decades (Gelezeau 2013). Unlike other border regions in the world that allow some form of active exchange and movement of people and goods (Martinez 1994), South Korea's border region had limited interaction of people because of the tension between the two Koreas, which largely qualifies it to be called alienated borderlands according to Martinez's classification. Alienated borderlands can transition to coexistent borderlands, which allow a limited opening of the border for cooperative development (Martinez 1994). Such transitions have been linked to changes in governance regimes in either or both countries, resulting in borderlands somewhat uncertain and dynamic places (Newman 2006). In recent years South Korea's border region became a dynamic place, as did other regions of the country, as South Korea's political and economic regimes have shifted. In particular, for the past couple of decades, the change from liberal to conservative governments in South Korea resulted in different national economic and environmental policies, which also set up different pathways for development in the border region. We seek to test whether the existing border theories could be applicable or not suitable for explaining the changing geography of South Korea's border regions.

Although there were some previous studies about Korea's border region in the geographic literature (Kim and Lee 2004), most of them focused on limited aspects of geography. For example, S. O. Park (2005) studied the changing economic and social geography of the border region, and Jin (2005) focused on cultural–political geography around South Koreans' Kumgangsan tour in North Korea. Physical geographers and environmental scientists focused on the border region's protected area (north of the civilian control zone) to investigate biodiversity (Kim 1997), forest ecology (Seo et al. 2017), and land-cover change in North Korea including the border area (Kang and Choi 2014). They suggested that the border region be protected for its pristine environment. As such, there has been no attempt to understand the complex interplay between the

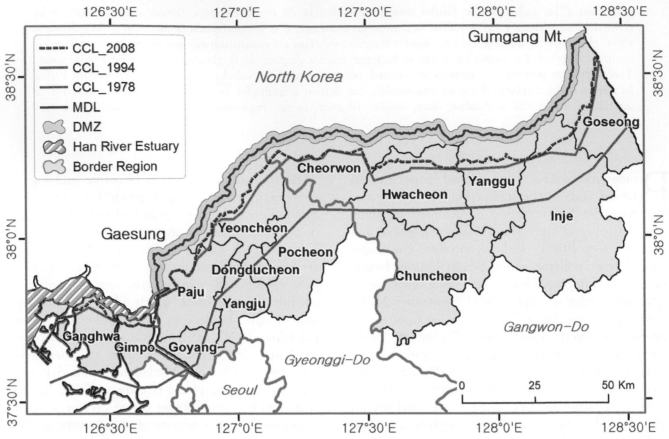

Figure 1. South Korea's border region and changes in CCL over time. *Note.* CCL = Civilian Control Line; MDL = Military Demarcation Line; DMZ = Demilitarized Zone.

changing human and physical geography collectively. Additionally, previous studies did not pay enough attention to smaller scale changes in the border region driven by multiple actors, failing to address the complex dynamics and interaction between the state and local government that help transform the space.

Our current research seeks to fill this gap by examining how national-scale political regime shifts and the associated land policy changes affect the local-scale socioeconomy and environment within the border region. We attempt to unravel how locals respond to such changes and examine under what social and political conditions their dreams might be realized or evaporate. In other words, what conditions trigger active interaction across scales in the border region? To investigate these complex and hierarchical relationships, we draw methods from various subdisciplines of geography: from landscape ecology to population geography to resilience theory. Specifically, we seek to answer the following research questions.

1. What is the spatial pattern of sociodemographic and economic changes within the border region over time?
2. How does the political regime shift affect land-cover change and landscape fragmentation in South Korea's border region? Is there an intraborder regional difference?
3. How does local environmental change relate to the changes in their socioeconomy and landscape resulting from policy shifts in the border region?

We first describe major social, economic, and environmental issues in South Korea's border region in general and two gateway regions in particular. This is followed by South Korea's land use policy changes with the shifts in governments and how they affect local places focusing on two gateway regions. Next, we discuss the primary data and methods used in our analysis. Results of the analysis and the interpretation of the results are reported next. The article concludes with a summary of the main findings of the study and recommendations for future research. We test whether new borders in motion and political ecological theories help unravel the coupled dynamics of a political regime and landscape changes in South Korea's border region.

South Korea's Border Region

The 7,000-km² border region (7 percent of South Korea's total area) encompasses fifteen *Si* (cities) and *Gun* (counties) that are adjacent to the border between North and South Korea (Figure 1). The border region was severely affected politically, economically, and environmentally by the Korean War (1950–1953) and thereafter. During the Korean War, the border region was the fierce battleground between the two Koreas, with most structures destroyed. After the 1953 Armistice Agreement that established the Military Demarcation Line (MDL), the 4-km-wide Demilitarized Zone (DMZ) was created to buffer any military actions and human settlement. The Civilian Control Zone (CCZ), which restricts civilian access, lies approximately 10 to 15 km south of the MDL. Farther south at a distance up to 50 km from the MLD are Military Installations Protection Districts (MIPDs), which severely restrict residential and economic development thanks to the presence of the military. As MIPDs occupy approximately three quarters of the border region, the border region remained largely underdeveloped until the mid-1990s (S. O. Park 2005). As a result of many regulations that prohibit the area from further development, the border region has been one of the most impoverished regions in South Korea. The border region has a population of approximately 1 million, with its population density (153 people/km²) being much lower than the average population density in South Korea (509 people/km²) in 2015.

Since the mid-1990s, however, the CCZ has shrunk, as the local government attempted to loosen regulations to accommodate the requests of its inhabitants, who wanted to engage in more economic activities within the border region. This shrinkage of the CCZ coincided with the rising citizen movements and the autonomy of the local government after South Korea adopted a new policy to transfer more power from the central government to local governments. Additionally, in 1997 with the rise of the liberal government (Kim Dae Jung) that promoted the sunshine policy toward North Korea, the border region was opened for new development and movement. The two most representative examples were Paju and Goseong, which served as gateway regions to North Korea, namely, to Kaesong Industrial Complex (for Paju) and Kumgangsan tour (for Goseong) in 2003 and 2004, respectively. These two regions have symbolic meanings, geopolitically and socioeconomically. The Kaesong industrial complex, which was a result of the first inter-Korean summit in 2000, once had more than 53,000 North Koreans working at 124 South Korea–owned factories in the North Korean territory. The industrial complex was temporarily closed by the conservative South Korean government in 2016 partly in response to continued North Korean provocations. The Kumgangsan tour, a tour to the scenic Kumgang Mountain in North Korea, was initiated by Jung Joo Young, who was a Hyundai chaebol who migrated from North Korea. The tour became the first place in North Korea that was opened for South Korean tourists in 1998, and it lasted until 2008 when the conservative government closed the tour after the death of a South Korean tourist. Many South Koreans who had migrated from North Korea visited the mountain and, as such, the tour was a symbol of the reunification of divided families. With the arrival of a new liberal government in 2017, these two cities also received attention in 2018; South and North Korea are currently exploring the possibility of constructing new roads and other infrastructure around these two places.

Although the two areas are similar in size, Paju, taking advantage of its proximity to Seoul, the capital of South Korea, had a higher population density (625 people/km²) than Goseong (46 people/km²) in 2015 (Statistics Korea 2016). Additionally, Paju's main industry is manufacturing, whereas Goseong's main industry is services mostly associated with tourism. Financial independence was much higher in Paju (44.0 percent) than in Goseong (13.4 percent). Comparing these two gateway regions serves an as excellent laboratory for understanding the spatial transformation of the border region as political regime shifts and how landscape conditions and specific historical geographical context played a role in shaping the transformation (Table 1).

Land Use Policy and Impacts on the Border Region in South Korea

South Korea's land development plans have changed since the country pursued industrialization. The first comprehensive national territorial plan

Table 1. Similarities and differences between Paju and Goseong

	Similarities Both regions	Differences	
		Paju	Goseong
Geography	Close to North Korean border; served as gateway cities	Closer to Seoul (27 km), drawing more domestic and international visitors (approximately one-eighth foreign tourists)	Distant from Seoul (160 km), drawing fewer visitors, mostly domestic (less than 2% foreign tourists)
Regional economy	Traditionally below national average regional income; affected by national and regional land policies	Good mix of manufacturing and service industries; flux of capital flow from outside; high financial independence (39.6%)	Heavily dependent on service industry (e.g., tourism); limited flow of capital from outside; low financial independence (7.8%)
Population	Aging population over time, reflecting national trend	Population increasing (3% growth per year; proportion of population over 65 in 2015 = 16.7%)	Population decreasing (0.8% decline per year; proportion of population over 65 in 2015 = 33.4%)
Land use	Military presence with limited land development; mostly forest, small fraction of urban areas	Urban areas are 11.8% of the region	Urban areas are 3.2% of the region
Environment	Relatively well protected in undeveloped areas near Demilitarized Zone	Relatively flat (mean slope = 7 degrees; elevation difference = 649 m), prone to flooding	Complex terrain (mean slope = 14.5 degrees; elevation difference = 1,293 m), prone to landslides

(CNTP; 1972–1981) focused on the industrial development of Busan and its surrounding southeastern part of South Korea to increase its export-driven economy. This CNTP has resulted in uneven development between the so-called the Seoul–Busan axis and other parts of the country. The second CNTP (1982–1991) attempted to balance regional economic disparities by investing in three other major cities—Daegu, Gwangju, and Daejeon—while expanding the existing industrial complex to the industrial complex belt in the southeastern region. The third CNTP (1992–2001), which coincided with a power transfer from central government to local government, further promoted the creation of the multicentered belt. Although the border region was first mentioned in the third national development plan as the basis of South–North cooperation, it was not until the fourth CNTP (2002–2011) that the spatial policy of the border region was explicitly mentioned. With the rise of South Korea's economic status in the world economy, the government designed four national supereconomic network zones that enhance financial flow to world regions. The west coastal region was designed as a new industrial belt to promote economic trade between China and other Southeastern Asian countries, whereas the east coastal

region was designated as an energy and tourism belt to promote economic trade with distant countries. The border belt reflects these two different interests in the west and east regions. The western part of the border belt (to which Paju belongs) was included in the industrial economic area, whereas the eastern part of the border belt (to which Goseong belongs) was included in the coastal tourism area. The border region's spatial policy was specifically designed to identify the types of industries and development potentials. For example, Paju was designated as an exchange and cooperation zone and Goseong was designated as an ecosphere reserve zone. These somewhat contrasting designations set up potentially different pathways of development and environment for each region. As a result, Paju was able to attract manufacturing industries such as book printing and electronic display manufacturing that were tied to the Kaesong industrial complex as well as national consumption and exports. Goseong's tourism industry, however, primarily relied on the Kumgangsan tour and thus became sensitive to changes in the political regime.

Another notable policy is the Seoul Metropolitan Area readjustment plan, which affected Paju's land development. The first plan (1982–1996) was characterized as an era of strong regulation that limited growth in the border region. The second plan (1997–2006) was a readjustment era that allowed some form of new development in the southern section of the border region. This period was aligned with the arrival of the liberal government (Kim Dae Jung and Noh Mu Hyun) that implemented the sunshine policy, which opened active political and economic engagement with North Korea. The third plan (2006–2020) has a clear multicentered network, with Paju becoming a north–south exchange industrial belt in the growth management zone. The Paju local government took full advantage of such national and regional policies that allowed it to host people and industry within its jurisdictional boundary. In contrast, no substantial resources were allocated to Goseong during the conservative regime to attract alternative tourist populations after the closure of the Kumgangsan tour, and unfinished developed land was largely abandoned.

To better examine the borderland evolution in a coupled social and environmental system framework, political ecological approaches, which have been used to understand the links between environmental governance and land-use and land-cover change and subsequent environmental change, could offer useful insights. In the United States, exurban political ecologists have stressed the importance of places that have unique social and political economic relations with nature, which continuously evolve with changing institutions and power dynamics (Hurley, Maccaroni, and Williams 2017). For example, Walker and Hurley (2004) identified that local pro-growth activists' protest against conservation resulted in the derailment of community-based collaborative natural resource management in Nevada County, California, emphasizing that understanding local political contexts and histories is essential for the success of collaborative environmental management. In a similar vein, Hiner (2015) remarked that the divergent views of environmental management in different stakeholders along the rural–urban interface in Calaveras County, California, were associated with different political and environmental ideologies and preferences. Additionally, Hurley, Maccaroni, and Williams (2017) stressed the importance of land use histories and landscape qualities as well as local political dynamics between long-term residents and in-migrants for understanding the new patterns of conservation development in suburban southeastern Pennsylvania. With regard to local environmental management, the increasing use of scientific information (e.g., monitoring of environmental quality) is often associated with citizens' protest against environmental degradation in the local environment (Chang et al. 2014). Using the theories and empirical findings of such case studies, we seek to identify which mechanisms might be relevant for explaining the socioeconomic and environmental changes in South Korea's borderland.

Data and Methods

Data

We used population (1995–2015), employment rate in specific industries (1993–2014; Statistics Korea 2016), land cover (1990–2010; Ministry of Environment 2017), and water quality (1990–2016; National Institute of Environment Research 2017) to understand the spatiotemporal characteristics of changes in sociodemographic and environmental conditions. We also used national policy documents as they relate to land and industrial development and environmental policy obtained from the Korea Research Institute for Human Settlement (2017).

GIS and Landscape Analysis

To detect the direction of land-cover change from one type to another, we calculated the percentage of land cover for each region each year using zonal statistics in ArcGIS 10.4 (Esri, Inc., Redlands, CA, USA). To evaluate the degree of landscape fragmentation, we used four landscape metrics: patch density (PD), area-weighted mean patch shape index (SHAPE_AM), connectance (Connect), and Shannon's diversity index (SHDI) as ways to gauge landscape change. These metrics were chosen as they represent landscape composition (PD), configuration (SHAPE_AM), connectivity (Connect), and diversity (SHDI) and were used in previous studies to detect landscape homogenization or heterogenization over time (Su et al. 2012; Cabral and Costa 2017). They were calculated using the aforementioned land-cover maps for each year in FRAGSTATS software (McGarigal et al. 2002).

Statistical Analysis

We used descriptive statistics to identify the change in growth rates in economy and population

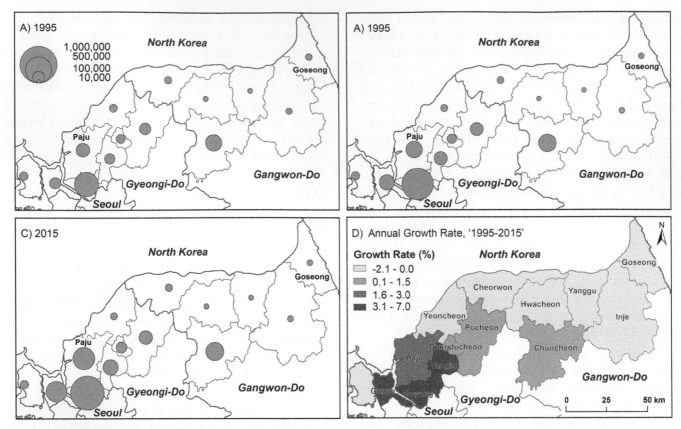

Figure 2. Change in population, 1995 to 2015, and annual growth rate in the border region. (Color figure available online.)

in each Gun within the border region. For detecting trends in water quality by year, we used the Mann–Kendall test for each station (Kendall 1975), which has been widely used in identifying significant changes in water resource data (Mainali and Chang 2018). Considering relatively fewer years of sampling ($n < 22$), we used the 10 percent significance level to detect significant trends of Sen slope estimates. A significant positive Sen slope indicates an increasing concentration of water quality parameters over time.

Regional Resilience

Regional resilience in this study is defined as a social, economic, and ecological resilience of a region, which commonly includes buffering capacity to maintain stability, learning and innovation, and capacity to adapt (Hotelling 2001; Peng et al. 2017) in response to pressure and pulse of the system. In our case, shifts in governmental regime (ten-year period cycle) and the associated land and environmental policies are considered pressure, whereas abrupt changes such as the closure of Kumgangsan tour and Kaesong industrial complex

are regarded as the pulse of the system. Although we do not attempt to include all potential variables that are associated with regional resilience, we strive to include key factors that represent each of the three pillars of resilience. Economic resilience was assessed based on annual growth rate in population and employees as these are two of the most common variables used in evolutionary economic geography (Simmie and Martin 2010; Di Caro and Fratesi 2018), and social resilience was assessed by demographic structure and suicide rate, which represent the diversity and stability of a community (Keck and Sakdapolrak 2013; Reeves et al. 2014). Ecological resilience was addressed by using the degree of landscape fragmentation (Su et al. 2012) and water quality change (Beck 2005; Chang et al. 2014).

Results

Sociodemographic Change by *Si/Gun*

As shown in Figure 2, during the period between 1995 and 2015, population growth rate was

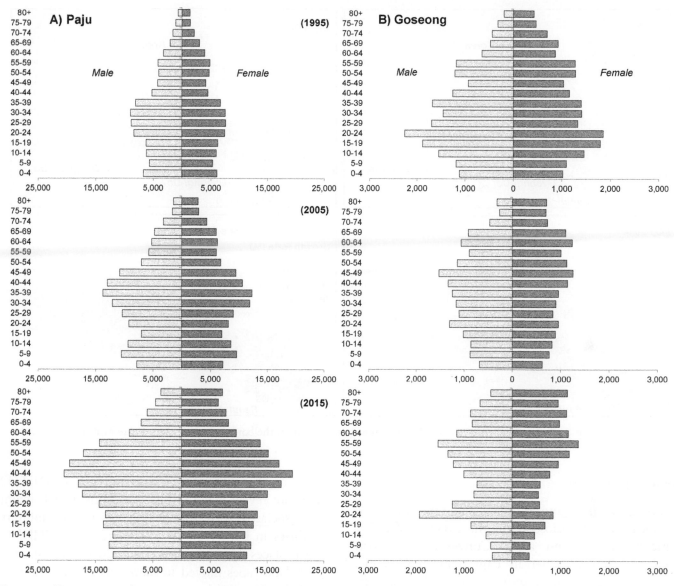

Figure 3. Change in population structure, 1995 to 2015: (A) Paju and (B) Goseong.

substantially higher in the western part of the study area, namely, Gyeonggi-Do, than the eastern part, Gangwon-Do. Together with the suburban areas of Seoul (Goyang and Gimpo), Paju had one of the highest population growth rates (around 3 percent). In contrast, Goseong's population declined by 23 percent. This is further explained by changes in demographic structure (Figure 3). Between 1995 and 2015, Paju's population increased in all age groups. Although the highest age group increased by ten years (from thirty to thirty-four to forty to forty-four) during the period, those age groups are active participants in various economic sectors within the region. In contrast, Goseong's absolute population declined substantially in all age groups

below fifty-five. As a result, like most rural areas, Goseong had a rapidly aging population. The highest number of males for the age group between twenty and twenty-four is associated with the number of soldiers in that Gun. Changes in employment showed spatial patterns similar to those for population (Figure 4). Although the number of employees grew in all regions during the study period, Paju has one of the three highest growth rates (over 21 percent), whereas Goseong had one of the lowest annual growth rates (below 5 percent). These statistics show the uneven pattern of spatial development within the border region, and the regional disparity between the west and east increased over time.

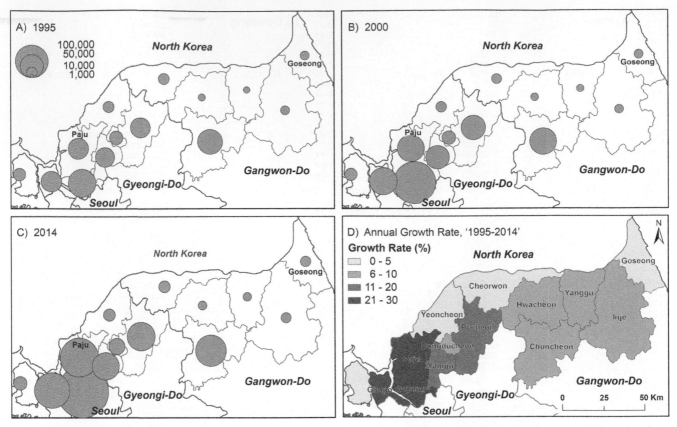

Figure 4. Change in employee number, 1995 to 2014, and annual growth rate in the border region. (Color figure available online.)

Landscape Change

As shown in Figure 5, urban land cover increased in both Paju and Goseong at the expense of either agricultural or forest areas. Conversion to urban areas was much higher in Paju than in Goseong. Additionally, spatial patterns of urban development differed between the two regions. Whereas Paju's development was more dispersed with the development of new urban centers in the south-central part of the *Gun* over time, Goseong's urban expansion was confined to the existing developed areas along the east coast. Such different patterns of landscape changes between the two regions were more clearly explained by examining landscape fragmentation (Table 2). In Paju, PD increased rapidly between 1990 and 2000, whereas it continuously declined in Goseong during the study period. The shape index (SHAPE_AM) initially declined in Paju but increased from 2000 to 2010, indicating that the landscape became more irregular in shape in the later period. In contrast, SHAPE_AM continuously declined in Goseong during the entire study period, suggesting that the landscape of Goseong became

more regular. The connectivity index (CONNECT) does not reveal hypothesized directions, which might be associated with the heavy presence of military facilities in both regions. The increases in SHDI in both places between 1990 and 2000 suggest that both landscapes moved to more mixed land-use patterns in this period.

Water Quality Monitoring and Change in Water Quality

The number of water quality monitoring stations increased in Paju during the study period. In contrast, Goseong had only one station throughout the study period. The increasing number of stations in Paju coincided with the development of land areas as shown in Figure 5. It is notable that different agencies across different levels started water quality monitoring beginning in the mid-1990s and into the late 2000s. Environmental awareness and movement rapidly increased in South Korea after 1990 when the Ministry of Environment was established. The regional and local governments appeared to respond

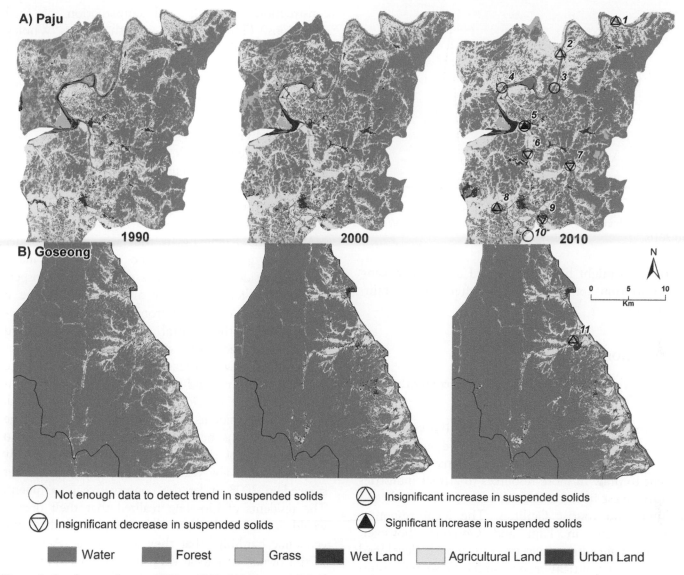

Figure 5. Land-cover change, 1990 to 2010: (A) Paju and (B) Goseong; changes in water quality, 1995 to 2016.

Table 2. Landscape pattern change between 1990 and 2010 in Paju and Goseong

| Year | 1990 | | 2000 | | 2010 | |
Place	Paju	Goseong	Paju	Goseong	Paju	Goseong
PD	37.27	20.53	44.66	18.22	37.31	15.76
SHAPE_AM	24.63	28.1	19.18	21.81	21.05	21.64
CONNECT	0.27	0.37	0.21	0.46	0.27	0.44
SHDI	1.19	0.60	1.27	0.62	1.22	0.62

Note. PD = patch density; SHAPE_AM = area-weighted mean patch shape index; CONNECT = connectance; SHDI = Shannon's diversity index. *Source:* McGarigal et al. (2002).

to these movements to make their land more sustainable both economically and environmentally. As a result, even though more land was converted to residential and industrial development in Paju, its water quality did not necessarily deteriorate much during the same period, except one station located in the downstream section of a tributary (Station 5) that has experienced rapid urban growth near the monitoring station (Figure 6). On the other hand, in Goseong, water quality as measured by total solids showed abrupt rises in the years 2003, 2004, and 2008. These years followed heavy rainfall brought by typhoons (2002) and wildfires (2008). Together with these extreme events, there were some construction activities associated with the Kumgangsan tour between 2003 and 2008. Because land development was concentrated around the city area (Figure 5), it is likely that the construction of resorts and

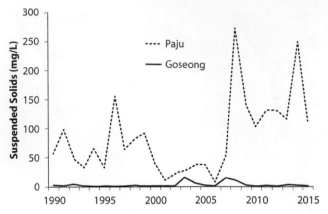

Figure 6. Change in water quality in Paju and Goseong, 1990 to 2016.

restaurants might have contributed to the increasing concentration of total solids after heavy rainfall events.

Discussion

Local Responses to the Central Government's Policy Changes

In response to shifts in the central government's policies, such as a new comprehensive support plan for the border area, local politicians aggressively sought to draw in more resources to attract industries and to invest in infrastructure such as new roads and cultural and tourist facilities. The local responses were different in Paju and Goseong, however. Whereas Paju's infrastructure continuously developed regardless of political regime shifts, Goseong's infrastructure projects mostly ended after the closure of the Kumgangsan tour. For example, large-scale road projects and construction of industrial complexes were continuously developed in the whole study period in Paju, irrespective of the government regime. In contrast, the Hyundai-Asan company, the chaebol that initiated the Kumgangsan tour, stopped building a major tourist hotel in Goseong. In search of more capital that could flow into the borderlands in Gangwon, Kim Jinsun, who was the mayor of Gangwon province in 2008, stated, "Gangwon has a bigger portion of the DMZ than any other province in Korea. I hope Gangwon will be the foundation stone for unification. We want to send a message of peace to North Korea, and the world" (Voice of America 2008). Similarly, another politician remarked, "I think the residents of the DMZ need an economic approach, which should include

construction of roads and construction of road extensions" (This is America with Dennis Wholey 2011a). A local city mayor stated, "It (DMZ) is a place where people are still living alongside the military. If the government can come up with a new law to improve the quality of the lives of the residents, then this area will not only be preserved but it will become a living heritage" (This is America with Dennis Wholey 2011b). Whereas local politicians used the rhetoric of reunification and support for development in anticipation of reelection in the local governments, local residents' responses were more revealing.

When the Kumgangsan tour was opened under the liberal government in the early 2000s, many Goseong residents had dreams of becoming prosperous as they felt that their area would become a major gateway for tourists from South Korea or elsewhere. When the central government announced the symbolic reopening of roads and railways that could link the two Koreas in 2002, many real estate offices were opened and new hotel and restaurant construction started in Goseong in a very short period. Some local residents unreasonably borrowed large amounts of loans beyond their financial capacity to join this line of dreamers, but their dreams largely evaporated in a few years when the Kumgangsan tour stopped in 2008 under the conservative government. The residents of Goseong realized that their dreams would not materialize and thus complained about the economic hardships that they were facing. According to the owner of a local shop in Goseong, "After the Kumgangsan tour program had ended, many business owners closed their offices and left the area, turning the area into a ghost town" (Joogangilbo 2017). Another local restaurant owner remarked, "When the Kumgangsan tour was in flux, many guests lined up for eating, but when the tour program became unavailable, no guests came and I had to close the restaurant. I just reopened the restaurant last year." Yoon Seong Geun, the mayor of Goseong, stated, "I wish we could reopen the Kumgangsan tour as soon as possible so that our regional economy could revive. All Goseong residents would like the road to Kumgangsan to be reopened as quickly as possible" (Joogangilbo 2017). These sentiments reflect the local residents' displeasure with the conservative central government's tourism policy that disrupted the local economy and their lives. In reality, there were several local protests against the conservative government's policy, lobbying for the reopening of the Kumgangsan tour. These points, although the context is quite different, reiterate how local society could gain or deny access to the use of local land resources with changing government regimes (Eilenberg 2016). In contrast, there has been no complaint report about any social and economic damages among Paju residents after the closure of the Kaesong industrial complex under the conservative government, suggesting that Paju's responses were largely independent of changes in governmental policies. These findings reinforce the recent theories of borders in motion, which view borders as spaces of mobility and uncertainty (Sheller and Urry 2006; Konrad 2015) that are continuously evolving with their own geography and dynamic power relationships across spatial scales (Newman 2006).

Landscape Change and Environmental Quality

According to our landscape analysis, the increase in PD and SHAPE_AM in Paju suggests that the area became more fragmented and the overall configuration of the land became more complex than before. In contrast, the continuous decline of SHAPE_AM in Goseong indicates that the overall landscape became more homogeneous. These contrasting trends can be explained by multiple social, economic, and political factors that are unique to the study areas that also have different proximity to the Seoul metropolitan area. Paju's rapid landscape fragmentation was associated with the new development of industrial complexes and related infrastructure such as housing, roads, and public facilities around the city. For example, Paju printing industry, a reunification park, and an LG display high-technology complex are now functioning as new nuclei in the urban spatial structure. As a result of these new developments, agricultural areas, which were once protected, continuously declined to attract more industries and people to Paju. On the other hand, near the North Korean border within the CCZ, forested areas have been converted into agricultural areas for growing cash crops such as Korean ginseng (E.-J. Park and Nam 2013). Together with different governmental regimes, Paju's master plans have been revised several times under different mayors since the local government took office in 1995. By updating insufficient items (mostly environmental-related items) in previous plans, Paju's new master plan included the identification of both developable areas (nearly 10 percent of the entire *Gun*) and conservation need areas based on scientific ecological assessment (City of Paju 2017). This local government's conservation efforts, while promoting sustainable development, reflect the increasing use of scientific information to monitor changes in environmental quality and to make decisions on environmental management. For example, Gyonggi Research Institute, a research center that employs more than 100 biophysical and policy scientists, performed a feasibility study on ecotourism (Seong et al. 2007). Additionally, the new residents, who were typically well educated and migrated from congested and densely developed Seoul, also sought the amenities and associated environmental quality (clean air and water) in Paju. When new citizens perceived or identified ecologically harmful development projects, they protested against new developers and the county government, often halting development projects (K.-M. Park 2017). In contrast, the economically underdeveloped Goseong merely used the short window of opportunity under the liberal government to further develop land near existing developed areas, resulting in maintaining a homogeneous landscape. Although there had been somewhat haphazard small-scale development, there were no reports of environmental complaints by residents in Goseong during the construction. There was also no additional effort to increase environmental monitoring by the local government or by residents. This could be associated with the fact that Goseong, with naturally complex terrain and the presence of military camps, has historically had many environmental regulations that prohibited newer development. Local governments and residents had been constantly looking to ease such environmental regulations for land development in hopes of economic prosperity. Our findings are in agreement with other borderland studies in Senegal and Guinea Bissau (Cabral and Costa 2017), in Austria and the Czech Republic (Sklenicka et al. 2014), and in Poland and Lithuania (Senetra et al. 2013), which indicated that the rapid conversion of borderlands was associated with local landscape quality and history and political and economic transformation, leading to the increase in spatial heterogeneity within the border area. Local power dynamics that were identified in the U.S. exurban political ecology literature appear to be relevant to understand the different pathways of land development and environmental management in response to the abrupt and gradual policy changes in the two border regions in South Korea.

Implications for Regional Resilience

A region's resilience can be examined by its socioeconomic status, landscape fragmentation, and environmental governance and quality occurring at multiple scales within the coupled social–ecological systems (Folke et al. 2010). How each region is able to absorb external shocks and internal stresses and adapt to these changes is dependent on multiple factors. After short-term shocks or long-term stresses, some regions might completely transform and move to a new state, whereas others might either collapse or remain resilient. Our study supports the notion that social systems could increase in complexity in response to disturbance (Weichselgartner and Kelman 2015; Wilson 2017). Although changes in national governmental regimes from a liberal to a conservative government affected the national economic and environmental policies and thus outcomes at the local scale by changing the roles of gateway areas, the various responses of regional and local governments should also be noted. Paju and Goseong illustrate such representative cases.

In Paju, regardless of the diminished flow to the Kaesong Industrial Complex and eventual shutdown of the complex under the conservative government in 2016, the region remained resilient as its economy and population continuously grew while generally maintaining environmental quality. Movement of goods and people to the region continuously flourished in both liberal and conservative government eras. With its fast economic growth, Paju has become the most financially independent Gun in the border region. During this development era, however, Paju's environment became somewhat mixed. Its landscape has become more fragmented than before as new development centers were formed to accommodate the needs of growing industries in the area. Although water quality was somewhat degraded in some places during the development stage, additional water quality monitoring stations were installed to better monitor and maintain the environment in developing areas by the local government. Similar local responses were found in the study of environmental governance and water quality change in the Portland-Vancouver metropolitan area

(Chang et al. 2014). This complex feedback between social and environmental systems suggests that the relationship between ecological resilience and social resilience in the region is not always linear, as stated by Adger (2000).

In contrast, Goseong's economic boom was much more ephemeral, as its economy was at its peak during the liberal government that initiated the Kumgangsan tour. The short period of prosperity ended with unfulfilled dreams and environmental costs. During the short period between 2003 and 2008, most new developments were concentrated in the surrounding areas of the existing town, mainly along local highways, and, as a result, with the spatially concentrated newer construction, water quality deteriorated when heavy rainfall hit the region. After 2008, when construction projects associated with the tourism stopped, it left a financial burden for residents and the local government and further lowered the financial independence rate in Goseong. After the closure of the Kumgangsan tour, between 2008 and 2010, the area's suicide rate increased by more than 50 percent from the previous years (before 2008). Similar to the case of Greece, where the suicide rate surged after the financial crisis of 2008, economic hardships with unfulfilled dreams have left some households with higher suicide rates in Goseong. Between 2008 and 2017, 241 restaurants were closed, and low-income or single-parent families increased dramatically, disrupting the local economy and society (Gangwon CBS 2017). Goseong's resilience was weak as its regional pathway was heavily dependent on somewhat competitive and external resources with a focus on a single industry. Considering that a suicide rate is one of the best indicators of social resilience (Reeves et al. 2014), this alarming statistic showed how the region was devastated by the end of tourism. The region was not able to rebound after the political and economic shock in 2008, but its environment might become resilient, as water quality has been restored since the Kumgangsan tour program ended.

Conclusions

We investigated the changing landscape of South Korea's border region with a focus on two gateway regions—Paju and Goseong—and how both regions underwent socioeconomic and environmental transformation during the past two decades when the

Table 3. Cross-scale factors affecting resilience of the border region in South Korea

Scale	National	Regional	Local
Political/economic	Governmental regime shift, cooperation between North and South Korea	Regional economy, provincial leadership, financial independence rate	Job opportunity, personal finance, city leadership
Sociodemographic	National population trend	Migration, % highly educated people	Local population structure; suicide rate
Environmental	National environmental standards and regulation	Regional environmental policy and monitoring effort	Environmental perception; grassroots organization

political regime of South Korea shifted from liberal to conservative governments. Although the border region was once considered a stagnant place, many sociodemographic and environmental shifts occurred during this period. Under the liberal government that implemented the sunshine policy toward North Korea, the border region was open to new development, allowing the movement of people and goods. The two gateway regions—Paju and Goseong—exemplified how the border region in South Korea transformed into different states as both places revealed different trajectories of regional resilience. While Paju's landscape became more fragmented as a result of urban expansion, different subcenters appear to have contributed to the increasing economic and environmental resilience of the region. In contrast, with continuously declining population, Goseong's landscape became less fragmented with one remaining main urban center, but its economy and environment became fragile after the closure of the Kumgangsan tour during the conservative regime. Local responses to government policy changes were distinct between the two regions, as Goseong, with an aging and declining population, was primarily interested in developing the tourism-based fragile industry, whereas Paju, with its well-educated and increasing population, diversified its economy and adopted environmental management policies in its long-term plan, which contributed to the resilience of the region. These different trajectories of regional resilience can be fully understood by considering various social, economic, and environmental factors acting on multiple scales that helped shape the region's stability (Table 3). Our results indicate that regional resilience is tightly coupled with internal and extraregional political and economic pressures and pulses that contribute to regional stability and innovation (Christopherson, Michie, and Tyler 2010). Thus, understanding how past political and economic legacies produced the current landscape patterns is crucial to a comprehensive

understanding of the borderland dynamics over the long term. Future studies could include systematic interviews of key officials and practitioners who were involved in land-use planning and policy as well as environmental organizations and residents who participated in environmental governance. Given the recent dialogues between the two Koreas, local politicians and residents are already responding to such movement in anticipation of reopening of the border. The integrated approach conceptualized in our study could be further expanded to understand the dynamics of the border region in relation to the rapidly changing international and national politics and how locals would respond to such changes in the coming years.

Acknowledgments

We appreciate Dr. Junghoon Lee of Kyonggi Research Institute who provided necessary documents used in the article. We also thank the two anonymous reviewers whose comments helped refine many points of the article.

Funding

This work was partially supported by the Korea Institute of Planning and Evaluation for Technology in Food, Agriculture, Forestry and Fisheries (IPET) through (Animal Disease Management Technology Development Program), funded by the Ministry of Agriculture, Food and Rural Affairs (MAFRA) (grant number 315038-2).

ORCID

Heejun Chang ⓘ http://orcid.org/0000-0002-5605-6500

References

Adger, W. N. 2000. Social and ecological resilience: Are they related? *Progress in Human Geography* 24 (3):347–64.

Apostolopoulou, E., D. Bormpoudakis, R. Paloniemi, J. Cent, M. Grodzińska-Jurczak, A. Pietrzyk-Kaszyńska, and J. D. Pantis. 2014. Governance rescaling and the neoliberalization of nature: The case of biodiversity conservation in four EU countries. *International Journal of Sustainable Development & World Ecology* 21 (6):481–94.

Beck, M. B. 2005. Vulnerability of water quality in intensively developing urban watersheds. *Environmental Modelling & Software* 20 (4):381–400.

Cabral, A. I. R., and F. L. Costa. 2017. Land cover changes and landscape pattern dynamics in Senegal and Guinea Bissau borderland. *Applied Geography* 82:115–28.

Chang, H., P. Thiers, N. R. Netusil, J. A. Yeakley, G. Rollwagen-Bollens, S. M. Bollens, and S. Singh. 2014. Relationships between environmental governance and water quality in growing metropolitan areas: A synthetic view through the coupled natural and human. *Hydrology and Earth System Sciences* 18 (4): 1383–95.

Christopherson, S., J. Michie, and P. Tyler. 2010. Regional resilience: Theoretical and empirical perspectives. *Cambridge Journal of Regions, Economy and Society* 3 (1):3–10.

City of Paju. 2017. City Master plan 2030. Paju: City of Paju, South Korea.

Di Caro, P., and U. Fratesi. 2018. Regional determinants of economic resilience. *The Annals of Regional Science* 60 (2):235–40.

Eilenberg, M. 2016. A state of fragmentation: Enhancing sovereignty and citizenship at the edge of the Indonesian state. *Development and Change* 47 (6): 1338–60.

Folke, C., S. R. Carpenter, B. Walker, M. Sheffer, T. Chapin, and J. Rockstrom. 2010. Resilience thinking: Integrating resilience, adaptability and transformability. *Ecology and Society* 15 (4):1–20.

Gangwon CBS. 2017. Status of Goseonggun damage. Accessed June 25, 2018. http://www.nocutnews.co.kr/news/4813585#csidx184bb4bc82e9f60b66ba9d6edb8504b.

Gelezeau, V. 2013. Life on the lines: People and places of the Korean border. In *De-bordering Korea: Tangible and intangible legacies of the sunshine policy*, ed. V. Gelezeau, 13–33. London and New York: Routledge.

Hiner, C. 2015. (False) dichotomies, political ideologies, and preferences for environmental management along the rural–urban interface in Calaveras County, California. *Applied Geography* 65:13–28.

Hotelling, C. S. 2001. Understanding the complexity of economic, ecological and social system. *Ecosystem* 4:390–405.

Hurley, P. T., M. Maccaroni, and A. Williams. 2017. Resistant actors, resistant landscapes? A historical political ecology of a forested conservation object in exurban southeastern Pennsylvania. *Landscape Research* 42 (3):291–306.

Jin, J. H. 2005. The experiences and discourses of the Kumkangsan tour: "Tourist gaze" and social construction of nature. *Cultural and Historical Geography* 17 (1):31–46.

Joogangilbo. 2017. The 10-year-old ruined town, The wish for revival through the reopening of Kumgangsan tour: The hope of residents in the northernmost part of the east coast. Accessed June 20, 2018. http://news.joins.com/article/21603605

Kang, S., and W. Choi. 2014. Forest cover changes in North Korea since the 1980s. *Regional Environmental Change* 14 (1):347–54.

Keck, M., and P. Sakdapolrak. 2013. What is social resilience? Lessons learned and ways forward. *Erdkunde* 67 (1):5–19.

Kendall, M. G. 1975. *Rank correlation methods.* 4th ed. London: Charles Griffin.

Kim, K. C. 1997. Preserving biodiversity in Korea's demilitarized zone. *Science* 278 (5336):242–43.

Kim, S. B., and W. H. Lee. 2004. Theoretical models and research trends of border region research. *Journal of the Economic Geographical Society of Korea* 7 (2): 117–36.

Konrad, V. 2015. Toward a theory of borders in motion. *Journal of Borderlands Studies* 30 (1):1–17.

Korea Research Institute for Human Settlement. 2017. *National territorial planning.* Anyang, South Korea: KRIHS.

Mainali, J., and H. Chang. 2018. Landscape and anthropogenic factors affecting spatial patterns of water quality trends in a large river basin, South Korea. *Journal of Hydrology* 564:26–40.

Martinez, O. J. 1994. The dynamics of border interaction. In *Global boundaries, world boundaries*, ed. D. H. Schofield, vol. 1, 1–15. London and New York: Routledge.

McGarigal, K., S. A. Cushman, M. C. Neel, and E. Ene. 2002. FRAGSTATS: Spatial pattern analysis program for categorical maps. Accessed June 10, 2018. http://www.umass.edu/landeco/research/fragstats/fragstats.html.

Medzini, A. 2016. Life on the border: The impact of the separation barrier on the residents of the Barta'a enclave demilitarized zone. *Journal of Borderlands Studies* 31 (4):401–25.

Ministry of Environment. 2017. Environmental geographic information services. Accessed January 15, 2017. http://egis.me.go.kr/req/intro.do.

National Institute of Environment Research. 2017. Water information system. Accessed July 7, 2017. http://water.nier.go.kr/waterMeasurement.

Newman, D. 2006. The lines that continue to separate us: Borders in our borderless world. *Progress in Human Geography* 30 (2):143–61.

Park, E.-J., and M.-A. Nam. 2013. Changes in land cover and the cultivation area of ginseng in the civilian control zone: Paju city and Yeoncheon county. *Korean Journal of Environment and Ecology* 27 (4): 507–15.

Park, K.-M. 2017. Protect Eurasian eagle-owl before the Paju Jangdankong project. Accessed July 3, 2018. http://www.hani.co.kr/arti/society/area/819188.html.

Park, S. O., ed. 2005. *Economic and social geography of the border zone: Aspects and change of backwardness and alienation.* Seoul, South Korea: SNU Press.

Peng, C., M. Yuan, C. Gu, Z. Peng, and T. Ming. 2017. A review of the theory and practice of regional resilience. *Sustainable Cities and Society* 29:86–96.

Pérez-Nieto, E. G. 2016. Centralization as a barrier to cross-border cooperation? Some preliminary notes from an Iberian approach. *Journal of Borderlands Studies* 31 (4):481–95.

Reeves, A., M. McKee, D. Gunnell, S. S. Chang, S. Basu, B. Barr, and D. Stuckler. 2014. Economic shocks, resilience, and male suicides in the great recession: Cross-national analysis of 20 EU countries. *European Journal of Public Health* 25 (3):404–9.

Scott, C. A., and A. L. Ley. 2016. Enhancing water governance for climate resilience: Arizona, USA–Sonora, Mexico comparative assessment of the role of reservoirs in adaptive management for water security. In *Increasing resilience to climate variability and change: The role of infrastructure and governance in the context of adaptation,* ed. C. Tortajada, 15–40. Berlin: Springer.

Senetra, A., A. Szczepańska, D. Veteikis, M. Wasilewicz-Pszczółkowska, R. Simanauskienė, and J. Volungevičius. 2013. Changes of the land use patterns in Polish and Lithuanian trans-border rural area Adam Senetra. *Baltica* 26 (2):157–68.

Seo, Y., D. Lee, D. Cha, and J. Choi. 2017. Comparing forest structures in the border areas of the demilitarized zone of South Korea. *Forest Science and Technology* 13 (1):47–53.

Seong, H.-C., H.-S. Kim, J.-A. Ok, S.-W. Jun, and M.-Y. Shin. 2007. A study on the conservation of national environment and ecotourism in DMZ. *Policy Study* 1:3–5.

Sheller, M., and J. Urry. 2006. The new mobilities paradigm. *Environment and Planning A* 38 (2):207–26.

Simmie, J., and R. Martin. 2010. The economic resilience of regions: Towards an evolutionary approach. *Cambridge Journal of Regions, Economy and Society* 3 (1):27–42.

Sklenicka, P., P. Simova, K. Hrdinova, and M. Salek. 2014. Changing rural landscapes along the border of Austria and the Czech Republic between 1952 and 2009: Roles of political, socioeconomic and environmental factors. *Applied Geography* 47:89–98.

Statistics Korea. 2016. Population and employment statistics. Accessed October 8, 2016. http://kostat.go.kr/portal/korea/index.action.

Su, S., R. Xiao, Z. Jiang, and Y. Zhang. 2012. Characterizing landscape pattern and ecosystem service value changes for urbanization impacts at an eco-regional scale. *Applied Geography* 34:295–305.

Tagliacozzo, E. 2016. Jagged landscapes: Conceptualizing borders and boundaries in the history of human societies. *Journal of Borderlands Studies* 31 (1):1–21.

This is America with Dennis Wholey. 2011a. Interview with Governor Lee kwang-jae, Gangwon Province, South Korea; pofessor Kim Chang-Hwan; Congressman Han Ki-ho. Accessed April 18, 2018. https://archive.org/details/WHUT_20110109_230000_This_Is_America_With_Dennis_Wholey/start/540/end/600

———. 2011b. Interview with Governor Lee kwang-jae, Gangwon Province, South Korea; pofessor Kim Chang-Hwan; Congressman Han Ki-ho. Accessed April 18, 2018. https://archive.org/details/WHUT_20110109_230000_This_Is_America_With_Dennis_Wholey/start/540/end/600

Voice of America. 2008. South Korea offers a kinder, gentler DMZ. Accessed June 13, 2018. https://www.voanews.com/a/a-13-2008-05-13-voa17/401775.html

Walker, P. A., and P. T. Hurley. 2004. Collaboration derailed: The politics of "community-based" resource management in Nevada County. *Society and Natural Resources* 17 (8):735–51.

Weichselgartner, J., and I. Kelman. 2015. Geographies of resilience: Challenges and opportunities of a descriptive concept. *Progress in Human Geography* 39 (3):249–67.

Wilson, G. A. 2017. Resilience and human geography. In *The international encyclopedia of geography,* ed. R. Richardson, N. Castree, M. Goodchild, A. Kobayashi, and R. A. Martson, 1–10. Chichester, UK: Wiley-Blackwell.

HEEJUN CHANG is a Professor and Chair of the Department of Geography and a Fellow of the Institute for Sustainable Solutions at Portland State University, Portland, OR 97201. E-mail: changh@pdx.edu. His research interests include combined impacts of climate change and urbanization on water resources, spatial analysis and modeling, and community resilience in coupled social and ecological systems.

SUNHAK BAE is an Associate Professor in the Department of Geography Education, Kangwon National University, Chuncheon-Si, Gangwon-Do, South Korea. E-mail: gis119@kangwon.ac.kr. His research interests include disease epidemiology, the DMZ of North and South Korea, geographic information system–based space analysis methods, and machine learning.

KYUNGHYUN PARK is an Associate Research Fellow and a Manager of Research Planning and Evaluation Team at the Korea Research Institute for Human Settlements, Sejong, South Korea. E-mail: khpark@krihs.re.kr. His research interests include urban and economic geography, industrial agglomeration, cultural and creative industries, and uneven development.

Afro-Brazilian Resistance to Extractivism in the Bay of Aratu

Adam Bledsoe

This article analyzes environmental governance and black geographies to explore the connections between Brazil's erstwhile populist government and President Michel Temer's conservative administration. Although on the surface Temer's austere approach appears to put him at fundamental odds with the Workers' Party's populist emphasis on social welfare and wealth redistribution, this article argues that Brazilian populism and conservatism contain striking similarities vis-à-vis the environment and racialized violence. I examine the ways in which natural resource extraction was a central component of governance under the Workers' Party and persists under Temer. By analyzing the struggles of three black communities in the state of Bahia, I draw particular attention to the ways in which a reliance on extractivism contributes to racialized landscapes, because these communities' autonomous territories remain grievously threatened. This article points out that the environmental tendencies of the new conservative government are not novel so much as they are a fulfillment of a trend propagated under the auspices of populism. This is not, however, the final word on the topic, because affected communities resist the environmental effects of extractive industry. Although extractive measures remain central to Brazilian governance, social movements like those in Bahia nonetheless enact a politics and counternotion of the environment that establish alternative ways of life.

本文分析环境治理与黑色地理学来探讨巴西过往的民粹政府和总统米歇尔．特梅尔的保守政府之间的连结。尽管表面上特梅尔的撙节政策似乎使其与工党强调社会福利与财富重分配的民粹诉求呈现根本上的对立，但本文主张，巴西的民粹主义和保守主义在面对环境与种族暴力上，包含了惊人的相似性。我检视自然资源搾取的方式作为工党政府的核心构成要件，并在特梅尔执政下持续如此。我通过分析巴伊亚州内三大黑人社群的抗争，特别关注对搾取主义的依赖如何导致种族化的地景，因为这些社群的自治领土仍然悲惨地受到威胁。本文指出，新保守主义政府的环境倾向并不新颖，而是体现民粹主义兴盛下普及的趋势。但这并不是该议题的最终结论，因为受影响的社群正在反抗搾取产业的环境影响。尽管搾取措施仍然是巴西治理的核心，诸如在巴伊亚的社会运动，仍然启动了能够建立另类生活方式的政治及反抗的环境概念。关键词: 黑色地理学，巴西，环境种族主义，工党。

Este artículo analiza la gobernanza ambiental y las geografías negras para explorar las conexiones entre el anterior gobierno populista y la nueva administración conservadora del presidente Michel Temer. Si bien en la superficie el austero enfoque de Temer pareciera colocarlo en desacuerdo fundamental con el énfasis populista del Partido de los Trabajadores, en bienestar social y redistribución de la riqueza, este artículo arguye que el populismo y el conservatismo brasileños muestran notables semejanzas, con respecto al medio ambiente y la violencia racializada. Examino los modos como la extracción de recursos naturales fue un componente central de la gobernanza bajo el Partido de los Trabajadores, lo cual persiste bajo Temer. Analizando las luchas de tres comunidades negras en el estado de Bahía, pongo particular atención a las maneras como una confianza en el extractivismo contribuye a los paisajes racializados, debido a que los territorios autónomos de estas comunidades siguen seriamente amenazados. Este artículo indica que las tendencias ambientales del nuevo gobierno conservador no son mayormente novedosas en cuanto que son la culminación de una tendencia propagada con los auspicios del populismo. No obstante, esta no es la última palabra sobre el tópico, porque las comunidades afectadas oponen resistencia a los efectos ambientales de la industria extractiva. Aunque las medidas extractivas siguen siendo centrales a la gobernanza brasileña, movimientos sociales como los que ocurren en Bahía promueven, sin embargo, una política y una contra-intención del medio ambiente que establecen medios de vida alternativos. *Palabras clave: Brasil, geografías negras, racismo ambiental, Partido de los Trabajadores.*

The shift in Brazil from the "populist" administration of ex-President Dilma Rousseff to the government of the "conservative" Michel Temer is a useful case for studying how changes in government relate to environmental racism. Although Temer's austere approach appears to put him at odds with the Workers' Party's populist emphasis on social welfare and wealth redistribution, an examination of Afro-Brazilian territorial struggles demonstrates how Brazilian populism and conservatism contain striking similarities vis-à-vis the environment and racialized violence. This article analyzes black geographies literature on anti-blackness (McKittrick and Woods 2007; McKittrick 2013; Eaves 2016) and political ecology literature on capitalist resource extraction (Sundberg 2008; Perreault 2013; Valdivia 2015; Pulido 2017) to evidence how extractivism is inherently tied to the spatial marginalization of Afro-descendant communities (Mollett 2011; Loperena 2017). Moreover, I show how this anti-black extractivism is perpetrated by seemingly distinct political parties.

The case of three Afro-Brazilian communities in the Bay of Aratu in the state of Bahia evidences how populist and conservative political forces have perpetuated capitalist accumulation and anti-black environmental practices through shipping and militarization. As processes underpinning natural resource extraction, shipping and militarization result in the spatial displacement and poisoning of the communities in the Bay of Aratu. Extractivism—as a component of global capital accumulation—thus depends on the denial of the spatial legitimacy of Afro-descendant communities (Loperena 2017). Extractivism also evidences how the coconstitution of race and the environment inform the spatial and social organization of the Bay of Aratu (Sundberg 2008). In this article, I draw on four years of participant observation and personal conversations with these communities to evidence how the extractive agenda of Brazil's progressive and conservative governments is made possible via the spatial subordination of the Afro-Brazilian communities in the Bay of Aratu. Furthermore, I examine alternatives to this agenda by foregrounding the politics of the Afro-Brazilian communities in the Bay of Aratu. This article highlights how anti-blackness remains central to extractivism in Brazil, regardless of shifts in governance. It also evidences how the communities in the Bay of Aratu enact subjectivities and counternotions of the environment that establish alternative ways of life.

The Rise and Fall of "Progressivism"

Brazil's Workers' Party (PT) officially formed in the 1980s and gained national prominence with the 2002 presidential election of Luiz Inácio "Lula" da Silva (Branford and Rocha 2015). Neopopulist PT measures like minimum wage increases and cash transfer programs for needy families served to pull more than 20 million Brazilians out of poverty, thereby establishing a public image of state benevolence (Oliveira 2006; Anderson 2011). In August 2016, the Brazilian Senate removed president Dilma Rousseff of the PT from power for violation of federal budget protocol. Michel Temer of the center-right Brazilian Democratic Movement replaced Rousseff. Temer immediately implemented fiscally and socially conservative measures, dissolving, defunding, and merging different government agencies dedicated to serving indigenous and Afro-descendant communities. He also moved to facilitate natural resource extraction (Arsenault 2017).

To many onlookers, attempts at increasing resource extraction and defunding organs committed to helping racialized communities demonstrate Temer's lack of concern for the environment and marginalized groups. Thus, Temer is often cast as a radical departure from the "progressive" agenda of the PT. An engagement with the Afro-Brazilian communities in the Bay of Aratu, however, reveals the centrality of extractivism and environmental racism to both progressive and conservative administrations.

The Bay of Aratu

The Bay of Aratu is a small bay in the state of Bahia, located around 25 km from the city-center of the state capital of Salvador. In this area reside the communities of Ilha de Maré, Tororó, and Rio dos Macacos. These communities exist within 7 km of one another. All three communities have more than 200 years of history in the region. Prior to the 1888 abolition of Brazilian slavery, the enslaved ancestors of all three communities worked on, and escaped from, plantations located in the region, establishing a measure of autonomy from dominant society. Thus, all three communities have demonstrated a generations-long commitment to remaining autonomous from the always shifting structural violence present in Bahia, effectively creating "the pillars of a parallel social order" in the Bay of Aratu (Woods 2017, 13).

These three communities are physically proximate to one another and maintain close political and personal connections with each other. Still, their histories are unique and deserve individual investigations.[1] Ilha de Maré is an island collectively composed of the communities of Santana, Bananeiras, Nevis, Botelho, Maracanã, Itamoabo, Porto dos Cavalos, Caquende, Martelo, Ponta Grossa, and Praia Grande. It is home to between 10,000 and 12,000 individuals. Ilha de Maré traces its history back to Africans who escaped slave ships and established maroon (fugitive slave) communities on the island. Tororó lays on the mainland shores of the Bay of Aratu and is home to around 130 families. Today's inhabitants are the descendants of enslaved Afro-descendant populations from the plantations Pombau, Bela Vista, Muribêca, Gameleira, Ponto de Areia, and Sapoca; the indigenous Tupinambá people; and escaped Afro-descendant maroons who lived in the area prior to the establishment of the plantations. Rio dos Macacos exists roughly 5 km inland from Tororó and is currently home to just over twenty families. This community traces its history back to the plantations Fazenda Macaco, Fazenda Mereles, Fazenda Carne Verde, and Fazenda Martins. The ancestors of today's community members worked on these plantations while also maintaining autonomous territories independent of the plantations (author's field notes 2014).

Despite having unique histories, these communities share the common characteristic of subsistence production. All three communities have historically relied on fishing and the collection of crabs, shrimp, oysters, and mussels from the sea, mangroves, and rivers. In addition to fishing, residents are skilled cultivators of a variety of crops, such as beans, okra, mangoes, watermelons, corn, manioc, guava, and cucumbers. They also forage for wild crops like jackfruit, limes, and honey. The communities' knowledge of, and commitment to, producing their own crops is best summarized by Rio dos Macacos's oldest resident, who succinctly told me, "I know how to plant everything" (author's field notes 2014). Drawing on the natural environment to provide for themselves has, for centuries, allowed the communities in the Bay of Aratu to maintain an autonomous way of life.

The communities' autonomy is, in part, based on their relationship to the environment. While discussing their environmental practices at a 2014 community meeting, Ilha de Maré's leaders insisted that the communities in the Bay of Aratu "have a different understanding of nature," based on "dependence and respect." This unique relationship leads to the communities seeing themselves as a part of the environment, instead of separated from it (author's field notes 2014). Rather than dominating the environment, these communities interact with the environment in a mutually sustaining manner. This relation to the natural environment has made community autonomy possible. Autonomy is vital, given the history of Brazilian anti-blackness (Bledsoe 2015) and the fact that life in Bahia remains informed by premature black death (Gilmore 2002).

Afro-Brazilian practices of environmental sustainability are the results of sociospatial practices that supersede a commitment to securing means of subsistence (see Voeks 1990; Werneck 2010). Nonetheless, subsistence and material autonomy remain central to the environmental approach of the communities in the Bay of Aratu. By providing Ilha de Maré, Tororó, and Rio dos Macacos with means of subsistence, the ocean, rivers, and forests of the Bay have provided the communities with alternatives to the highly exploitative labor practices immanent to mainstream society, such as low-paid, informal, and domestic labor (Neto and Azzoni 2011; Harrington 2015). Instead of remaining beholden to the (often abusive) demands of bosses (Perry 2013), the communities' subsistence practices allow members to spend time with each other and work in a manner befitting them and their communal needs (author's field notes 2014). Moreover, the communities' reliance on the environment means that they do not have to live near Salvador's city core to work. Instead, as explained by leaders from Ilha de Maré during a public audience with municipal officials, the communities intentionally avoid urban Salvador because of "how blacks ... are treated in the city" (author's field notes 2014). Conditions in urban Brazil are, generally, rife with multiple forms of state and nonstate violence (Garmany 2011). In urban Salvador, police violence is particularly ubiquitous. Police murders of Afro-descendant residents of the city are an all-too-common phenomenon, having increased 212 percent between 1995 and 2005 (Smith 2013). Although no way of life is perfect, these communities have preserved autonomy for more than 200 years by constantly analyzing their life conditions and adjusting to the various permutations of Bahia's anti-blackness.

Some of the most recent manifestations of anti-black violence in the Bay of Aratu are effects of extraction—specifically embodied in commercial shipping and militarization. These phenomena have taken place under both populist and conservative regimes, as both administrations have clearly committed themselves to participating in the global capitalist economy. This participation is made possible through logics of anti-blackness, which see black populations and their geographies as "unknowable and unseeable" (McKittrick and Woods 2007, 3). These logics lead to everyday, racialized events (Mollett 2011) like processes of pollution and routine violence, described later. By highlighting the ways in which Brazilian extractivism remains imbricated with anti-blackness, the case of the Bay of Aratu evidences that presumably distinct political administrations reify capitalism through structural, environmentally racist mechanisms (Pulido 2000; Pulido 2017).

Extraction and Shipping in the Bay of Aratu

The Bay of Aratu has been a site of development projects since the 1960s. Under Brazil's military dictatorship (1964–1985), the state government embarked on a journey to "modernize" the state of Bahia. Baiano elites believed that, through government-led development, the state of Bahia—nationally perceived as underdeveloped, impoverished, backward, and black—could begin to compete with more economically prosperous, "whiter" southern cities like Rio de Janeiro and São Paulo (Viana Filho 1984). To combat Brazil's apparent regional hierarchies (Weinstein 2015), elites in Bahia undertook a number of projects. In addition to the expansion of the military in the Bay of Aratu—discussed in more detail here—one of the largest projects undertaken during this time was the Port of Aratu, constructed in 1971 (Viana Filho 1984). Whereas Baiano politicians cast the port as a boon for the state's economy, the communities in the Bay of Aratu critique the port's construction. Tororó community leaders who lived through this land grab maintain that nobody consulted the communities in the Bay prior to the building of the port. These same leaders maintain that neither public nor private actors have displayed

any interest in the effects this construction has subsequently had for the communities (author's field notes 2017).

In the decades since its original construction, the port has expanded and currently represents a site of global commodity transport. Today, the Port of Aratu is home to three terminals in which freighters dispense and receive shipments of various goods made from extracted natural resources. In many ways, the port has lived up to elite aspirations of connecting Bahia to the wider national and global economy. Ships from all around the world come to the port to drop off and pick up commodities like aluminum, copper, ammonium, diesel, and ethanol. The present circulation of minerals and natural resources through the port is part of Brazil's political economic commitment to extractivism. The "progressive" PT played a significant role in building Brazil's current extractive capabilities, as it heavily financed extractive industries in the country. Under the PT, state-run corporations like BNDES, Previ, and Petros invested billions of dollars in extractive endeavors (Zibechi 2014). The subsequent extraction of natural resources, minerals, petroleum-based products, and agro-fuels made Brazil the largest extractive power in South America by 2011 (Gudynas 2013). This PT-led extraction of natural resources adversely affected the indigenous and Afro-descendant communities that depend on the environment.

Although the PT cast itself as a progressive, "populist" party, their commitment to "extractivism maintain[ed] a style of development based on the appropriation of Nature" and created "negative environmental and social impacts" (Gudynas 2010, 1). These impacts took various forms. Natural resource extraction physically changed the environment, as evidenced in the fact that more than 24 million hectares of land in Brazil were under commercial agricultural cultivation as of 2014. Such cultivation drastically altered the country's ecological diversity (Gudynas 2013). This extractive agenda has led to irreversible environmental changes for the communities present in the Bay of Aratu.

In the Bay of Aratu, pollution from the shipping of copper, aluminum, propane, butane, gasoline, and coal, among other commodities, has led to the destruction of the communities' means of subsistence. Increased freighter traffic to the Port of Aratu has led to the spillage of metals and toxic materials, which poisons marine life and reduces the number of fish and shellfish species on which the communities depend. In discussing the environmental changes her community has experienced, an experienced fisherwoman from Tororó explained to me that "in the past, we could get between ten and fifteen kilos of shrimp from the Bay. Today we only get one kilo or maybe eighty grams." In addition to this, it is all too common for "the shellfish in the mangroves to come out of the mud dead and soft" from pollution (author's field notes 2014). In addition to the long-term effects of shipping, singular events—occurring while the PT was in power—have had catastrophic results in Ilha de Maré.

In December 2013, a cargo of natural gas on the Singaporean vessel *Golden Miller* caught fire. To prevent the flames from spreading, the crew jettisoned the ship's fuel directly into the Bay of Aratu. In the following months, fish and shellfish began disappearing at an alarming rate in and around Ilha de Maré. Environmental effects expanded beyond marine life, as well. At a meeting with state officials and environmental scientists, a member of Ilha de Maré tearfully discussed how, following the spill, mangrove mud, which "had always been a source of health" in the community was now leaving them with skin rashes (author's field notes 2015). In subsequent years, things worsened. Residents of Ilha de Maré describe odd sicknesses like bone pains and vertigo. Perhaps even more troubling, cancers have begun setting in among the community's youth, killing at least three people below the age of thirty in the past three years. "In the past," exclaimed a leader from Ilha de Maré at an audience with a state attorney general, "people would live past one hundred. Today, people aren't making it to thirty!" (author's field notes 2017).

Community appeals to the "progressive" federal government immediately following the petrol spill elicited no response. Follow-up demands, presented to both the progressive and conservative governments over the past several years, have resulted in meetings with low-level government representatives unable to issue any kind of environmental or monetary relief. Leaders from Ilha de Maré unwaveringly maintain that government regimes—both progressive and conservative—have failed to respond to them because the administrations do not see Ilha de Maré as an area of production. As such, neither administration has had a problem sacrificing the

community's lived space to the capitalist agenda of economic progress. It is, in short, the cost of doing business (author's field notes 2017). Despite the fact that the environment of the Bay of Aratu is a constitutive part of Ilha de Maré's community, state actors resolve to perpetuate capital accumulation that erases the community's territorial practices. "It's as if we don't exist!" remains a common refrain among the members of Ilha de Maré when describing their treatment at the hands of powerful actors in the Bay. The environmental approaches of the Brazilian government and international capital clash antagonistically with those of the more than 200-year-old community of Ilha de Maré. In this situation, capital accumulation, embodied in the physical transportation of extracted natural resources, is possible due to the structural denial of black spatial legitimacy.

Shipping in the Bay of Aratu results in the erasure of the autonomous ways of life practiced by Ilha de Maré. Whereas the community sees the environment as something they must maintain, state and capitalist interests present in the Bay of Aratu continue to poison and upset the environmental balance in the name of facilitating commodity distribution. Meanwhile, the Brazilian government—both the erstwhile populist PT and the current conservative administration—refuses to intervene on behalf of Ilha de Maré's environment, thereby underwriting the disasters the community suffers. Unfortunately for the communities in the Bay of Aratu, shipping is not the only phenomenon supporting extractivism. Military potency walks hand in hand with extraction in Brazil.

Militarism in the Bay of Aratu

An intense focus on military buildup occurred during Lula's presidency. Under the PT, military expansion was apparently necessary to protect the "national development" embodied in state-sponsored extractive industries and manufacturing (Zibechi 2014). Military buildup signals a second form of state intervention aimed at protecting and increasing extractive capabilities in Brazil. The Bay of Aratu is a major site of this military expansion.

In 1969, the Brazilian navy completed construction of a naval base near Tororó. As the navy built and expanded the base, the traditional territory of Tororó became smaller due to enclosures committed by the navy. In addition to the base, the navy built a dam in the mid-1960s and a villa in the late 1970s to provide energy for the base and house its officers, respectively. "[The navy] came to me and said, 'We need this land to build a dam for energy.' So, we moved," a nonagenarian matriarch from Rio dos Macacos explained to me as she recounted how the navy forced her, her family, and the wider community to leave their original homes (author's field notes 2014).

If the origins of the Brazilian navy in the Bay of Aratu truncated the respective territories of Rio dos Macacos and Tororó, the treatment of community members by the navy in the decades following the conclusion of the naval base has been nothing short of catastrophic. The navy has established state-sponsored spaces of "fear, intimidation, and spatial isolation" that spatially incapacitate and displace the communities (Shabazz 2015, 6). Enforced expulsions of members of Rio dos Macacos began to take place in the 1980s. During this time the navy continued its practices of enclosing the territory of Tororó. When, in the early 2000s, the navy decided that it needed more land to expand the villa and create military training zones, conditions worsened. The accounts of normalized, routine brutality suffered by Rio dos Macacos at the hands of the navy evidence the continuity of black geographic dispossession, inaugurated during slavery (McKittrick 2013). Members of Rio dos Macacos recognize this fact. While protesting the navy's destruction of a community member's house in Rio dos Macacos, one of the community's leaders angrily described this violence, stating, "The navy only thinks of beating women, of beating poor people, and crushing black communities!" (author's field notes 2014). Beatings, rapes, arson, the destruction of property, refusal of education, and daily intimidation comprise the treatment of both Tororó and Rio dos Macacos at the hands of the navy, including during the PT's time in power. At a 2014 public audience with federal officials, an elder of Rio dos Macacos wept while insisting that the federal government recognize, "Life for us is rape! It's gunshots! It's shameful!" (author's field notes 2014). When the brutal treatment of Tororó and Rio dos Macacos did not succeed in displacing the communities, the navy employed environmental destruction.

Members of Rio dos Macacos relate stories of discovering naval soldiers spreading poisonous liquids on their fruit trees, setting corn patches on fire, prematurely ripping up manioc, and crushing nascent bean plants, actively curtailing the community's ability to draw on the natural environment for sustenance (author's field notes 2014). In addition to interfering with Rio dos Macacos's traditional poly-cultural way of life, the navy prevents community members from accessing water sources for fishing. Traditionally combining their foraging and farming activities with fishing, Rio dos Macacos relied heavily on the local rivers to provide access to this vital fauna. "[When I was young], my mother would make nets out of sacks, take two nets, leave for the night, and come back in the morning with a huge quantity of fish … we would spread the fish out on the ground and members of the community would fill their buckets [with the fish]" a middle-aged member of the community told me, reflecting on the historical importance of fishing in the community (author's field notes 2014). In building the naval dam, the navy not only changed the trajectory of local rivers but also prohibited members of Rio dos Macacos from accessing the dam and its tributaries by beating them and taking them to the naval jail for fishing there (author's field notes 2013). When questioned by state and federal officials about such behavior during public audiences, naval officials justify themselves by arguing that they cannot simply allow individuals to come and go to the dam as they please. They claim that the dam is a site of national security to which only they should have access (author's field notes 2014). These phenomena show that the Brazilian navy sees the natural environment as something they can dominate and manipulate for the proliferation of their own existence and for the protection of fixed capital. Their actions also show that they view and treat the environmental practices of Tororó and Rio dos Macacos as illegitimate, as they actively hinder the communities' abilities to reproduce themselves.

Persistent Anti-Blackness

Structural notions of black spaces as "emptied out of life" and "lands of no one" (McKittrick 2013, 7) underwrite the continued accumulation of capital via extractivism, specifically in the guise of shipping and militarism. For shipping interests and the navy, land in the Bay of Aratu is appropriate for the propagation of capital—in the guise of ports and commodity circulation—as well as the implementation of sovereign state power—in the guise of the navy's built environment. Rio dos Macacos and Tororó's engagement with the natural environment remains subordinated to this extractive agenda, as their environmentalist geographies elicit no recognition of legitimacy from the state or shipping interests. When I discussed the navy's complete disregard for the community's territory with one of Rio dos Macacos's leaders, she cogently diagnosed the situation, stating, "We're not seen as people [by the state]" (author's field notes 2014). State and capitalist commitments to the domination of the natural environment qua extractivism remain legitimate precisely because the communities' geographies receive no recognition in dominant renderings of space. Extractivism thus entails the "forcible removal *and* the elimination" of Afro-descendant communities like those in the Bay of Aratu (Loperena 2017, 806, italics in original). Both progressive and conservative administrations are responsible for this commitment to extractivism.

As mentioned earlier, the progressive PT had a central role in funding extractive industries, bolstering militarization to protect extractive endeavors, and supporting the circulation of extracted resources. According to members of each community, the conservative government has continued these trends. Temer's administration has deepened the commitment to extractivism in the Bay of Aratu through attempts at displacing Rio dos Macacos via new rounds of arson, as well as the planned construction of a new pier for Brazilian petrochemical corporation Braskem in the Port of Aratu (author's field notes 2017; personal communication 2018).

The navy, state actors running the Port of Aratu, and private shipping interests enact a politics that manipulates the environment at the same time that it erases the livelihoods of the communities in the Bay of Aratu. Because progressive and conservative administrations remain committed to a capitalist agenda of natural resource extraction, both reinforce anti-black violence, because shipping and militarization are dependent on the destruction of Afro-Brazilian communal life. Capitalist ideas of development and progress—to which both progressive and conservative regimes adhere—are possible via the erasure of Afro-Brazilian communities. In this way, the interworking of the Brazilian state and capital results in the death and degradation of Afro-

descendant populations in the Bay of Aratu (Pulido 2017). Nonetheless, through cultivating black geographies rooted in "numerous layers of organizational, familial, and cultural activities" (Woods 2002), these communities have found ways to resist the extractive agenda of the progressive PT and the current conservative regime.

Resistance to Extractivism

The communities in the Bay of Aratu experience multifaceted forms of environmental racism. Nonetheless, they find ways to combat these deleterious phenomena. Through their struggles, the communities show themselves part of a wider tendency among Afro-Brazilians to refuse structural displacement and demand dignified material living conditions (Perry 2004). Like many Latin American social movements, the kinds of resistance put forward by the communities take form in actions like protests and spatial occupation (Zibechi 2012; Lopes de Souza 2016). More than resistance, however, the communities in the Bay of Aratu actively create alternative ways of life by relying on their environment to remain autonomous from the violent erasure inherent to the universalizing agenda of capitalism and its many purveyors (Escobar 2016). Resistance to the widespread effects of extraction and the ability to create alternative environmental praxes make the communities in the Bay of Aratu important examples of how oppressed populations enact distinct subjectivities.

Despite the devastating effects of *Golden Miller*'s oil spill in Ilha de Maré, neither the state government of Bahia nor the PT-led federal government made any effort to acknowledge the community's plight. After reaching out to municipal-, state-, and federal-level officials regarding their quickly deteriorating situation, the members of Ilha de Maré decided to take matters into their own hands. On 20 February 2014, community members staged a blockade of the Port of Aratu, blocking the road used by trucks to pick up and drop off cargo at the port. "It is for our lives we are protesting. ... Because of the neglect of the governing bodies and the irresponsibility of the business owners," stated one of the community's leaders, explaining the reasons for the protest (author's field notes 2014). Ilha de Maré's leadership insisted that the circulation of extracted natural resources would not continue until Ilha de Maré could parlay with the Docks Company of Bahia (CODEBA)—the government organ responsible for the port. After protestors obstructed the port from 5:00 a.m. until 2:00 p.m., the president of CODEBA agreed to meet with community leaders. Although this event did not lead to an immediate resolution for Ilha de Maré, it was the first step in forcing the Brazilian government to recognize the disaster that had befallen the community. It was also the initial moment of negotiations between Ilha de Maré and the government—a process that continues today.

Like Ilha de Maré, Rio dos Macacos and Tororó have also employed practices of obstruction and sabotage to protest their mistreatment by the navy. Perhaps even more noteworthy, however, are the ways in which both Rio dos Macacos and Tororó have succeeded in continuing subsistence practices despite the navy's attempts to erase them from the landscape of the Bay of Aratu. Despite the navy's environmental terrorism, Rio dos Macacos and Tororó have managed to overcome these violent measures, stubbornly continuing to plant, cultivate, and harvest subsistence crops like corn, beans, mangoes, jackfruit, guava, okra, watermelons, and squash. This perseverance is possible due to the collective efforts in the communities. "We share seeds with one another and help clear each other's land" explained a member of Rio dos Macacos as she showed me around her planted fields. Members also share foodstuffs like African palm oil and fruit juice with one another (author's field notes 2016). Members of Tororó, on the other hand, find locations in which the navy has not built walls around their community to plant crops and forage fruits, vegetables, and herbs (author's field notes 2014). In pushing forward with their centuries-old practices of autonomy and subsistence, Rio dos Macacos and Tororó illustrate the ways in which alternative visions of the environment lead to concrete spatial practices that protect the lives of Afro-descendant communities. As evidenced in the communities' subsistence practices, the environment can provide a long-term way of life. This way of life is not dependent on rampant extraction and degradation, however. Rather, it establishes a radical humanism and propagates geographies in which "liberation is evident, visible, and available to all" (McCutcheon 2016, 21).

Conclusion

Although many in Brazil and around the world rightly critique Temer's austerity measures, the situation in the Bay of Aratu shows the parallels between the progressive and conservative governments. It is true that Temer's relaxation of environmental laws and his defunding and dissolving of government organs dedicated to working with racialized groups have the potential to further the environmental and racialized violence already taking place in the country. These measures are not aberrations, however; they are a deepening of the capitalist tendency previously espoused by the PT.

Despite claims of progressive populism, in reality the PT implemented and enacted death-dealing arrangements that gravely affected racialized communities—particularly Afro-Brazilian communities, as evidenced in the case of the Bay of Aratu. A progressive devotion to capital accumulation qua extractivism meant an inherent commitment to anti-black violence in the Bay. Temer and his government have essentially inherited these practices and expanded them materially and discursively. Under Temer's administration, the extractive agenda propagated by the PT continues. The Port of Aratu remains a vital location for the shipment of natural resources. As the environmental effects of shipping in the Bay continue to poison Ilha de Maré, Temer's government seeks to expand the functional and physical capacities of the Port of Aratu. At the same time, Rio dos Macacos and Tororó continue to face the environmental outcomes of the Brazilian navy's occupation of their territories and the naval insistence that their militarized landscape expand into the communities' territories.

By examining the relational nature of space under both the PT and Temer's administration and by taking seriously the geographical experiences and expressions of Afro-Brazilian communities, it becomes clear that what has occurred in Brazil vis-à-vis the environment and Afro-descendant peoples is not new governance. Rather, new actors are continuing established "concrete and epistemic actions and structural patterns [that] harm, kill, or coerce a particular grouping of people" (McKittrick 2011, 947). Analyzing the environmental experiences of Afro-Brazilian communities in this case offers a sense of the limitations of liberal state power, because two supposedly different approaches to governance actually have clear, macabre connections. Nonetheless, the spatial politics of the communities in the Bay of Aratu offer hope as the communities refuse to mask the structural, environmental racism they face (Eaves 2016). These communities acknowledge and seek to mediate the effects of structural racism and environmental domination and destruction. "Our struggle is a long, difficult one," acknowledged a member of Rio dos Macacos, whose family constantly receives threats from naval soldiers. "[W]e're not afraid to die," she continued. "Those of us here have [hundreds] of years in this place," and "all we want is our land and to be left alone" (author's field notes 2014). This commitment to rejecting anti-blackness and creating a life-affirming environment in the Bay of Aratu typifies these communities' politics and offers blueprints for global black struggle.

Acknowledgments

The author thanks attendee of Indiana University's Workshop on Race, Ethnicity, and Migration who commented on this work when it was in its initial stages.

Note

1. The histories described here were told to me by leaders and elders from all three communities. These histories are passed down orally within the communities and I have chosen to privilege such communal knowledge in this article.

References

Anderson, P. 2011. Lula's Brazil. *London Review of Books* 33 (7):3–12.

Arsenault, C. 2017. Brazil, home of Amazon, rolls back environmental protection. *Reuters*, May 15. https://www.reuters.com/article/us-brazil-politics-environment-idUSKCN18B21P.

Bledsoe, A. 2015. The negation and reassertion of black geographies in Brazil. *ACME: An International E-Journal for Critical Geographies* 14 (1):324–43.

Branford, S., and J. Rocha. 2015. *Brazil under the Workers' Party: From euphoria to despair*. Warwickshire, UK: Practical Action Publishing.

Eaves, L. 2016. We wear the mask. *Southeastern Geographer* 56 (1):22–28.

Escobar, A. 2016. Thinking-feeling with the earth: Territorial struggles and the ontological dimension of the epistemologies of the South. *Aibr, Revista de Antropología Iberoamericana* 11 (1):11–32.

Garmany, J. 2011. Drugs, violence, fear, and death: The necro- and narco-geographies of contemporary urban space. *Urban Geography* 32 (8):1148–66.

Gilmore, R. 2002. Fatal couplings of power and difference: Notes on racism and geography. *The Professional Geographer* 54 (1):15–24.

Gudynas, E. 2010. The new extractivism of the 21st century: Ten urgent theses about extractivism in relation to current South American progressivism. *Americas Program Report*, January 21. Washington, DC: Center for International Policy.

———. 2013. Brazil: The biggest extractivist in South America. *Latin America in Movement*. Accessed February 7, 2018. https://www.alainet.org/es/node/75975.

Harrington, J. 2015. A place of their own: Black feminist leadership and economic and educational justice in São Paulo and Rio de Janeiro, Brazil. *Latin American and Caribbean Ethnic Studies* 10 (3):271–87.

Loperena, C. 2017. Settler violence? Race and emergent frontiers of progress in Honduras. *American Quarterly* 69 (4):801–7.

Lopes de Souza, M. 2016. Lessons from praxis: Autonomy and spatiality in contemporary Latin American social movements. *Antipode* 48 (5):1292–1316.

McCutcheon, P. 2016. The "radical" welcome table: Faith, social justice, and the spiritual geography of Mother Emanuel in Charleston, South Carolina. *Southeastern Geographer* 56 (1):16–21.

McKittrick, K. 2011. On plantations, prisons, and a black sense of place. *Social & Cultural Geography* 12 (8):947–63.

———. 2013. Plantation futures. *Small Axe: A Caribbean Journal of Criticism* 17 (3):1–15.

McKittrick, K., and C. Woods. 2007. *Black geographies and the politics of place*. Cambridge, MA: South End.

Mollett, S. 2011. Racial narratives: Miskito and Colono land struggles in the Honduran mosquitia. *Cultural Geographies* 18 (1):43–62.

Neto, R., and C. Azzoni. 2011. Non-spatial government policies and regional income inequality in Brazil. *Regional Studies* 45 (4):453–61.

Oliveira, F. 2006. Lula in the labyrinth. *New Left Review* 42:5–22.

Perreault, T. 2013. Dispossession by accumulation? Mining, water and the nature of enclosure on the Bolivian Altiplano. *Antipode* 45 (5):69.

Perry, K. 2004. The roots of black resistance: Race, gender and the struggle for urban land rights in Salvador, Bahia, Brazil. *Social Identities* 10 (6):811–31.

———. 2013. *Black women against the land grab: The fight for racial justice in Brazil*. Minneapolis: University of Minnesota Press.

Pulido, L. 2000. Rethinking environmental racism: White privilege and urban development in Southern California. *Annals of the Association of American Geographers* 90 (1):12–40.

———. 2017. Geographies of race and ethnicity II: Environmental racism, racial capitalism and state-sanctioned violence. *Progress in Human Geography* 41 (4):524–33.

Shabazz, R. 2015. *Spatializing blackness: Architectures of confinement and blackmasculinity in Chicago*. Urbana: University of Illinois Press.

Smith, C. 2013. Strange fruit: Brazil, necropolitics, and the transnational resonance of torture and death. *Souls* 15 (3):177–98.

Sundberg, J. 2008. Placing race in environmental justice research in Latin America. *Society & Natural Resources* 21 (7):569–82.

Valdivia, G. 2015. Oil frictions and the subterranean geopolitics of energy regionalisms. *Environment and Planning A: Economy and Space* 47 (7):1422–39.

Viana Filho, L. 1984. *Petroquímica e Industrialização da Bahia, 1967–1971* [Petrochemicals and industrialization in Bahia, 1967–1971]. Brasilia: Senado Federal, Centro Gráfico.

Voeks, R. 1990. Sacred leaves of Brazilian candomble. *Geographical Review* 80 (2):118–31.

Weinstein, B. 2015. *The color of modernity: Sao Paulo and the making of race and nation in Brazil.* Durham, NC: Duke University Press.

Werneck, J. 2010. Nossos passos vêm de longe! Movimentos de mulheres negras e estratégias políticas contra o sexismo e o racismo [Our steps come from far! Black women's movements a political strategies against sexism and racism]. *Revista Da ABPN* 1 (1):8–17.

Woods, C. 2002. Life after death. *The Professional Geographer* 54 (1):62–66.

———. 2017. *Development drowned and reborn: The blues and bourbon restorations in post-Katrina New Orleans.* Athens: University of Georgia Press.

Zibechi, R. 2012. *Territories in resistance: A cartography of Latin American social movements.* Oakland, CA: AK Press.

———. 2014. *The new Brazil: Regional imperialism and the new democracy.* Oakland, CA: AK Press.

ADAM BLEDSOE is an Assistant Professor of Geography and African American Studies at Florida State University, Tallahassee, FL 32306. E-mail: abledsoe@fsu.edu. His research interests include black geographies and black social movements in the context of the Americas.

Infrastructure and Authoritarianism in the Land of Waters: A Genealogy of Flood Control in Guyana

Joshua Mullenite ⓘD

Although often viewed as serving as a public good, infrastructure can have important political effects resulting from the way in which it is designed, built, and managed that preexist its stated or implied technical goals. It acts as a mediator and enforcer of state interests, defining the ways in which the state can enter everyday life and, in turn, it shapes the possibilities of life around the goals of the state. Although this politics of infrastructure has seen renewed interest from geographers, anthropologists, and other social scientists concerned with the power of artifacts, the role that infrastructure plays in defining and characterizing the particularly nationalist and racialized state remains undertheorized. Through a genealogy of water control infrastructure in Guyana, I show how apparently banal aspects of everyday life, such as infrastructure, can play an important role in the rise of an authoritarian government, first colonial and later postcolonial. Because 90 percent of Guyana's population and most of the nonmineral economic resources are below sea level, water control infrastructure plays an important functional role in the country. Rather than just a means for preventing coastal flooding and irrigating the patchwork of sugar and rice fields that define the economy, however, I argue that this infrastructure played a key role in driving ethnic divisions between laborers in the colonial era that undermined anticolonial sentiment and laid the groundwork for the creation and perpetuation of an ethnic nationalist and authoritarian postcolonial regime.

尽管基础建设常被视为服务公众之用，但先于其所宣称或意味的技术目的存在的设计、建造和管理方式，则可能造成重大的政治影响。基础建设作为国家利益的中介物和执行者，决定国家进入日常生活的方式，并回头塑造围绕着国家目标的生命可能。尽管此般基础建设政治已重获关注权力构造的地理学者、人类学者和其他社会科学家的兴趣，基础建设在定义并描绘特定国族和种族化的国家中所扮演的角色却尚未充分理论化。我通过圭亚那水资源控制基础建设的系谱研究，展现诸如基础建设的每日生活平庸面向，如何能够先后在殖民与后殖民的威权政体的兴起中扮演要角。由于圭亚那人口的百分之九十、以及多半的非矿物经济资源皆低于海平面，水资源控制的基础建设因而在该国扮演重要的功能性角色。我主张，此一基础建设不仅只是预防海岸淹水以及灌溉定义该国经济的甘蔗田与稻米田相间的工具，而是在殖民时期驱动劳工间的族裔分野中扮演关键角色，并减损了反殖民的态度，且为族裔国族主义和威权后殖民政体的创造与续延奠定了基础。关键词：殖民主义，洪水，基础建设，种族，水。

Aunque con frecuencia es vista como algo que sirve como bien público, la infraestructura puede tener efectos políticos importantes que surgen de la manera como se la diseña, construye y maneja anticipando sus metas técnicas declaradas o implícitas. La infraestructura actúa como mediador y ejecutante de los intereses del estado, definiendo los modos como éste puede meterse en la vida cotidiana, en tanto que aquélla configura las posibilidades de vida alrededor de las metas del estado. Aunque esta política de infraestructura ha recibido un renovado interés de parte de geógrafos, antropólogos y otros científicos sociales preocupados con el poder de los artefactos, el papel que juega la infraestructura para definir y caracterizar al estado particularmente nacionalista y racializado sigue escasamente teorizado. Por medio de una genealogía de la infraestructura para el control del agua en Guyana, muestro el modo como aspectos aparentemente banales de la vida cotidiana, tales como la infraestructura, pueden desempeñar un papel importante en el encumbramiento de un gobierno autoritario, primero colonial y más tarde poscolonial. Debido a que el 90 por ciento de la población de Guyana y la mayoría de los recursos económicos no minerales se hallan debajo del nivel del mar, la infraestructura del control hídrico juega un importante rol funcional en el país. Sin embargo, más que un simple modo de prevenir las inundaciones de la costa y la colcha de retazos de campos irrigados de caña de azúcar y arroz que definen la economía, arguyo que esta infraestructura jugó un rol clave en el impulso de divisiones étnicas entre los trabajadores de la era colonial que socavaron el sentimiento anticolonialista y aportó el trabajo preliminar para la creación y perpetuación de un régimen étnico poscolonial nacionalista y autoritario. *Palabras clave: agua, colonialismo, infraestructura, inundación, raza.*

The small South American country of Guyana is still dealing with its authoritarian past. A century and a half of British colonial rule in which the needs and desires of the plantocracy were of paramount importance and the needs of the laboring populations remained ignored was followed by nearly three decades of control under a postcolonial regime that would quickly become dictatorial, rigging at least seven local and national elections and referenda and growing the military to maintain and tighten their power (Committee of Concerned Citizens 1978; Working People's Alliance 1983; Collymore 1990; Kanhai 2015).

Although election rigging is one of the more visible ways of establishing an authoritarian regime in an ostensibly democratic country (Thomas 1984), I want to suggest that there are more subtle methods at the disposal of governments seeking to consolidate power that involve the imposition of economic and political barriers that limit individuals' abilities in ways that foster state power (see Foucault 1982; Deleuze 1992). Beeson (2010) argued, for example, that the Leviathanesque threat posed by climate change could result in the growth of a new form of environmental authoritarianism, especially in countries seeking economic development (see also Wainwright and Mann 2013). Beeson (2010) contended that "the emergence of an environmentally-conscious, politically-savvy, effective civil society that can transform environ-mental practices [is] obviated by uncertain economic development and inequality" (277). Key to Beeson's argument is that environmental management and economic stability go hand in hand, thus making environmental management a crucial tool of governments seeking stability in times of crisis. To what extent, though, are these conditions unique to climate change? What existing technologies of government allow authoritarian politics to develop? Social and economic inequality, uncertain economic futures, and social conflict bred by unequal access to resources are long-standing issues, especially in postcolonial societies where authoritarianism has a long history (Thomas 1984).

In this article, I demonstrate how apparently banal aspects of everyday life such as flood control infrastructure can play a significant role in the rise and persistence of authoritarian regimes. Guyana serves as a valuable site from which these questions can be approached and through which entanglements between infrastructure—including its development, funding, construction, and maintenance—and authoritarian governmentalities can be articulated. I draw my understanding of authoritarianism from the work of Guyanese political economist Thomas (1984), who argued that, in the postcolonial world, authoritarian regimes take on a unique form arising from the consolidation of political power under a petty bourgeois ruling class. For Thomas, this class gains and maintains its power through clientelism, manipulation of electoral processes, and terror that are explicitly learned from histories of colonial violence they experienced. In addition to rigged elections and the assassination of political opponents (see Kanhai 2015), I argue that Forbes Burnham—Guyana's first postcolonial leader—drew on experiences garnered in the country's colonial past and exploited the geographical segregation of Afro- and Indo-Guyanese groups to economically disenfranchise opposition supporters through negligence of the flood control infrastructure needed to support their local economies.

With 90 percent of the country's population and most of the nonmineral economic resources lying on a coastal plain that sits below sea level, flood control infrastructure in the form of sea walls, canals, levees, and conservancies plays a significant functional role in the country's political economy and serves as a key technology through which governmental projects can be expressed and enforced (Kooy and Bakker 2008; Meehan 2014). Specifically, as Meehan (2014) argued, the growth and development of infrastructural works and the administrative commitments they entail provide new means through which the state can govern a population while alternatives to governmental schemes serve as acts of resistance that limit state power. I draw on archival research conducted between 2015 and 2017 to argue that the construction and maintenance of flood control infrastructure during the colonial and early postcolonial period in the country has been intimately entwined with the state's governance structures. Through a

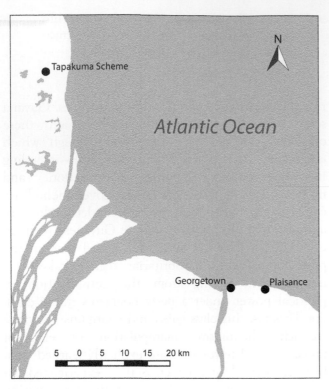

Figure 1. Map showing the relative areas of locations discussed in this article.

genealogy of flood control, I show how the legislation, construction, and maintenance of infrastructural works empowered the colonial plantocracy and played a key role in driving ethnic divisions between laborers in the colonial era that undermined a unified anticolonial sentiment while laying the groundwork for the creation and perpetuation of a nationalist authoritarian postcolonial regime. Rather than acting as background on which political, economic, and racial struggles took place, I argue that flood control was an integral part of colonial governance, used to maintain and govern the marginalized populations of the colony both through its construction and maintenance and through its neglect. I likewise argue that the maintenance (or lack thereof) of these works is upheld as a key governance structure in the racial politics of the postcolonial period, particularly between 1966 and 1985 under the authoritarian rule of Linden Forbes Burnham.

Sugar and the Political Economy of Guyana's Water Control Infrastructure

First established as a series of Dutch colonial trading outposts in the sixteenth century, the area now called Guyana rose to colonial prominence as a

British sugar producer in the nineteenth century. Describing the political conditions in 1871, Jenkins called the young colony "a mild despotism of sugar" due in no small part to the fact that estate owners effectively controlled the legislature and manipulated the colony to meet their demands (Jenkins 1871, 58). From the late eighteenth century until 1834, the colony was home to a large number of plantations that relied exclusively on the labor of enslaved Africans. After emancipation in 1834, a number of plantations found themselves insolvent and abandoned their lands, and others turned to the Colonial Office to provide a new source of cheap labor in the form of indentured workers from South and Southeast Asia, many of whom would themselves become permanent settlers when the British government could no longer afford to pay for their return passage as had previously been promised (Rodney 1981).

During the four-year period of postemancipation apprenticeship common in the British Caribbean, black workers were offered meager wages in exchange for their labor. In the years that followed, a portion of this newly independent laboring population pooled their resources and purchased the abandoned plots to establish new lives outside of the plantation economy (Young 1958; Rodney 1981; Moore 1987). This was the case in the village of Plaisance, located approximately six miles east of Georgetown in the East Coast Demerara Region (see Figure 1). In 1842, four years after the end of postslavery apprenticeship, sixty-five families of formerly enslaved people purchased the 500-acre former cotton plantation with the intention of forming a village independent of the plantation economy (Haynes 2016). Within a year, 210 acres of ground crops including staple foods such as plantains and cassava were under cultivation in the village.

In 1849, Plaisance experienced a breech in a portion of its front sea dam, causing severe inundation. Then-Governor Harold Barclay levied a large fine against the villagers for failure to maintain their sea defense infrastructure. The money raised from the fine was used to pay for a large drainage pump meant to improve drainage not only for the village but also for the sugar estates that sandwiched it. The villagers did not feel that they were responsible for the flooding and should not have to pay for the pump, eventually petitioning the governor to reconsider but ultimately failing to persuade him (Haynes 2016). With the pump in place, villagers began paying rates to the Department of Public Works for its maintenance and upkeep. This imposition marked the beginning of a series of legislative changes in which the colonial government gained increasing control over the day-to-day governmental functions of the villages and thus limited their options in terms of livelihood practices. By 1882, under then-Governor Irving, it was argued that the villages should have no right to self-determination or government but that they should fall instead under the colonial bureaucracy, especially with regard to the ability of villages to build and maintain appropriate infrastructure (Rodney 1981). Irving's position was that it was unjust to have villages, which have little income, pay the same rates for maintenance as the profitable sugar plantations. Rather than allowing villages to develop flood control systems more suitable to their needs, however, his recommendation was to end the de facto black self-government of the village system to ensure that the system of drainage, irrigation, and sea defenses continued to work in the interests of the sugar estates. Irving's plans were never realized and villages continued to pay high rates to maintain the works necessary for the continuation of coastal sugar production.

Despite paying the rates for pump maintenance and the increasing administrative presence of the colonial government in the villages, the pump failed in 1886, causing severe flooding to inundate village crops, forcing villagers to labor in nearby sugar plantations for income. One villager, venting his frustrations in the *Daily Chronicle*, asked, "How must the people undertake again to put their labour in the ground when the white man allows negroes [sic] plantains and cassava to be inundated and thus suffer?" (Ratepayer 1886). The following year, flooding struck again, this time from the collapse of the back dam. The villagers of Plaisance this time suggested that this was the direct responsibility of the estate manager and the neighboring sugar plantation of Goedverwagting, who they accused of modifying an aspect of their shared drainage infrastructure to extend the cultivating season for sugar and resulting in the inundation of their fields and village lands (Truth 1887). This sparked a long debate in the daily paper regarding the relationship among sugar, villages, and the economy, culminating with Garnett (1887) of Plantation Nonpariel stating that the colony were "one and all dependent on sugar" and suggesting that villagers should

abandon their efforts for autonomy and get back in line with the industry.

A clear link had been established in the colony to this point between colonial white supremacy and the economic supremacy of sugar. Jenkins's (1871) understanding of the "despotism" of sugar was clearly accepted by the planters, but villagers continuously sought to challenge it through their attempts to develop other ways of living on the coast, such as subsistence cropping and aquaculture, only to be constantly frustrated by floods resulting from infrastructural failure that forced them back into the plantation system (Ratepayer 1886; Truth 1887; Young 1958; for a discussion of the causes of flooding in Guyana, see Richardson [1973]; Strachan [1980]; Lakhan [1994]). This pattern continued well into the twentieth century. In this way, villages acted as spaces for reserve laborers for the estates. As Hintzen (2018) argued, this is in line with a broader project of managing capitalist accumulation in peripheral countries. Here the failures of the pumps and dams in Plaisance could be "deployed to control, discipline, and regulate those located outside of the space of capitalist formation in order to ensure they did not disrupt the process of capital accumulation. The latter also became organized into 'segments' of 'surplus labor' to be made available for capitalist production when needed" (Hintzen 2018, 42). Thus, despite attempts to stay outside of the primary motivator for capitalist accumulation in the colony (i.e., the plantation system), flooding and the failures of flood control infrastructure could be manipulated and legislated in such a way that failures in the villages were all but guaranteed, forcing villagers to come back into the system and accept nearly any wage, particularly during periods where additional labor was needed on the plantations (Greenidge 2001). Rather than experimental, anticolonial democracies in the context of a sugar colony, the villages became further sites in which the despotic rule of the plantocracy persisted and through which colonial authoritarianism could be practiced (Thomas 1984; see also Rodney 1981).

Flooding, Infrastructure, and the Rise of Anticolonialism

At the turn of the twentieth century, the situation for Guyanese workers was bleak. Costs of living in the colony were rising and Colonial Office surveys of working conditions showed that wages were stagnant or falling. In 1905, facing an unregulated industrial situation in which laborers were expected to take whatever wages they were offered, workers from across the coast went on strike, demanding higher wages and better working conditions and giving rise to the first organized labor movement in British Guiana (Woolford 1998). Over the next two decades, this labor movement grew rapidly, with workers organizing and striking not only against the plantocracy but also against the role of the government in propping it up through the subservience of the police and military to the plantocracy (Woolford 1998). This was accomplished by organizing and action through the British Guiana Labor Union (BGLU) and the Guyana Industrial Workers Union (GIWU), representing Guyana's dock workers and sugar workers, respectively (Greenidge 2001).

In the 1920s and 1930s and in the context of this nascent labor movement, planters sought new ways to exert their control over the laborers, particularly those in the black villages on whom they relied for sugar production. Through several acts of legislation, representatives from the Bookers McConnell and Sandbach Parker sugar firms used their influence in the colonial office to force villages to again pay the same tax rates for flood control infrastructure construction and maintenance as sugar estates by recommending and eventually legislating a flat tax rate per acre of land rather than those based on income or assessed value. This happened despite the much lower economic productivity of the villages due to their subsistence nature. As with Irving's earlier contestation, the issue was raised with no input from the villages themselves.

What differed between this new flat tax plan and those established in the 1880s was that the low market costs of sugar were causing estates some level of financial distress. Planters continuously claimed to be unable to afford to pay laborers for maintenance of defense works, requesting grants-in-aid from the Colonial Office. In at least one instance, the sugar firm Sandbach Parker loaned the government money for its own grant-in-aid, receiving interest on repayment and thus profiting off of free, predominantly black labor provided by the state to repair the sea walls, levees, and canals necessary for continued sugar production (Letter to Lord Passfield, 12 March 1930, CO 111/683/6, National Archives UK).

Meanwhile, the economic burden posed by flat tax rates and continued need for maintenance and construction of these same infrastructural works in the villages caused them to suffer further and pushed many villagers back into the plantation economy to keep their villages afloat. A 1930 report on the economic situation in the colony demonstrated the persistence of previous methods of labor pooling by describing elevated levels of unemployment in the capital city of Georgetown due to the introduction of beet sugar in the European market while the sugar plantations were hiring villagers to work in their estates in greater numbers than previous years, but with lower total wages paid ("Report of the Investigative Committee," 1930). A fundamental part of this control and regulation was through the nondemocratic legislation of flood control systems against the interests of the colonized populations of the coast.

While planters were working with the local colonial government to ensure a steady supply of low-cost labor during the global depression of the 1930s, the Colonial Office was seeking to develop a new market that could provide a source for additional food in the West Indies and for its rapidly increasing population to reduce the reliance on imports (Greenidge 2001). Rice had proven a minor success since its introduction in 1886 and as early as 1917, the year Indian immigration ended, the colony had established a viable export market for rice products with output doubling every two decades since its introduction.

After World War II and under the auspices of the Colonial Development and Welfare Act of 1945, the Colonial Officse sought to expand rice production through the development of a number of empolderment schemes. Under these schemes, new systems of drainage, irrigation, and coastal defense were designed, built, and funded by the colonial office with minimal financial input from the villages themselves, allowing rice villages to focus on production without the debt associated with flood control maintenance in African[1] villages. Although some water management schemes were established in majority black areas, this was primarily due to the preexistence of African villages there, meaning that many villagers were still grappling with the legacies of debt associated with earlier legislative acts. At the same time, however, water management schemes in less populated spaces were specifically designed to create rice-producing regions, including new villages for Indian laboring populations who would be granted housing and land for both cultivation and subsistence in lieu of the return passage promised in their indentureship contracts (Greenidge 2001). This allowed for the establishment of large Indian-majority populations socially, economically, and geographically separate from the African community and without the infrastructural debts associated with the earlier, autonomous settlement patterns (Lakhan 1994; Greenidge 2001).

With the number of agricultural workers increasing, and with wages continuing to remain stagnant in the face of the increased costs of living associated with the postwar period, the labor movement grew not only in numbers but also in its willingness to make demands on the government and began to take on an anticolonial character. Using this sentiment as a means for decolonization, in November 1946 Guyanese dentist Cheddi Jagan, along with others, formed the Political Affairs Committee (PAC; Premdas 1974). In PAC's analysis, the colonial state was used rhetorically as a common enemy of all Guyanese working people, existing primarily as a tool of class oppression at the hands of the colonizers (Premdas 1974; Hintzen 1989). This analysis, combined with relentless propaganda from the PAC, led to Jagan's election to parliament in the 1947 general election.

Having captured leadership positions in Guyana's growing union movement, the PAC reformed themselves as a new labor-backed, multiracial, and anticolonial political party, combining with the BGLU in 1950 to form the People's Progressive Party (PPP). Although working among unions with different racial characteristics, Jagan's popularity rested primarily with Indo-Guyanese, who saw in him one of their own (Premdas 1974). To unite the Afro-Guyanese and Indo-Guyanese working classes under a single party, Jagan sought out a charismatic Afro-Guyanese counterpart in Forbes Burnham, a London-educated lawyer, master orator, and then-leader of the BGLU (Premdas 1974). Because the union movement and its political arms were divided by industrial sector and racial characteristics, the PPP sought to establish a party that would provide a platform for all workers, keeping the working classes from remaining divided in the colonial system. It was under this united banner that the PPP would advocate for self-government and universal adult

suffrage, succeeding in the latter and bringing about a new general election in Guyana in 1953, which they won. Despite the victory of the PPP in the country's first mass election, the party was deposed by British armed forces after only six months, reverting the political system to an autarchic form of colonial governance (Premdas 1974; Hintzen 1989). During this time, British Guiana's new constitution was suspended, Jagan was jailed, and Burnham was placed under house arrest after agreeing to cease any political activity (Hintzen 1989). Now separated, the two leaders of the PPP began to espouse different tactics and goals for the party going forward, leading to their eventual split and the creation of a two-party, racially bifurcated political system that would persist for decades (see Hintzen 1989).

The split in the PPP did not kill the independence movement in Guyana, but it did change the terms on which it occurred. With African and Indian villagers now separated spatially, socially, economically, and politically, the U.S. government via the Central Intelligence Agency (CIA) began to promote Burnham's rise to power (Rabe 2005). Although a socialist, Burnham lacked Jagan's ties to Cuba and the Soviet Union and the CIA felt that he could be more easily manipulated to follow U.S. hemispheric interests and an anticommunist agenda (Crosthwait 1966; Rabe 2005). Because Indo-Guyanese populations far outnumbered Afro-Guyanese, the first-past-the-post system of elections similar to that of the United States and advocated for within the colony would not allow Jagan's defeat. The U.S. and UK governments pushed for and were successful in implementing a parliamentary system that allowed the People's National Congress (PNC) to form a coalition government with the liberal capitalist United Force (UF) and gain a parliamentary majority (Hintzen 1989). In 1965, this coalition was able to take control of parliament and establish the first postcolonial government when the country received independence the following year, but UF ministers were quickly removed from power and replaced by individuals loyal to Burnham.

The Politics of Flooding in the Postcolonial Period

Once in power, the PNC-led government removed many of the administrative and financial programs meant to maintain the rice growing (and largely Indian-focused) schemes, at the same time consolidating power around the party and its leader, Forbes Burnham (Lakhan 1994). The focus of flood control and the infrastructural works associated with it shifted from these rice-producing and Indian-dominated villages to the urban centers and later, through its nationalization in 1976, to the sugar industry, where a small number of black workers still had a home. The governmental abandonment of rice-producing regions also allowed for the reentry of an imperialist and antisocialist politics of flood management to occur along racialized lines—as demonstrated by UK and U.S. aid programs focused on the rehabilitation and development of the Tapakuma Scheme Project in the 1970s.

Tapakuma has long been of interest as a rice-producing region, but even as other areas were being developed for Indo-Guyanese settlement, the project was frequently placed on the back burner by a colonial regime focused on quick returns. Located along Guyana's Essequibo coast and inhabited by a primarily Indo-Guyanese population, the area remains somewhat isolated from the capital city but has a large and successful agricultural industry. As early as 1829, planters worked together to provide a reservoir and canal system for this area that remained in place until the 1960s (Strachan 1980). In the 1930s, the region was hit hard by the recession and falling sugar prices, causing many of the estates to be abandoned and later converted to rice production. The system of water management for sugar was not satisfactory for the rice industry, leading to a number of engineering projects being recommended until the Tapakuma conservancy project, centered around Tapakuma Lake (see Figure 1), was finally approved in the 1950s (Strachan 1980). With the major hydraulic works completed by 1963, land clearing in 1964, and settlement by a large contingent of Indo-Guyanese who would work the land in 1965, the postcolonial government needed to offer little more than maintenance and support to provide for the region's success (Vining 1977). With cuts to spending on infrastructural maintenance in Indian majority schemes, however, this is not what occurred.

By the late 1970s, a decade and a half of neglect from the PNC regime meant that the Tapakuma scheme was in a state of disrepair. Sea defenses were crumbling, roads were impassible at times, and the rice industry was under constant threat of catastrophic inundation (*Guyana Tapakuma Irrigation Scheme* 1976–1978; Vining 1977). In considering projects to fund in 1978, the UK Ministry of Overseas Development (OD) strongly considered and eventually funded large-scale infrastructure projects meant not only to solve the issues stemming from the lack of maintenance but also to increase the region's economic productivity, which they felt would threaten Burnham's growing authoritarian power (*Guyana Tapakuma Irrigation Scheme* 1976–1978; *Guyana - Sea Defences (Essquibo)* 1976–1978). Their justification for taking on this project over others was simple: "The vast majority of the beneficiaries of the scheme will be of Indian race. The present Government in Guyana has tended to favour the Afro-Guyanese rather than the Indian element, and this could help to constitute an argument for supporting the Tapakuma scheme" ("Guyana: Tapakuma Irrigation Scheme and Sea Defence" 1978). Frustrated by Burnham's focus on growing the military to keep order in an increasingly degenerating regime, the OD sought to undermine his centralization of power by providing services to the base of his main political opposition (Minister of State 1978).

This British interference into flood control infrastructure maintenance appears to have been successful in reducing the economic burdens imposed by flooding on Indian villages. During fieldwork in the Tapakuma region in 2017, an Indo-Guyanese man described coastal flooding as "a black issue," saying that it was a concern for Georgetown and the African villages but not along the Essequibo coast. According to a number of coastal residents, the current government—a multiracial coalition party led by the remnants of the PNC—had seemingly moved past the racial politics of flood control and was focusing equally on all areas, perhaps to the extent of spreading themselves thin. One woman, an Afro-Guyanese teacher in the village of Uivlught, expressed concern that this was effectively colorblind and ignored the long-standing inequalities and hardships faced by both groups in favor a "feel good" policy (see Hardy, Milligan, and Heynen 2017). Regardless of these specific positions and experiences, however, the historical analysis I have presented here demonstrates how flood control has continued to serve as a technology of government in the postcolonial era.

Conclusion

The financial and legislative mechanisms over flood control enacted by colonial and postcolonial regimes served as a central means by which the government could regulate the lives and livelihoods of Guyana's population. These projects corresponded to the political, economic, and racial dynamics of the country through authoritarianisms rooted in both the hegemony of the plantocracy in the colonial period and in the racially driven dictatorship of Forbes Burnham and the PNC. Beyond strict electoral control (through the refusal of adult suffrage under colonialism and blatant election rigging under Burnham), these regimes were able to use flood control and its related infrastructural commitments to maintain and grow their economic and political power. Equally important to these projects were the tax burdens imposed on Afro-Guyanese populations during colonial rule and the willful neglect of flood control infrastructure in both the colonial and postcolonial era as a means of increasing the vulnerability of populations to ensure they could not actively resist the authoritarian state. Additionally, in the former case, it could be argued that taxation itself served as an act of governance separate from its role in funding the colony by making colonized populations governable (Bush and Maltby 2004). During the colonial period, new means of legislating and funding flood control infrastructure ensured that alternatives would not develop and that the laboring population would remain reliant on the plantation system for its survival (see also Young 1958). In the postcolonial period, infrastructural neglect was used to economically disenfranchise opposition groups who would otherwise have threatened Burnham's authoritarian regime. In this way, environmental management in the form of flood control and the infrastructure necessary to support it served as an important site for authoritarian governance that is often overlooked, echoing recent arguments from Barnes (2017) that "the bureaucracy and technologies of water management can be seen as one of the conduits through which [state] authority may be both forged and contested" (149). Beyond the potential for future authoritarianisms posed by climate change (Beeson 2010), these everyday practices render authoritarian governance practices a banal part of daily life, allowing for new authoritarianisms to creep in largely unnoticed.

Funding

This research was supported by a Morris and Anita Broad Research Fellowship from the School of International and Public Affairs and Doctoral Evidence Acquisition and Dissertation Year Fellowships from the University Graduate School at Florida International University.

ORCID

Joshua Mullenite http://orcid.org/0000-0001-8477-4190

Note

1. I use the term African in the Caribbean sense to refer broadly to the African-descended population of the country. In Guyana, African, black, and Afro-Guyanese are used interchangeably.

References

Barnes, J. 2017. States of maintenance: Power, politics, and Egypt's irrigation infrastructure. *Environment and Planning D: Society and Space* 35 (1):146–64.

Beeson, M. 2010. The coming of environmental authoritarianism. *Environmental Politics* 19 (2):276–94.

Bush, B., and J. Maltby. 2004. Taxation in West Africa: Transforming the colonial subject into the governable person. *Critical Perspectives on Accounting* 15 (1):5–34.

Collymore, C. 1990. *A quarter century of failure.* Georgetown, Guyana: People's Progressive Party.

Committee of Concerned Citizens. 1978. *A report on the referendum held in Guyana, July 10th 1978.* Georgetown, Guyana: CEDAR Press and Cole's Printery.

Crosthwait, T. L. 1966. "Despatch No. 3 – Guyana: 100 Days After Independence." Guyana – Internal Political Situation, DO 200/200. The National Archives, London.

Deleuze, G. 1992. Postscript on the societies of control. *October* 59 (1):3–7.

Douglas-Jones, C. 1930. Letter to Lord Passfield. March 12. *British Guiana – Sea Defences,* CO 111/683/6. The National Archives, London.

Foucault, M. 1982. The subject and power. *Critical Inquiry* 8 (4):777–95.

Garnett, H. 1887. The planters and the floods. *Daily Chronicle* (Georgetown, British Guiana), February 15.

Greenidge, C. B. 2001. *Empowering a peasantry in a Caribbean context: The case of land settlement schemes in Guyana, 1865–1985.* Kingston, Jamaica: University of the West Indies Press.

Guyana – Sea Defences (Essquibo). 1976–1978. OD 50/53. The National Archives, London.

Guyana Tapakuma Irrigation Scheme. 1976–1978. OD 50/52. The National Archives, London.

"Guyana: Tapakuma Irrigation Scheme and Sea Defence." Report. Guyana Tapakuma Irrigation Scheme, OD 50/52. The National Archives, London.

Hardy, R. D., R. A. Milligan, and N. Heynen. 2017. Racial coastal formation: The environmental injustice of colorblind adaptation planning for sea-level rise. *Geoforum* 87:62–72.

Haynes, B. A. 2016. *Plaisance: From emancipation to independence and beyond.* Georgetown, Guyana: F&H Printing.

Hintzen, P. C. 1989. *The costs of regime survival.* Cambridge, UK: Cambridge University Press.

———. 2018. Rethinking identity, national sovereignty, and the state: Reviewing some critical contributions. *Social Identities* 24 (1):39–47.

Jenkins, E. 1871. *The coolie: His rights and wrongs.* New York: Routledge.

Kanhai, R. 2015. What "context" can justify Walter Rodney's assassination? *Groundings* 2 (2):25–39.

Kooy, M., and K. Bakker. 2008. Technologies of government: Constituting subjectivities, spaces, and infrastructures in colonial and contemporary Jakarta. *International Journal of Urban and Regional Research* 32 (2):375–91.

Lakhan, V. C. 1994. Planning and development experiences in the coastal zone of Guyana. *Ocean & Coastal Management* 22 (3):169–86.

Meehan, K. M. 2014. Tool-power: Water infrastructure as wellsprings of state power. *Geoforum* 57:215–24.

Minister of State. 1978. Draft Letter to Mr. R.G. Taylor. n.d. *Guyana – Sea Defences (Essquibo),* OD 50/53. The National Archives, London.

Moore, B. L. 1987. *Race, power, and social segmentation in colonial society.* New York: Gordon and Breach.

Premdas, R. R. 1974. The rise of the first mass-based multi-racial party in Guyana. *Caribbean Quarterly* 20 (3–4):5–20.

Rabe, S. G. 2005. *U.S. intervention in British Guiana: A cold war story.* Chapel Hill: University of North Carolina Press.

Ratepayer. 1886. Plaisance village. *Daily Chronicle* (Georgetown, British Guiana), February 2.

"Report of the Investigative Committee." 1930. *British Guiana – Sugar Industry: Relief Works.* CO 111/688/8. The National Archives, London.

Richardson, B. C. 1973. Spatial determinants of rural livelihood in coastal Guyana. *The Professional Geographer* 25 (4):363–68.

Rodney, W. 1981. *A history of the Guyanese working people, 1881–1905.* Baltimore, MD: Johns Hopkins University Press.

Strachan, A. J. 1980. Water control in Guyana. *Geography* 65 (4):297–304.

Thomas, C. Y. 1984. *The rise of the authoritarian state in peripheral societies.* London: Monthly Review Press.

Truth. 1887. Sparendaam and Plaisance. *Daily Chronicle* (Georgetown, British Guiana), February 9.

Vining, J. W. 1977. Presettlement planning in Guyana. *Geographical Review* 67 (4):469–80.

Wainwright, J., and G. Mann. 2013. Climate Leviathan. *Antipode* 45 (1):1–22.

Woolford, H. M. 1998. The origins of the labour movement. In *Themes in African-Guyanese history,* ed. W. F. McGowan, J. G. Rose, and D. A. Granger, 277–95. Georgetown, Guyana: Free Press.

Working People's Alliance. 1983. *Arguments for unity against the dictatorship in Guyana.* Georgetown, Guyana: Working People's Alliance.

Young, A. 1958. *The approaches to local self-government in British Guiana.* London: Longmans.

JOSHUA MULLENITE is a Doctoral Candidate in the Department of Global and Sociocultural Studies at Florida International University, Miami, FL 33199. Beginning in fall 2018 he will be Visiting Assistant Professor in the Department of Anthropology at Wagner College, Staten Island, NY 10301. E-mail: joshua.mullenite@wagner.edu. His research interests include the social, economic, and political factors that shape flood control decision making.

Border Thinking, Borderland Diversity, and Trump's Wall

Melissa W. Wright

Donald Trump's agenda to build a "big" and "beautiful" border wall continues to raise alarms for anyone concerned with social justice and environmental well-being throughout the Mexico–U.S. borderlands. In this article, I examine how the border wall and its surrounding debates raise multiple issues central to political ecological and human geographic scholarship into governance across the organic spectrum. I focus particularly on a comparison of the different kinds of "border thinking" that frame these debates and that provide synergy for those coalitions dedicated to the preservation of diversity throughout the ecological and social landscapes of the Mexico–U.S. borderlands.

唐纳德．特朗普在边境筑起一座"大而美"的高墙之议程, 对关注美墨边境的社会正义与环境福祉的任何人而言持续发出警报。我于本文中检视边境城墙及其相关辩论, 如何将政治生态学与人文地理学中的多重核心议题, 带入横跨生物光谱的治理议题之中。我特别聚焦比较不同种类的"围墙思考", 这些思考框架了上述辩论, 并对致力于保存美墨边境的生态与社会地景之联盟提供了协同作用。 关键词: 生物多样性, 去殖民, 女权主义, 美墨边境, 新自由主义。

La agenda de Donald Trump para levantar un muro fronterizo "grande" y "hermoso" sigue prendiendo las alarmas entre quienes se preocupan por la justicia social y el bienestar ambiental a lo largo de la frontera México-EE.UU. En este artículo examino el modo como el muro fronterizo y los debates que lo rodean despiertan múltiples interrogantes que son centrales en la erudición político-ecológica y humana a través del espectro orgánico en términos de gobernanza. Me enfoco en particular en la comparación de las diferentes clases de "pensamiento fronterizo" que enmarcan estos debates y que suministran sinergia a las coaliciones dedicadas a preservar la diversidad a través de todos los paisajes ecológicos y sociales de las áreas limítrofes entre México y EE.UU. *Palabras clave: biodiversidad, descolonial, feminista, neoliberal, tierras fronterizas.*

Beyond jeopardizing wildlife, endangered species, and public lands, the U.S.–Mexico border wall is part of a larger strategy of ongoing border militarization that damages human rights, civil liberties, native lands, local businesses, and international relations. The border wall impedes the natural migrations of people and wildlife that are essential to healthy diversity.

—The Center for Biological Diversity (2017)

The border wall (the wall) of "hardened concrete, rebar, and steel" that Donald Trump promises to build along the southern U.S. border with Mexico continues to raise alarms for social and environmental justice groups on both sides (Noel 2017). In November 2017, the eight wall prototypes unveiled by private contractors offered a preview of what to expect: thick concrete and steel structures that stand thirty feet high and six feet deep, some fully blocking visibility and any animal migrations along with wind and water flows, and all designed to withstand attacks by "sledgehammer, car jack, pick ax, chisel, battery-operated impact tools, battery-operated cutting tools, oxy/acetylene torch or other similar hand-held tools" (Hawthorne 2017). Prototypes for digital barriers and surveillance are also in development. Currently, the wall represents one of the most costly border infrastructure projects in U.S. history and is a continuum of policies over the last century that have supported the building of barriers and militarized surveillance of the southern U.S. land border. Indeed, the Trump Wall agenda calls for the rebuilding of the most recent border barrier construction project—the 2006 Secure Fence Act, which had already erected barriers along some 650 miles of the almost 2,000-mile divide. This wall will also revive formerly abandoned efforts to create a "digital" barrier along many stretches. Through a fortification of these preexisting barriers and their extension into areas previously deemed inappropriate for further infrastructure, the current wall project is a defining feature of the Trump administration's promise to "make America great again."

Opposition to the wall and the policies behind it has gained steam through environmental and social justice coalitions that warn of its damaging consequences to the borderland landscape. Through their alliances that merge advocacies for human rights, immigrant well-being, and environmental stewardship, antiwall coalitions have formed powerful campaigns that have eroded public support for the project. These alliances emphasize three intersecting concerns. One is the irreversible ecological damage guaranteed by the wall's building. The barrier will slice through several national parks, conservation areas, and wildlife refuges and will create a hazard for some ninety-three endangered animal species with the potential of directly contributing to the extinction of the jaguar and pygmy owl. As one environmental activist warns, what Trump affectionately refers to as his "big" and "beautiful" border wall "will choke off life from both sides" (Silva and Gamboa 2017). A second antiwall emphasis is a demonstration of the inefficacy of border walls as technologies for controlling immigration, managing land use, or diminishing criminal activity in border regions (see Brookings Institute 2017). As a preponderance of evidence indicates, regardless of whether formed of physical or digital elements, border walls are inefficient and often counterproductive instruments for deterring criminal activity or controlling border flows. Furthermore, and given that most illegal immigration occurs through the overstaying of visas, they are not effective means for stopping the principle sources of undocumented immigration

(Barone and Tweeten 2017). As a director for the Center for Biological Diversity summarized, "The border wall will do many things—but it will not be effective at stopping people seeking a better life from getting to this country, rendering it useless for the purposes Trump says it will serve" (Greenwald 2017). A third area of emphasis for opposing the wall is that of its racist message along with its incalculable price tag and associated profits for the handful of companies that will build and maintain it (see Sierra Club 2018). The Sierra Club refers to this project as an "18 billion dollar boondoggle" (a conservative estimate by most accounts) that will line the pockets of the handful of firms lucky enough to be awarded the lucrative government contracts for building and maintaining the wall that will stand as a monument to racism and environmental degradation (Sierra Club 2018). By focusing on these principal issues through myriad political and community education campaigns across the United States, the activist networks of social–environmental–immigrant alliances have jeopardized congressional funding and broader public support. The Trump administration is fighting back.

In this article, I focus on how these successes in turning the public tide against the wall expose how the coalitions' commitments to diversity across the social–biological–physical landscape reflect a powerful legacy of border thinking within their strategies. "Border thinking," as border scholar Saldívar (2006) wrote, "emerges from critical reflections of (undocumented) immigrants, migrants, bracero/a workers, refugees, campesinos, women, and children on the major structures of dominance and subordination of our times. Thus envisaged, border thinking is the name for a new geopolitically located thinking of epistemology from both the internal and external borders of the modern (colonial) world-system." Rooted in the border thinking inspired by Latina border scholar Anzaldúa and the engagements with her work generated by feminist, decolonial, critical race, and indigenous studies in multiple contexts, the scholarly-activist dialogues against Trump's wall celebrate the borderlands as a place not characterized by binaries in need of more segregation but rather, to use the words of Juarense-border thinker Rosario Sanmiguel (2001, as cited in Castillo and Tabuenca Córdoba 2001, 8), as a place where "we get all confused and all mixed up." In my reflections here, I argue that the recent successes of the antiwall coalitions reflect the promise of such "mixed up" border thinking for fortifying the broad-based alliances so necessary to struggles against the intersected capitalist and racist interests that equate the wall's building with national security (see CNN 2017; Fox News 2018). Deepening and expanding on border thinking is essential, I maintain, for strengthening these alliances particularly throughout the borderlands most directly affected by the wall and its devastating impact on daily life across the organic spectrum.

Understanding the stakes in such struggles is crucial, I believe, for providing the analytical support to the coalitions that promote diversity in this part of the world and as part of a broader commitment to planetary sustainability. Their alliances entail constant shifts in strategy along with intransigent commitments to working together, across a wide array of different priorities, some of which (e.g., immigrant rights and environmental protection) have histories of tense or even unfriendly politics that still involve careful negotiation (see Solnit 2017).[1] They also require analytical support for the intransigent challenges of forming the cross-border collaborations that can coordinate and sustain action in the places most directly threatened by border wall construction and its lasting devastation throughout the environment. With such challenges in mind, I draw from dialogues between the critical geography literature that emphasizes the political force of alliances across the biosocial spectrum and the "border thinking" literature out of Latinx, Mexico–U.S. border, feminist political ecology, and decolonial studies (see Ortega 2016). I also draw from my own studies and experiences as a critical scholar of Mexico and the Mexican–U.S. border region who has directly experienced and documented competing perspectives over the meaning of the border for daily life and across the many scales of geographic belonging in a region with clear significance for transnational politics and global economic processes (Wright 1998, 2004, 2011, 2018).[2]

Border Thinking and the Wall

As a starting point, the border thinking most commonly articulated by the prowall coalitions regards the wall as a practical means for securing the border against illegal activity, such as smuggling and undocumented immigration, simultaneously distinguishing an Anglo-dominated U.S. territory from its Latin American others (see Zelizer 2018). As one wall supporter proclaimed on a fundraising Web site: "BUILD THE TRUMP WALL— REPUBLICANS MUST SUPPORT THIS PROJECT! KEEP ALL AMERICAN CHILDREN & AMERICANS SAFE—KEEP AMERICAN JOBS— KEEP AMERICA DRUG FREE—KEEP OUT DANGEROUS GANGS" (Build the Trump Wall Foundation 2018). This kind of border thinking, and its easily identifiable roots in nativist Anglo perspectives regarding national belonging (Zelizer 2018), explains that the wall is essential to establishing the rule of law and protecting national security from the undocumented immigrants coming from the southern Americas who represent unnamed threats to U.S. children, to the U.S. economy, and to U.S. public health and safety in relation to the drug trade. Although this prowall contingent does not overtly applaud the foreseeable environmental damage presented by its construction, they rarely raise this inevitable consequence. Moreover, their loyalty to Trump and his wall agenda does not falter when confronted with the overwhelming evidence indicating the impracticality of a wall for addressing the very concerns they raise. A recent *Newsweek* article based on interviews with wall supporters points to this contradiction: "In a world where most illegal immigrants don't walk across the border and not a single terrorist ever has, where small businesses rely on immigrant labor, and where average Americans could be hurt by increased prices or violations of their civil liberties, does it make sense to spend tens of billions of dollars on this wall?" (Monticello and Weissmuller 2018). As the journalists who wrote this article concluded, Trump and his wall supporters do not flinch at these facts.

By contrast, the border thinking most common to the antiwall contingent embraces a commitment to fostering the social and biological diversity fundamental to the thriving of life throughout the borderlands and its varied ecosystems. This perspective on diversity includes intersected collaborations across academic border studies, border environmental coalitions, and immigrant labor campaigns. With many connections to the farmworker activism of the 1960s and 1970s along with the first articulations, in the early 1980s, of "border thinking" by feminist Chicana lesbian essayist, activist, and scholar Gloria Anzaldúa and her benchmark writings in

"*Borderlands/La Frontera: The New Mestiza*" (Anzaldúa 2012), border studies emerged through the intersection of academic and activist struggles that connected myriad social campaigns regarding civil rights, feminism, labor, racism, and immigrant experiences (see Ortega 2016). Since then, the border thinking launched by this earlier work has inspired a proliferation of decolonial scholarship, with commitments in anti-imperialist, indigenous, and feminist movements in the southern Americas. As the decolonial and feminist scholar Lugones (2008) wrote, border thinking is essential to mobilizing "*los entrecruzamientos*" (cross-linkages) of academic and activist synergies that challenge the ongoing legacies of colonial or gendered and racialized systems of power such as those endemic to a militarization of borders and the criminalization of immigrant communities around the world (see also Quijano 2000; Mignolo 2011). This expansion of border thinking across the Americas over the last twenty years has yielded fruitful collaborations across the anti-imperial and anticolonial movements throughout the Global South.

As the interventions of border thinking have reshaped transnational politics and the epistemologies of borderlands studies into the present day, their influence has also spread to environmental justice advocacy as an embrace of social–environmental connections throughout the physical, animal, and social landscape (Ybarra 2009; Sundberg 2014; Yang 2017). As border thinker Ybarra (2009) wrote, in her reflection on border thinking's application to environmental justice: "[I]f we approach the borderlands as bioregion, we gain even more from Anzaldúa's landmark text" (175) for understanding the challenges of protecting the borderlands as a place sustained through diverse encounters. Such merging of environmental and social concepts of well-being has proliferated throughout the southern Americas particularly out of *indigena* (women's indigenous) border thinking that advocates, as geographer Radcliffe (2015) described, "a solidarity economy, recognition of unpaid—including reproductive—labor, cultural diversity, viewing nature as constitutive of and intrinsically valuable as social life, and environmental sustainability." The resonance with the current antiwall campaign is evident throughout the multiplying of rebuttals to Trump's border thinking and the wall that it pledges to produce. As the Brookings Institute (2017) warned,

"Rather than a line of separation, the border should be conceived of as a membrane, connecting the tissues of communities on both sides, enabling mutually beneficial trade, manufacturing, ecosystem improvements, and security, while enhancing inter-cultural exchanges." This vision of a diverse border region—across the biosocial landscapes of two countries that have emerged always and inevitably in relation to their shared border—exposes the productive engagements currently supporting the antiwall coalitions as they fight for life, in all of its manifestations, against the impending threat of Trump's wall and related political agenda.

To fully understand the stakes in the contrasting approaches to border diversity that characterize the pro- and anti-border wall camps, it is important to discuss the specific measures that the Trump administration is taking to set up the conditions for the wall's building. As a starting point and to initiate any form of wall construction, the government has to acquire all of the land, both "public" and "private," needed to establish the literal groundwork of the wall and then limit access to all of this territory whose sole legal function will be to sustain the wall and related border security measures. Toward this end, and building on the 2006 Secure Fence Act, the Trump administration has petitioned Congress to extend previous policies for waiving every possible legal obstacle to acquiring the remaining tracts necessary to complete the wall. Included within such obstacles are environmental regulations, treaties with Native American tribes and their rights to move freely throughout the border region, international environmental accords, and respect for public and private access to land and other resources. If Trump succeeds with his border plans, these, among other uses and ways of interacting with the land specifically required for the wall's building, will be eliminated or rendered illegal by the closure of public access to the affected territory. The extent of this transformation of borderlands into a place whose sole purpose is to sustain the wall is not yet clear. As a report issued by a U.S. congressional committee explains: "[T]he administration cannot provide the committee with any definitive real estate costs or requirements, cannot tell the committee how many American citizens will have their land seized, and has no timeline for completing land acquisition efforts necessary to build the wall that President Trump has ordered" (Kopan 2017). In other words, the U.S. government is prepared to litigate, in perpetuity, to seize both public and private lands and abolish, by outlawing, any possible uses that could compete with the dedication of lands needed for the foundation of a wall. Such deals expose, as Prudham (2013) wrote, the socio-natural transformations characteristic of neoliberal governance as the state mediates the transfer of public resources through the forced sale of private lands (bought with public funds) or the closure of public lands, in addition to other entities. This transaction and its associated transfer of public goods to a privileged few in the private sector remain, in orthodox neoliberal fashion, protected by laws that criminalize any challenge to these measures even as the public sector assumes the costs and risks, including the burdens of the inevitable environmental hazards, in the process (see also Harvey 1996).

At issue here is a double-pronged form of privatization. One is the restriction of public access to public resources (e.g., parks and conservation lands) via their condemnation and then subsequent management by private firms that reap enormous profits to build barriers and maintain them. Another is via the U.S. government's role as a kind of broker that oversees the transfer of some private property and Native American communal lands to the public sphere where it will be accessible to those private firms that, again, stand to profit handsomely from construction and maintenance contracts (Monticello and Weissmuller 2018). Meanwhile, the vitriolic distrust, even hatred, of immigrants expressed by many wall proponents fuels a concomitant political agenda that generates distrust and cruel policies, such as the forcible separation of children from their families and their subsequent detention in isolated tent cities (Whittaker 2017; Molloy 2018).

The stakes in this battle for the ecological and social health of the border region are difficult to exaggerate. As a spokesperson for the Center for Biological Diversity described, "Beyond jeopardizing wildlife, endangered species and public lands, the U.S.–Mexico border wall is part of a larger strategy of ongoing border militarization that damages human rights, civil liberties, native lands, local businesses and international relations. The border wall impedes the natural migrations of people and wildlife that are essential to healthy diversity" (Center for Biological Diversity 2017). According to numerous analysts and reports, the proportions of this "looming tragedy" for environmental and social diversity, and for different ways of thinking and living the border, are so great that they regard the wall as a kind of warfare that will wreak incalculable damage from which the social and ecological communities that make up the borderlands might never recover (Brookings Insititute 2017; Center for Biological Diversity 2017). As Juanita Sundberg (2017) wrote, the political agenda behind the wall "frames the borderlands as yet another theater of war." The warfare here, as is made clear both by the coalitions opposed to the wall as well as by the decolonial and geographic scholarship that engages with them, is against diversity in border thought, land use, or actions in support of sustaining varied habitats throughout the region.

Walls as Warfare

As antiwall activists publicize the connections linking the project to a cruel politics that reveals a startling disdain for human and environmental well-being, congressional and broader public support has stalled. In response, Trump has vowed to use whatever means he can, including a shutdown of the federal government (by refusing to sign the federal budget), to force congress to fund his wall (Burnett 2018). His most loyal supporters continue this fight in other ways as well and have turned their collective vitriol on the intertwined immigrant and environmental movements that form the platform of wall opposition. This politics of division emerged noticeably when, in early 2018, Trump declared that any hope for his extension of protections to those immigrants known as "the Dreamers," the mostly young adults and teens who were accepted into the Obama-era Deferred Action for Childhood Arrivals (DACA) program, depended on congressional support for the wall. Numbering around 800,000, the Dreamers are now in legal limbo as their fate is now linked to that of Trump's wall. The embrace of progressive border thinking within environmental activism, however, has provided a powerful platform for rebuking, as one antiwall organizer in Austin recently put it, "the greening of hate" (personal communication, June 2018; see also Hartmann 2010). Through an embrace of a commitment to diversity as an intertwined environmental and social value that the binational borderlands must sustain, environmental activists are refusing to reopen the old fault lines of racism and anti-immigrant politics that once split their own coalitions. For instance, at another 2018 antiwall rally in Washington, DC, environmental and immigrant rights' activists joined forces to lobby congress to support a "Clean Dream Act" that would stop using the Dreamers as "bargaining chips for Trump's cruel anti-immigrant agenda and racist border wall" (League of Conservation Voters 2018). As one of the participants from Sustainable Practices explained at that event, "We will stand with and support those who need their voice amplified for the net outcome of promoting a global society that understands the interconnectedness of the planet with a foundation in social justice, environmental justice and economic equity. We are Sustainable Practices and our mission is the promotion of a culture of sustainability" (League of Conservation Voters 2017).

Such alliances across the nexus of diversity expose ways of border thinking that echo with, as geographers Collard, Dempsey, and Sundberg (2015) wrote, feminist–decolonial–political ecological commitments to "manifestos for more abundant futures"; that is, of "futures with more diverse and autonomous forms of life and ways of living together" (324). To use the words of an environmental activist in the Tijuana River Estuary in Southern California who, when explaining to me in 2008 his own political evolution and embrace of coalition politics with immigrant and human rights groups, "This border is not only the place of a division. It is the place where all come together. Where all kinds of life come together. This is not the place of a wall."

Although the current success of biosocial diversity coalitions to derail Trump's wall offer valuable insights into the political possibilities of putting socially and environmentally progressive "bordering thinking" into action, they also reveal the need for expanding resilient commitments to diversity across and along borders. The fight is hardly won given that Trump and his political allies have much political power and many tools in their arsenal for turning a wall into a weapon aimed against those who promote diversity as a value and way of life. As border thinkers and their allied activists continue to face big money and bellicose anti-immigrant stances in the United States, the need for increased solidarity across border geographies is paramount. Perhaps one of the most strategically pressing matters is the imperative for strengthening coalitions across the

diversely experienced realities of the Mexico–U.S. border and the myriad approaches to environmental and social justice across the divide. For instance, and in relation to the wall, many on the Mexican side of the line explain how even though the project (despite Trump's bluster that Mexico would "pay" for it) is a decidedly "U.S. issue," its impact across the borderlands means that it has a connection to events in both countries. Yet, those in the southern neighbor are rarely raised in relation to it. As one resident of the Mexican border city Ciudad Acuña (Coahuila), explained to me in 2008 and in reference to the 2006 border barrier construction, "It's ugly, but it's on their side. The real issues here are jobs, drugs, and water. That affects everyone on this border. Both sides. You should be asking questions about those things. Not about what I think of this stupid wall."

Such sentiments expose the constant challenges in efforts to strengthen cross-border alliances within broad coalitions committed to expansive commitments to diversity and border ways of life. As many border thinkers and scholars explain, these challenges also reveal themselves as such within border thinking where the experiences of those on the Mexican side often fail to emerge as priorities in Anglophone scholarship oriented to publics more familiar with epistemologies and political priorities rooted within the United States or the Global North (see, e.g., Castillo and Tabuenca Córdoba 2001). As literary scholars and border thinkers Castillo and Tabuenca Córdoba (2001) wrote: "[I]f we cross the border literally as well as metaphorically, we would have to note that because of the comparatively fewer social, economic, and political advantages enjoyed by Mexico's northern border states" (16), we can trace how Mexican perspectives are often missing from even progressive accounts of border life. Certainly with regard to the ongoing debates regarding Trump's wall and related issues of rights, well-being, and sustainability, many issues most acutely felt on the "Mexican side" of the divide such as the violence of a U.S.-funded Mexican "drug war," endless economic crises, increasing salinity rates in water tables, and poor air quality have yet to emerge as prominent priorities in cross-border coalitional politics (see Expansión 2017). Rather than regard such problems as "failures," Castillo and Tabuenca Córdoba (2001), among other border thinkers from the *southern edge*, call for creative attention to the urgencies of bridging such gaps in the "evolving study of border theory and border culture" and border life (32). As a border scholar and activist in Ciudad Juárez said to me in 2017, shortly after Trump's election and the possibility of the border wall loomed large over both countries, "To fight this wall, we have to fight all of the thinking behind it. We can only do that if people from both sides come together. We are the people of the borderlands. We are not the people of a place that is divided by a wall. We are not from that place."

Concluding Thoughts

As border thinkers in collaboration with critical geographic literature indicate, the synergy linking understandings of social and ecological mutual reliance with political activism dedicated to protecting this synergy is vital for the sustaining of a healthy border landscape. The Trump wall is not an isolated case of how the modern state colludes with the private sector to destroy social and natural diversity in the name of national security and at great benefit to a very privileged few; nor is it an isolated case of the political and economic warfare waged against those who value diversity as an interconnected relationship fundamental to the well-being of the planet. In fact, the border thinking that reveals this warfare behind the wall's building as a monument to nativist concepts of place and belonging has succeeded in generating broader public recognition of the interconnectedness—the *entrecruzamientos*—that is required for healthy border living. This recognition has undermined the wall's support to such an extent that its defeat seems possible if the coalitions behind its opposition expand the solidarity against the warfare of this wall and the devastation that it promises for the borderlands and for planetary well-being. As border thinking, in its most collaborative spirit, leads us to see, fighting the wall and the punishing values behind it requires such fortifications across the epistemologies that remain committed to justice as a daily and geographically grounded practice and across the political divide. Critical geographers have the tools to further this project. The Trump wall demonstrates, yet again, the urgency for doing so now.

Acknowledgments

I am especially grateful to Guadalupe D'Anda for her insights over the years on the topics I discuss here. I also owe enormous thanks to Dr. Rosalba Robles and other participants in a 2017 border studies seminar at the Universidad Autónoma de Ciudad Juárez and also to the participants in my border geographies seminar at the Centro de Estudios Superiores de México y Centroamérica (CESMECA), in 2014, in San Cristóbal de las Casas. My understanding of border thinking across the Americas deepened through these important encounters. I am also indebted to Dr. Hector Padilla, Dr. Juanita Sundberg, and Leobardo Alvarado, with whom I worked on a collaborative project on militarization along the Mexico–U.S. border from 2010 to 2016. I am solely responsible for any errors.

Funding

This project has received funding from the National Science Foundation under award number 1023266. Any opinions, findings, and conclusions or recommendations expressed in this material are those of the author and do not necessarily reflect the views of the National Science Foundation. I am also grateful to the Penn State College of Earth and Mineral Sciences for funding the initial study.

Notes

1. References to this evolving history were common to my discussions with environmental and immigrant rights organizations throughout the borderlands in 2008.
2. I have intermittently studied or been in involved in a variety of community efforts in El Paso, Texas, and Ciudad Juárez, Chihuahua, over the last thirty years, which are interests stemming from the political convictions of my youth and family in the area. Most specifically, from 1993 to 1996, I was active in community efforts in El Paso to oppose the building of barriers and further militarization of the borderlands under the Clinton administration. Then, in 2006 and 2007, during the initial implementation of the 2006 Secure Fence Act, I conducted a study into the meaning of this project for border residents in both Ciudad Juárez, where I was living that year, and in El Paso, Texas. In the subsequent year, I conducted a several-week field study, with funding from Penn State's College of Earth and Mineral Sciences and with companionship from Guadalupe D'Anda (and our young daughter), into the perspectives of border residents regarding the 2006 Act and who were involved in progressive social and environmental coalitions in the borderlands. We initiated the project in San Diego and Tijuana in

May 2008 and conducted interviews in key urban areas along both sides of the border until concluding the fieldwork in the Brownsville and Matamoros area. From 2010 to 2016, I collaborated on a project on border militarization with Dr. Hector Padilla, Dr. Juanita Sundberg, and Leobardo Alvarado. Although our focus was not directly on the border walls or barriers, our discussions regarding border governance and militarization have helped me in innumerable ways as I work through these ideas. More recently, I have chaired seminars in border studies at both the Universidad Autónoma de Ciudad Juárez (in 2017) and at the Centro de Estudios Superiores de México y Centroamérica (CESMECA), in 2014, in San Cristóbal de las Casas. My understanding of border thinking and decolonial feminist thought across the Americas deepened through these important encounters.

References

Anzaldúa, G. 2012. *Borderlands/La frontera: The new mestiza*. 4th ed. San Francisco: Aunt Lute Books.

Barone, A., and L. Tweeten. 2017. This graphic shows why President Trump's border wall won't stop immigrants from crossing. *Time*, May 16. Accessed January 14, 2019. http://time.com/4729470/mexico-border-wall-trump-undocumented-immigrants/.

Brookings Institute. 2017. The wall: The real costs of a barrier between the United States and Mexico. Accessed July 31, 2018. https://www.brookings.edu/essay/the-wall-the-real-costs-of-a-barrier-between-the-united-states-and-mexico/.

Build the Trump Wall Foundation. 2018. Fundly page. Accessed October 26, 2018. https://fundly.com/build-the-trump-wall-foundation.

Burnett, J. 2017. Borderland Trump supporters welcome a wall in their own backyard. *National Public Radio*, August 12. Accessed January 14, 2018. https://www.npr.org/2016/08/12/489350032/borderland-trump-supporters-welcome-a-wall-in-their-own-backyard.

———. 2018. NPR poll: 2 in 3 support legal status for DREAMers; majority oppose building a wall. NPR, February 6. Accessed January 14, 2019. https://www.npr.org/2018/02/06/583402634/npr-poll-2-in-3-support-legal-status-for-dreamers-majority-oppose-building-a-wall.

Castillo, D., and M.-S. Tabuenca Córdoba. 2001. *Border women: Writing from La frontera*. Minneapolis: University of Minnesota Press.

Center for Biological Diversity. 2017. No border wall. Accessed January 14, 2019. https://www.biologicaldiversity.org/campaigns/border_wall/.

CNN. 2017. How Republicans came to support Trump's wall. Accessed July 31, 2018. https://www.cnn.com/2017/01/28/politics/donald-trump-republicans-support-border-wall/index.html.

Collard, R.-C., J. Dempsey, and J. Sundberg. 2015. A manifesto for abundant futures. *The American Association of American Geographers* 105 (2):322–30.

Expansión. 2017. Opinion: ¿Cómo afecta al ecosistema la construcción de un muro fronterizo? Accessed July 31, 2018. https://expansion.mx/opinion/2017/02/17/opinion-como-afecta-al-ecosistema-la-construccion-de-un-muro-fronterizo.

Fox News. 2018. Trump's border wall: A look at the numbers. Accessed July 31, 2018. http://www.foxnews.com/politics/2018/03/14/trumps-border-wall-look-at-numbers.html.

Greenwald, N. 2017. Trump's border wall not just a disaster for people. Accessed July 31, 2018. https://www.huffingtonpost.com/entry/trumps-border-wall-not-just-a-disaster-for-people_us_5927504fe4b065b396c06b79.

Hartmann, B. 2010. Betsy Hartmann: An environmentalist essay on the greening of hate. Accessed October 26, 2018. https://climateandcapitalism.com/2010/08/31/the-greening-of-hate-an-environmentalists-essay/.

Harvey, D. 1996. *Justice, nature, and the geography of difference*. Oxford, UK: Basil Blackwell.

Hawthorne, C. 2017. Trump's border wall through the eyes of an architecture critic. *Los Angeles Times*, December 30. Accessed January 14, 2019. http://www.latimes.com/entertainment/arts/la-ca-cm-building-type-border-wall-20171231-html story.html.

Kopan, T. 2017. Trump Admin taking quiet steps on seizing border land. CNN, November 13. Accessed January 14, 2018. http://www.cnn.com/2017/11/13/politics/border-wall-eminent-domain/index.html.

League of Conservation Voters. 2017. Environmental coalitions stand with dreamers. http://origin.lcv.org/article/environmental-coalition-stands-dreamers-greens-clean-dream-act-visual-demonstration/. Accessed January 17, 2019.

———. 2018. Home page. Accessed July 31, 2018. http://origin.lcv.org.

Lugones, M. 2008. Colonialidad y género [Coloniality and gender]. *Tabula Rasa* 9: 73–101.

Mignolo, W. 2011. *The darker side of Western modernity: Global futures, decolonial options*. Durham, NC: Duke University Press.

Monticello, J., and Z. Weissmuller. 2018. Should we build the wall? We asked Trump supporters. Accessed October 26, 2018. https://reason.com/reasontv/2018/03/23/trump-wall-immigration-borders-illegal.

Noel, A. 2017. Here's why the great wall of Trump is an alternative fact. *The Daily Beast*, February 19. Accessed January 14, 2018. https://www.thedailybeast.com/heres-why-the-great-wall-of-trump-is-an-alternative-fact.

Ortega, M. 2016. *In-between: Latina feminist phenomenology, multiplicity, and the self*. Albany: SUNY Press.

Prudham, S. 2013. Men and things: Karl Polanyi, primitive accumulation, and their relevance to a radical green political economy. *Environment and Planning A* 45 (7):1569–87.

Quijano, A. 2000. Coloniality of power, ethnocentrism, and Latin America. *Nepantla* 1 (3):533–80.

Radcliffe, S. 2015. *Dilemmas of difference*. Durham, NC: Duke University Press.

Saldívar, J. D. 2006 Unsettling race, coloniality, and caste. *Cultural Studies* 21 (2):339–67.

Sierra Club. 2018. Trump's border wall: An 18 billion dollar boondoggle. Accessed July 31, 2018. https://www.sierraclub.org/press-releases/2018/01/trump-s-border-wall-18-billion-dollar-boondoggle.

Silva, D., and S. Gamboa. 2017. Trump's border wall "catastrophic" for environment, endangered species: Activists. NBC News, April 11. Accessed January 14, 2018. https://www.nbcnews.com/science/environment/trump-s-border-wall-catastrophic-environment-endangered-species-activists-n748446.

Solnit, R. 2017. The Sierra Club's 125-year history has been a story of evolution. Accessed July 31, 2018. https://www.sierraclub.org/sierra/2017-3-may-june/feature/sierra-clubs-125-year-history-has-been-story-evolution.

Sundberg, J. 2014. Decolonizing posthumanist geographies. Cultural Geography 21 (1):33–47.

———. 2017. The nature of border control. Accessed January 14, 2018. https://nacla.org/blog/2017/05/12/nature-border-control.

Whittaker, M. 2017. 8 companies that can profit from a border wall with Mexico. U.S. News & World Report, March 10. Accessed July 31, 2018. https://money.usnews.com/investing/articles/2017-03-10/8-companies-that-can-profit-from-a-border-wall-with-mexico.

Wright, M. W. 1998. Maquiladora mestizas, and a feminist borderpolitics: Revisiting Anzaldúa. Hypatia 13 (3):114–31.

———. 2004. From protests to politics: Sex work, women's worth and Ciudad Juárez modernity. Annals of the Association of American Geographers 94:369–86.

———. 2011. Necropolitics, narcopolitics and femicide: Gendered violence on the Mexico–U.S. border. Signs: Journal of Women in Culture and Society 36:707–31.

———. 2018. Against the evils of democracy: Fighting forced disappearance and neoliberal terror in Mexico. Annals of the American Association of Geographers 108 (2):327–36. doi:10.1080/24694452.2017.1365584.

Yang, M. 2017. The Trump wall: A cultural wall and a cultural war. Lateral 6 (2). Accessed January 14, 2019. http://csalateral.org/issue/6-2/trump-wall-cultural-war-yang/.

Ybarra, P. 2009. Borderlands as bioregion: Jovita González, Gloria Anzaldúa, and the twentieth-century ecological revolution in the Rio Grande valley. MELUS 34 (2):175–89.

Zelizer, J. 2018. America's mirror on the wall. The Atlantic, January 28. Accessed July 31, 2018. https://www.theatlantic.com/politics/archive/2018/01/americas-mirror-on-the-wall/551165/.

MELISSA W. WRIGHT is Professor in the Department of Geography and in the Department of Women's, Gender, and Sexuality Studies, The Pennsylvania State University, University Park, PA 16802. E-mail: mww11@psu.edu. Her research interests include social movements, feminism, resistance to state terror, and transnational coalitions throughout Mexico and the Mexico–U.S. borderlands.

Environmental Deregulation, Spectacular Racism, and White Nationalism in the Trump Era

Laura Pulido, Tianna Bruno, Cristina Faiver-Serna, and Cassandra Galentine

This article examines the relationship between racism and environmental deregulation in President Trump's first year in office. We collected data on all environmental events, such as executive actions at the federal level or Trump's tweets. Likewise, we documented racist events targeting indigenous people, people of color, Muslims, and South Asians or Arabs. We found important differences in how these agendas unfolded: Environmental events were more likely to be concrete actions, whereas racist events were more likely to involve "noisy" rhetoric. The differing forms are not associated with particular levels of harm; rather, they suggest the unanticipated and complex ways in which racism intersects with environmental governance under neoliberal, authoritarian regimes. We argue that Trump's "spectacular racism," characterized by sensational visibility, helps obscure the profound deregulation underway. The white nation plays a critical role, as Trump uses spectacular racism to nurture his base, consolidate his power, and implement his agenda. Such an analysis expands how environmental racism is typically conceptualized.

本文检视特朗普总统上任第一年间, 种族主义和环境去管制之间的关系。我们搜集所有环境事件的数据, 诸如联邦层级的行政行动或特朗普的推文。我们同样记录针对原住民族、有色人种、穆斯林、以及南亚或阿拉伯人的种族歧视事件。我们发现, 这些议程开展的方式具有重要的差异: 环境事件更倾向是具体的行动, 而种族歧视事件则更可能涉及 "嘈杂的" 修辞。不同的形式并非关乎特定程度的伤害; 反之, 它们指向威权新自由主义政体下, 种族主义和环境治理交叉的非预期与复杂的方式。我们主张, 特朗普以轰动的可见度为特征的 "奇观种族主义", 有助于掩盖正在进行中的深刻去管制。 特朗普运用奇观种族主义来培养其基层、巩固其权力并执行其议程时, 白人国族扮演了关键角色。此般分析扩展了环境种族主义一般被概念化的方式。关键词: 环境去管制, 奇观种族主义, 特朗普, 白人国族。

Este artículo examina la relación entre racismo y desregularización ambiental durante el primer año de gobierno del presidente Trump. Recabamos datos de todos los eventos ambientales, tales como las acciones ejecutivas a nivel federal, o los tuits de Trump. También, documentamos eventos racistas enfocados contra gente indígena, gente de color, musulmanes y asiáticos del sur, o árabes. Encontramos diferencias importantes sobre la manera como estas agendas fueron desplegadas: Seguramente, los eventos ambientales fueron acciones concretas, en tanto que los eventos racistas muy probablemente se revistieron de retórica "ruidosa". Las formas discrepantes no están asociadas con particulares niveles de daño; más que eso, sugieren el modo imprevisto y complejo como el racismo intersecta con la gobernanza ambiental bajo regímenes neoliberales y autoritarios. Sostenemos que el "espectacular racismo" de Trump, caracterizado por su despliegue sensacionalista, ayuda a ocultar la profunda desregulación que está en marcha. La nación blanca juega un rol crítico a medida que Trump usa el racismo espectacular para nutrir su base, consolidar su poder e implementar su agenda. Tal tipo de análisis amplía la gama de maneras como el racismo ambiental es típicamente conceptualizado. *Palabras clave: desregulación ambiental, nación blanca, racismo espectacular, Trump.*

This article examines the relationship between racism and environmental deregulation in President Trump's first year. A hallmark of the Trump era (defined as his campaign and presidency) is his use of transgressive racism, such as declaring Mexicans rapists and introducing a Muslim ban. Geographers have analyzed how and why he has employed this strategy and its impacts (Gokariksel and Smith 2018; Inwood 2018; Page and Dittmer 2018). The profundity of transgressive racism, which we call spectacular racism, is akin to Nixon's (2011) "spectacular violence" in its visible and sensational nature (6, 13). Not surprisingly, Trump's spectacular racism has shifted the U.S. racial formation. Racial formation is "the process by which social, economic and political forces determine the content and importance of racial categories, and by which they are in turn shaped by racial meanings" (Omi and Winant 1994, 61). Until recently, hegemonic racial culture was marked by political correctness and "neoliberal multiculturalism" (Melamed 2011), but it now coexists with overt white supremacy. Because the United States is a deeply racialized society, not only are all sites racialized but racial processes can produce a multitude of consequences—including unanticipated and seemingly unrelated ones. Thus, we explore how the shift in racial formation, specifically the ascent of spectacular racism, might have affected environmental governance. We do this by comparing how Trump's environmental and racist agendas unfolded during his first year in office. The data, which include speeches, proposed legislation, and appointments, indicate that Trump's spectacular racism drew massive media attention because of its transgressive nature, whereas his environmental agenda attracted far less scrutiny. Indeed, Trump's environmental agenda was extremely well-orchestrated and unfolded quietly. Regardless of intent, we suggest that Trump's spectacular racism overshadowed environmental deregulation in his first year. Consequently, one of the many outcomes of spectacular racism has been to facilitate deregulation and a larger neoliberal agenda.

Such a reframing is important because it expands how racism is conceptualized within the discipline of geography and environmental justice. Within geography, racial analyses often focus on how racism affects people of color, rather than attending to the larger political culture (for exceptions, see Kobayashi and Peake 2000; Gilmore 2002; Inwood 2015; Bonds and Inwood 2016; Inwood and Bonds 2016; Pulido 2017, 2018). Within environmental justice, racism is usually conceptualized as the spatial relationship between environmental hazards and marginalized communities, including procedural justice (Schlosberg and Carruthers 2010; Yen-Kohl and The Newtown Florist Club Writing Collective 2016). Our analysis builds on critical environmental justice scholars seeking to expand the contours of environmental racism (Heynen 2016; Pulido 2016, 2017; Pellow 2017). We argue that Trump's spectacular racism transformed the political environment by facilitating an authoritarian neoliberal regime, with profound consequences for the natural environment through deregulation. We do not claim that this was planned, although possible; rather, we trace two parallel processes to uncover unanticipated linkages.

We ask the following questions: What has been the role of spectacular racism during Trump's first year? What has been the relationship between spectacular racism and environmental deregulation? Which populations have been targeted, and why? How has environmental deregulation occurred? We answer these questions by comparing two interconnected data sets: one list of environmental events and one of racist/anti-indigenous, anti-immigrant, and anti-Muslim/Arabs events (hereafter referred to as racist). Although we distinguish racist events from environmental ones for the sake of data collection and analysis, we acknowledge their somewhat fictitious nature. Despite such challenges, however, the data indicate that the two agendas have unfolded distinctly and that U.S. white supremacy is so pervasive that it affects seemingly "nonracial" phenomena.

Spectacular Racism and the White Nation in the Trump Era

The Trump era marks a shift in the racial formation. According to Melamed (2011), the early twenty-first century was characterized by neoliberal multiculturalism, a form of antiracism in which race is dematerialized and severed from capitalism's violence. Neoliberal multiculturalism, including political correctness, was delegitimized by the Trump campaign. Trump's racist rhetoric, which affirms whiteness and connected emotionally, inspired intense media coverage precisely because of its transgressive nature (Hochschild 2016). His transgressive racism accomplishes numerous political objectives, including dehumanizing his targets, consolidating his power, eroding democratic norms, and distracting from policy and legal changes.

The spectacle of Trump's racism—the incessant tweets, outrageous statements, and dehumanizing immigration policies—cannot be divorced from his neoliberal environmental agenda. Indeed, a growing number of scholars have shown how deep historicization is necessary to understand how racism informs contemporary economic structures and processes (Wilson 2000; Gilmore 2002; Baptist 2016; Woods 2017). Recently, Inwood (2018) argued that whiteness, which is a form of white supremacy, is a powerful "counter-revolutionary" force that "impede[s] progressive and racial reconfigurations of the US racial state" (3). This is an important move that foregrounds the deeply problematic nature of whiteness itself, as it readily slips into white nationalism and other forms of white supremacy. One reason that whiteness impedes progressive change, including environmentalism, is because it is, by definition, antidemocratic, as it seeks to exclude and subordinate. Never in U.S. history has mobilizing along whiteness led to greater democracy and equality. Indeed, its opposite—the black radical tradition (BRT)—consistently leans toward greater freedom and inclusion (Robinson 2000). The BRT refers to the centuries-long struggle of black people to resist racial capitalism, a system that "not only extracts life from black bodies, but dehumanizes all workers while colonizing indigenous lands and incarcerating surplus bodies" (G. T. Johnson and Lubin 2017, 12). The BRT, which is not limited to black people, includes abolition geographies (Gilmore 2017) and abolition ecologies (Heynen 2016) and is generally associated with progressive change.

Inwood (2015) noted a second problem with whiteness: its usage as a "racial fix." Similar to a spatial fix, racial fixes are employed to resolve a crisis, however temporarily. In the United States, crises are routinely resolved by deploying racism to either blame people of color or otherwise appeal to the white nation (Hochschild 2016). Solving crises through preexisting relations, including racism, is fundamental to U.S. politics (Gilmore 1999; Woods 2017).

We see both dynamics occurring in the Trump era. Trump employs a racial fix by blaming racial others and immigrants to offer the white nation a psychological wage (Du Bois 2014). This wage does not merely validate white people's superiority; rather, it addresses their emotional dislocation, fear, and resentment of a changing world (Hochschild 2016), affirming their status as the true nation (Thobani 2007). Clearly, this works against progressive change by aligning the white working class with capital (Du Bois 2014), but it also has implications for environmental governance (Hochschild 2016).

The current wave of white supremacy has been brewing for decades as Republicans attracted southern whites through racial resentment (Hajnal and Rivera 2014; Inwood 2015; Tesler 2016; Woods 2017). The election of Barack Obama, growing economic precarity, changing demographics, and multiculturalism all contributed to a deep resentment on the part of many whites, which candidate Trump tapped into.

If a nation is an "imagined political community" (Anderson 1983), the U.S. white nation is a political community constituted by whiteness and Christianity. It is not defined by the exclusion of non-whites and non-Christians but rather by the valorization of whiteness and Christianity. Not only does it equate white Christianity with the essence of the U.S. nation, but it considers it superior to those others deemed threats or intruders.[1] Moreton-Robinson (2016) argued that white nationalism in settler societies is largely defined by entitlement and possession, especially in terms of territory and state (see also Harris 1993). Although whiteness has varied geographically in the United States (Vanderbeck 2006; see also C. Johnson and Coleman 2012), contemporary white nationalism has been consolidated and strengthened through populism and white supremacy (Inwood 2015), although it might contain diverse racist ideologies.

By nurturing the white nation via spectacular racism, Trump has shifted the racial formation so that overt white supremacy is increasingly normalized (Page and Dittmer 2018). This, in turn, paves the way for dehumanizing policy. For instance, when he declared that there "were fine people on both sides" ("Full Transcript and Video" 2017), after the 2017 Unite the Right rally in Charlottesville, Trump legitimized white supremacist violence, thereby normalizing it. Such acts contribute to a climate in which Latinx children were separated from their parents at the U.S.–Mexico border, as happened in 2018. Such rhetoric and actions not only nurture and embolden the white nation, but they dehumanize racial others: They are not worthy of full legal and moral consideration. Indeed, they are arguably mere pawns in Trump's political machinations.

Another important shift in the racial formation is spectacular racism's ability to deflect attention away from political and economic crises. This is especially significant early in new regimes, when accepted norms are being violated and spectacular racism devours both media attention and individual energy.

Relatedly, because the white nation is being nurtured, the possibility of structural critique is foreclosed. Instead, the white nation is offered false explanations, further dividing the working class. At the time of this writing, for example, during midterm elections, Trump hoped to send 15,000 troops to the southern U.S. border to block Central American immigrants, who he has demonized and blames for U.S. problems (Gonzales 2018).

Spectacular racism fuels and coexists with other manifestations of racism, including institutional racism and white privilege (Pulido 2000). It functions to enhance authoritarian and populist power by solidifying and empowering a political base that is partially animated by white supremacy and xenophobia (Zeskind 2012; McElwee and McDaniel 2017). It generates loyalty to an individual, rather than to political ideas or institutions (Frum 2017; see also Taub 2016). Spectacular racism, white nationalism, and authoritarianism are all distinct but work together in the Trump era. Portions of the U.S. electorate have always supported authoritarian figures (Hetherington and Suhay 2011; MacWilliams 2016; Koch 2017), who offer easy solutions to insecurity and change (Taub 2016). They rise to power through elections by aligning themselves with establishment politics and appealing to the public (Levitsky and Ziblatt 2018). Although certainly not Trump's only means of attracting support, spectacular racism has been central to his strategy. Trump's calculated use of spectacular racism has channeled diffuse anger and anxiety into a ferocious wave that is the white nation. Thus, the white nation is foundational to Trump's power. Despite violating social norms, disregarding laws, and maintaining a chaotic administration, Trump has retained a loyal base (Graham 2017). Trump has successfully reshaped the Republican Party (Malone, Enten, and Nield 2016; Isenstadt 2017; Thompson 2018), so that it no longer even espouses racial equality (Page and Dittmer 2018). Trump understands the power of his base and seeks to nurture it, as it allows him to continue to function as an authoritarian. The white nation is the fulcrum that enables Trump's agenda, which clearly favors elites and capital.

Method

Data were collected via a class-based research project at the University of Oregon in the winter of 2018.[2] Students were divided into teams focused on

Table 1. Delineation of events, action, and discourse

Events	
Actions	Discourse
Budgets	President tweets
Court rulings	Agency administrator tweets
Observed acts of censorship	President speeches
Executive orders/memoranda	
Policy: new, changes, delays	
(Non)appointments	

environmental and racial issues. Environmental teams studied water, land, climate, air, toxins, and environmental justice, and the racial teams studied black people, Indigenous people, Jews and Palestinians, Mexicans and Latinxs, South Asians, Muslims and Arabs, whites, China, and the Koreas. Each group sought to find all events related to their topic that emanated from the Trump administration, federal legislature, and judiciary between 20 January 2017 and 20 January 2018. Initially, we drew from extant lists, such as Columbia and Harvard law schools' environmental trackers and the American Civil Liberties Union's lists, and then we branched out. We followed news leads, the *Congressional Record*, the White House's Presidential Actions, and the like.

As Table 1 shows, an event could be rhetoric, a policy, or anything in between. Specifically, an event was defined as new policy, policy changes, policy delays, observed acts of censorship, executive actions, appointments (including nonappointments), budgets, court rulings, tweets, and speeches. We included all legislative action, including proposed bills. We included all tweets from Trump, as the Department of Justice views them as official presidential statements, as well as tweets from then-Environmental Protection Agency (EPA) Director Scott Pruitt. We struggled to decide which racial groups to study and how to categorize them. This became especially difficult in terms of groups that clearly transcend the domestic racial formation, such as Koreans. Ultimately, we included as many identifiable groups as possible. Events that are both racial and environmental, including the Presidential Memorandum Regarding Construction of the Dakota Access Pipeline, appear on both lists.

We acknowledge the fictitious nature of our categories. Indeed, environmental justice scholars have advocated for collapsing such boundaries to understand how environmental injustice is produced

(Pulido 2000, 2016). Our objective, however, is to document state actions. Although it is true that universal environmental regulations have uneven consequences, only rarely is the current federal government connecting racism and environmental governance (Holifield 2012; Newkirk 2018). Recall that under neoliberal multiculturalism the state has "dematerialized" race, and whereas previous administrations were invested in contracting the conception of racism, the Trump administration is committed to disavowing its existence. Thus, with a few exceptions, such as the poisoned water of Flint, Michigan, pesticide deregulation, and public lands involving Indigenous people, the state treats race and the environment as separate spheres.

For each event we noted the date, responsible entity, publicity level, kind (e.g., judiciary vs. budget), justification, push-back (if any), and potential impacts across space and time. We reviewed the data, addressed duplication, and analyzed it according to the patterns and themes detailed in the tables included here. The data set is available at www.laurapulido.org.

Spectacular Racism and Environmental Deregulation in the Trump Era

The data indicate that Trump's environmental and racial agendas have unfolded in distinct fashions, which we believe are meaningful. Despite significant chaos, we view Trump as a strategist and environmental deregulation as part of a larger neoliberal agenda embraced by the Republican Party. This is important because agendas that fully align with the GOP have been more successfully implemented than those that do not. It is important to recall that Republican opposition to Trump existed early on, although it has since vanished (Jacobson 2017; Chait 2018). Thus, these data capture a more fractious GOP—one that fully supported environmental rollbacks but had some unease with racism. Given these dynamics, it is not surprising that spectacular racism has helped obscure the relatively smooth and devastating deregulation. This obscuring has occurred through numbers and "noise."

Numbers

Numbers refers to the frequency of events. We found 195 environmental and 354 racial events (Table 2). A first glance suggests that more resources were invested in racial rather than environmental events. There is an important distinction, however, between types of events: concrete actions versus rhetoric. Specifically, we found that concrete actions were more likely to be environmentally related, whereas rhetoric was more likely to be racist. These differences are further illuminated in Figures 1 and 2.

Figure 1 indicates that 68 percent of all concrete actions were environmentally related, compared to 32 percent that were racial. This divergence is even more pronounced in Figure 2, which shows that 92.5 percent of all discursive events were racial, whereas only 7.5 percent were environmental. Closer inspection of the environmental actions is revealing. Table 3 lists the range of environmental events. Topping the list are policy, appointments, and executive actions. These three categories account for almost 70 percent of all events and embody a broad neoliberal agenda. Policy actions included rollbacks from the Obama era, the Department of Interior's (DOI) "streamlining" of National Environmental Policy Act reviews, and reconsidering fuel economy standards. As many have documented, the easing of regulatory requirements is central to neoliberalism (Holifield 2004; Heynen, McCarthy, and Robbins 2007; Faber 2008; Himley 2008; Castree 2010). We were also interested in how deregulation was narrated, so we analyzed the justifications offered. The Trump administration was quite

Table 2. Frequency and kind of environmental and racial events: 20 January 2017 to 20 January 2018

Event	Frequency	Percentage
Environmental actions	173	88.72% of environmental events
Environmental discourse	22	11.28% of environmental events
Total	195	100.00%
Racial actions	83	23.45% of racial events
Racial discourse	271	76.55% of racial events
Total	354	100.00%

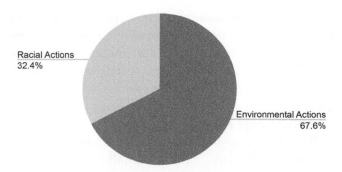

Figure 1. All actions during the first year of Trump's presidency (N = 256) broken into environmental and racial arenas.

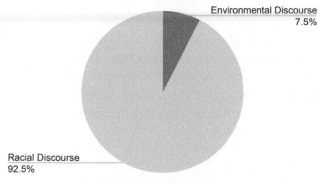

Figure 2. All discursive events during the first year of Trump's presidency (N = 293) broken into environmental and racial arenas.

Table 3. Mechanisms for environmental events: 20 January 2017 to 20 January 2018

Mechanism	Count	Percentage
Policy	59	29.44
Appointments	43	21.83
Executive	32	16.24
Discourse	22	11.17
Delays	16	9.64
Budget	9	4.57
Censorship	5	2.54
Legislative	5	2.54
Judiciary	3	1.52
Omission or denial	1	0.51
Total	195	100

Table 4. Justifications for environmental events: 20 January 2017 to 20 January 2018

Justification	Count	Percentage
Undetermined	57	29.23
Business interests or competitiveness	42	21.54
Efficiency	40	20.51
National interests or security	24	12.31
Energy	17	8.72
Science	8	4.10
Core mission	7	3.59
Total	195	100

frank in justifying its actions. Table 4 lists explanations given by the responsible party.

Significantly, there was no explanation for almost one third of all environmental actions. This silence registers as an absence of noise, which we explore later. The next two categories, which total just over 40 percent, are business interests or competitiveness and efficiency. Although it is sometimes necessary to decipher coded language, this is not the case. We believe that the justifications accurately reflect the Republican Party's stated priorities. Efficiency justified 20 percent of all environmental actions. Here, efficiency is an effort to make the regulatory state leaner and more agile. Efficiency portends to save taxpayers money and make regulation less burdensome for industry, but efficiency also embodies what Peck (2001, 447) called the "hollowing-out" of regulatory capacity. This was exemplified by thirty environmental nonappointments.

Appointments serve multiple purposes. Some, like Ben Carson, are random choices for departments that are not priorities. Others are carefully chosen for their experience in undermining regulatory agencies, such as the appointment of Nancy Beck to the

EPA's toxic chemicals unit. Beck came from the American Chemistry Council, where she made tracking toxins more difficult. In a major sweep, Pruitt barred scientists who received EPA grants from serving on the agency's advisory boards (Cornwall 2017). He justified his actions as creating a level playing field for industry. Indeed, between 2017 and 2018 the percentage of private consultants and industry scientists grew from 7 percent to 32 percent (Gustin 2018). These are textbook examples of the "polluter-industrial complex," which includes polluters capturing regulatory agencies (Faber 2008).

A third form of neoliberal governance is privatization and creating opportunities for industry. This was most pronounced in terms of fossil fuels. One of the first things Trump did was issue a Presidential Memorandum supporting the Keystone XL and Dakota Access pipelines. This was followed by expanding offshore drilling (Department of Interior Secretarial Order 3354 2017), and reviewing national monuments. The breadth, depth, and speed of environmental deregulation suggests that—unlike other parts of the Trump agenda, such as the Muslim ban, which was characterized by chaos and massive opposition—this wave has been relatively smooth and carefully crafted. One reason for this is because environmental governance has been envisioned as a sequential process. Consider that in February 2017 Pruitt was appointed EPA director. On 28 March Trump signed the Presidential Executive Order on Promoting Energy Independence and Economic Growth. On 29 March the DOI lifted the ban on coal mining. The administration explained, "Given the critical importance of the Federal coal leasing program to energy security, job creation, and proper conservation stewardship, this Order directs efforts to enhance and improve the Federal coal leasing program" (Secretary of the Interior, Order No. 3348). These events should be seen as linked. Despite having to resign over ethics violations, Pruitt was a skilled administrator who previously sued the EPA numerous times. Unlike other Trump appointees, he understood the regulatory bureaucracy and set to work.

Executive actions are also part of this sequencing. They provide a different but complementary way to enact policy, which is essentially circular: Executive actions drove agency agendas; Pruitt implemented changes, which were then justified by executive actions.

Returning to Table 4 and the justifications for environmental events, we see that business interests, efficiency, and national security comprised almost 50 percent of the total. These are core values codified in early executive orders. Hence, when the EPA limited power plants' toxic metal emissions in April 2017, the justification was that it was an "inordinate cost to industry."

The administration's environmental deregulation has been so exceptionally fast and smooth that it suggests some level of preplanning. Like the Federalist Society, which vetted judicial nominees for Trump (Savage 2017), conservative organizations like the American Legislative Exchange Council (ALEC) have served a similar function (Center for Media and Democracy 2018). ALEC has written probusiness legislation intended to proliferate and eventually reach the federal level. Although we do not claim that ALEC influenced the Trump administration, we assume that something comparable was at work.

In contrast, there were far fewer concrete racist actions than environmental ones. Table 2 indicates that less than 25 percent of all racial events were actions. Although there were many more racial events (354) than environmental ones (195), there were still ninety more environmental actions than racial ones. Of course, numbers do not indicate impact. The Muslim ban, for instance, has profoundly affected peoples' lives, as has the narrowing of asylum claims. It bears repeating: Racist rhetoric creates the conditions for extraordinary dehumanization and should never be dismissed.

Racial actions include seeking to dismantle the Deferred Action for Childhood Arrivals (DACA) legislation, building a southern U.S. border wall, not responding to a Congressional Black Caucus letter, and banning Muslims. Such efforts are intended to both insult or harm particular groups, while affirming the white nation. Likewise, the appointment of a disproportionate number of white men actively signals to whites that they are rightful owners of the nation (Hochschild 2016). Such appointments, given their conservative orientation, are then likely to prioritize the interests of the white nation and capital.

Other efforts seek to "exalt" the white nation (Thobani 2007). Trump withdrew funding for Life After Hate, an organization offering alternatives to white supremacy and redirected $400,000 toward fighting anti-Islamic extremism. Similarly, the Global War on Terrorism Memorial Act was passed while the National Park Service cut $98,000 for a project honoring the Black Panther Party. These actions exemplify Trump's refusal to criticize white supremacist violence while honoring the white nation. These actions must be seen as relational—there is no white nation without racial and colonized others.

Noise

Noise highlights the attention associated with specific events. Let us return to Figure 2, which shows all discursive events. The fact that environmental issues comprised less than 10 percent of all discursive events indicates that the Trump regime was more likely to be silent on environmental issues and relatively "noisy" on racial ones. With some exceptions, such as the Paris Accord, pipelines, and national monuments, most environmental actions have been unfolding in silence. As one student observed, "Trump is not tweeting about air." These silences are meaningful, especially in light of the number of environmental actions. When Trump does make environmental noise, such as his tweet in

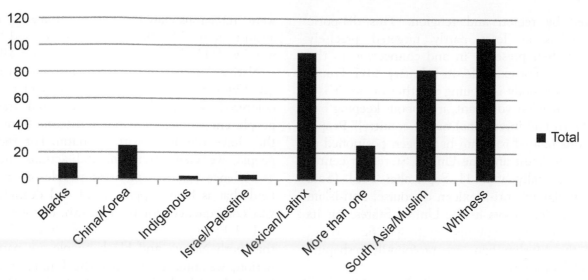

Figure 3. All racial tweets during the first year of Trump's presidency (N = 355) broken down by racial group.

Table 5. Number of discursive events by group: 20 January 2017 to 20 January 2018

Group	Count	Percentage
Mexican and Latinx	94	34.69
Muslim, South Asian, and Arab	87	32.10
International: China and North Korea and Israel and Palestine	34	12.55
Multiple groups	27	9.96
Black people	26	9.59
Native Americans	3	1.11
Total	271	100

Note. Multiple groups refers to events that addressed more than one targeted group.

November 2017, it is symbolic: "It is finally happening for our great clean coal miners!"

We categorized racist rhetoric by all discursive events and by tweets alone. On Twitter, whites were the most frequently referenced group (Figure 3). Despite Trump's attempt to equate the protesters in Charlottesville's Unite the Right Rally, most white-directed comments were subtle. The vast majority were references to "Make America Great Again," which we interpreted as hailing the white nation (Ngo 2017; Weems 2017).

Table 5 details how racist rhetoric was distributed by group. Mexicans and Latinxs led the way with ninety-four racist comments. This rhetoric was extremely noisy, as seen in the tweet data in Figure 3. Recall that Trump initiated his campaign with transgressive remarks about Mexicans and the 2018 midterm elections were marked by the vilification of Central Americans. Although these events lie outside the data set, they provide important context. Despite most of Trump's comments targeting

immigrants, it is impossible to conceive of Mexico and Mexican immigrants outside of the U.S. racial formation. As Rivera (2006) argued, Mexico has served as a racial and national other to the United States since the Mexican–American War, contributing to Mexicans being ideal fodder for Trump's spectacular racism. Because of the deep U.S. history of anti-Mexican racism, pejorative meanings are well established and pervasive. One need not "produce" new meanings or connections—one can simply draw on hegemonic anti-Mexican ideology and sentiment. Although Trump attacked Mexico and Mexican-Americans, he reserved most of his bile for immigrants, most of whom cannot vote. Indeed, Mexican immigrants are complex racial subjects, as they are seen as posing racial and national threats to the white nation (Huntington 2004; Chávez 2008).

Next is the category of Muslims, South Asians, and Arabs, with eighty-seven discursive events. As a transnational population, the rhetoric is less directed against any state but rather at multiple peoples connected by region and religion. This diasporic population is at least partly targeted precisely because of their presence in and connection to the United States. For example, in February 2017 Trump tweeted: "Everybody is arguing whether or not it is a BAN. Call it what you want, it is about keeping bad people (with bad intentions) out of country!" People and regions related to Islam have been positioned as others to the West and the United States for centuries but especially after 11 September 2001 (Said 2004). Similar to anti-Mexican discourse, anti-Islam rhetoric has deep roots in the United States, but its spectacular nature combined with specific actions, such as the Muslim ban, are profoundly reshaping the racial formation.

Significantly, black people and Native Americans were targeted far less. We used the term black people to describe Trump's racism because although most remarks refer to African Americans, some were aimed at black people in Africa and the Caribbean. We saw these as distinct from other international groupings because they were rarely directed toward another country but rather black people per se. Further, such racist rhetoric reflects a global blackness that is universally subordinate (Sharpe 2016). We were initially surprised by the relatively low levels of antiblack racism but attributed this to two things.[3] First, African Americans, despite their secondary status, are more accepted as part of the U.S. nation in that their racialization is not tied to an immigrant status. To be clear, black people are certainly not part of the white nation, but nationalism is a driving force behind Trump's racism. Many Americans recognize, however begrudgingly, that African Americans are part of U.S. history through slavery, even if they refuse to acknowledge its afterlife that shapes contemporary black experiences (Hartman 2008; Sharpe 2016).

Black peoples' position as the ultimate racial other within the U.S. racial formation might also influence their treatment by Trump. Because they are considered the leaders of the civil rights movement and have influenced how the United States frames race, African Americans serve as a litmus test for what constitutes racism. Thus, whereas it has been acceptable to vilify Mexicans and Muslims, it is more transgressive to attack African Americans. Indeed, Trump has a long history of antiblack racism, but he limited his explicit antiblack rhetoric during his first year in office. Nonetheless, there were numerous law-and-order comments directed against groups like Black Lives Matter and "Twitter wars" with black athletes protesting police killings.

Native Americans, who were also infrequently targeted by racist rhetoric, occupy a distinct position as colonized people. Although black and Indigenous people encountered less racist rhetoric from Trump, they have not been without harm. For Indigenous people, we view environmental actions themselves as violence against colonized people, as it is their land that is being appropriated and degraded, with vast consequences for their health, cultures, nations, and ability to survive. Although it can be argued that both Native and black people are part of the nation, we must recall that the United States was forged through slavery and colonization. Thus, their exclusion, domination, and eradication were central to the formation of the United States (Harris 1993; Smith 2012). The fact that Mexicans and Latinx and Muslims and Arabs are currently more targeted by Trump's racism reflects the complexity and fluidity of the U.S. racial formation (Bonilla-Silva 2004). Alternatively, attacking Latinxs might prepare the path for more concentrated assaults on black people.

Multiple racial groups were also targeted simultaneously, at a frequency comparable to that of black and Native people. In July Trump tweeted, "At some point the Fake News will be forced to discuss our great jobs numbers, strong economy, success with ISIS, the border & so much else!" Here, he brings together ISIS (Islam) and the border (Mexicans), while emphasizing national security and affirming the white Christian nation.

Although we have distinguished between rhetoric and actions, we are cognizant of how the two inform each other. Consider Trump's proposed border wall. Clearly, the wall maintains the borders of the white nation both metaphorically and literally. It nurtures Trump's base, thereby securing continued support. Repeated reference to the border wall has also pushed the realm of political possibility to the right. In the winter of 2018 the fate of unaccompanied minors, DACA recipients, was held hostage to Trump's desire for the wall. Finally, such tactics, both spectacular racism and using DACA recipients as bargaining chips, further erode the norms and humanity that are essential to a functioning democracy. As such, Trump's relentless racist rhetoric normalizes a spectacularly racist policy agenda and obscures vast environmental deregulation.

Concluding Thoughts

We have explored the patterns between racist and environmental events during the Trump administration's first year. During this period, environmental issues were far more likely to be concrete actions, whereas racism was more likely to be rhetorical. Accordingly, we suggest that spectacular racism, regardless of intent, obscures environmental deregulation.

Perhaps more significant are the larger implications for the racial formation and environmental governance, which we can only gesture to. The ascent of the white nation contributes to the abandonment of neoliberal multiculturalism and environmental protections, as capital and whiteness are protected. This raises several questions regarding the relationship between the two, including why millions of the white working class would vote against their material interests. Such a question assumes, of course, that voters are rationale beings, when, in fact, the evidence suggests that people supported Trump for emotional reasons (Hochschild 2016; Koch 2018), whether fear, anger, or hope.

If much of the vote was emotionally driven, is environmental deregulation simply collateral damage? Hochschild (2016), in her study of Louisiana Tea Party members, examined the relationship between conservative voters and environmental regulation and painted a more complex picture. She identified two reasons why whites resented the government, even amidst severe pollution. First, whites were deeply resentful because they felt displaced by racial others who had "cut in line," through affirmative action, equal rights, and immigration, and, second, they did not believe that regulations served them (Hochschild 2016, 137). Instead, they believed that regulators targeted small businesses and individuals, which could not readily afford compliance, whereas big corporations were free to pollute. It is true that larger polluters are lightly regulated (Pellow 2004; Sze 2006; Faber 2008; Pulido, Kohl, and Cotton 2016; Dillon 2018), but Trump argued that deregulation would "Make America Great Again." Hence, deregulation became a tool to build both the white nation and economic prosperity. Here, economic populism resonated and intersected with the grievances of the white nation.

Both our data and Hochschild's (2016) findings seem to affirm Inwood's (2018) contention that white supremacy impedes progressive change—albeit at two different scales—the national and the local.

Clearly, more research is necessary to clarify these dynamics. What is evident, however, is that overt white supremacy can never be isolated from other spheres of society, including seemingly unrelated ones. Its regressive nature can only be challenged by drawing on such formations as the BRT (Robinson 2000), of which the environmental justice, antiracist, and even mainstream environmental movements are a part.

Acknowledgments

We gratefully acknowledge the research assistance of Cristina Zepeda-Yanez and Aakash Upraity. Versions of this article were presented at the University of Uppsala and Willamette University, where we received very helpful feedback. We alone remain responsible for all shortcomings.

Notes

1. For different articulations of the white nation, see Belew (2018).
2. The students were Tianna Bruno, Bob Craven, Fiona de los Rios-McCucheon, Shiloh Deitz, Cristina Faiver-Serna, Lisa Fink, Cassandra Galentine, Theodore Godfrey, Ben Hinde, Nick Machuca, Katya Reyna, Derek Robinson, Kate Shields, Michael Skaja, Aakash Upraity, Adriana Uscanga Castillo, Shianne Walker, Claire Williams, Olivia Wing, Sara Worl, Jordan Wyant, Cristina Zepeda-Yanez, Holly Moulton, and Natalie Mosman.
3. For evidence on the continued toll that racism takes on African Americans, see Levine et al. (2001), Phelan and Link (2015), and Cunningham et al. (2017).

References

Anderson, B. 1983. *Imagined communities: Reflections on the origins and spread of nationalism.* London: Verso.

Baptist, E. E. 2016. *The half has never been told: Slavery and the making of American capitalism.* New York: Basic Books.

Belew, K. 2018. *Bring the war home: The white power movement and paramilitary America.* Cambridge, MA: Harvard University Press.

Bonds, A., and J. Inwood. 2016. Beyond white privilege: Geographies of white supremacy and settler colonialism. *Progress in Human Geography* 40 (6):715–33. doi:10.1177/0309132515613166.

Bonilla-Silva, E. 2004. From bi-racial to tri-racial: Towards a new system of racial stratification in the USA. *Ethnic and Racial Studies* 27 (6):931–50. doi:10.1080/0141987042000268530.

Castree, N. 2010. Neoliberalism and the biophysical environment 1: What "neoliberalism" is, and what difference nature makes to it. *Geography Compass* 4 (12):1726–33. doi:10.3167/ares2010010102.

Center for Media and Democracy. 2018. ALEC exposed. Accessed July 20, 2018. http://www.alecexposed.org/wiki/ALEC_Exposed.

Chait, J. 2018. The anti-Trump right has become Trump's base. *New York Magazine*, July 3. Accessed July 18, 2018. http://nymag.com/intelligencer/2018/07/anti-trump-conservatives-have-become-trumps-base.html?gtm=top&gtm=bottom.

Chávez, L. 2008. *The Latino threat.* Stanford, CA: Stanford University Press.

Cornwall, W. 2017. Trump's EPA has blocked agency grantees from serving on science advisory panels. Here is what it means. *Science*, October 31. Accessed March 30, 2018. http://www.sciencemag.org/news/2017/10/trump-s-epa-has-blocked-agency-grantees-serving-science-advisory-panels-here-what-it.

Cunningham, T. J., J. B. Croft, Y. Liu, H. Lu, P. I. Eke, and W. H. Giles. 2017. Vital signs: Racial disparities in age-specific mortality among blacks or African Americans—United States, 1999–2015. *Morbidity and Mortality Weekly Report* 66 (17):444–56. doi:10.15585mmwrmm6617e1.

Department of Interior. 2017. Supporting and improving the federal onshore oil and gas leasing program and federal solid mineral leasing program. Secretarial Order 3354. July 6. Accessed July 18, 2018. https://www.doi.gov/sites/doi.gov/files/uploads/doi-so-3354.pdf.

Dillon, L. 2018. The breathers of Bayview Hill: Redevelopment and environmental justice in southeast San Francisco. *Hastings Environmental Law Journal* 24 (2):227–36.

Du Bois, W. E. B. 2014. [2014]. *Black reconstruction in America.* New York: Oxford University Press.

Faber, D. 2008. *Capitalizing on environmental injustice: The polluter-industrial complex in the age of globalization.* Lanham, MD: Rowman & Littlefield.

Frum, D. 2017. How to build an autocracy. *The Atlantic*, January 30. Accessed March 20, 2018. https://www.theatlantic.com/press-releases/archive/2017/01/how-to-build-an-autocracy-the-atlantics-march-cover-story-online-now/515115/.

Full transcript and video: Trump's news conference in New York. 2017. *New York Times*, August 15. Accessed February 3, 2019. https://www.nytimes.com/2017/08/15/us/politics/trump-press-conference-transcript.html?action=click&module=RelatedCoverage&pgtype=Article®ion=Footer

Gilmore, R. W. 1999. Globalisation and U.S. prison growth: From military Keynesianism to post-Keynesian militarism. *Race & Class* 40 (2–3):171–88. doi:10.1177/030639689904000212.

———. 2002. Fatal couplings of power and difference: Notes on racism and geography. *The Professional Geographer* 54 (1):15–24.

———. 2017. Abolition geography and the problem of innocence. In *Futures of black radicalism*, ed. T. G. Johnson and A. Lubin, 225–40. New York: Verso.

Gokariksel, B., and S. Smith. 2018. Tiny hands, tiki torches: Embodied white male supremacy and its politics of exclusion. *Political Geography* 62:209–11. doi:10.1016/polgeo201710010.

Gonzales, R. 2018. Trump says he'll send as many as 15,000 troops to the southern border. National Public Radio, November 1. Accessed November 5, 2018. http://www.npr.org/2018/10/31/662735242/trump-says-hell-send-as-many-as-15-000-troops-to-the-southern-border.

Graham, D. 2017. Trump's shrinking, energized base. *The Atlantic*, September 8. Accessed March 30, 2018. https://www.theatlantic.com/politics/archive/2017/09/trumps-shrinking-impassioned-base/539160/.

Gustin, G. 2018. Trump administration deserts science advisory boards across agencies. *Inside Climate News*, January 18. Accessed November 6, 2018. https://insideclimatenews.org/news/18012018/science-climate-change-advisory-board-epa-interior-trump-administration.

Hajnal, Z., and M. U. Rivera. 2014. Immigration, Latinos, and white partisan politics: The new democratic defection. *American Journal of Political Science* 58 (4):773–89. doi:10.1111/ajps12101.

Harris, C. 1993. Whiteness as property. *Harvard Law Review* 106 (8):1710–91.

Hartman, S. 2008. *Lose your mother: A journey along the Atlantic slave route.* New York: Farrar, Straus and Giroux.

Hetherington, M., and E. Suhay. 2011. Authoritarianism, threat, and Americans' support for the war on terror. *American Journal of Political Science* 55 (3):546–60. doi:10.1111/15405907201100514.

Heynen, N. 2016. Urban political ecology II: The abolitionist century. *Progress in Human Geography* 40 (6):839–45. doi:10.1177/0309132515617394.

Heynen, N., J. McCarthy, and P. Robbins. 2007. *Neoliberal environments: False promises and unnatural consequences.* London and New York: Routledge.

Himley, M. 2008. Geographies of environmental governance: The nexus of nature and neoliberalism. *Geography Compass* 2 (2):433–51. doi:10.1111/17498198200800094.

Hochschild, A. 2016. *Strangers in their own land.* New York: New Press.

Holifield, R. 2004. Neoliberalism and environmental justice in the United States Environmental Protection Agency: Translating policy into managerial practice in hazardous waste remediation. *Geoforum* 35 (3):285–97. doi:10.1016/geoforum200311003.

———. 2012. Environmental justice as recognition and participation in risk assessment: Negotiating and translating health risk at a superfund site in Indian country. *Annals of the American Association of Geographers* 102 (3):591–613. doi:10.1080/000456082011641892.

Huntington, S. 2004. *Who are we? The challenges to America's national identity.* New York: Simon & Schuster.

Inwood, J. 2015. Neoliberal racism: The "Southern strategy" and expanding geographies of white supremacy. *Social & Cultural Geography* 16 (4):407–23. doi:10.1080/146493652014994670.

———. 2018. White supremacy, white counter-revolutionary politics, and the rise of Donald Trump. *Environment and Planning C: Politics and Space.* Advance online publication. doi:10.1177/2399654418789949.

Inwood, J., and A. Bonds. 2016. Confronting white supremacy and a militaristic pedagogy in the U.S. settler colonial state. *Annals of the Association of American Geographers* 106 (3):521–29. doi:10.1080/2469445220161145510.

Isenstadt, A. 2017. Trump purges enemies and reshapes party in his image. *Politico*, October 24. Accessed March 30, 2018. https://www.politico.com/story/2017/10/24/trump-republicans-corker-flake-purge-244139.

Jacobson, G. C. 2017. The triumph of polarized partisanship in 2016: Donald Trump's improbable victory. *Political Science Quarterly* 132 (1):9–41. doi:10.1002/polq12572.

Johnson, C., and A. Coleman. 2012. The internal other: Exploring the dialectical relationship between religious exclusion and the construction of national identity. *Annals of the Association of American Geographers* 102 (4):863–80. doi:10.1080/0004560820411602934.

Johnson, G. T., and A. Lubin. 2017. Introduction. In *Futures of black radicalism*, ed. G. T. Johnson and A. Lubin, 9–18. London: Verso.

Kobayashi, A., and L. Peake. 2000. Racism out of place: Thoughts on whiteness and an antiracist geography in the new millennium. *Annals of the Association of American Geographers* 90 (2):392–403. doi:10.1111/0004560800202.

Koch, N. 2017. Orientalizing authoritarianism: Narrating U.S. exceptionalism in popular reactions to the Trump election and presidency. *Political Geography* 58:145–47. doi:10.1016/polgeo201703001.

———. 2018. Trump one year later: Three myths of liberalism exposed. *Political Geography* 62:212–14. doi:10.1016/polgeo201710010.

Levine R. S., J. E. Foster, R. E. Fullilove, M. T. Fullilove, N. C. Briggs, P. C. Hull, B. A. Husaini, and C. H. Hennekens. 2001. Black–white inequalities in mortality and life expectancy, 1933–1999. Implications for healthy people 2010. *Public Health Reports* 116 (5):474–83. doi:10.1093/phr/116.5.474

Levitsky, S., and D. Ziblatt. 2018. *How democracies die.* New York: Crown.

MacWilliams, M. C. 2016. Who decides when the party doesn't? Authoritarian voters and the rise of Donald Trump. *PS: Political Science & Politics* 49 (4):716–21. doi:10.1017/S1049096516001463.

Malone, C., H. Enten, and D. Nield. 2016. The end of a republican party. Accessed March 30, 2018. https://fivethirtyeight.com/features/the-end-of-a-republican-party/.

McElwee, S., and J. McDaniel. 2017. Economic anxiety didn't make people vote for Trump, racism did. *The Nation*, May 8. Accessed March 20, 2018. https://www.thenation.com/article/economic-anxiety-didnt-make-people-vote-trump-racism-did/.

Melamed, J. 2011. *Represent and destroy: Rationalizing violence in the new racial capitalism.* Minneapolis: University of Minnesota Press.

Moreton-Robinson, A. 2016. *The white possessive: Property, power and indigenous sovereignty.* Minneapolis: University of Minnesota Press.

Newkirk, V. R., II. 2018. Trump's EPA concludes environmental racism is real. *The Atlantic*, February 28. Accessed March 30, 2018. https://www.theatlantic.com/politics/archive/2018/02/the-trump-administration-finds-that-environmental-racism-is-real/554315/.

Ngo, B. 2017. Immigrant education against the backdrop of "Make America great again." *Educational Studies* 53 (5):429–32. doi:10.1080/00131946.2017.1355800.

Nixon, R. 2011. *Slow violence and the environmentalism of the poor.* Cambridge, MA: Harvard University Press.

Omi, M., and H. Winant. 1994. *Racial formation in the United States: From the 1960s to the 1990s.* London and New York: Routledge.

Page, S., and J. Dittmer. 2018. Mea culpa. *Political Geography* 62:207–9. doi:10.1016/polgeo201710010.

Peck, J. 2001. Neoliberalizing states: Thin policies/hard outcomes. *Progress in Human Geography* 25 (3):445–55.

Pellow, D. N. 2004. *Garbage wars: The struggle for environmental justice in Chicago.* Cambridge, MA: The MIT Press.

———. 2017. *What is critical environmental justice?* Cambridge, UK: Polity.

Phelan J. C., and B. G. Link. 2015. Is racism a fundamental cause of inequalities in health? *Annual Review of Sociology* 41:311–30. doi:10.11146/annurev-soc-073014-112305.

Pulido, L. 2000. Rethinking environmental racism: White privilege and urban development in Southern California. *Annals of the Association of American Geographers* 90 (1):12–40.

———. 2015. Geographies of race and ethnicity I: White supremacy vs. white privilege in environmental racism research. *Progress in Human Geography* 39 (6):809–17. doi:10.1177/0309132514563008.

———. 2016. Flint Michigan, environmental racism and racial capitalism. *Capitalism Nature Socialism* 27 (3):1–16. doi:10.1080/1045575220161213013.

———. 2017. Geographies of race and ethnicity II: Environmental racism, racial capitalism and state-sanctioned violence. *Progress in Human Geography* 41 (4):524–33. doi:10.1177/0309132516646495.

———. 2018. Geographies of race and ethnicity III: Settler colonialism and nonnative people of color. *Progress in Human Geography* 42 (2):309–18. doi:10.1177/0309132516686011.

Pulido, L., E. Kohl, and N. Cotton. 2016. State regulation and environmental justice: The need for strategy reassessment. *Capital Nature Socialism* 27 (2):12–31. doi:10.1080/10455752.2016.1146782.

Rivera, J. 2006. *The emergence of Mexican America: Recovering stories of Mexican peoplehood in U.S. culture.* New York: New York University Press.

Robinson, C. J. 2000. *Black Marxism: The making of the black radical tradition.* 2nd ed. Chapel Hill: University of North Carolina Press.

Said, E. 2004. Orientalism once more. *Development and Change* 35 (5):869–79. doi:10.1111/146776602 00400383.

Savage, C. 2017. Trump is rapidly changing the judiciary. Here's how. *New York Times*, November 11. Accessed March 30, 2018. https://www.nytimes.com/2017/11/11/us/politics/trump-judiciary-appeals-courts-conservatives.html.

Schlosberg, D., and D. Carruthers. 2010. Indigenous struggles, environmental justice, and community capabilities. *Global Environmental Politics* 10 (4): 12–35. doi:10.1162/GLEPa00029.

Secretary of the Interior, Order No. 3348. Concerning the federal coal moratorium. Accessed February 3, 2019. https://www.doi.gov/sites/doi.gov/files/uploads/so_3348_coal_moratorium.pdf

Sharpe, C. 2016. *In the wake: On blackness and being.* Durham, NC: Duke University Press.

Smith, A. 2012. Indigeneity, settler colonialism, White supremacy. In *Racial formation in the twenty-first century*, ed. D. Martinez Hosang, O. LaBennett, and L. Pulido, 66–90. Berkeley and Los Angeles: University of California Press.

Sze, J. 2006. *Noxious New York: The racial politics of urban health and environmental justice.* Cambridge, MA: The MIT Press.

Taub, A. 2016. The rise of American authoritarianism. Accessed March 30, 2018. https://www.vox.com/2016/3/1/11127424/trump-authoritarianism.

Tesler, M. 2016. *Post-racial or most racial? Race and politics in the Obama era.* Chicago: University of Chicago Press.

Thobani, S. 2007. *Exalted subjects.* Toronto: University of Toronto.

Thompson, D. 2018. Donald Trump's language is reshaping American politics. *The Atlantic*, February 15. Accessed March 30, 2018. https://www.theatlantic.com/politics/archive/2018/02/donald-trumps-language-is-reshaping-american-politics/553349/.

Vanderbeck, R. 2006. Vermont and the imaginative geographies of American whiteness. *Annals of the Association of American Geographers* 96 (3):641–59. doi:10.1111/14678306200600710.

Weems, M. E. 2017. Make America great again? *Qualitative Inquiry* 23 (2):168–70. doi:10.1177/1077800416674752.

Wilson, B. M. 2000. *America's Johannesburg: Industrialization and racial transformation in Birmingham.* Lanham, MD: Rowman & Littlefield.

Woods, C. 2017. *Development drowned and reborn*, ed. L. Pulido and J. Camp. Athens: University of Georgia Press.

Wright, M. 2015. *Physics of blackness: Beyond the middle passage epistemology.* Minneapolis: University of Minnesota.

Yen-Kohl, E., and The Newtown Florist Club Writing Collective. 2016. "We've been studied to death, we ain't gotten anything": (Re) claiming environmental knowledge production through the praxis of writing collectives. *Capitalism Nature Socialism* 27 (1):52–67. doi:10.1080/1045575220151104705.

Zeskind, L. 2012. A nation dispossessed: The Tea Party movement and race. *Critical Sociology* 38 (4):495–509. doi:10.1177/0896920511431852.

LAURA PULIDO is Professor of Ethnic Studies and Geography at the University of Oregon, Eugene, OR 97403. E-mail: lpulido@uoregon.edu. Her research interests include race, comparative ethnic studies, environmental justice, and critical human geography.

TIANNA BRUNO is a Doctoral Student in the Geography Department at the University of Oregon, Eugene, OR 97403. E-mail: tbruno@uoregon.edu. Her research interests include critical environmental justice, black geographies, political ecology, and critical physical geography.

CRISTINA FAIVER-SERNA is a Doctoral Candidate in the Geography Department at the University of Oregon, Eugene, OR 97403. E-mail: cfaiver@uoregon.edu. Her research interests include critical environmental studies, critical race studies, science and technology studies, Latinx geographies, and women of color feminist theory and praxis.

CASSANDRA GALENTINE is a Doctoral Student in the English Department at the University of Oregon, Eugene, OR 97403. E-mail: cassieg@uoregon.edu. Her research interests include working-class literature, critical environmental justice studies, and new materialism.

Reaction, Resilience, and the Trumpist Behemoth: Environmental Risk Management from "Hoax" to Technique of Domination

Matthew Sparke and Daniel Bessner

The election of Donald Trump to the U.S. presidency has led to significant changes in environmental governance, unleashing an authoritarian, nationalistic, and business-deregulating juggernaut aimed at destroying various forms of environmental protection. We seek to name and explain this juggernaut as the Trumpist Behemoth. Using this terminology, we show how Neumann's classic 1942 study of the Nazi Behemoth can be used to build a critique of the Trump administration's approach to governance. We join this with the contemporary critique of Climate Leviathan by Mann and Wainwright along with diverse critical literatures on resilience to argue that the Trumpist Behemoth is further distinguished by its anti-Leviathan reactionary appropriation of the politics and practices of resilience. This creates a regime that retains certain neoliberal commitments to market rule but rearticulates and reterritorializes them nationalistically. Connecting business interests with a border-building vision of "America First," it simultaneously reterritorializes nature as national in ways that obscure the global ecosystems and contradictions of the Anthropocene cum capitalocene out of which the Trumpist Behemoth has been birthed.

唐纳德．特朗普当选美国总统，带来了环境治理的显着改变，释放了一股威权主义、国族主义且商业去管制的毁灭力量，旨在摧毁各种形式的环境保护。我们企图将此般毁灭力量命名为"川普巨兽"（Trumpist Behemoth）并对此进行解释。我们运用此一术语，展现诺伊曼1942年对纳粹巨兽的经典研究，如何能够用来发展对特朗普政权治理方式的批判。此外，我们加入曼恩与温赖特对于气候利维坦的当代批判，以及有关恢复力的多样批判文献，以此主张特朗普巨兽进一步由其反利维坦式的反动政治与恢复力实践进行辨识。它创造了保有若干新自由主义对市场规则的承诺，但却以国族主义的方式重新接合并将之再领域化的政体。它连结了商业利益与"美国优先"的边境打造计画，同时以掩盖川普巨兽从中诞生的全球生态系统和人类世与资本世的矛盾之方式，将自然再领域化为国族的。 关键词：威权主义，气候巨兽，绿色新自由主义，反动，恢复力。

La elección de Donald Trump a la presidencia de los EE.UU. ha producido cambios significativos en la gobernanza ambiental, desatando una endemoniada fuerza autoritaria, nacionalista y comercialmente caótica, que enfila a la destrucción de varias formas de protección ambiental. Buscamos nombrar y explicar tan descomunal despropósito como el Monstruo Trumpista. Usando esta terminología, indicamos cómo el clásico estudio de 1942 de Neumann sobre el Monstruo Nazista puede utilizarse para elaborar una crítica al enfoque de gobernanza de la Administración Trump. Unimos esto a la crítica contemporánea del Leviatán Climático de Mann y Wainwright, junto con diversas literaturas críticas sobre resiliencia, para argüir que el Monstruo Trumpista se distingue principalmente por su apropiación reaccionaria anti-Leviatán de las políticas y prácticas de la resiliencia. Esto genera un régimen que retiene ciertos compromisos neoliberales con el poder del mercado, aunque los rearticula y reterritorializa con criterio nacionalista. Al conectar los intereses de los negocios con la visión constructora de talanqueras de "América primero", simultáneamente reterritorializa la naturaleza como bien nacional de formas que oscurecen los ecosistemas y contradicciones globales del Antropoceno, o Capitaloceno, de donde nació el Monstruo Trumpista. *Palabras clave: autoritarismo, Monstruo climático, neoliberalismo verde, reacción, resiliencia.*

We should be focused on clean and beautiful air—not expensive and business closing GLOBAL WARMING—a total hoax!

—Donald Trump (2013)

National Socialism has no political theory of its own, and … the ideologies it uses or discards are mere *arcana dominationis*, techniques of domination.

—Franz Neumann (1942, 459)

How can we best come to terms with the impacts of the Trump administration on environmental governance? In this article, we argue that theories of authoritarian domination under what Neumann (1942) once described as the Nazi Behemoth can be adapted to theorize the reactionary reworking of environmental policymaking by a new Trumpist Behemoth. Although we do not believe that Donald J. Trump himself is literally a neo-Nazi, we suggest that theorizing his administration as a Behemoth highlights underlying tendencies that are often overlooked in the numbing 24/7 focus on Trump's tweets, impulsive personality, and chaotic administration. Ultimately, we argue in this way that a defining feature of the Trumpist Behemoth is an approach to the environment that selectively reuses and reterritorializes Obama-era resilience thinking. This damaging approach, we want to underline, illustrates especially clearly the neoliberal hegemony and political quiescence that many critics have already identified in appeals to resilience (MacKinnon and Derickson 2013; Neocleous 2013; Sparke 2013; Nelson 2014; Watts 2015; Leitner et al. 2018; McKeown and Glenn 2018; Swyngedouw and Ernstson 2018). It also takes the associated trends in disaster capitalism, including the dispossession of those long disenfranchised by colonialism and racism, in still more devastating directions. Although some of the most destructive of Trump's plans were stalled during the start of his tenure by constitutional checks and organized resistance (Bomber 2017), we anticipate that the rest of his time in office will nevertheless enable extractive and polluting industries to expand their dominance at the expense of sustainable human–environment relations. The end result of these dynamics still remains to be seen, but as resilience rhetoric is incorporated into the Trump administration's *arcana dominationis*, we offer this provisional account of the Trumpist Behemoth as part of the effort to theorize its authoritarian approach to environmental governance.

Our article builds on an essay we published early in Trump's presidency in which we suggested that the new administration's underlying governmental contradictions could be understood in terms of "a monstrous merging of Nazi and neoliberal tendencies" (Bessner and Sparke 2017b, 1214). Here we seek to develop this argument further by drawing on an additional series of theories about both Nazism and the neoliberalization of environmental governance. We therefore remain interested in the contemporary remixing of Nazism and neoliberalism (Giroux 2018), but our specific focus here is instead on how the mixed-up monster progeny that results—the Trumpist Behemoth—is defined in its governmental effects by a distinctly reactionary response to "green neoliberalism" and by the resulting reworking and reterritorializaton of resilience. This means more than just emphasizing the point that "resilience thinking tends to be reactive in nature" (McKeown and Glenn 2018, 205). What we see the Trumpist Behemoth as doing differently involves making this reactive tendency truly reactionary by turning the more inclusive long-term disaster management approaches of green neoliberalism into territorially and racially exclusionary innovations in disaster capitalism.

Our use of the term *green neoliberalism* is inspired by earlier critical geographical explorations of its many variations (McCarthy and Prudham 2004; Bakker 2010; Watts 2015). We are interested in turn in how the Trumpist Behemoth's identity politics relates to green neoliberalism's signature emphasis on resilience. One important inspiration in this regard is Fraser's (2017) argument about the way in which the emphasis on recognition (rather than redistribution) in Hillary Clinton's progressive neoliberalism became the target for Trump's own reactionary brand of recognition concerned with the resentments of working-class whites. We see the politics of environmental resilience in green neoliberalism playing an analogous role in creating a kind of takeover target for Trump. We want to suggest in this way that the identity politics of this takeover is racist as well as hypernationalist, illustrating anew the racist logics that postcolonial scholars have already argued lie latent in assumptions about insurability and security in geopolitical renditions of resilience thinking more generally (Chaturvedi and Doyle 2015; Baldwin 2016; Bracke 2016). Our point is that the Trumpist Behemoth is effectively surfacing these racist logics, connecting them to capitalist concerns with accumulation by dispossession that are thereby increasingly imagined in fascistic terms. Although the administration is undoubtedly bent on increasing the business opportunities of capitalist elites and their rights to grab land and resources, it also therefore seems to be seeking to secure wealthier and whiter subpopulations at the expense of others deemed foreign.

For all of these reasons, resilience thinking plays a contradictory role for the Trumpist Behemoth. On the one side it appears as a series of Obama-era policymaking commitments to climate change adaptation that have been targeted for rollback and defeat. On the other side, though, it is reactivated as a national security code word for managing environmental crises selectively and preferentially, radically reterritorializing the imagined community of resilience while also refusing any acknowledgment of the causes of the environmental crises for which this exceptional national community's resilience is to be prepared. The administration has thereby defined itself against Obama-era resilience understood as environmentally conscious risk management that acknowledges global climate change, prepares communities to endure disaster, and envisions forms of business sustainability that might manage the negative environmental externalities of global capitalism. Instead of this framework, Trumpian resilience is distinguished by a redlining of risk, its protection of privileged risk managers, and its class- and race-based abandonment of the more vulnerable.

The evidence of all of these tendencies is widespread and we can only profile a small set of examples in the space provided here. A key conclusion that we nevertheless want to defend is that the result involves a contradictory combination of calculation and candor in relation to environmental governance. Thus, alongside the Behemoth's basic bureaucratic work of deregulating drilling, mining, logging, and other environmentally damaging industries, there is the concurrent—albeit inadvertent—acknowledgment that a real alternative to global environmental catastrophe demands fundamental changes to global capitalism and real controls on privileged exploiters of the environment (Foster 2017b). Clearly the administration has no desire to restructure global capitalism, unless one counts its ad hoc experiments in imposing tariffs on trade (Bessner and Sparke 2017a). Nonetheless, by adding resilience to the *arcana dominationis* of a fascistic ruling class, the Trumpist Behemoth makes manifest the power relations of the neoliberal Anthropocene (as what critics have variously renamed the Capitalocene, Plantationocene, and Anthropo-obscene) in a way that the green neoliberalism of an earlier era tended to green-wash and depoliticize (Perkins 2009; Vergès 2017; Swyngedouw and Ernstson 2018).

In the end we are suggesting that the Trumpist Behemoth is both attacking and assimilating green neoliberalism in the course of reproducing neoliberal rule for and by privileged elites. By defending this unsustainable environmental agenda in the name of authoritarian nationalism, of U.S. global energy dominance, and of freedom for domestic fossil fuel extraction and pollution, the administration's reactionary tendencies have led to the reuse of resilience rhetoric as a technique of domination. We do not mean to suggest that the nationalist authoritarianism of the Trumpist Behemoth is the only reason its environmental agenda is so damaging. Traditional business interests dedicated to deregulated environmental exploitation remain an overwhelming influence in this regard. To the extent that its authoritarian impulses and arguments have enabled the Trumpist Behemoth to increase the deregulation of industry in the name of national freedom, however, and to the extent that this has involved a reactionary and thus identitarian reworking of resilience, it has also highlighted how ineffectual green neoliberalism is in the face of capitalist interests that are simultaneously allied to authoritarianism.

To conceptualize how the Trumpist Behemoth has reclaimed resilience as an authoritarian technique of domination, we draw on three interventions in the political theory of governance: Neumann's *Behemoth*, the classic 1942 critique of the structure and practice of national socialism; the contemporary arguments about *Climate Leviathan* made by Mann and Wainwright (Wainwright and Mann 2013, 2015; Mann and Wainwright 2018); and the many recent critical writings on resilience cited earlier. Combining these diverse theoretical approaches enables us to suggest that the Trumpist Behemoth is creating a regime of environmental governance that is committed to corporate profitability and elite insurability at the same time as it imposes blame and disposability on everybody else.

Materialism and Expansionism from the Nazi to the Trumpist Behemoth

Franz Neumann (1900–1954), a social democratic lawyer and political theorist forced to flee Nazi

Germany after Adolf Hitler assumed power in 1933, provided in his book *Behemoth* one of the most comprehensive accounts of national socialist governance. The book addresses in detail Nazism's authoritarian elevation of the *Führer*, its corporatist ties to industrialists, its construction of a monopolistic economy, its militaristic pursuit of racial imperialism, and its racialized assumptions about nation, blood, and belonging in a *Grossdeutsche Reich*. In the next section, we explore how Neumann's analyses of Nazism must be adapted to come to terms with the Trumpist Behemoth. First, though, we here pursue the questions of political sovereignty and state-making that are central to Neumann's account of Nazism and that offer an entrée into the tensional space of national sovereignty versus global sovereignty that have been taken up by Mann and Wainwright in their account of Climate Leviathan.

Although the racial and religious geopolitics of the Trump regime are undoubtedly different from those of the Nazis, we believe that there are some important political geographical arguments in Neumann's analysis of the *Grossdeutsche Reich* that help us understand the implications of the Trumpist Behemoth. Specifically, through a critique of Schmitt's geostrategic discourse, Neumann offered an analysis of Nazi expansionism that provides insights into Trumpist geopolitics, especially the administration's declared national security strategy of global energy dominance. In *Behemoth*, Neumann argued that Schmitt's justification for Nazi expansionism was at base about what we would now call—following David Harvey—the need for a spatial fix for German capital. He thereby summarized Schmitt as arguing that "[l]arge-space economics, precedes large-space politics" (Neumann 1942, 156–57). Neumann argued in this way that Schmitt's justifications for German *Lebensraum* were premised on the economic interests of German industry.

Neumann's materialist critique of Schmitt connects in turn to Wainwright and Mann's critical political–economic arguments. They also focus on the economic tendencies pushing state-making in transnational and expansionist directions, thereby producing the effects of a "planetary sovereign" that they termed, reworking Schmitt, Climate Leviathan. Unlike Schmitt's own authoritarian investment in *Lebensraum*, however, and more in the spirit of Neumann's critique, Mann and Wainwright argue that such a sovereign is likely to emerge when the

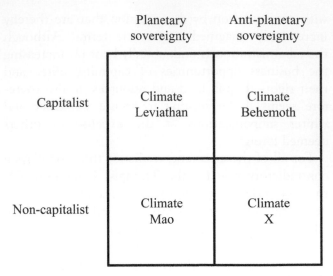

Figure 1. Four potential political futures. *Source:* Wainwright and Mann (2013); reproduced with permission.

urgency of "climate-induced disruptions of accumulation and political stability" force "the dominant capitalist nation-states" to establish a power structure able to manage an increasingly interrupted global capitalist system (Wainwright and Mann 2013, 13). Under Climate Leviathan, Wainwright and Mann maintain, capitalism itself comes to be seen as the best means to end climate change, inevitably giving rise to new varieties of green neoliberalism. Just as Neumann connect the expansion of the Nazi Behemoth to capital's spatial expansionism, Wainwright and Mann connected the contemporary transnational expansion of sovereignty to these very same tendencies.

Nevertheless, Wainwright and Mann do not argue that the creation of a procapitalist Climate Leviathan is guaranteed. Rather, they insist that there are three additional alternatives that could emerge in response to global climate change: Climate Behemoth (composed chiefly of reactionary yet capitalist refusals of planetary sovereignty), Climate Mao (consisting of anticapitalist adaptation through planetary sovereignty), and Climate X (imagined as an anticapitalist and anti-Leviathan adaptation in which the political is no longer organized by sovereign exceptions). Wainwright and Mann diagram the four possible political responses to global climate change as a 2 × 2 table (see Figure 1).

Although there are undoubtedly dangers of radical reductionism involved in any attempt to map political possibilities on such a grand scale with so simple a grid (see Lothman 2012; Braun 2015), we nevertheless think that the tensional oppositions diagrammed by Wainwright and Mann provide a useful window into the Trumpist Behemoth. Most important, they compel us to analyze this formation as emergent in and through opposition to the other identified tendencies. It is in this way that we are conceptualizing the Trumpist Behemoth as organized in opposition to the cosmopolitan globalism of green neoliberalism.

Wainwright and Mann themselves posit the anti-planetary sovereignty, procapitalist Behemoth as an important potential challenge to the proplanetary sovereignty, procapitalist Leviathan. "Behemoth is Leviathan's greatest immediate threat," they declare, "and, while unlikely to become hegemonic, may well remain disruptive enough to prevent Leviathan from achieving a new hegemonic order" (Wainwright and Mann 2013). Writing well before Trump announced his presidential candidacy, they nonetheless anticipate something like the Trumpist Behemoth:

> Consider the persistence of a more-or-less conspiracist climate denialism in mainstream political discourse, especially in the USA. ... The disproportionate influence of this proudly unreasonable minority, agitated by the ill-gotten riches of a handful, will persist. (Wainwright and Mann 2013, 13)

Insofar as this observation foresees the Trumpist Behemoth, it also suggests that Trump himself is not *sui generis*, as so many liberal and conservative commentators have claimed, but instead embodies reactionary tendencies long present in U.S. culture. Taking this wider and longer range view, we next evaluate how older models of authoritarianism and corporatism identified by Neumann can also be seen as organizing the Trumpist Behemoth.

On the Behemoth's Techniques of Domination

Two particular political patterns critiqued in Neumann's *Behemoth* distinguish the Trumpist Behemoth as well: authoritarianism and corporatism. Given the fast-paced 1930s history of *Gleichschaltung* (the Nazi "bringing into line" or "synchronization" of all federal, state, and municipal affairs), we must here state that there has not been, and likely will not be, a corresponding Trumpist *Gleichschaltung*

(Neumann 1942). Indeed, the U.S. federal system combined with the commitments to states' rights by U.S. conservatives and white supremacists create significant barriers to Trumpist national synchronization. Nevertheless, following Foster (2017a), the comparisons are still important to note.

Authoritarianism and the Leader

The authoritarianism outlined in Neumann's account of Nazism had three distinguishing features. First was the embrace of radical opportunism, with pronouncements being made and jettisoned as fast as the historical context evolved (Neumann 1942). Second was the consolidation of absolute authority in the hands of the *Führer* (Neumann 1942). Third was the privileging of action and reaction over deliberation and evaluation, legitimated after Schmitt in terms of the inherent legitimacy of the leader's decisions (Neumann 1942).

Each of these aspects of Nazi authoritarianism offers some purchase on the Trumpist Behemoth. Although U.S. courts have so far halted Trump's attempt to consolidate all governmental authority in the executive, the unwillingness of Republicans in Congress to challenge the president means that the presidential bully pulpit has become ever more bullying, amplifying forms of intimidation already long established in business (Reid 2017). Since his assumption of office in January 2017, Trump's authoritarian inclinations have been increasingly apparent, with the president and his surrogates regularly raging against the mainstream press, "fake news," and enemies real and imagined. In so doing, Trump and his administration have inflamed old U.S. traditions of racist, masculinist, and xenophobic cultural politics (Rosa and Bonilla 2017). Indeed, as Giroux (2017) argued, Trump can be thought of as "both a symptom and enabler of this culture, one that enables him to delight in taunting black athletes, defending neo-Nazis in Charlottesville and mocking anyone who disagrees with him."

Following Neumann, we can further trace how the Trumpist Behemoth has been characterized by a vulgar decisionism, especially in terms of environmental governance. The most egregious example of this is Trump's denial of evidence of climate change and his repeated insistence on it being a hoax (De Pryck and Gemenne 2017). Early on, in an interview with the *New York Times* given in late November 2016, Trump unabashedly defended his denialism in a rambling, stream-of-consciousness rant that also illustrated his narcissistic, self-sanitizing, and extraordinarily privileged albeit bunkered worldview (Donald Trump's *New York Times* Interview 2016). His avowal of interest in "clean air" and "crystal clean water" in the interview and elsewhere might further be interpreted as a twenty-first-century example of older fascist professions of desire to defend the purity of the *Heimat*'s cleanliness and ecology (Theweleit 1987). Certainly, as feminist geographers have underlined, when these sorts of assertions of identity are intertwined with Trump's masculinism and white supremacism they articulate a fascist body politics (Gökariksel and Smith 2018). Yet unlike the Nazis, who imagined that their racialized and sexualized national identity would transcend traditional class formations, Trump presents his environmental credentials in an explicit class form. Specifically, Trump tends, as in the *New York Times* interview, to tie his putative environmental interests to the privileged landscapes of expensive golf courses and, through these, to his larger—which is to say, smaller and ultraenclaved—corporate *Weltanschauung*. In the same way, regulation is imagined by administrators of the Trumpist Behemoth as necessarily a restriction on the freedoms of affluent executives to make decisions (Talbot 2018).

These ideological commitments to a CEO worldview in turn inform the Trumpist Behemoth's more practical deregulatory agenda such that every new reaction to an environmental crisis leads the president and his administration to turn the associated disaster into capitalist class opportunity. Most egregious (at least at the time of revising this article in the summer of 2018), Trump and his interior secretary Ryan Zinke responded to the largest ever wildfires in California history with demands for new business freedoms from environmental regulations. While Zinke propounded the timber industry line that more logging would help, Trump tweeted that it was water controls that also needed deregulating. "California wildfires are being magnified & made so much worse by the bad environmental laws," he asserted, "which aren't allowing massive amounts of readily available water to be properly utilized. It is being diverted into the Pacific Ocean. Must also tree clear to stop fire from spreading!" (Brown 2018). This ignorant argument

was also clearly combined with an interested refusal to consider what scientists have shown to be the real contributing role of climate change to the underlying environmental risk (Abatzoglou and Williams 2016). Notable, too, was the knee-jerk enthusiasm for a policy response involving more corporate deregulation. It is to all of the other associated corporate collusions of the Trumpist Behemoth that we now turn.

Corporatism and the Monopolistic Machinery of Rationalization

In *Behemoth*, Neumann argued that one of Schmitt's most important contributions to Nazi rule was to accommodate corporate business interests in Hitler's regime legally by declaring that the ideal "Germanic totality" was one in which "a strong and powerful state [had] full political control but left economic activities unrestricted" (Neumann 1942, 49). Here, Neumann usefully underlined the laissez-faire elements of the Nazi regime before going on to highlight how the resulting mix of political authoritarianism and economic liberalism led Hitler and his cronies to leave corporate monopolies intact and create spaces for big industry to expand profit making (Neumann 1942). Neumann's account thus rejects the notion that Nazi rule was premised on complete state control of capital, which accords with the emphasis other critical scholars have put on the Nazi interest in creating a legal process for so-called reprivatization (e.g., Poulantzas 1974).

There are also clear concurrences between Nazi reprivatization and the contemporary neoliberal push to reprivatize public services, spaces, and lands. In the last year, representatives of the natural resource extractive industries and members of the finance, real estate, and insurance industries have supported the Trumpist Behemoth in its rejection of global environmental crisis management by taking advantage of Trump's own instincts as a real estate mogul. This corporate phalanx has thereby turned the Trumpist Behemoth into a battering ram designed to demolish the administrative state or, to be more precise, the administrative state of green neoliberalism bequeathed by the Obama administration.

The executive machinery assembled by Trump to direct this battering ram is quite different from the Nazi *Führerstaat*, which, as Neumann described, placed the entire German economy on a scientific-ally planned, productive, and well-organized war footing (Neumann 1942). In lieu of such rational-ized planning, Trump has adopted a science-deny-ing, destructive, and disorganizing approach in his war against the administrative state. Yet, because the Trumpist Behemoth is composed of a mix of industry and government leaders who have spent their careers fighting federal environmental protec-tion, it is an effective demolition machine that is effectively using climate change denialism to under-mine environmental protections (Lipton and Ivory 2017). As well as the aforementioned interior sec-retary Zinke, other protagonists involved include Mick Mulvaney, the director of the Office of Management and Budget, who has said that he is "not yet convinced that it is a direct correlation between man-made activity and the change in the climate" and who has sought to eliminate funding for climate research and green energy programs; Rick Perry, the Secretary of Energy, who has said that "the science is out" on climate change and who has also restructured the Department of Energy to focus less on technologies aimed at reducing carbon emissions; Sonny Purdue, the Secretary of Agriculture, who has said that "liberals have lost all credibility when it comes to climate science" and whose staffers at the U.S. Department of Agriculture have told employees to avoid refer-ence to climate change; Jeff Sessions, Trump's ini-tial Attorney General who has said that CO_2 is "really not a pollutant. It's a plant food"; Kirstjen Nielsen, Homeland Security Secretary, who has said of climate change that she "can't unequivocally state it's caused by humans" and who nevertheless has oversight over the Federal Emergency Management Agency (FEMA); and Mike Pompeo, the CIA director, who has equivocated: "There's some who think we're warming, there's some who think we're cooling" and who has contributed to Trump's national security strategy documents in ways that, as we review later, turn the terminology of "resilience" into a euphemism for avoiding any mention of climate change (Holden and Lin 2018). There are many other personalities involved, of course, including the disgraced Scott Pruitt, who had to resign from directing the Environmental Protection Agency (EPA) due to personal scandals largely separate from his pro-polluter agenda, and Trump's Supreme Court Justice appointees Neil

Gorsuch and Brett Kavanaugh, who are widely expected to take a conservative approach to review-ing suits brought against federal agencies for failed enforcement of environmental law (McClammer 2018). Even this brief review of its leading person-alities, however, highlights how the Trumpist Behemoth is staffed with skeptics of climate science who are ready to rationalize rollbacks of environ-mental protection with denialist discourse. As a result, scientists across a wide range of federal agen-cies have also been prevented from measuring or even mentioning climate change and related chal-lenges (Columbia Law School 2018; Dillon et al. this issue).

Although Trump has not fully adhered to a cam-paign boast that he would eliminate the EPA (which he told Fox News was really the "Department of Employment Prevention"), his administration has still imposed huge budget cuts on the agency and freed corporate polluters that the EPA is supposed to regulate from any rigorous oversight and control (Talbot 2018). For related reasons, the EPA has experienced a huge loss of more than 1,500 staff, including 260 scientists, 185 "environmental protec-tion specialists," and 106 engineers, while making less than 400 new hires (Dennis, Elperin, and Tran 2018). Before he resigned, Pruitt also worked to start reducing Obama-era clean car standards and to slash support for the EPA's Ann Arbor laboratory, where vehicles are tested for emissions (Talbot 2018). Subsequently, in August 2018, the administration announced its intent to freeze fuel efficiency stand-ards for cars and contest the right of states such as California to set more stringent requirements to reduce carbon emissions (Davenport 2018). In September 2018, it proposed to roll back Obama-era rules intended to reduce leaks of methane from oil and gas facilities.

More widely Trump's team has worked with con-gressional Republicans to create a remarkable "machinery of rationalization," to borrow Neumann's term, that has systematically rolled back the limited environmental protections that had been advanced under the resilience initiatives of the Obama years. More than fifty federal environmental rules have been identified for elimination (Popovic and Albeck-Ripka 2017). Tax cuts for developers, deregulation of polluters, and reprivatizations of pub-lic land are all being advanced with arrogant haste on an unprecedented scale. National monuments designed to protect fragile environments have been shrunk, formerly protected spaces such as the Arctic National Wildlife Refuge have been targeted for oil drilling, Obama's Clean Power Plan has been reversed, maritime oil exploitation off both the West and East coasts has been set in motion, and new oil pipeline approvals that had been held in check during the Obama years are being green-lighted with alacrity. Trump's record on global climate initiatives is similarly destructive. He has stopped funding both the United Nations Framework Convention on Climate Change and the Green Climate Fund and, in his showpiece nationalist reaction to Obama's green neoliberalism, the president also initiated the withdrawal of the United States from the Paris Climate Agreement in the name of putting America first.

The Reactionary Reworking and Reterritorialization of Resilience

Notwithstanding its appeals to nationalist exceptionalism, the contradictions of the Trumpist Behemoth repeatedly raise the question of how to reconcile the promises of increased homeland security with the deregulatory fight against environmental protection (Mann and Wainwright 2018). To put the contradictions in truly Trumpist language, how can America win climate change? (Conway 2017). Trump himself, perhaps unsurprising, does not seem particularly bothered by these contradictions. Instead of taking the environmental risks seriously, the president seems to think that national security is protected if one increases jobs in the newly deregulated "hard man" polluting sectors of the economy while building up border defenses. As his 2016 *New York Times* interview also illustrated, in Trump's imagination the wealthy will go on living free from the worries of climate change in oases of affluence protected by hardened barricades ("Donald Trump's *New York Times* interview" 2016). The federal managers working under Trump cannot afford to be so detached, however, especially given the fact that the environmental crises produced by climate change, from hurricanes to floods to fires to droughts, have already destabilized both the United States and the world. The Department of Defense (DoD) has been especially concerned with the resulting destabilization. Yet while military chiefs deliberate their "responsibility to prepare," invoke "whole of

government" responses, and plan for hardening bases, building sea walls, and adapting to a global geopolitical landscape transformed by environmental crises, they now only talk about adaptive security responses rather than wrestling with the causes of catastrophe (Klare 2017; Climate Security and Advisory Group 2018). This is where the reworking of resilience has proved so useful.

One clear example of the resulting reworking of resilience has come in the form of the administration's self-described "America First" U.S. National Security Strategy (NSS) published in December 2017 (The White House 2017). In lock step with the denialism previously reviewed, not a word is said in the NSS about the threats posed to national security by global climate change. The word *climate* itself is in fact only used four times and, revealing of the Trumpist Behemoth's enduring neoliberal norms and nomenclature, three of these references are to the need for "transparent" and "investor-friendly" "business climates" (The White House 2017, 20, 21, 22). Only once in the NSS is climate used in an environmental sense. "Climate policies will continue to shape the global system," it says, before proceeding to insist that this demands "U.S. leadership" to counter an "anti-growth energy agenda" associated with said climate policies. By contrast, the words *resilience* and *resilient* are used repeatedly throughout the NSS, including in a special section commandingly entitled "Promote American Resilience" (The White House 2017, 14). The section explains how the U.S. government should "help Americans remain resilient in the face of adversity" by improving risk management, building a culture of preparedness, improving planning, and incentivizing information sharing. The information to be shared is clearly not meant to include any scientific information about the threats posed by climate change, however. Instead, it is supposed to be information about foreign threats and anything else that might jeopardize America's so-called energy dominance. This dominance is in turn made so central to the overall argument that at one point the NSS insists that it is U.S. energy dominance itself that "ensures that markets are free and U.S. infrastructure is resilient and secure" (The White House 2017, 22). Thus is resilience reimagined and reterritorialized in radically exceptionalist America First terms. Not a response to the climate change dangers created by an overreliance on fossil fuels, it is re-presented as actually being dependent on U.S. energy dominance and allied forms of military "overmatch" (The White House 2017, 28). At odds thus with green neoliberalism's more globalist geoeconomic concern with the environmental sustainability of global capitalism, the NSS nevertheless recycles resilience rhetoric geopolitically in a way that covers for its climate change denialism while conveying at least a little enduring attention to systemic risk.

To be sure, none of these patterns are entirely unique to the times of Trump. The 2002 NSS crafted by Condoleezza Rice for the Bush administration also mixed exceptionalist geopolitical assertions with the reworking of geoeconomic terminology in its attempts to conjugate U.S. dominance with concerns about global capitalist coordination (Sparke 2005). In another way, DoD leaders previously deployed the language of resilience during the Obama presidency to talk about foreign policy (Center for Climate and Security 2018). In a novel departure, though, what we see with the Trumpist Behemoth is other federal agencies, including the EPA, FEMA, the Department of Homeland Security (DHS), the Department of Agriculture, and the Federal Highway Administration, also all now recycling resilience rhetoric domestically as a strategic euphemism (Green 2017; Milman 2017; Mooney and Rein 2017; Talbot 2018). The whole administration has found it useful in this way to rework resilience as a technique of domination that canalizes concern with global dangers and secures public acquiescence by focusing on disaster management rather than on the causes of environmental disasters. To illustrate this pattern, let us first review how administration leaders scrambled to respond to the epic hurricanes of 2017 without mentioning climate change.

In the lead-up to Hurricane Irma in September 2017, Pruitt asserted that it was "insensitive" and "misplaced" to talk about the environmental causes of such storms when the focus should be on helping people recover (Diaz 2017). Similarly, in November a FEMA manager responded to a reporter's questions about the connections between climate change and natural disasters by affirming that although "[t]here are plenty of people who want to debate the vocabulary," his mandate was merely to "reduce the costs of future disasters" and contribute to local "resilience" (as cited in Plumer 2017). Other government officials have also employed this kind of euphemistic phraseology. For instance, a DHS bureaucrat told CNN that "[r]egardless of what causes disasters, it's our job within the Department of Homeland Security and FEMA to manage the consequences." A different FEMA administrator likewise avowed that "we always have to look at not just the response, but the preparedness and the resilience" of organizations in the face of disaster (Green 2017). As these remarks suggest, resilience is used repeatedly in this way as a euphemism that enables agents of the Trumpist Behemoth to manage environmental disasters without mentioning climate change, reterritorializing connected phenomena as a series of local crises that can be tackled with a show of quick and highly targeted interventions.

In practice, however, not everyone receives the government help required to defend against or recover from climate catastrophes. There is instead a racist logic embedded in the Trumpist Behemoth's resilience speak that is linked to ideas of who truly deserves to be protected from environmental crisis (Miller 2017). To reuse the language of the NSS in a way that further illustrates its reterritorialization of resilience, "In difficult times, the true character of the [Trumpist Behemoth's vision of the] American people emerges" (The White House 2017, 14). It needs noting that this type of racializing approach to apportioning resilience and risk in the Anthropocene is global in scope (Vergès 2017). As Baldwin argued, there is a racist rationality running right through resilience thinking, especially when it comes to anticipating the challenges of climate change adaptation. Specifically, Baldwin suggested that resilience thinking imports a form of redlining of risk, race, and responsibility common in insurance and real estate business: "If insurability is an index of adaptability and thus a key trait of a valued life under changing climatic conditions, then insurability must also be understood to imply its opposite, uninsurability, where uninsurability signifies unvalued or devalued life" (Baldwin 2016, 4). Not surprising, though, Trump—who is himself no stranger to real estate redlining—appears especially drawn to these racist ways of justifying abandonment in the face of environmental crisis. This was especially clear in his response to Hurricane Maria, which wrecked Puerto Rico in the autumn of 2017.

Less than two weeks after Maria devastated communities across the island, Trump took to Twitter to claim that Puerto Rico had already faced "a financial crisis ... largely of their own making." Trump immediately followed this tweet with two others, the second of which declared that "[w]e [i.e., the federal government] cannot keep FEMA, the Military & the First Responders, who have been amazing ... in P.R. forever!" By juxtaposing his victim-blaming comment with the assertion that the United States would soon abandon the island, Trump implied that Puerto Rico's supposed fiscal irresponsibility justified his neglect of its people. Adding a neoliberal-turned-neopaternalist twist to the U.S. colonial relationship with the island, Trump indicated that Puerto Ricans would need to prove their resilience by surviving Maria's aftermath without official U.S. assistance. Trump had no interest in considering the deep-rooted and ongoing neocolonial causes of Puerto Rico's precarity (Arbasetti et al. 2017).

The Puerto Rican example reveals that in the Trumpist Behemoth resilience-speak reinforces a form of subcitizenship that is tied not only to the global inequalities produced by neoliberalism but also to the racialized and neopaternalistic responses to these inequalities (Adams 2012; Mitchell and Sparke 2016; Sparke 2017). Viewed from a critical long-term perspective, this subcitizenship was clearly a continuation of enduring colonial subjugation in what Bonilla has termed the "archipelago of neglect" (Bonilla 2018; see also Font-Guzman 2017; Rodríguez Soto 2017). By blaming Puerto Rico's debt-encumbered denizens for their own suffering, the administration's response also indicated how the Trumpist Behemoth effectively red-lined resilience, dividing those deemed deserving of recovery from racialized others who are ignored and abandoned or, worse, retargeted for more dispossession (Klein 2018).

Conclusion

We have argued that the Trumpist Behemoth is a mixed-up monster regime that is in the process of rejecting the green neoliberalism of the Obama era by reworking its rationalities and rhetorics of resilience as *arcana dominationis*. Although both scholars and pundits have repeatedly—and correctly—noted that Trump and his coterie are undisciplined, this by no means assures a speedy collapse for the regime. Indeed, in some respects the Trumpist Behemoth thrives on chaos. The general air of scandal that surrounds the president has not prevented him from appointing dedicated officials who enact his

reactionary preferences. In this way the Trumpist Behemoth is institutionalizing reactive, short-term, and exclusionary approaches to climate management that protect the wealthy and the white at the expense of everybody else. Yet it is important to emphasize that these Trumpist tendencies are an apotheosis of previous trends in disaster capitalism that have already disenfranchised large masses of people in both the United States and the world. At the same time people continue to resist these trends, including in Puerto Rico (Werner 2017; Klein 2018). For this reason, if we want to bring this fossil-fueled Behemoth to justice and imagine alternatives to its neoliberal-neopaternalist world order, we also desperately need to learn from the resistance—as well as the resilience—of all who have already endured its cruelties and indignities in years past.

References

Abatzoglou, J. T., and A. P. Williams. 2016. Impact of anthropogenic climate change on wildfire across Western U.S. forests. *Proceedings of the National Academy of Sciences* 113 (42):11770–75.

Adams, V. 2012. The other road to serfdom: Recovery by the market and the affect economy in New Orleans. *Public Culture* 24 (1):185–216.

Arbasetti, J., C. Minet, A. V. Hernandez, and J. Stites. 2017. 100 years of colonialism: How Puerto Rico became easy prey for profiteers. *In These Times*, November 12. Accessed January 16, 2019. http://inthesetimes.com/features/puerto_rico_colonialism_hurricane_vulture_funds.html

Bakker, K. 2010. The limits of "neoliberal natures": Debating green neoliberalism. *Progress in Human Geography* 34 (6):715–35.

Baldwin, A. 2016. Resilience and race, or climate change and the uninsurable migrant: Towards an anthroporacial reading of "race." *Resilience* 5 (2):129–143.

Bessner, D., and M. Sparke. 2017a. Don't let his trade policy fool you: Trump is a neoliberal. *Washington Post*, March 22. Accessed January 16, 2019. https://www.washingtonpost.com/posteverything/wp/2017/03/22/dont-let-his-trade-policy-fool-you-trump-is-a-neoliberal/?utm_term=.9dadaee861dd

———. 2017b. Nazism, neoliberalism, and the Trumpist challenge to democracy. *Environment and Planning A* 49 (6):1214–23.

Bomber, E. 2017. Environmental politics in the Trump era: An early assessment. *Environmental Politics* 26 (5):956–63.

Bonilla, Y. 2018. The wait of disaster: Hurricanes and the politics of recovery in Puerto Rico. Lecture delivered at the University of California Santa Cruz, Santa Cruz, CA, January 31.

Bracke, S. 2016. Is the subaltern resilient? Notes on agency and neoliberal subjects. *Cultural Studies* 30 (5):839–55.

Braun, B. 2015. New materialisms and neoliberal natures. *Antipode* 47 (1):1–14.

Brown, A. 2018. Ryan Zinke uses climate-fueled wildfires to boost the timber industry. *The Intercept.* Accessed August 30, 2018. https://theintercept.com/2018/08/18/ryan-zinke-wildfires-timber-industry/.

Center for Climate and Security. 2018. *Chronology of U.S. military leadership on climate change and security: 2017–2018.* Washington, DC: Center for Climate and Security.

Chaturvedi, S., and T. Doyle. 2015. *Climate terror: A critical geopolitics of climate change.* Basingstoke, UK: Palgrave Macmillan.

Climate Security and Advisory Group. 2018. A responsibility to prepare—Strengthening national and homeland security in the face of a changing climate. Climate Security and Advisory Group. Accessed August 30, 2018. https://climateandsecurity.org/csagrecommendations2018.

Columbia Law School. 2018. Silencing science tracker. Accessed September 20, 2018. http://columbiaclimatelaw.com/resources/silencing-science-tracker/silencing-climate-science/.

Conway, P. 2017. Dismay, dissembly and geocide: Ways through the maze of Trumpist geopolitics. *Law and Critique* 28 (2):111–18.

Davenport, C. 2018. Trump administration unveils its plan to relax car pollution rules. *New York Times,* August 2. Accessed January 16, 2019. https://www.nytimes.com/2018/08/02/climate/trump-auto-emissions-california.html

Dennis, E., J. Elperin, and A. B. Tran. 2018. With a shrinking EPA, Trump delivers on his promise to cut government. *Washington Post,* September 8. Accessed January 16, 2019. https://www.washingtonpost.com/national/health-science/with-a-shrinking-epa-trump-delivers-on-his-promise-to-cut-government/2018/09/08/6b058f9e-b143-11e8-a20b-5f4f84429666_story.html?utm_term=.e6f314d73f00

De Pryck, K., and F. Gemenne. 2017. The Denier-in-Chief: Climate change, science and the election of Donald J. Trump. *Law and Critique* 28 (2):119–26.

Diaz, D. 2017. EPA chief on Irma: The time to talk climate change isn't now. CNN, September 7. Accessed December 1, 2017. https://www.cnn.com/2017/09/07/politics/scott-pruitt-hurricanes-climate-change-interview/index.html

Donald Trump's *New York Times* interview. 2016. *New York Times,* November 23. Accessed February 10, 2018. https://www.nytimes.com/2016/11/23/us/politics/trump-new-york-times-interview-transcript.html

Font-Guzman, J. 2017. Puerto Ricans are hardly U.S. citizens: They are colonial subjects. *Washington Post,* December 13.

Foster, J. B. 2017a. Neofascism in the White House. *Monthly Review* 68 (11). Accessed March 12, 2018. https://monthlyreview.org/2017/04/01/neofascism-in-the-white-house/

———. 2017b. Trump and climate catastrophe. *Monthly Review* 68 (9):1. Accessed March 12, 2018. https://monthlyreview.org/2017/02/01/trump-and-climate-catastrophe/

Fraser, N. 2017. The end of progressive neoliberalism. *Dissent,* January 2. Accessed May 10, 2018. https://www.dissentmagazine.org/online_articles/progressive-neoliberalism-reactionary-populism-nancy-fraser.

Giroux, H. 2017. Gangster capitalism and nostalgic authoritarianism. Accessed March 1, 2018. http://www.tikkun.org/nextgen/gangster-capitalism-and-nostalgic-authoritarianism.

———. 2018. Neoliberal fascism and the twilight of the social. Accessed September 12, 2018. https://truthout.org/articles/neoliberal-fascism-and-the-twilight-of-the-social/.

Gökariksel, B., and S. Smith. 2018. Tiny hands, tiki torches: Embodied white male supremacy and its politics of exclusion. *Political Geography* 62:207–15.

Green, M. 2017. Trump administration swaps "climate change" for "resilience." Accessed February 15, 2018. http://www.cnn.com/2017/09/30/politics/resilience-climate-change/index.html.

Holden, E., and J. Lin. 2018. Trump's climate science doubters. *Politico.* Accessed August 10, 2018. https://www.politico.com/interactives/2018/climate-science-doubters/

Klare, M. 2017. This department is the last hideout of climate change believers in Donald Trump's government. *The Nation,* September 18. Accessed January 16, 2019. https://www.thenation.com/article/this-department-is-the-last-hideout-of-climate-change-believers-in-donald-trumps-government/

Klein, N. 2018. *The battle for paradise: Puerto Rico takes on the disaster capitalists.* New York: Haymarket.

Leitner, H., E. Sheppard, S. Webber, and E. Colven. 2018. Globalizing urban resilience. *Urban Geography* 39 (8):1276–84.

Lipton, E., and D. Ivory. 2017. Under Trump, E.P.A. has slowed actions against polluters, and put limits on enforcement officers. *New York Times,* December 10. Accessed January 16, 2019. https://www.nytimes.com/2017/12/10/us/politics/pollution-epa-regulations.html

Lothman, L. 2012. Commentary on "Climate Leviathan." *Antipode.* Accessed July 1, 2017. https://wp.me/p16RPC-rh

MacKinnon, D., and K. D. Derickson. 2013. From resilience to resourcefulness: A critique of resilience policy and activism. *Progress in Human Geography* 37 (2):253–70.

Mann, G., and J. Wainwright. 2018. *Climate Leviathan: A political theory of our planetary future.* New York: Verso.

McCarthy, J., and S. Prudham. 2004. Neoliberal nature and the nature of neoliberalism. *Geoforum* 35 (3):275–83.

McClammer, J. 2018. Facing the behemoth: Gorsuch's implications for environmental law. *The Legal Intelligencer* 255 (39):1–2.

Mckeown, A., and J. Glenn. 2018. The rise of resilience after the financial crises: A case of neoliberalism rebooted? *Review of International Studies* 44 (2): 193–214.

Miller, T. 2017. *Storming the wall: Climate change, migration and homeland security.* San Francisco: City Lights.

Milman, O. 2017. U.S. federal department is censoring use of term "climate change," emails reveal. *The Guardian*, August 7.

Mitchell, K., and M. Sparke. 2016. The new Washington consensus: Millennial philanthropy and the making of global market subjects. *Antipode* 48 (3):724–49.

Mooney, C., and L. Rein. 2017. Don't call it "Climate Change": How the government is rebranding in the age of Trump. *Washington Post*, May 26. Accessed January 16, 2019. https://www.washingtonpost.com/news/energy-environment/wp/2017/05/26/just-dont-call-it-climate-change-rebranding-government-in-the-age-of-trump/?utm_term=.9d0f82326a16

Nelson, S. H. 2014. Resilience and the neoliberal counterrevolution: From ecologies of control to production of the common. *Resilience* 2 (1):1–17.

Neocleous, M. 2013. Resisting resilience. *Radical Philosophy* 178:3–7.

Neumann, F. 1942. *Behemoth: The practice and structure of national socialism.* New York: Oxford University Press.

Perkins, H. 2009. Out from the (Green) shadow? Neoliberal hegemony through the market logic of shared urban environmental governance. *Political Geography* 28 (7):395–405.

Plumer, B. 2017. Trump ignores climate change. That's very bad for disaster planners. *New York Times*, November 9. Accessed March 20, 2018. https://www.nytimes.com/2017/11/09/climate/fema-flooding-trump.html

Popovic, N., and L. Albeck-Ripka. 2017. 52 Environmental rules on the way out under Trump. *New York Times*, October 6. Accessed January 16, 2019. https://www.nytimes.com/interactive/2017/10/05/climate/trump-environment-rules-reversed.html

Poulantzas, N. 1974. *Fascism and dictatorship.* London: Verso.

Reid, R. 2017. *Confronting political intimidation and public bullying: A citizen's handbook for the Trump era and beyond.* San Bernardino, CA: Amazon ebook.

Rodríguez Soto, I. 2017. Colonialism's orchestrated disasters in Puerto Rico. *Anthropology News*, November 27. Accessed January 16, 2019. https://anthrosource.onlinelibrary.wiley.com/doi/10.1111/AN.711

Rosa, J., and Y. Bonilla. 2017. Deprovincializing Trump, decolonizing diversity, and unsettling anthropology. *American Ethnologist* 44 (2):201–8.

Sparke, M. 2005. *In the space of theory: Postfoundational geographies of the nation-state.* Minneapolis: University of Minnesota Press.

———. 2013. *Introducing globalization: Ties, tensions and uneven development.* New York: Wiley-Blackwell.

———. 2017. Austerity and the embodiment of neoliberalism as ill-health: Towards a theory of biological sub-citizenship. *Social Science & Medicine* 187:287–95.

Swyngedouw, E., and H. Ernstson. 2018. Interrupting the Anthropo-obScene: Immuno-biopolitics and depoliticizing ontologies in the Anthropocene. *Theory, Culture & Society* 35 (6):3–30.

Talbot, M. 2018. Dirty politics: Scott Pruitt's EPA is giving ostentatious polluters a reprieve. *The New Yorker*, April 2:38–51.

Theweleit, K. 1987. *Male fantasies: Women, floods, bodies, history.* Vol. 1. Minneapolis: University of Minnesota Press.

Vergès, F. 2017. Racial Capitalocene: Is the Anthropocene racial? In *Futures of black radicalism*, ed. G. Johnson and A. Lubin, 24–35. Brooklyn, NY: Verso.

Wainwright, J., and G. Mann. 2013. Climate leviathan. *Antipode* 45 (1):1–22.

———. 2015. Climate change and the adaptation of the political. *Annals of the Association of American Geographers* 105 (2):313–21.

Watts, M. 2015. Adapting to the Anthropocene: Some reflections on development and climate in the West African Sahel. *Geographical Research* 53 (3):288–97.

Werner, M. 2017. "We are citizens!": Puerto Rico and the Caribbean from Hurricane Katrina to Maria. *Society and Space.* Accessed March 30, 2018. http://societyandspace.org/2017/10/06/we-are-citizens-puerto-rico-and-the-caribbean-from-hurricane-katrina-to-maria/#

The White House. 2017. *National security strategy of the United States of America.* Washington, DC: White House. https://www.whitehouse.gov/wp-content/uploads/2017/12/NSS-Final-12-18-2017-0905.pdf.

MATTHEW SPARKE is the Director of Graduate Studies in the Politics Department at the University of California Santa Cruz, Santa Cruz, CA 95064. E-mail: msparke@ucsc.edu. The author of *In the Space of Theory* and *Introducing Globalization*, his most recent work has focused on neoliberalism, subcitizenship, and global health.

DANIEL BESSNER is the Anne H. H. and Kenneth B. Pyle Assistant Professor in American Foreign Policy at the University of Washington, Seattle, WA 98195. E-mail: dbessner@uw.edu. The author of *Democracy in Exile: Hans Speier and the Rise of the Defense Intellectual*, he works on intellectual and cultural history, U.S. foreign relations, and the history of democratic thought and the social sciences.

Situating Data in a Trumpian Era: The Environmental Data and Governance Initiative

Lindsey Dillon, Rebecca Lave, Becky Mansfield, Sara Wylie, Nicholas Shapiro, Anita Say Chan, and Michelle Murphy

The Trump administration's antienvironmental policies and its proclivity to dismiss evidence-based claims creates challenges for environmental politics in a warming world. This article offers the Environmental Data and Governance Initiative (EDGI) as a case study of one way to respond to this political moment. EDGI was started by a small group of Science and Technology Studies and environmental justice researchers and activists in the United States and Canada immediately after the November 2016 elections. Since then, EDGI has engaged in four primary activities: archiving Web pages and online scientific data from federal environmental agencies; monitoring changes to these agencies' Web sites; interviewing career staff at the Environmental Protection Agency and the Occupational Safety and Health Administration as a means of tracking changes within those agencies; and analyzing shifts in environmental policy. Through these projects and practices, EDGI members developed the concept of environmental data justice. Environmental data justice is deeply informed by feminist approaches to the politics of knowledge, especially in relation to critical data and archival studies. In this article we establish the theoretical basis for environmental data justice and demonstrate how EDGI enacts this framework in practice.

特朗普政府的反环境保护政策，及其轻视根据证据的主张之倾向，为世界暖化中的环境政治带来了挑战。本文提供环境数据与治理行动 (EDGI) 之案例研究，作为回应此般政治时刻的方式。2016年十一月大选过后，美国与加拿大的一小群科学与科技研究和环境正义的研究者与社会运动者，随即发起EDGI。EDGI涉入四大主要活动：对联邦环境局的网页和网路科学数据进行建档；监测这些单位的网站改变；访问环境保护局和职业安全与健康管理局的从业工作者，作为追踪这些局处内部变迁的方式；以及分析环境政策的改变。EDGI的成员透过这些计画与实践，发展出环境数据正义的概念。环境数据正义深爱女权主义之于知识政治的方法所影响，特别是有关批判数据与档案研究方面。我们于本文中建立环境数据正义的理论基础，并展现EDGI如何将此一架构付诸实践。 *关键词: 批判数据研究, 环境数据正义, 女权主义科学研究, 知识的政治, 社会实践。*

Las políticas anti-ambientalistas de la administración Trump y su proclividad para descartar reclamos respaldados con evidencia da lugar a desafíos a la política ambiental en un mundo en proceso de calentamiento. Este artículo presenta la Iniciativa de Gobernanza y Datos Ambientales (EDGI) como estudio de caso sobre una manera de responder al momento político actual. EDGI empezó como iniciativa de un pequeño grupo de los Estudios de Ciencia y Tecnología, investigadores de justicia ambiental y activistas de Estados Unidos y Canadá, inmediatamente después de las elecciones de noviembre de 2016. Desde entonces, EDGI se ha involucrado en cuatro actividades primarias: el archivo de páginas Web y datos científicos online de las agencias ambientales federales; monitoreo de los cambios en los sitios Web de estas agencias; entrevistas a los funcionarios de carrera de la Agencia de Protección Ambiental y de la Administración de la Salud y Seguridad Ocupacional, como medio de seguimiento a los cambios que ocurran en esas agencias: y analizar cambios en la política ambiental. A través de estos proyectos y prácticas, los miembros de EDGI desarrollaron el concepto de justicia de los datos ambientales. La justicia de los datos ambientales está profundamente imbuida de enfoques feministas hacia la política del conocimiento, especialmente en relación con los estudios de datos críticos y archivos. En este artículo ponemos las bases teóricas de la justicia de los datos ambientales y demostramos cómo EDGI promueve este marco en la práctica. *Palabras clave: estudios de datos críticos, justicia de datos ambientales, estudios de ciencia feminista, política del conocimiento, práctica social.*

The point is to make a difference in the world, to cast our lot for some ways and not others. To do that, one must be in the action, be finite and dirty, not transcendent and clean.

—Haraway (1997, 36)

The Environmental Data and Governance Initiative (EDGI 2018) was started by a small group of Science and Technology Studies (STS) and environmental justice researchers and activists in the United States and Canada immediately after the November 2016 U.S. elections.[1] Like many people at the time, EDGI's founders were concerned about the potential impact of the Trump administration on the environment and human health. Given EDGI's collective expertise on this topic and the all too recent memory of Prime Minister Stephen Harper's attack on public science in Canada (Turner 2013), our concern had a particular focus: the future of environmental science, data, and policy in the face of a virulently antiscience and antienvironment administration. In its first year, EDGI concentrated on four primary activities: archiving Web pages and online scientific data from federal environmental agencies[2]; monitoring changes to these agencies' Web sites; interviewing career staff at the Environmental Protection Agency (EPA) and the Occupational Safety and Health Administration as a means of tracking changes within those agencies; and analyzing shifts in environmental policy (Sellers et al. 2017, Underhill et al. 2017; Rinberg et al. 2018).

Given that many members of EDGI, including the authors of this article, are long-standing critics of state knowledge production and regulatory practice, it is ironic that EDGI's initial activities centered on "rescuing" what we saw as vulnerable federal data and protecting state regulations.[3] EDGI's governmental accountability and oversight work could be seen as advocating a return to Obama-era liberalism, based on the idea that state environmental science and data represent an unambiguous form of "truth" in contrast to the Trump administration's "fictions." The stakes in this political moment are indeed high, particularly as many of the EPA's political appointees are on the record denying both anthropogenic climate change and the harmful effects of pollution (Lipton 2017). Still, as scholars theoretically grounded in feminist STS and environmental justice research, we find a political strategy of uncritically defending facts and data untenable.

EDGI's work attempts to reconcile the need to preserve publicly accessible environmental data and protect state environmental agencies with our shared conviction that it would be a mistake to simply reinstate normal science and state regulation. We argue for the importance of continuing to critique state science, even under an administration that seeks to dismantle state agencies and undermine their scientific work. Critiquing the regulatory state is not enough, though; we also must work to change it. What follows is a reflection on the transformative potential of EDGI, with a focus on the emergent framework and set of practices around *environmental data justice*, a term coined by EDGI member and University of Toronto professor Michelle Murphy to encourage the work of building alternative social and technical data infrastructures and more just socio-environmental futures.

The framework of environmental data justice is deeply informed by feminist approaches to the politics of knowledge, especially in relation to critical data and archival studies. Environmental data justice is also informed by decolonial approaches to knowledge practices, particularly Tuck's (2009, 416) call for "desire-based" rather than "damage-centered" research. Desire-based research does not pathologize communities by merely documenting harm; rather, it emphasizes capacities, multiplicities, and hope and actively works toward building a better world.

In this article we establish a provisional theoretical basis for environmental data justice and demonstrate how EDGI enacts this framework in practice, through specific projects and working groups. We argue that environmental data justice can be of use for all of us in geography and beyond who face the seemingly contradictory imperatives to both defend and critique environmental data and its role in the liberal state. EDGI offers one way of imagining and building alternative, justice-oriented knowledge practices and forms of environmental governance, while creating new modes of counting and accountability.

Environmental Governance and "Alternative Facts"

A consistent characteristic of the Trump administration has been the blurring of any distinction between fact and falsehood. Trump advisor Kellyanne Conway famously used the term "alternative facts" in January 2017 to justify purposeful lies about the size of Trump's inaugural crowd (Sinterbrand 2017). Meanwhile, Trump consistently targets news outlets like CNN and *The New York Times* as "fake" news (Schwartz 2018). This disregard for any kind of adherence to factual statements or evidence-based claims, combined with the administration's white supremacist and xenophobic policies and rhetoric, led public commentators in 2017 to return to Arendt's (1951) book, *The Origins of Totalitarianism* (Berkowitz 2017; Harnett 2017).[4] For Arendt (1973), a defining component of totalitarian movements is their "contempt for facts":

> The ideal subject of totalitarian rule is not the convinced Nazi or the convinced Communist, but people for whom the distinction between fact and fiction (i.e., the reality of experience) and the distinction between true and false (i.e., the standards of thought) no longer exist. (474)

The proclivity of Trump and many of his supporters to undermine these sorts of distinctions has generated widespread dismay across the United States, leading to social protests in defense of truth and facts, and especially environmental facts. For example, signs displayed at the Women's March, March for Science, and related events in the months after Trump took office included "Pro-facts" and "Climate Change Is Real."

Nevertheless, the narrative of Trump as a populist or authoritarian leader leaves out the pervasive corporate influence on his administration—particularly from the petrochemical industry—which deeply shapes its approach to environmental policies (Roberts 2017; Tabuchi and Lipton 2017; Dillon et al. 2018). Former EPA administrator Scott Pruitt[5] and current Interior Secretary Ryan Zinke are openly sympathetic to fossil fuel interests.[6] As Zinke told the National Petroleum Council in September 2017,

"We're now in the business of being partners, rather than adversaries" (Venook 2017). Pruitt actively sought to unravel the EPA through steep budget and staff cuts, deregulation, and a reluctance to enforce environmental laws (Sellers et al. 2017; Irfan 2018). He also restructured the EPA's Science Advisory Board and its Board of Scientific Counselors, prevented EPA scientists from serving on those boards, and (for the first time in the agency's history) allowed lobbyists on EPA science advisory boards (Kimm and Rafferty 2017; Millman 2018). This is in keeping with many of Trump's early Executive Orders as U.S. president, which took aim at environmental agencies and policies. For example, Executive Order 13783 withdrew from former President Barack Obama's Clean Power Plan and reversed other climate change policies. Significantly, Trump signed this Executive Order at EPA Headquarters as part of a highly publicized event, during which coal miners were brought on stage to demonstrate the agency's new political commitments (Sellers et al. 2017).

Grassroots projects such as EDGI represent a form of political resistance and academic research in a moment when a pervasive corporate influence over environmental policy and federal agencies has called the integrity of state environmental data and the cultures of state scientific research into question. The problem of environmental governance under the new administration is not only Trump's cavalier, authoritarian-like relationship with facts (e.g., his well-known comment that climate change is a "hoax") but also the corporate capture of regulatory agencies by companies with a long history of actively "manufacturing doubt" about climate change (Oreskes and Conway 2011). From this latter perspective, the politics of what counts as an environmental fact is not new, nor is corporate influence over environmental agencies like the EPA (Wylie 2018). EDGI's projects have sought to challenge the immediate threat to environmental agencies and policies—Pruitt's dismantling of the EPA, for example—while maintaining a longer, historical view and critique that goes beyond the Trump administration.

One of the ways in which we have sought to challenge the Trump administration and advance a broader critique of the liberal state and its forms of knowledge production is through the concept and practices of environmental data justice. Environmental data justice builds from several analytical frameworks, including feminist STS and critical data studies. In what follows, we put these two scholarly literatures in conversation through the notion of "situating data."

Situating Data and the Politics of Knowledge

A tension that EDGI confronts, and the underlying question of this article, is how to account for the social construction of knowledge when environmental facts and data are also vital to any hope of state and corporate accountability for environmental harms. Does arguing for the social construction of knowledge inadvertently align with a political agenda supporting "alternative facts"? Does it enable the fossil fuel industry's co-optation of environmental agencies by encouraging further doubt about climate change? Although such questions are particularly salient today, similar questions have been of long-standing interest in academic fields such as science studies and political ecology (see Neimark et al. this issue). Here, we turn to foundational feminist STS scholarship that we find especially useful for engaging environmental data in the current political moment. We focus on Haraway's (1988) concept of *situated knowledge* and on Harding's (1992) framework for *strong objectivity*.

Haraway (1988) argued that all knowledge claims are partial, produced in and through practices that are corporeally, socially, geographically, and technically situated. Thus, conventional understandings of universalist knowledge and objectivity, which she described as the "god trick of seeing everything from nowhere," are fundamentally inaccurate descriptions of scientific research. Haraway (1988) wrote, "I am arguing for politics and epistemologies of location, positioning, and situating, where partiality and not universality is the condition of being heard to make rational knowledge claims. ... I am arguing for a view from a body" (589). A politics of location involves acknowledging and theorizing the conditions of knowledge production, rather than claiming a transcendent universality or the "view from nowhere." Acknowledging these conditions also entails taking greater responsibility and accountability for knowledge claims.

Haraway (1988) developed the concept of situated knowledges as a way out of the "two poles" of absolutist objectivity on the one hand and the strong social constructivist argument (in which all knowledge claims could be reduced to a play of power) on the other. In this sense, the concept of situated knowledges is particularly relevant to EDGI's work because Haraway, writing in the 1980s, was also responding to the political stakes of the time: She worried that strong social constructivism encouraged an ethical relativism, leaving any claim to the "real world" to political blocs, such as the Christian fundamentalists who supported then-President Reagan's militaristic policies. Haraway found both absolutist objectivity and strong social constructivism untenable in a world demanding social change. Her description of the needle that must be threaded is prescient of our current dilemma:

> I think my problem, and "our" problem, is how to have *simultaneously* an account of radical historical contingency for all knowledge claims and knowing subjects, a critical practice for recognizing our own "semiotic technologies" for making meanings, *and* a no-nonsense commitment to faithful accounts of a "real" world. (579)

Feminist philosopher of science Harding's (1992) theory of "strong objectivity" offers one response to Haraway's dilemma. Harding argued that "[o]bjectivity has not been 'operationalized'" (440) in scientific practice, because common scientific methods do not identify the collective biases of scientists. In conversation with Haraway, she wrote, "It is a delusion—and a historically identifiable one—to think that human thought could completely erase the fingerprints that reveal its production process." Therefore, "culturewide assumptions [drawn from 'racist, sexist, heteronormative beliefs'] *that have not been criticized within the scientific research process* are transported into the results of research" (Harding 1992, 446).

To counter the unacknowledged politics of scientific research, Harding (1992) argued that we must recognize how knowledge practices—including scientific practices such as developing a hypothesis, developing research tools, selecting methods of data collection and analysis, and reporting results—are inescapably shaped by their social and (we would add) geographical conditions. To develop *strong objectivity*, we must practice *strong reflexivity* by foregrounding the sociospatial positionalities and the historical specificities of knowledge claims and knowledge-producing social systems. Although Harding and Haraway disagree about whether or not there are better, or more adequate, standpoints from which to produce knowledge, they share a commitment to reflexive, inclusive, and participatory knowledge making. Perfect reflexivity might be impossible, because it requires being able to perform the god-trick on ourselves (Rose 1997). We can still strengthen the rigor and reflexivity of our work, though—and our evaluations of others' work—by engaging more with the context of knowledge production, not less.

One point we draw from Haraway and Harding—as we develop EDGI's approach to theory and practice—is that critique of state science is fundamental to justice-centered approaches to environmental data and policy. Foregrounding the socially situated character of knowledge—for us, environmental data—does not automatically lead to more accountable and responsible knowledge or data, but it is a necessary beginning of this process. Situating data requires us to ask questions such as these: Where does the research funding come from? What sorts of information and ideas are included and excluded? What assumptions are embedded in knowledge claims? Who does and does not gets to make valid knowledge claims and through what processes and institutions?

Critical data scholars have taken up similar questions in highlighting the nontransparency of data and digital archival practices. Much of this work addresses commercial and state surveillance practices, as well as corporate structures of data collection—one of EDGI's primary concerns. For example, digital media companies extract large amounts of data on individual users, at the same time keeping those data sets, and the larger data ecologies in which they lie, inaccessible to users, as exemplified by Cambridge Analytica's use of Facebook data to develop voter profiles and personalize political ads for the 2016 Trump campaign. Along the lines of Haraway's call for responsible knowledge claims, critical data scholars have called for greater "algorithmic transparency" and "audits" as new strategies to demand greater user access to the underlying code of commercial software systems (Graham 2005; Sandvig et al. 2014). Other scholars call attention to the cultural and political struggles underlying the production of algorithms and the expansion of large data archives on and around individuals, whose online activities and digital traces are continuously tracked and analyzed for commercial and state profiling projects (and hybrid versions of the two; Dourish 2016; Noble 2018).

Against the dominant idea within mainstream archival studies that practitioners should remain neutral, critical archival scholars have drawn from feminist, critical race, decolonial, and Indigenous studies scholarship to argue for deepened forms of accountability to and collaboration with marginalized communities that are the subjects of data collection (Christen 2011; Punzalan and Caswell 2016). As one example, Kukutai and Taylor (2016) advanced the notion of "Indigenous data sovereignty," which addresses the extractive relationship between Indigenous people and the state, such as the history of biopiracy and misuse of Indigenous knowledge, and the simultaneous absence of reliable data collection on Indigenous people, making it difficult for Indigenous communities to make justice claims through the state. Kukutai and Taylor (2016) argued that there should be "effective participation in data gathering and research" and that "Indigenous people should control these data" (xxii). Projects like the Inuvialuit Living History (2018), the Plateau Peoples' Web Portal (2018), and Mukurtu (2018) have likewise sought to support Indigenous knowledge systems and values through digital access to the archival holdings of various institutions (also see Duarte 2017).[7] Indigenous data sovereignty also includes the refusal to be researched and objectified through scholarship and other data collection projects (Tuck and Yang 2014).

Similar to Haraway's and Harding's calls for situated and contextualized knowledge practices, critical data and archival scholars can be understood to call for "situating data." Strategies of situating data—from opening up the practices of data collection to rethinking infrastructures of data stewardship—are pivotal to developing more responsible, accountable relationships with data. They also move us in the direction of a desire-based approach to data, oriented toward building more habitable relations that acknowledge and nurture the "complexity, contradiction, and the self-determination of lived lives" (Tuck 2009, 416). Situating data undermines any claim to absolute objectivity—for example, the notion that state data simply reflect "what is"— while also acknowledging the power and potential of grassroots data collection projects ("The Counted" 2016; Maharawal and McElroy 2017).[8]

Feminist STS and critical data science aspire to build more responsible and accountable forms of knowledge. These scholarly and activist traditions offer a theoretical approach to environmental data that is not merely deconstructive but seek to build more just sociotechnical data infrastructures and new relationships with data. We turn now to the ways in which EDGI has worked in conversation with these scholarly and activist insights.

Toward Environmental Data Justice

The concept of environmental data justice developed (and continues to develop) through EDGI's projects and practices. Here we explain its initial contours and key concerns and show how it offers an analytical framework and set of practices oriented toward transforming environmental data and governance. EDGI theorizes environmental data justice as a "desire-based framework" (Tuck 2009) that seeks to foster justice, inclusion, and accountability in environmental knowledge practices (Paris et al. 2017; Walker 2017). By this, we mean that environmental data justice is explicitly proactive about creating practices, technologies, governance, forms of community, and infrastructures aimed at bringing about a more just world.

EDGI's mission is to "document current changes to environmental data and governance practices and to foster stewardship, participatory civic technologies, and new communities of practice to make data more accessible and governments and industry more accountable." Our work also aims to make justice and equity central to environmental, climate, and data governance, reflecting Harding's (1986) claim that "commitments to anti-authoritarian, anti-elitist, participatory, and emancipatory values and projects … increase the objectivity of science" (27). We developed this mission statement through working group exercises and a collaborative writing process.

EDGI's political and theoretical commitments inform its internal organizational practices. EDGI structured itself as a consensus-based, horizontal organization. Its Member Protocol is inspired by feminist values, drawing from do-it-yourself science organization Public Lab (in which some EDGI members are also involved), Civic Tech Toronto's Code of Conduct, and the Geek Feminism Wiki.[9] EDGI is also interdisciplinary: Its members include social scientists, physical and life scientists, lawyers, librarians, archivists, artists, and open-access technology communities dedicated to public access to scientific data and analysis. Through its interdisciplinary and

horizontal organization, EDGI brings together different perspectives and forms of expertise. Valuing different forms of knowledge and expertise is another way EDGI puts feminist principles into practice.

In its first few months of existence, the most visible aspect of EDGI was the DataRescue project, coordinated with a partner organization, DataRefuge (2018), and through a collaboration with the Wayback Machine at the Internet Archive. DataRescue crowdsourced the archiving of Web sites and data sets from federal environmental agencies, to maintain the public accessibility of those data sets in the context of a profound uncertainty about the future of online data and other environmental resources. DataRescue unfolded through a series of grassroots events (many at university libraries), with the first event at the University of Toronto in December 2016. The location is significant: Canada had only recently emerged from the administration of Prime Minister Stephen Harper (2006–2015). Harper undermined many of Canada's environmental science programs, policies, and agencies; censored federal scientists from speaking publicly; deleted content from federal environmental Web sites; and closed and destroyed materials from environmental libraries (Sellers et al. 2017). Under Harper's administration, concerned Canadians mobilized around evidence-based environmental policies through "Death of Evidence" rallies and the Right2Know network, affirming the value of science in the public interest (Bell 2012). After the first DataRescue event in Toronto, EDGI worked with DataRefuge to coordinate almost fifty DataRescue events in cities across the United States and Canada. The project received extensive coverage in news outlets including *The Washington Post* and the BBC (BBC 2016; A. Brown 2017).[10]

In some ways, DataRescue appeared as an uncritical form of activism. In part, it reacted to the immediate political moment through rhetoric of "saving" government data from the new administration. The notion of saving environmental data was understood by many, including the news media and some of the event participants, as an effort to rescue or save the liberal state. Arguably, the popularity of DataRescue stemmed from the notion that the Trump administration's environmental policies represent a political anomaly, rather than the extension of a well-funded and long-standing effort to undermine environmental regulation. Here, we explain how DataRescue, from its inception, also involved feminist practices and a critique of the liberal state—therefore enacting a form of environmental data justice. Moreover, projects that have emerged from DataRescue reflect EDGI's commitment to building new social and technical infrastructures and capacities.

The first DataRescue event in Toronto (which did not yet bear the name "DataRescue") was not simply about archiving as many EPA Web sites as possible; it was also about empowering a broad community to work together to copy and preserve data that they cared about. To this end, a key part of the event was producing a toolkit to enable communities in other cities or institutions to replicate and build on the process developed in Toronto.[11] The toolkit included technical information about how to archive Web pages and data sets, as well as documents such as a code of conduct, with an antiharassment policy, aimed at fostering an inclusive and enabling work environment.[12] The toolkit also included a "Code for Crediting, Licensing, and Acknowledgement," which shared EDGI-developed documentation under a "Creative Commons Attribution-Sharealike" license (with coding tools shared GLP 3 and MIT 2017).[13] The code also asked subsequent data archiving events to "generously credit local and nonlocal collaborators on the development of your tools, events, and social media." In extending credit and acknowledging many forms of labor, the environmental data archiving project aimed to create inclusive communities of concern around environmental data. That is, DataRescue was never merely a technical project of saving data. Rather, through this toolkit and other practices, EDGI also sought to create communities to care for data and for each other.

Through the popularity of DataRescue, EDGI was able to raise questions about the stewardship and potential vulnerabilities of state-produced data. DataRescue also literally resituated public environmental data by moving copies of it into alternative archives like DataRefuge (2018). In building the tools of a distributed, community-based archiving, DataRescue created alternative infrastructures to care for public data, guided by feminist practices in the minutia of the project's details. In this way, we understand DataRescue to have both encouraged a tendency to fetishize facts and reify the state and to

have mobilized a critical and expansive politics of data care and justice.

DataRescue ended in June 2017, but it led to a new project, Data Together. Data Together emerged in part through conversations on the potentials and limitations of DataRescue. The project is a collaboration between EDGI and two companies, Protocol Labs (which builds open-source protocols, systems, and technologies of data stewardship) and qri.io (which develops research tools for the distributed Web). Data Together aims to develop community-based and decentralized models of data management and stewardship. Notably, it relies on a system of peer-to-peer data storage and retrieval (which Protocol Labs has been instrumental in developing) that allows communities to hold copies of data. This is in line with the community-centered digital capacities that Kukutai and Taylor (2016) advocated for in their notion of Indigenous data sovereignty. Decentralized models of data stewardship limit the state's power to control and disappear data sets (as Canadian scientists had experienced under the Harper administration). Data Together thus represents a shift within EDGI from a politics of preserving existing data sets toward building new open-source social and technical infrastructures to enable alternative relationships to data (Walker forthcoming).

Along with DataRescue, in January 2017, EDGI began to monitor changes to tens of thousands of federal environmental Web sites. In the process it has developed new tools and methods to hold the federal government accountable for censorship and reduced access to environmental data and information. As of December 2017, EDGI has issued twenty reports on significant changes in wording and public access to environmental information and resources, resulting in more than 100 media reports in venues including *ProPublica*, the *Washington Post*, and the *New York Times*.[14] By documenting Web page changes over time—such as the EPA's removal of resources on climate change—EDGI revealed particular vulnerabilities of online environmental information to shifts in political power (see Friedman 2017). Nevertheless, EDGI's Web site monitoring project is not merely a government oversight project; we have also developed an open-source Web site monitoring platform to make this process more financially accessible (both for EDGI and for other community groups). This open-source platform represents another effort to build new sociotechnical infrastructures.

The concept of environmental data justice emerged through these projects and practices, and also through theoretical reflection. EDGI formed in large part through the merging of data justice and environmental justice communities, and over the past year we have begun to systematically examine the tensions and overlaps between the ways these communities relate to environmental knowledge and data (see Walker 2017). As discussed earlier, critical data studies and data activism have tended to focus on issues of state and corporate data collection practices and demanded less surveillance by the state and corporations. In contrast, environmental justice scholars and activists generally demand more surveillance from the state, especially better monitoring and data collection of industrial emissions and toxic exposure. Environmental justice activists have mobilized to collect their own data on industrial pollution, in the absence of reliable monitoring and data collection by the state. As one example, the Louisiana Bucket Brigades constructed inexpensive air pollution monitors to register peak emissions from a Shell chemical factory, because the state's data collection practices failed to do so (Ottinger 2010). We argue that bringing critical data studies into conversation with environmental justice expands the latter's traditional focus on toxic exposure to include questions of data stewardship, the politics of technical infrastructures, and coding tools. Likewise, data justice activism can engage with desires for more information and greater access to large-scale environmental data sets, particularly having to do with climate change.

In an EDGI working group on environmental data justice, at conferences, and in our collective writing, we continue to explore what environmental data justice means, where it exists in practice already, and how we can foster it more widely in EDGI projects. Some initial ideas, published in our report *Pursuing a Toxic Agenda: Environmental Injustice in the Early Trump Administration* (Paris et al. 2017), include the following:

- Holding the state, corporations, and other polluters responsible for environmental harms. This includes drawing attention to the state's pervasive use of industry-produced data.
- Fostering social, political, and technical infrastructures in which communities can determine what

kinds of data are collected about their own conditions, including offering forms of consent to participate in data collection frameworks, building from the United Nations Declaration on the Rights of Indigenous Peoples. Within this framework consent includes the possibility of refusal.

- Opposing surveillance practices that oppress, dispossess, and marginalize.
- Supporting practices that avoid damage-based research—frameworks that represent communities as damaged and that do not alleviate environmental harms.
- Rethinking the ways we organize, steward, and distribute data.

In reflecting on EDGI's environmental data activism to date, we are drawn to a quote by Haraway (1997), which is the epigraph for this article: "The point is to make a difference in the world, to cast our lot for some ways and not others. To do that, one must be in the action, be finite and dirty, not transcendent and clean" (36). DataRescue, Data Together, EDGI's Web monitoring project, and our work of theory building around environmental data justice represent some of our efforts to engage with the environmental politics of the moment—to be in the mix. None of these projects are free from critique. We emphasize, however, that they have all included utopian, desire-based elements, and have sought to create inclusive communities of concern and envision alternative environmental knowledge and data practices. From its inception, EDGI's conversations included what we called "positive visioning," naming this desire to build something new—and signaling that our scholarship and activism is not simply a reaction to the Trump administration. In developing these projects and reflecting on them, we strive to enact a form of environmental data justice.

Conclusion

In this article we have explored some of the ways in which EDGI combines critique with political engagement, toward the goal of building alternative social and technical infrastructures and pursuing what we call environmental data justice. It does so in a moment when petrochemical interests dominate federal environmental agencies and therefore also the regulatory and data infrastructures that many vulnerable communities rely on to mitigate industrial-environmental harms. In developing environmental data justice as a desire-based framework, we respond to this situation by asking this: What forms of environmental governance, data, and justice could be built to meet the needs of the world we would like to see come into being? Feminist science studies and critical data studies offer important conceptual tools in answering these questions.

We have not resolved tensions between the critical theories that inspire our work and the practical ways we have sought to address the environmental and health threats of the Trump administration; indeed, these might not be fully resolvable. We are committed, however, to the ongoing practice of reflection and change, rather than an easy resolution. In these and other ways, we think that EDGI's work offers intellectual resources and social practices for geographers and other social scientists conducting research on environmental knowledge, justice, and politics in the years to come.

Acknowledgments

The authors are grateful for very helpful feedback from three anonymous reviewers as well as from Becky Mansfield's Space and Sovereignty Working Group. This article builds on the collective work of EDGI as a whole, as well as our partners on several of the initiatives discussed here, Data Refuge and Internet Archive.

Notes

1. EDGI is now a network of more than 150 people, including all of the authors of this article.
2. Including the EPA, Department of Energy, Department of the Interior, NASA, National Oceanic and Atmospheric Administration, Occupational Safety and Health Administration, the U.S. Department of Agriculture, and U.S. Geological Survey.
3. For example, see Murphy (2006), P. Brown (2007), Mansfield (2012), Dillon (2014), Shapiro (2014), and Harrison (2015).
4. The online bookseller Amazon briefly ran out of stock of Arendt's book after Trump's election (Harnett 2017).
5. This article was accepted prior to Pruitt's resignation in June 2018.
6. Most of Trump's political appointees to the EPA have previously worked for climate change–doubting think tanks or petrochemical industries, including Senior Deputy General Counsel Edward Baptist (formerly with the influential American Petroleum Institute) and Deputy Assistant Administrator for

the Office of Chemical Safety and Pollution Prevention Nancy Beck (formerly with the American Chemistry Council; Center for Public Integrity 2018).
7. See Inunialuit Living History (http://www. inuvialuitlivinghistory.ca/), Plateau People's Web Portal (https://plateauportal.libraries.wsu.edu/), and Murkurtu (http://mukurtu.org/about/).
8. See also the Data for Black Lives Conference, 2017 (http://d4bl.org/conference.html).
9. All can be viewed at https://envirodatagov.org/about/mission-vision-values/ and https://github.com/edgi-govdata-archiving/overview/blob/master/CONDUCT.md. EDGI's Member Protocol (similar to a code of conduct) is important for an organization like EDGI, in which member's primary interactions are through online platforms.
10. DataRescue was also mentioned by Klein as an example of resistance in her 2017 book, *No Is Not Enough: Resisting the New Shock Politics and Winning the World We Need*. Full coverage of EDGI's work can be viewed at https://envirodatagov.org/press/#coverage.
11. The Data Rescue Code of Conduct can be viewed at https://docs.google.com/document/d/1bmMTOCgzZslkQwy03NoqX4pEFFDFyMoEQDro7h35E7c/edit#.
12. Materials from DataRescue Toolkit, included the Code for Crediting, Licensing, and Acknowledgment document, can be viewed at https://envirodatagov.org/datarescue/.
13. EDGI also uses MIT license.
14. See https://envirodatagov.org/website-monitoring/ for the full list of reports and media coverage.

References

Arendt, H. 1973. *The origins of totalitarianism*. Boston: Houghton Mifflin Harcourt.

Bell, A. 2012. Why Canada's scientists need our support. *The Guardian*, July 11. Accessed May 5, 2017. https://www.theguardian.com/commentisfree/2012/jul/11/canada-scientists-strike-protests.

Berkowitz, R. 2017. What Arendt matters: Revisiting "The origins of totalitarianism." *Los Angeles Review of Books*, March 18. Accessed May 5, 2018. https://lareviewofbooks.org/article/arendt-matters-revisiting-origins-totalitarianism/#!.

Brown, A. 2017. A coalition of scientists keeps watch on the U.S. government's climate data. *The Intercept*, January 26. Accessed May 5, 2018. https://theintercept.com/2017/01/27/a-coalition-of-scientists-keeps-watch-on-the-u-s-governments-climate-data/.

Brown, P. 2007. *Toxic exposures: Contested illnesses and the environmental health movement*. New York: Columbia University Press.

Christen, K. 2011. Opening archives: Respectful repatriation. *American Archivist* 74 (1):185–210.

The counted: People killed by police in the U.S. 2016. *The Guardian*. Accessed May 5, 2018. https://www.theguardian.com/us-news/ng-interactive/2015/jun/01/the-counted-map-us-police-killings.

DataRefuge. 2018. Accessed May 5, 2018. https://www.datarefuge.org.

Dillon, L., C. Sellers, V. Underhill, N. Shapiro, J. L. Ohayon, M. Sullivan, P. Brown, J. L. Harrison, S. Wylie, and EPA Under Siege Writing Group. 2018. The Environmental Protection Agency in the early Trump administration: Prelude to regulatory capture. *American Journal of Public Health* 108 (S2):S89–S94.

Dourish, P. 2016. Algorithms and their others: Algorithmic culture in context. *Big Data and Society* 3 (2):1–11.

Duarte, M. E. 2017. *Network sovereignty: Building the Internet across Indian country*. Seattle: University of Washington Press.

Environmental Data and Governance Initiative. 2018. Mission, vision, and values. About. Accessed May 5, 2018. https://envirodatagov.org/about/mission-vision-values.

Friedman, L. 2017. E.P.A. scrubs a climate change website of "climate change." *New York Times*, October 20. Accessed May 5, 2018. https://www.nytimes.com/2017/10/20/climate/epa-climate-change.html?_r=1.

Graham, S. 2005. Software-sorted geographies. *Progress in Human Geography* 29 (5):562–80.

Haraway, D. 1988. Situated knowledges: The science question in feminism and the privilege of partial perspective. *Feminist Studies* 14 (3):575–99.

———. 1997. *Modest − witness@second − millennium. FemaleMan − meets − OncoMouse: Feminism and technoscience*. London: Psychology Press.

Harding, S. G. 1986. *The science question in feminism*. New York: Cornell University Press.

———. 1992. Rethinking standpoint epistemology: What is "strong objectivity?" *The Centennial Review* 36 (3):437–70.

Harnett, S. 2017. Trump election spurs sales of books about white working class and totalitarianism. KQED News, January 19. Accessed May 5, 2018. https://www.kqed.org/news/11275396/trump-election-spurs-sales-of-books-about-white-working-class-and-totalitarianism.

Harrison, J. L. 2015. Coopted environmental justice? Activists' roles in shaping EJ policy implementation. *Environmental Sociology* 1 (4):241–55.

Inuvialuit Living History. 2018. Accessed May 5, 2018. http://www.inuvialuitlivinghistory.ca

Irfan, U. 2018. Scott Pruitt is slowly strangling the EPA. *Vox*, March 8. Accessed May 5, 2018. https://www.vox.com/energy-and-environment/2018/1/29/16684952/epa-scott-pruitt-director-regulations.

Kimm, S., and A. Rafferty. 2017. Pruitt makes EPA science board more industry friendly. *NBC News*, November 3. Accessed March 21, 2018. https://www.nbcnews.com/politics/white-house/pruitt-makes-epa-science-board-more-industry-friendly-n817276.

Kukutai, T., and J. Taylor. 2016. *Indigenous data sovereignty: Toward an agenda*. Canberra, Australia: Australian National University Press.

Lipton, E. 2017. Why has the E.P.A. shifted on toxic chemicals? An industry insider helps call the shots. *New York Times*, October 21. Accessed May 5, 2018. https://www.nytimes.com/2017/10/21/us/trump-epa-chemicals-regulations.html.

Maharawal, M. M., and E. McElroy. 2017. The anti-eviction mapping project: Counter mapping and oral history toward bay area housing justice. *Annals of the American Association of Geographers* 108 (2):380–89.

Mansfield, B. 2012. Race and the new epigenetic biopolitics of environmental health. *BioSocieties* 7 (4):352–72.

Millman, O. 2018. EPA head Scott Pruitt says global warming may help "humans flourish." *The Guardian*, February 7. Accessed May 5, 2018. https://www.theguardian.com/environment/2018/feb/07/epa-head-scott-pruitt-says-global-warming-may-help-humans-flourish.

Massachusetts Institute of Technology. 2017. Data for Black Lives Conference, MIT. Accessed December 15, 2017. http://d4bl.org/conference.html

Mukurtu. 2018. Accessed May 5, 2018. http://mukurtu.org/.

Murphy, M. 2006. *Sick building syndrome and the problem of uncertainty: Environmental politics, technoscience, and women workers*. Durham, NC: Duke University Press.

Oreskes, N., and E. Conway. 2011. *Merchants of doubt: How a handful of scientists obscured the truth on issues from tobacco smoke to global warming*. New York: Bloomsbury.

Ottinger, G. 2010. Buckets of resistance: Standards and the effectiveness of citizen science. *Science, Technology & Human Values* 35 (2):244–70.

Paris, B. S., L. Dillon, J. Pierre, I. R. Pasquetto, E. Marquez, S. Wylie, M. Murphy, et al. 2017. Pursuing a toxic agenda: Environmental injustice in the early trump administration. Accessed May 5, 2018. https://envirodatagov.org/publication/pursuing-toxic-agenda.

Peoples' Web Portal. 2018. Accessed May 5, 2018. http://plateauportal.wsulibs.wsu.edu.

Punzalan, R. L., and M. Caswell. 2016. Critical directions for archival approaches to social justice. *The Library Quarterly* 86 (1):25–42.

Rinberg, T., M. Anjur-Dietrich, M. Beck, A. Bergman, J. Derry, L. Dillon, G. Gehrke, et al. 2018. Changing the digital climate: How climate change web content is being censored under the Trump administration. Accessed May 5, 2018. https://envirodatagov.org/publication/changing-digital-climate.

Roberts, D. 2017. Donald Trump is handing the federal government over to fossil fuel interests. *Vox*, June 14. Accessed May 5, 2018. https://www.vox.com/energy-and-environment/2017/6/13/15681498/trump-government-fossil-fuels.

Rose, G. 1997. Situating knowledges: Positionality, reflexivities and other tactics. *Progress in Human Geography* 21 (3):305–20.

Sandvig, C., K. Hamilton, K. Karahalios, and C. Langbort. 2014. An algorithm audit. In *Data and discrimination: Collected essays*, ed. S. P. Gangadharan, V. Eubanks, and S. Barocas, Part I. Washington, DC: New America Foundation.

Schwartz, J. 2018. Trump gives out "Fake News Awards" to CNN, N.Y. Times, Wash Post. *Politico*, January 17. Accessed May 5, 2018. https://www.politico.com/story/2018/01/17/trump-fake-news-awards-345482.

Sellers, C., L. Dillon, J. L. Ohayon, N. Shapiro, M. Sullivan, C. Amoss, S. Bocking, et al. 2017. The

EPA under siege. Accessed May 5, 2018. https://envirodatagov.org/publication/the-epa-under-siege.

Shapiro, N. 2014. Un-knowing exposure: Toxic emergency housing, strategic inconclusivity and governance in the U.S. Gulf South. In *Knowledge, technology and law*, ed. E. Cloatre and M. Pickersgill. Los Angeles: The UCLA Institute for Society and Genetics.

Sinterbrand, R. 2017. How Kellyanne Conway ushered in an era of "alternative facts." *The Washington Post*, January 22. Accessed May 5, 2018. https://www.washingtonpost.com/news/the-fix/wp/2017/01/22/how-kellyanne-conway-ushered-in-the-era-of-alternative-facts/?utm_term=.76f8386a613b.

Tabuchi, H., and E. Lipton. 2017. How rollbacks at Scott Pruitt's EPA are a boon to oil and gas. *The New York Times*, May 20. Accessed May 5, 2018. https://www.nytimes.com/2017/05/20/business/energy-environment/devon-energy.html.

Tuck, E. 2009. Suspending damage: A letter to communities. *Harvard Educational Review* 79 (3):409–28.

Tuck, E., and K. W. Yang. 2014. Unbecoming claims: Pedagogies of refusal in qualitative research. *Qualitative Inquiry* 20 (6):811–18.

Turner, C. 2013. *The war on science: Muzzled scientists and willful blindness in Stephen Harper's Canada*. Vancouver: Greystone Books.

Underhill, V., M. Martenyi, S. Lamdam, and A. Bergman. 2017. Public protections under threat at the EPA: Examining safeguards and programs that would have been blocked by H.R. 1430. Accessed May 5, 2018. https://envirodatagov.org/publication/public-protections-under-threat/.

Venook, J. 2017. The Trump administration's conflict of interest: A crib sheet. *The Atlantic*, January 18. Accessed May 5, 2018. https://www.theatlantic.com/business/archive/2017/01/trumps-appointees-conflicts-of-interest-a-crib-sheet/512711/.

Walker, D. 2017. Towards environmental data justice: Initial thoughts. Environmental Data and Governance Initiative. Accessed May 5, 2018. https://envirodatagov.org/towards-edj-statement/.

Walker, D., E. Nost, A. Lemlin, R. Lave, and L. Dillon. Forthcoming. Practicing environmental data justice: From DataRescue to Data Together. *Geo: Geography and Environment*.

Wylie, S. A. 2018. *Fractivism: Corporate bodies and chemical bonds*. Durham, NC: Duke University Press.

LINDSEY DILLON is an Assistant Professor in the Department of Sociology at the University of California, Santa Cruz, Santa Cruz, CA 95062. E-mail: lidillon@ucsc.edu. Her current research looks at the political ecologies of urban redevelopment, focusing on the entanglements of race and toxic waste.

REBECCA LAVE is an Associate Professor in the Department of Geography at Indiana University, Bloomington, IN 47405. E-mail: rlave@indiana.edu. Her critical physical geography research combines political economy, STS, and fluvial geomorphology to analyze water regulation, the construction of scientific expertise, and the ecosocial consequences of market-based environmental management.

BECKY MANSFIELD is a Professor in the Department of Geography at the Ohio State University, Columbus, OH 43210. E-mail: mansfield.32@osu.edu. Her current research centers on political ecology of environmental health, with a focus on the politics of science of hazardous exposures.

SARA WYLIE is an Assistant Professor of Sociology/Anthropology and Health Sciences in the Northeastern University Social Science Environmental Health Research Institute (SSEHRI), Boston, MA 02115. E-mail: s.wylie@northeastern.edu. She studies the environmental and human health impacts of oil and gas extraction through community-based participatory research and citizen science.

NICHOLAS SHAPIRO is an Assistant Professor of Biology and Society at UCLA, Institute for Society and Genetics, Los Angeles, CA 90095. Email: nick shapiro@ucla.edu. His work revolves around the multiple ways we attempt to detoxify our collective atmospheres.

ANITA SAY CHAN is an Associate Professor in the School of Information Sciences and the Department of Media and Cinema Studies and a Fiddler Innovation Faculty Fellow at the National Center for Supercomputing Applications at the University of Illinois at Urbana–Champaign, Urbana, IL 61801. E-mail: achan@illinois.edu. Her research interests include innovation networks and the "periphery," science and technology studies in Latin America, and collaborative data cultures and practice.

MICHELLE MURPHY is Professor of History and Women and Gender Studies and Director of the Technoscience Research Unit at the University of Toronto, Toronto, ON M6H3H9, Canada. E-mail: michelle.murphy@utoronto.ca. Her research concerns decolonial approaches to environmental violence, chemicals, environmental data, and colonialism in the Great Lakes area.

Rocket Wastelands in Kazakhstan: Scientific Authoritarianism and the Baikonur Cosmodrome

Robert A. Kopack

In this article, I examine how the authoritarian control of scientific research with regard to the Russian space program and the Baikonur Cosmodrome sustains toxic geographies and an information void in Kazakhstan. Baikonur is the oldest, largest, and now busiest space complex in the world, operating continuously since the clandestine Soviet program began in 1957. After 1991, Baikonur became part of a global services industry. Since 2007, a string of violent explosions of Proton class rocket engines, littering designated "fall zones" in central Kazakhstan with toxic debris, have revealed public concern over the use of unsymmetrical dimethyl-hydrazine (heptyl) fuel. When activists' opposition to the use of Proton engines is not squelched as an irrational fear of the cosmos or *cosmophobia*, Russian and Kazakh authorities resort to censorship, intimidation, and imprisonment. Although located in Kazakhstan, Baikonur's launch facilities, the adjacent closed city of the same name, and rocket "fall zones" are administered by the Russian Federation through several post-Soviet techno-diplomatic leasing agreements. All environmental assessment or remediation related to Baikonur is channeled through the Russian Space Agency (RosCosmos), rendering access, publishing, and independent scientific research outside of public scrutiny. Based on twenty months of field research and key interviews with Russian space industry actors, Kazakh state officials, environmental groups, environmental consultants, and local citizens, I examine how the post-Soviet privatization of Baikonur and a legally binding lease agreement facilitate the emergence of authoritarian forms of environmental governance that normalize pollution and block activist interventions.

我于本文中检视对于俄罗斯太空计画和贝康诺太空无人机的科学研究之威权控制，如何维系有毒地理以及哈萨克斯坦的资讯真空。贝康诺是世界上最悠久、最大型，且目前最为繁忙的太空复合体，并从1957年苏联秘密进行的计画开始持续运作至今。1991年后，贝康诺成为全球服务产业的一环。自2007年起，质子火箭引擎的一连串爆炸事件，将有毒残骸弃置于哈萨克斯坦中部的指定"坠落区"，揭发了公众对于使用偏二甲基肼（heptyl）燃料的忧虑。当社会运动人士反对使用质子引擎之诉求无法压制成为对宇宙的非理性恐惧抑或宇宙恐惧症时，俄罗斯和哈萨克政府便诉诸审查、恫吓，以及囚禁。尽管位于哈萨克斯坦，贝康诺的发射设施——其邻近的封闭城市亦以此为名——以及火箭"坠落区"，是由俄罗斯联邦通过若干后苏维埃科技外交的契约协议进行管理。所有有关贝康诺的环境评估或矫正皆通过俄罗斯太空局（RosCosmos）传达，使得取得管道、出版和独立科学研究无法受到公共监督。我根据二十个月的田野研究，以及对俄罗斯太空产业参与者、哈萨克政府官员、环保团体、环境顾问、以及在地公民的关键访谈，检视后苏维埃时期贝康诺的私有化，以及合法的契约协议，如何促成常态化污染、并阻碍社会运动人士介入的威权式环境治理的诞生。关键词：威权主义，环境治理，哈萨克斯坦，空间，有毒地理。

En este artículo examino cómo con el control autoritario de la investigación científica, en lo que concierne al programa espacial ruso y el Cosmódromo de Baikonur, se mantienen en Kazakstán unas geografías tóxicas y un vacío de información. Baikonur es el complejo espacial más antiguo, más grande y ahora más atareado del mundo, que ha operado de manera continua desde que el programa soviético clandestino empezó en 1957. Después de 1991, Baikonur pasó a ser parte de una industria global de servicios. Desde el 2007, una serie de violentas explosiones de motores de cohete de la clase Protón, que contaminan la designadas "zonas de precipitación" de la parte central de Kazakstán con desechos tóxicos, han revelado la preocupación pública por el uso del combustible dimetil-hidracina asimétrica (heptilo). Cuando la oposición de activistas sobre uso de motores de Protón no es reprimida como temor irracional hacia el cosmos, o cosmofobia, las autoridades rusas y kazajas recurren a la censura, la intimidación y la prisión. Aunque ubicadas en Kazakstán, las instalaciones de lanzamiento de Baikonur y la ciudad cerrada adyacente del mismo nombre, lo mismo que las "zonas de precipitación" de cohetes, son administradas por la Federación Rusa, gracias a varios acuerdos tecno-diplomáticos pos-soviéticos de arrendamiento. Toda evaluación o remedio ambiental relacionados con

Baikonur se canaliza a través de la Agencia Espacial Rusa (RosCosmos), colocando el acceso, publicaciones e investigación científica independiente fuera del escrutinio público. Con base en veinte meses de trabajo de campo y entrevistas claves con actores de la industria espacial rusa, funcionarios estatales kazajos, grupos ambientalistas, consultores ambientales y ciudadanos locales, yo examino cómo la privatización pos-soviética de Baikonur y un acuerdo de arrendamiento de obligatoriedad legal facilitan la emergencia de formas autoritarias de gobernanza ambiental que normalizan la contaminación y bloquean las intervenciones activistas. *Palabras clave: autoritario, espacio, geografías tóxicas, gobernanza ambiental, Kazakstán.*

On the morning of 6 September 2007, 135 seconds after liftoff (T + 135) from the Russian-leased Baikonur Cosmodrome (space launch facility), a Proton M commercial grade rocket exploded over the steppe in central Kazakhstan. A failure in the second-stage engine sent a $200 million, Lockheed Martin-built, Japanese telecommunication satellite (JCSAT 11) plummeting to the ground. Segments weighing as much as 800 pounds carved out a series of blackened craters, some as large as sixty feet across and forty-five feet deep, near the Ulitau farming district in the Karaganda *oblast'* (region; see Figure 1).[1] The descent of fire brigades and hazardous material crews revealed the space industry's shadowy assemblage of mobile wreckage managers and the jurisdiction they command over the Kazakh steppe. In the days following the event and to the chagrin of anti-Russian, anti-space industry activists, Kazakhstan's Space Agency (KasCosmos) and their contractor Garysh Ecologiya investigated the damages produced by the crash. As for the hundreds of tons of spilled and burning heptyl fuel (1,1-dimethyl hydrazine or unsymmetrical dimethyl hydrazine [UDMH]) that the rocket was carrying, Garysh Ecologiya concluded with familiar watchwords: "sparsely populated" and "within allowable limits."[2]

For nongovernmental organizations (NGOs) in Kazakhstan, most notably Antigeptil and Baikonur for Human Rights, Garysh Ecologiya's findings were deeply problematic for two reasons. First, the areas down range from Baikonur were populated. Second, they questioned the legitimacy of the official scientific conclusions, insisting that state environmental agencies were minimizing the seriousness of human harm caused by heptyl fuel. According to the activists I spoke with, this was "a common tactic," one that protects the lucrative Russian space industry from public scrutiny. A series of frightening spectacles in 1999, 2004, 2006, and most recently in 2013 brought entire rockets to the ground with propellants, engines, and millions of dollars of satellite debris falling into inhabited areas. Like the activists, Garysh Ecologiya found these accidents to be dangerous and claimed as much publicly. This quasi-public–private contractor, however, also found general fears of the space industry to be "irrational" and "exaggerated"—a manifestation of a psychoemotional disorder that the firm explained to me during a 2015 interview was simply "cosmophobia." Such dismissals

Figure 1. Ulitau crater from 2007 crash site, Karaganda region, Kazakhstan. *Photo:* EcoMuseum. (Color figure available online.)

have in recent years been outweighed by a more martial response to activism from the Kazakh state and its unforgiving penal code. Small protests (some as few as one person) within the last decade in the country's largest cities, Astana and Almaty, have been disbanded, with activists harassed, fined, and arrested by the State Security Bureau (KNB), a descendant of the Soviet-era KGB.

A single rocket accident in Kazakhstan offers a window into the methods and techniques of environmental governance underwriting an authoritarian power structure, its institutional legitimacies, and the particular "social-spatial relations" (Bulkeley 2012, 2428) that a profitable space industry depends on. During the Soviet era, the geography of the space program was a closed militarized zone without environmental obligations or public input about its operation. With the fall of the Soviet Union in 1991, the Baikonur Cosmodrome (hereafter Baikonur) and its historically polluted landscapes are "new objects" of management, rescaled into parcels of governance that are both within and yet outside of Kazakhstan (Cohen and McCarthy 2015). Baikonur's management by a split cast of space industry actors shows how people and places are rendered disposable, and how the logics of maintaining secrecy about risk expose the optics of overlapping state security and economic interests that depoliticize a landscape enrolled in the service of global capital flows (see United Nations Development Program [UNDP] 2004; Swyngedouw 2013). The structures that maintain Baikonur as an "industrial enclave" (Ferguson 2006; Yessenova 2012) provide insulation and jurisdictional guarantees for the Russian Federation to do business in Kazakhstan, ensuring that the troubling socioenvironmental and political dynamics remain internal to the ex-Soviet satellite.

In what follows, I examine the post-Soviet life of Baikonur through the lens of its obligatory wreckage and toxicity. I show how legally binding land lease agreements that include a wealth of infrastructure and down-range areas to catch falling rocket debris depend on authoritarian forms of environmental governance that normalize pollution and block activist interventions in Kazakhstan. The transformation of a Cold War space complex into its present commercial form connects historical geographies to a nascent global space industry that is protected by the control of scientific data and social dissent.

Like elsewhere in the former Soviet Union, most notably Chernobyl in Ukraine and Kazakhstan's own Semipalatinsk nuclear test site, the question of environmental contamination is mired within the collapse of the Soviet Union. The management by new states such as Kazakhstan of the environmental legacies of the Soviet period has largely demonstrated neglect for basic human care and remediation attempts. Across much of the entire post-Soviet region, scientists and government officials have resorted to labeling dissidents as having psychological "phobias" (see Petryna 2013; Goldstein and Hall 2015; Stawkowski 2017).[3] Thus, in places such as Chernobyl and the Semipalatinsk nuclear test site the problem frequently is considered to be people and not environmental toxicity. What makes Baikonur unique among post-Soviet cases is that this former military asset with a deep ecological footprint is not abandoned. Ongoing rocket pollution at Baikonur is accommodated within the open steppe landscapes that continue a historical disposability inherited from the Soviet period. Current access rights cloak obligatory waste products that underpin the success of the Russian space industry and make Kazakhstan politically, environmentally, and economically vulnerable to an ecological problem that is "too big to fix" (S. Davis and Hayes-Conroy 2017, 18).

This article draws on twenty months of fieldwork conducted in 2015 and 2016 to explore how the space industry works, what kinds of information are produced about it, and how state actors deal with environmental toxicity and resistance. I conducted interviews with representatives of Russian and Kazakh space industry firms, state officials, environmental activists, consultants, and citizens in Almaty, the city of Baikonur, and Karaganda. These data show the range of opinions, scientific conclusions, and conversations among public and private agencies. I also consulted published local and international news sources, industry materials, and environmental reports prepared by the UNDP and others.[4] The story of Baikonur, today, is a winding narrative through a courageous history of Soviet cosmic triumphs of the Cold War, leading to a booming commercial enterprise built on old military infrastructures and landscapes in the Kazakh steppe after the Soviet collapse in 1991. As I show here, however, restricted access to key sites and people, environmental governance without public oversight, and an authoritarian grasp on the media and public spaces are pivotal to Baikonur's operation.

Baikonur Then and Now

News about Kazakhstan can be sparse. When astronaut crews blast off to the International Space Station (ISS) or parachute back to Earth followed by Russian rescue helicopters, the vast Eurasian steppe appears, if only to vanish again. A long political history has made Baikonur difficult to know. In 1955, the arid, sparsely settled Kazakh steppe east of the Aral Sea was selected for developing long-distance telemetry systems to guide rockets, to undertake experiments, and to receive falling debris (Siddiqi 2000; Chertok 2009). Over several decades, Sputnik 1, space dogs, Yuri Gagarin, Valentina Tereshkova and other cosmonauts, generations of intercontinental ballistic missiles (ICBMs), space shuttles, and space stations became key elements of Soviet modernity linked to the secret site called Baikonur (Andrews 2009; Gerovitch 2011).

In Kazakhstan, Cold War mentalities from the 1950s envisioned a "sacrifice area" for rocket testing, similar to those that led to land condemnations for military–industrial purposes seen in the U.S.-produced "nuclear landscapes" in the U.S. West, the South Pacific, and other areas (Kirsch 1997; Masco 2002; J. Davis 2005; Voyles 2015). An underwriting "wasteland discourse" (Kuletz 1998; see also Pitkanen and Farish 2017) allowed these territories to be suspended from environmental governance. The process has produced demilitarized, toxic post–Cold War landscapes with contamination issues that are legion and far from being solved (Krupar 2012; Havlick, Hourdequin and John 2014; Alexander 2016; Stawkowski 2016). Whereas many such U.S. sites are "closed" and under state or federal management as wildlife refuges, the Baikonur launch complex and the wasteland discourse is, for the most part, operating today as it was during the Cold War.

Baikonur and its adjacent residential and administrative city were arguably among the most strategic military possessions developed in the Soviet Union. When formally dissolved in 1991, the collage of territory became part of Kazakhstan and there was neither a clear governance road map nor funding structure ready for the largest space complex on earth (Alexandrov 1999; Laruelle and Peyrose 2015). It was evident in the early 1990s that this suite of military infrastructure or "residual assets" (Cooley 2000) had remarkable use value. With fifteen active launch pads for different rocket engines and ballistic missiles, eleven vehicle assembly and

testing buildings, three fueling stations, a cryogenic oxygen–nitrogen plant, a 60-MW combined heat and power station, two runways (one over a mile long), almost 300 miles of rail lines, more than 4,000 miles of power lines, and almost 800 miles of roads, Baikonur was not going to sit dormant (Kazakhstan Ministry of Energy 2014, 202).

A ten-year interstate agreement between the Russian Federation and Kazakhstan in 1994 for the price of $115 million per year (renewed in 2004 until 2050) secured a future for the former Soviet space complex. This comprehensive lease (*dogovor arendy*) between two sovereign nations grants the Russian Federation full use and responsibility for all existing launch infrastructure, municipal facilities, the environmental costs of operation, and liberty to pursue commercial ventures. The demand by the United States, the European Space Agency (ESA), and other foreign national space programs in the 1990s generated contracts for Russian rocket engines, expertise, and launch infrastructure at Baikonur. (House Committee on Science, Space, and Technology 1994) Since then, Russia, thanks partly to the Baikonur facilities, has played a key service role in a competitive and profitable global market. Currently, all piloted missions to the ISS are aboard Russian-made Soyuz rockets launched from Baikonur. As of 2015, the United States pays over $80 million for each of their seats aboard one (Klotz 2015). Contracts on nonpiloted Proton carrier engines (offered only through Baikonur) are priced at as much as $65 million (Federal Aviation Administration 2016).

Today, the Russian Federation operates one of the busiest space programs in the world through Baikonur, with their national agency RosCosmos at the helm. Nearly 300 of Russia's 428 launches between January 2001 and October 2017 have left from Baikonur.[5] Facilitating this commercial activity are large multistage rockets (weighing in excess of 700 tons) that perform routine separations as crew or cargo are delivered into orbit (International Launch Services 2017). At roughly the two-minute mark, the first in a series of engines and fuel tanks plummet to the ground with significant portions of the rocket's hull (see Figure 2). Decade's worth of jettisoned segments of Proton, Zenith, Dnepr, and Soyuz rockets and ICBM chassis can be found today as barns, fencing, or other building materials of down-range residents (see Figure 3). The ecological

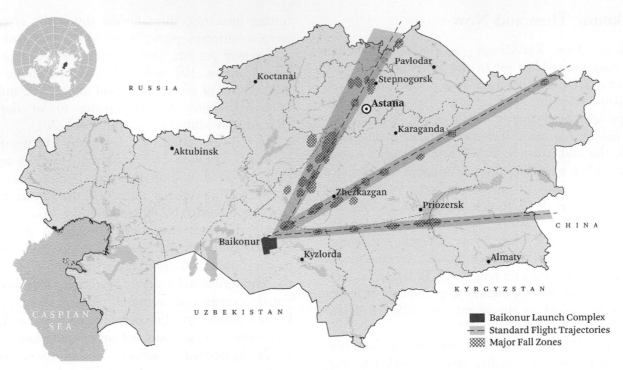

Figure 2. Map of Baikonur's flight trajectories and fall zones in Kazakhstan. *Map:* Travis Bost, Department of Geography and Planning, University of Toronto.

Figure 3. Rocket barn, Karganda region, Kazakhstan. *Photo:* Konstantin Yudin. (Color figure available online.)

damage of this toxic fallout over the last fifty years can at best be estimated. Of the thirty fall zones that are known, some are as large as 3,000 square miles, covering parts of the Karaganda, Pavlodar, Akmola, and East Kazakhstan regions (Baikonur Rental Agreement 2005, appendix 2). With more than 2,000 total launches and counting, more than 27,000 square miles of land have been described by international observers as "zones of ecological crisis" or "zones of ecological disaster" (UNDP 2004, 51).

The post-Soviet life of Baikonur, driven by privatization and commercial ventures, revolves around a complex set of privileges and use rights that define the relationship between the Russian Federation and Kazakhstan that directly shapes the relationship between the Kazakh state and its citizens. Authoritarian governance and the control of information in this dynamic have several levels. As outlined in the interstate lease agreement, only "authorized parties" from their two national space programs—RosCosmos and

KazCosmos—can assess the environmental outcomes of the space industry (Baikonur Rental Agreement 2005, 8.4b). Scientific work thereafter occurs through consultant firms from both agencies, the Center for the Control and Operation of Ground Space Infrastructure or TSENKI for the Russian Federation and Garysh Ecologiya for Kazakhstan, who make remediation decisions together. Although the burden and costs of cleanup ultimately fall on RosCosmos as per the lease agreement, Garysh Ecologiya has consulted on 137 launches of Proton, Soyuz, and Zenith rockets in addition to eighteen ICBMs from Baikonur since 2002 in addition to serving as a public intermediary. Following five major Proton accidents over the last twenty years, they have featured repeatedly in the news as Kazakhstan's lead watchdog for Baikonur. Their team of self-proclaimed *kosmicheskiye ekologi* (space ecologists) includes chemists, biologists, several medical doctors, and two four-wheel-drive mobile laboratory units for on-site environmental sampling. During several interviews I conducted in their Almaty offices in 2015, their director was proud to "have an important job to do" and be, as he put it, "on the front lines."

Founded in 2001 by state decree to "implement applied scientific research on environmental safety related to rocket and space activities in Kazakhstan," Garysh Ecologiya's ability to provide independent, objective oversight is limited in several ways (Garysh Ecologiya n.d.). First, Garysh Ecologiya is not permitted to survey or work independent of the Russian Federation, which reserves the right to restrict research access or toxicity assessments unless a collaborative project is "deemed necessary" (Baikonur Rental Agreement 2005, Article 5). Between 2011 and 2013, for example, there were twenty-four launches of Soyuz rockets (piloted and supply missions to the ISS) during which Garysh Ecologiya did not participate, because they were not "invited" to do so (Kazakhstan Ministry of Energy 2014, 203–05). Second, any information they gather becomes proprietary data of the Russian Federation. The firm's newly developed methods for heptyl fuel detection in the soil, for example, an ongoing ecological challenge that some have deemed a crisis (Kenessov et al. 2008; Carlsen, Kenessov, and Batyrbekova 2009; Tovassarov et al. 2016), could only be shared with RosCosmos or filed away. "Regretfully, our scientists can't publish it, so better that they [the Russians] use it—and know that we developed it," the director explained with a mixture of pride and resignation. He was adamant to remind me on several occasions, "We are for a clean cosmos and care

Figure 4. Rocket wreckage assessment, Karaganda region, Kazakhstan. *Photo*: EcoMuseum.

deeply for the environment, but our relationship with Russia is a deal that we cannot change."

The management of data and space between Kazakhstan and the Russian Federation is best described by what I call *sporadic sovereignty*. The collective infrastructures of Baikonur (launch facilities and the adjacent residential city) are technically within Kazakhstan, although under the jurisdiction of the Russian Federation. Assigned steppe fall zones, required for each and every launch, represent the temporary usage of Kazakh territory that becomes the jurisdiction of the Russian Federation when physical wreckage, planned or otherwise, crashes to the Earth as property. During every launch, the Russian contractor NPO mashinostroyeniya mobilizes in ground units to impound rocket debris and TSENKI secures wreckage sites and proceeds with the impact assessment. Afterward, Garysh Ecologiya might conduct its own investigation, if permitted. During an interview in 2016, a decorated RosCosmos veteran likened the Kazakh contribution to Baikonur as one of "janitors (*dvorniki*)—coming around once the work has already been done" (see Figure 4).

Within this legal governance structure or sporadic sovereignty, the concise activities of Russia's contractors under the umbrella of RosCosmos cannot be entirely known. Many activist groups share frustration, together with private environmental consultants, about the effects of heptyl fuel and falling space debris. "Whom do we protest against? Where do we turn?" asked the director of one Karaganda-based NGO in 2016. This careful environmental activist also noted that Garysh Ecologiya's very public and official role is one that gives "the appearance of a functioning state where there isn't one … who else do you know in the world that lives under the paths of rockets, where pieces are designed to fall from the sky, or blow up entirely, and are told that things are okay?" Similar questions were common across a range of informants I spoke with, most preferring to remain anonymous and still more declining to be recorded.

Recent scholarship has shown how scientific expertise is enrolled in the service of business interests and governments, "rendering the toxic visible and in making the resulting issues public" (Boudia and Jas 2014, 2). At the same time, and this is the case with Baikonur, expertise can create "situations of invisibility and accommodation" (Boudia and Jas 2014, 23). In further interviews with scientists from

Garysh Ecologiya, Kazakhstan's role in the governance of Baikonur's toxic waste showed to be more than an edifice that minimized the dangers of the Russian-controlled space industry as many have suggested. Given the Russian Federation's express right to deny outside oversight and their history of doing so, the "cosmic ecologists" of Garysh Ecologiya navigate the boundaries of overlapping authoritarian regimes that they themselves are weary of but pivotal to. To the detriment of many activists, a wealth of information about Baikonur's environmental footprint, and especially heptyl fuel, is not a matter of transparent public debate.

Normalizing Pollution and Blocking Activist Response

The environmental effects of past Cold War and recent market-driven rocket launches are cumulative. Whereas other major spaceports in the world, like Cape Canaveral (U.S.) or French Guyana (ESA), are positioned on the coasts, meaning that rocket debris and fuels fall into the ocean and out of sight, the situation in Kazakhstan is one of in-land fall zones and rocket trajectories above populations. Vast areas of pasturelands, forests, reservoirs, rivers, power stations, and populous cities like Zhezkazgan and Karaganda, as well as hundreds of other settlements are down range (UNDP 2004). Space industry experts from Garysh Ecologiya negotiate a difficult jurisdiction as official environmental stewards, given their subservient position. Yet all the while, phrases like "sparsely inhabited" and "within allowable limits" suggest dutiful oversight and containment to processes that would seem to evade both. During a typical launch, hundreds of cubic miles of air are contaminated by fuel wakes and up to four tons of unburned propellant is released when engines separate from heights of thirty to sixty miles (Carlsen, Kenessov, and Batyrbekova 2008; Kolumbayeva et al. 2014; Liao, Feng, and Wang 2016). Heptyl—the staple fuel for Russia's commercial Proton rockets—has been found in "remarkable concentrations" at more than 1,000 sites (Carlsen, Kenesova, and Batyrbekova 2007, 1115).

Across the global space industry, heptyl fuels are increasingly under scrutiny for their risks. Studies have shown hydrazines (of which heptyl is one) to be "local irritants, convulsants, hepatoxins, hemolytic agents, cardiac depressants, neurotoxicants, and suspected carcinogens" (Choudhary and Hansen 1998,

820). The European Commission's Registration, Evaluation, Authorization and Restriction of Chemical Substances classifies heptyl as a toxin "of very high concern" that the ESA is looking to "sunset" (ESA 2015; see also European Chemicals Agency 2011, 2). As one engineer from the European Space Research and Technology Center darkly quipped, "One tablespoon of hydrazine in a swimming pool would kill anyone who drank the water" (Giles 2005, 95). In 2008 the United States, equally alarmed, shot down one of its own wayward reconnaissance satellites; it was carrying 1,000 pounds of heptyl and posed "an unacceptable risk to human life on the ground" (Shanker 2008; see also Missile Defense Agency 2010).

At the Garysh Ecologiya Almaty laboratory, heptyl keeps scientists busy. During a series of interviews in 2015 I was greeted by numerous color images: maps of fall zones and accident sites; twisted rocket remains; technicians working mass spectrometers and peering through microscopes; racks of vials and test tubes with dates and coordinates; and field scientists on the steppe in full chemical suits wielding augers, surrounded by soil darkened by spilled fuel. When I asked the director about public health and the space industry, he explained:

> There are many things I cannot speak of because the information technically doesn't belong to us, because of our relationship with the Russians, even though we work on the problems. But I can tell you this, heptyl is extremely dangerous [*chrezvychayno opasnyy*] and we've been working with the Russian Federation for years to manage it. But fall zones are chosen because they're sparsely populated. This is done on purpose to prevent public harm. Accidents are exceptions, and we respond to these. But *cosmophobia* can create real medical conditions or exacerbate existing ones. People who live in stress, or anticipation, especially in rural areas, can develop anxieties—these can turn into real physical conditions. They [the general public] are drawn to rumors. The media or activist groups only inflame peoples' worries. Our technicians can look pretty scary in the field. One of our jobs is to educate.

Recently Garysh Ecologiya penned several articles and press releases explaining a "psycho-emotional condition" that their researchers identified following a longitudinal study in the aftermath of the 2007 Proton crash (Asherbekov and Akhmedova 2012; Posdniakova et al. 2014; Posdniakova, Permenev, and Astanin 2014). The agency concluded that

illnesses of the reproductive system, blood and liver disorders, digestive troubles, cancers, and low birth weights reported by residents (and the UNDP 2004) near wreckage sites were afflictions common throughout rural Kazakhstan where poverty, poor air and water quality, and ignorance were endemic issues. Even with an archive of toxic samples, cross-referenced with detailed maps outlining their location, Garysh Ecologiya's experts insisted that public fear was an irrational response, more dangerous than the space industry itself.

The timing of Garysh Ecologiya's publications coincided with both the emergence of activist protest in Kazakhstan and the uncompromising state response under Article 174 (Penal Code of the Republic of Kazakhstan 2014). This sweeping legal code prohibits the incitement of social, national, racial, class, or religious discord through mass or social media, protest, or the distribution of literatures. With virtually any public statement or confrontation treated as potentially provocative, many grassroots organization members, most notably Antigeptil, have been threatened and arrested. Amnesty International (2017) responded recently: "Long subjected to severe restriction, the rights to freedom of expression and peaceful assembly came under renewed attack in Kazakhstan in 2015 and 2016" (4). Similarly, in 2015, the UN Office for the High Commissioner on Human Rights made appeals to the Kazakh government over "preemptive arrests" and overbroad legal injunctions when a well-known anti-Baikonur activist was seized from his flat in Almaty for posting on Facebook (UN 2015).

As the aftermath of recent rocket accidents demonstrates, public organizations play no effective role in official environmental assessments or governance. RosCosmos and TSENKI have no obligations to communicate environmental harms to the public in Kazakhstan. The near unreachability of the Russian Federation has made Garysh Ecologiya the official voice. The policing of public space and information have created a situation in which the work of any disinterested environmental consultants is unwelcome.

Prior to the formation of Garysh Ecologiya, the Kazakh Ministry of the Environment responded to public outcry by soliciting the expertise of a Karaganda-based NGO to explore the socioenvironmental impacts at an accident site. One event, a Proton explosion in 1999 that rained heptyl and rocket debris across several farming communities, cast the realities of the space industry and the question of Kazakhstan's sovereignty vis-à-vis the Russian Federation into the public consciousness. "We were severely limited in what we could do and what we could study," explained a Soviet-trained geochemist and director of the NGO during an interview in 2015:

> We looked too closely at the actual populations, their animals, their water sources, the vegetation, types of soil, the wind patterns, and so on—like a real environmental assessment would. Heptyl soaks into the ground, you know, becomes an aerosol, mixes with water. It chemically transforms into other substances so you don't know what you're looking for or how to find it. The problem was enormous. And what about the people that were breathing anywhere near the accident that day? These were questions we tried to work together on, but we weren't asked to collaborate after that.

The NGO's findings were seen as a threat. Early in 2001 the geochemist was "invited to relocate," as he said, his Karaganda offices to a state-owned building whose neighbor is still the KNB. This form of "authoritarian persuasion" shows how "naked coercion remains important, but is less common, more targeted, and more rationed than in its hard authoritarian counterparts" (Schatz and Maltseva 2012, 46). Today, the NGO offers a public, historical–environmental venue space about the industrial landscape of Kazakhstan where they "carefully" educate the public about pollution. "Mostly what we can suggest—without getting into

too much trouble—is that Baikonur needs more regulation. Then we're okay." Indeed, the existence of Article 174 shapes the message of his community-based organization and the kinds of educational opportunities it offers to the public curious to look at the pieces of rockets that he displays. As it would seem, at least in the museum, the space industry is not an environmental risk. It is a part of the historical and contemporary landscape of Kazakhstan that is managed by experts.

Conclusions: Authoritarian Governance and Toxicity's Routine

In the summer of 2016 I visited the closed city of Baikonur to meet with a civil rights attorney and social advocate who, unbeknownst to either of us at the time, would face criminal charges of slander in the coming year. A lifelong resident of Baikonur, this man is not up against the Kazakh legal system but that of the Russian Federation. As he explained, "Once you come through the checkpoint, you've left Kazakhstan."

Today, the city of Baikonur retains much of the Soviet material form: a military–industrial enclave administered from Moscow (see Brown 2013). Access is controlled at three guarded points where all vehicles are stopped. A barren highway leads a short distance north to another gate to the largest launch complex on Earth. The combined territory of this former Soviet asset from where Yuri Gagarin made history in 1961 is more than 4,000 square miles. Inside the walls sit the offices of the key

Figure 5. Young children next to rocket chassis, Karaganda region, Kazakhstan. *Photo*: Konstantin Yudin.

Russian rocket manufacturers Khurnischev, Progress, in addition to others like TSENKI, RosCosmos, and Garysh Ecologiya. In the city where only the ruble is accepted, the attorney made the mistake of defaming RosCosmos, citing the political and scientific cover-up of toxic accidents involving heptyl. For the crime he faces a prison sentence and fines of more than $10,000, more than the sum of his annual salary. As of 2017 his grassroots organization Baikonur for Civil Rights has been declared illegal in the city and his office was closed.

Looking at Baikonur brings to light an underexamined picture of the global space industry and particularly the "background conditions of possibility" (Fraser 2014, 57) that are made visible by a rocket accident. Building on the collapse of the Soviet Union and the subsequent signing of a land and infrastructure lease and the emergence of competitive global services market, the appraisal from Moscow and Astana of Kazakhstan is still one of a wasteland for rocket debris (see Figure 5). Baikonur might as well demonstrate a "global pattern of deterritoriality" and the "loss of commitment by modern nation-states (and even the international community) to particular lands or regions" (Kuletz 1998, 7). Although NASA's celebrated Green Propellant Infusion Mission plans to phase out the use of heptyl altogether at U.S. launch sites, they and many others who hire services through Baikonur for piloted missions to the ISS or for placing satellites into orbit have accepted no responsibility for what happens in Kazakhstan (Giles 2005). With clear industry entrenchment, it might be more accurate to conclude that the global space industry depends not on a pattern of deterritoriality as much as an undertheorized, complex form of land usage that blurs the boundaries of state sovereignty and environmental governance.

Although policed and clandestine, the space industry in Kazakhstan is nevertheless a "social institution," putting citizens at risk (S. Davis and Hayes-Conroy 2017, 1). From the casual remarks from residents in Karaganda about headaches and nausea to the more serious concerns from women about miscarriages, Baikonur is a narrative trope that brings to mind rockets in the air trailed by fuel wakes and more often burning wreckage in the open steppe. The political and environmental governance structure that keeps Baikonur running, however, speaks to an authoritarian era of post-truth, where information is proprietary, landscapes are still secret, and resistance is neutralized as a "phobia" or a threat to order. Facts in this case show Kazakhstan to be a "commercial partner" to the Russian Federation compelled to protect stipulations in a fifty-year lease agreement at all costs (Yessenova 2012, 98). The future holds the real possibility that operations at Baikonur will expand, including a new line of heavy-lift commercial rockets that could replace the Proton, new carriers that, hopefully, will run on less hazardous fuels. This has been repeatedly postponed.

Acknowledgments

Much gratitude to Robert Lewis and Matt Farish for helping this article come together with their unmatched thoroughness to detail, insight, and patience. Many thanks to Magdalena Stawkowski for her invaluable comments and intellectual community. My sincere appreciation to Lynne Viola and the terrific reading group that she sponsors who closely read this article in Toronto; to the Department of Geography at the University of North Carolina at Chapel Hill, Scott Kirsch and John Pickles, who sponsored a guest lecture to present this work; to my colleagues at the University of Toronto, Kilian McCormack, Lazar Konforti, and Travis Bost. My deepest thanks to Dima and Yulia Kalmykov, Dana Yermalyonok, and other colleagues in Karaganda, without whom this research would never have been possible. To all of those I spoke with in Kazakhstan about Baikonur whose names have been changed, thank you so much. This work is a testament to your time, generosity, and trust. My sincere thanks to the special issue editor James McCarthy, managing editor Jennifer Cassidento, and the anonymous peer reviewers for their work.

Notes

1. "They almost didn't get the satellite." Industrial Karaganda. 7 September 2007:5–6 (archive).
2. "This is becoming ordinary." Industrial Karaganda. 15 September 2007:3–5 (archive).
3. With Chernobyl in post-Soviet Ukraine, biological citizenship has become a way for citizens to secure social welfare "based on medical, scientific, and legal criteria that both acknowledge biological injury and compensate for it" (Petryna 2013, 6). In Kazakhstan, even though a very nominal compensation program for environmental harm exists, it is geographically

restricted to the Aral Sea and the Semipalatinsk nuclear test site.

4. All translations and interviews were done by the author. To protect the identities of informants vital to this research, all of their names have been removed.

5. Although Baikonur is the busiest launch complex, the Russian Federation has other two other launch sites at Plesetsk and Dombarovsky.

References

Alexander, C. 2016. Cleaning up and moving on: Kazakhstan's nuclear renaissance. In *Les Chantiers du Nucléaire*, ed. R. Gracier and F. Lafaye, 1–26. Paris: Archives Contemporaines.

Alexandrov, M. 1999. *An uneasy alliance: Relations between Russia and Kazakhstan in the post Soviet era, 1992–1997*. Westport, CT: Greenwood.

Amnesty International. 2017. *Think before you text: Closing down social media space in Kazakhstan*. London: Amnesty International.

Andrews, J. 2009. *Red cosmos K.E. Tsiolkovskii, grandfather of Soviet rocketry*. College Station: Texas A&M University Press.

Asherbekov, G., and G. Akhmedova. 2012. The psychogenic effects on populations living in the flight trajectory of rockets. *Journal of the Almaty Institute for the Enhancement of Doctors* 629.7 (504):15–17.

Boudia, S., and N. Jas. 2014. *Powerless science? Science and politics in a toxic world*. New York: Berghahn.

Brown, K. 2013. *Plutopia: Nuclear families, atomic cities, and the great Soviet and American plutonium disasters*. Oxford, UK: Oxford University Press.

Bulkeley, H. 2012. Governance and the geography of authority: Modalities of authorization and the transnational governing of climate change. *Environment and Planning A* 44:2428–44.

Carlsen, L. 2009. A QSAR/QSTR study on the human health impact of the rocket fuel 1,1-dimethyl hydrazine and its transformation products: Multicriteria hazard ranking based on partial order methodologies. *Environmental Toxicology and Pharmacology* 27 (3):415–23.

Carlsen, L., O. Kenesova, and S. Batyrbekova. 2007. A preliminary assessment of the potential environmental and human health impact of unsymmetrical dimethylhydrazine as a result of space activities. *Chemosphere* 67:1108–16.

Carlsen, L., B. Kenessov, and S. Batyrbekova. 2008. A QSAR/QSTR study on the environmental health impact by the rocket fuel 1,1-dimethyl hydrazine and its transformation products. *Environmental Health Insights* 1:11–20.

Chertok, B. 2009. *Rockets and people: Vol. III. Hot days of the Cold War*. Washington, DC: National Aeronautics and Space Administration.

Choudhary, G., and H. Hansen. 1998. Human health perspective on environmental exposure to hydrazines: A review. *Chemosphere* 37 (5):801–43.

Cohen, A., and J. McCarthy. 2015. Reviewing rescaling: Strengthening the case for environmental considerations. *Progress in Human Geography* 39 (1):3–25.

Cooley, A. 2000. Imperial wreckage: Property rights, sovereignty, and security in the post-Soviet space. *International Security* 25 (3):100–27.

Davis, J. 2005. Representing place: "Deserted isles" and the reproduction of Bikini Atoll. *Annals of the Association of American Geographers* 95 (3):607–25.

Davis, S., and J. Hayes-Conroy. 2017. Invisible radiations reveals who we are as people: Environmental complexity, engendered risk, and biopolitics after the Fukushima disaster. *Social and Cultural Geography* 18:1–21.

European Chemicals Agency. 2011. *Agreement of the member state committee support on the identification of hydrazine as a substance of very high concern*. EC No. 206-114-9, Helsinki: European Chemicals Agency.

Federal Aviation Administration. 2016. The annual compendium of commercial space transportation: 2016. Accessed October 5, 2018. https://www.faa.gov/about/office_org/headquarters_offices/ast/media/2016_Compendium.pdf

Ferguson, J. 2006. *Global shadows: Africa in the neoliberal world order*. Durham, NC: Duke University Press.

Fraser, N. 2014. Behind Marx's hidden abode: For an expanded conception of capitalism. *New Left Review* 86:55–72.

Gerovitch, S. 2011. Why are we telling lies: The creation of Soviet space history myths. *The Russian Review* 70:460–84.

Giles, J. 2005. Study links sickness to Russian launch site. *Nature* 433:95.

Goldstein, D., and K. Hall. 2015. Mass hysteria in Le Roy, New York: How brain experts materialized truth and outscienced environmental inquiry. *American Ethnologist* 42 (4):640–57.

Havlick, D., M. Hourdequin, and M. John. 2014. Examining restoration goals at a former military site. *Nature and Culture* 9 (3):288–315.

International Launch Services. 2017. Proton M Brochure. Accessed November 18, 2018. https://mk0ilslaunchupbj5chy.kinstacdn.com/wp-content/uploads/2018/07/Proton-Breeze-MBrochure.pdf.

Kenessov, B., S. Batyrbekova, M. Nauryzbayev, T. Bekbassov, M. Alimzhanova, and L. Carlsen. 2008. GC-MS determination of 1-methyl-1H1,2,4-triazole in soils affected by rocket fuel spills in central Kazakhstan. *Chromatographia* 67 (5–6):421–24.

Kirsch, S. 1997. Watching the bombs go off: Photography, nuclear landscapes, and spectator democracy. *Antipode* 29 (3):227–55.

Klotz, I. 2015. NASA extends contract with Russia for rides to Space Station. Reuters. 5 August. Accessed November 18, 2018. https://www.reuters.com/article/us-space-nasa-russia/nasa-extends-contract-with-russia-for-rides-to-space-station-idUSKCN0QA2P920150805.

Kolumbayeva, S., D. Begimbetova, T. Shalakhmetova, T. Saliev, A. Lovinskaya, and B. Zhunusbekova. 2014. Chromosomal instability of rodents caused by pollution from Baikonur Cosmodrome. *Exotoxicology* 23:1283–91.

Krupar, S. 2012. Transnatural ethics: Revisiting the nuclear cleanup of Rocky Flats, CO, through the queer ecology of nuclear waste. *Cultural Geographies* 19 (3):303–27.

Kuletz, V. 1998. *The tainted desert: Environmental ruin in the American West.* London and New York: Routledge.

Laruelle, M., and S. Peyrose. 2015. *Globalizing central Asia: Geopolitics and the challenges of economic development.* London and New York: Routledge.

Liao, Q., C. Feng, and L. Wang. 2016. Biodegredation of unsymmetrical dimethylhydrazine in solution and soil by bacteria isolated from activated sludge. *Applied Sciences* 6 (95):2–19.

Masco, J. 2002. Lie detectors: On secrets and hypersecurity in Los Alamos. *Public Culture* 14 (3):441–67.

Ministry of Defense and Aerospace Industry of the Republic of Kazakhstan Aerospace Committee. Republican State Enterprise Research Center "Garysh-Ecology." Accessed November 18, 2018. http://ghecology.kz/ru/o-predpriyatii.

Ministry of Energy, Kazakhstan. 2014. National report on the state of the environment and the use of natural resources 2011–2014. Accessed June 11, 2017. http://doklad2014.ecogosfond.kz/государственное-управление-охраноЙ.

Office of the High Commissioner for Human Rights (UN). 2015. *Mandates of the working group on arbitrary detention; The Special Rapporteur on the promotion and protection of the right to freedom of opinion and expression; The Special Rapporteur on the rights to freedom of peaceful assembly and of association; and the Special Rapporteur on the situation of human rights defenders,* Geneva, Switzerland, Office of the High Commissioner for Human Rights.

Organisation for Economic Co-Operation and Development. 2017. *Multi-dimensional review of Kazakhstan: Vol. 2. In-depth analysis and recommendations. OECD Development Pathways.* Paris: OECD Publishing.

Penal Code of the Republic of Kazakhstan. 2014. Article 15. Persons subject to criminal responsibility. Accessed November 18, 2018. http://adilet.zan.kz/eng/docs/K1400000226.

Petryna, A. 2013. *Life exposed: Biological citizens after Chernobyl.* Princeton, NJ: Princeton University Press.

Pitkanen, L., and M. Farish. 2017. Nuclear landscapes. *Progress in Human Geography.* Advance online publication. doi:10.1177/0309132517725808.

Posdniakova, A., A Galaiva, Z. Adeelgireli, and G. Asherbekov. 2014. The specificity of illness suffered in the Ulitau District of the Karaganda Region after the Proton M accident of 2007. *Kazakh National Medical University* 3 (1):5–14.

Posdniakova, A., Y. Permenev, and D. Astanin. 2014. Relation to space-rocket activity of the population on the territories adjacent to the place of accident of the Proton launch vehicle in 2007 in the Karaganda region. *News of the National Academy of Sciences of the Republic of Kazakhstan* 6 (306):42–46.

Rental Agreement of the Baikonur Cosmodrome Between the Government of the Republic of Kazakhstan and the Government of the Russian Federation. 2004.

Schatz, E., and E. Maltseva. 2012. Kazakhstan's "authoritarian" persuasion. *Post-Soviet Affairs* 28 (1):45–65.

Shanker, T. 2008. U.S. to attempt to shoot down faulty satellite. *The New York Times,* February 15. Accessed July 2, 2018. https://www.nytimes.com/2008/02/15/us/15satellite.html.

Siddiqi, A. 2000. *Sputnik and the Soviet space challenge.* Gainesville: University Press of Florida.

Stawkowski, M. 2016. "I am a radioactive mutant": Emerging biological subjectivities at the Semipalatinsk Nuclear Test Site. *American Ethnologist* 43 (1):144–57.

———. 2017. Radiophobia had to be reinvented. *Culture, Theory and Critique* 58 (4):357–74.

Swyngedouw, E. 2013. Apocalypse now! Fear and doomsday pleasures. *Capitalism, Nature, Socialism* 240 (1):9–18.

Tovassarov, A., S. Bissarieyeva, M. Nursultanov, and N. Kassymova. 2016. Combined environmentally safe method of detoxification of soils contaminated with unsymmetrical dimethylhydrazine and its toxic derivatives. *IRASCT—Engineering Science and Technology: An International Journal* 6 (3):1–5.

United Nations Development Program (UNDP). 2004. *Environment and development nexus in Kazakhstan.* Almaty, Kazakhstan: United Nations Development Program.

United Nations Human Rights Council. 2015. *Report of the Special Rapporteur on the implications for human rights of the environmentally sound management and disposal of hazardous substances and wastes: Mission to Kazakhstan.* Geneva, Switzerland: United Nations.

U.S. Department of Defense, Missile Defense Agency. 2010. Missile Defense Agency Presents Ronald W. Reagan and Technology Achievement Awards. *MDA News Release.* Accessed November 18, 2018. https://www.mda.mil/news/10news0003.html.

U.S. Government. Chairman's Report of the Committee on Science, Space and Technology. House of Representatives. 1994. 103rd Congress, 2nd Session. House Report 03-451. Oversight Visit Baikonur Cosmodrome.

Voyles, T. 2015. *Wastelanding: Legacies of uranium mining in Navajo country.* Minneapolis: University of Minnesota Press.

Yessenova, S. 2012. The Tengiz oil enclave: Labor, business, and the state. *Political and Legal Anthropology Review* 35 (1):94–114.

ROBERT A. KOPACK is a Doctoral Candidate in the Department of Geography and Planning at the University of Toronto, Toronto, ON M5S3G3, Canada. E-mail: robert.kopack@mail.utoronto.ca. His research interests are the political economies of military and postmilitary landscapes, the global space industry, and the former Soviet Union.

Avoiding Climate Change: "Agnostic Adaptation" and the Politics of Public Silence

Liz Koslov

What does it mean to adapt to climate change without talking about climate change? The term *agnostic adaptation* has emerged to refer to actions that address climate change's effects without acknowledging its existence or human causes. Although prevalent, agnostic adaptation has yet to be the focus of significant empirical research. Most studies of climate silence and denial examine the absence of action rather than its paradoxical presence. This article, by contrast, explores how action and silence coexist and even serve to reinforce each other. It draws on fieldwork in Staten Island, New York City's most politically conservative and only predominantly white borough, where residents mobilized after Hurricane Sandy in favor of government buyouts of their damaged homes that would pay them to relocate rather than rebuild in place. The areas that received buyouts have been lauded from afar as exemplary sites of community-led climate adaptation in one of its most radical forms, managed retreat. On the ground, however, those who participated in the push for retreat were largely silent on the topic of climate change, which was not seen as politically enabling or efficacious to discuss. Agnostic adaptation minimized conflict, made for more tractable claims, and maintained relations of power but in so doing offered protection to only a select few. These findings point to the practical effects of climate silence as it exists in relation to climate talk, both of which share omissions, erasures, and forms of agnosticism that narrow the space for transformative action.

适应气候变迁、却不谈论气候变迁意味着什麼？"不可知论的调适"措辞的浮现, 指涉在不承认气候变迁存在或其人类导因之下, 应对气候变迁效应的行动。尽管"不可知论的调适"相当盛行, 但却尚未成为显着的经验研究焦点。研究气候缄默与否认, 多半检视行动的缺乏, 而非其矛盾的存在。反之, 本文探讨行动与缄默如何同时存在, 甚至相互强化。本研究运用在纽约市史泰登岛这个在政治上最为保守、且唯一一个白人佔优势的自治市所进行的田野工作。该地居民在珊蒂飓风过后进行动员, 支持政府买断其损坏的房屋以偿付异地重新安置之费用, 而非就地进行重建。接受买断的地区, 被外界誉为由社区主导的最为激进的气候调适形式之一之案例——安排撤离。但实际上, 参与推动撤离的人们, 却大半对气候变迁的议题维持缄默, 并且在政治上无法视为具培力作用或有效而论之。不可知论的调适最小化冲突, 导致顺从的主张, 并维持权力关系, 但这麼做, 却仅为获选的少数提供保护。这些研究发现, 说明了其存在关乎气候讨论的气候缄默的实际效应, 两者共享忽略、抹除, 以及不可知论主义的形式, 因而窄化了转变行动的空间。
关键词: 调适, 气候变迁, 否认, 灾害, 环境政治。

¿Qué significa adaptarse al cambio climático sin que se discuta sobre el mismo? Ha aparecido el término adaptación agnóstica para referirse a las acciones que enfrentan los efectos del cambio climático sin reconocer su existencia o las causas humanas del mismo. A pesar de su prevalencia, la adaptación agnóstica aún tiene que ser objeto de investigación empírica significativa. La mayoría de los estudios del soslayo y la denegación del problema climático examinan la falta de acción más que lo paradójico de su presencia. Por contraste, este artículo explora el modo como la acción y el silencio coexisten e incluso sirven para reforzarse el uno con el otro. El artículo se apoya en trabajo de campo realizado en Staten Island, el único sector predominantemente blanco y políticamente más conservador de la Ciudad de Nueva York, donde los residentes se movilizaron después del Huracán Sandy a favor del programa de adquisición de sus casas afectadas por parte del gobierno como mecanismo de relocalización en vez de reconstruir en el mismo sitio. Las áreas que se beneficiaron con adquisiciones por el gobierno han sido elogiadas desde lejos como sitios ejemplares de adaptación climática de orientación comunitaria a través de una de las alternativas más radicales, la retirada dirigida. En el terreno, sin embargo, quienes participaron en la presión por la retirada se mantuvieron en gran medida silenciosos sobre el tópico del cambio climático, que no fue visto como políticamente habilitante o conveniente para discutir. La adaptación agnóstica minimizó el conflicto, sirvió para hacer más manejables las reclamaciones y mantuvo las relaciones de poder, aunque al hacerlo brindó protección a tan solo una minoría selecta. Estos hallazgos puntualizan los efectos prácticos del silencio climático como se dan dentro de la discusión del clima, compartiendo omisiones, enmendaduras y formas de agnosticismo que estrechan el espacio para la acción transformadora. *Palabras clave: adaptación, cambio climático, denegación, desastre, política ambiental.*

In December 2015, as the United Nations Climate Change Conference (COP21) got underway in Paris, Don sent me an e-mail.[1] "Looks like I may [be] becoming a climate change advocate," he wrote, linking to a recent news article. The article related how Don and a group of fellow homeowners in Staten Island, New York, fought for the demolition of their waterfront neighborhood, Oakwood Beach, after Hurricane Sandy in 2012. Rather than rebuild, they pressed the government to buy out their damaged houses, restore wetlands in their place, and prohibit future development. In response to their plea, New York Governor Andrew Cuomo declared a portion of Oakwood Beach the pilot site of a new state buyout program, which later expanded to include homes in two other Staten Island neighborhoods. The program's success in Oakwood Beach meant, the article noted, that at least part of New York was prepared for climate change. Thanks to local leaders like Don, retreat from rising seas and stronger storms had begun. Replying to Don, a self-described conservative Republican, I asked whether he ever anticipated becoming, or being portrayed as, a climate change activist when he began advocating buyouts. "Nope!!!" he wrote. In fact, he thought, climate change had not really entered into it at all.

The piece Don shared was just one of many to cast Staten Island's buyouts as an exemplary case of community-led adaptation to climate change, a heartening counterpoint to the usual denial and inaction. For Jacob (2015), a member of the New York City Panel on Climate Change, what happened in Oakwood Beach marked a rare exception to the "incremental changes to reduce risk" (42) that predominated after Sandy despite admonitions from him and other scientists to rebuild—and in some places *unbuild*—with a warmer, wetter future in mind. "We need more resilient development, to be sure," wrote coastal geologist Pilkey (2012) in one post-Sandy op-ed. "But we also need to begin to retreat from the ocean's edge." Many, like New York City Mayor Michael Bloomberg (2013), pushed back on such prescriptions. People in Oakwood Beach, however, appeared to embrace them, making it possible "to see what a future of managed retreat might look like" once sea level rise and other effects of climate change grew sufficiently undeniable to provoke adaptation on a grander scale (A. Rice 2016).

I began spending time in Oakwood Beach and other low-lying neighborhoods along Staten Island's east and south shores in the months after Sandy. Interested in following the process of retreat from an ethnographic perspective, I conducted fieldwork and interviews over the next four years, observing daily life in affected areas, participating in community meetings and events, and speaking with residents as they decided whether to stay or to go. As block after block of households beyond Oakwood Beach mobilized in favor of buyouts, demand for which came to span at least eight neighborhoods, I expected there to be ample talk of climate change. Yet as Don's e-mail suggested, the topic hardly arose—unless I brought it up myself, which I often waited to do until the end of an interview or conversation. "Are people talking about climate change?" I would ask. "No," I was consistently told. "Not in this neighborhood." "I'm sure 90 percent of people who had a buyout didn't believe in climate change," said Danny, who organized a push for buyouts on the south shore of Staten Island. Regardless of its accuracy, Danny's assessment reflected what soon became readily apparent: Talk of climate change, prominent in much outside discussion of Staten Island's buyouts, was scant among participants themselves. Whatever people's private views on the topic, in public, silence prevailed.

This article examines what it means to adapt to climate change without talking about climate change. From the perspective of the Staten Island homeowners whose collective action precipitated the buyouts, retreat proceeded—and succeeded—largely as what legal theorist Kuh (2015) termed *agnostic adaptation*, or "adaptation without the why—the divorce of adaptation from knowledge or acceptance of climate change being humans' fault" (10046). Acts of adaptation can, of course, coexist even with overt climate change denial; look no further than the Trump Organization's attempt to gain permitting for a seawall to protect a golf course while its chairman and president, Donald J. Trump, denounced global warming as a hoax (Schreckinger 2016). Among individuals, agnostic adaptation is "natural and ubiquitous," Kuh (2015) wrote, because one can easily take action to adjust to a particular weather pattern (e.g., by turning on an air conditioner) without considering that pattern's context or root causes (10047). On a larger scale, however, it takes work to sustain public silence on an issue, particularly one of climate change's enormity and ever-more-apparent impact (Norgaard 2011). As Eliasoph (1998) noted in *Avoiding Politics: How Americans Produce Apathy in Everyday Life*, "It can be as difficult to ignore a problem as to try to solve it" (6). Simultaneously ignoring a problem and trying to solve it, or at least evade its effects, might be more difficult still. What does such an approach afford? In what ways does silence serve to facilitate rather than hinder action? What does the seeming success of agnostic adaptation in this case reveal about the broader politics of climate change and the stakes of collective action in its name?

In the sections that follow, I first situate these questions in relation to research on the forms of public silence that have come to characterize climate discourse and (in)action. Turning to agnostic adaptation in post-Sandy Staten Island, I analyze what climate silence meant in this context, both how it was interpreted and what it worked to accomplish. By focusing on silence's shared meanings and practical effects, which hold regardless of whether intent exists to minimize or evade talk of climate change, I do not aim to ascertain the presence of "denial" or assess motivations for people's silence. Rather, I identify ways in which not talking about climate change was perceived to be more politically enabling and efficacious than engaging with dominant climate change discourse. By way of conclusion, I discuss what these findings imply for climate action given that both weather and politics are rapidly becoming more extreme.

From Denial to Action: Manifestations of Climate Silence

Much scholarship on the social life of climate change seeks to explain inaction. Why, it asks, has so little been done to rectify global warming and respond to its impacts when there is long-standing scientific agreement on its causes and threats? The efforts by powerful special interests, namely, fossil fuel companies that spent decades funding right-wing think tanks, politicians, and contrarian "experts," to sow doubt about established science in hopes of stymieing regulation are by now well known (Oreskes and Conway 2010). This disinformation and lobbying campaign has helped inflame partisan divides and elect vocal deniers to public office, but it remains the case that the majority of people in the United States accept the scientific consensus that global warming is happening and human activity is the cause (Leiserowitz et al. 2018). Still, conversation about climate change is rare, suggesting a "climate 'spiral of silence'" that leads people who do not hear about the topic in daily life to avoid discussing it themselves (Maibach et al. 2016, citing Noelle-Neumann). It is in this context that the "puzzle of climate change" persists: an apparent dearth of grassroots action despite reported levels of individual awareness, belief, and concern (McAdam 2017, 192).

Even as warming begins to have profound and palpable local effects, everyday silence on the subject prevails. Beyond the polarized setting of the United States, Norgaard (2006, 2011) showed how populations both knowledgeable and alarmed about climate change can act little differently than those who would deny the problem's existence or urgency. Silence, Norgaard showed, can be symptomatic of "implicatory denial" (S. Cohen 2001), which differs from literal denial by entailing "not in most cases a rejection of information per se, but the failure to integrate this knowledge into everyday life or to transform it into social action" (Norgaard 2011, 10–11). In the Norwegian village Norgaard studied, where flooding and reduced snowfall threatened the local economy, feelings of fear, helplessness, and guilt about climate change fostered reluctance to talk or even think too much about it. These feelings violated "emotion norms," shared expectations for how people should feel, or at least act: optimistic, in control, and proud of Norway, a country whose high standard of living rested on oil wealth and high per-capita carbon emissions (Norgaard 2006, 390–91). Norway's political economy not only added to the dissonance residents felt. It also raised the stakes of their silence. "Citizens of wealthy nations," Norgaard (2011) noted, "benefit from their denial in economic terms" (72). Avoiding the topic buffered the status quo, offering stability even amidst decidedly unsettling effects.

Yet not only members of privileged groups poised to benefit most, or be harmed least, from climate change are wary of its mention. Climate silence also exists among those who bear the brunt of environmental destruction, past and present. These groups, historically oppressed, have been further marginalized by the dominant discourses and practices of climate science, policy, and planning, fields that harbor their own forms of silence—they tend, studies show, to elide the concerns and complexities of everyday life and to disregard the role of non-elites in producing knowledge of climate change and its possible solutions (J. L. Rice, Burke, and Heynen 2015; Hardy, Milligan, and Heynen 2017; Paprocki 2018). Gaillard (2012) termed this the "climate gap," showing how modes of addressing climate change "are disconnected from local realities, including people's needs, the cultural fabric and the traditional system of governance" (262). In the Gullah/ Geechee Nation along the southeastern U.S. coast, the climate gap manifests in the prevailing sentiment "that scientific knowledge and discourse about climate change and sea-level rise comes from outside Geechee life, and, conversely, that knowledge from the [local] community will not be considered valid, valued, or welcome in such fora" (Hardy, Milligan, and Heynen 2017, 70). Here, the climate gap is racialized, reinforced both by the underrepresentation of African Americans in science and by "colorblind adaptation planning" that ignores the role of racism in producing environmental risk and vulnerability (Hardy, Milligan, and Heynen 2017, 70). It is thus that talk of climate change can come to seem irrelevant, ignorant of people's lived experience and incapable of addressing the full range of challenges they face. Facilitating climate silence is the silencing of certain voices and perspectives in climate talk.

Silence and inaction might appear to go hand in hand, but powerful climate-related action can proceed in the absence of climate talk. D. A. Cohen (2016, 2017) illustrated this point by following those he termed "accidental" or "other" "low-carbon protagonists," namely, participants in urban social movements for affordable density and public transit. In what might be considered a case of agnostic mitigation, participants in these movements fight for aims that align with efforts to reduce carbon emissions but do not tend to regard themselves as climate activists or receive recognition as such. Contributing to the silence on the climate ramifications of such movements' work are several factors: the appropriation of environmentalist rhetoric by elites favoring luxury redevelopment; lack of time, resources, and technical knowledge necessary for "the elaboration of a whole other political and discursive framework"; and belief that it makes strategic sense to foreground issues seen as more pressing than a slowly warming planet (D. A. Cohen 2017, 153–56). Nonetheless, to miss these climate actors for their lack of climate talk is to enjoy at best a partial view of the breadth and possibilities of collective action in the present moment.

Although perhaps more explicitly oriented to another end, practices and forms of mobilization that promote mitigation and adaptation exist beyond the expected sites and styles of climate action. Addressing the "puzzle of climate change" is therefore not only about accounting for the absence of action. It also requires attending more closely to its paradoxical presence—to action that emerges amidst climate silence and even outright denial. To this end, the following section turns to examine what talking—and not talking—about climate change came to mean in the context of post-Sandy Staten Island, as people mobilized for buyouts while remaining, for the most part, publicly agnostic about the relationship between retreat and anthropogenic warming.

Putting Silence to Work: Agnostic Adaptation in Staten Island

Staten Island, known for being New York City's most conservative borough, is not the place one might expect to find at the forefront of collective adaptation to climate change. This is particularly true of the east and south shores, where the mobilization for buyouts took shape. In these waterfront neighborhoods, the majority of voters cast their ballots for Donald Trump in the 2016 presidential election. Homeowners outnumber renters here, and many residents boast multigenerational ties to the shore, where extended families often live in close proximity. Careers in public service are common, with a sizable share of Staten Islanders working as police officers, firefighters, sanitation workers, or in other unionized blue-collar occupations. The only predominantly white borough, Staten Island is also

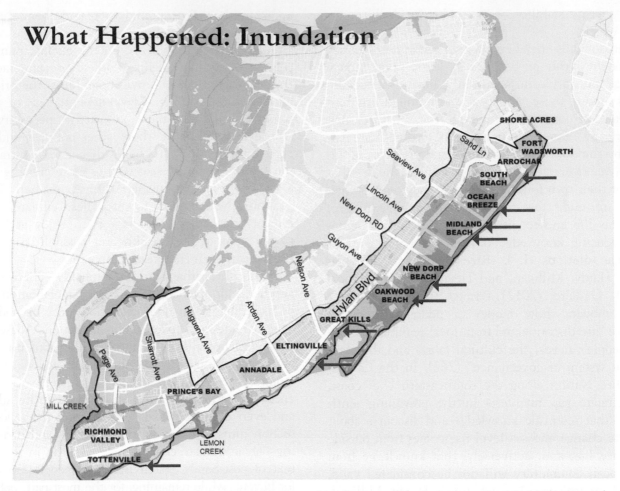

Figure 1. Map showing the extent of Sandy's flooding with Staten Island's east and south shores outlined in black. Arrows have been added to show neighborhoods where groups of residents petitioned for buyouts, which were granted in portions of Oakwood Beach, Ocean Breeze, and Graham Beach (between Ocean Breeze and Midland Beach). *Source:* Retrieved from the Web site of the New York Storm Recovery Resources Center (Governor's Office of Storm Recovery n.d).

highly segregated. Traveling south from the racially diverse north shore, divided from the rest of the island by the Staten Island Expressway, neighborhoods become whiter and wealthier. In the years leading up to Sandy, more recent immigrants hailing from parts of Central America, Eastern Europe, Russia, and the Middle East made homes on the east shore. Still, longer-time residents attest, the shore's prolific fruit and vegetable gardens continue to recall the old country of Italy to which many trace their roots. An unusual sight elsewhere in the city, these gardens spoke to how people, regardless of their tenure, rooted themselves in the "forgotten borough"—prior, at least, to the disruption of the storm.

The worst of Sandy's surge was felt on the low-lying east and south shores, which were the site of the most deaths citywide and an outsize number of

buildings substantially damaged or destroyed. Many of these neighborhoods have suffered recurrent flooding over the years. It was a Nor'easter in 1992 that first spurred a group of residents in Oakwood Beach to organize in response to flooding. They formed a Flood Victims Committee, which lobbied for protection and a halt to new housing development in the neighborhood's wetlands. The committee enjoyed little success, however, and when Sandy struck two decades later disillusionment was rife. Members of the original committee organized a neighborhood meeting and found overwhelming interest in moving rather than rebuilding should there be support to do so in the form of buyouts. After the meeting, a new committee formed: The Oakwood Beach Buyout Committee, which drew up a list of interested homeowners and renewed efforts to garner government support.

This time, with federal disaster recovery funds flowing in, the resident-led committee met with success. Less than four months after the storm, in February 2013, New York State's Democratic Governor Andrew Cuomo traveled down to Staten Island, where he gave a State of the State address to an audience that included invitees from Oakwood Beach. "Climate change is real," he said. Skipping over any mention of climate change's human causes, the governor omitted sentences from an earlier version of the address given in Albany about the need to cut greenhouse gas emissions and increase the use of renewable energy. Instead, he went straight to introducing the state's new buyout program. "There are some places Mother Nature owns," the governor declared. "I want to be there for people and communities who want to say, 'I want to give this parcel back to Mother Nature'" (NYGovCuomo 2013). To rousing applause, he announced a portion of Oakwood Beach as the program's pilot site, where homeowners who opted to participate would receive the pre-storm value of their damaged houses plus incentives meant to encourage widespread uptake and ease the cost of relocating to a less risky area.

After the governor's announcement, households beyond the pilot site in Oakwood Beach began seeking buyouts of their own. All along the shore, from South Beach down to Tottenville, storm-affected Staten Islanders organized buyout groups and pressed state officials to include them, too (see Figure 1; Governor's Office of Storm Recovery n.d.). It was at this point that I began fieldwork, following these groups as they mobilized and debated the merits of moving away from the shore. The data I draw on here come from field notes taken over a four-year period and from interviews I conducted with forty-seven Staten Island residents plus a selection of others involved in the buyout and rebuilding process. Most people I spoke with were white, working- or middle-class homeowners eligible, or potentially eligible, for a buyout. To gain a deeper understanding of retreat from multiple perspectives, I also reached out to community leaders, renters and other nearby residents, architects and urban planners, local journalists, and city, state, and federal government officials. As efforts to secure buyouts ramped up in Staten Island, talk of climate change played what seemed a surprisingly small role given the compelling rationale rising seas and stronger storms present for planned retreat. Over the course of my fieldwork, however, I came to understand how not

talking about climate change also played a role in facilitating adaptation in this context. What emerged was a tension; talking about climate change could prove both politically enabling and disabling. In the remainder of the section I draw out aspects of this tension, focusing on the role of climate silence in minimizing conflict, making claims, and maintaining relations of power.

Minimizing Conflict

Staten Island might possess a significant portion of New York City's right-leaning voters, but the borough is by no means politically homogenous. To organize and sustain neighborhood-wide support for buyouts, crucial to winning backing from the state, required coordinating action across party lines among residents who held different views and even, in some cases, ardently disliked each other. Minimizing conflict was paramount, discord an ever-present risk. Climate change, meanwhile, was widely understood to be a polarizing issue, prone to provoke the very disagreements that people were seeking to avoid. "You don't want my own view on climate change," one woman warned me. "It's become just totally a political football." She made clear that airing one's position on climate change was not the norm in her east shore neighborhood. "I have not heard people discuss it at all, if you want to know the truth," she said. "It's not something that they would throw out there in conversation." Her comments resonated with Danny's assessment of climate change as "the third rail," "the 800-pound gorilla" that people were prone to evade or ignore. To broach the matter was to take a stance, potentially dividing those who needed to unite.

When I raised the topic of climate change, some responded with avowed denial, others alarmed concern. Most responded cautiously, hesitating to assign responsibility to humans even if they agreed that the climate was changing. When I asked Gloria, a member of the Oakwood Beach Buyout Committee, whether she and her neighbors wanted to move because they thought a storm like Sandy would happen again, she equivocated. "Um," she said, "it seems to be a global *trend*. Um, not that it's global *warming*—I don't want to be, you know, jump—I— I'm not jumping on any theory bandwagon. It seems to be a trend regardless of what is causing it." Even people who expressed stronger views in private seemed reluctant to stake out a position in public.

In one interview, a high-ranking staffer for a local democrat told me that he hoped Sandy spurred more talk about climate change and more government action to address it. "Katrina was a wakeup call," he said, "and Irene was a wakeup call. And if Sandy didn't wake you up, you're dead." Yet when I saw him speak at a community meeting, he hedged. "I don't care what you think the cause of climate change is," he told the crowd, "but now we're getting more storms."

In public meetings, local media, and everyday conversation, buyouts were rarely framed as a response to climate change, which might have encouraged residents and officials of differing political stripes to band together in their support. Proponents included not just State Senators Diane Savino, a democrat, and Andrew Lanza, a republican, but also Congressman Michael Grimm, a Tea Party favorite, and other members of the local Republican Party who, one post-Sandy article noted, "have rejected the scientific consensus about climate change, and … routinely voted against measures meant to combat it" (Knafo and Shapiro 2012). The exception to this rule was Danny, who told me he "couldn't live with the contradiction" of staying silent on climate change as he pressed for buyouts in his own south shore neighborhood. One of the only people I saw raise the topic in public, Danny circulated a clip on rapid global warming to his buyout group's e-mail list and included climate-themed cartoons and facts about sea-level rise on posters and in a binder he distributed to make the case for retreat. His general outspokenness, however, fed into the reputation that he acquired among some other local leaders as "a rogue elephant," "a smart guy" but one who lacked political savvy and wound up "pissing everyone off" instead of cultivating the broad base of support needed to get a neighborhood bought out.

Making Claims

Whereas climate change could be a divisive topic, there were other, more unifying ways to talk about environmental transformation that also served to justify claims for buyout funding. The governor might have prefaced his introduction of the buyout program with, "Climate change is real," but it was the latter portion of his remarks that echoed in the months to come, as people asserted their wish to return to Mother Nature land they said should never

have been built on in the first place. "Gov. Cuomo, 'Mother Nature Wants Her Land Back.' BUY US OUT and Give It Back," read one sign erected in Ocean Breeze. There, as elsewhere, I heard little talk of climate change but many laments about flooding. "Climate change never seemed to be an issue here," an Oakwood Beach man named Jack told me. "As part of any sort of casual conversation, no. *Flooding*, yeah. But not talking about, like, if you mean, like, man-made climate change or anything, no." By contrast to talk of climate change, talk of flooding drew on a long history of local activism. It provided ready targets of blame and what seemed a more tractable path to redress, which climate change, with its diffuse causes and widespread effects, did not.

Far from agnostic when discussing the role of development in worsening flooding, the Staten Islanders I met spoke frequently of decades spent watching new housing encroach on nearby wetlands. "Over the years the city allowed development," Pauline in Ocean Breeze said, sharing a story I heard versions of countless times. "They allowed them to fill in the creeks and the wetlands and build on them. So now the water's a problem because it doesn't have its natural places to go anymore." "I *know*," she said, "that I should be grateful that they're offering to buy us out and get us out of this situation—but a big part of me says, they did this. This is a problem that was *created*." Jack likewise faulted "greedy developers" and the "politicians [that] let them get away with it" for worsening flooding. Attributing responsibility for Sandy's damage in this way helped resolve any dissonance that residents might have felt between the conservative and libertarian views that many held and their desire for government intervention in the form of a buyout. It conveyed a sense that their claims were legitimate, a means of righting past wrongs regardless of what the future had in store.

In conversations among residents, attention to past flooding channeled anger at powerful actors who some felt risked evading responsibility should climate change become the primary focus. A reporter at WNYC, the city's public radio station, voiced this concern as he explained why he tended not to talk about climate change in his coverage of Sandy:

I actually think if you say this is a probabilistic problem, that Sandy was due to happen at some point and we totally underprepared for it, it's a real indictment of government, and lack of preparation, and of what we've been doing, whereas actually, if we say it's climate change, we never experienced something like this before, we never could have expected it before, it actually lets current government office-holders off the hook.[2]

In granting buyouts, however, a focus on Mother Nature and recurrent flooding also allowed government officials to appear responsive without admitting responsibility or providing those at risk with comprehensive support. Governor Cuomo might have invoked climate change to underscore the importance of piloting retreat, but he went on to position buyouts as an investment to restore what Mother Nature already owned—land that it could be argued was naturally subject to flooding. This made no admission of culpability and narrowed the scope of possible claims. Mayor Bloomberg, for his part, used the threat of climate change to argue for more—albeit more "resilient"—waterfront development. Calling for emissions cuts to mitigate warming and avert the direst scenarios, the mayor positioned retreat as infeasible and premature. The future was still within human control, the costs of transformation too high given the scale of people and assets potentially at risk. Uncoupling retreat from climate change made for not only more limited but also more targeted and politically palatable claims, from the perspective of those who made them as well as those called on to respond. "It's hard to say we made a mistake, we should have maintained the infrastructure, we should have done it the way that we were supposed to do it," noted Gloria. "It's much easier for *them* to say let's just let them go away." A focus on flooding, which felt possible to contain and, if not solve, at least avoid by moving away from the water, offered an out, one whose efficacy and ethics talk of climate change risked throwing into dispute.

Maintaining Power

When it came to petitioning for resources to retreat after Sandy, the homeowners who participated in Staten Island's buyout groups occupied positions of relative power. Compared to many of their neighbors and people in other parts of the city, including the low-income renters who made up a majority of those affected or immigrants whose status left them ineligible for government aid, they could expect to benefit from greater political access and influence. As an audience member at one post-

Sandy event noted of the neighborhood whose pleas prompted the buyout program's launch, "Oakwood Beach is a very white area with a very high concentration of uniformed city employees, and the city and the state *take note* when such people start pushing back." To make their voices heard, there was little need for buyout groups to align themselves with a broader or more diverse coalition of actors. Moreover, their political leanings made for an uneasy fit with groups that put climate change higher on the agenda, such as the city's environmental justice organizations centered on advancing equity for marginalized people whose environments prove deadly on a daily basis, not just in the event of flooding or a storm. The mobilization for buyouts sought to reestablish the security that its participants possessed prior to Sandy; its aim was not to redistribute resources to those not afforded such security in the first place. It could thus remain silent on climate change, a problem with impacts that fall disproportionately on others.

At the same time, Sandy served as a reminder of just how precarious power could be. It wiped out people's homes, for many their primary asset, and pushed those already on the edge after the financial crisis even deeper into debt. With post-disaster assistance slow to arrive and widely understood to be in short supply, the feeling was often that of a zero-sum game. Gloria expressed this when she described the Oakwood Beach Buyout Committee's strategy: "We knew—I knew—from day one we were going to have to be the fastest and the loudest," she said. "We wanted to get to the governor first ... because look at everybody else." Gloria referred to the other groups of similarly situated homeowners who also sought buyouts once they became public knowledge (as Don recalled, there was pressure within the committee to keep their request quiet and out of the local paper, due to fears that demands from others would dilute their own claims). After the governor's announcement, subsequent groups found solace in the fact that interest in the program was largely limited to Staten Island. "One thing that's helping us—nobody in Queens wants the buyout," a man in Ocean Breeze told his neighbors. "*Here*, that's good for *us*. It means there's more money." Still, it was months before state officials admitted any of these additional groups to the program. Even then, the program wound up including a fraction of interested households. Many if not most who mobilized were

like Jack and Danny, whose blocks were not bought out despite their dedicated efforts. In the meantime, worry abounded that what little funding there was would be diverted elsewhere, not least toward measures that might increase future climate resiliency but not help people suffering at present recover from the storm.

It was in this context that acknowledging climate change could be seen as not just a politically inefficacious distraction but also a threat. Both the causes of climate change and its likely solutions seemed to implicate Sandy survivors themselves, particularly those who held some power but were fearful of losing it. Admitting climate change meant, first of all, reckoning with one's own role and contribution. As Maria, a young woman who emigrated from Colombia and rented a house in South Beach, told me, "I think I was the only one talking about [climate change]. ... Because they don't—I—I haven't heard anyone blame, like, [say], you know, 'that happened because it's *our* fault, we all contribute to the climate change somehow, a little bit.'" Jack, in Oakwood Beach, expressed concern that should climate change be taken seriously it was people like himself who would shoulder the costs. He told me that he did not believe in global warming but conceded, "I know there's loads of people out there that believe in it. And I—I just—okay, you can believe in it, but so what is the answer? To tax the people more? To make us pay more?" "And," he said, "the other thing is, like, okay, it's a big deal in the United States of America, right? But, what about India? What about China? They've got more people than us ... and they're not going to be held to account to the same—so, you know, we're going to be paying more to protect the atmosphere, and they're just going to keep dumping in it." Whether it was a carbon tax passed down to consumers or the sharp increase in flood insurance rates already underway, climate change seemed to provide cover and justification for a host of new potential disasters. Despite the rising emissions of developing countries, it remained the case that the United States, with its dominant historical and per capita contributions—not to mention high consumption of goods produced via other countries' emissions—bore substantial responsibility. How such responsibility would be distributed, though, was an open question; it was easy to imagine how the costs of responding to climate change could fall on everyday people, including

those who, while reaping diminishing benefits from the system as it stood, also felt that they could ill afford its dismantling. Unable to personally pay for or leverage the ability to be protected in place, they were nevertheless at an advantage when it came to accessing resources to otherwise adapt—an advantage that coming to terms with climate change might undermine and even undo.

Discussion and Conclusion

Through a case of agnostic adaptation, this article has explored how climate action can proceed in the absence of climate talk and what it means for it to do so. In post-Sandy Staten Island, not talking about climate change was paradoxically seen to facilitate rather than hinder residents' mobilization for buyouts. Agnostic adaptation minimized conflict, helping to transcend the partisanship and polarization that mire much political action. Reluctance to broach the topic reflected not only the conservative bent of Staten Island but also a broader tendency in the United States to distance oneself and one's civic engagement from the negative connotations of politics as usual (Bennett et al. 2013). What Bennett (2013) and her colleagues termed "disavowal of the political" spans the ideological spectrum and has become a common "cultural idiom" at a time of widespread mistrust in government and cynicism about the political process (518). In an argument that extends to what I observed of climate agnosticism and public silence, "disavowal," they found, "can be *productive* of civic engagement. However, divorcing the concept of 'politics' from the everyday work of active citizenship involves trade-offs, such as excluding marginalized groups and minimizing the value of conflict in democratic debate" (Bennett et al. 2013, 520).

This is not to say that agnostic adaptation was *apolitical*. To the contrary, minimal talk of climate change was accompanied by ample discussion of the politics of development and land use. Such themes drew on a long history of local activism and offered a unifying narrative in parts of the shore where residents shared firsthand experience watching wetlands give way to housing and impermeable surfaces that worsened flooding in readily observable ways over the years. A focus on Mother Nature wanting her land back begged the question of who had taken it from her in the first place, channeling anger at government for aiding and abetting development in ways that contributed to the production of flood risk. It legitimated claims for buyouts and made retreat feel like a meaningful act; giving back to Mother Nature was very different than giving in to rising seas or giving up in the face of climate change. Agnostic adaptation about the role of climate change allowed for the kind of bottom-up claims and action that dominant technocratic and "post-political" approaches to climate change are often critiqued for foreclosing (Swyngedouw 2010; Eriksen, Nightingale, and Eakin 2015; Lindegaard 2018). Not talking about climate change can thus reflect "not simply an unwillingness to face an 'inconvenient truth,' but a political reaction against those who would use truth to eliminate politics" (Brown 2014, 141).[3]

At the same time, while rallying residents to act, talk of recurrent flooding gave more powerful figures an out, a way to redress past wrongs without taking responsibility for the full scope of risk and systemic contributions to its production. Rather than acknowledge and attend to uneven vulnerability, differential responsibility, and the power disparities that underlie and perpetuate the existence of global warming, agnostic adaptation offered security only to a select few, evading difficult questions of justice and redistribution—questions that overt climate talk has also tended to elide. Those able to access resources to adapt to climate change without *talking* about climate change were generally those who already possessed the power to have their demands heard and recognized. Agnostic adaptation thus acted as a form of facilitation (Collins 2010), which has typically enabled more privileged groups to remain in hazardous but desirable areas, including on the waterfront. As the effects of climate change grow more severe, however, facilitation can be expected to increasingly entail the kind of assisted relocation offered by buyouts, a form of "adaptation privilege" premised on property ownership and by extension on whiteness (Marino 2018) that risks reproducing and compounding racialized inequality through the process of adaptation, agnostic or not.

Future research could do more to interrogate the connections between climate silence and other forms of disavowal such as racism denial and evasiveness (Nelson 2013; Beeman 2015; Underhill 2018). There are salient parallels between the affordances and limits of agnostic adaptation and those presented by strategies of racism evasiveness in the context of color-blind ideology (Beeman 2015). Meanwhile, scholars such as Pulido (2018) point out the failure of Anthropocene discourse, scholarship, and policy to grapple with the role of racism. Examples of climate action such as international agreements to limit warming to 2 °C are lauded, she wrote, despite it being "fully understood that two degrees would eliminate some island states and be absolutely disastrous for much of Africa. This is key: *knowingly* allowing large swathes of nonwhite, mostly poor people to die" (Pulido 2018, 121, citing Klein). Such indifference, Pulido (2018) argued, constitutes and must be analyzed as "a particular form of widespread contemporary racism" (121). Condemning agnostic adaptation without grappling with the erasures, omissions, and violent forms of indifference that characterize explicit ways of conceptualizing and responding to climate change has consequences that bear noting in relation to Nelson's (2013) caution on denial and racism: "Locating the problem of racism in ordinary, non-elite, often socioeconomically disadvantaged, white people can also be considered a form of denial," Nelson argued. "The implication of shifting responsibility for racism onto this group is that white privilege broadly remains unquestioned and protected" (92). Further research would do well to consider climate silence and denial not simply as oppositional to climate talk but also as occurring in relation to it, even sharing patterns of silence and disavowal.

Following a string of devastating hurricanes, it was reported in November 2017 that the Trump White House had requested $12 billion for measures to promote resilience to flooding, including "large-scale buyouts" (Natter 2017). At the same time, the administration proceeded with efforts to roll back climate policy and remove mentions of climate change and related terms from government Web sites (Environmental Data & Governance Initiative 2018). It also continued ramping up anti-immigrant and anti-refugee sentiment, pursuing various and often violent means to block people entering the country and to expel those already within its borders—actions that amount to an unacknowledged but "particularly brutal form of climate change adaptation" as disasters and environmental destruction contribute to driving growing numbers from their homes (Klein 2017; see also Miller 2017).

With cities and counties across the country confronting ever more costly problems due to the failure to reduce emissions and stem dangerous warming, some localities have begun suing fossil fuel companies for the financial support needed to adapt, while some respond by rejecting any mention of greenhouse gases or predictions of future change. The stakes of talking—and not talking—about climate change are high. This makes it crucial to gain a deeper understanding of the meaning and implications that talk and silence hold in various contexts, not just at the highest levels but also in everyday life.

Acknowledgments

This article benefited immensely from the comments of three anonymous reviewers and Editor James McCarthy. I owe thanks to them and to a number of other colleagues and readers, including Daniel Aldana Cohen, Rebecca Elliott, Katherine Gottschalk, Dorothy Huey, Stephen King, Alexis Merdjanoff, Kasia Paprocki, John Lyon Paul, Caitlin Petre, Shelly Ronen, Ariel Schwartz, Raka Sen, and Elana Sulakshana, all of whom generously offered their feedback and ideas. I presented earlier versions of this article at Rising Waters: A Workshop on Urban Waterscapes hosted by the Penn Program in the Environmental Humanities and at annual meetings of the American Anthropological Association and American Sociological Association. My thanks to participants in these events, especially organizers Nikhil Anand, Bethany Wiggin, Sophie Bjork-James, and Kari Norgaard and discussant Adriana Petryna. Additional thanks to Jennifer Cassidento for her work throughout the review process, to Michael Rosch for his help transcribing interviews, and, as always, to the many interlocutors who shared their stories and insights with me over the course of this research. Mistakes are mine alone.

Funding

Funding for this research was provided by a Mellon/ACLS Dissertation Completion Fellowship and by New York University's Institute for Public Knowledge and Department of Media, Culture, and Communication. The writing was supported by a Mellon Postdoctoral Fellowship in the Humanities at the Massachusetts Institute of Technology. See https://shass.mit.edu/academics/graduate/mellon/post-doctoral-fellows

Notes

1. All names are pseudonyms.
2. In an apt illustration of this theory, the cover of Mayor Michael Bloomberg's eponymous magazine, *Bloomberg Businessweek*, featured "IT'S GLOBAL WARMING, STUPID" in all caps on its first poststorm issue.
3. As J. L. Rice, Burke, and Heynen (2015) insightfully noted, "expert-only politics runs the risk of excluding the knowledge of individuals who do not prioritize scientific explanations, who in some cases might also be the most vulnerable. Insisting on 'climate literacy' might actually be a way of working *on* these communities rather than working *with* them" (260).

References

Beeman, A. 2015. Walk the walk but don't talk the talk: The strategic use of color-blind ideology in an interracial social movement organization. *Sociological Forum* 30 (1):127–47. doi: 10.1111/socf.12148.

Bennett, E. A., A. Cordner, P. T. Klein, S. Savell, and G. Baiocchi. 2013. Disavowing politics: Civic engagement in an era of political skepticism. *American Journal of Sociology* 119 (2):518–48. doi: 10.1086/674006.

Bloomberg, M. R. 2013. Mayor Bloomberg presents the city's long-term plan to further prepare for the impacts of a changing climate. Accessed September 29, 2018. http://www1.nyc.gov/office-of-the-mayor/news/200-13/mayor-bloomberg-presents-city-s-long-term-plan-further-prepare-the-impacts-a-changing.

Brown, M. B. 2014. Climate science, populism, and the democracy of rejection. In *Culture, politics and climate change: How information shapes our common future*, ed. D. A. Crow and M. T. Boykoff, 129–45. London and New York: Routledge.

Cohen, D. A. 2016. Petro Gotham, people's Gotham. In *Nonstop metropolis: A New York atlas*, ed. R. Solnit and J. Jelly-Schapiro, 47–54. Berkeley: University of California Press.

———. 2017. The other low-carbon protagonists: Poor people's movements and climate politics in São Paulo. In *The city is the factory: New solidarities and spatial strategies in an urban age*, ed. M. Greenberg and P. Lewis, 140–57. Ithaca, NY: Cornell University Press.

Cohen, S. 2001. *States of denial: Knowing about atrocities and suffering*. Cambridge, UK: Polity Press.

Collins, T. W. 2010. Marginalization, facilitation, and the production of unequal risk: The 2006 *Paso del Norte* floods. *Antipode* 42 (2):258–88. doi: 10.1111/j.1467-8330.2009.00755.x.

Eliasoph, N. 1998. *Avoiding politics: How Americans produce apathy in everyday life*. Cambridge, UK: Cambridge University Press.

Environmental Data & Governance Initiative. 2018. Changing the digital climate: How climate change web content is being censored under the Trump administration. Accessed October 18, 2018. https://envirodatagov.org/wp-content/uploads/2018/01/Part-3-Changing-the-Digital-Climate-1.pdf.

Eriksen, S. H., A. J. Nightingale, and H. Eakin. 2015. Reframing adaptation: The political nature of climate change adaptation. *Global Environmental Change* 35:523–33. doi: 10.1016/j.gloenvcha.2015.09.014.

Gaillard, J. C. 2012. The climate gap. *Climate and Development* 4 (4):261–64. doi: 10.1080/17565529.2012.742846.

Governor's Office of Storm Recovery. n.d. Staten Island Committee Meeting #2—Presentation. NY Rising Community Reconstruction Plan. Accessed November 12, 2018. https://stormrecovery.ny.gov/regional-communities/staten-island.

Hardy, R. D., R. A. Milligan, and N. Heynen. 2017. Racial coastal formation: The environmental injustice of color-blind adaptation planning for sea-level rise. *Geoforum* 87:62–72. doi: 10.1016/j.geoforum.2017.10.005.

Jacob, K. H. 2015. Sea level rise, storm risk, denial, and the future of coastal cities. *Bulletin of the Atomic Scientists* 71 (5):40–50. doi: 10.1177/0096340215599777.

Klein, N. 2014. Why #BlackLivesMatter should transform the climate debate: What would governments do if Black and Brown lives counted as much as White lives? The *Nation*, December 12. Accessed October 18, 2018. https://www.thenation.com/article/what-does-blacklivesmatter-have-do-climate-change/.

———. 2017. Canada prepares for a new wave of refugees as Haitians flee Trump's America. *The Intercept*, November 22. Accessed October 18, 2018. https://theintercept.com/2017/11/22/canada-prepares-for-a-new-wave-of-refugees-as-haitians-flee-trumps-america/.

Knafo, S., and L. Shapiro. 2012. Staten Island's Hurricane Sandy damage sheds light on complicated political battle. *Huffington Post*, December 6. Accessed December 17, 2017. http://www.huffingtonpost.com/2012/12/06/staten-island-hurricane-sandy_n_2245523.html.

Kuh, K. F. 2015. Agnostic adaptation. *Environmental Law Reporter* 45 (1):10027–48.

Leiserowitz, A., E. Maibach, C. Roser-Renouf, S. Rosenthal, M. Cutler, and J. Kotcher. 2018. *Climate change in the American mind*. New Haven, CT: Yale Program on Climate Change Communication.

Lindegaard, L. S. 2018. Adaptation as a political arena: Interrogating sedentarization as climate change adaptation in central Vietnam. *Global Environmental Change* 49:166–74. doi: 10.1016/j.gloenvcha.2018.02.012.

Maibach, E., A. Leiserowitz, S. Rosenthal, C. Roser-Renouf, and M. Cutler. 2016. *Is there a climate "Spiral of Silence" in America*. New Haven, CT: Yale Program on Climate Change Communication.

Marino, E. 2018. Adaptation privilege and voluntary buy-outs: Perspectives on ethnocentrism in sea level rise relocation and retreat policies in the U.S. *Global Environmental Change* 49:10–13. doi: 10.1016/j.gloenvcha.2018.01.002.

McAdam, D. 2017. Social movement theory and the prospects for climate change activism in the United States. *Annual Review of Political Science* 20 (1):189–208. doi: 10.1146/annurev-polisci-052615-025801.

Miller, T. 2017. *Storming the wall: Climate change, migration, and homeland security*. San Francisco: City Lights.

Natter, A. 2017. Trump seeks $12 billion to fight flooding tied to climate change. Accessed December 17, 2017. https://www.bloomberg.com/news/articles/2017-11-22/trump-seeks-12-billion-to-fight-flooding-tied-to-climate-change.

Nelson, J. K. 2013. Denial of racism and its implications for local action. *Discourse & Society* 24 (1):89–109. doi: 10.1177/0957926512463635.

Noelle-Neumann, E. 1993. *The spiral of silence: Public opinion—Our social skin*. 2nd ed. Chicago: University of Chicago Press.

Norgaard, K. M. 2006. "People want to protect themselves a little bit": Emotions, denial, and social movement nonparticipation. *Sociological Inquiry* 76 (3):372–96. (doi: 10.1111/j.1475-682X.2006.00160.x.

———. 2011. *Living in denial: Climate change, emotions, and everyday life*. Cambridge, MA: MIT Press.

NYGovCuomo. 2013. State of the state in Staten Island. YouTube, posted August 23, 2014. Accessed October 19, 2018. http://www.youtube.com/watch?v=BesAWi2SJng.

Oreskes, N., and E. M. Conway. 2010. *Merchants of doubt*. New York: Bloomsbury.

Paprocki, K. 2018. Threatening dystopias: Development and adaptation regimes in Bangladesh. *Annals of the American Association of Geographers* 108 (4):955–73. doi: 10.1080/24694452.2017.1406330.

Pilkey, O. H. 2012. We need to retreat from the beach. *New York Times*, November 14. Accessed October 19, 2018. http://www.nytimes.com/2012/11/15/opinion/a-beachfront-retreat.html.

Pulido, L. 2018. Racism and the Anthropocene. In *Future remains: A cabinet of curiosities for the Anthropocene*, ed. G. Mitman, M. Armiero, and R. S. Emmett, 116–28. Chicago: University of Chicago Press.

Rice, A. 2016. When will New York City sink? *New York Magazine*, September 7. Accessed October 19, 2018. http://nymag.com/daily/intelligencer/2016/09/new-york-future-flooding-climate-change.html.

Rice, J. L., B. J. Burke, and N. Heynen. 2015. Knowing climate change, embodying climate praxis: Experiential knowledge in Southern Appalachia. *Annals of the Association of American Geographers* 105 (2):253–62. doi: 10.1080/00045608.2014.985628.

Schreckinger, B. 2016. Trump acknowledges climate change: At his golf course. *Politico*, May 23. Accessed October 19, 2018. https://www.politico.com/story/2016/05/donald-trump-climate-change-golf-course-223436.

Swyngedouw, E. 2010. Apocalypse forever? Post-political populism and the spectre of climate change. *Theory, Culture & Society* 27 (2–3):213–32. doi: 10.1177/0263276409358728.

Underhill, M. R. 2018. Parenting during Ferguson: Making sense of white parents' silence. *Ethnic and Racial Studies* 41 (11):1934–51. doi: 10.1080/01419870.2017.1375132.

LIZ KOSLOV is an Assistant Professor in the Department of Urban Planning and the Institute of the Environment and Sustainability at the University of California, Los Angeles, CA 90095-1656. E-mail: koslov@ucla.edu. Her research explores questions of justice and the environment, the social dimensions of climate change, and the cultural and environmental politics of cities.

The People Know Best: Situating the Counterexpertise of Populist Pipeline Opposition Movements

Kai Bosworth

Critical scholarship suggests that environmental populism is either an expression of radical democracy beyond the paternalistic liberalism of mainstream environmentalism (Meyer 2008) or that it is paranoid, irrational, and merely reactive to elite technocratic governance (Swyngedouw 2010). Because both frameworks take populism to instrumentalize knowledge production, they miss how practices of counterexpertise might condition the emergence of left-populist oppositional identities. I argue that counterexpertise is a political activity not by producing an alternative epistemology but as a minor science that contests science from within and in the process shapes left-populist political coalitions. This is illustrated through research on populist responses to the Keystone XL and Dakota Access pipelines in the Great Plains region of North America, where environmentalists, landowners, and grassroots organizers sought to position themselves as experts. Through public participation in environmental review, pipeline mapping projects, and construction monitoring, environmental populists created an educational campaign concerning topics as diverse as hydrology, economics, and archaeology. Developing counterexpertise not only contested the evidence produced by oil infrastructure firms and the state but also consolidated the oppositional identity of "the people." By examining populist knowledge production within the broader field of contentious politics, I argue that we can better understand it as neither an irrational reaction nor transparently democratic but as part of a processual production of identities of resentment and resistance. One implication is that climate change denial and disinformation spread by the oil industry might be challenged by resituating science for political ends rather than renewing neutral objectivity.

批判研究主张，环保民粹主义不是超越温和专制的自由主义下的主流环境保护主义之基进民主的展现 (Meyer 2008)，便是仅只是针对精英官僚治理的偏执、非理性之反动 (Swyngedouw 2010) 。上述两种架构皆运用民粹主义操作知识生产，因而忽略了反专家的实践如何可能成为左翼民粹主义的反抗性身份认同的浮现之条件。我主张，反专家作为一种政治活动，并非透过生产另类的认识论，而是在科学内部进行争夺的微科学，并在过程中塑造左翼民粹主义的政治联盟。此一论点通过研究北美大平原区域中的基斯顿输油管 (Keystone XL) 和达科他输油管 (Dakota Access pipelines) 之民粹反应进行阐述，其中环境专家、土地所有者和草根组织者寻求将自身置于专家的位置。通过环境审查、输油管製图计画、以及工程监督的公众参与，环保民粹主义者创造了考量水文、经济和考古等多样主题的教育倡议。发展反专家运动不仅对石油基础建设公司和国家所生产的证据进行争夺，同时巩固了"人民"作为反对者的身份认同。我通过检视更广泛的争议政治领域中的民粹知识生产，主张不将其视为不理性的反动或显而易见的民主，而是更佳地将其理解为生产愤怒与抵抗的身份认同的过程中的一部分。其中一个意涵便是，气候变迁否认主义和石油产业所传播的虚假信息，或可通过将科学至于政治端、而非重拾客观中立性来进行挑战。
关键词: 环境保护主义，专家，输油管，民粹主义。

La erudición crítica sugiere que el populismo ambiental es, o una expresión de la democracia radical que trasciende el liberalismo paternalista de la principal corriente del ambientalismo (Meyer 2008), o paranoico, irracional y meramente reactivo a la gobernanza tecnocrática de la élite (Swyngedouw 2010). Debido a que ambos marcos toman al populismo para instrumentalizar la producción de conocimiento, ellos no captan cómo las prácticas de contraexperticia podrían condicionar la aparición de identidades opositoras izquierdo-populistas. Sostengo que la contraexperticia es una actividad política no productora de una epistemología alternativa, sino como una ciencia menor que cuestiona la ciencia desde dentro, proceso en el cual configura coaliciones políticas izquierdo-populistas. Esto se ilustra por medio de investigación sobre las respuestas populistas a los oleoductos Keystone XL y Dakota Access en la región de los Grandes Llanos de América del Norte, donde los ambientalistas, propietarios de la tierra y organizadores de las bases buscan posicionarse como expertos. A través de la participación pública en la revisión ambiental, proyectos de mapeo de los oleoductos y monitoreo de las construcciones, los populistas ambientales crearon una campaña educativa relacionada con tópicos tan diversos como hidrología, economía y arqueología. El desarrollar contraexperticia no solo cuestionó la evidencia producida por las firmas de infraestructura de petróleos y el estado, sino que también consolidó la identidad opositora de "el pueblo". Al examinar la producción populista de conocimiento dentro del campo más amplio de la política de confrontación, sostengo que es posible entenderla mejor si no la consideramos como reacción irracional ni transparentemente democrática, sino como parte de una producción de proceso de identidades de resentimiento y resistencia. Una implicación es que la denegación del cambio climático y la desinformación difundida por la industria petrolera podrían retarse resituando la ciencia más para fines políticos que para renovar la objetividad neutral. *Palabras clave: ambientalismo, experticia, oleoductos, populismo.*

Although recent scholarship largely associates populism with demagoguery, authoritarianism, and reactionary illiberalism (Müller 2016; Mudde and Kaltwasser 2017; Scoones et al. 2018), this assessment has been countered by a "persistent counter-refrain" (Grattan 2016, 19) that understands progressive or left populism as a counterhegemonic performative construction of "the people." Variously understood as "grassroots populism," "everyday populism," or "democratic populism" (Grattan 2016, 33), such emergent "environmental, pro-democracy, and anti-corruption mobilisations" (Gerbaudo 2017, 6) are distinguished from right-populisms through their desires to actualize an ideal of popular sovereignty. Such progressive populism has consistently animated leftist and radical movements in the U.S. Great Plains since the 1890s to the extent that the Marxist historian Pollack (1976) approvingly claimed that in the Midwest, "populism described the results of ideology, and Marx its causation" (72). I argue that this genre of oppositional, cross-class populism is at work in aspects of some contemporary oil pipeline opposition movements.

Even narrowly defined, progressive, grassroots populism has been incredibly divisive for the political left. On the one hand, some theorists suggest that "racism is essential" (Rancière 2016, 102) in the creation of the collective subject of populism, such that populism "harbors ... a long term proto-fascist tendency" (Žižek 2008, 280). Such a position is shared by the political center and Keynesian liberals, for whom "every populism, right or left, is equally suspect, because each one represents the pathologically unhinged demos that the existing institutional order seeks to moderate, filter, and contain" (Riofrancos 2017; see also Mann 2017). On the other hand, Laclau and Mouffe contended that populism "must be conceived as a 'radical reformism' which strives to recover and deepen democracy" (Mouffe 2016) and that it is "the royal road to understanding something about the ontological constitution of the political as such" (Laclau 2005, 67). Assessments of populist politics in the global climate justice movement are also deeply split. From the World People's Conference on Climate Change and the Rights of Mother Earth at Cochabamba, Bolivia, to the People's Climate March in New York City, scholars and activists disagree. Some suggest that "low-carbon populism" (Huber 2017), "a sustained and populist climate movement" (Klein 2014, 157), or "a popular movement for climate justice ... is a necessary condition for more radical actions" (Smucker and Premo 2014). Others assert that such strategies smack more of a "corporate PR campaign" (Gupta 2014) "which, because its demands are amorphous, can be joined by anyone" (Hedges 2014).

For the progressive environmentalists, landowners, and community organizers fighting the Keystone XL and Dakota Access pipelines in the U.S. Great Plains, environmental populism unfolded precisely through an iterative politics of scientific counterexpertise. To be clear, not all pipeline opposition is populist in character, nor did it necessarily uphold counterexpertise. Struggles for decolonization and Native sovereignty, for example, did not emphasize retrieving a supposedly lost U.S. American democracy or popular sovereignty. Nonetheless, many progressive citizens' groups opposing pipelines were decidedly populist, and some explicitly called themselves populists. This fact that should give those conflating populism and the political right some pause. In this article, I contend that contesting the scientific process of environmental review through counterexpertise was one important condition that consolidated a collective identity of "the people." To pipeline opponents, the failure of environmental review demonstrated that fossil fuel industries had so deeply influenced third-party contractors and state agencies that they could no longer grasp the truth. This consolidated their opposition to both the state and fossil fuel firms, creating conditions for more radical political possibilities. This research further suggests that both climate change denial and technocratic liberalism might be challenged by resituating scientific knowledge production toward clear political ends.

Populism and Environmental Expertise

Populism is a contested concept. At its most general level, it is defined as the performative political act that constructs "the people" as a unified, collective body in opposition to the perceived corrupt power of institutionalized elites or outsiders (Canovan 1981; Laclau 2005). As a colloquial signifier and political discourse, "populism" is frequently used to symmetrically equate extreme positions on both the political left and right, both of which are said to express grievances against institutionalized liberalism. This recently common use of populism has its roots in denunciations of agrarian politics in modernization theory and, most famously, the work of Hofstadter (1960), who understood populism as paranoid, anti-intellectual, and antidemocratic.[1]

This use of populism cannot be upheld when applied to left populisms, because no symmetry exists in the political discourse or social formation of left- and right-wing populisms and the manner in which they construct the people (Sibertin-Blanc 2013). For the political right, the language of the people substantializes nationalism, nativism, and reactionary politics. By contrast, a growing body of political theory argues that left populism can be distinguished by its desire to enact democracy as popular sovereignty, against the lip service it is paid by elites, elected politicians, and the liberal state more generally (Grattan 2016; Gerbaudo 2017). With its roots in the agrarian and producerist movements of the Farmer's Alliance and the People's Party (or Populist Party) of the late 1800s, "democratic populism" or "grassroots populism" could seem like a regionally specific U.S. understanding. Yet Gerbaudo (2017)

showed how this definition of left populism can have broad application to social movements around the world fighting for justice, equality, and a deeper democracy.

Analyses of left-populist discursive strategy are still fundamentally split. The position upheld by Laclau (2005) and Mouffe (2016) claims that populism is an authentic expression of radical democracy with the flexibility and creativity to counter institutionalized postpolitics. By contrast, many Marxists uphold the position that "populism places too little emphasis on class" (Dean 2017, S44). By refusing to name a particular, properly political subject (e.g., the proletariat), populism is too vague a political identity to enact justice (Swyngedouw 2010). Such normative dismissals have some merit but hardly explain populism's ongoing persuasive abilities (Kazin 1998). Through what processes does left populism enroll its subjects, and with what effects? It is my contention that, in the arena of environmental politics, disputes over expertise play an important role unacknowledged by contemporary scholarship.

Meyer (2008) argued that U.S. environmentalist discourse is split between a paternalistic and a populist persuasion. Paternalistic environmentalism consolidates elite power through a white, upper-class orientation and demonizes the poor and marginalized peoples as mindless, antiecological masses. The close relationship between some elements of science and environmental governance has undoubtedly contributed to further consolidation of paternalistic power. In this situation, "politics more and more becomes a struggle between those who have expertise and those who do not" (Fischer 2000, 23). This is evidenced in the United States by the manner in which "Big Green" nongovernmental organizations attempt to retain this exclusive power through their supposed expertise or counterexpertise (Eden 1996; Klein 2014).

On the other hand, environmental populism, like contemporary anti-extraction movements worldwide, "perhaps ... shouldn't be referred to as an environmental movement at all, since it is primarily driven by a desire for a deeper form of democracy" (Klein 2014, 295). Yet environmental populism adds a new valence to left populism through its emphasis on "local knowledge rooted in the particularities of place and community" (Meyer 2008, 225). Because many contemporary North American environmentalisms forefront expert knowledge as a site of struggle, they can provide an important case of the understudied manner in which contestation of expertise can generate populism.

Swyngedouw (2010) claimed that any sense that "the people know best" is upheld by their investment in evidence emerging from a "scientific technocracy assumed to be neutral" (223). For this reason, he argued that environmental populism is "inherently non-political and non-partisan" (Swyngedouw 2010, 223). This argument runs counter to Meyer's sympathetic view, which sees the populist persuasion elaborating not a faith in technocracy but instead in experiential, nonscientific knowledges. In my assessment, both of these positions see the role of environmental expertise in environmental populism as too instrumental.[2] Each suggests populism does not actually hinge on practices and processes of scientific knowledge production but only claims-making based on contesting scientific results. For its critics, populists suspiciously subordinate ecological expertise to the conspiracy theories of the people. For its champions, populists already have all of the knowledge they need in their lived experiences and thus need no supplementary scientific expertise.

What remains scarcely explained by such detractors of the contemporary consolidation of expert knowledges is how the development of practices of counterexpertise could condition the emergence of environmental populism. Fischer (2000) noted that reactions against expertise can engender "both right- and left-wing populisms, [which] hold out a return to grassroots democracy as the key to revitalizing American society" (28). Although it is clear that divisions in types of knowledge can engender resentment against elites, I demonstrate how the process of developing counterexpertise can contribute to the populist political form.

With what concepts can we understand the construction of the people through rather than only against expertise? First, I would argue that we must take the postfoundationalist stance that there is no essential identity to the people, that in a phrase Deleuze (1989) drew from Paul Klee, "the people are missing" (216). This position counters the dismissive thesis that populisms are merely reactionary movements concerned with "'THE' Environment and 'THE' People, Humanity as a whole" (Swyngedouw 2010, 221). Contrary to this claim, much of environmental populist discourse is characterized by an intense attention to place-based, open-ended constructions of "a people," constructed through provisional alliances (Iveson 2014; Hébert 2016; Grossman 2017; Andreucci 2018). In such formulations, the people is not assumed as a given nation or population, let alone all of humanity, but instead must be carefully and provisionally assembled, always with the risk of failure.

Second, through attempts to mobilize expert knowledge, environmental populisms are frequently constructed through minor sciences that leak from or cut at the edge of elite or "major" science. A minor science could be understood to be involved in the never-finished, always-processual construction of an oppositional sense or tone that composes a people through alliance or affinity (Katz 1996; Thoburn 2016; Barry 2017). Importantly, minor science takes part in "the organization of the social field," the latter being immanently "a part of that science itself" (Deleuze and Guattari 1987, 368–69). Counterexpertise could be considered minor science, insofar as it is an iterative process of scientific contestation through *bricolage*, a "taking up of whatever is at hand" (Secor and Linz 2017, 568) that goes beyond common sense, lay, or experiential knowledge toward developing new, scientific skills among the people. Seeing counterexpertise as a minor science confounds the assumed division between elite science and popular or lay knowledge. Both counter-expert and lay knowledges are capable of engendering or being captured by either "paternalistic" or "populist" environmentalisms. Thus, a situation in which counterexpertise congeals a collective subject of the people can teach us much about contemporary populism. Indeed, because environmental populism decomposes and recomposes scientific knowledge precisely as if it were not neutral, it is capable of grounding a distinctly political (rather than depoliticizing) science.

Briefly contrasting such a position from contemporary Gramscian political analyses of both populism and experiential knowledge (Mann 2009; Hart 2012; Crehan 2016) can elucidate the specific relationship between science and politics at stake, which otherwise has much in common with an analysis of minor science (see Keeling 2007; Featherstone 2011). Gramsci understood the construction of a people through the counterhegemonic process of unraveling "common sense" to contest the hegemonic consolidation of knowledge and national identity. Although

Gramsci's analysis should not be understood as economically or class reductionist, he was undoubtedly drawn toward understanding counterhegemonic knowledge production that is emergent from subalternity, marginality, or class struggle. By contrast, the portion of the antipipeline movement I examined did not, in my assessment, hinge on a pedagogy, knowledge, or category of "the oppressed." This is not to say that it could not eventually lead to a praxis-oriented politics (see Carter and Kruzic 2017) but that open-ended, performative construction of the people elaborated in pipeline opposition cut across various class positions, social identities, and spaces.

Second, whereas Gramsci paid close attention to popular culture and knowledge, vigorously contesting positivist theories of knowledge, he devoted less attention to science and expertise as a field of struggle. Gramscian analysis has instead paid more attention to how "experiential, placed-based, and nonscientific knowledge" (Rice, Burke, and Heynen 2015, 254) exceeds and challenges science, taking the latter to be evidence of depoliticization. By contrast, Wainwright and Mercer's (2009) understanding of a Gramscian elaboration of situated science as a social process of iterability is closer to the concept of minor science. The minor for Deleuze and Guattari (1987) emerges not in outright opposition to the major but from within "a scientific field" (367). Thus, counterexpertise as minor science does not elaborate an alternative epistemology based in common sense, popular culture, or lay knowledge but is constructed by augmenting scientific practices. Insofar as minor science strategically affiliates, it has further affinity with Haraway's (1990) political tendency toward a feminist coalitional or "united front politics" (151).

This conceptual framework further reiterates that left and right populisms are not at all symmetrical in form despite the fact that both construct the people against "elites." On the political right, petro-populism and conservative skepticism of climate science hinge on fear of institutionalized elites and government interference in the market (Huber 2013) to consolidate the normative force of a determinate, substantialized, majoritarian people, namely, white Americans. On the other hand, minor science exposes and unravels the majoritarian people through staging its own performative assembly toward a utopian and not-yet-existent popular

sovereignty (Sibertin-Blanc 2013; Butler 2016). There is no reason to be especially romantic about minor science. In the case of pipeline opposition, it was partial, fragmentary, and largely unsuccessful in constructing a durable political subject. Yet, important, the development of expertise as a minor science and subsequent populist social movements also created the conditions of possibility for deeper resentment toward state and corporate forces as well as opening possibilities for more politically radical forms of pipeline opposition.

The People versus the Pipelines

A number of climate activists recall that "in the mid- to late-2000s, the US climate movement was flailing and fractured, and had not unified around common opponents" (Russell et al. 2014, 167). Focusing on climate policy at a national level seemed to reach a final death knell with the failure of the American Clean Energy and Security Act in 2009. On the international stage, the Copenhagen Summit was a disappointment. Yet at the same time, on the Great Plains of the upper Midwest, a new and different kind of environmental movement was forming. Antipipeline sentiment had been bubbling in the Dakotas and Nebraska, where farmers, ranchers, Native nations, users of public parks, and drinkers of water were increasingly disgruntled by the sudden appearance of TransCanada's plans for the Keystone XL pipeline. As these emerging antipipeline sentiments coalesced into organized opposition, mainstream climate activists began to see this movement as "more capable of keeping carbon in the ground than lobbying efforts" (Russell et al. 2014, 168). The strategy and discourse of populist opposition would have a transformational effect on U.S. environmentalism.

TransCanada's Keystone pipeline system is a network of proposed and completed oil pipelines designed to bring diluted bitumen over 2,000 miles from the Canadian tar sands near Hardisty, Alberta, across the continental United States to storage facilities in Cushing, Oklahoma, and refineries near Port Arthur, Texas, and Patoka and Wood River, Illinois. The Keystone XL phase of the system was proposed in 2008 and included a route from Hardisty to Steele City, Nebraska, traversing Montana and South Dakota to also interlink to the Bakken oil field in North Dakota. Although Keystone I, an earlier TransCanada pipeline routed through the eastern Dakotas, was finished in 2009, Keystone XL became mired in controversy as it crossed hundreds of parcels of private land, ecologically sensitive wetlands, and the historic land base of the Oceti Ŝakowiŋ Oyate—the Lakota, Dakota, and Nakota people—as well as several other Native nations. Dozens of public comment sessions, public scoping meetings, and evidentiary hearings served as hotbeds for opposition from 2010 to 2014. Just when it appeared likely that the permit for Keystone XL would be ultimately rejected by the Obama administration in 2015, the Dakota Access Pipeline (DAPL) was being permitted by Energy Transfer Partners (ETP) to connect increasingly desperate Bakken producers to Patoka by way of South Dakota and Iowa. Because DAPL was determined to require no federal environmental impact statement (EIS), it was pushed through state public utilities commissions at breakneck speed before its rise to international prominence due to the blockade near the Standing Rock Sioux reservation. In January 2017, President Trump reversed the former administration's decision on Keystone XL as his first act of office, symbolically demonstrating the significance the project holds for the political right.

The pipeline buildout provoked numerous political actions at different scales, including testimony at hearings, public protest, concerts, cookouts, and blockades. From 2014 to 2016, I acted as a participant-observer at more than forty of these events, recording public discourse of the people, helping build counterexpertise where I could and tracking grievances and successes throughout the Dakotas, Nebraska, and Iowa. My object of analysis was left-populist responses to the pipelines; this then excluded mainstream Big Greens on the one hand and radical autonomist, anarchist, or decolonial activism on the other. I further conducted twenty-three semistructured interviews with community organizers, landowners, and activists who saw themselves as part of this movement. Finally, I coded and analyzed a sample of 700 unique written and oral public comments recorded during the Keystone XL EIS permitting process (2008–2015) and the South Dakota Public Utilities Commission evidentiary hearings for both Keystone XL (2009 and 2015) and DAPL (2014) for language of the people and its connection to grievances against expertise.

Among various political activities taken against the pipelines, populist discourse and activity emerged in a manner oblique to the mainstream environmentalism of Big Greens and the radical tactics of direct action and sabotage. Rather than appeal to the power of policymakers, the people was taken to be the principle subject capable of enacting democracy and defending the land. Protest signs and public testimony frequently displayed slogans such as "people power," "people > pipelines," and "we the people …" as grounds for opposition. Comments on the EIS were replete with the sentiment that, in the words of one commenter, "We have sent a clear message to President Barack Obama, Transcanada, and the U.S. Congress. They need to listen to us, because We Are The People" (U.S. Department of State 2013a, 84). Due to the history of successes of the Farmer's Alliance and the People's Party (or Populist Party) in the region, populism has long played a role in the cultural identities of the upper Midwest and has resulted in its identification at times with grassroots environmentalism (Ostler 1993; Husmann 2011; Lee 2011; Ferguson 2015). One nonprofit leader involved reflected on the evident desire for "less establishment [and] more populism," claiming that "a movement of We the People, in the Heartland of America, is one of the big reasons we stopped a pipeline" (Kleeb 2016). Importantly, then, populism became not just a political ascription from the outside but also an identity of pipeline opponents.

Yet this populist identity or political formation did not precede a politics of knowledge that was then instrumentally organized in support of the campaign against the pipelines. Instead, I argue that populist discourse emerged from struggles over expertise. Many pipeline opponents took offense to the sense that TransCanada, ETP, or the federal government considered the people to be unintelligent. As one public commenter put it, "It shocks me to think that those people who work in government (on the dime of We the People) seem to believe that we, out here in the rest of the country, are morons" (U.S. Department of State 2013a, 13). To properly understand the emergence of populist politics in the contestation of the pipelines, we need to examine how individuals and groups had to position themselves as knowing more than these outside entities.

Counterexpertise Conditions the Emergence of the People

In response to the overwhelming amount of evidence under review in the EIS, individuals and groups took it upon themselves to educate each other about subjects as diverse as environmental law, pipeline spill cleanup regulations, soil science, economics, and hydrology. The purpose of developing counterexpertise was initially to contest the evidence presented in environmental review. As these attempts failed, however, this minor science frequently began to serve another function: to consolidate the oppositional identity of the people. It was through demonstrating that the traditional legal mechanisms for adjudicating knowledge were exhausted that the people emerged as a subject of identification.

Scoping meetings, public comment sessions, and evidentiary hearings served as initial sites of the coherence of a populist subject against pipelines. These participatory governance mechanisms were historically created with the idea that the recognition of public knowledge could enhance success of development projects. Interfacing with the public might also create broader acceptance of industrial change. At early public meetings, many pipeline opponents embraced the opportunity to testify, enthusiastically building a case that the pipeline would result in negative impacts to tribal land and water, farm and ranchland, and sensitive ecological areas. Early testimony served to ground opposition in experience, local knowledge, and long-term heritage and frequently contested the scientific understandings of land and water described in the EIS.

For example, citing their long-term life and labor on the land in question, ranchers testified that the high water table in south central South Dakota and northern Nebraska was not adequately considered in the EIS. They argued that a pipeline leak in this area could result in contamination of the Ogalala Aquifer. Lay opponents and scientists both testified that the "boundaries" of the Sandhills bioregion corresponding to this unique hydrogeology as mapped by the Nebraska Department of Environmental Quality (NDEQ) did not correspond to its actual extent. Opponents made such a good case that Keystone XL was rerouted around the new boundaries of the Sandhills. From the perspective of the environmental review process, this might appear to be a prime example of the importance of public testimony. Yet, of course, a simple rerouting of the pipeline was deeply unsatisfying to opponents, who were forced to develop new evidence against the pipeline. Understanding the limits of lay knowledge, they instead began conducting their own research to try to demonstrate what the boundaries of the Sandhills were, taking soil samples, and remapping the Sandhills region based on this evidence. The mobilization of expert evidence, the failure of lay knowledge, and the subsequent development of counterexpertise only strengthened the resolve of pipeline opponents in Nebraska.

State-scale evidentiary hearings put expertise on trial, as landowners, Native nations, and environmental and community groups honed their arguments while attempting to discredit those of TransCanada. Differences in performance, professionalism, dress, knowledge of the law, and argumentation between paid experts and unpaid lay people augmented perceptions of knowledge and expertise. At the first South Dakota Public Utilities Commission (SD PUC) evidentiary hearings in 2009, the parade of expert witnesses took on a near-absurdist quality. One commenter captured this sense particularly well in suggesting that "TransCanada cannot even get their lies straight between their own expert witnesses. They have to bring in an expert witness to refute what other expert witnesses say when it does not fit their agenda" (Public Utilities Commission of South Dakota 2009b, 27). Another individual involved in the second set of evidentiary hearings in 2015 later described to me how TransCanada "continually called people as experts [who were in fact] senior company people who knew virtually nothing about any of the things that were going on ostensibly under their direction." When these experts were less than convincing, TransCanada simply "tried to distance themselves as far as possible from anything that could have given them a fault."

Like many of my interlocutors, I found it difficult to find a real differential in expertise on display in evidentiary hearings. Just like the various kinds of experiential, lay, and local knowledge laid out against the pipeline, emergent scientific counterexpertise tended also to be dismissed. Climate change was not allowed to be discussed in the state-level review process in South Dakota. When scientific evidence was brought from expert witnesses called by pipeline opponents, it was often unclear how it was being judged. Independent paleontologist and *Tyrannosaurus rex* expert Peter Larson testified that Keystone XL construction could harm fossils in the rich Hell Creek Formation in northwestern South Dakota that date from the late Cretaceous. He argued that this would further affect the ability to understand prehistoric mass extinction caused by drastic climate change, a rather important set of evidence given contemporary global warming. Later, his testimony was disputed when a state witness suggested that although they were "not an expert," they did not expect paleontological resources to be harmed because "if you go to any museum that has fossils, you'll see them in pieces" (The Public Utilities Commission of South Dakota 2009a, 259). The dispute around paleontology did not provide enough evidence for the SD PUC. Instances such as this, in which evidence and expertise seemed to actually favor those against the pipeline, left many with the feeling that it was not the evidence itself but the stakes of scientific inquiry that were on trial.

Grievances toward the "major" mode of expertise leveraged in federal environmental review coalesced into the construction of the populist subject position. The connection between scientific expertise and will of the people is perfectly captured by one public comment:

> The State Department statements regarding the Keystone XL review are incomprehensible and an outrage against the concept of scientifically robust analysis—even American democracy itself. If I am to take reported comments and analysis seriously, there is a dramatically evident disconnect between what State Department looks at on one hand, and what any competent evaluator would look at to judge the long term safety, health, environmental and economic merits of the project. We The People who care about this and related issues devote tremendous time and energy to pursuing fact-based information upon which we rely to make our decisions as "informed citizens." … The State Department must go back to the beginning and do a competent review and report that will withstand the scrutiny of the scientific community and We The People. (U.S. Department of State 2013b, 469)

As this comment demonstrates, expertise was often part of the composition of the subject position of "We The People" as well as what that subject took to be at stake in environmental permitting. The commenter demanded recognition of the complementarity of scientific and popular authority. Yet federal environmental review would respond to an abridged version of this comment by omitting its political content, instead arguing simply that the EIS sufficiently "presents information and analyses regarding indirect cumulative impacts and lifecycle GHG emissions, including the potential impact of further development of the oil sands on climate change" (U.S. Department of State 2013c, 181).

Pipeline opponents found that the minor strategies of public testimony and counterexpertise were even less successful in arguing against DAPL. Many felt forced to go beyond the established political process, which had quickly approved the pipeline. Organized in part through the Science and Environmental Health Network, Indigenous Environmental Network, and Dakota Rural Action, the Bakken pipeline watchdogs network was one new strategy. The group began monitoring the nascent DAPL construction process, using the law to delay construction while legal cases and blockades escalated opposition elsewhere. This strategy required not only that opponents understand environmental laws and regulations but also that they cultivate the skills to see violations. The pipeline watchdogs held trainings that helped attune themselves to violations of the law, basic surveillance skills, and the fortitude to follow and observe construction crews. The construction watchdogs shared images of legal violations from all along the 1,200-mile pipeline route. Countermapping was a crucial aspect of monitoring, as a public map that displayed active work sites and completed portions of the pipeline was frequently updated. In addition to actions that took place within the rule of law, some pipeline opponents began to cultivate their knowledge of how to sabotage construction equipment. Crucial to such actions were also a range of scientific and technical knowledges but, departing from populist strategy, these skills did not circulate in public until long after the pipeline's completion.

The failure of acts of counterexpertise to prevent pipeline permitting or construction might seem like an example of the depoliticizing effect of technocratic politics and a political dead end. I was surprised, however, to find that many pipeline opponents, reflecting on their participation in practices of counterexpertise, disagreed with this sentiment. The belief that "the people know best" grounded their opposition well beyond whatever form of expertise the state recognized. The problem was not that they lacked expertise but that they lacked financial resources comparable to those of TransCanada to hire experts to give testimony in evidentiary review. Contrasting the populist movement in South Dakota with the strategy of the Big Greens, one organizer suggested that "if we could get 350.org [to] give us 1 percent of their public relations budget, we could downright pay the lawyers and pay the experts." Another landowner told me, "I have nine boxes of evidence printed out back home, but none of it matters." She felt that her testimony would not make a difference.

Nonetheless, opponents found that the development and performance of their counterexpertise reinforced their commitment. The disheartening experience of going through the environmental review process and losing despite the obvious truth of their position reinforced the identities of resentment and resistance that composed populist politics. One community organizer attested to both of these sentiments while also taking a characteristic trust of the people and skepticism toward elite environmentalisms.

> You know TransCanada didn't have any problem paying for its so-called experts and all the PUC with our money. And we … could not call or it was very difficult to call [upon experts] because you know just trying to raise resources to do that. … The so-called Big Greens are so caught up with their multimillion-dollar projects and, and, just trying to play nice, um, and they had no time let alone any willingness to invest resources—a fraction of what they're using on their full page ads in the *Washington Post* or whatever—to help us with experts or anything like that. It's a little bit disconcerting but it's all educational. We know that ultimately any protection of our water resources is gonna come from the people here and that's the only place it'll come from. … Sometimes you have to push some of these agencies to do their jobs and if you get them to do it, great, that's what we want them to do, and if you can't, hopefully people will learn that you need to try something else.

Although the position of pipeline opponents was increasingly cynical about the role of counterexpertise in environmental review, they did believe that the minor science of self-education was crucial. Through such acts, they exhausted the political potential of contesting the pipeline through official channels and demonstrated that the people ultimately need to take power themselves. When the organizer earlier noted that "it's all educational," they are suggesting that the people are learning how to contest through expertise and that politics actually emerges beyond that very venue. The failure of minor science to actually stop DAPL should not obscure its success in enabling a collective of political subjects increasingly capable of moving beyond that sphere to do "something else"—communicate, inform, organize, blockade, or even sabotage.

Conclusion

It is important to attest to the wide range of expert knowledges that were contested by pipeline opponents. These further included disputes about aquifer boundaries and communication, diluent chemical composition, cultural resource surveys, flow rates of heavy crude in water systems, the economic impact of pipeline construction and oil export, and several other micro- to macro-antagonisms surrounding the supposed "national interest" in constructing new oil pipelines. Through this engagement with struggles over and within expertise, pipeline opponents came to understand a fundamental split—not between elite knowledge and local or lay experience but between a science in the interests of the state and capital and a minor science—what we might call a science for the people. This understanding led to disaffection with traditional routes of political contestation and eventually a path more open to radical politics.

Since the initial rounds of public review, pipeline politics has become even more polarized. Fossil fuel–funded public relations firms attempt to dispel any counterexpertise through "transparent fact-checking" Web sites. In response to the supposed success of "fake news" in capturing the rural masses, many political analysts have doubled down on the liberal distrust of populism, left or right. Others on the political left believe that in forming their identities as an alternative to elites, populists are doomed to subordinate proper politics to unprincipled argumentation with experts. These uncharitable views, I have argued, miss the ways in which populism can incrementally construct itself out of a minor science. Although it might be insufficient in itself, populism thus can produce conditions of possibility for subjects willing to go beyond the status quo, intriguingly by maintaining a ground in practices of scientific counterexpertise that were precisely interested rather than objective. Pipeline opposition demonstrates that common people are keen at developing expertise in a wide range of knowledges, including the art of politics. Given that no political collective is born with a ready-made critique of the state and capital in hand, these minor sciences offer a glimpse at the cultivation of radical opposition without recourse to a messianic event to come.

The strategies of pipeline opponents have a final implication for combating the depoliticization of climate science, suggesting that climate denial and postpolitical governance might be more effectively challenged if scientific and expert practices are not understood to be modes of depoliticization opposed to local experience. Minor sciences split scientific epistemology from within, creating points of alliance and leverage through which the hold of the state and fossil fuel industry on the scientific field can be severed. This strategy offers no guarantees. The risk of even minimal left populism activating nationalisms or other reactionary or authoritarian mobilizations in the United States is very real. The deeply American understanding of popular sovereignty relies on either forgetting settler coloniality or appropriating indigenous resistance (Bosworth forthcoming). Nonetheless, taking scientific counterexpertise as a fundamentally depoliticizing aspect of populism runs counter to the lessons of the struggle against Keystone XL and DAPL. If a mass mobilization is indeed necessary for any chance at climate justice, we will have to learn from activists and organizers that perhaps belief in the rule of expertise could lead affirmatively to a science for the people.

Acknowledgments

The author thanks three anonymous reviewers and James McCarthy for their generous comments.

The author is now a Visiting Assistant Professor at Brown University, Providence, RI.

Funding

Aspects of this research were supported by funding from the Social Science Research Council's Dissertation Proposal Development Fellowship.

Notes

1. Prior to the 1950s, populism in the United States was largely understood as a left-wing—even vaguely socialist—form of politics. Yet emergent forms of modernization theory and postwar Keynesian fears of the people attempted to redefine populism as an irrational and anti-Semitic form of politics. This meaning was challenged by U.S. historians, especially Pollack (1976) and Goodwyn (1976, 1978). In Europe, however, populism continued to be associated with the political right such that something like left populism could seem an oxymoron. Jäger (2017) meticulously traced this "semantic drift" in the meaning of populism, with the conclusion that "recent conceptualizations [of populism] may lack an awareness of the implications of the vocabulary it deploys" (311).
2. A similar bifurcation has structured historians' assessments of the Farmer's Alliance and the People's Party. As Postel (2009) wrote, "Historians have tended to cast academic experts in the role of modernizers battling to overcome the inertia of 'reluctant farmers,' who were mired in tradition and unconvinced of the value of education" (47). Postel challenged this thesis through evidence of a massive campaign of counterexpertise that fought not against the modernizing ideals of agricultural science but against the method and ends to which they were used.

References

Andreucci, D. 2018. Populism, hegemony, and the politics of natural resource extraction in Evo Morales's Bolivia. *Antipode* 50 (4):825–45.

Barry, A. 2017. Minor political geographies. *Environment and Planning D: Society and Space* 35 (4):589–92.

Bosworth, K. Forthcoming. "They're treating us like Indians!": Political ecologies of property and race in North American pipeline populism. *Antipode.* doi:10.1111/anti.12426

Butler, J. 2016. "We, the people": Thoughts on freedom of assembly. In *What is a people?*, ed. A. Badiou, J. Butler, G. Didi-Huberman, S. Khiari, J. Rancière, P. Bourdieu, B. Bosteels, and K. Olson, 49–64. New York: Columbia University Press.

Canovan, M. 1981. *Populism.* New York: Harcourt Brace Jovanovich.

Carter, A., and A. Kruzic. 2017. Centering the commons, creating space for the collective: Ecofeminist #NoDAPL praxis in Iowa. *Journal of Social Justice* 7:1–22.

Crehan, K. 2016. *Gramsci's common sense: Inequality and its narratives.* Durham, NC: Duke University Press.

Dean, J. 2017. Not him, us (and we aren't populists). *Theory & Event* 20 (1):S38–S44.

Deleuze, G. 1989. *Cinema 2: The time image,* trans. H. Tomlinson and R. Galeta. Minneapolis: University of Minnesota Press.

Deleuze, G., and F. Guattari. 1987. *A thousand plateaus: Capitalism and schizophrenia,* trans. B. Massumi. Minneapolis: University of Minnesota Press.

Eden, S. 1996. Public participation in environmental policy: Considering scientific, counter-scientific and non-scientific contributions. *Public Understanding of Science* 5 (3):183–204.

Featherstone, D. 2011. On assemblage and articulation. *Area* 43 (2):139–42.

Ferguson, C. 2015. *This is our land: Grassroots environmentalism in the late twentieth century.* New Brunswick, NJ: Rutgers University Press.

Fischer, F. 2000. *Citizens, experts, and the environment: The politics of local knowledge.* Durham, NC: Duke University Press.

Gerbaudo, P. 2017. *The mask and the flag: Populism, citizenism, and global protest.* New York: Oxford University Press.

Goodwyn, L. 1976. *Democratic promise: The populist moment in America.* New York: Oxford University Press.

———. 1978. *The populist moment: A short history of the agrarian revolt in America.* New York: Oxford University Press.

Grattan, L. 2016. *Populism's power: Radical grassroots democracy in America.* Oxford, UK: Oxford University Press.

Grossman, Z. 2017. *Unlikely alliances: Native nations and white communities join to defend rural lands.* Seattle: University of Washington Press.

Gupta, A. 2014. How the People's Climate March became a corporate PR campaign. Accessed June 24, 2018. http://www.counterpunch.org/2014/09/19/how-the-peoples-climate-march-became-a-corporate-pr-campaign/.

Haraway, D. 1990. *Simians, cyborgs, and women: The reinvention of nature.* London and New York: Routledge.

Hart, G. 2012. Gramsci, geography, and the languages of populism. In *Gramsci: Space, nature, politics,* ed. E. Michael, H. Gillian, K. Stefan, and L. Alex, 301–20. West Sussex, UK: Wiley.

Hébert, K. 2016. Chronicle of a disaster foretold: Scientific risk assessment, public participation, and the politics of imperilment in Bristol Bay, Alaska. *Journal of the Royal Anthropological Institute* 22 (S1):108–26.

Hedges, C. 2014. The last gasp of climate change liberals. Accessed June 24, 2018. https://www.truthdig.com/articles/the-last-gasp-of-climate-change-liberals/.

Hofstadter, R. 1960. *The age of reform.* New York: Vintage.

Huber, M. T. 2013. *Lifeblood: Oil, freedom, and the forces of capital.* Minneapolis: University of Minnesota Press.

———. 2017. A climate policy for the people. *The American Prospect,* November 16. Accessed September 20, 2018. http://prospect.org/article/climate-policy-people.

Husmann, J. 2011. Environmentalism in South Dakota: A grassroots approach. In *The Plains political tradition: Essays on South Dakota political culture,* ed. J. K. Lauck, J. E. Miller, and D. C. Simmons, 239–66. Pierre: South Dakota State Historical Society.

Iveson, K. 2014. Building a city for "the people": The politics of alliance-building in the Sydney green ban movement. *Antipode* 46 (4):992–1013.

Jäger, A. 2017. The semantic drift: Images of populism in post-war American historiography and their relevance for (European) political science. *Constellations* 24 (3):310–23.

Katz, C. 1996. Towards minor theory. *Environment and Planning D: Society and Space* 14 (4):487–99.

Kazin, M. 1998. *The populist persuasion: An American history.* Ithaca, NY: Cornell University Press.

Keeling, K. 2007. *The witch's flight: The cinematic, the black femme, and the image of common sense.* Durham, NC: Duke University Press.

Kleeb, J. F. 2016. Let's get rural: Middle America wants less establishment, more populism. Accessed June 24, 2018. https://medium.com/@janekleeb/lets-get-rural-middle-america-wants-less-establishment-more-populism-c182224adca3.

Klein, N. 2014. *This changes everything: Capitalism vs. the climate.* New York: Simon & Schuster.

Laclau, E. 2005. *On populist reason.* London: Verso.

Lee, R. A. 2011. *Principle over party: The Farmers' Alliance and populism in South Dakota, 1880–1900.* Pierre: South Dakota State Historical Society.

Mann, G. 2009. Should political ecology be Marxist? A case for Gramsci's historical materialism. *Geoforum* 40 (3):335–44.

———. 2017. *In the long run we are all dead: Keynesianism, political economy, and revolution.* London: Verso.

Meyer, J. M. 2008. Populism, paternalism and the state of environmentalism in the U.S. *Environmental Politics* 17 (2):219–36.

Mouffe, C. 2016. The populist challenge. Accessed June 24, 2018. https://www.opendemocracy.net/democraciaabierta/chantal-mouffe/populist-challenge.

Mudde, C., and C. R. Kaltwasser. 2017. *Populism: A very short introduction.* Oxford, UK: Oxford University Press.

Müller, J.-W. 2016. *What is populism?* Philadelphia: University of Pennsylvania Press.

Ostler, J. 1993. *Prairie populism: The fate of agrarian radicalism in Kansas, Nebraska, and Iowa, 1880–1892.* Lawrence: University Press of Kansas.

Pollack, N. 1976. *The populist response to industrial America: Midwestern populist thought.* Cambridge, MA: Harvard University Press.

Postel, C. 2009. *The populist vision.* New York: Oxford University Press.

The Public Utilities Commission of South Dakota. 2009a. HP09-001 transcript of proceedings, Vol II. Accessed June 24, 2018. https://puc.sd.gov/commission/minutes/2009/hp09-001/110309vol2.pdf.

———. 2009b. HP09-001 transcript of public input hearing. Accessed June 24, 2018. https://puc.sd.gov/commission/minutes/2009/hp09-001/110309public.pdf.

Rancière, J. 2016. The populism that is not to be found. In *What is a people?,* ed. A. Badiou, J. Butler, G. Didi-Huberman, S. Khiari, J. Rancière, P. Bourdieu, B. Bosteels, and K. Olson, trans. J. Gladding, 101–6. New York: Columbia University Press.

Rice, J. L., B. J. Burke, and N. Heynen. 2015. Knowing climate change, embodying climate praxis: Experiential knowledge in southern Appalachia. *Annals of the American Association of Geographers* 105 (2):253–62.

Riofrancos, T. 2017. Democracy without the people. *N + 1.* Accessed June 24, 2018. https://nplusonemag.

com/online-only/online-only/democracy-without-the-people/.

Russell, J. K., L. Capato, M. Leonard, and R. Breaux. 2014. Lessons from direct action at the White House to stop the Keystone XL pipeline. In *A line in the tar sands: Struggles for environmental justice.* ed. T. Black, S. D'Arcy, T. Weis, and J. K. Russell, 167–180. Oakland, CA: PM Press.

Scoones, I., M. Edelman, S. M. Borras, R. Hall, W. Wolford, and B. White. 2018. Emancipatory rural politics: Confronting authoritarian populism. *The Journal of Peasant Studies* 45 (1):1–20.

Secor, A., and J. Linz. 2017. Becoming minor. *Environment and Planning D: Society and Space* 35 (4):568–73.

Sibertin-Blanc, G. 2013. From democratic simulacrum to the fabulation of the people: Minority populism. *Actuel Marx* 54 (2):71–85.

Smucker, J., and M. Premo. 2014. What's wrong with the radical critique of the people's climate march. *The Nation,* September 30. Accessed June 24, 2018. https://www.thenation.com/article/whats-wrong-radical-critique-peoples-climate-march/.

Swyngedouw, E. 2010. Apocalypse forever? Post-political populism and the spectre of climate change. *Theory, Culture & Society* 27 (2–3):213–32.

Thoburn, N. 2016. The people are missing: Cramped space, social relations, and the mediators of politics.

International Journal of Politics, Culture, and Society 29 (4):367–81.

U.S. Department of State. 2013a. *Keystone XL draft environmental impact statement: Public comment unique submissions, Part 01.* Washington, DC: U.S. Department of State.

———. 2013b. *Keystone XL draft environmental impact statement: Public comment unique submissions, Part 03.* Washington, DC: U.S. Department of State.

———. 2013c. *Summary of public comments and responses to the Keystone XL Project draft supplemental environmental impact statement.* Accessed June 24, 2018. https://2012-keystonepipeline-xl.state.gov/documents/organization/221210.pdf.

Wainwright, J., and K. Mercer. 2009. The dilemma of decontamination: A Gramscian analysis of the Mexican transgenic maize dispute. *Geoforum* 40 (3):345–54.

Žižek, S. 2008. *In defense of lost causes.* London: Verso.

KAI BOSWORTH is a Visiting Assistant Professor at the Institute at Brown for Environment & Society at Brown University, Providence, RI 02912. Email: kai_bosworth@brown.edu. His research interests include populism's possibilities and limits for left political organizing in the context of fossil fuel economies and North American settler colonialism.

Beyond Narratives: Civic Epistemologies and the Coproduction of Environmental Knowledge and Popular Environmentalism in Thailand

Tim Forsyth

Popular environmentalism can have limited democratic outcomes if it reproduces structures of social order. This article seeks to advance understandings of environmental democratization by examining the analytical framework of civic epistemologies as a complement to the current use of environmental narratives in political ecology and science and technology studies. Civic epistemologies are the preexisting dimensions of political order that the state and other actors seek to maintain as unchallengeable. They add to current analysis because they show the structures around which narratives form, as well as how knowledge and political agencies of different actors are coproduced in reductive ways. The article applies this analysis to popular environmentalism in Thailand and especially concerning community forests and logging from 1968 to present. Using a combination of interviews and content analysis of historic newspaper reporting, the article shows how diverse actors—including state, elite conservationists, and peasant activists—have organized political activism and ecological claims about forests according to unchallenged norms of appropriate community culture and behavior. These actions have kept narratives about forests and society in place and worked against alternative and arguably more empowering visions of communities and forests in recent years. The article argues that revealing civic epistemologies can contribute to a deeper form of environmental democratization than engaging in environmental politics based on existing narratives or analyzing the limitations of narratives alone.

大众环境主义若再生产社会秩序结构的话, 则可能产生有限的民主成果。本文通过检视公民认识论的分析架构, 补充政治生态学和科学与技术研究的环境叙事的当前使用, 寻求推进我们对于环境民主化的理解。公民认识论是先于国家及其他行动者企图维持不容挑战的政治秩序的存在面向。它们以还原的方式显示叙事形成的结构, 以及不同行动者的知识与政治主体如何共同生产, 因而扩充当前的分析。本文将此一分析运用至泰国的大众环境主义, 并特别关照1968年至今的社区森林与伐木。本文结合访谈与历史新闻报导的内容分析, 展现多样的行动者——包括国家、环境保育菁英, 以及农民运动倡议者——如何遵循适当的社区文化与行为的未受挑战之常规, 组织有关森林的政治倡议与生态宣称。近年来, 这些行动维护有关森林与社会的叙事, 同时反对对社区和森林而言可说更具培力远见的另类方案。本文主张, 揭露公民认识论, 能够较根据既有叙事涉入环境政治、抑或是单独分析叙事的限制而言, 对更深刻的环境民主化形式做出贡献。关键词: 威权主义, 环境主义, 政治生态学, 科学与技术研究, 泰国。

El ambientalismo popular puede generar resultados democráticos limitados si reproduce estructuras del orden social. Este artículo busca avanzar en el entendimiento de la democratización ambiental examinando el marco analítico de epistemologías cívicas como complemento del actual uso de las narrativas ambientales en ecología y ciencia políticas, y en estudios de tecnología. Las epistemologías cívicas son las dimensiones preexistentes del orden político que el estado y otros actores buscan mantener como indisputable. Aquellas contribuyen al análisis actual porque muestran las estructuras alrededor de las cuales se forman las narrativas, así como el conocimiento y las agencias políticas de diferentes actores son coproducidos de maneras reductivas. El artículo aplica este análisis al ambientalismo popular de Tailandia y especialmente con respecto a comunidades forestales y madereras, desde 1968 hasta el presente. Usando una combinación de entrevistas y análisis de contenido de reportajes históricos de periódico, el artículo muestra el modo como diversos actores—incluyendo el estado, las élites conservacionistas y los activistas campesinos—han organizado el activismo político y los reclamos ecológicos sobre los bosques de acuerdo con normas no disputadas de cultura y comportamiento comunitarios apropiados. Estas acciones han mantenido en su lugar las narrativas acerca de los bosques y la sociedad, y trabajaron contra visiones de empoderamiento de comunidades y bosques supuestamente mayores, en años recientes. El artículo sostiene que las epistemologías cívicas reveladoras pueden contribuir a una forma más profunda de democratización ambiental que comprometerse con una política ambiental basada en las narrativas existentes, o mediante el análisis de las limitaciones de las propias narrativas. *Palabras clave: ambientalismo, autoritarismo, ecología política, estudios de ciencia y tecnología, Tailandia.*

A persistent concern at the interface of political ecology and science and technology studies (STS) is the democratizing potential of popular environmentalism under conditions of authoritarianism. Frequently, environmentalism is presented as a democratizing force. Yet, increasingly, scholars acknowledge that it does not always empower marginalized people but instead can reproduce narratives, which are "devices through which actors are positioned, and through which specific ideas of 'blame' and 'responsibility' and 'urgency' and 'responsible behavior' are attributed" (Hajer 1995, 64–65). Various studies have shown how narratives project simplistic explanations of complex environmental problems while ordering social actors into blame-worthy or responsible roles (Leach and Mearns 1996; Peet and Watts 2004; Goldman, Nadasdy, and Turner 2011; Lejano, Ingram, and Ingram 2013). Yet, despite the popularity of narratives-based approaches, there is growing concern that these analyses might say too little about how narratives remain powerful and what can be done to make them more governable. Do narratives gain power because they were established in history and remain unchallenged? How do contemporary politics bring authority to different configurations of knowledge and activism?

This article aims to advance understandings of environmental democratization by examining how the analytical framework of civic epistemologies can give insights to how narratives retain political and epistemic authority in contemporary politics. Civic epistemologies have been defined as "the institutionalized practices by which members of a given society test and deploy knowledge claims used as a basis for making collective choices" (Jasanoff 2005, 255; C. Miller 2005, 2008). As such, civic epistemologies offer important ways to identify tacit connections between different configurations of evidence, actors, and styles of contestation and so provide insights for how narratives are made and can be governed. Simultaneously, this article also seeks to advance debates about civic epistemologies by analyzing the informal, civil society–based sources of knowledge and expertise within developing countries and by

examining how current political debates adopt, and are influenced by, narratives established in the past.

The article applies this analysis to Thailand, a country with significant histories of authoritarianism and popular environmentalism. In particular, it focuses on the emergence of communities as a specific category imbued with meaning and agency within forest politics. Much research has examined themes of territorialization and land titling concerning forests and community forests in Thailand (Vandergeest and Peluso 1995; Chusak and Baird 2018) or the significance of popular environmental protests in resisting oppressive state policy (Hirsch 1996; Fahn 2003; Pye 2005b). There has been relatively less attention, though, to the configuration and meaning of communities simultaneously with environmental narratives relating to forests (Pinkaew 2005; Atchara 2009; Bencharat 2014). The article adopts three main methods: First, it considers the key characteristics of social order that might shape civic epistemologies relating to communities and forests in Thailand and how these have influenced environmental narratives about forests. Second, it draws on interviews with activists and observers concerned with civil society and forest politics from the 1990s to the present. Third, it uses content analysis of historic newspaper reporting about environmentalism as an indicator of how narratives have formed over time with the selective involvement of different actors, values, and framings. These methods show how popular environmentalism concerning communities and forests has acted as a proxy for democratization under authoritarianism in ways that coproduced simplistic representations of both forests and communities. They also provide insights for understanding civic epistemologies as a template for the coproduction of environmental knowledge and agency and for ways of making these political processes more socially inclusive.

Narratives, Coproduction, and Civic Epistemologies

It is now widely accepted in geography and environmental social science that explanations of environmental problems and politics sometimes fit convenient patterns that misrepresent complex realities. An alleged example was the tendency of some historic political ecology to adopt an analytical framework that studied the how resource-dependent communities resisted destructive state development projects or unregulated capitalism (Cockburn and Ridgeway 1979; Bryant and Bailey 1997). Later studies, influenced by discourse analysis and STS, sometimes cheekily caricatured this framework as "the good, the bad, and the ugly" for using predefined normative positions to represent communities as good, states bad, and transnational corporations ugly (Béné 2005; Dwyer and Minnegal 2006). This form of predetermined environmental analysis silences knowledge and social identities inconvenient to those norms (Wynne 1996).

Many analysts in political ecology and STS now seek to indicate how the connections between political agency and knowledge are contextual, coproduced, and nonessential (Jasanoff 2004b; Latour 2005). As part of this analysis, scholars have used the framework of environmental narratives or storylines to show how nature and society are ordered together into convenient, but usually misrepresentative, statements of cause and effect (Hajer 1995). Narratives are problematic because, according to Roe (1991), "[they] tell scenarios not so much about what should happen as about what will happen according to their tellers—if the events or positions are carried out as described" (288). The concept has been used especially within critical political ecology to show how "received wisdom" (Leach and Mearns 1996) about problems such as desertification or deforestation have been shaped by historic social and political influences on the generation of knowledge (Thompson, Warburton, and Hatley 1986; Forsyth 2003; Bassett, Crummey, and Beusekom 2004; Benjaminsen 2009; Beymer-Farris and Bassett 2013). More generally, however, the framework of narratives has been a link between discourse analysis and STS to show how the identity and expected agency of social actors are shaped simultaneously with authoritative knowledge about environmental problems (Ku and Tian 2002; H. T. Miller 2012).[1]

Over time, STS scholars have expanded this approach to the related frameworks of assemblages and actor networks, which, according to Latour (2005) has "borrowed from narrative theories" (Rodríguez-Giralt, Marrero-Guillamón, and Milstein

2018, 257). Narratives, actor networks, and assemblages share the common purpose of showing how configurations of contexts, actors, and knowledge—often forged through historic entanglements—have become stabilized in common discourse and scientific and institutional practices as unquestioned truths. Narrative analysis shows how environmental problems are represented in terms of overly simple statements of cause and effect. Actor network theory refers to this process as *purification* but makes the additional point that both problems and actors exist because of their relationship to each other rather than their essential qualities (Akrich and Latour 1992). Indeed, insights from assemblage theory have been used in studies of forest history (Peluso and Vandergeest 2011) and agroforestry policies (Smith and Dressler 2017) in Southeast Asia.

Yet, despite the popularity of these frameworks, there is also growing concern that current approaches to narratives undertheorize the role of contemporary politics in coproducing political and epistemic authority (Jasanoff 2004a; Hajer 2009). First, critics have argued that narratives should not just be seen as ordering devices for interconnections between society and environment. Instead, there should also be an attempt to explain how these interconnections arise (Jones and Radaelli 2015; Lejano 2015). This argument reflects a broader debate about whether discourse analysis in political research should show how language facilitates different arguments, versus its role in shaping supposedly "real" visions of the world (Feindt and Oels 2005; Hajer and Versteeg 2005).

Second, other scholars have asked whether assemblages and actor networks have tended to overemphasize the fixity of historical events, interests, and networks in shaping knowledge and social actors (Latour 2005; Müller and Schurr 2016). Doing this might undertheorize the coproduction of knowledge and social agency because it implies that narratives act in a historically deterministic way on current politics, rather than examining how contemporary politics use, or indeed enhance, narratives in selective ways. Accordingly, critics prefer to analyze how knowledge and actors gain or lose political saliency dynamically, according to how they benefit from or invoke sources of authority that lie outside of the specific events, interests, and networks outlined in narratives (Law and Singleton 2005; Anderson et al. 2012). Indeed, Jasanoff (2004a) stated, "Coproductionist accounts are not content simply to ask what *is*; they seek to understand how particular states of knowledge are arrived at and held in place, or abandoned" (19, italics in original).

Civic epistemologies are one analytical framework that aims to show how knowledge and actors gain saliency within narratives (Jasanoff 2005; C. Miller 2005, 2008). These describe

> ways of knowing and reasoning about policy problems intertwined with ways of organizing political order. These knowledge orders are reasonably stable, in that they persist over relatively long periods of time, often embedded in institutionalized epistemic, social, and political practices. But, they are also dynamic: open to change through novel processes of co-production that link epistemic, social, and political contestation and innovation. (C. Miller 2008, 1898)

Civic epistemologies add to the debate by indicating the rules or structures of authority that make both knowledge and activism influential and hence allow the analysis of current practices that make or reproduce narratives. They also refer to the political influences on these structures, the so-called knowledge orders that create norms of appropriate discussion (Jung, Korinek, and Straßheim 2014), or "the dimensions of political order that each state seeks to immunize or hold beyond question" (Jasanoff 2012, 10). These unchallenged themes within political debate influence how actors form common identities; adopt different standards of evidence, knowledge, and activism; and maintain or challenge social hierarchies. For example, civic epistemologies contribute to shared visions of reality through helping to facilitate discourse coalitions among different political actors who disagree on cognitive points of interest but who share the same perspective on other themes (Hajer 1995; Hajer, van den Brink, and Metze 2006). Discourse coalitions often result in unseen acts of world-making (Goodman 1978) or what Law (2011) called "collateral realities" that impose unquestioned visions of cause and effect for environmental problems in exclusionary ways. They also influence which events or actions are considered as acceptable evidence of cause and effect in practice.

Yet, civic epistemologies can also be questioned. First, much analysis so far has focused on "national cultures of rationality" (Winickoff 2012, x) that refer to the formal means of resolving conflicts such as lawsuits in the United States or spokespeople from trusted institutions in Germany (Jasanoff [2005]).

(The political debate to hold the British European referendum as beyond reproach is possibly another example.) Yet, critics have questioned whether these national epistemologies are too general (Barry 2012) or whether they leave sufficient space for less formal or nonstate forms of civic epistemologies (Beck and Forsyth 2015). These concerns are especially relevant for popular environmentalism, which is usually characterized by social divisions and different values, as well as being a proxy for other forms of democratization under authoritarianism (Tickle and Welsh 1998).

Second, there is also a need to understand how civic epistemologies can be reframed through popular activism. Some research has shown, for example, how alliances between social movements and scientific institutions in the United States have succeeded in diversifying treatments for breast cancer (Batt 1994; Ley 2009) or reframing HIV research from slowing the spread of the virus to improving the lives of people with it (Epstein 1996). These examples of social movements are different from others because they have sought to transform the way in which formal scientific knowledge is generated. There has been relatively little research, however, on how social movements engage with civic epistemologies in developing world contexts and therefore might transform (rather than enact) narratives.

This Study

This article presents research on popular environmentalism concerning communities and forests in Thailand from the late 1960s to present. It asks the following questions:

- How has environmentalism expressed narratives about the problems, role, and impacts of communities concerning forest policies?
- How do these narratives indicate the influence of civic epistemologies on environmental activism?
- What are the insights for understanding civic epistemologies as an academic framework and for contributing to environmental democratization?

The article first reviews the challenges of understanding community in Thailand and evidence for civic epistemologies within this debate. It then analyzes how environmental activism about communities and forests in Thailand has been influenced by civic epistemologies. It then discusses evidence for how civic epistemologies can be reframed.

As the article seeks to investigate civic epistemologies, the research analyzes the development of activism over time rather than an in-depth or ethnographic analysis of one or more cases. The analysis starts in the late 1960s because this is when environmentalism became an identifiable political concern in Thailand. It initially refers to logging as the main concern of the 1970s and 1980s, which led to a national logging ban in 1989. After this period, the article refers mainly to the debate about community forests as the legal framework determining local rights over forest land. As noted before, there is much existing literature on the history and challenges of community forests and their relationship with state-making and land titling (Vandergeest 1996; Sato 2003; Salam, Noguchi, and Pothitan 2006; Vandergeest and Peluso 2006; Atchara 2009; Usher 2009; Ting et al. 2011; Bencharat 2014; Forsyth and Walker 2014; Chusak and Baird 2018). This article, instead, focuses more on the meaning and representation of communities within environmentalism about forests as a way to understand civic epistemologies.

The article's primary empirical content is a combination of original interviews with environmental activists and analysts in Thailand and content analysis of historic newspapers as a way to indicate how narratives about communities and forests emerged. Indeed, it has been widely noted that "journalism is history's first draft,"[2] and so tracing narratives in newspapers offers a useful insight for how the values, framings, and actors evolved. In particular, the study compared the two key search terms "logging" and "community forest bill(s)" for different periods: 1968 to 2000 for logging and 1993 to 2017 for community forest bill(s). These different terms and dates were selected because the Bangkok Post database did not refer to community forests before 1993 and the physical (paper-based) records for logging ended in 2000, requiring a different system of online searching for news after 2000.[3] Before 1993, all news about forests, including concerns about communities, conservation, and ecology, were accumulated under the single label of logging. The later focus on community forest bill(s) referred to all reports about events and developments relating to community action concerning forests, as well as the national discussions about appropriate forms of community forests legislation. Comparison of these two themes over different, but overlapping, time periods was the most effective

way to show how narratives about communities and forests changed over time.

The Bangkok Post was selected because it is Thailand's oldest broadsheet; it is also written in English for Thai readers, which facilitated fast reading.[4] It also has an archive collection of historic news reports going back to the 1960s, which were already organized into different subject themes and thus followed the newspaper's own logic of organization. At the same time, using the Bangkok Post posed dilemmas: It is by no means a neutral or uncensored publication (McCargo 2000); its reports also reflect the interests of the elite Thai journalists working for it.[5] Yet, at the same time, the purpose of narrative analysis is to study what is reported, rather than to pretend that there is a clear and unquestioned version of each storyline. It should be noted that it was relatively more difficult to use the second English-language Thai broadsheet, The Nation, because this newspaper did not have the same accessible archive; it has a smaller circulation; and its environment section was edited by a U.S. citizen during the 1990s (Fahn 2003), whereas the Bangkok Post was almost totally written by Thais. On the balance of these factors, the Bangkok Post was a valuable historical resource to use alongside interviews to trace public debate and how community activism was linked to different knowledge claims.

The method for conducting the content analysis is described in Forsyth (2007). Newspaper reports were grouped together using the Bangkok Post's own classification system and then each report was analyzed to identify the number and identity of actors reported (classified into broad categories of community representatives, national conservation groups, formal expert organizations, government, etc.) and then to record the underlying frames or discourses contained in each story (e.g., whether the report emphasized ecological fragility, risks to local people and livelihoods, strength of community, democratization, state failure, etc.).

Authoritarianism and Community Politics in Thailand

Thailand has become both a byword for environmental concern and democratic worries. Since its establishment as a modern state in 1932, Thailand has experienced twelve coups, seven attempted coups, twenty-five general elections, and four major occasions when soldiers or paramilitaries shot and killed protestors urging greater democracy. Meanwhile, the World Bank has called Thailand "one of the great development success stories" (World Bank 2018), due to its rapid economic growth. Environmentalism has blossomed, partly out of concern for the degradation of forests, coasts, and city environments (Anat 1988) but also because environmentalism has acted as a proxy for democratic activism under military regimes (Hirsch 1996; Fahn 2003). Authoritarianism still persists: In 2006 and 2014, military forces overthrew democratically elected governments. Between these coups, Thailand also became embroiled in conflicts between different political alliances called Red Shirts and Yellow Shirts, who represented, respectively, supporters of the deposed prime minister, Thaksin Shinawatra, and more royalist critics of Thaksin (Pasuk and Baker 2010). In 2016, the widely respected King Bhumibhol died after a reign of seventy years. Criticisms of the monarchy are severely curtailed: One man who made a sarcastic Facebook comment about one of the King's dogs was arrested in 2015 with the prospect of a long sentence (Political Prisoners in Thailand 2017). Another man accused of sending four offensive text messages about the royalty was denied bail on eight occasions and died in prison of liver cancer in 2012 while serving a twenty-year sentence (Fuller 2012). Since the 2014 coup, all public political protests have been restricted.

Some of the factors underlying authoritarianism can be described as forms of civic epistemologies, in both formal and informal ways. During the Cold War, anti-insurgency concerns led Prime Minister Sarit Thanarat to insist on a form of democratic debate based on traditional Thai society, and policy decisions centralized within central government (Saneh 2006; Bencharat 2014). These factors contributed, according to Haberkorn (2011), to the failure of revolutionary movements during the 1960s. In October 1973, the military government cracked down by shooting student democracy activists in Bangkok. This event led to the fall of the government and the installation of a more progressive regime, with experiments in local democratization. In October 1976, though, paramilitary groups again killed students in Bangkok, a further coup happened, and Thailand experienced authoritarian military rule until 1988. In 1976, a well-known Buddhist monk,

Phra Kittivuddho, pronounced that "killing communists is not demeritorious" (Haberkorn 2011, 131). During this period, many Thais feared that Thailand, too, would become communist, following the conversion of Vietnam, Laos, and Cambodia. Indeed, various analysts have argued that this fear contributed to many Thais accepting the coup and its resulting authoritarianism (Thongchai 1994; Glassman 2004; Ji 2006).

Against the background of war and the fear of revolution, the concept of community has had a contested history. In particular, the long-standing philosophy of "community culture" in Thailand organized the perception of rural communities along moral grounds of "sharing, taking turn to work for each other, non-exploitation of each other, and righteousness" (Bamrung 1984, 238; Bencharat 2014, 124; see also Chatthip 1991a, 1991b). Some analysts have argued that community culture is empowering to villagers (Seri 1986; Hewison 1993). Critics, however, have called the philosophy essentialist and nostalgic and have alleged it represents a form of conservative nationalism that opposes democratization or engagement with capitalism by villagers (Rigg 1991; Thongchai 2008; Reynolds 2013). A related framework is the sufficiency economy promoted by King Bhumibhol that discusses ways of engaging in modern living without greed (Isager and Ivarsson 2010; Prasopchoke 2010).

These ideas about communities have also affected the relationships or, indeed, alliances between rural villagers and urban or elite members of Thai society. Chatthip Nartsupha, one philosopher most associated with community culture, called for "a progressive bourgeoisie" in the 1980s and 1990s to work with the peasantry to "overthrow the parasitic capitalism that develops from exploitation of the countryside, develop industrial capitalism, and allow the countryside to remain in its old state" (Chatthip 1991a, 58). Critics, however, have argued that community development is not a gift but a right (Vandergeest 1991) and that the kind of assistance offered under community culture was a form of anti-insurgency control (Kanok 1981; Quinn 1997; Atchara 2014; Bencharat 2014).

These debates suggest that community politics reflected class-based divisions in Thai society, but class remains a controversial theme. Some analysts have argued that the clashes between Red Shirts and Yellow Shirts since 2005 was "a growing class war between the urban and rural poor and the old elites" (Ji 2009, 83) or a "script deftly managed by the conservative 'royalist' establishment" to block populist voting (Bello 2014, 1; see also Glassman 2010). Others have preferred to use specifically Thai concepts of hierarchy—for example, the word *phrai* was used under times of the absolute monarchy to mean "commoner," in contrast to *amartaya* ("elite"; Ferrara 2014, 35). A similar, informal phrase, *siwilai* (literally, "civilize") also differentiated less-civilized *chao bannok* (rural people) or *pu-noi* ("little" people) and the more advanced *chao muang* (city people; Anan 1984; Thongchai 2000). Nonetheless, as the next section shows, questions of social difference have influenced how communities have been represented in environmental politics.

Communities and Forest-Based Activism

For some analysts, rural communities have played a key role in environmentalism in Thailand, partly because these activists greatly outnumber middle classes (Fahn 2003). These statements, however, hide the symbolism and tensions involved in both communities and forests that have affected the relationship of environmentalism and democratization. First, as discussed earlier, the term *community* invokes images of appropriate social order and tradition. Second, forests also carry concerns about historic security threats and political instability. Since the late 1950s, Thai governments demarcated conservation forests as part of their anti-insurgency measures (Peluso and Vandergeest 2011). During the final years of the Vietnam War, and afterward, in the 1970s and 1980s, successive governments closed public access to forest zones to protect against the spread of communism. Indeed, after the 1976 coup, opponents to the military regime, including the Communist Party of Thailand (CPT), established camps in forest zones. Forests therefore came to signify resistance and political insecurity. According to one ex-senator and campaigner for social development: "I don't see the words 'fleeing to the jungle' to mean 'forest' ... the jungle was a euphemism for the CPT stronghold" (Jon Ungpakorn interview, March 2008).

The combined histories of communities and forests in Thailand therefore created templates for the later development of community-based forest activism. At first, communities in zones demarcated as forest sought to gain extra recognition for social and economic development. Over time, though, two competing narratives developed. On one hand, communities could benefit strategically by framing themselves as traditional and appropriate forest users (Bencharat 2014). On the other hand, they could also be accused of encroaching into protected areas, and rejecting the appropriate community role expected of them (Pye 2005a). Predefined visions of community culture and forests therefore provided forms of civic epistemologies against which villagers' activities were seen and judged.

These tensions were shown in the years between 1973 and 1976 when Thailand's military rule relaxed. During this time, a progressive organization called the Campaign for Dissemination of Democracy project encouraged some 3,000 students to visit rural areas (this group arguably constituted Chatthip's "progressive bourgeoisie"), and the Farmers' Federation of Thailand (FFT) was established in 1974. These groups worked to demand rural development and to support villagers who had reoccupied farming land previously claimed for conservation.[6] Critics, however, claimed that students were assisting farmers to become encroachers ("NSCT Denies Urging Farmers" 1975). Meanwhile, between March 1974 and September 1979, thirty-three leaders of the FFT were assassinated, with a further eight seriously injured, and five more disappeared (Haberkorn 2011).

Environmental conflicts, however, became more prominent during the 1980s and were driven by a wider range of concerns. First, there was continued opposition to government-based expansion of tree plantations (especially eucalyptus) in the northeast of Thailand.[7] Second, there was growing public concern about unregulated logging, especially by the army, politicians, or connected businesses.[8] Third, there was an international campaign to halt the construction of a World Bank–financed dam in the Nam Choan rainforests of western Thailand (Hirsch and Lohmann 1989; Hirsch 1993; Chusak 2008; Chusak and Baird 2018). Activism about forests therefore became to represent various aspects of democratization, and this public dissent led to a period of significant change in the late 1980s: the postponement of the Nam Choan dam (1988), the passing of a national ban on logging (1989), and the reintroduction of general elections (1988, until a further coup in 1991).

The period following the logging ban, however, demonstrated a return to the tensions concerning communities and forest. In the early 1990s the apparent alliance between community activists and conservation groups continued in opposition to the so-called Khor Jor Kor program of reforestation and enforced resettlement in the northeast of Thailand (Pye 2005b). During the later 1990s, though, divisions emerged between activities seeking to establish new legislation to define how local people can use forests in the wake of the logging ban and other campaigns to protect forests against alleged irresponsible encroachment or degradation from commercial activities such as mining or tourism (Pinkaew and Rajesh 1992). These tensions were seen most prominently in discussions about legislation for community forests. Various analysts trace the origin of debates about community forests to research started by Saneh and Yos (1993), which emphasized ideas dating back to community culture (Attajak 2005; Bencharat 2014). Simultaneously, community activists began to form social movement organizations such as the Assembly of Isaan (northeast) Farmers for Land Rights and Improvement of Natural Resources, the Northern Farmers Movement, and, by 1995, an overarching Assembly of the Poor (AOP; Missingham 2003). The AOP promoted various concerns including community rights, opposition to dams (including, most prominently, the Pak Mul dam near the Thai–Laos border) and the health and welfare of factory workers and slum dwellers (Prapart 1998; Baker 2000). Thailand's new constitution of 1997 added to this movement by confirming the rights of local people to participate in decisions about infrastructure and natural resources (later constitutions in 2007, 2014, and 2017 reduced this statement).

The debate about community forests legislation included various draft bills. A "People's Version" of the bill was proposed in 1993, based on consultations with the Local Development Institute and Project for Ecological Recovery (two nongovernmental organizations [NGOs] linked to community rights); the NGO coordinating committee (NGO-CORD); and various village representatives, academics, and foresters (Anonymous n.d.). This bill was then followed by alternative drafts from the Royal Forest Department (RFD) in 1995 and two further versions before a draft was approved by the Cabinet in 1997. This bill stated that villagers living in conservation forest areas could only request a community forest if they had lived in the area at least five years before the Community Forest Act and could "possess behavior which indicates a culture or harmony with forest conservation." A bill was passed in 2007, but stated that community forest rights do not also confer land rights. Moreover, the government delayed the final passing of the bill into law in 2008 (RFD 2007; Regional Community Forestry Training Center 2011; Bencharat 2014, 192, 213). The implications of these terms for environmental democratization and civic epistemologies are discussed in the next sections.

Coproducing Knowledge and Agency

Despite the activism described earlier and the claims that environmentalism in Thailand is socially inclusive (Hirsch 1997; Fahn 2003), there is one shocking fact: "Every community forest bill drafted, including the two so-called people's version bills, prohibits the occupation of, and farming and living on community forest land" (Bencharat 2014, 212). How did this exclusionary outcome happen?

This section presents information from the content analysis of historic *Bangkok Post* reporting for logging and community forests. As discussed earlier, this analysis summarizes how different news stories relating to logging and community forests were framed in terms of democratization in general (including the success of community activism), worries about the state (e.g., corruption), or public approval of the state (e.g., success in implementing laws). Figure 1 shows activities related to logging (as defined by the newspaper itself) between 1968 and 2000, which includes political disputes, community action, conflicts over encroachment and conservation, and the timber trade. Figure 2 shows reports specifically mentioning community forest bills (including debates and activism) between 1993 and 2017.[9] These figures show changes in the narratives concerning communities and forest and can be used alongside other information such as interviews.

Figure 1A shows how news about logging between 1968 and 2000 was framed in terms of democratization, state failure, or state success. This chart shows a sudden increase in themes relating to democratization and state success during the period 1973 to 1976 when military rule was relaxed. After the 1976

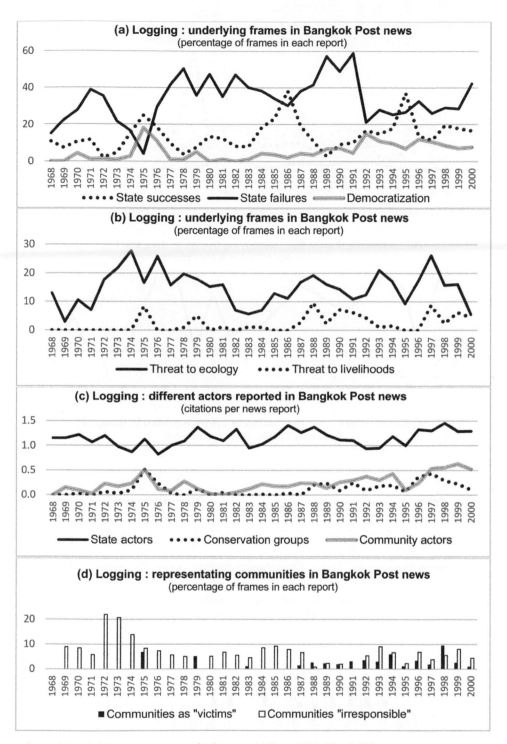

Figure 1. Content analysis of *Bangkok Post* news reports for logging, 1968 to 2000. $N = 1{,}518$ reports.
Source: Bangkok Post library archive.

crackdown, newspaper reporting returned to criticizing the government for alleged failures, although this trend declined immediately after the 1991 coup (possibly because of censorship). Newspapers reported increases in public approval of the state ("state success") in the mid-1980s and mid-1990s

largely because of reports about state action to stamp out corruption or illegal logging at the time. (Of course, this reporting did not mean that illegal logging stopped during these periods.)

Figure 1B shows that logging in general, however, was represented during this period as more of a threat

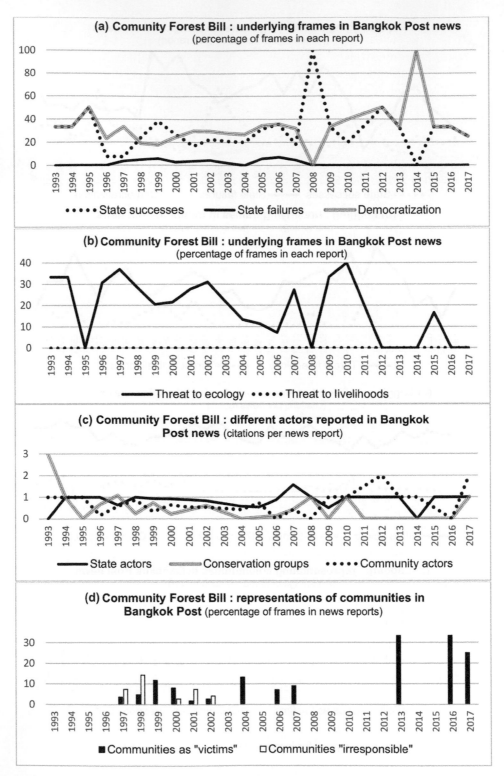

Figure 2. Content analysis of *Bangkok Post* news reports for community forest bill, 1993 to 2017. *N* = 165 reports. *Source:* Factiva database.

to ecology (meaning impact on water, wildlife, and heritage) than livelihoods or community rights. Despite this framing, Figure 1C shows that the actual number of community actors reported in the news remained more or less equivalent to elite or national conservation groups during this time, although state actors (e.g., the RFD or ministers) had the greatest prominence in day-to-day reporting about logging.

Figure 1D, however, focuses on how communities were represented. In keeping with the structure of narratives (Hajer 1995), communities were sometimes portrayed as irresponsible (e.g., when they were blamed for illegal encroachment) or as victims (when they were affected by insensitive state resettlement policies, or by illegal logging). Figure 1D shows that the representation of communities as irresponsible was highest during the 1970s, including between 1973 and 1976. This finding is in contrast to Figure 1A, which showed an increase in themes of democratization during 1973 to 1976. This apparent tension probably demonstrates the effect indicated in Figure 1B: Newspapers framed community-based activism as democratizing when it sought to protect forests, but activism to increase access to forest land was reported as irresponsible. During the 1980s and 1990s, however, news reports increasingly referred to communities as victims and criticisms of illegal logging and the state in general were growing.

Figure 2 shows similar information for debates about community forest bills between 1993 and 2017, but it shows a different set of trends. Most striking, the diagrams in Figure 2 show the community forests narrative as a story of successful state action and democratization (Figure 2A). For example, the final agreement of community forest legislation was framed 100 percent as a success for the state in 2008; and in 2014 community forest use was framed 100 percent as democratizing following the decisions of the new military government to resettle some villages and install commercial plantations (discussed below). The main challenge to be addressed was threats to forest ecology; indeed, threats to livelihoods were not discussed (Figure 2B). Yet, despite the lack of attention to local livelihoods, community actors were reported more or less as frequently as state actors and conservation groups (Figure 2C). Throughout this time, communities were predominantly (and increasingly) represented as victims rather than irresponsible (Figure 2D). Together, Figures 2B and 2D indicate that communities were only represented in news reporting about community forests in terms of victims and as part of a narrative about protecting forest ecology and traditional lifestyles rather than access to land for agriculture. Even when the new military government in 2014 began to re-install top-down plantation forestry, journalists portrayed

this as an assault on traditional villages living sensitively in fragile ecological zones, rather than as a threat to livelihoods and rights in general (e.g., "Forest Anger" 2014).

A comparison of Figures 1 and 2 shows that logging during the 1970s and 1980s was a source of diverse concerns, but by the 1990s and 2000s, the discussion of community and forests had become largely defined by predefined framings of what forests and communities meant. In turn, these definitions framed environmental democratization in terms of protecting a certain type of community in an allegedly common objective of protesting forests, rather than according to a wider range of democratic objectives or options for forest landscapes. There is also evidence that different actors also contributed to this narrative about communities and forests.

First, it is worth noting that various conflicts where villagers asserted rights to livelihoods, or sought to occupy conservation forest land, were not reported as relevant to debates about community forests legislation. For example, in the Dongyai forest of Buriram province in northeastern Thailand, villagers engaged in various acts of resistance against state-enforced eucalyptus plantations on land used by local people from the mid-1980s to the mid-1990s. Most famously, villagers, led by a monk, Phra Phrachak Khuttachitto, cut down 200,000 plantation saplings and set fire to the offices of the RFD (Rajesh 1992; Magagnini 1994; Pye 2005a). In March 1994, the *Bangkok Post* published twenty-one stories about government actions to resettle people in this district. The most common frame adopted in these reports were the tensions between seeing these people as either illegal or as victims (roughly 30 percent of all frames). At no point, however, was the forest policy itself questioned; indeed, the second most common frame was of the state successfully implementing policy (15 percent). In another example, villagers were accused of encroaching on protected forest land in Dong Larn, also in the northeast, which had been demarcated as reserved forest land since 1964 (Ayuwat 1993). In the 1990s, the government asked these people to leave. Between 1996 and 1999, the *Bangkok Post* published 242 stories on this topic, where the main framing was tensions between successful implementation of government resettlement policy (27 percent of all frames) versus whether the state was overly forceful (17 percent). None of these reports was connected to debates about the community forests bill(s). These and other examples provide evidence that the narrative of communities and forests excluded communities considered to be acting unlawfully. This preordering of cases of community activism is further evidence of a civic epistemology that excludes these kinds of cases from debate about community forests and whether plantation forests on agricultural land is indeed an appropriate form of forest policy.

Second, there is also evidence that community activists fed the narrative themselves by framing their activities strategically to emphasize the role of communities as forest protectors (Lohmann 1995; Forsyth and Walker 2008). For example, an NGO called the Community Love Forest Project was formed in northern Thailand in 1995 with the objective of changing the perception of middle-class groups (Bencharat 2014). Communities also participated in forest ordination, where trees were wrapped in saffron robes of Buddhist monks to dissuade loggers. In 1996, the Northern Farmer Federation launched a campaign to ordain 50 million trees in honor of King Bhumibhol's fiftieth anniversary (Darlington 2012). Sometimes community actors also used these actions to legitimize alternative demands for development. Indeed, at a protest outside Chiang Mai provincial hall in 1999 organized by the AOP, community activists held banners showing images of villagers demanding the right to participate in decisions about natural resources with their fists raised. In the background, meanwhile, the banner carefully added pictures of undisturbed forest and an ordained tree as a nod to more orthodox concerns about conservation (personal observation May 1999). Other posters and murals at the time also emphasized the royal initiatives for reforestation and the alleged connections between forests and lowland water supply.

Third, other activists also represented communities and forests in this way and sometimes in terms akin to the broader philosophies of community culture and sufficiency economy (as discussed earlier). In particular, these viewpoints rejected capitalism and championed traditional values and local ecological wisdom. For example, Wanida Tantiwittayapitak was a prominent member of the AOP. In one interview, she justified the protection of both forests and community on the basis of counterposing them to capitalist investment:

The poor people can live with the forest. Poor people eat, but not a lot. They must respect nature and trees. … If they live in the forest margins, they can use the forest for food. It is better to give the forest to the poor people than to investors from Singapore, Japan, Taiwanese, or Farang [westerners]. … OK, villagers may use the forest to plant rice, but you must ask why, but the investors plant cabbage and this is a problem. (Interview 1999, author's translation)

Similarly, Baramee Chiyarat is a current representative of the AOP. He was keen to state that community statements about forests are not simply tactics:

It is very normal that people who live with nature take care of nature. For example, making offerings to spirits of forest and water. It is a long tradition. Caring for nature is not a tactic but the way of communicating this is the tactic. (Interview 2008, author's translation)

Nonetheless, some critics have identified a trend in reporting communities in these terms. In particular, Walker (2001) argued that there is a "consensus" about representing the Karen ethnic minority on the Thai–Myanmar borders that portrays this group as closer to nature because they adopt traditional shifting cultivation that allows forest regeneration and make offerings to spirits of forest and water. Indeed, the Bangkok Post once referred to Karen people as "naturally peace-loving and docile." ("Editorial: A Prime Example" 1998). Walker proposed that this representation arose from the debate about community forests, where both conservation and development groups wish to find examples of people who can live in forest zones yet also show appropriate appreciation of forest conservation (Walker 2004; Usher 2009). This representation, however, does not reflect the majority of Karen livelihoods today that rely more on commercialized agriculture and migration than these ideas suggest.[10]

The effect of these representations, however, is to reinforce two discourse coalitions about communities and forests that produce their own exclusionary or collateral realities (Law 2011). First, there is an overriding representation of ecology as fragile, especially concerning the role of forests (and reforestation) as necessary for ecosystem services such as rainfall and the avoidance of droughts, despite evidence from research in and outside Thailand that shows that these statements are more complex or even challengeable (Alford 1992; Calder 1999; Bruijnzeel 2004; Forsyth and Walker 2008). Second, rural people tend to be represented as communities only in terms of tradition, or an existence outside commercialization and modernity (e.g., Fahn 2003).[11] These two perceptions can sometimes feed each other: For example, in one interview with the conservation group, the Seub Nakhasathien Foundation in 1999, the director stated that the entire territory of Thailand had been covered with forest in the early twentieth century and that it was the job of conservationists to restore this cover.[12]

Together, these factors reduce debate about potential options for agriculture or commercial advancement by rural villagers and create a new form of "good, bad, and ugly" structure to allocate social roles under environmentalism. As Baker (2000) noted, "Civil society is a slippery term, and in the late 1990s [in Thailand] was often appropriated for a particular urban and middle-class view of political change" (6).

Following and Reforming Civic Epistemologies

The implication of the preceding discussion is that the representation of communities and forests in Thailand has followed narratives but that these narratives have been structured by preexisting cultural norms that have acted as civic epistemologies because they have predetermined which kinds of knowledge, actors, and forms of activism carry political and epistemic authority. In particular, these norms have related to ideas about community culture, which reflect social hierarchies between elites and commoners, and these have reinforced traditional ideas about communities and their relationship to nature. They also exclude certain kinds of community activism as either appropriate forms of environmentalism or as counting toward the negotiation of community forests.

Various analysts have pointed out that the community forests debate has "reflected the social reality of division between urban-based elite classes and rural-based farmer classes in Thai society" (Chusak and Baird 2018, 322). This statement is too simple: It hides the ways in which community activists have used these norms as short-term tactics to gain political advantage but that doing so has also reinforced the underlying ideas that limit rural development. In other words, civic epistemologies are shared by different actors and shape both the knowledge and agency expressed by them.

Can civic epistemologies be reformed? Since the community forests debates of the 1990s, various significant changes have occurred in Thailand. Between 2001 and 2006, a new prime minister, Thaksin Shinawatra, introduced a new style of populist politics based on harnessing rural voters. Thaksin was then deposed by a coup in 2006, which was followed by conflicts between Red Shirts and Yellow Shirts until a further coup in 2014. During this time, the AOP also declined, partly because Thaksin's regime offered an alternative arena for politics but also because of internal divisions and the death of leading activists in the AOP (Baramee Chiyarat interview 2017).

Thaksin's policies have been described as challenging the traditional vision of community culture because they introduced economic opportunities and liberalization to farming communities (McCargo and Ukrit 2005). For example, the 1 million baht fund for village development and the One Tambol One Product (OTOP) program gave villagers money for infrastructure development and encouraged local districts to trade directly with markets (Baker 2016). At the same time, these policies inspired resistance from critics apparently motivated by traditional ideas of community culture. For example, Thaksin was accused of "pushing capitalism into rural areas" (Pasuk 2004) or, as noted by one Bangkok Post editor, "As a baby of ruthless globalization, Mr. Thaksin's victory spells doom for the environment and the rural poor. Believing that money has no nationality, he will continue to sell the natural environment to the highest bidder at the cost of the villagers' livelihood and environmental destruction" (Sanitsuda 2006). Various analysts have linked Thaksin to financial corruption (McCargo and Ukrit 2005; Pasuk and Baker 2010).

Reforming civic epistemologies is likely to be difficult because, as noted earlier, they are knowledge orders stabilized in institutionalized epistemic and political practices (C. Miller 2008). For example, one social commentator in Thailand explained that the legacy of mistrust against a "leftist" agenda in Thailand made it difficult to define environmentalism in terms of poverty and access to livelihoods (Chayan Vaddhanaputhi interview 2008). Yet, seeking to diversify knowledge production might also reproduce old structures. For example, the Thai Baan (or Thai Village) research initiative was introduced during the 2000s to increase the space for villagers to participate in knowledge generation about local ecology, as a way to influence policy, and to empower community development. It has especially been used in researching fish species in sites where dams have been proposed (Friend 2009; Living River Siam Association 2017). This work has clearly created new knowledge and represented community members as local experts about the environment. Yet, simultaneously, Thai Baan has been criticized for allegedly being insensitive to how its practices also coproduce traditional "subsistence narratives" about community livelihoods, similar to the Karen consensus, rather than fighting for access to markets or other socioeconomic change (Lamb 2018). The desire to resist projects such as dams can therefore reproduce narratives about appropriate community life.

It is also important to emphasize that authoritarianism is still strong in Thailand and has grown since the 2014 coup. The 2007 constitution rolled back many of the public consultation provisions in the 1997 constitution. The new constitution of 2014 went further by introducing a new clause called Section 44, which gives the government executive freedom to issue any order for the sake of reform or social progress. In addition, the government has reasserted its objective to restore Thailand's forest reserve land to 25 percent of national territory, leading to evictions. Many specific evictions, however, are not reported in national newspapers (Kongpob 2017). Under these conditions, it is not surprising that rural actors and NGOs might resort to older, safer, roles. For example, when the government proposed to build the Kaeng Sua Ten dam in the northern province of Phrae in 2012, activists held a rally where villagers ordained trees and used both red and yellow colors to portray themselves as traditional and separate from national politics (personal observation September 2012).

Some critics, however, adopt a more critical tone. According to some observers, Thai NGOs are now mainly "tools of the junta" (Pinkaew 2017) or indeed "NGOs just want the government to solve their problems and don't mind if the military regime takes over the country" (Baramee Chiyarat interview 2017). This is fighting talk, yet these statements actually refer to long-standing worries that the Thai concept of community culture encourages people to see community development as a gift rather than a right (Vandergeest 1991; Chatthip 1991a, 1991b). It could be, with the sudden increase in authoritarianism following the 2014 coup, that public discussions of communities and forests have resorted to familiar narratives rather than seeking to change deeper social hierarchies.

Conclusion: Interpreting Civic Epistemologies

This article has sought to advance debates about environmental democratization by showing the influence of civic epistemologies as a way to show how narratives produce simplified and exclusionary explanations of environmental politics. The article has shown that, in Thailand, civic epistemologies are constituted by long-standing ideas of community culture and social hierarchies that have defined visions of appropriate rural life. These visions have been strengthened through worries about national security in rural areas and by a perception of forests and traditional lifestyles as threatened by modern economic growth. These epistemologies act as knowledge orders by qualifying which kinds of environmentalism or community activism are seen as relevant to debates about forests and which kinds of knowledge claims are considered appropriate. Accordingly, popular environmentalism frequently results in reinforcing predefined ideas about the ecological functions of forests and the characteristics of communities that can be considered simplistic and exclusionary of alternative, and more flexible understandings.

What lessons can this article show for implementing civic epistemologies in environmental analysis? Three conclusions can be made.

First, this article adds to the literature that argues that environmental democratization should not be understood in terms of clashes between actors but on what actors create together (Hajer 1995; Wynne 1996). For this reason, there should be caution about seeing community-based activism as necessarily progressive and instead more attention to how activism contributes to existing social orders. The analysis of civic epistemologies shows the templates and areas of commonality that are often not apparent but which influence how community activism proceeds. This article has shown that this is important for Thailand. As Bowie (1992) and Walker (2014) noted, there is an "uncanny resemblance between leftist and royalist prescriptions for rural society, both drawn to images of an authentically Thai village in which local production systems and local culture and mutually reinforcing" (Walker 2014, 203).

Second, the analysis of civic epistemologies demonstrates a larger role for contemporary politics in shaping and using narratives. There is a tendency to see narratives (especially if interpreted as assemblages) as historically deterministic. Under this framework, current politics are shaped by hybrid facts and norms melded in the past (Law and Singleton 2005; Müller and Schurr 2016); for example, concerning the role of counterinsurgency in shaping definitions or demarcation of forests (Peluso and Vandergeest 2011). A focus on civic epistemologies, however, opens space to consider how current politics might select and repeat narratives more dynamically; for example, how contemporary actors use history selectively to create modern myths or stereotypical interpretations of the past, such as Thailand's frameworks of community culture or sufficiency economy. Focusing more on the present also allows researchers to ask whether narratives are summaries of how alleged truth claims were made in the past or whether they are indications of how meaning is projected onto diverse and contested events in the past and present. If so, then it is important to ask who controls narratives. This article has assumed that newspapers shape narratives because (so the saying goes) journalism is history's first draft, and indeed in many cases it is the only written record. It remains a testing question, however, how far the received wisdom of narratives (Leach and Mearns 1996) is made and how it becomes unchallenged.

Third, there is also a need to consider academics' own roles in maintaining civic epistemologies and narratives. Many of the problems of the so-called good, bad, and ugly approach to environmental politics arise from an unexamined application of normative agendas to complex problems, which give rise to narratives. Indeed, in Thailand, some academics have resisted calls to deconstruct narratives, asking whether these constitute "an attack" on community rights (Pinkaew 2009). Indeed, Latour (2005) stated, "Social scientists have too often confused their role of analyst with some sort of political call for discipline and emancipation" (61).

The objective of understanding civic epistemologies is to investigate the formal or informal political processes that lead to dimensions of political order being held beyond question (C. Miller 2005; Jasanoff 2012). Accordingly, one starting point is for analysts to consider how their own normative positions coproduce different observations and vice versa. Highlighting civic epistemologies is a way to shift the analysis of environmental democratization away from tacit narratives, or on the narratives themselves, toward an understanding of why narratives emerge and remain unchallenged.

Notes

1. Indeed, Spicer (2013) noted, "The ideographs and narratives that we use are not simply words. They constitute who we are, and they can have consequences for us, both good and bad" (771). It should be noted that this socially constructivist application of narratives is different from the more cognitive "analytical narratives" espoused by Bates et al. (1998).
2. This statement is commonly attributed to Philip Graham, the publisher of the *Washington Post* from 1946 to 1963.
3. Online searching was conducted through the Factiva database (https://www.dowjones.com/products/factiva/).
4. The researcher speaks and writes Thai, although he works faster in English.
5. In particular, some especially notable long-term specialists on environmentalism at the *Bangkok Post* include Sanitsuda Ekichai, Supara Janchitfah, and Wasant Techawongtham, all of whom have been interviewed in relation to this work.
6. Alleged encroachment especially took place in Chaiayphum, Phitsanulok, Si Saket, and Chantaburi provinces ("Trespassers Into National Parks" 1973; "Villagers Can Settle in Forest" 1975; "Students Ask Government" 1976).
7. Government reforestation during 1986 and 1987 became known as "Green Isaan" partly because of the color of army uniforms. Isaan is the Thai name for the northeast. Similar reforestation and resettlement in Isaan was attempted during the Khor Jor Kor program of the early 1990s (Pye 2005b).
8. Two important early conservation groups were the Association for Conservation of Wildlife and the Society for the Conservation of Treasure and the Environment.
9. Other environmental topics such as dams, pollution, agriculture, tourism development, and so on, are not included in these figures.
10. A further example is resistance to government plans to relocate Karen villagers from the Kaeng Krachan forest in western Thailand, which has repeated the usual representation of the Karen. See "Respect Rights of the Karen" (2018).
11. In particular, Fahn pointed to the corrupting influence of General Chavalit Yongchaiyudh, who was prime minister in 1996–1997, as an example of a politician who supported business interests that threatened forest conservation.
12. Interview with Rathaya Janthien, 1999. This statement was based on the record of an aviator who flew from Bangkok northeast to the Mekong River and claimed to have seen no gaps in the forest.

References

Akrich, M., and B. Latour. 1992. A summary of a convenient vocabulary for the semiotics of human and nonhuman assemblies. In *Shaping technology/building society: Studies in sociotechnical change*, ed. E. W. Bijker and J. Law, 259–64. Cambridge MA: MIT Press.

Alford, D. 1992. Streamflow and sediment transport from mountain watersheds of the Chao Phraya basin, northern Thailand: A reconnaissance study. *Mountain Research and Development* 12 (3):237–68.

Anan, G. 1984. The idiom of Phii Ka: Peasant conception of class differentiation in northern Thailand. *Mankind* 14 (4):325–9.

Anat, A. 1988. *Thailand: Natural resources profile*. Singapore and Oxford, UK: Thailand Development Research Institute and Oxford University Press.

Anderson, B., M. Kearnes, C. McFarlane, and D. Swanton. 2012. On assemblages and geography. *Dialogues in Human Geography* 2 (2):171–89. doi: 10.1177/2043820612449261.

Anonymous. n.d. *Draft community forest laws: Differences between the people's version and the government versions*. Bangkok, Thailand: Project for Ecological Recovery.

Atchara, R. 2009. *Constructing the meanings of land, resource, and a community in the context of globalization*. Chiang Mai, Thailand: Chiang Mai University.

———. 2014. *Kanmuang khong kanlodthon khwampen kanmuang: kanprab khwam samphan rawang angkorn phathana achon rath lae prachachon* [The politics of depoliticization: The transformation of NGOs: State–people relations]. Bangkok, Thailand: Office of the Higher Education Commission Research Fund, Department of Social Sciences, Silpakorn University.

Attajak, S. 2005. Raboksapsin nai chon bot phark nua khong prathait thai [Property regime in rural northern Thailand]. In *Prawathisart khwarmkhit thai kap naew khit chumchon*, ed. N. Chatthip and B. Vanvipa, 123–72. Bangkok, Thailand: Srangsan.

Ayuwat, D. 1993. Effects of migration patterns on forest use and forestry projects in a Thai village. *Society & Natural Resources* 6 (2):195–202. doi: 10.1080/08941929309380819.

Baker, C. 2000. Thailand's Assembly of the Poor: Background, drama, reaction. *South East Asia Research* 8 (1):5–29. doi: 10.5367/000000000101297208.

——— 2016. The 2014 Thai coup and some roots of authoritarianism. *Journal of Contemporary Asia* 46 (3):388–404.

Bamrung, B. 1984. Wathanatham chumchon [Community culture approach]. In *Approaches in Thai social development: The proceedings from the seminar on development experiences*. Bangkok, Thailand: Thai Volunteer Service Project.

Barry, A. 2012. Political situations: Knowledge controversies in transnational governance. *Critical Policy Studies* 6 (3):324–26. doi: 10.1080/19460171.2012.699234.

Bassett, T. J., D. Crummey, and M. M. V Beusekom. 2004. African savannas: Global narratives and local knowledge of environmental change. *African Studies Review* 47 (3):209–11. doi: 10.1017/S0002020600030584.

Bates, R. H., A. Greif, M. Levi, J.-L. Rosenthal, and B. R. Weingast. 1998. *Analytic narratives*. Princeton, NJ: Princeton University Press.

Batt, S. 1994. *Patient no more: The politics of breast cancer*. Edinburgh, UK: Scarlet.

Beck, S., and T. Forsyth. 2015. Co-production and democratizing global environmental expertise: The IPCC and adaptation to climate change. In *Science and democracy: Making knowledge and making power in the biosciences and beyond*, ed. R. Hagendijk, S. Hilgartner, and C. Miller, 113–32. London and New York: Routledge.

Bello, W. 2014. Class war: Thailand's military coup. *Foreign Policy in Focus*, May 27. Accessed January 6, 2019. http://fpif.org/class-war-thailands-military-coup/.

Bencharat, S. C. 2014. *Redefining citizenship rights: The community forest movement in Thailand and strategic rights claim*. Bundoora, Australia: School of Social Sciences and Communications, Faculty of Humanities and Social Sciences, LaTrobe University.

Béné, C. 2005. The good, the bad and the ugly: Discourse, policy controversies and the role of science in the politics of shrimp farming development. *Development Policy Review* 23 (5):585–614. doi: 10.1111/j.1467-7679.2005.00304.x.

Benjaminsen, T. A. 2009. Climate change and conflicts in the Sahel—Politics versus science. *Internasjonal Politikk* 67 (2):151–72.

Beymer-Farris, B. A., and T. J. Bassett. 2013. Environmental narratives and politics in Tanzania's Rufiji Delta: A reply to Burgess et al. *Global Environmental Change* 23 (5):1355–58. doi: 10.1016/j.gloenvcha.2013.06.007.

Bowie, K. A. 1992. Unraveling the myth of the subsistence economy: The case of textile production in nineteenth century northern Thailand. *Journal of Asian Studies* 51 (4):797–823. doi: 10.2307/2059037.

Bruijnzeel, L. A. 2004. Hydrological functions of tropical forests: Not seeing the soil for the trees? *Agriculture, Ecosystems and Environment* 104 (1):185–228. doi: 10.1016/j.agee.2004.01.015.

Bryant, R. L., and S. A Bailey. 1997. *Third world political ecology*. London and New York: Routledge.

Calder, I. 1999. *The blue revolution: Land use and integrated resource management*. London: Earthscan.

Chatthip, N. 1991a. The community culture school of thought. In *Thai constructions of knowledge*, ed. M. Chitakasem and A. Turton, 118–41. London: School of Oriental and African Studies.

———. 1991b. *Watthanatham thai kap khabuan kanplianpleang sangkhom* [Thai culture and social-change movements]. Bangkok, Thailand: Chulalongkorn University Press.

Chusak, W. 2008. History and geography of identifications related to resource conflicts and ethnic violence in northern Thailand. *Social Europe* 49 (1):111–27.

Chusak, W., and I. G. Baird. 2018. Communal land titling dilemmas in northern Thailand: From community forestry to beneficial yet risky and uncertain options. *Land Use Policy* 71:320–28. doi: 10.1016/j.landusepol.2017.12.019.

Cockburn, A., and J. Ridgeway. 1979. *Political ecology: An activist's reader on energy, land, food, technology, health, and the economics and politics of social change.* New York: Times Books.

Darlington, S. M. 2012. *The ordination of a tree: The Thai Buddhist environmental movement.* Albany, NY: SUNY Press.

Dwyer, P., and M. Minnegal. 2006. The good, the bad and the ugly: Risk, uncertainty and decision-making by Victorian fishers. *Journal of Political Ecology* 13 (1):1–23. doi: 10.2458/v13i1.21675.

Editorial: A prime example of what must stop. 1998. *Bangkok Post*, April 29.

Epstein, S. 1996. *Impure science: AIDS activism and the politics of knowledge.* Los Angeles: University of California Press.

Fahn, J. 2003. *A land on fire: The environmental consequences of the Southeast Asian boom.* Boulder, CO: Westview.

Feindt, P. H., and A. Oels. 2005. Does discourse matter? Discourse analysis in environmental policy making. *Journal of Environmental Policy & Planning* 7 (3):161–73. doi: 10.1080/15239080500339638.

Ferrara, F. 2014. Unfinished business: The contagion of conflict over a century of Thai political development. In *Good coup gone bad: Thailand's political development since Thaksin's downfall*, ed. P. Chachavalpongpun, 17–46. Singapore: Institute of Southeast Asian Studies.

Forest anger runs deep. 2014. *Bangkok Post*, November 1.

Forsyth, T. 2003. *Critical political ecology: The politics of environmental science.* London and New York: Routledge.

———. 2007. Are environmental social movements socially exclusive? An historical study from Thailand. *World Development* 35 (12):2110–30.

Forsyth, T., and A. Walker. 2008. *Forest guardians, forest destroyers: The politics of environmental knowledge in northern Thailand.* Seattle: University of Washington Press.

———. 2014. Hidden alliances: Rethinking environmentality and the politics of knowledge in Thailand's campaign for community forestry. *Conservation and Society* 12 (4):408–17. doi: 10.4103/0972-4923.155584.

Friend, R. M. 2009. Fishing for influence: Fisheries science and evidence in water resources development in the Mekong basin. *Water Alternatives* 2 (2):167–82.

Fuller, T. 2012. Thai man jailed for insulting king dies in detention. *New York Times*, May 8. Accessed December 19, 2018. http://www.nytimes.com/2012/05/09/world/asia/thai-man-jailed-for-insulting-king-dies-in-detention.html.

Glassman, J. 2004. *Thailand at the margins: Internationalization of the state and the transformation of labour.* Oxford, UK: Oxford University Press.

——— 2010. The provinces elect governments, Bangkok overthrows them. Urbanity, class and post-democracy in Thailand. *Urban Studies* 47 (6):1301–23.

Goldman, M., P. Nadasdy, and M. Turner. 2011. *Knowing nature: Conversations at the intersection of political ecology and science studies.* Chicago: University of Chicago Press.

Goodman, N. 1978. *Ways of worldmaking.* Hassocks, UK: Harvester.

Haberkorn, T. 2011. *Revolution interrupted: Farmers, students, law, and violence in northern Thailand.* Madison: University of Wisconsin Press.

Hajer, M. A. 1995. *The politics of environmental discourse: Ecological modernization and the policy process.* Oxford, UK: Clarendon.

———. 2009. *Authoritative governance: Policy-making in the age of mediatization.* Oxford, UK: Oxford University Press.

Hajer, M. A., M. van den Brink, and T. Metze. 2006. Doing discourse analysis: Coalitions, practices, meaning. *Nederlandse Geografische Studies* 344:65–76.

Hajer, M. A., and W. Versteeg. 2005. A decade of discourse analysis of environmental politics: Achievements, challenges, perspectives. *Journal of Environmental Policy & Planning* 7 (3):175–84. doi: 10.1080/15239080500339646.

Hewison, K. 1993. Nongovernmental organizations and the cultural development perspective in Thailand: A comment on Rigg (1991). *World Development* 21 (10):1699–1708. doi: 10.1016/0305-750X(93)90103-G.

Hirsch, P. 1993. *Political economy of environment in Thailand.* Manila, Philippines: Journal of Contemporary Asia Publishers.

———. 1996. *Seeing forests for trees: Environment and environmentalism in Thailand.* Chiang Mai, Thailand: Silkworm.

———. 1997. The politics of environment: Opposition and legitimacy. In *Political change in Thailand: Democracy and participation*, ed. K. Hewison, 179–95. London and New York: Routledge.

Hirsch, P., and L. Lohmann. 1989. The contemporary politics of environment in Thailand. *Asian Survey* XXIX (4):439–51. doi: 10.2307/2644886.

Isager, L., and S. Ivarsson. 2010. Strengthening the moral fibre of the nation: The King's Sufficiency Economy as etho-politics. In *Saying the unsayable: Monarchy and democracy in Thailand*, ed. S. Ivarsson and L. Isager, 223–39. Copenhagen: NIAS Press.

Jasanoff, S. 2004a. Ordering knowledge, ordering society. In *States of knowledge: The coproduction of science and social order*, ed. S. Jasanoff, 13–45. London and New York: Routledge.

———. 2004b. *States of knowledge: The co-production of science and social order.* London and New York: Routledge.

———. 2005. *Designs on nature: Science and democracy in Europe and the United States.* Princeton, NJ: Princeton University Press.

———. 2012. *Science and public reason.* London and New York: Routledge.

Ji, U. 2006. The impact of the Thai "sixties" on the peoples movement today. *Inter-Asia Cultural Studies* 7 (4):570–88.

———. 2009. Class struggle between the colourful T-shirts in Thailand. *Journal of Asia Pacific Studies* 1 (1):76–100.

Jones, M. D., and C. M. Radaelli. 2015. The narrative policy framework: Child or monster? *Critical Policy Studies* 9 (3):339–55. doi: 10.1080/19460171.2015.1053959.

Jung, A., R.-L. Korinek, and H. Straßheim. 2014. Embedded expertise: A conceptual framework for reconstructing knowledge orders, their transformation and local specificities. *Innovation: The European Journal of Social Science Research* 27 (4):398–419. doi: 10.1080/13511610.2014.892425.

Kanok, W. 1981. Communist revolutionary process: A study of the communist party of Thailand. PhD thesis, Department of Political Science, The Johns Hopkins University.

Kongpob, A. 2017. Investors in, poor out: Junta's land policy after 3 years in power. *Prachatai*, May 29. Accessed January 6, 2019. https://prachatai.com/english/node/7169

Ku, A. S. M., and H. Tian. 2002. Narratives, politics, and the public sphere: Struggles over political reform in the final transitional years in Hong Kong (1992–1994). *Discourse and Society* 13 (3):412–14.

Lamb, V. 2018. Who knows the river? Gender, expertise, and the politics of local ecological knowledge production of the Salween River, Thai–Myanmar border. *Gender, Place & Culture* 1–16. Advance online publication. doi: 10.1080/0966369X.2018.1481018.

Latour, B. 2005. *Reassembling the social: An introduction to actor-network-theory.* Oxford, UK: Oxford University Press.

Law, J. 2011. Collateral realities. In *The politics of knowledge*, ed. F. Dominguez Rubio and P. Baert, 156–78. London and New York: Routledge.

Law, J., and V. Singleton. 2005. Object lessons. *Organization* 12 (3):331–55. doi: 10.1177/1350508405051270.

Leach, M., and R. Mearns, eds. 1996. *The lie of the land: Challenging received wisdom on the African environment.* Oxford, UK: James Currey.

Lejano, R. P. 2015. Narrative disenchantment. *Critical Policy Studies* 9 (3):368–71. doi: 10.1080/19460171.2015.1075736.

Lejano, R. P., M. Ingram, and H. Ingram. 2013. *The power of narrative in environmental networks.* Cambridge, MA: MIT Press.

Ley, B. L. 2009. *From pink to green: Disease prevention and the environmental breast cancer movement.* New Brunswick, NJ: Rutgers University Press.

Living River Siam Association. 2017. The Mekong River basin. Accessed December 10, 2018. http://www.livingriversiam.org/index-eng.html.

Lohmann, L. 1995. No rules of engagement: Interest groups, centralization and the creative politics of "environment" in Thailand. In *Counting the costs: Economic growth and environmental change in Thailand*, ed. J. Rigg, 211–34. Singapore: Institute of Southeast Asian Studies.

Magagnini, S. 1994. If a tree falls …. a monk's blessing for Thailand's forest. *The Amicus Journal* 16 (2):12–14.

McCargo, D. 2000. *Politics and the press in Thailand: Media machinations.* London and New York: Routledge.

McCargo, D., and P. Ukrit. 2005. *The Thaksinization of Thailand.* Copenhagen: NIAS Press.

Miller, C. 2005. New civic epistemologies of quantification: Making sense of indicators of local and global sustainability. *Science, Technology & Human Values* 30:403–32. doi: 10.1177/0162243904273448.

———. 2008. Civic epistemologies: Constituting knowledge and order in political communities. *Sociology Compass* 2 (6):1896–1919.

Miller, H. T. 2012. *Governing narratives: Symbolic politics and policy change.* Tuscaloosa: University of Alabama Press.

Missingham, B. 2003. *The Assembly of the Poor in Thailand: From local struggles to national protest movement.* Bangkok, Thailand: Silkworm.

Müller, M., and C. Schurr. 2016. Assemblage thinking and actor-network theory: Conjunctions, disjunctions, cross-fertilisations. *Transactions of the Institute of British Geographers* 41 (3):217–29. doi: 10.1111/tran.12117.

NSCT Denies Urging Farmers to Violence. 1975. *Bangkok Post*, June 10.

Pasuk, P. 2004. Address delivered at April 2 seminar "Statesman or manager? Image and reality of leadership in Southeast Asia." *Bangkok Post*, April 21.

Pasuk, P., and C. Baker. 2010. *Thaksin.* 2nd ed. Bangkok, Thailand: Silkworm.

Peet, R., and M. Watts. 2004. *Liberation ecologies: Environment, development, social movements.* 2nd ed. London and New York: Routledge.

Peluso, N. L., and P. Vandergeest. 2011. Political ecologies of war and forests: Counterinsurgencies and the making of national natures. *Annals of the Association of American Geographers* 101 (3):587–608. doi: 10.1080/00045608.2011.560064.

Pinkaew, L. 2005. On the politics of nature conservation in Thailand. In *After the logging ban: Politics of forest management in Thailand*, ed. N. Rajesh, 48–67. Bangkok, Thailand: Foundation for Ecological Recovery.

———. 2009. Botkhwarm prasit: Amnart niyom thongthin niyom lae seri niyom mai kap botphanna wa duay sithi chumchon [Review article: Popular and new liberalism and depicting community rights]. *Journal of the Social Sciences, Chiang Mai University* 21 (2):277–304.

———. 2017. Are Thai NGOs tools of the junta? *Pratichai*, August 2. Accessed January 6, 2019. https://prachatai.com/english/node/7304

Pinkaew, L., and N. Rajesh, eds. 1992. *The future of people and forests in Thailand after the logging ban.* Bangkok, Thailand: Project for Ecological Recovery.

Political Prisoners in Thailand. 2017. Thanakorn Siripaiboon. Accessed December 10, 2018. https://thaipoliticalprisoners.wordpress.com/pendingcases/thanakorn-siripaiboon/.

Prapart, P. 1998. *Kan muang bon thong thanon: 99 wan samatcha khon chon* [Politics on the street: 99 Days of the Assembly of the Poor]. Bangkok, Thailand: Krirk University.

Prasopchoke, M. 2010. The philosophy of the sufficiency economy: A contribution to the theory of development. *Asia-Pacific Development Journal* 17 (1):123–43.

Pye, O. 2005a. Forest policy and strategic groups in Thailand. *Internationales Asienforum* 36 (3–4):311–36.

———. 2005b. *Khor Jor Kor: Forest politics in Thailand.* Bangkok, Thailand: White Lotus.

Quinn, R. 1997. *NGOs, peasants and the state: Transformation and intervention in rural Thailand, 1970–1990.* Canberra, Australia: Department of Politics, Australian National University.

Rajesh, N. 1992. Monks battle to save forests. *Down to Earth*, September 15. Accessed January 15, 2019. https://www.downtoearth.org.in/news/monks-battle-to-save-forest-30134

Regional Community Forestry Training Centre. 2011. *Community forestry in Thailand.* Bangkok: Regional Community Forestry Training Centre.

Respect rights of the Karen. 2018. *Bangkok Post*, 16 June.

Reynolds, C. J. 2013. Chatthip Nartsupha, his critics, and more criticism. In *Essays on Thailand's economy and society for Professor Chatthip Nartsupha at 72*, ed. P. Phongpaichit and C. Baker, 1–22. Bangkok, Thailand: Sangsan.

Rigg, J. 1991. Grass-roots development in rural Thailand: A lost cause? *World Development* 19 (2–3):199–211. doi: 10.1016/0305-750X(91)90255-G.

Rodríguez-Giralt, I., I. Marrero-Guillamón, and D. Milstein. 2018. Reassembling activism, activating assemblages: An introduction. *Social Movement Studies* 17 (3):257–68. doi: 10.1080/14742837.2018.1459299.

Roe, E. M. 1991. Development narratives, or making the best of blueprint development. *World Development* 19 (4):287–300. doi: 10.1016/0305-750X(91)90177-J.

Royal Forest Department (RFD). 2007. *Draft community forest bill.* Bangkok: Royal Forest Department.

Salam, M. A., T. Noguchi, and R. Pothitan. 2006. Community forest management in Thailand: Current situation and dynamics in the context of sustainable development. *New Forests* 31 (2):273–91. doi: 10.1007/s11056-005-7483-8.

Saneh, C. 2006. *Kanmuang thai kap phathanakan rat thum nun* [Thai politics and constitutional development]. Bangkok, Thailand: The Foundation for the Promotion of Social Sciences and Humanities.

Saneh, C., and S. Yos, eds. 1993. *Pa chumchon nai prathet thai: Naeo thang kan phatana* [Community forest in Thailand: Development guidelines]. Bangkok, Thailand: Rural Development Institute.

Sanitsuda, E. 2006. Commentary: Caretaker making sure he does not sink alone. *Bangkok Post*, July 20.

Sato, J. 2003. Public land for the people: Institutional basis of community forestry in Thailand. *Journal of Southeast Asian Studies* 32 (2):329–46.

Seri, P., ed. 1986. *Back to the roots: Village and self-reliance in a Thai context.* Bangkok, Thailand: Rural Development Documentation Centre.

Smith, W., and W. H. Dressler. 2017. Rooted in place? The coproduction of knowledge and space in agroforestry assemblages. *Annals of the American Association of Geographers* 107 (4):897–914. doi: 10.1080/24694452.2016.1270186.

Spicer, M. W. 2013. Contesting narratives in politics: Governing narratives: Symbolic politics and policy change. *Public Administration Review* 73 (5):768–71. doi: 10.1111/puar.12106.

Students ask government to save forests. 1976. *Bangkok Post*, 11 June.

Thompson, M., M. Warburton, and T. Hatley. 1986. *Uncertainty on a Himalayan scale: An institutional theory of environmental perception and a strategic framework for the sustainable development of the Himalaya.* London: Enthographica, Milton Ash Editions.

Thongchai, W. 1994. *Siam mapped: A history of the geo-body of a nation.* Honolulu: University of Hawaii Press. doi: 10.1086/ahr/100.2.477.

———. 2000. The quest for "Siwilai": A geographical discourse of civilizational thinking in the late nineteenth and early twentieth-century Siam. *The Journal of Asian Studies* 59 (3):528–49.

———. 2008. Nationalism and the radical intelligentsia in Thailand. *Third World Quarterly* 29 (3):575–91.

Tickle, A., and I. Welsh, eds. 1998. *Environment and society in Eastern Europe.* Harlow, UK: Longman.

Ting, Z., C. Haiyun, G. P. Shivakoti, R. Cochard, and K. Homcha-Aim. 2011. Revisit to community forest in northeast of Thailand: Changes in status and utilization. *Environment, Development and Sustainability* 13 (2):385–402. doi: 10.1007/s10668-010-9267-3.

Trespassers into national park swell to 10,000. 1973. *Bangkok Post*, September 15.

Usher, A. D. 2009. *Thai forestry: A critical history.* Chiang Mai, Thailand: Silkworm.

Vandergeest, P. 1991. Gifts and rights—Cautionary notes on community self-help in Thailand. *Development and Change* 22 (3):421–43. doi: 10.1111/j.1467-7660.1991.tb00420.x.

———. 1996. Mapping nature: Territorialization of forest rights in Thailand. *Society and Natural Resources* 9 (2):159–75. doi: 10.1080/08941929609380962.

Vandergeest, P., and N. Peluso. 1995. Territorialization and state power in Thailand. *Theory and Society* 24 (3):385–426. doi: 10.1007/BF00993352.

———. 2006. Empires of forestry: Professional forestry and state power in Southeast Asia, part 1. *Environment and History* 12 (1):31–64. doi: 10.3197/096734006776026809.

Villagers can settle in forest area. 1975. *Bangkok Post*, 16 June.

Walker, A. 2001. The "Karen Consensus:" Ethnic politics and resource-use legitimacy in northern Thailand. *Asian Ethnicity* 2 (2):145–62. doi: 10.1080/14631360120058839.

———. 2004. Seeing farmers for the trees: Community forestry and the arborealisation of agriculture in northern Thailand. *Asia Pacific Viewpoint* 45 (3):311–24. doi: 10.1111/j.1467-8373.2004.00250.x.

———. 2014. Is peasant politics in Thailand civil? In *"Good coup gone bad": Thailand's political development since Thaksin's downfall*, ed. P. Chachavalpongpun, 199–215. Singapore: Institute of Southeast Asian Studies.

Winickoff, D. 2012. Preface. In *Science and public reason*, ed. S. Jasanoff, ix–xii. London and New York: Routledge.

World Bank. 2018. The World Bank in Thailand. Accessed December 10, 2018. http://www.worldbank.org/en/country/thailand/overview.

Wynne, B. 1996. SSK's identity parade: Signing-up, off-and-on. *Social Studies of Science* 26 (2):357–91. doi: 10.1177/030631296026002007.

TIM FORSYTH is a Professor in the Department of International Development at the London School of Economics and Political Science, London WC2A 2AE, UK. E-mail: t.j.forsyth@lse.ac.uk. His research interests include environment and development, especially relating to the governance of contested knowledge claims and expertise, particularly in Asia.

Speaking Power to "Post-Truth": Critical Political Ecology and the New Authoritarianism

Benjamin Neimark, ⓘ John Childs, ⓘ Andrea J. Nightingale, Connor Joseph Cavanagh, Sian Sullivan, ⓘ Tor A. Benjaminsen, ⓘ Simon Batterbury, ⓘ Stasja Koot, ⓘ and Wendy Harcourt

Given a history in political ecology of challenging hegemonic "scientific" narratives concerning environmental problems, the current political moment presents a potent conundrum: how to (continue to) critically engage with narratives of environmental change while confronting the "populist" promotion of "alternative facts." We ask how political ecologists might situate themselves vis-à-vis the presently growing power of contemporary authoritarian forms, highlighting how the latter operates through sociopolitical domains and beyond-human natures. We argue for a clear and conscious strategy of speaking power to post-truth, to enable two things. The first is to come to terms with an internal paradox of addressing those seeking to obfuscate or deny environmental degradation and social injustice, while retaining political ecology's own historical critique of the privileged role of Western science and expert knowledge in determining dominant forms of environmental governance. This involves understanding post-truth, and its twin pillars of alternative facts and fake news, as operating politically by those regimes looking to shore up power, rather than as embodying a coherent mode of ontological reasoning regarding the nature of reality. Second, we differentiate post-truth from analyses affirming diversity in both knowledge and reality (i.e., epistemology and ontology, respectively) regarding the drivers of environmental change. This enables a critical confrontation of contemporary authoritarianism and still allows for a relevant and accessible political ecology that engages with marginalized populations likely to suffer most from the proliferation of post-truth politics.

有鉴于政治生态学挑战环境问题的霸权"科学"叙事之历史，当前的政治时刻面临了强大的难题：如何（持续）批判性地涉入环境变迁叙事，同时面对"民粹主义"所提倡的"另类事实"。我们质问政治生态学者如何能够置身于今日威权主义形式增长中的力量，并强调该力量如何通过社会政治领域和超越人类的自然进行运作。我们主张对后事实的话语权需要有清晰且有意识的策略，以促成以下两件事：首先是接受应对企图模糊或否认环境恶化与社会不公者时的内部矛盾，同时保留政治生态学者在面对自身对于西方科学与专家知识决定环境治理的主流形式上所拥有的优势角色时的历史性批判。此一涉及对后事实及其孪生的另类事实与假新闻之理解，它们是由企图巩固权力的政体所进行的政治运作，而非体出对现事实本质的一致本体论理模式。再者，我们区辨后事实与断言环境变迁的导因体现知识与现实的多样性之分析（例如分别就认识论与本体论而言）。这麼做，使得对当代威权主义的批判性对抗成为可能，同时考量有意义且具可及性的政治生态学，该学问涉入可能因后事实政治的盛行而受害最深的边缘人口。关键词: 关键词: 威权主义, 环境政策, 政治生态学, 后事实, 科学。

Dada una historia en ecología política que reta las narrativas hegemónicas "científicas" en lo que concierne a los problemas ambientales, el momento político actual pone de presente un enigma portentoso: cómo (seguir) involucrándose críticamente con narrativas del cambio ambiental al tiempo que se confronta la promoción "populista" de "los hechos alternativos". Nos preguntamos cómo podrían situarse los ecologos políticos en relación con el creciente poder que registran las formas autoritarias contemporáneas, destacando el modo como opera el segundo a través de los dominios sociopolíticos y las naturalezas que están más allá de lo humano. Inquirimos por una estrategia clara y consciente del poder de la palabra ante la pos-verdad, para habilitar dos cosas. La primera es llegar a un acuerdo con una paradoja interna de hablarle a quienes buscan ofuscar o negar la degradación ambiental y la injusticia social, en tanto se retiene la propia historia crítica de la ecología política sobre el papel privilegiado de la ciencia occidental y el conocimiento experto para determinar las formas dominantes de la gobernanza ambiental. Esto implica entender la pos-verdad y sus pilares gemelos de hechos alternativos y noticias falsas, como si estuviesen siendo operados políticamente por aquellos regímenes que buscan respaldar el poder más que personificar un modo coherente de razonamiento ontológico en relación con la naturaleza de la realidad. Segundo, diferenciamos la pos-verdad de los análisis que afirman la diversidad tanto en conocimiento como en la realidad (i.e., epistemología y ontología, respectivamente) en relación con los controladores del cambio ambiental. Esto permite una confrontación crítica del autoritarismo contemporáneo y deja campo todavía para una ecología política relevante y accesible que se comprometa con las poblaciones marginales, más propensas a sufrir los efectos de la proliferación de políticas pos-verdad. *Palabras clave: autoritarismo, ciencia, ecología política, política ambiental, pos-verdad.*

Post-truth is the latest manifestation of a long, troubled history in the relation between truth, politics, and power. Indeed, it is hardly a revelation that politicians selectively choose (or construct) their facts to serve particular ends. Yet, the current political moment has also managed to provoke a heightened level of anxiety about the nature of truth in science and politics that has emerged as particularly disruptive (Dillon et al. 2019). This anxiety has ushered in new language with terms such as *alternative facts* and *fake news* becoming part of an everyday vocabulary.[1] For geographers, and in particular political ecologists, post-truth presents a familiar yet intensified challenge. Post-truth provokes questions for scholars critical of scientific institutions and their knowledge-making practices that shape environmental policy, given that these same institutions are now under attack from populist authoritarian discourse and policies.

A paradox thereby emerges between working with, while also problematizing, the production of knowledge associated with positivist science—a paradox that demands both reflection and action from critical political ecologists and activists alike (Robbins 2015). How can political ecologists mount an effective challenge against the propagation of alternative facts in service of populist authoritarian agendas, while also embracing multiple knowledges and realities associated with cultural and linguistic diversity (de la Cadena 2010; Burman 2017)? How can we defend this stance against charges that our dismay with post-truth politics stems from an elite, liberal "chagrin at the fact that the wrong kinds of people are suddenly claiming authority and having their say?" (Mair 2017,

3). Finally, how can political ecologists, many of whom have long insisted on the need to analyze the politics of knowledge production within science, work with science to show that the form of critical engagement we advocate and practice is different from that propounded by the authoritarian right?

Both political ecology and post-truth politics take issue with certain hegemonic types of truth making.[2] It is political ecology, however, that concerns itself with the epistemological violence effected through the coloniality of reality that subjugates cultural, and especially indigenous, diversity in relation to ecological knowledges and praxis (Burman 2017; Sullivan 2017). Our main contribution in response to this is to affirm the necessity of speaking power to post-truth (Collingridge and Reeve 1986): by amplifying an inclusive, effective, and publicly accessible political ecology that both refracts populist (re)framings of socioenvironmental concerns—at times mobilizing and allying with positivist science to do so (King 2010; Brannstrom and Vadjunec 2013)—and organizes to contest mechanisms of authoritarian power.

This strategy, first, situates political ecology as a useful bridge to a diversity of approaches that probe the co-constitutive relationship between environmental politics and scientific truth making (Jasanoff 2006). It recognizes and welcomes the conceptual convergence between, for example, political ecology, science and technology studies (STS), and anthropology (Rocheleau 2008; Goldman, Nadasdy, and Turner 2011; Dillon et al. 2019). Combining perspectives across these approaches means accepting that knowledges do not necessarily become authoritative because they more accurately portray "the truth." Rather, they become paradigmatic as the truth in part through their generation and endorsement in politically empowered networks as the best means of uncovering the truth (cf. Kuhn 1970; Foucault 1980; Guthman and Mansfield 2013). Foregrounding (once again) these relationships between political power and truth claims makes it possible to clarify mechanisms of knowledge production and exclusion and thereby to clarify possibilities for contestation (Hulme 2010).

Second, as well as having an established history of critically analyzing environmental truth making, political ecologists are experienced and motivated in acting and collaborating beyond the academy, to speak power to post-truth through new knowledge coalitions and action. Coalitions beyond the academy are about creating an accessible political ecology that can empower a politically engaged and informed resistance to current post-truth narratives. We argue that political ecology and cognate disciplines can combine with reflexive scientific knowledge production to offer collective responses within this eco-political moment. This sort of critical political ecology (Forsyth 2003) contributes to broader public discourse and builds on recent attempts to decolonize knowledge production inside and outside the academy not by creating a geographic and academic silo, but rather to be united against a reductive and regressive post-truth debate.[3]

In what follows, we provide a brief genealogy of political ecology in relation to post-truth. We proceed by offering three interrelated areas for intervention that, taken together, may articulate a political ecology counternarrative to truth making while remaining critical of authoritarian attacks on knowledge production. We insist throughout that it is possible to retain our critical stance toward scientific knowledge production through careful positioning of it within the circuits of its own production. When this same critical approach is applied to alternative facts, we can show that these are not new ways of knowing but rather new mechanisms of deploying power within an erstwhile and reductive ontology that colonizes other ways of knowing.

Political Ecology beyond Post-truth

Political ecology has long been concerned with authoritarian forms of power and politics in relation to environmental knowledges, policies, and infrastructures, as well as to understandings of the

materiality of nature itself.[4] At its core, early political ecology analyzed historically and spatially situated (and differentiated) powers to access and control natural resources, originally seen through class and later through other forms of social difference such as gender, ethnicity, age, and, sexuality. Political ecology thereby brought into focus how "the environment is an arena of contested entitlements, a theatre of which conflicts or claims over property, assets, labor, and politics of recognition play themselves out" (Peluso and Watts 2001, 25; Rocheleau, Thomas-Slayter, and Wangari 1996). A second related dimension of political ecology soon emerged that involved a more poststructuralist understanding of the politics of environmental knowledge production and its material-discursive interplay with environmental governance (Escobar 1995; Peet and Watts 1996; Stott and Sullivan 2000). Reflecting the influence of Foucault, a key emphasis has been on the institutional and other societal structures through which environments and environmental truths are defined, known, and therefore controlled and managed (Peet and Watts 1996; Robertson 2006; Burke and Heynen 2014).

A series of early empirical studies showed how local ecological problems have origins in trans-scalar political and economic contexts, rather than merely the allegedly maladaptive behaviors of local land users (Watts 1983; Blaikie and Brookfield 1987). Environmental processes were presented by apolitical (and Malthusian) ecological analyses as caused by small-scale producers, while research in political ecology demonstrated how these problems were incorrectly explained, or largely exaggerated, thereby challenging received wisdom on environmental degradation (Fairhead and Leach 1996). An outcome of these local(ized) studies was that there were different ways of knowing and managing environments which were frequently bypassed by mainstream environmental policies. For Forsyth (2003; see also Benjaminsen, Aune, and Sidibé 2010), this also meant linking political economy and epistemologies of environmental change to empirically challenge dominating environmental policies.

Although certainly critical, such challenges to dominant narratives and theories are—as the explicitly antiauthoritarian The Open Society and its Enemies (Popper 1971) observed—simply an integral feature of good (social) scientific inquiry. A certain degree of skepticism toward knowledge claims and findings is part of conventional scientific practice. As such, political ecology's relationship to environmental science has over the years been complex. Playing the "trickster," political ecology both engages and borrows methodology from mainstream science regarding land use change, hazards, and environmental health, only "to undermine them, demonstrating power-laden implications in any such foundational account of human/environmental relationships" (Robbins 2015, 93).

Recently, political ecology has been shaped more explicitly by postcolonial, subaltern, feminist, and queer critiques, opening up new avenues to counter "universalizing dimensions" of knowledge production associated with Western science and modernity (e.g., Nightingale 2006; Burman 2017; Sullivan 2017). Political ecologists have also found fertile ground in debates emanating from assemblage theorists in actor network theory (ANT) within STS, emphasizing how environmental phenomena and governance are mediated by technology and materiality (Bennett 2010) and the roles of beyond-human actants in socio-techno-natural assemblages (Castree and Braun 2001; Kosek 2006; Goldman, Nadasdy, and Turner 2011).[5]

Equipped with these new epistemological and ontological tools, political ecology has the ability both to distinguish itself vis-à-vis power, especially in its contemporary authoritarian forms and to push similar work to explore how forms of power operate through sociopolitical domains and nonhuman natures. Therefore, in echoing contemporary calls to scrutinize alternative facts, political ecology's attention to power-laden scientific claims is well equipped to examine differing environmental representations to expose the multiple ways in which power operates to produce, maintain, and privilege particular "truths" about the environment.

The openness and fluidity of poststructuralist approaches to knowledge production, however, lend themselves both to a seeming "overcomplexification" of socioecological circumstances and to cooptation by far right agendas. The latter have knowingly borrowed tactics and strategies used by left-leaning activists and scholars to highlight the politics of knowledge production, to push for the acceptance of alternative facts and to relativize the views of scientists and right-wing ideologues (Nagel 2017). Thus, the awkward conceptual resemblance between alternative facts and academic debates about the politics of knowledge production is not mere coincidence.

Yet, there are crucial distinctions to be drawn between critical approaches of scientific practice and the tactics now adopted by the alt-right. A critical approach to the environmental sciences underscores the ways in which power constitutes, moves within, and reproduces sociomaterial relations to shape which knowledges, social relations, and practices (and corresponding ecologies) are hegemonic. For example, although not always accomplished, many political ecologists attempt to challenge dominant environmental narratives and recognize multiple non-Western knowledge perspectives to analyze the production of uneven environmental outcomes for diverse individuals and populations (Burman 2017). Such groups and individuals are stratified by differences and inequalities of—*inter alia*—class, ethnicity, and gender and are commonly those most vulnerable to socioecological shocks or stressors. Difference and inequality in turn shapes and are shaped by environmental change processes themselves (Nightingale 2006). Moreover, by observing everyday and mundane forms of authoritarian power and governmental control, critical political ecologists have sought to take account of how knowledge and governance of resources are actively resisted and have been a focal point for empowerment of marginalized groups through both individual and collective agency (Li 2007; Wolford 2010).

Future political engagement by political ecologists and others therefore requires a sharpened focus on knowledge production and who holds the power to define truth (Gramsci 1971; Foucault 1980).[6] This ontological politics probes the values, relations, and practices through which some forms of knowledge (epistemologies) come to be accepted as more true than others. One way forward could be to carefully distinguish between the ontological and epistemological politics of asserting that there are many ways of knowing, measuring, and relating to or being in "different" worlds (ontology). If we accept the notion of multiple ontologies (that what the world is can be different across communities of knowing), political ecologists have much to say about the sociomaterial relations through which different ontologies arise and are sustained. There is an accompanying epistemological politics of asserting the truth about how one ostensibly should know or live in a single world. This latter stance largely rejects the notion of multiple ontologies and rather probe how asserting a single epistemology (how we can know the world) is inextricably bound up in claims to authority.

One role for political ecologists is to illuminate how the privileging of alternative facts exacerbates tensions between different ontologies and thereby claims space for competing knowledge claims. Some take the position that feminist political ecologists' engagement with power and privileged forms of environmental knowledge construction could help guide us to navigate the paradox of post-truth politics, while some others prefer seeing power through the lens of structuralism and/or post-structuralism. Nevertheless, we thereby advocate that political ecology, in all its forms, be made more relevant, accessible, and engaging to (newly) marginalized populations while we work to bridge the binary of science and activism closely with social movements toward new "liberation ecologies" (Peet and Watts 1996) and alternative sustainablilities (Cavanagh and Benjaminsen 2017).

Speaking Power to Post-truth

A constructive and critical political ecology, then, is about meeting power with power, mobilizing not only the discourses and social networks of critical scholarship, which at times can be just as universalizing in their own right, but also publicly informed elements, such as collective action and activism, or what we define throughout as to speak power to post-truth. Taken together, we argue that we can effectively counter the purveyors of post-truth and their inventive uses of environmental messages. This requires not only exposing the workings of power in the generation of alternative facts but also in consolidating an alternative edifice of knowledge production, policies, institutions, and relationships that can counter authoritarian politics with new social (and socionatural) relations. This is not only about building a better, more nuanced version of science via the practice of political–ecological research but also about harnessing more-than-scientific resources in ways that seek to change rather than merely describe the world (Castree, Chatterton, and Heynen 2010).

We call for a sensitivity to the power of both ontological and epistemological politics through which environmental issues are defined and known, and that thereby shape conflicts (Blaser 2013; Escobar 2016). We put forward three pathways—expose, teach and learn, and engage—to show what an effective political–ecological critique in the post-truth moment might look. Our aim is to inspire a response that counters post-truth, to think about

how to engage with the public that form enduring resistance networks to authoritarian power. We caution, however, that this should not be read as a singular prescriptive solution; rather, we advocate for multiple emerging pathways to counter and resist the onslaught of authoritarian post-truth narratives.

Expose

The power of political ecology is that it cuts through post-truth to *expose* it. Political ecology is not alone in this, as there have been many other fruitful attempts to deconstruct science debates in STS.[7] Political ecology, however, has been at the forefront of calling out the role of powerful authoritarian states, individuals, and corporations who link post-truth discourse to policy and take shortcuts with democratic rights, especially with territorially based and indigenous communities, but also with global planetary health (Batterbury 2016). "Alternative facts" are often central to such efforts. This perhaps involves political ecology's role as the "trickster," both mimicking and calling out hegemonic science and political discourse (Robbins 2015), but more its willing to use this science to critically think about how truth claims emerge and can be judged.

For example, the framing of climate change brings powerful actors, institutions, and capital together in shaping the political economy of oil (Bridge and Le Billon 2017). This kind of culturally, historically, and politically contextual analysis shows that alternative facts on climate change emerge from within the same relations and logics that perpetuate current capitalist projects, rather than existing as an alternative to a capitalist worldview. This needs to be distinguished from the kinds of alternative ontologies that sit outside of capitalist structuring, such as those that may be practiced by indigenous peoples (Valdivia 2009; Sundberg 2010; Theriault 2017; Anthias 2018). Exposing unsubstantiated "alt. facts" will not suffice, though. The role of political ecologists is to expose power, profit making, and threats to the environment and social justice (Martinez-Alier et al. 2016; Nightingale 2017). This is reflected in the work of environmental justice organizations and other nongovernmental organizations, like the EJOLT project (see http://www.ejolt.org/project) and *Accion Ecológica* in Quito, Ecuador, which brave personal risks to expose environmental injustices and make essential links between scholars and environmental justice activist networks.

Power that coalesces through exposure is not singular but can take many forms. Examples include the Environmental Justice Atlas, or the growing Political Ecology Network initiative (see https://politicalecologynetwork.org/), which links academic output to social media and political journalism (see http://www.aljazeera.com/profile/william-g-moseley.html). Another way to expose is through collaborative attempts, such as the ENTITLE writing collective, which mainstreams critical environmental scholarship through less-known public and activist stories. It aims to link policymakers, scientific researchers, and activists, "through engagement in movements and institutions" (see https://entitleblog.org/). Meanwhile, the network of academics and nonprofits working under the Environmental Data & Governance Initiative (EDGI; see https://envirodatagov.org/) are on the front lines exposing authoritarianism threats to progressive U.S. "federal environmental and energy policy, and to the scientific research infrastructure" meant to "investigate, inform, and enforce them" (Dillon et al. 2019).

These efforts are a small sampling of the initiatives taken by political ecologists to link across communities of knowledge. A question that emerges in these efforts is this: Whose voices are privileged and whose are marginalized, even within collaborative projects? It is arguably more important than ever, in an era of post-truth, to use the counternarratives and explanations generated by political ecology offer much by way of evaluating environmental 'post-truths' asserted in domains of populist authoritarian politics.

Teach and Learn

Going beyond exposure, political ecology *teaching* and *learning* can expand the impact of our critique of alternative facts. Geographers are learning fast that effective communication can challenge authoritarianism through deliberately networking, publishing, increased social media presence, and, moreover, mobilizing this effectively to students and the broader public. For example, political ecologists have been at the forefront of recent attempts at "decolonizing" how ecology and the Anthropocene (Schulz 2017) are delivered in the classroom and approached by the institutions that structure them

(e.g., Fletcher 2017; Meek and Lloro-Bidart 2017; Meyerhoff and Thompsett 2017; Osborne 2017). These efforts serve to decenter some forms of science as hegemonic ways of knowing, at the same time providing students with the critical skills to place all ways of knowing within the power relations that perpetuate them.

Feminist political ecologists have been at the forefront of the coproduction of knowledge with people outside academia and how values and facts that drive outside involvement combine in everyday politics. Harcourt and colleagues, for example, have overseen a movement to engage feminist political ecology with grassroots organizations worldwide that brought forward insights into how smaller scale, localized resistances to hegemonic economic and political relations can succeed (Harcourt and Nelson 2015). The recently formed WEGO (Well-being, Ecology, Gender and Community) network will collect together knowledge of local communities' own understandings of strategies to build resilient and equitable futures. This work highlights the coproduction of knowledge to help community and network activists better understand the institutional, economic, and political contexts that serve to support or inhibit their efforts. Scholars engaging in these practices also gain experiential and in-depth understanding of alternative ontologies and visions for a better world. These efforts have shown the importance of scholarship in not only exposing but also learning from community efforts at challenging hegemonic relations of power.

Other efforts at coproduction of knowledge through teaching and learning include the ENTITLE collective's political ecology syllabus (see http://www.politicalecology.eu/) and also POLLEN's online teaching resources (https://politicalecologynetwork.org/political-ecology-syllabi/) that produces scholarship through community building and stimulating dialogue among "diverse communities" (Harcourt and Nelson 2015), albeit ones that are most likely to use Web-based resources for learning. Political ecologists can learn from recent decolonizing efforts that call for new forms of "epistemic disobedience"—political and epistemological delinking of one's colonial past (Mignolo 2011, 4; Hawthorne and Meché 2016). A good example of this learning in practice through disobedience is the historical problematizing of neoliberal or market conservation that has displaced local practices and knowledges (e.g., Igoe, Sullivan, and

Brockington 2010). The key, however, is to not only bring to light meaningful political ecology research but to integrate this learning, both within the academy (Sundberg 2014) and through broader networks of resistance (Dillon et al. 2019).

Engage

Some political ecologists have taken the notion of learning to another level by trying to translate it directly into policy arenas. For example, Ojha, Paudel, and Dipak (2013) experimented with policy labs in the forestry sector (earlier called *Ban Chautari* but now used beyond the forestry sector to deal with climate and water issues) to generate critical thinking about environmental governance questions for which conventional expertise is inadequate. Policy labs bring together political actors and sectoral specialists (i.e., hydrologists, agricultural officers, and forestry officers) to tackle environmental governance problems. Using Chatham House rules, policy labs are designed to create safe spaces of ignorance, encouraging people to ask questions rather than providing answers. A core concern is to show how different sectors are linked together, the histories surrounding how and why that is the case, and where their agendas are conflicting. This helps to place the issues at stake within a wider contextual frame and can offer opportunities for everyone involved to learn and generate new critical ideas about action.

A renewed focus on rights infuses geographical work, faced with threats that are existential and real and geographical—from border policing to reneging on international treaties and agreements (Sundberg 2010). Social scientists have a particular duty to call out the broader publics'—from civil society groups and individuals to those marginal or invisible—rights to participation (Neimark and Vermeylen 2017). For instance, the "Political Ecology for Civil Society" open access publication by the ENTITLE group is an excellent example of bridging the gap between activist groups and critical social science (ENTITLE Fellows 2016). Also relevant is the Emancipatory Rural Politics Initiative (ERPI) work on authoritarian populism that looks to provoke debate and action among scholars, activists, practitioners, and policymakers on how "exclusionary politics are deepening inequalities," through issues of growth, climate disruptions, and social division and focused on generating alternatives to regressive, authoritarian

politics (Scoones et al. 2017). There are even more overt political campaigns that require new alliances and coalitions (de Vrieze 2017) around antifracking, food sovereignty movements, and pollution cleanup (Hudgins and Poole 2014; Cambell and Veteto 2015; D'Alisa et al. 2017).

Yet, new opportunities beyond academia have also opened up. These are particularly in settings less examined by political ecologists but nonetheless at the heart of current political dynamics around post-truth. They include rural white working-class communities who are generally (mis)represented as "conservative, xenophobic, and reactionary" (Van Sant and Bosworth 2017) but that many times also share experiences of marginality and forms of local knowledge with some of the subjects conventionally focused on in political ecology studies (McCarthy 2002). Although political ecology is effective in highlighting political activism and social movements, if anything, it has been historically less successful at delivering its research results in ways that are easily mobilized to diverse political coalitions. It is these diverse political coalitions where we argue that political ecology research if delivered to nonacademic settings can gain traction in countering post-truth narratives.

Public outreach beyond academia is therefore vital. Political ecology's Public Political Ecology Lab (PPEL) is one important public outreach project (see http://ppel.arizona.edu). It narrates the need for practical and political engagement through academic work, providing training on research methods (participatory action research) and pragmatic media and communication skills to activist-minded students and the wider public. It also provides an online forum to make vital connections between community organizations and graduate students for direct impact. Similarly, the rapidly growing Political Ecology Network is now reaching beyond Europe to facilitate exchanges with a number of "nodes" consisting of non-Western institutions, academics, and civil society organizations. As Martinez-Alier et al. (2014) showed, there is a "reverse movement" of concepts and ideas coming from environmental justice organizations to academic political ecology, thereby, "favor[ing] cooperation between activist and academics because they do not compete for the same turf" (49). This demonstrates the potential for scientists, political ecologists, and activists to form essential alliances to counter post-truth discourse and new forms of authoritarianism.

Conclusion

If anything, political ecologists are responding to contemporary authoritarianism, drawing attention not only to injustice but also to social and political resistance through collective action around the world. To be effective, though, we need to move beyond just illustrating obvious tensions that exist within our own practice and praxis. We must question "truth" based on empirically based natural and social science through multiple perspectives, also explicitly amplifying an inclusive, effective, and publicly accessible political ecology that speaks power to post-truth. Crucially, we must continue to explore links between knowledge and authority, in our own scholarship and in other very relevant cognate studies and also with and as we evaluate knowledge claims emanating from different communities globally.

If anything, our collective response to this post-truth moment is to call out the dominant hegemonic discourses that accompany alternative facts through exposure of the links between power and knowledge and through seeding new counterinitiatives. As those on the political far right successfully adopt poststructuralist ideas and techniques and methods of grassroots activism to maintain authoritarianism, political ecologists need once again to reappropriate these methods of public engagement and civil action. This is a long and difficult project and by no means do we pose a single solution here. Yet, our collective goal is to add tactics and analysis, making our scholarship more relevant, accessible, and engaging to populations most likely to suffer from the proliferation of post-truth politics, notably around the denial of climate change and its impacts.

Acknowledgments

This article represents work conducted as part of the Political Ecology Network (POLLEN). We thank Rob Fletcher and Bram Büscher for help with earlier drafts of this article. Special thanks to James McCarthy, Jennifer Cassidento, and three anonymous reviewers from the *Annals of the American Association of Geographers* for suggestions.

ORCID

Benjamin Neimark ⓘ https://orcid.org/0000-0003-3229-0869

John Childs ⓘ http://orcid.org/0000-0003-3293-9517
Sian Sullivan ⓘ http://orcid.org/0000-0002-0522-8843
Tor A. Benjaminsen ⓘ https://orcid.org/0000-0003-0192-833X
Simon Batterbury ⓘ https://orcid.org/0000-0002-2801-7483
Stasja Koot ⓘ https://orcid.org/0000-0001-8625-7525

Notes

1. Since being used by the U.S. president's special counsel to defend demonstrably false statements by the White House Press Office, the term *alternative facts* has been invoked widely in the media to question the relationship between science and truth. Similarly, President Donald Trump makes personal and repeated dismissals of major international media and research outlets as "fake news."
2. Although used somewhat interchangeably, we recognize that hegemony and dominant forms of science, and knowledge, are not necessarily always the same (see Guha 1997).
3. Critical political ecology is an open-ended and empirically based approach that combines deconstruction with a realist belief in science as a means to achieve a more accurate description and understanding of environmental realities. This is not the only attempt to do this. In fact, there is a long history of previous work in "critical realism" to integrate sociopolitical values with positivism (see Bhaskar [1975] 1997) and also to some degree in sustainability science (see Clark et al. 2016).
4. We do not provide a review of political ecology but rather a snapshot of some examples of its breadth; for fuller reviews, see Robbins (2011), Bryant (2015), and Perreault et al. (2015).
5. Albeit a key theme in earlier political ecology, our hope is that given the particular political climate of post-truth, more studies today can reemphasize the importance of the emergence of facts simultaneously with values and structure.
6. From this perspective, truth making is more about establishing an effective hegemony (understood as the articulation of different interests around a common cause) than trying to champion a particular constellation of facts.
7. Although STS does include debates around positivist science and many, particularly those geographers and others adopting the language of assemblage, claim that their frameworks do explain the entanglement of facts simultaneously with values and structures, it is critical political ecology that has been more willing to adopt positivist science as a tool to counter dominant scientific claims.

References

Anthias, P. 2018. *Limits to decolonization: Indigeneity, territory, and hydrocarbon politics in the Bolivian Chaco.* Ithaca, NY: Cornell University Press.

Batterbury, S. P. J. 2016. Ecología política: Relevancia, activismo y posibilidades de cambio [Political ecology: Relevance, activism and possibilities for change]. Ecología Política 50:45–54.

Benjaminsen, T. A., J. B. Aune, and D. Sidibé. 2010. A critical political ecology of cotton and soil fertility in Mali. Geoforum 41 (4):647–56.

Bennett, J. 2010. Vibrant matter: The political life of things. Durham, NC: Duke University Press.

Bhaskar, R. A. [1975] 1997. A realist theory of science. London: Verso.

Blaikie, P., and H. Brookfield. 1987. Land degradation and society. London and New York: Routledge.

Blaser, M. 2013. Notes towards a political ontology of "environmental" conflicts. In Contested ecologies: Dialogues in the South on nature and knowledge, ed. L. Green, 13–27. Cape Town, South Africa: Human Sciences Research Council Press.

Brannstrom, C., and J. M. Vadjunec. 2013. Notes for avoiding a missed opportunity in sustainability science: Integrating land change science and political ecology. In Land change science, political ecology, and sustainability synergies and divergences, ed. C. Brannstrom and M. Vadjunec, 1–23. Abingdon, UK: Routledge.

Bridge, G., and P. Le Billon. 2017. Oil. Oxford, UK: Wiley.

Bryant, R. L., ed. 2015. The international handbook of political ecology. Cheltenham, UK: Edward Elgar.

Burke, B. J., and N. Heynen. 2014. Transforming participatory science into socioecological praxis: Valuing marginalized environmental knowledges in the face of the neoliberalization of nature and science. Environment and Society 5 (1):7–27.

Burman, A. 2017. The political ontology of climate change: Moral meteorology, climate justice, and the coloniality of reality in the Bolivian Andes. Journal of Political Ecology 24 (1):921–38.

Campbell, B. C., and J. R. Veteto. 2015. Free seeds and food sovereignty: Anthropology and grassroots agrobiodiversity conservation strategies in the U.S. South. Journal of Political Ecology 22 (1):445–65.

Castree, N., and B. Braun. 2001. Social nature theory, practice, and politics. Malden, MA: Blackwell.

Castree, N., P. A. Chatterton, and N. Heynen, eds. 2010. The point is to change it: Geographies of hope and survival in an age of crisis. Oxford, UK: Wiley.

Cavanagh, C., and T. A. Benjaminsen. 2017. Political ecology, variegated green economies, and the foreclosure of alternative sustainabilities. Journal of Political Ecology 24 (1):200–216.

Clark, W. C., T. P. Tomich, M. Van Noordwijk, D. Guston, D. Catacutan, N. M. Dickson, and E. McNie. 2016. Boundary work for sustainable development: Natural resource management at the Consultative Group on International Agricultural Research (CGIAR). Proceedings of the National Academy of Sciences 113 (17):4615–22.

Collingridge, D., and C. Reeve. 1986. Science speaks to power: The role of experts in policy making. New York: St. Martin's.

D'Alisa, G., A. R. Germani, P. M. Falcone, and P. Morone. 2017. Political ecology of health in the land of fires: A hotspot of environmental crimes in the south of Italy. Journal of Political Ecology 24:59–86.

de la Cadena, M. 2010. Indigenous cosmopolitics in the Andes: Conceptual reflections beyond politics. Cultural Anthropology 25 (2):334–70.

de Vrieze, J. 2017. "Science wars" veteran has a new mission. Science 358 (6360):159.

Dillon, L., R. Lave, B. Mansfield, S. Wylie, N. Shapiro, A. S. Chan, and M. Murphy. 2019. Situating data in a Trumpian era: The environmental data and governance initiative. Annals of the American Association of Geographers. doi:10.1080/24694452.2018.1511410

ENTITLE Fellows. 2016. Political ecology for civil society. Accessed January 15, 2016. http://www.politicalecology.eu/documents/events/94-entitle-manual-may-2016/file.

Escobar, A. 1995. Imagining a post-development era. In Power of development, ed. J. Crush, 211–27. New York: Routledge.

———. 2016. Thinking-feeling with the earth: Territorial struggles and the ontological dimension of the epistemologies of the South. AIBR, Revista de Antropología Iberoamericana 11 (1):11–32.

Fairhead, J., and M. Leach. 1996. Misreading the African landscape: Society and ecology in a forest–savanna mosaic. Cambridge, UK: Cambridge University Press.

Fletcher, R. 2017. Connection with nature is an oxymoron: A political ecology of nature-deficit disorder. The Journal of Environmental Education 48 (4):226–33.

Forsyth, T. J. 2003. Critical political ecology: The politics of environmental science. London and New York: Routledge.

Foucault, M. 1980. Power/knowledge: Selected interviews and other writings, 1972–1977, ed. C. Gordon. London: Harvester Wheatsheaf.

Goldman, M. J., P. Nadasdy, and M. D. Turner, eds. 2011. Knowing nature: Conversations at the intersection of political ecology and science studies. Chicago: University of Chicago Press.

Gramsci, A. 1971. Selections from the prison notebooks of Antonio Gramsci, ed. and trans. Q. Hoare and G. N. Smith. New York: International.

Guha, R. 1997. Dominance without hegemony: History and power in colonial India. Cambridge, MA: Harvard University Press.

Guthman, J., and B. Mansfield. 2013. The implications of environmental epigenetics: A new direction for geographic inquiry on health, space, and nature–society relations. Progress in Human Geography 37 (4):486–504.

Harcourt, W., and L. L. Nelson, eds. 2015. Practicing feminist political ecologies: Moving beyond the "green economy." London: Zed.

Harding, S. 2008. Sciences from below: Feminisms, postcolonialities, and modernities. Durham, NC: Duke University Press.

Hawthorne, C., and B. Meché. 2016. Making room for black feminist praxis in geography. Society and Space. Accessed January 15, 2017. http://societyandspace.

org/2016/09/30/making-room-for-black-feminist-praxis-in-geography/.

Hudgins, A., and A. Poole. 2014. Framing fracking: Private property, common resources, and regimes of governance. *Journal of Political Ecology* 21 (1):303–19.

Hulme, M. 2010. Problems with making and governing global kinds of knowledge. *Global Environmental Change* 20 (4):558–64.

Igoe, J., S. Sullivan, and D. Brockington. 2010. Problematizing neoliberal biodiversity conservation: Displaced and disobedient knowledge. *Current Conservation* 3 (3):4–7.

Jasanoff, S. 2006. Just evidence: The limits of science in the legal process. *The Journal of Law, Medicine & Ethics: A Journal of the American Society of Law, Medicine & Ethics* 34 (2):328–41.

King, B. 2010. Political ecologies of health. *Progress in Human Geography* 34 (1):38–55.

Kosek, J. 2006. *Understories: The political life of forests in northern New Mexico.* Durham, NC: Duke University Press.

Kuhn, T. S. 1970. *The structure of scientific revolutions.* 2nd ed. Chicago: University of Chicago Press.

Mair, J. 2017. Post-truth anthropology. *Anthropology Today* 33 (3):3–4.

Martinez-Alier, J., I. Anguelovski, P. Bond, D. Del Bene, and F. Demaria. 2014. Between activism and science: Grassroots concepts for sustainability coined by environmental justice organizations. *Journal of Political Ecology* 21 (1):20–60.

Martinez-Alier, J., L. Temper, D. Del Bene, and A. Scheidel. 2016. Is there a global environmental justice movement? *The Journal of Peasant Studies* 43 (3):731–55.

McCarthy, J. 2002. First world political ecology: Lessons from the wise use movement. *Environment and Planning A* 34 (7):1281–1302.

Meek, D., and T. Lloro-Bidart. 2017. Introduction: Synthesizing a political ecology of education. *The Journal of Environmental Education* 48 (4):213–25.

Meyerhoff, E., and F. Thompsett. 2017. Decolonizing study: Free universities in more-than-humanist accompliceships with Indigenous movements. *The Journal of Environmental Education* 48 (4):234–47.

Mignolo, W. 2011. *The darker side of Western modernity: Global futures, decolonial options.* Durham, NC: Duke University Press.

Li, T. M. 2007. *The will to improve: Governmentality, development, and the practice of politics.* Durham, NC: Duke University Press.

Nagel, A. 2017. *Kill all normies: Online culture wars from 4chan and Tumblr to Trump and the alt-right.* Winchester, UK: Zero Books.

Neimark, B. D., and S. Vermeylen. 2017. A human right to science?: Precarious labor and basic rights in science and bioprospecting. *Annals of the American Association of Geographers* 107 (1):167–82.

Nightingale, A. 2006. The nature of gender: Work, gender, and environment. *Environment and Planning D: Society and Space* 24 (2):165–85.

———. 2017. Power and politics in climate change adaptation efforts: Struggles over authority and recognition in the context of political instability. *Geoforum* 84:11–20.

Ojha, H. R., N. S. Paudel, and B. K. Dipak. 2013. Can policy learning be catalyzed? Ban Chautari experiment in Nepal's forestry sector. *Journal of Forest and Livelihood* 10 (1):1–27.

Osborne, T. 2017. Public political ecology: A community of praxis for earth stewardship. *Journal of Political Ecology* 24 (1):843–60.

Peet, R., and M. Watts. 1996. *Liberation ecologies: Environment, development, social movements.* London and New York: Routledge.

Peluso, N. L., and M. Watts, eds. 2001. *Violent environments.* New York: Cornell University Press.

Perreault, T., G. Bridge, and J. McCarthy, eds. 2015. *The Routledge handbook of political ecology.* London and New York: Routledge.

Popper, K. R. 1971. *The open society and its enemies.* London and New York: Routledge.

Robbins, P. 2011. *Political ecology: A critical introduction.* Oxford, UK: Wiley.

———. 2015. The trickster science. In *The Routledge handbook of political ecology,* ed. T. Perreault, G. Bridge, and J. McCarthy, 89–101. London and New York: Routledge.

Robertson, M. M. 2006. The nature that capital can see: Science, state, and market in the commodification of ecosystem services. *Environment and Planning D: Society and Space* 24 (3):367–87.

Rocheleau, D. E. 2008. Political ecology in the key of policy: From chains of explanation to webs of relation. *Geoforum* 39 (2):716–27.

Rocheleau, D., B. Thomas-Slayter, and E. Wangari. 1996. Gender and environment: A feminist political ecology perspective. In *Feminist political ecology: Global issues and local experiences,* ed. D. Rocheleau, B. Thomas-Slayter, and E. Wangari, 3–26. New York: Routledge.

Schulz, K. A. 2017. Decolonizing political ecology: Ontology, technology and "critical" enchantment. *Journal of Political Ecology* 24 (1):125–43.

Scoones, I., M. Edelman, S. M. Borras, Jr., R. Hall, W. Wolford, and B. White. 2017. Emancipatory rural politics: Confronting authoritarian populism. *The Journal of Peasant Studies* 45 (1):1–20.

Stott, P., and S. Sullivan. 2000. *Political ecology: Science, myth and power.* London: Edward Arnold.

Sullivan, S. 2017. What's ontology got to do with it? On nature and knowledge in a political ecology of "the green economy." *Journal of Political Ecology* 24 (1):217–42.

Sundberg, J. 2010. Diabolic caminos in the desert and cat fights on the Río: A posthumanist political ecology of boundary enforcement in the United States–Mexico. *Annals of the Association of American Geographers* 101 (2):318–36.

———. 2014. Decolonizing posthumanist geographies. *Cultural Geographies* 21 (1):33–47.

Theriault, N. 2017. A forest of dreams: Ontological multiplicity and the fantasies of environmental government in the Philippines. *Political Geography* 58 (58):114–27.

Valdivia, G. 2009. Indigenous bodies, indigenous minds? Towards an understanding of indigeneity in the Ecuadorian Amazon. *Gender, Place & Culture* 16 (5):535–51.

Van Sant, L., and K. Bosworth. 2017. Intervention—Race, rurality, and radical geography in the U.S. *Antipode Online.* Accessed September 1, 2017. https://wp.me/p16RPC-1Da.

Watts, M. 1983. *Silent violence.* Berkeley: University of California Press.

Wolford, W. 2010. *This land is ours now: Social mobilization and the meanings of land in Brazil.* Durham, NC: Duke University Press.

BENJAMIN NEIMARK is Senior Lecturer of Human Geography in the Lancaster Environment Centre, Lancaster University. Library Avenue, Lancaster University, Lancaster LA1 4YQ, UK. E-mail: b.neimark@lancaster.ac.uk. His research interests include the political ecology and political economy of bio- and green economy interventions, uneven development, and labor and global commodity chains in Madagascar and Africa.

JOHN CHILDS is in the Lancaster Environment Centre, Lancaster University, Library Avenue, Lancaster University, Lancaster LA1 4YQ, UK. E-mail: j.childs@lancaster.ac.uk. His research interests include the political ecology of resource extraction in the Global South, particularly focused on mining and its various forms, geographies, and effects.

ANDREA J. NIGHTINGALE is a Professor in the Department of Sociology and Human Geography at the University of Oslo, Norway, and the Department of Urban and Rural Development at the Swedish University of Agricultural Sciences (SLU), Sweden. Email: a.j.nightingale@sosgeo.uio.no. Her current research interests include the nature–society nexus; feminist theorizations of emotion and subjectivity in relation to development, transformation, collective action, and the commons; political violence and climate change; and public authority, collective action, and state formation.

CONNOR JOSEPH CAVANAGH is a Post-Doctoral Research Fellow in the Department of International Environment and Development Studies (Noragric), Norwegian University of Life Sciences, 1433 As, Norway. E-mail: connor.cavanagh@nmbu.no. His research and publications explore the political ecology of conservation and development interventions, with a focus on land and resource tenure conflicts and the institutional evolution of laws, regulations, and policies for governing both ecosystems and rural populations.

SIAN SULLIVAN is Professor of Environment and Culture at Bath Spa University, Newton Park, Bath BA2 9BN, UK, and Associate of Gobabeb Research and Training Centre, Namibia. E-mail: s.sullivan@bathspa.ac.uk. Her research interests include cultural landscapes, political ecology, and the financialization of nature.

TOR A. BENJAMINSEN is a Professor in the Department of International Environment and Development Studies, Faculty of Landscape and Society, Norwegian University of Life Sciences, 1432 As, Norway. E-mail: torbe@nmbu.no. He works on issues of environmental change and conservation, pastoralism, land rights, resistance, and justice in Mali and Tanzania, as well as in Arctic Norway.

SIMON BATTERBURY is a Professor of Political Ecology at the Lancaster Environment Centre, Lancaster University, Library Avenue, Lancaster University, Lancaster LA1 4YQ, UK. E-mail: simonpjb@unimelb.edu.au. He is also Principal Fellow in the School of Geography at the University of Melbourne. His research interests include the political ecology of natural resources in West Africa and Oceania.

STASJA KOOT is an Assistant Professor in the Department of Sociology and Anthropology of Development at Wageningen University, Wageningen, 6700 EW Wageningen, The Netherlands. E-mail: kootwork@gmail.com. His research interests are predominantly in Southern Africa, including nature conservation, tourism, wildlife crime, capitalism, indigenous people, land, and philanthropy.

WENDY HARCOURT is Professor of Gender, Diversity & Sustainable Development, Westerdijk Professor at the International Institute of Social Studies of Erasmus University, Rotterdam, ISS, 2518 AX The Hague, The Netherlands. E-mail: harcourt@iss.nl. Her research interests include feminist political ecology, feminist theory, and postdevelopment.

Populism, Emancipation, and Environmental Governance: Insights from Bolivia

Diego Andreucci (iD)

The rise of the populist right and the concomitant crisis of progressive neoliberalism have reactivated debates about the possibility and desirability of a left populism. Through an engagement with the work of Ernesto Laclau, and drawing insights from contentions around resource governance in Evo Morales's Bolivia, this article addresses the question of whether and how populism can be a valid strategy to achieve emancipatory transformations in environmental governance. In Bolivia, the construction of a collective identity out of indigenous–popular mobilizations facilitated a counterhegemonic articulation capable of subverting the neoliberal order and achieving progressive changes in the governance of natural resources. Yet, following the electoral victory of Morales in 2005, this counterhegemonic project turned into a passive revolution that frustrated its most genuinely transformative political aspirations. Reflecting on the Bolivian experience, I make three interrelated claims. First, the main strength of populism lies in enabling socioenvironmental movements to transcend their particularistic struggles and, through the (re)definition of a collective identity, build a broader counterhegemonic bloc capable of subverting the dominant institutional order. Second, for populism to be conducive of emancipatory transformation, the process of articulation should emerge out of subaltern socioenvironmental struggles and revendications and have radical, egalitarian-democratic ambitions transcending the horizon of the state. Third, short of a full social reordering, counterhegemonic projects are likely to be reabsorbed within the dominant institutional configuration and yet they remain necessary to challenge the socioenvironmentally regressive tendencies of capitalist domination and enable progressive transformations in environmental governance.

右翼民粹主义的兴起，以及同时发生的激进新自由主义之危机，重新燃起了有关左翼民粹主义的可能性与可欲性之辩论。通过涉入．拉克劳的理论，并运用玻利维亚的埃沃．莫拉莱斯政权下的资源治理争议之洞见，本文应对民粹主义是否能够作为在环境治理中取得解放性转变的有效策略。在玻利维亚，从本土大众动员中建构而成的集体身份认同，促进能够颠覆新自由主义秩序并在自然资源治理中取得激进变迁的反霸权接合。但在 2005 年莫拉莱斯选举胜利之后，此一反霸权计画转变成为消极革命，并使真正的转型政治期待落空。我将做出三大相关宣称来反应玻利维亚的经验。首先，民粹主义的主要长处在于让社会环境运动能够超越其特定的斗争，并且通过（重新）定义共同的身份认同，建构能够颠覆宰制的制度次序之更为广泛的反霸权集团。再者，为使民粹主义得以贡献解放性的转型，接合的过程应从从属的社会环境斗争和收复失地的要求中浮现，并具有超越国家水平的激进、自主的民主抱负。第三，由于缺乏完整的社会再次序化，反霸权计画很可能被重新吸纳进支配的制度构造，但它们仍然是挑战资本主义支配下社会环境的倒退倾向、并促发环境治理激进转型的必要条件。关键词：玻利维亚，反霸权，环境治理，政治生态学，民粹主义。

El ascenso de la derecha populista y la crisis concomitante del neoliberalismo progresista ha reactivado los debates acerca de la posibilidad y el atractivo de un populismo de izquierda. A través de un compromiso con el trabajo de Ernesto Laclau, y derivando perspicacias de las disputas sobre la gobernanza de los recursos en la Bolivia de Evo Morales, este artículo aborda la cuestión de si el populismo puede ser una estrategia válida para lograr transformaciones emancipadoras en gobernanza ambiental, y cómo puede serlo. En Bolivia, la construcción de una identidad colectiva a partir de movilizaciones indígeno-populares facilitó una articulación contra-hegemónica capaz de subvertir el orden neoliberal y alcanzar cambios progresivos en la gobernanza de los recursos naturales. No obstante, después de la victoria electoral de Morales en 2005, este proyecto contra-hegemónico se convirtió en una revolución pasiva que frustró sus aspiraciones políticas más genuinamente transformadoras. Reflexionando sobre la experiencia boliviana, formulo tres reclamaciones interrelacionadas. Primera, la principal fuerza del populismo descansa en capacitar los movimientos socioambientales para trascender sus luchas particularistas y, a través de la (re)definición de una identidad colectiva, construir un bloque contra-hegemónico más amplio capaz de subvertir el orden institucional dominante. Segunda, para que el populismo sea propicio a la transformación emancipadora, el proceso de articulación debe surgir de las luchas socioambientales y reivindicaciones subalternas y tener ambiciones radicales e igualitario-democráticas que trasciendan el horizonte del estado. Tercera, cortos de un reordenamiento social pleno, los proyectos contra-hegemónicos quedan propensos a ser reabsorbidos dentro de la configuración institucional dominante y aun así siguen siendo necesarios para retar tendencias socioambientalmente reaccionarias de la dominación capitalista y activan transformaciones progresivas en gobernanza ambiental. *Palabras clave: Bolivia, contra-hegemonía, ecología política, gobernanza ambiental, populismo.*

Commenting on the electoral victory of Donald Trump in the United States, Fraser (2017) argued that the left should reject the false choice between reactionary populism and "progressive neoliberalism" and mobilize its own transformative populism. As Mouffe (2005, 51) noted, it is precisely the (neo)liberal reduction of politics to a "postpolitical" consensus that favors the emergence of right-wing populism. Mainstream environmental politics is also caught in a false choice between increasingly authoritarian forms of nature's enclosure and resource extractivism and the false promises of a financialized, "green" capitalism (Wainwright and Mann 2015). Yet, despite its normative commitment to promoting social justice and structural political change (Bridge et al. 2015), political ecology has so far dedicated comparatively little effort to conceptualizing and reflecting on what emancipatory socioenvironmental change could look like and how it might take place.

The overall purpose of this article is to advance debates in political ecology around emancipatory strategy. I endorse Fraser's argument that populism might offer a way of advancing a counterhegemonic project. Populism, however, is a controversial strategy: It is associated with widely diverse political projects and can be conducive of progressive as well as regressive social change, depending on historically and geographically specific circumstances. In critically appraising its political potential, the key question is therefore this: How might populism be a valid strategy to achieve emancipatory transformations in environmental governance?

To address this question, I critically engage with the work of the Argentinian philosopher Ernesto Laclau, the most influential advocate and theorist of populism on the left. I also draw insights from qualitative interviews and secondary literatures delving into the case of struggles around environmental governance in Bolivia.[1] Over the last fifteen years, Bolivia's current governing party, Evo Morales's Movement Towards Socialism (MAS), has gone from being celebrated for its socially and environmentally radical political agenda—which it drew from indigenous–popular mobilizations (Collins 2014)—to being criticized as promoting a form of authoritarian, neoextractivist populism (Svampa 2015). The case of Bolivia therefore offers a valid entry point to both (1) analyze how populism might favor emancipatory change and (2) identify and reflect on some of the challenges involved in constructing a counterhegemonic strategy through populism.

The article is structured as follows. In the next section, I argue that the main strength of populism lies in enabling socioenvironmental movements to transcend their particular struggles and build broader counterhegemonic ambitions with the potential to subvert existing institutional frameworks for governing environments and resources. I also point out some of the pitfalls of populism as a counterhegemonic strategy, particularly the risk of passive revolution and the correspondent possibility of ideological and political degeneration. Following that, I consider some instances of how the potentialities and perils of populism played out in the Bolivian case, particularly how the indigenous– popular counterhegemonic project—which emerged out of struggles over resource governance— was reabsorbed into the existing order through the "electoralization" and subsequent institutionalization of its political agenda. The last section, reflecting on the Bolivian experience, discusses the implications of the theory of populism for the politics of environmental governance.

Environmental Governance, Hegemony, and Populism

The notion of environmental or resource governance refers to a multiscalar ensemble of (broadly defined) institutional frameworks through which environments and resources are governed (Perreault 2008). Political ecologists have distanced themselves from mainstream analyses of governance and, drawing on the regulation approach among others, have analyzed spatial and scalar restructurings associated with post-Fordism and related processes of "nature's neoliberalization" (see Bakker 2015). Additionally, they have shown how reconfigurations in environmental governance relate to the stabilization of accumulation and the need to stave off crises in nature-facing industries (Bridge and Perreault 2009).

This work has considered the agency of social movements and organized class interests in challenging or reproducing such institutional alignments (Himley 2013). In the Bolivian case considered in this article, for instance, political ecologists have analyzed, among other aspects, the role of popular protest in contesting neoliberal resource governance (Perreault 2006, 2008), hegemonic strategies by landed capital and elite groups resisting agrarian reform (Valdivia 2010), the role of mine workers' unions and cooperatives in reproducing the hegemony of mining (Kaup 2014; Marston and Perreault 2017), and state repression of indigenous groups struggling over the governance of extraction (Andreucci and Radhuber 2017). I suggest that this work might fruitfully be put into dialog with Gramscian-inspired political ecologies of (counter) hegemony.

Political ecologists have found in Gramsci's philosophy of praxis a productive way of examining the discursive and the "ethico-political" moments in hegemonic struggles, without renouncing the materiality of class and of relations of (re)production. Gramsci's dialectical understanding of nature has facilitated its adoption by political ecologists, who have placed analytical emphasis on how the environment is mobilized, materially and ideologically, in processes of hegemonic formation and contestation (Ekers et al. 2009). Within this subfield, reflections on emancipatory politics have not taken center stage. As Glassman (2013) recognized, in recent years, Gramsci's ideas have primarily been adopted by researchers for analyzing the reproduction of ruling class hegemony, whereas those concerned with transformation and alternatives have often turned to other conceptual–political frameworks (e.g., anarchist, autonomist, or post-Marxist approaches). There have been important attempts to advance debates around the politics (and political ecology) of counterhegemony, however.

Counterhegemony refers to the ability of a dominated class or faction to exercise leadership over other subaltern groups, to build a social bloc capable of challenging the hegemony of dominant classes. In Gramscian theory, such a leadership role is assigned, of course, to the working-class movement (Thomas 2013). Yet, political ecologists have rightly argued that other actors and groups—including peasant, indigenous, and other socioenvironmental movements—could become counterhegemonic actors (Karriem 2009). They have explored the ways in which, to exercise leadership, these movements develop the ability to transcend their particularistic, "economic-corporate" interests and engage other subaltern groups. For instance, Karriem (2009) stressed the importance of critiquing the received "common sense" and of the material and cultural practices through which alternative "conceptions of the world" and of nature are elaborated, highlighting the role of "organic intellectuals." Similarly, Calvário et al. (2017) underscored the politics of solidarity through which subaltern groups engage in relations of mutual transformation and alliance, thereby building hegemony from below.

I suggest that a critical and selective adoption of Laclau's ideas on populism might help to advance these emerging conversations.[2] Populism, in Laclau, refers precisely to the process through which a subaltern class or group could transcend the particularism of localized struggles and engage in counterhegemonic politics. To struggle for hegemony, for Laclau (1977, 173), a dominated class or faction must engage with other subaltern groups on the class-neutral terrain of "popular-democratic" struggles. In this way, it constructs a common identity—a shared understanding of "the people"—pitted against the dominant ideology. In his book *On Populist Reason*, Laclau (2005a, 73–87) expanded on the mechanisms through which such a common identity is constructed. He argued that the basic unit of analysis is that of "social demands." When the demands of multiple social groups go unmet, solidarity can arise among them: They share a common source of frustration, through which they form a "chain of equivalences" and develop shared "signifiers." A sense of collective identity thus emerges in opposition to an "other" or "enemy"—typically "those in power"—that functions as a "constitutive outside" and divides society into two opposed blocs.

Liberal theories treat populism as a discursive strategy employed by actors in a position of power, such as an established political party or leader, to mobilize mass support. Yet, the most significant aspect of Laclau's theory is, in my view, his treating populism as a transformative process grounded in socioenvironmental struggles and revendications (Collins 2014). Some critics on the left consider that Laclau's populism is itself postpolitical, in that it reinforces the need for an elite to which demands are addressed and displaces social antagonism onto an external "other" (Žižek 2006; Swyngedouw 2010). For Laclau, however, it is only through an accumulation of unmet demands, and the resulting oppositional construction of a collective identity, that otherwise separate subaltern struggles can join up to challenge and subvert the existing institutional configuration.

A problem that this framework leaves open is that of the challenges associated with the realization of transformative ambitions. A key concept developed by Gramsci to analyze the process of absorption of transformative, counterhegemonic ambitions into an established institutional order is that of "passive revolution," a political process that draws on subaltern demands to produce changes in the existing social order but with the ultimate goal of neutralizing or diffusing popular pressures from below. A passive revolution institutionalizes transformative ambitions, containing their subversive potential. This might, in some cases, result in progressive changes, but it also entails the risk of repression for recalcitrant subaltern groups (Jessop 1990). Laclau (2005b, 47) suggested that, in the institutionalization of a transformative political process, the new power bloc might appropriate the subaltern "signifiers," repressing the social demands from which they emerged. This leaves us therefore with another central issue; that is, to understand how and why counterhegemonic populism might lose its transformative character or even acquire reactionary traits.

Struggles over Resource Governance in Bolivia

Revolutionary Epoch

From 2000 to 2005, Bolivia underwent a cycle of popular struggles that precipitated a crisis of legitimacy of the neoliberal order and prepared the terrain for the political rise of Evo Morales and the MAS.

Webber (2011) referred to this period as a "revolutionary epoch." I argue that, in this conjuncture, a discursive redefinition of the Bolivian people made it possible for distinct socioenvironmental mobilizations to develop into a counterhegemonic bloc with a transformative horizon (Collins 2014). A collective identity emerged out of conflicts such as the Cochabamba Water War of April 2000, which inaugurated this cycle of upheavals (Perreault 2006). After years of defensive and largely ineffective antineoliberal struggles, the Water War marked the beginning of popular counteroffensive (Tapia 2011). The Coordinator (*Coordinadora*) for the Defense of Water and Life—the main political actor in the mobilization—included movements as diverse as factory workers, irrigators, environmentalists, and neighborhood associations. The mobilization began as a struggle against the private, transnational control of water services and it quickly developed a broader set of revendications directed against the neoliberal government (Gutiérrez-Aguilar 2008).

The sense of subaltern solidarity that emerged in this conjuncture facilitated the development of a transformative political ambition with an "egalitarian-democratic" horizon (Swyngedouw 2014). It began to delineate a new understanding of "the people" as those dispossessed of both access to natural wealth and of meaningful political participation. According to Oscar Olivera, one of the *Coordinadora* leaders and organic intellectuals, the two main axes of struggle of the movement were the "social reappropriation of the common good" and the "revendication of direct, popular democracy" (author interview,[3] 11 March 2014). Despite claims for a Constituent Assembly since the 1990s, the *Coordinadora* was the first Bolivian movement to force it into the national political agenda and debate. In this conjuncture, significantly, the proposal for a Constitutent Assembly was not conceived as a way of demanding state reform but rather as an autonomous space of subaltern political sovereignty, outside and against the state itself (Gutiérrez-Aguilar 2008).

This cycle of mobilizations reached its most intense point in 2003. The center of dispute shifted toward hydrocarbon governance. Since the 1990s, hydrocarbons, and particularly natural gas, had become Bolivia's main export; due to the neoliberal governance regime in place, however, transnational firms controlled gas production, commercialization, and, most significant, rents (Kaup 2013). In 2003, there were other intense social conflicts that caused confrontations with the state and violent repression (Gutiérrez-Aguilar 2008). Yet, gas emerged as a central focus of popular mobilizations, whose epicenter was the city of El Alto. Slogans such as "Gas Is Ours" or "Gas for the Bolivian People" became common in demonstrations (Perreault 2006). In Laclau's (2005a) terms, this demand emerged as a (tendentially empty) signifier of the whole chain of equivalences, including a host of other indigenous–popular revendications around natural resource governance as well as issues of democracy and citizenship. The mobilizations culminated in a victory for popular organizations and the resignation of President Sánchez de Lozada.

The issue of indigeneity was a key axis in the remaking of "the people" (Canessa 2006). In this conjuncture, whereby popular struggles were conceived as in opposition to white elites and their "imperial" allies, indigeneity emerged as a shared marker of subalternity among several social sectors. In the 2001 census, 62 percent of Bolivians self-identified as indigenous (Canessa 2006), which signals a politicization of the category and its adoption by many urban and rural poor, not necessarily affiliated with any indigenous ethnicity or organization (Schavelzon 2014). Such a politicization of indigeneity had its roots in different historical trajectories through which the radical Indianist tradition in the Bolivian highlands and the process of indigenous political organization in the lowlands converged with urban indigenous and working-class struggles (García-Linera et al. 2010). This assertion of indigenous identity functioned, too, as a partially empty signifier, standing in for a chain of demands around the subaltern reappropriation of the social wealth—including access to natural resources and rents—and the radical redefinition of citizenship and democracy. It gave this political process an anticolonial and anti-imperialist character, central to the construction of a distinctly indigenous–popular counterhegemony (Postero 2017).

The proposal for a Constituent Assembly had itself originated in the indigenous mobilizations that took place since the early 1990s (Tapia 2011). After 2000, it became a key unifying agenda of the indigenous movement, which gained relevance as a national political actor and created an alliance with *campesino* and coca grower unions. In 2004, this led to the formation of a "Unity Pact," which, in the conjuncture of struggle against neoliberalism, functioned as "the collective 'organic intellectual' of the *campesino* and indigenous peoples" (Tapia 2011, 93). In 2006, the Unity Pact presented a first constitutional proposal; this appropriated and re-elaborated demands for redistribution and political participation from the preceding cycle of struggles—such as the nationalization of natural resources—alongside other historical revendications, including land redistribution and indigenous–*campesino* territorial self-government (Schavelzon 2012). The idea of a "plurinational state" was proposed as the political horizon for a process of decolonization of the state and the partial decommodification of society–nature relations in the country, via the promotion of indigenous–communitarian political and economic practices.

Restoration and "Electoralization"

Today, Bolivia is no longer considered a virtuous example of indigenous-led, emancipatory socioenvironmental transformation. The record of change of the MAS administration is mixed. With regard to environmental governance, the reach of reform has been significantly lesser than demanded by indigenous–popular mobilizations in key areas. Indigenous demands for including principles of territorial self-determination in the governance of mineral and hydrocarbon extraction, for instance, were only partly recognized and then gradually sidelined or reverted (Andreucci and Radhuber 2017). The expansion of resource extractivism, moreover, caused increasing tensions between the government and the indigenous movement. After 2011, such tensions turned into a phase of open conflict, inaugurated by the dispute over the construction of a highway cutting through the Isiboro-Sécure National Park and Indigenous Territory (TIPNIS; McNeish 2013). The TIPNIS conflict pitted indigenous organizations, opposed to the project, against (largely pro-government) *campesino* unions and led to the breakup of the Unity Pact. In the context of increasing state repression of indigenous organizations, nongovernmental organizations and media critical of the government also came under attack, accused of aiding a conservative comeback (Svampa 2015).

Several critics have argued that the institutionalization of the MAS's political project from 2006 on represents an instance of passive revolution and "transformism" (Tapia 2011; Modonesi 2013; Webber 2016; Andreucci 2017) and have drawn parallels with the conservative turn taken by Bolivian politics after the 1952 revolution (Hesketh and Morton 2014). Increased capture and redistribution of resource rents has allowed for improvements in the material conditions of the country's poor. This created broad legitimacy around resource-based accumulation, thereby diffusing popular opposition and marginalizing recalcitrant subaltern groups. Yet, critics argue, the Morales government has not attacked structural causes of inequality and it has increased dependence on transnationally controlled resource extraction (Webber 2016). Over the course of Morales's first term (2006–2009), the government already moderated its positions in some important respects. For instance, during the negotiations with right-wing parties that preceded the approval of the 2009 constitution, the radical character of the Unity Pact's proposal was limited in key aspects, such as land reform and indigenous political participation (Schavelzon 2012). This, according to Unity Pact activist Fernando Garcés, marked a key point of rupture between the indigenous movement and the MAS (author interview, 20 October 2014) and prepared the terrain for the subsequent phase of conflict.

It can be argued, however, that the roots of political degeneration were already present before 2006, particularly in what I would call the MAS's electoralization of indigenous–popular counterhegemony after 2003 and the appropriation of its signifiers. In the 2000 to 2003 conjuncture, the organizations mobilizing in Cochabamba, El Alto, and elsewhere were largely independent from formal institutions as well as (with the exception of the cocalero movement) political parties. For Carlos Crespo, for instance—a Bolivian sociologist and activist of the Water War—the struggle of the Coordinadora was "as foundational as Occupy Wall Street or the Indignados movement … in its strongly anti-party posture; that is, in its strong direct-democratic posture, and also in its strong antistate posture" (author interview, 13 March 2014). In the words of Oscar Olivera, in the 2000 to 2003 period, "there wasn't a political party that led the mobilizations … we tried to put in place a horizontal articulation, participative and assembly-based, free from any interference from party politics. … The orders didn't come from above, they came from below" (author interview, 11 March 2014). The

mobilizations and counterhegemonic articulations playing out in this conjuncture still had a bottom-up, autonomist character (Gutiérrez-Aguilar 2008). At the same time, however, Morales's MAS, which rose in parallel to the development of the indigenous-popular mobilization, gradually articulated with its counterhegemonic project and subsumed it under a statist and reformist program.

The passive revolutionary character of the MAS government therefore can be seen as a continuation of its earlier trajectory. Created as a political instrument of the cocalero movement, by 2002 the MAS had become the main opposition party. According to Raúl Prada—a member of the Constituent Assembly and a vice-minister in the first Morales government—the MAS drew its program and discourse from the street mobilizations, capitalizing politically on the construction of counterhegemony already articulating from below (author interview, 16 October 2014). For sympathetic commentators such as Errejón and Guijarro (2016), channeling popular–indigenous aspirations toward an electoral contest was a great merit of the MAS and was precisely what made its project hegemonic. Critics, however—like the Bolivian political theorist Luis Tapia—argue that the MAS was never a hegemonic actor in a Gramscian sense: "It was never at the forefront of the indigenous-popular articulation, but always behind it" (author interview, 13 October 2014). During the government of Carlos Mesa (2003–2005), the MAS maintained a moderate position aimed at winning over the middle-class vote and had an ambiguous posture toward social mobilizations (Webber 2011). This ambivalent relationship of the MAS with indigenous–popular mobilizations helps to explain the relative ease with which it later turned into a "disarticulator" of civil society (Tapia 2014; Andreucci 2018).

Implications for Environmental Politics

The Bolivian experience shows that there are at least two main features that characterize what we might call emancipatory populism. First, a key element of a genuinely transformative populism is the bottom-up character of the process of articulation. This is not an argument about the appropriate scale of counterhegemonic projects: Environmental movements adopt complex and often contradictory scalar politics (McCarthy 2005; Neumann 2009), whereby both their demands and strategies transcend clear dualisms of bottom-up versus top-down. Rather, I mean to suggest that, to be conducive of emancipatory change, the construction of a collective identity should emerge from a movement's own political practices and "organic intellectuals," instead of being imposed from outside (or "above") the movement itself. Indeed, discursive constructions of "the people" (and of a people–enemy opposition) from political actors in a position of power are often associated with authoritarian or postpolitical populism (Swyngedouw 2010; Hart 2013). Even when it might have a progressive character, as in the case of institutionalized left-populist parties, a construction of the people from outside the movement can hardly have a transformative character; it can, of course, constitute a progressive articulation of diverse social struggles and demands, but it remains, at best, within a reformist horizon (see, e.g., Kioupkiolis and Katsambekis 2018).

The second central element regards the political character of the social demands and ambitions that constitute a counterhegemonic project. In his early writings, Laclau (1977) defined populism as a way of articulating a certain class project within the non-class terrain of popular-democratic struggles. I have suggested that the political horizon of struggles in Bolivia during the 2000 to 2005 revolutionary epoch resulted precisely from articulating a subaltern class-based, material struggle with radically democratic ambitions. It was, in other words, an egalitarian-democratic horizon, whereby demands around the socialization of the country's natural resources such as water and gas in favor of traditionally marginalized and dispossessed groups were articulated with principles of direct democracy and autonomous self-government. This counterhegemonic project, enabled by the construction of an indigenous-popular collective identity, constituted a radical alternative to the dominant, neoliberal mode of resource governance (Perreault 2006) and overflowed into ambitions for fundamental rethinking of democracy, sovereignty, and citizenship.

What are the implications of such a conceptualization of populism for the politics of environmental governance? I noted earlier that, under capitalist arrangements, institutional frameworks for governing environments and resources are associated with the goal of stabilizing accumulation (Bridge and Perreault 2009). This article has argued that a theory

of emancipatory populism provides a productive way of analyzing how environmental movements might challenge and subvert such institutional configurations. This might help to identify the political and ideological strategies that facilitate the articulation of transformative, counterhegemonic ambitions out of otherwise disconnected socioenvironmental struggles. Such a perspective points to the necessity for socioenvironmental movements to engage in struggles for hegemony (Glassman 2013). Although this is not meant to diminish the importance of struggles taking place at a physical distance from the state—as is the case for many socioenvironmental struggles and instances of "resistance"—it also posits that state power and capitalist hegemony cannot be simply contained or countered "from the outside" (Poulantzas 2008, 370–72). Indeed, emancipatory change in the dominant political configuration and mode of governing socionatural relations implies a deep transformation of the capitalist state (Poulantzas 2008; Thomas 2013). Populism, as a vehicle for the articulation of such an emancipatory project, contributes to achieving this goal.

At the same time, I have also suggested that, when engaging on the "battleground" of the state, the threat of reformist degeneration is ever present (Poulantzas 2008, 376). Another important insight gained from the Bolivian case is that, despite the genuinely transformative character of a counterhegemonic project, short of a fully revolutionary outcome, it risks being reabsorbed within the dominant institutional order. This might result from a combination of various factors (and historically and place-specific circumstances), including the actuality or threat of a conservative reaction and the tendency for social mobilizations to be subsumed under reformist electoral programs.

Passive revolution need not be conservative in character (Modonesi 2013). Although the most radical demands are inevitably filtered out—typically, those that pose a threat to the economic or ideological base of the new power bloc's hegemony—others are selectively adopted and institutionalized, although in "domesticated" form, potentially giving the overall political dynamic a progressive character. Nevertheless, the most genuinely emancipatory possibilities are likely to be frustrated, whereas the oppositional ideological strategy might degenerate into mystification and the legitimation of authoritarianism.

Although the idea of passive revolution is not meant to inspire political fatalism (Thomas 2013), the risk of reformist and statist degeneration is always present. This, I would argue, is the aporetic character of counterhegemony: the fact that it is in a sense likely to "fail" in its most radical ambitions, but it is nonetheless necessary to challenge the regressive tendencies of capitalist domination, with its corollary of creative destruction, inequality, and socioenvironmental injustice. Understanding this contradictory dialectic could perhaps help us to "fail better"—to identify the conditions under which an egalitarian-democratic articulation might come into being, while being prepared to anticipate the pitfalls that it will inevitably encounter.

Acknowledgments

Thanks to those who participated in and supported my research in Bolivia and to Melissa García Lamarca, Irmak Ertör, the Special Issue Editor James McCarthy, and three anonymous reviewers for their comments on previous versions of this article. All mistakes remain my own.

Funding

Research for this article benefited from the People Programme (Marie Curie Actions) of the European Union's Seventh Framework Programme, under REA agreement No. 289374, "ENTITLE."

ORCID

Diego Andreucci http://orcid.org/0000-0002-5411-4578

Notes

1. This article is part of a broader research project, for which I carried out twelve months of fieldwork in Bolivia in 2013 and 2014, conducting eighty-one semistructured interviews with members of communities affected by mineral and hydrocarbon extraction, indigenous and environmentalist organizations, state and extractive industry representatives, and political analysts and commentators. Specifically, the arguments presented in this article are based on a subset of fourteen expert interviews with activists and intellectuals in Bolivian political, social, and environmental organizations.
2. I am aware that Gramscian scholars, including political ecologists, are strongly critical of some of Laclau's positions (e.g., Loftus 2014). Most notable, perhaps, they take issue with the discursive turn taken by Laclau since the 1980s in his interpretation of hegemony, whereby class (and the materiality of socionatural relations more generally) is displaced from the center analysis. Although I agree with this critique, following Hart (2013), I propose to read Laclau's more recent contributions in light of his earlier work, which develops a theory of populism within a materialist framework, explicitly framed as a way of articulating class politics with ideological strategy.
3. All translations from the Spanish are my own.

References

Andreucci, D. 2018. Populism, hegemony, and the politics of natural resource extraction in Evo Morales's Bolivia. *Antipode* 50:825–45.

———. 2017. Resources, regulation and the state: Struggles over gas extraction and passive revolution in Evo Morales's Bolivia. *Political Geography* 61:170–80. https://doi.org/10.1016/j.polgeo.2017.09.003.

Andreucci, D., and I. M. Radhuber. 2017. Limits to "counter-neoliberal" reform: Mining expansion and the marginalisation of post-extractivist forces in Evo Morales's Bolivia. *Geoforum* 84:280–91. https://doi.org/10.1016/j.geoforum.2015.09.002.

Bakker, K. 2015. Neoliberalization of nature. In *The Routledge handbook of political ecology*, ed. T. Perreault, G. Bridge, and J. McCarthy, 446–56. London and New York: Routledge.

Bridge, G., J. McCarthy, and T. Perreault. 2015. Editors' introduction. In *The Routledge handbook of political ecology*, ed. T. Perreault, G. Bridge, and J. McCarthy, 3–18. London and New York: Routledge.

Bridge, G., and T. Perreault. 2009. Environmental governance. In *A companion to environmental geography*, ed. N. Castree, D. Demeritt, D. Liverman, and B. Rhoads, 442–60. Malden, MA: Wiley.

Calvário, R., G. Velegrakis, and M. Kaika. 2017. The political ecology of austerity: An analysis of socio-environmental conflict under crisis in Greece. *Capitalism Nature Socialism* 28 (3):69–87. https://doi.org/10.1080/10455752.2016.1260147.

Canessa, A. 2006. *Todos Somos Indígenas*: Towards a new language of national political identity. *Bulletin of Latin American Research* 25 (2):241–63. https://doi.org/10.1111/j.0261-3050.2006.00162.x.

Collins, J. N. 2014. New left experiences in Bolivia and Ecuador and the challenge to theories of populism. *Journal of Latin American Studies* 46 (1):59–86. https://doi.org/10.1017/S0022216X13001569.

Ekers, M., A. Loftus, and G. Mann. 2009. Gramsci lives! Themed issue: Gramscian political ecologies. *Geoforum* 40:287–91. https://doi.org/10.1016/j.geoforum.2009.04.007.

Errejón, I., and J. Guijarro. 2016. Post-neoliberalism's difficult hegemonic consolidation: A comparative analysis of the Ecuadorean and Bolivian processes. *Latin American Perspectives* 43 (1):34–52. https://doi.org/10.1177/0094582X15579901.

Fraser, N. 2017. Against progressive neoliberalism, a new progressive populism. *Dissent Magazine*, January 28. Accessed October 10, 2018. https://www.dissentmagazine.org/online_articles/nancy-fraser-against-progressive-neoliberalism-progressive-populism.

García-Linera, A., M. Chávez-León, and P. Costas-Monje. 2010. *Sociología de los Movimientos Sociales en Bolivia: Estructuras de movilización, repertorios culturales y acción política*. [Sociology of social movements in Bolivia: Structures of mobilization, cultural repertoires, and political action]. 4th ed. La Paz, Bolivia: Plural Editores.

Glassman, J. 2013. Cracking hegemony: Gramsci and the dialectics of rebellion. In *Gramsci: Space, nature, politics*, ed. M. Ekers, G. Hart, S. Kipfer, and A. Loftus, 241–57. Malden, MA: Wiley-Blackwell.

Gutiérrez-Aguilar, R. 2008. *Los ritmos del Pachakuti: Movilización y levantamiento popular-indígena en Bolivia (2000–2005)* [Rhythms of the Pachakuti: Indigenous uprising and state power in Bolivia]. Buenos Aires, Argentina: Tinta Limón.

Hart, G. 2013. Gramsci, geography, and the languages of populism. In *Gramsci: Space, nature, politics*, ed. M. Ekers, G. Hart, S. Kipfer and A. Loftus, 301–20. Malden, MA: Wiley-Blackwell.

Hesketh, C., and A. D. Morton. 2014. Spaces of uneven development and class struggle in Bolivia: Transformation or *trasformismo? Antipode* 46 (1): 149–69.

Himley, M. 2013. Regularizing extraction in Andean Peru: Mining and social mobilization in an age of corporate social responsibility. *Antipode* 45 (2):394–416.

Jessop, B. 1990. *State theory: Putting the capitalist state in its place*. Cambridge, UK: John Wiley & Sons.

Karriem, A. 2009. The rise and transformation of the Brazilian landless movement into a counter-hegemonic political actor: A Gramscian analysis. *Geoforum, Themed Issue: Gramscian Political Ecologies* 40 (3):316–25. https://doi.org/10.1016/j.geoforum.2008.10.005.

Kaup, B. Z. 2013. *Market justice: Political economic struggle in Bolivia*. Cambridge, UK: Cambridge University Press.

———. 2014. Divergent paths of counter-neoliberalization: Materiality and the labor process in Bolivia's natural resource sectors. *Environment and Planning A* 46 (8):1836–51.

Kioupkiolis, A., and G. Katsambekis. 2018. Radical left populism from the margins to the mainstream: A comparison of Syriza and Podemos, In *Podemos and the new political cycle left-wing populism and anti-establishment politics*, ed. G. A. Óscar and M. Briziarelli, 201–26. London: Palgrave Macmillan.

Laclau, E. 1977. *Politics and ideology in Marxist theory: Capitalism, fascism, populism*. London: New Left Books.

———. 2005a. *On populist reason*. London: Verso.

———. 2005b. Populism: What's in a name? In *Populism and the mirror of democracy*, ed. F. Panizza, 32–49. London: Verso.

Loftus, A. 2014. Against a speculative leftism. In *The post-political and its discontents: Spaces of depoliticization, spectres of radical politics*, ed J. Wilson and E. Swyngedouw, 229–43. Edinburgh, UK: Edinburgh University Press.

Marston, A., and T. Perreault. 2017. Consent, coercion and cooperativismo: Mining cooperatives and resource regimes in Bolivia. *Environment and Planning A: Economy and Space* 49 (2):252–72. https://doi.org/10.1177/0308518X16674008.

McCarthy, J. 2005. Scale, sovereignty, and strategy in environmental governance. *Antipode* 37 (4):731–53. https://doi.org/10.1111/j.0066-4812.2005.00523.x.

McNeish, J.-A. 2013. Extraction, protest and indigeneity in Bolivia: The TIPNIS effect. *Latin American and Caribbean Ethnic Studies* 8 (2):221–42. https://doi.org/10.1080/17442222.2013.808495.

Modonesi, M. 2013. Revoluciones pasivas en América Latina: Una aproximación gramsciana a la caracterización de los gobiernos progresistas de inicio de siglo [Passive revolutions in Latin America: A Gramscian perspective on the progressive governments of the beginning of the century]. In *Horizontes Gramscianos: Estudios en torno al pensamiento de Antonio Gramsci*, ed. M. Modonesi, 209–36. Mexico City: Facultad de Ciencias Políticas y Sociales, UNAM.

Mouffe, C. 2005. The "end of politics" and the challenge of right-wing populism. In *Populism and the mirror of democracy*, ed. F. Panizza, 50–71. London: Verso.

Neumann, R. P. 2009. Political ecology: Theorizing scale. *Progress in Human Geography* 33 (3):398–406. https://doi.org/10.1177/0309132508096353.

Perreault, T. 2006. From the Guerra del Agua to the Guerra del Gas: Resource governance, neoliberalism and popular protest in Bolivia. *Antipode* 38 (1):150–72. https://doi.org/10.1111/j.0066-4812.2006.00569.x.

———. 2008. Custom and contradiction: Rural water governance and the politics of *usos y costumbres* in Bolivia's irrigators' movement. *Annals of the Association of American Geographers* 98 (4):834–54. https://doi.org/10.1080/00045600802013502.

Postero, N. 2017. *The indigenous state: Race, politics, and performance in plurinational Bolivia*. Oakland: University of California Press.

Poulantzas, N. 2008. Towards a democratic socialism. In *The Poulantzas reader: Marxism, law and the state*, ed. J. Martin, 361–76. London: Verso.

Schavelzon, S. 2012. *El nacimiento del Estado Plurinacional de Bolivia: Etnografía de una Asamblea Constituyente* [The birth of the plurinational state of Bolivia: Ethnography of a constituent assembly]. La Paz, Bolivia: Plural Editores.

———. 2014. Mutaciones de la identificación indígena durante el debate del censo 2012 en Bolivia: Mestizaje abandonado, indigeneidad estatal y proliferación minoritaria [Mutations of indigenous identification during the debate over the 2012 census in Bolivia: Abandoned miscegenation, state indigeneity and minority proliferation]. *Journal of Iberian and Latin American Research* 20 (3):328–54. https://doi.org/10.1080/13260219.2014.995872.

Svampa, M. 2015. América Latina: De nuevas izquierdas a populismos de alta intensidad [Latin America: From new Left governments to high-intensity populisms]. *Contrapunto* 7:83–93.

Swyngedouw, E. 2010. Apocalypse forever? Post-political populism and the spectre of climate change. *Theory Culture Society* 27 (2–3):213–32. https://doi.org/10.1177/0263276409358728.

———. 2014. Insurgent architects, radical cities, and the promise of the political. In *The post-political and its discontents: Spaces of depoliticization, spectres of radical politics*, ed. J. Wilson and E. Swyngedouw, 169–88. Edinburgh, UK: Edinburgh University Press.

Tapia, L. 2011. *El estado de derecho como tiranía* [The rule of law as tyranny]. La Paz, Bolivia: Autodeterminación.

———. 2014. *La sustitución del pueblo* [The substitution of the people]. La Paz, Bolivia: Autodeterminación.

Thomas, P. D. 2013. Hegemony, passive revolution and the modern prince. *Thesis Eleven* 117 (1):20–39. https://doi.org/10.1177/0725513613493991.

Valdivia, G. 2010. Agrarian capitalism and struggles over hegemony in the Bolivian lowlands. *Latin American Perspectives* 37 (4):67–87. https://doi.org/10.1177/0094582X10373354.

Wainwright, J., and G. Mann. 2015. Climate change and the adaptation of the political. *Annals of the Association of American Geographers* 105 (2):313–21. https://doi.org/10.1080/00045608.2014.973807.

Webber, J. R. 2011. *From rebellion to reform in Bolivia: Class struggle, indigenous liberation, and the politics of Evo Morales*. Chicago: Haymarket.

———. 2016. Evo Morales and the political economy of passive revolution in Bolivia, 2006–15. *Third World Quarterly* 37 (10):1855–76. https://doi.org/10.1080/01436597.2016.1175296.

Žižek, S. 2006. Against the populist temptation. *Critical Inquiry* 32 (3):551–74. https://doi.org/10.1086/505378.

DIEGO ANDREUCCI is a Juan de la Cierva Postdoctoral Researcher in the Department of Political and Social Science at Pompeu Fabra University, Barcelona, Catalunya, Spain. E-mail: diego.andreucci@gmail.com. His research interests include political ecologies of development and of natural resource governance in Latin America.

Whatever Happened to Green Collar Jobs? Populism and Clean Energy Transition

Sarah Knuth

In today's populist moment, climate change response has become anything but "postpolitical." The project to decarbonize energy supplies is generating ongoing political clashes today, including between competing forms of capital/ism. In the United States, rising renewable energy industries in places like California contend with fossil fuel blocs and their regional bases. Such confrontations are sparking populist organizing on the right and left. I argue that critical geography must further consider left populist movements' role in these politics of clean energy transition, grievance, and reparation and openings for collectively advancing more liberatory futures. I survey a wave of coalition-building that has evolved in the United States since the beginnings of the New Economy, allying U.S. environmentalists, organized labor, and, more recently, racial and community justice organizers. This movement became most visible as it built networks around calls for national "green collar" job creation during the late 2000s financial crisis and 2008 presidential campaign. Its organizing shaped noteworthy, if ultimately limited Obama administration programs and continues to influence clean energy rollout in regions such as California, particularly campaigns for job quality and racial diversity in green construction. I consider here both these successes and their limits in a turbulent clean-tech sector: the need for farther reaching transformations in energy–industrial policy and democratic participation in shaping them.

在当今的民粹时刻，气候变迁回应绝非已成为"后政治"。去碳化的能源供给计画，在今日持续产生政治冲突，包括相互竞争的资本／主义形式之间的冲突。在美国，加州等地成长中的可再生能源产业，与石化集团及其区域基础相互竞争。此般冲突正激起左翼与右翼的民粹组织。我主张，批判地理学必须进一步考量左翼民粹运动在这些乾淨能源变迁、不满和修復政治中的角色，以及为集体促进更具解放性的未来起头。我调查美国自新经济开始以降演化中的联盟建构浪潮，其中美国环保主义者、有组织的劳工，以及更为晚近的种族和社区正义组织者相互结盟。此一运动打造2000年代晚期金融风暴和2008年总统选举竞选时号召全国"绿领"工作创造的网络，进而成为最引人注目的运动。其组织形塑了显着的但最终受限的欧巴马政府之计画，并持续影响诸如加州等区域乾淨能源的推出，特别是对绿色工程中的工作平等与种族多样化之倡议。我于此同时考量其在动盪的清洁技术部门中的成功与限制：能源产业政策必须要有更为广泛的变迁，以及形塑这些变迁的民主参与。 关键词: 乾淨能源转换，气候变迁，绿领工作，绿色经济，民粹主义。

En el momento populista que nos acompaña, la respuesta al cambio climático se ha convertido en algo menos que "pospolítico". El proyecto de descarbonificar los abastos energéticos está generando confrontaciones políticas continuas, incluyendo las de las formas competitivas del capital/ismo. En los Estados Unidos, emergentes industrias de energía renovable en lugares como California se enfrentan a bloques de combustibles fósiles y a sus bases regionales. Tales confrontaciones están desencadenando la organización populista a derecha e izquierda. Yo sostengo que la geografía crítica debe dar mayor consideración al papel de los movimientos populistas de izquierda en estas políticas de transición a la energía limpia, quejas y reparación; y a las aperturas para promover colectivamente futuros más libertarios. Examino una ola de construcción de coaliciones que ha evolucionado en los Estados Unidos desde los comienzos de la Nueva Economía, que alía los ambientalistas americanos, el trabajo organizado y más recientemente los organizadores de la justicia racial y comunitaria. Este movimiento adquirió mayor visibilidad a medida que construyó redes alrededor de llamados por la creación nacional del tipo de empleo de "cuello verde" durante la crisis financiera de finales del 2000 y la campaña presidencial del 2008. Su organización configuró los notables, aunque a la postre limitados, programas de la administración Obama, y siguen influyendo la promoción de la energía limpia en regiones como California en particular, en campañas por calidad del empleo y diversidad racial en la construcción verde. Aquí considero estos éxitos y sus limitaciones en un turbulento sector de tecnolimpieza: la necesidad de transformaciones de mayor alcance en la política industrial energética y participación democrática en su conformación. *Palabras clave: cambio climático, economía verde, empleos de cuello verde, populismo, transición a la energía limpia.*

In today's moment of surging populisms, climate response has become anything but "postpolitical"—if, indeed, it ever was so (Swyngedouw 2010; McCarthy 2013). Behind Swyngedouw's "socio-chemical" enemy lie livelihoods, accumulation regimes, and entrenched power relations, many constituencies for whom climate action appears more dangerous and damaging than inaction. The project to decarbonize energy supplies shapes many forms of this in/action, and critical energy geography is now illuminating many such struggles. Beyond the field's ongoing examination of extractive economies and petropolitics (too extensive to review here), a wave of recent research is taking on global geographies of clean energy rollout and the green capitalist programs typically propelling it (see, e.g., Pasqualetti 2011; Bridge et al. 2013; Huber and McCarthy 2017; McEwan 2017). This work is revealing new forms and articulations of grievance and new claims for justice, remedy, and reparation—key lineaments in the production of political formations, including left- and right-wing populisms. Given the political ecological and agrarian political economic bent of much of this scholarship, it is not surprising that struggles over rural land and livelihoods feature prominently. Indeed, it would be startling to find otherwise: Solar and wind energy infrastructures are now being rapidly deployed in many contexts and producing major land transformations (joining parallel conversions for biofuels, notably in land grabs after the 2008 financial collapse; Baka 2013).

I argue that this research, although necessary to expand geography's frontiers of scholarship and praxis on energy transition, is not sufficient to confront current political movements, on (at least) two levels. Like political ecology in general, it better captures capitalism's imperialist moment in resource (neo-)peripheries than its dialectically entangled struggles "at home." Particularly, it insufficiently addresses capital's tendency to fracture into competing blocs, techno-industrial and regional, ones that wage bitter struggles for supremacy and ongoing accumulation. Such zero-sum confrontations and competitive devaluations are a classic concern of geographical political economy, as the field has tracked this creative destruction in the conjoined rise and fall of regional and urban economies (Harvey 1982; Storper and Walker 1989; Markusen et al. 1991). In the U.S. context that I discuss here, such clashes have helped spark new populist formations, as cross-class alliances and blocs assemble to boost regions or combat their decline—often through competition to capture the federal state apparatus, with its powers over development policy and the geographic redistribution of wealth (Fraser and Gerstle 1989). Crucially, the U.S. energy transition (and climate policy) is now being fought out in such sectional battles and prospective devaluations, increasingly openly (Knuth 2017).[1] Rising renewable energy industries in places like California now contend with entrenched and new fossil fuel production regions. In the long rightward shift in U.S. populism since (especially) the Reagan Revolution (Kazin 1998; McGirr 2002; Frank 2007), right-wing movement-builders have drawn on this sharpening geographical division multiple times—most recently, in the Tea Party's "drill, baby, drill" advocacy for domestic oil and gas (amid broader invocations to defend "fly-over country" and its hazily imagined true America from coastal elites) and Donald Trump's 2016 campaign promises to roll back Obama-era climate policy and "save" Appalachian coal.[2]

In deepening energy geography's analysis of these confrontations, scholars must expand political ecology's work on conservative populist movements (McCarthy 2002) and continue to bridge political ecology and economy on industrial questions (e.g., Huber 2017). More particular, I argue, energy geographers must further consider left populisms. We are still struggling to conceptualize their role, existing and potential, in political blocs for clean energy transition, particularly in urban and techno-industrial capitalist centers. Such ambiguities often bleed into our own critical praxis. Green economic development programs dreamed up in places like Silicon Valley present optimistic techno-futurist visions of new surplus. This turn away from Anthropocene millenarianism and neo-Malthusianism has been attractive to some political ecologists, notwithstanding its naivety, genuine or strategic, about the failings of green capitalism. (For example, see conflicting and shifting takes from Latour [2011, 2015]; Collard, Dempsey, and Sundberg [2015]; and Robbins and Moore [2015] on how to interact with the Breakthrough Institute, the particularly aggressive ecomodernist advocate based in the San Francisco Bay Area.) Today's radical cheapening and expansion of renewable energy presents undeniable opportunities for a climate movement long confounded by the difficulties of organizing around shared scarcity. An already dubious prospect for winning broad allegiance, in the U.S. context such a project has been further encumbered by cultural suspicions (from many directions) of liberal-elite "voluntary simplicity." With the prospect of affordable clean energy and new jobs producing it, transition supporters might more aggressively and successively contest the fossil fuel bloc—overcoming postpolitical conciliations like Obama-era "all of the above" energy policy.[3] This new hope conceals crucial uncertainties, though: How many clean energy jobs will there actually be, and for whom? Will they be good and stable ones? Will there be enough of them in the right places? If not, are there other options for making remedy or reparation to fossil fuel industry workers and regions? Such dilemmas are matters of both justice and political urgency for a movement that confronts substantial populist resistance. They suggest a further, crucial one: Who gets to participate in debating and deciding these questions?

In the remainder of this article, I briefly consider one U.S. attempt to answer the last question and, in the process, many of the others. I survey a wave of left populist coalition building that has evolved over the last twenty years.[4] Among its other outcomes, this organizing has helped shape major clean energy development programs in California and nationally under the Obama administration. As policy advocacy, it became most visible in calls for national "green collar" job creation during the late 2000s financial crisis and stimulus, prominently articulated by V. Jones (2008) and advanced during the 2008 presidential campaign by entities such as the Apollo Alliance, Green for All, the Center for American Progress, and the Center on Wisconsin Strategy (COWS). In these calls, blue–green alliances of labor unions and environmentalists joined with racial and community justice organizers in new ways. This movement has deeper roots, though. Its vision reflects the distinctive political tensions of places like California and the Pacific Northwest in the

1980s and 1990s. In the New Economy, these regions rose as "postindustrial" technological leaders and environmentalist hotbeds, even as their older resource extraction and manufacturing job bases declined—a conjoining that prompted both enduring grievance and new visions for remedy. In the 2010s, California regional advocates continue to push for good, diverse jobs in clean energy, as the state experiments with what kind of "clean-tech" innovator it might become and who might share in the planning and proceeds of such development. Since the late 2000s, many movement proposals have centered around building a clean energy economy in the United States, in a quite literal sense: They focus their visions for working-class clean energy jobs on weatherization and energy efficiency retrofitting, rooftop solar panel installation, and infrastructure development for utility-scale solar and wind power plants. All resemble traditional construction and building trades work. This sector presents a problematic U.S. political inheritance—notably, if not exclusively, in its racial politics—that populist alliances have recently sought to reform into self-consciously progressive race–class programs. The movement's strategy raises deeper questions, however. In a clean energy economy jostled by would-be tech visionaries and destabilizing global restructurings, must left populisms (among other prospective left political formations) demand farther reaching transformations in energy-industrial policy and the power to shape them?

The Movement for Green Collar Jobs: Building a/s Solution?

Understanding the roots of green collar jobs alliances requires looking also to right-wing populism. As the New Economy began to reshape Western resource economies and cultures, it sharpened existing antienvironmentalist resistance—for example, from rural right-wing movements like Wise Use (McCarthy 2002). New Economy in-migrants, including environmentalists and wealthy rural home-buyers-turned-preservationists, drew fresh ire (e.g., Walker 2003). Through such protests, the right constructed a persistent commonsense (after Gramsci) grievance of "jobs versus the environment." It now recurrently deploys this trope to flog environmentalists for supposed elitism and cluelessness about the costs of environmental regulation to workers, particularly in already embattled resource industries—echoing elements of political ecologists' own critiques of mainstream conservation, sans their nuance on white settler colonialisms. The Pacific Northwest became one important proving ground for such confrontations, in nationally prominent clashes between the timber industry, workers, and conservationists. The term *green collar jobs* was coined in this regional political moment, in a book by Durning (1999). In it, Durning attempted to grapple with the Pacific Northwest's declining rural jobs, amid its simultaneous transformation into a hub of New Economic technology growth. The volume argued that newly emerging working-class jobs in fields such as environmental restoration had the potential to offset resource industry jobs lost. It offered an alternative to both the timber industry's zero-sum framing and the New Economy's bifurcated job structure: high-wage tech employment for the highly educated, particularly white men, combined with low-wage, insecure service work (or no work) for most others.

Notions of green collar employment reflected the period's broader ecological modernization proposals: nascent green capitalist visions of economic decoupling (growth detached from its energy and resource metabolisms) and a rejuvenated postindustrial economy. The idea's subsequent travels, however, were facilitated by other new imaginaries of the era, ones that required more deliberate political work to construct. Through the 1990s, progressive and left populist organizers pioneered the notion of blue–green alliances, a project that sought to unite disparate strands in the embattled U.S. left. This alliance-building brought together progressive wings of established institutions: the Democratic Party's mid-twentieth-century base of industrial labor unions and the environmental advocacy interests that rose in the party from the 1970s onward. At the grassroots level, environmental activists and workers developed these blue–green politics through shared resistance to neoliberal free trade rollout, with the 1999 anti–World Trade Organization (WTO) protests in Seattle a crystallizing moment (Cockburn and Clair 2000).

Through the 2000s, this movement-building gained national prominence and saw mainstream Democratic Party uptake, in mobilizations around successive Democratic defeats in presidential elections and George W. Bush–era energy and environmental policies. Newly formed blue–green advocates like the Apollo Alliance embraced novel kinds of

ecomodernism and technological futurism. Increasingly, these efforts opened up potential collaborations with Silicon Valley venture capitalists and tech interests. From the tech side, this green turn proved attractive to a host of entrepreneurs and investors, then casting about for new opportunities after the collapse of the first New Economy boom in 2001 (Caprotti 2017; Knuth 2017). Collectively, these efforts helped kick off a wave of U.S. clean-tech investment centered in Silicon Valley and the Bay Area (although see Goldstein 2018). As this nascent boom gathered steam from the mid-2000s, would-be tech visionaries imagined new fundamental breakthroughs in clean energy technologies and business models—a fresh source of regional superprofits and surplus from a national and global economy (Knuth 2018a).

From the mid-2000s, urban organizers expressed growing concerns about the exclusions and costs of this new green growth. Racial and community justice institutions in places like West Oakland warned of its power to exacerbate an already highly uneven New Economy and its problems—to once again pass impoverished communities by in terms of stable employment, atop existing harms from institutionalized racism and the carceral state (Gilmore 2007), environmental injustice, deindustrialization, and long-term urban disinvestment. In 2005, V. Jones questioned, "Will the green wave lift all boats?" He was skeptical: "Right now we have eco-apartheid. Look at Marin; they've got solar this, and bio this, and organic the other, and fifteen minutes away by car, you're in Oakland with cancer clusters, asthma, and pollution" (V. Jones, as cited in Strickland 2005). These criticisms echoed similar calls for "just sustainabilities" (Agyeman and Evans 2003), which as the green wave spread—increasingly as a project for urban (re)development and branding as much as clean energy transition—have expanded into protests against new "green" gentrification (e.g., Checker 2011; Knuth 2016, 2018b).

In the late 2000s, however, a raft of organizers in Oakland and elsewhere simultaneously questioned whether the green economy opened up opportunities to do technologically advanced economic development differently. Pinderhughes helped translate Durning's green collar jobs concept to the urban Bay Area context around 2004 (e.g., Pinderhughes 2006). By 2005, V. Jones, who became green collar jobs' most visible U.S. champion, was working to disseminate the term and its organizing vision regionally and jumping scale (Smith 1992) to reframe it as a national economic strategy. A raft of efforts in the late 2000s fleshed out this call (e.g., Gordon et al. 2008; V. Jones 2008). In a 2009 interview in *Antipode* (Mirpuri, Feldman, and Roberts 2009), V. Jones articulated this strategy using similar Gramscian framing to that informing this article as one for galvanizing the U.S. left while building better energy and climate policy: "There is a struggle going on within the upper echelons of US capital … between the military-petroleum complex that's still the dominant bloc of capital, and greener less polluting forms of capital … we'd probably call it eco-apartheid, because it would be, left to its own devices, just as unjust and just as exploitative as gray capitalism." In the political moment of the late 2000s, he argued that left organizers should nonetheless ally strategically with the green bloc: "I don't believe everything works out better when it all falls apart, and certainly not from the position of relative organizational ideological weakness that the Left is in right now" (V. Jones, as cited in Mirpuri, Feldman, and Roberts 2009, 405–6). These kinds of efforts from V. Jones and others brought new prominence for community and racial justice organizers within the blue–green coalition—including a seat for V. Jones on the board of the Apollo Alliance and, following Obama's 2008 election, an administration position as green jobs advisor.

The 2008 presidential election, amid the popular anger and openings of the financial crisis, gave new force to the green collar jobs movement. Its organizing was fueled by a broader wave of left populism that included calls for punishing bankers and condemnations of financial-sector parasitism on the "real economy," on a spectrum between radical protest and (ultimately dominant) technocratic quasifixes. Green collar jobs discourses articulated commonalities between U.S. interests, sectors, places, and populations dispossessed and damaged in the crisis, including black and minority communities particularly targeted for exploitative subprime loans (and see Coates 2014). Moreover, they spoke to the longer term abandonments, ravages, and political undermining of decades of neoliberal political rollout: of polluted and disinvested urban neighborhoods like West Oakland but also labor unions, disinvested industrial regions like the Rust Belt, and frustrated climate activists. This organizing aided in the consolidation of a significant constituency in Obama's successful 2008 presidential campaign (one whose

commonalities and political potential remain apparent if embattled in the 2010s, as Bernie Sanders's 2016 primary run and subsequent democratic socialist organizing have sounded many similar notes).

As Obama's election gave the green collar jobs call a level of national policy support in the late 2000s, it encountered new pressures and dilemmas. In the United States, its organizing vision was increasingly translated into programs for a new, "green" New Deal and revived Keynesianism (see Tienhaara [2014] on similar visions abroad)—with considerable ambiguities in how the United States' nascent green economy would support such broadbased employment. Manufacturing experienced a moment of popular cultural enthusiasm—a (temporary) turn from accustomed lionization of postindustrialism since the New Economy. None of a stillyoung crop of clean-tech startups, however, provided a clear path to breaking through to a new U.S. Golden Age: of global green manufacturing leadership and rents but with national surplus to be this time shared and redistributed beyond the white working class. Certainly the stimulus and subsequent Obama administration programs contained stabs at resurrecting U.S. industrial policy. (This proved an uphill battle for a cascade of tech-sector, domestic political, and global economic reasons too extensive to treat here, but see Block and Keller [2011]; Caprotti [2015, 2017]; Mazzucato [2015]; Mulvaney [2016]; Knuth [2017, 2018a]; and Goldstein [2018] for various elements.) Quickly, however, it became clear that the construction sector was to provide the bulk of immediate green collar jobs in programs such as the crisis-era stimulus package. This was so for reasons besides the uncertain future of U.S. manufacturing. The collapse of the U.S. housing bubble fueled a keen construction lobby and clear political sell: "[putting] construction workers back to work with good jobs that can't be outsourced" (White House 2013). Moreover, already existing technologies for rooftop solar, building energy retrofits, and utility-scale renewable energy needed no blue-sky clean-tech breakthroughs to deploy. Some funding candidates like the federal Weatherization Assistance Program (WAP) already possessed institutional infrastructure and seemed to need only an influx of money to become eminently "shovel-ready" (WAP received $5 billion in the stimulus, then nearly equal to its cumulative funding over its thirty-plus-year history; U.S. Department of Energy 2009; Tonn et al. 2011).

Construction jobs represented a significant opportunity for additional reasons. One notable green collar jobs call during the 2008 campaign came from the Emerald Cities Partnership, a collaboration of entities including COWS and the Service Employees International Union (Grabelsky and Thompson 2010). Proponents sold green construction as an opportunity to effect multiple levels of redress and reparation simultaneously: to a working class hurting from the financial crisis and particularly to black and minority workers. U.S. building trades unions were historically notorious for their racial exclusions (Kazin 1988; Sugrue 2004). Now, minority workers were to be ushered in with union leaders' eager welcome and open acknowledgment of past wrongs (see, e.g., Ayers in Grabelsky 2010). Of course, there was political calculation here but one with broad potential benefit. The sector promised substantial job numbers if successful—although a host of economic modelers debated precisely how many and how in any case to define a green economy and labor within it. This prospective employment did not benefit only cities and workers of color. For decades, programs like WAP have worked equally to ameliorate white rural poverty (Harrison and Popke 2011). To try to safeguard the quality of jobs created, administration programs used both union tools like Project Labor Agreements (PLAs; which the George W. Bush administration had banned on federally funded projects) and other federal instruments like prevailing wage rates, via the Davis–Bacon Act. When the latter was applied to WAP funding, some unions speculated that it might help reverse the U.S. residential construction sector's long decline in union density, pay, and job quality (Fine 2011; Osterman and Chimienti 2012).

Reflecting on these early Obama administration years, it is easy to find classic problems of populism. Supporters drawn to Obama's personal charisma and galvanizing but empty signifiers like shared "hope" were disappointed for multiple reasons. Genuine missed opportunities for transformative politics[5] combined with mundane institutional roadblocks—for example, WAP expansion was notoriously slow and proved limited in its ability to reform housing construction. Moreover, decades of neoliberal political restructuring, ideological undermining, and nonplanning presented a formidable challenge. Green collar organizing work from the late 2000s continues to bear fruit and evolve as a political strategy,

however. For example, as California rolls out ambitious clean energy and climate programs, labor and the building trades are forging a noteworthy advocacy and policy-shaping role. The prospect of green collar jobs continues to bolster policies like the state's renewable portfolio standard (RPS), expanded in 2015 to require state utilities to obtain 50 percent of their power from eligible renewable resources by 2030. Moreover, it has helped defend California's climate policies against fossil fuel industry attack for example, the fossil fuel industry–funded Proposition 23 in 2010.[6] Much of the state's new clean power comes from utility-scale solar deployment in rural areas. Because most of this infrastructure has been constructed with unionized building trades labor, job quality is protected by PLAs (which mandate union wages, benefits, and employer training support) and by state-certified union apprenticeship programs (B. Jones, Philips, and Zabin 2016). This labor infrastructure has had notable success in promoting a racially diverse workforce (Luke et al. 2017).[7] Meanwhile, labor concerns have influenced debates such as the priority of California state support for different fractions of its clean energy industry, as high-tech players and policymakers back competing visions for the state's clean energy economy (and see Caprotti 2015; Knuth 2018a). In 2015, building trades unions helped defeat the rooftop solar industry's lobbying for support in California's expanded RPS (Roth 2015).[8] As labor researchers argued, jobs in rooftop solar installation are generally nonunion, less well-paid, and lower quality, as with the residential construction work that they resemble in other ways (B. Jones and Zabin 2015). Such developments are suggestive rather than conclusive but present a compelling case for ongoing analytical and political attention.

Future Directions: From Green Populism to Green Industrial Policy?

Moving forward, the political experience discussed here suggests multiple insights for energy–environmental scholarship and praxis. First, it argues that as scholars, we need to think more carefully about how we theorize and examine notions like postpolitical populism, particularly as regards the role of left populisms in more overtly liberal-technocratic formations (as indeed Obama-era programs could appear and be). In part, this is a call for more empirical examination. In many ways, the concept of the post-political poorly reflects the material and political messiness and contestations of climate and energy politics on the ground. Nor does it capture the shifting and contingent alliances involved in dismantling a still-dominant fraction and form of capital/ism. As charges of left-elite condescension fly today, including against academics, we must take more care to engage what left populist strategy looks like in its own varying calculations.

Second, this discussion suggests a window into distinct forms of grievance and reparation arising in the experience of clean energy transition, painful as it will be on its losing end. That such an experience should stir populist anger should surprise no one. At the same time, such justified grievance is one among many forms competing for remedy and reparation today, dense indeed in the U.S. contemporary context. Critical energy scholars must cast an eye toward the latter side of this equation as well as the former. What forms of reparation are available in the difficult contexts we examine, and can we help imagine political strategies that more adequately conjoin pragmatism and restorative justice in their application? The 2016 presidential elections saw various alternatives mooted, which from the left included substantial reparations payments to U.S. coal regions. In tackling this thorny problem, the option of green collar jobs creation remains a significant one for conceptual and political attention.

Finally, this conclusion suggests a key task for geographical scholarship and the distinctive toolkit that political economic/ecological analysis brings to bear on energy transition as a problem, in and beyond the United States. Whatever the strategic achievements and failings of the left populist organization surveyed in this discussion, one element that it chronically lacked in the 2000s was a clear strategy for more comprehensive green economic development beyond strategic sectors. Left populist alliances might have successfully gotten a foot in the door in certain political forums, but sectors like construction remain troublingly dependent on higher level industrial strategies and choices for their long-term prospects. How can we translate a role such as now being experimented with in California into a more powerful say in not just who builds clean energy infrastructure but what that clean energy infrastructure consists of and means? Resistance from an intransigent fossil fuel bloc and its conservative

populist base at home or China's undeniable successes as a green industrial powerhouse abroad can only partially account for existing limitations here. In addition, we must look critically to U.S. industrial policy—or, rather, the country's long lack of an open one (Block and Keller 2011; Mazzucato 2015; Knuth 2018a). If any dimension of U.S. climate politics remains genuinely and stubbornly postpolitical, it is a techno-industrial culture that persistently subordinates both left populists and policy technocrats themselves to familiar New Economic fallacies: the need for entrepreneurial and venture capitalist "genius" and "breakthroughs" to solve problems like energy transition, the assumed inadequacies of government economic development planning, and the political irrelevance of most people affected in this decision making. All of these propositions present genuine opportunities for geographers' intervention: critical, practical, and imaginative.

Acknowledgments

Thanks very much to two anonymous reviewers for their constructive comments on a draft version of this article, as well as to James McCarthy for his editorial support and guidance. For thoughtful feedback on earlier versions of this argument, my deep thanks also to Peter Wissoker, John Stehlin, Noah Quastel, and participants in the "Biopolitics and Environmental Justice" workshop organized by Alida Cantor and Catherine Jampel at the 2016 Dimensions of Political Ecology Conference.

Notes

1. And in similar energy-producing contexts—internal conflicts different from the populist possibilities of an external enemy.
2. Ironically, threatened most directly by that same boom in unconventional oil and gas. These explanations remain important, alongside factors like cultural grievance and white racist revanchism.
3. Or to entice fossil fuel industry executives all too likely to abandon existing workers and regions once sufficiently attractive alternatives and exit strategies present themselves—more straightforward here than in places with more complex extractive industry politics (e.g., Andreucci 2018).
4. This discussion of populism and/as strategic alliance building draws primarily on Laclau's (1977, 2005) Gramsci-influenced theorization (particularly, following Hart [2012] and Andreucci [2018], Laclau's earlier, more political economic conceptualization)—

although with several returns to Swyngedouw's (2010) notion of the postpolitical. It builds on field-work conducted in the Bay Area between 2008 and 2013, as regional green collar jobs calls consolidated and were incorporated into U.S. federal policy. Besides participant observation and policy analysis in relevant forums (both for green collar jobs organizing and the clean-tech industry), this investigation and follow-ups have involved extensive engagement with contemporary archives—for example, tech industry news and blogs, think tank publications, government gray literature, and labor research published in policy and academic forums (including by geographers; e.g., Luke et al. 2017).

5. And other concessions, as when Jones was let go from his advisory role after attacks from Congressional Republicans.

6. Notably, Koch Industries. Made via the (unsuccessful in this case) use of the ballot initiative, California's quintessentially populist direct democracy instrument.

7. Less so in increasing gender diversity, an ongoing problem in construction employment.

8. Although rooftop solar is not included within generation sources eligible for California's RPS, it has been supported by other state policies.

References

Agyeman, J., and T. Evans. 2003. Toward just sustainability in urban communities: Building equity rights with sustainable solutions. *The Annals of the American Academy of Political and Social Science* 590 (1):35–53. doi:10.1177/0002716203256565.

Andreucci, D. 2018. Populism, hegemony, and the politics of natural resource extraction in Evo Morales's Bolivia. *Antipode* 50 (4):825–45. doi:10.1111/anti.12373.

Baka, J. 2013. The political construction of wasteland: Governmentality, land acquisition and social inequality in South India. *Development and Change* 44 (2):409–28. doi:10.1111/dech.12018.

Block, F., and M. R. Keller, eds. 2011. *State of innovation.* Boulder, CO: Paradigm.

Bridge, G., S. Bouzarovski, M. Bradshaw, and N. Eyre. 2013. Geographies of energy transition: Space, place and the low-carbon economy. *Energy Policy* 53:331–40. doi:10.1016/j.enpol.2012.10.066.

Caprotti, F. 2015. Golden sun, green economy: Market security and the U.S./EU–China "solar trade war." *Asian Geographer* 32 (2):99–115. doi:10.1080/10225706.2015.1057191.

———. 2017. Protecting innovative niches in the green economy: Investigating the rise and fall of Solyndra, 2005–2011. *GeoJournal* 82 (5):937–55. doi:10.1007/s10708-016-9722-2.

Checker, M. 2011. Wiped out by the "greenwave": Environmental gentrification and the paradoxical politics of urban sustainability. *City & Society* 23 (2):210–29. doi:10.1111/j.1548-744X.2011.01063.x.

Coates, T. N. 2014. The case for reparations. *The Atlantic* 313 (5):54–71.

Cockburn, A., and J. S. Clair. 2000. *Five days that shook the world: Seattle and beyond.* New York: Verso.

Collard, R. C., J. Dempsey, and J. Sundberg. 2015. A manifesto for abundant futures. *Annals of the Association of American Geographers* 105 (2):322–30. doi:10.1080/00045608.2014.973007.

Durning, A. T. 1999. *Green-collar jobs: Working in the new Northwest.* Seattle, WA: Sightline Institute.

Fine, J. 2011. When the rubber hits the high road: Labor and community complexities in the greening of the Garden State. *Labor Studies Journal* 36 (1):122–61. doi:10.1177/0160449X10397647.

Frank, T. 2007. *What's the matter with Kansas? How conservatives won the heart of America.* New York: Metropolitan Books.

Fraser, S., and G. Gerstle, ed. 1989. *The rise and fall of the New Deal order, 1930–1980.* Princeton, NJ: Princeton University Press.

Gilmore, R. W. 2007. *Golden gulag: Prisons, surplus, crisis, and opposition in globalizing California.* Berkeley: University of California Press.

Goldstein, J. 2018. *Planetary improvement: Cleantech entrepreneurship and the contradictions of green capitalism.* Cambridge, MA: MIT Press.

Gordon, K., J. Hays, J. Walsh, B. Hendricks, and S. White. 2008. *Green-collar jobs in America's cities: Building pathways out of poverty and careers in the clean energy economy.* San Francisco, CA, Oakland, CA, Washington, DC, and Madison, WI: Apollo Alliance and Green for All with the Center for American Progress and the Center on Wisconsin Strategy.

Grabelsky, J. 2010. "We're getting our country back": Reflections on politics, race, labor and community in the age of Obama. *WorkingUSA* 13 (4):545–59. doi:10.1111/j.1743-4580.2010.00311.x.

Grabelsky, J., and P. Thompson. 2010. Emerald cities in the age of Obama: A new social compact between labor and community. *Perspectives on Work* 13 (2):15–18.

Harrison, C., and J. Popke. 2011. "Because you got to have heat": The networked assemblage of energy poverty in eastern North Carolina. *Annals of the Association of American Geographers* 101 (4):949–61. doi:10.1080/00045608.2011.569659.

Hart, G. 2012. Gramsci, geography, and the languages of populism. In *Gramsci: Space, nature, politics,* ed. M. Ekers, G. Hart, S. Kipfer, and A. Loftus, 301–20. Hoboken, NJ: Wiley-Blackwell.

Harvey, D. 1982. *The limits to capital.* Oxford, UK: Blackwell.

Huber, M. T. 2017. Hidden abodes: Industrializing political ecology. *Annals of the American Association of Geographers* 107 (1):151–66. doi:10.1080/24694452.2016.1219249.

Huber, M. T., and J. McCarthy. 2017. Beyond the subterranean energy regime? Fuel, land use and the production of space. *Transactions of the Institute of British Geographers* 42 (4):655–68. doi:10.1111/tran.12182.

Jones, B., P. Philips, and C. Zabin. 2016. *The link between good jobs and a low carbon future: Evidence from California's renewables portfolio standard, 2002–2015.*

Berkeley: University of California Berkeley Labor Center.

Jones, B., and C. Zabin. 2015. UC Berkeley Labor Center Blog post: Are solar jobs good jobs? Accessed July 21, 2018. http://laborcenter.berkeley.edu/are-solar-energy-jobs-good-jobs/.

Jones, V. 2008. *The green collar economy.* New York: HarperCollins.

Kazin, M. 1988. *Barons of labor: The San Francisco building trades and union power in the progressive era.* Champaign: University of Illinois Press.

———. 1998. *The populist persuasion: An American history.* Ithaca, NY: Cornell University Press.

Knuth, S. 2016. Seeing green in San Francisco: City as resource frontier. *Antipode* 48 (3):626–44. doi:10.1111/anti.12205.

———. 2017. Green devaluation: Disruption, divestment, and decommodification for a green economy. *Capitalism Nature Socialism* 28 (1):98–117. doi:10.1080/10455752.2016.1266001.

———. 2018a. Breakthroughs for a green economy? Financialization and clean energy transition. *Energy Research & Social Science* 41:220–29. doi:10.1016/j.erss.2018.04.024.

———. 2018b. Cities and planetary repair: The problem with climate retrofitting. *Environment and Planning A: Economy and Space.* OnlineFirst. doi:10.1177/0308518X18793973.

Laclau, E. 1977. *Politics and ideology in Marxist theory.* New York: Verso.

———. 2005. *On populist reason.* New York: Verso.

Latour, B. 2011. Love your monsters. *Breakthrough Journal* 2 (11):21–28.

———. 2015. Fifty shades of green. *Environmental Humanities* 7 (1):219–25.

Luke, N., C. Zabin, D. Velasco, and R. Collier. 2017. *Diversity in California's clean energy workforce: Access to jobs for disadvantaged workers in renewable energy construction.* Berkeley: University of California Berkeley Labor Center.

Markusen, A. R., S. Campbell, S. Deitrick, and P. Hall. 1991. *The rise of the gunbelt: The military remapping of industrial America.* Oxford, UK: Oxford University Press.

Mazzucato, M. 2015. *The entrepreneurial state: Debunking public vs. private sector myths.* New York: Anthem.

McCarthy, J. 2002. First world political ecology: Lessons from the Wise Use Movement. *Environment and Planning A* 34 (7):1281–1302. doi:10.1068/a3526.

———. 2013. We have never been post-political. *Capitalism Nature Socialism* 24 (1):19–25. doi:10.1080/10455752.2012.759251.

McEwan, C. 2017. Spatial processes and politics of renewable energy transition: Land, zones and frictions in South Africa. *Political Geography* 56:1–12. doi:10.1016/j.polgeo.2016.10.001.

McGirr, L. 2002. *Suburban warriors: The origins of the new American right.* Princeton, NJ: Princeton University Press.

Mirpuri, A., K. P. Feldman, and G. M. Roberts. 2009. Antiracism and environmental justice in an age of neoliberalism: An interview with Van Jones. *Antipode* 41 (3):401–15. doi:10.1111/j.1467-8330.2009.00680.x.

Mulvaney, D. 2016. Energy and global production networks, In *The Palgrave handbook of the international political economy of energy,* ed. T. Van de Graaf, B. K. Sovacool, A. Ghosh, F. Kern, and M. T. Klare, 621–40. Basingstoke, UK: Palgrave Macmillan.

Osterman, P., and E. Chimienti. 2012. The politics of job quality: A case study of weatherization. *Work and Occupations* 39 (4):409–26. doi:10.1177/0730888412455155.

Pasqualetti, M. J. 2011. Opposing wind energy landscapes: A search for common cause. *Annals of the Association of American Geographers* 101 (4):907–17. doi:10.1080/00045608.2011.568879.

Pinderhughes, R. 2006. Green collar jobs: Work force opportunities in the growing green economy. *Race, Poverty & the Environment* 13 (1):62–63.

Robbins, P., and S. A. Moore. 2015. Love your symptoms: A sympathetic diagnosis of the Ecomodernist Manifesto. Accessed September 5, 2018. https://entitleblog.org/2015/06/19/love-your-symptoms-a-sympathetic-diagnosis-of-the-ecomodernist-manifesto/.

Roth, S. 2015. Rooftop solar battle brewing in Sacramento. *The Desert Sun,* July 10. Accessed January 16, 2019. https://eu.desertsun.com/story/tech/science/energy/2015/07/10/rooftop-solar-battle-brewing-sacramento/29995761/

Smith, N. 1992. Contours of a spatialized politics: Homeless vehicles and the production of geographical scale. *Social Text* 33:55–81. doi:10.2307/466434.

Storper, M., and R. Walker. 1989. *The capitalist imperative: Territory, technology, and industrial growth.* Oxford, UK: Blackwell.

Strickland, E. 2005. The new face of environmentalism. *East Bay Express,* November 2. Accessed January 16, 2019. https://www.eastbayexpress.com/oakland/the-new-face-of-environmentalism/Content?oid=1079539

Sugrue, T. J. 2004. Affirmative action from below: Civil rights, the building trades, and the politics of racial equality in the urban north, 1945–1969. *Journal of American History* 91 (1):145–73. doi:10.2307/3659618.

Swyngedouw, E. 2010. Apocalypse forever? *Theory, Culture & Society* 27 (2–3):213–32. doi:10.1177/0263276409358728.

Tienhaara, K. 2014. Varieties of green capitalism: Economy and environment in the wake of the global financial crisis. *Environmental Politics* 23 (2):187–204. doi:10.1080/09644016.2013.821828.

Tonn, B., E. Rose, R. Schmoyer, J. Eisenberg, M. Terns, M. Schweitzer, and T. Hendrick. 2011. Evaluation of the national weatherization assistance program during program years 2009–2011 (American Reinvestment and Recovery Act Period). ORNL/TM-2011/87, Oak Ridge National Laboratory, Oak Ridge, TN.

U.S. Department of Energy. 2009. *Weatherization assistance Program—The American Recovery and Reinvestment Act of 2009.* Washington, DC: Department of Energy Office of Energy Efficiency and Renewable Energy.

Walker, P. A. 2003. Reconsidering "regional" political ecologies: Toward a political ecology of the rural

American West. *Progress in Human Geography* 27 (1):7–24. doi:10.1191/0309132503ph410oa.

White House. 2013. Policy snapshot: Creating jobs. Accessed September 1, 2013. http://www.whitehouse.gov/snapshots/creating-american-jobs.

SARAH KNUTH is an Assistant Professor in the Department of Geography at Durham University, Durham DH1 3LE, UK. E-mail: sarah.e.knuth@durham.ac.uk. Her research interests span various topics in the geographical political economy and ecology of property, finance and financialization, urban and regional geographies, work, and technology, with a particular focus on the politics of climate change, clean energy transition, and the green economy.

Reparation Ecologies: Regimes of Repair in Populist Agroecology

Kirsten Valentine Cadieux, (iD) Stephen Carpenter, Alex Liebman, Renata Blumberg, and Bhaskar Upadhyay (iD)

Amidst the backdrop of attention to populism in general, it is instructive to understand populism through social movements focused on food and agriculture. Agrarian populism is particularly salient in agrifood movements. Agroecology has been widely identified as a domain of populist claims on environmental and social governance surrounding agricultural–ecological and political–economic systems. As authoritarian populist leaders gain power throughout the world at a time of expanding economic globalization and contingent socioecological crises, contests over populism in agrifood regimes can highlight current dynamics relevant for formative evaluation of alternative political agroecology strategies and of populist environmental governance more broadly. Can populism be harnessed by radical political agroecologies to simultaneously contest the hydra-headed nature of capitalism, authoritarianism, and pollution and implement forms of environmental governance based on repair? We argue that populist agroecology has untapped potential for repair and that the mechanism of focusing social movements on repair might help address some of the more problematic authoritarian tendencies of populism.

在关注民粹主义的普遍背景下，通过聚焦粮食与农业的社会运动理解民粹主义是具有启发性的。农业民粹主义在农粮运动中特别突出。农业生态已被大幅指认为民粹对于有关农业生态和政治经济系统的环境与社会治理之宣称的领域。当威权民粹主义领导者在扩张的经济全球化与耦合的社会生态危机中，于世界各地取得权力之时，农粮体制的民粹斗争，能够凸显关乎另类政治农业生态策略的形成评估的当下动态，以及更为广泛的民粹环境治理。激进的政治农业生态是否能够驾驭民粹主义，以同时和资本主义、威权主义与污染的多中心本质进行竞争，并施行以修复为基础的环境治理形式？我们主张，民粹农业生态具有尚未利用的修复潜力，而聚焦修复的社会运动之机制，或能有助于应对民粹主义更具有疑义的威权倾向。
关键词： 农业生态, 农业粮食行动主义, 解放的农村政治, 粮食运动, 民粹主义, 农村地理学。

En medio del trasfondo de atención al populismo en general, es instructivo entender el populismo a través de los movimientos sociales que se enfocan sobre la alimentación y la agricultura. El populismo agrario es particularmente saliente en los movimientos agroalimentarios. La agroecología ha sido ampliamente identificada como dominio de los reclamos populistas sobre gobernanza ambiental y social que rodea los sistemas agro-ecológicos y político-económicos. A medida que los líderes populistas autoritarios logran el poder alrededor del mundo en una época de la globalización económica en expansión y contingentes crisis socioecológicas, las disputas sobre el populismo en regímenes agroalimentarios pueden relievar la actual dinámica pertinente a la evaluación formativa de estrategias alternativas de agroecología política y, en términos más generales, de la gobernanza ambiental populista. ¿Puede llegar a controlarse el populismo por las agroecologías políticas radicales para simultáneamente enfrentar la naturaleza "cabeza de hidra" del capitalismo, al autoritarismo y la contaminación, e implementar formas de gobernanza ambiental basadas en la reparación? Argüimos que la agroecología populista tiene potencial sin utilizar para reparar, y que el mecanismo de enfocar los movimientos sociales en la reparación podría ayudar a abocar algunas de las tendencias autoritarias más problemáticas del populismo. Palabras clave: activismo agroalimentario, agroecología, geografía rural, movimiento alimentario, política rural emancipadora, populismo.

This article has been republished with minor changes. These changes do not impact the academic content of the article.

The basis of all wealth is the combination of land and labor, and to be self determining we must liberate both. We know the fight for the liberation of Black people will require us to build thriving movement hubs, to meet our basic needs, and to practice and engage in self governance. Access to land gives us the greatest opportunity to realize those steps towards liberation. ... Let us be clear that the value of suffering can never be calculated and the lives lost never returned. However, reparations is about *repairing our relations*.

—*Reparations for Black Land and Liberation Manifesto* (Black Land and Liberation Initiative 2017, italics in original)

Food and Farming Social Movements in a Populist and Agroecological Context

As a framework for social organization and political action, populism has considerable potential for engaging people in food system transformation and repair, especially repair of relations with food, land, and labor. A rich literature explores the impacts of various food regimes—organizing sets of principles and power relations and practices to enforce them (Friedmann 1987; Le Heron and Lewis 2009; Wittman 2009; Schneider and McMichael 2010; Grant 2017). We pick up themes of populism and repair in the agroecology turn toward food sovereignty in these analytic traditions (Altieri and Toledo 2011; Rosset and Martínez-Torres 2012; Timmermann and Félix 2015), particularly where they fit with what McMichael (2009) described as emerging regimes in tension with "the global food/fuel agricultural complex ... on the grounds of democracy, ecology and quality" (142). Populism is used to critique but also to erase, defend, and exacerbate exploitative food systems (Slocum et al. 2011; Dreher 2012; Holmes 2013; Beck and Bodur 2015, especially in light of the American Farm Bureau Foundation for Agriculture 2015). Popular responses to—and defenses of—genetically modified foods, for example, show that popular imagination about what food is, where it comes from, and how it is produced remains a powerful populist force in multiple ways. Beyond the role of temporary food relief in buying votes (Ullekh 2013; in contrast to commitments to structural change, Cadieux and Blumberg 2014), populist agrifood politics tend in predictable problematic directions:

toward agribusiness-aligned defenses of the virtue and necessity of food producers against unappreciative and ignorant urban elites (Murray 2018) and toward retreat to idyllic, local, and individualistic consumerism (Johnston and Baumann 2014). Both the well-rehearsed extractive populist "feeding the world" and idealized, foodie "defensive localist" versions of agrifood populism often fail to engage in systemic analysis of the intricate and linked processes of exploitation underpinning food systems (Aubrun, Brown, and Grady 2005; DuPuis and Goodman 2005; Guthman 2007; Slocum et al. 2016; Carpenter 2017; Patel and Moore 2017; Blumberg 2018). We identify some of the central populist features of agroecological social movements, in contrast, as focused on social and ecological repair. Repair is a recurring theme of an emerging food regime that we see as potentially corrective to the extractive regimes that have dominated agrifood-related environmental governance. We see the strong strand of repair-oriented agroecological regimes as operationalizing both literal repair and negotiative, collaborative governance processes that acknowledge harm and the need for repair, as we discuss in two cases later. While recognizing the dangers inherent in populism, such as a vulnerability to symbolic but empty political action and, worse, political demagoguery that vilifies largely powerless people, we argue here for understanding agroecological populism as a potential reparative food regime.

The multilayered accounts of populist agroecological "repair" we describe connect instances of regenerative efforts in diverse economies with systemic critiques of agrifood harms. Such repair can offer green infrastructure and community engagement while also, in the languages of the communities using repair to organize, functioning as a framework to build egalitarian grassroots solidarity and new forms of dispersed power, such as community-based land trusts (Davis 2010) and gardens centered around shifting narratives on racial justice. Given the slippages in this usage of repair narratives, it is important to acknowledge that rural and urban agricultural land is valued for many different reasons, which are often contested. As described in recent *Annals* articles by McClintock (2018) and Ekers and Prudham (2017), agrifood environments provide investment opportunities for many diverse arrangements, from formal circuits of capital investment in farmland and gentrification to everyday practices like kitchen gardening that enable social reproduction. The frameworks they and others provide are helpful in considering agrifood environmental governance in light of tensions between precarious fixes facilitating extractive, racialized investments and regenerative repair strategies that attempt to refigure agrifood political ecologies in terms of their values for circulating nourishment and supporting healing (Canty 2017). We consider reparative strains of populism, using a political agroecological reading of reparative agrifood practices that shape environmental governance, shifting state agroenvironmental policies and the social organization of food producers toward frameworks that, we argue, represent regimes of repair. We analyze the logic of repair that is mobilized in populist food strategies that contest elite domination of the governance of food environments. We focus on agroecology's provision of ecological understandings that contest extraction in both environmental and social terms. By linking agroecology and the right to food, La Via Campesina, the United Nations, and others highlight ways in which food regimes make claims to environmental governance (De Schutter 2011). Food sovereignty efforts, particularly, contest the extractive nature of dominant agroecologies and contrast populisms focused on repairing food, social, economic, and soil systems with extractive populisms reproducing agroexport regimes (Vía Campesina 2001; Wittman 2009; Schneider and McMichael 2010).

As scholar-practitioners who teach about and observe agrifood movements, as well as participate in them (both in the U.S. Upper Midwest and also much more broadly, nationally and internationally; Carpenter 2012, 2017; Cadieux et al. 2016; Upadhyay et al. 2017; Blumberg et al. 2018; see also the work of the Twin Cities Community Agricultural Land Trust), we have taken the opportunity of this broader conversation, concern over the demise of democracy with the rise of authoritarian populism, and ineffectual food movement activism to confer across our projects to identify *reparation ecologies*. We have used this construct over the past eight years to reflect and amplify the socioecological processes we have witnessed (e.g., Cadieux 2014), and we have appreciated and built on prior and subsequent uses of the construct of reparation ecology (e.g., Cairns 2003; Caney 2006; Hale et al. 2014; see particularly Patel and Moore 2017). We come to

this analysis of a potential food regime of repair through our work as farmers, organizers, and academics engaged in these dynamics during the past decade, during which time we have observed that one of the overarching dynamics characterizing Twin Cities agriculture is one of reparative populist formations.

We draw on historical and archival sources, along with our observations as participants, to trace the continuities in our contemporary case study of urban agriculture organizing in the legacy of the historic Farmers Holiday movement of the 1930s. All of the authors participated in public discourses and praxis around the contemporary case via the prominence of reparative agrarian populism in Twin Cities school gardens, food policy councils, land use conflicts, and other domains in which we were active in our research, teaching, professional, and volunteer capacities. We have been influenced considerably by the movement connected to the epigraph, including Grant's (2017) leadership of community food justice work here and by the case networks we describe. Further, Upadhyay's work on parallel issues in Nepal (Upadhyay et al. 2019), Liebman's in Chile and Columbia; Cadieux's in Canada and Aotearoa, New Zealand; and Blumberg's in Eastern Europe (Blumberg 2018) prompted us to also compare notes more systematically about parallel observations in our more geographically dispersed experiences and research relationships.

By comparing analyses that had emerged across the authors' research areas and identifying two case studies in which to explore agrarian populism as a regime of repair, we set out here to understand whether conflicts between community food systems and public–private food security practices are successfully able to mobilize populist agroecological strategies—as well as whether these can challenge extractive paradigms of food production by encouraging reparative environmental governance. We examined themes of reparative agroecological populism emerging from our widely divergent research programs (in Canada; Eastern Europe; Latin America; Aotearoa, New Zealand; South Asia; and with contemporary urban community farmers in the U.S. Midwest and mid-Atlantic regions, particularly in black, Indigenous, and immigrant communities) to focus on two case studies that explore how environments are known and governed through populist approaches to food regimes. Although the focus of this article is on two case studies, our methodological approach has involved weaving together conceptual insights gained through sustained scholarly engagement in diverse places. Following Massey (1994, 2005), we deploy a relational understanding of place, which underscores that places are open, unbounded, and forged by a multiplicity of material and immaterial flows, including the ongoing dialogue that forms the foundation of our collaborative effort of knowledge production (Blumberg et al. 2018). Like scholars of transnational feminist praxis (Katz 2001; Nagar and Ali 2003; Pratt and Yeoh 2003), our research process has involved crossing multiple, complex borders to trace connections between analogous processes experienced in diverse locales. Even as these locales are remade as sites of rupture by capitalist processes, we have analyzed how people contest these processes and forge collective efforts informed by regimes of repair and populist agroecology to take control over spatial flows and relations and remake their everyday places.

The following sections explain our use of *populism*, how we read populism in agrifood movements, and how we see reparative populist dynamics at work in two case studies from a region central to extractive agrifood practice. These cases, of the midwestern Farmers Holiday movement in the 1930s and the current community agricultural land movement, suggest ways in which the reparative focus of social movement agroecology might mitigate troubling aspects of populism, particularly in the domain of farmland governance.

On Agrarian Populisms

Part of what appears to fuel current reactive and authoritarian strains of populism is a reaction *against* calls for transformative change. The Coalition of Immokalee Workers (CIW 2017) calls out the Trump version of "populism" when describing the women's march: "Millions of men, women and children poured into the streets … to declare that people of the world would not allow the growing (and terribly misnamed) 'populist' movement, rooted in fear and repression, to turn back the clock on their civil, political, and fundamental human rights." Lost in this common representation is the notion of populism as a liberatory or emancipatory force, such as many U.S. agrarian struggles throughout the twentieth century, radical democracy populism, or

contemporary "indigenous populist" movements in Latin America (Brienen 2016; Grattan 2016; Bosworth 2019). In the contemporary moment, people seem vulnerable to what Judis (2016) described as right-wing triadic populism, which, rather than just rallying "people" against "elites," "sets up a triadic antagonism between the people, the elite, and a third segment of the population that is supposedly being coddled by the political establishment: Muslims, immigrants, effete intellectuals, and so on" (Mounk 2017). Mounk pointed out that Judis might be overly sanguine about the left's avoidance of the dangers of populism, as politicians claim to speak for the "real" people, with what Müller (2016) called a "moral monopoly of representation." In many cases this quickly devolves into scapegoating of others and performative gestures of provisioning seeking to signal leadership's allegiance to the people. We find Moffitt's (2015) article on the performance of populism useful in contextualizing the performative aspects of populism in agrifood movements and as we distinguish reparative characteristics of populism from other conceptual baggage that populism might bring. We use three characteristics of populism he reviews to establish the context of our argument that reparation ecologies might be an important populist aspect of agroecological social movements.

First, we use the term *popular* to distinguish movements made on behalf of a claimed "people," often in relation to land. In most of the cases we refer to, people distinguish themselves not against other people so much as against mechanisms of state and capital that they argue are dispossessing them, generally on behalf of finance. Following from this, it seems important to distinguish the modes of *crises*, *solutions*, and *claims for equivalencies in value* made in the agrifood populisms we discuss. The crisis in question in most of these cases is the extended crisis of extractive, racist (neo)colonialism, which is not the kind of "populism being an extraordinary phenomenon that only arises periodically during crisis" dramatically sweeping a population (Moffitt 2015, 193). In populism, Moffitt (2015) argued, "Actors actively participate in the 'spectacularization of failure' that underlies crisis, allowing them to pit 'the people' against a dangerous other, radically simplify the terms and terrain of political debate and advocate strong leadership and quick political action to stave off or solve the impending crisis" (190). In our cases, we see urgency in the calls to heed the harms of finance capital and neocolonial racism and reject the ongoing reproduction of the status quo but, contrary to authoritarian populisms, our cases likely understate crisis (e.g., of persistent rural or urban poverty, stress, etc.), often in favor of building community capacity to deal with the situation at hand. Central actors are less focused on leadership in a perpetuated state of crisis than working toward repair and changing the rules of procedural justice to reflect the populist principles they promote, as in the cases analyzed here.

Moffitt (2015) argued that "the 'slow politics' ... of consensus and negotiation are presented [by populists] as ineffectual, while strong and decisive political action, unencumbered by procedural checks and balances, are seen as desirable" (201). "Procedural simplification is evident in the often crude and immediate policy solutions offered by populist actors in the effort to stop crises" (Moffitt 2015, 205) he noted and, quoting Žižek's explanation of this formulation: "'The enemy is externalized or reified into a positive ontological entity (even if this entity is spectral) whose annihilation would restore balance and justice'. In such formulations, the cause of the crisis is not the system or general structure as such, but rather always the enemy" (206). In contrast, reparative populism identifies and addresses structural and systemic problems. These might be personified in simplified form as state or financial actors and might be countered with community process but generally not by eradication of the enemy. In the contemporary case we analyze, we see widespread efforts to connect interest in food in "the neighborhood" (the salient social scale of "the people") to complex issues of political economy, global finance, and structural racism, as we discuss later. This contrasts the often-critiqued representational poverty of populism and also speaks to the often deliberative and process orientation of reparative populism, particularly around establishing a plurality of operational value(s) of food system practices. Moffitt (2015, 199) summarized one of Laclau's key claims about the "emptiness" of the populist demand being key to populism's political saliency: "'the so-called "poverty" of the populist symbols is the condition of their political efficacy' (Laclau 2005: 40)"; however, although this emptiness might be characteristic in nationalist racist forms of authoritarian populism, we see reparative populist efforts often

hampering problematically efficient and not publicly accountable "progress" by interrogating exploitation and extraction and making procedural efforts to avoid them, as with antioppression and antiracism trainings.

Reparative Agroecological Populism as a Mode of Environmental Governance

Populism—disruption of elite power by mobilization of "the people" to redistribute that power—has, in the agrifood domain, tended to arise and be noticed during particularly acute moments of crises. Crisis points such as food shortages, natural disasters, economic depressions, and dispossession of land have prompted social movements with ephemeral success at protecting common interests against threats posed by perceived elites, particularly around the control of land, credit, infrastructure, and governing ideology and imagination (McMath 1995; Moffitt 2015). Agrifood and agrarian populisms often focus on tension between defending the interests of people facing dispossession and exploitation—farmers against creditors, the hungry against hoarding, gatherers against state conservationists—and the challenge of enrolling a larger populace in solidarity. This involves sociospatial strategies that we characterize as a form of ecology of repair, a regime we understand to have analytical, educational, and political–ethical functions (Campbell 2009).

The analytic category of a regime of repair focuses our attention on long-term, community-based efforts to build nourishing agroecologies and address land dispossession. Representations of family farming often provide the basis for reactionary populism in environmental governance and politics. Folksy farmers, and the populism they represent, are now staple images in the United States that are used to attack endangered species protection, promote subsidies for federal crop insurance that supports monocropping, defend agribusiness from a wide array of environmental regulations, support the end of estate tax, and even sell pricey pickup trucks, along with the erasure of racialized rural labor (Holmes 2013; Beck and Bodur 2015). Efforts to internalize the immense social costs in the agrifood sector are often opposed with agrarian populist imagery (Hollomon et al. 2017; Williams and Holt-Gimenez 2017). In the United States, at least, farm and food politics is populist politics (Murray 2018), often across the political spectrum. U.S. society in general has long responded to populist appeals (Phillips 1982), and nowhere has that logic been stronger than in the world of food and farming. Facing considerable co-optation of populist agrifood movement logic, questions about populist engagements are consequently centered around the kind of populism and populist methods used. An embrace of agrarian populism-focused ecological repair directly confronts, names, and challenges what we call "extractive populism."

We turn to agroecology, particularly as it has been understood through political ecology (Altieri and Toledo 2011; Méndez et al. 2013), as a salient domain for understanding social and ecological repair in agrifood systems. Agroecology, the study and practice of supporting ecological functions in agricultural ecosystems, is sometimes understood superficially as a technology of replicating ecological functions in agriculture. A more rigorous and socially embedded interpretation of agroecology sees institutional arrangements and political ecological formations that enable food production in regenerative socioecologies (Bawden et al. 1984; Rosset and Martínez-Torres 2014; Holt-Giménez and Altieri 2016; Montenegro de Wit and Iles 2016; Bezner Kerr et al. forthcoming)—what political ecologists might study as political agroecology (de Molina 2013; Méndez et al. 2013; Meek and Tarlau 2015). Peer-to-peer modes of knowledge sharing and respect for regenerative systemic perspectives, which decenter the dominance of extractive systems (e.g., as perpetuated by dependence on corporate input suppliers for farming knowledge and extension), are examples of agroecological priorities (Varghese and Hansen-Kuh 2013).

Analytically, in contrast to corrosive authoritarian blood-and-soil nationalist and socially xenophobic populisms based on eradication, exclusion, and narratives of scarcity, agroecology's social movement toward food sovereignty focuses significantly on what can be gained by agroecological methods, both mechanical and social, of food production (Bezner Kerr 2008; Wezel et al. 2011; Snipstal 2013; Bezner Kerr et al. forthcoming). This simultaneously critical and constructive framing enables broad public audiences to better understand the social and ecological consequences of agri-industrial externalization of the costs of production and exchange—as well as to understand a plurality of ways to internalize these costs, an important feature

of a nonhegemonic regime (Wark 2015). Educationally and ethically, by enrolling all eaters in agricultural and environmental relationships (Gussow 1991; Berry 1992) who bear responsibility for transforming agrifood systems to be less violent and more equitable (Thompson and Wiggins 2009; Holmes 2013; Reynolds and Cohen 2016; Alkon and Guthman 2017; CIW 2017; Marquis 2017), agroecological social movements move beyond direct participation by agrifood producers or laborers only to mobilize broader intersectional, reparative performances of populist solidarity.

Repair functions as a mode of approaching ecological dynamics with respect to the need to address harm and build regeneration into agricultural ecosystems. Rather than focusing on romanticized restorations, as environmental regime metaphors often encourage, reparative restoration ecology recognizes "that ecologies are always in flux," that "climate change disproportionately affects marginalized races, nationalities, genders, and classes of people, [so that] natures must be restored with the consent, participation, and design of those so affected," that "in the Anthropocene, there is no clean slate with which to begin; colonial and racist injustices have given rise to neocolonial injustices that climate change exacerbates," and that corporate support of environmentalism "might be an attempt to offer reparations for its history of plundering"; hence "restorative processes like native plantings, prairie burnings, or invasive removals aim to *redirect* ecological systems—in order to set them in motion again" and to "ask, what lessons have we inherited, and what skills can we hone, from our participation in both Earth-destroying, and Earth-regenerating, activities?" (Garvey 2016). Repair also functions as a regime of relationships and diffuse informal sanctions for preventing and dissipating concentrated power and resource control (Robinson and Tormey 2009), revealing some of the paradoxes around traditional conceptions of farming and environmentalist success, and disempowering capture of popular agrifood discourses and practices. As we hear particularly in arguments for addressing black and indigenous land loss in the United States, the need for repair of food systems is a constant refrain in contemporary community agrifood organizing—but there are very few well-established rubrics for the evaluation of repair (although see Anderson et al. 2009; Merkle 2013). Focusing on repair in engaging these narratives helps differentiate political ecologies of claims around land loss, vulnerability, and harm from losses suffered by privileged commodity farms and their investors. Agroecological framing of repair points advocates toward more socioecological, rather than merely symbolic, modes of repair work.

The Farmers Holiday Fight against Dispossession as Agrarian Populist Repair

A 1930s farmer movement, the Farmers Holiday, mobilized thousands of farmers and disrupted the capitalist consolidation of agriculture in the Great Depression. By the early 1930s, there were about 6 million farms in the United States. A large proportion of these farms were in the Midwest. Although continuing to be self-sustaining in some ways, these farms fit closely into an ideal type of household commodity production (Friedmann 1978). Midwest farmers had significant debt, paid substantial cash-based taxes, and sold commodities into a largely undifferentiated market. Midwestern tenant farmers were not under the day-to-day control of landlords. Whereas popular conceptions of farmer struggles during the Depression center around drought and the Dust Bowl, for the majority of farmers the central issue was dispossession.

In the early 1930s, Midwestern farm prices dropped dramatically, by about 75 percent. Farmers already earned far below average incomes (Rochester 1940). Rural Midwesterners had valued literacy and public education and borrowed for schools; that borrowing was repaid with property taxes paid by farmers. Tax delinquency and dispossession via tax sale became common. Farmers had borrowed to buy farm equipment and land, so mortgage foreclosure and the repossession of personal property, along with postforeclosure deficiency judgments, also became common. By 1932, no close observer could deny that dispossession threatened virtually all Midwestern farmers (Shover 1965; Dyson 1968).

Farmers turned to protest via the Farmers Holiday Association, a brief but influential populist movement (Saloutos and Hicks 1951; Kramer 1956; Shover 1965; Luoma 1967; Dyson 1968; Nass 1984). The name was a bitter nod to the bank holidays of the era in which banks closed and depositor savings were lost. Farmers Holiday was a quasi-national organization. The strongest presence was in the Midwest, but in every state from Pennsylvania to the West Coast,

farmers forcibly stopped foreclosures. The movement's ideology was populist in the sense that it was antielitist, in particular regarding class, but also in the sense that it remembered the original agrarian populism of the nineteenth century that had struggled against Gilded Age inequality (Goodwyn 1978; McMath 1990; Postel 2007).

The movement was populist in two further ways. First, it was an insurgency. The Holiday mobilized and forced attention on farm issues in hopes that government would respond—preferably with an increase in farm prices. In addition, farmers challenged state authority by creating a farmer-run set of rules that regulated the dispossession of farmers. Second, the Holiday movement was populist in its antielitism. Milo Reno, its most famous leader, had struggled as a farmer, was a part-time preacher, and played a fiddle at rallies. When founded in 1932, the Holiday's platform fit on a single page: ten paragraphs, and fewer than 400 words. A fair price for farmers and debt relief were the two main points. Political, intellectual, and journalistic elites, for their part, were unsympathetic to the movement, and the main farm organizations of the day—even the left-leaning Farmers Union—had no sympathy for the movement and its tactics. Even New Deal politicians who generally sympathized with farmers did not endorse the movement. The Holiday movement, for its part, berated New Dealers but refused to embrace the Klan-like organizations that expressed some interest in farm protest (Dyson 1968).

Farm protest took two main forms: movement mobilization and a quiet shadow system of debtor–creditor law that usurped state power. As an original tactic, farmers sought to strike to drive up prices. These actions largely failed to move prices. Then, Holiday farmers engaged in protests that made tax and foreclosure sales essentially impossible. At the outset, this meant a "penny auction," in which no one bid more than a nominal amount and the farm was returned to the owner. Writing about the Holiday, to the extent that it exists, tends to focus on dramatic confrontations. Farmers set up barricades and could seemingly mobilize hundreds at any time and anywhere to stop a foreclosure or tax sale. Sheriffs, judges, lawyers, and lenders were intimidated, and many chose not to proceed with creditor actions. Although the political history of the agricultural New Deal is complicated, there can be no doubt that the Farmers Holiday movement helped push political elites into reform—with both federal and state policies (Shover 1965; Dyson 1968).

Holiday farmers also created "councils of defense" that were "intended to adjudicate all disputes between creditor and debtor" and acted utterly outside of the legal system (Dyson 1968, 131). Each council had from five to eleven elected members and rarely included anyone who was not a farmer. Councils addressed mortgage foreclosures, chattel sales, and landlord–tenant disputes. They operated as a hybrid of mediation and court adjudication. The goal was to reach a peaceful accommodation. If the farmer wanted to avoid foreclosure, the council decided whether it should proceed. Some farmers agreed to foreclosure but hoped to avoid a deficiency judgment, and in these cases the council considered the deficiency judgments. Chattel sales were dealt with in a similar way. Councils also heard cases between landlords and tenants. For foreclosures, councils often recommended a moratorium on creditor action but called for the debtor to pay a fair rent for the land in the meantime. In rental cases, the result was generally reduced rent. Local councils heard thousands of cases and the practice extended across wide areas in the countryside. When the councils of defense failed to arrange an accommodation, farmers often blocked the forced sale. Journalists at the time wrote extensively about farmers halting foreclosure through direct action; the quiet work of the councils, however, stopped far more foreclosures (Shover 1965; Dyson 1968). The true extent of this effort will never be known, because of oaths of secrecy for council members and the secret existence of Holiday auxiliary organizations that were local, kept few records, and used pen names in correspondence. Farmers created a parallel state that prevented the official state from performing its perhaps most essential function in a capitalist economy—collecting taxes and enforcing debts. The populist repair here is one that moves utterly outside of legal, but not community, limits and creates a new, farmer-based control that eased the disruption of the worst economic crisis in agriculture in the country's history.

Depression-era farmers assumed that a market economy could function in a moral way but watched it stray into disaster and concluded that it was as up to farmers themselves to repair the rupture and hold fast until things improved. Holiday populism depended on organizing and, for a social base, on a relatively egalitarian rural social structure. With the

advent of the New Deal and various state reforms, the Farmers Holiday movement soon disappeared.

The agricultural New Deal was flawed in many ways. The programs created failed to protect the interests of struggling farmers over the long term, for example, and were often used effectively as a means of protecting class and especially racial inequality in the rural South. The Holiday triggered reform but was unable to shape the nature of the reform or to defend it over the long term. That said, real reform and real resources came with the New Deal and were in significant part due to the politics of farmer populist repair.

Midwestern Community Urban Agriculture Movements as a Mechanism of Repair

Because of the efforts of a diverse network of agrifood organizations to contest accumulation interests and to raise up community leaders, community urban agriculture in the Twin Cities metropolitan area of Minnesota provides a useful subsequent case study for exploring how populist agroecology movements can attend to repair, particularly in relation to the linked ecological and social composition of environmental governance challenges. The Twin Cities is home to significant communities of displaced Southeast Asian, East African, Latinx, and black farmers, in addition to significant communities of displaced Indigenous peoples, as well as rural-to-urban migrants whose farm families participated in egalitarian agrarian populist movements. Urban agriculture here has reproduced the agrarian question central to family farms' resistance to finance, as seen in the Farmers Holiday movement, and in movements the authors have all taken part in, as scholars, teachers sending service-learning students, and active members. Reparative agrarian populism can be seen in the way in which agrifood movements in the region have engaged racialized dynamics of repeated dispossession to work on repairing relations with food and land.

Despite shifts rightward in politics and challenges from state, financial, and structural forces, the Upper Midwest has retained an agrifood culture with foundations in populist values of equity and cooperation (e.g., the Land Stewardship Project). Grappling with structural barriers to equitable, successful agrifood and related environmental governance has refined the regional agrarian populist culture, adding more attention to solidarity with Indigenous peoples and people of color, often centered around questions of reparations, or at least repair of trauma, and relationally accountable, community-led action (LaDuke et al. 2010; Institute for Agriculture and Trade Policy 2012; Homegrown Minneapolis 2018). For example, the Hope Community Listening Project began: "We live in an era of food- and stress-related health crises, increasing disparities, and cultural and environmental erosion. We not only have to find solutions together; we must also honor each other so that we can work together to achieve them" (Hollomon et al. 2015, 5; also see how the Land Stewardship Project adds "racial justice" to its legacy of "keeping the land and people together"). The Twin Cities is proud of its history of adopting one of the first food policy councils in the nation in 1986 (the St. Paul–Ramsey County Food and Nutrition Commission) and of its active network of rent stabilization and affordable housing efforts working to address strong racial disparities in access to housing, healthy food, and other supportive infrastructure (Lindeke [2015] showed the tensions, particularly between food co-ops and housing needs; Burga [2016] provided a sampling of many organizations addressing these in integrated ways; see the Center for Urban and Regional Affairs for community-based research on many additional such projects). The integrated relationships, across topics and communities, of these efforts prompt us to argue for the value of considering repair ecologies as an emerging food regime. These ongoing efforts have contributed to residents—especially Indigenous and people of color—contesting development pressure in the Twin Cities and organizing to address environmental harms with specifically repair-focused food provisioning projects.

Over the past five years, following the introduction of the AB551 urban agriculture enterprise zone legislation in California, a large Twin Cities network was convened by the Council of Minnesotans of African Heritage to advocate for statewide urban agriculture support legislation (Project Sweetie Pie 2015). An extensive network of supporters collaborated over several years of community listening sessions. This effort attracted bipartisan support across the state, successfully recruiting rural districts as allies to urban agriculture. This involved recognizing

that rural livelihood strategies do not preempt experience of the need for repair (Shea 2013) and that (re)conciliative outreach to commodity producers asking for solidarity action that recognizes exclusions and harm can help to build reparative critical agricultural literacy in both the organizing and policy domains (Van Sant and Bosworth 2017). The urban agriculture legislation effort retained a broad platform of community development, positive environmental impact, and economic justice, promoting a progressive populist platform—without collapsing into single-leader or single-issue simplifications. Prioritizing a range of ways in which urban agriculture could benefit broad publics, the result of this legislative effort (e.g., Minnesota bills HF 1461 and HF 2076 in 2016; Minnesota Department of Agriculture 2018) was the assembling of a set of nineteen criteria that combined agroecological repair of degraded food systems (environments, bodies, livelihoods) with antiracist acknowledgment of the settler land dispossessions that have led Indigenous and people of color to need access to land for food while residing in urban spaces (cf. Williams and Holt-Gimenez 2017). Keeping a reparation frame in the forefront of these conversations as a way to understand how land access could repair harms continuing to be experienced in dispossessed communities led to the central prioritization in the resulting grants rubric of serving "communities of color or Native American tribal communities," despite the controversy of such language in the region. This also led Voices for Racial Justice to recognize these legislative efforts as some of the most progressive environmental justice work seen in the recent legislature (Racial Equity Tools 2016).

As with many disinvested metropolitan neighborhoods of the United States (and more globally), the urban agriculture and community food production being supported with this legislation in the Twin Cities has experienced a significant rise in visibility over the past few decades (Hollomon et al. 2017). Interest in inclusive agrifood movement politics has also grown (Union of Concerned Scientists, Center for Science and Democracy 2014, 2015; Hollomon et al. 2015), and institutional supports, particularly via foundations, nongovernmental organizations, and schools (Institute for Agriculture and Trade Policy 2012, 2014), have been variable, and land has been relatively abundant, especially due to disinvestment, foreclosure crises, and limited development throughout the urban core of the Twin Cities metropolitan region, although this last trend is now sharply reversing (Goetz, Damiano, and Hicks 2017; Orfield and Stancil 2017). As elsewhere in the United States, urban agriculture ranges from many backyard and informal vacant lot gardens to school gardens, church gardens, both new and many-decades-old community garden plots, intensive hydroponics and capital-intensive indoor agriculture, enclaves of immigrant growers, and small-scale urban farm businesses, often run by white-college graduates.

One such urban vegetable farm, Stone's Throw Urban Farm, a partnership in operation from 2011 to 2016 across three acres of dispersed sites, marshaled creative place making and small business legitimacy to change zoning codes in St. Paul while collaborating with a broader network to push for regionwide debates on land access for urban agriculture. This farm project and its supporting community served as a venue for exploring community food movement issues, such as the precarity of urban year-to-year land access, and experimenting with ways to contest gentrification. Stone's Throw's land access was gained through a variety of means—lease agreements through city council offices, contracts as part of landscaping for businesses, and private leases with landowners—and their tactics for using land centered around active and organized resistance to the association between urban agriculture and whitening ecogentrification. Working in close collaboration with the Twin Cities Community Agricultural Land Trust to pressure the metropolitan land use governance agencies to rethink a highest and best use policy in favor of indicators of success that address racially disproportionate stress and dispossession, the Stone's Throw farmers (like many others, e.g., Daftary-Steel 2015; see fooddignity.org) harnessed urban agriculture to education about and repair of structural harms. In addition to encouraging policy supports for agricultural land uses that meet community needs beyond conventionally recognizable garden plots (Phat Beets Produce 2012; Jacquemet 2016), they coordinated efforts to reject the appropriation of agriculture by growth coalitions and boosterism. They frequently intervened in education efforts for the networked Twin Cities land access community on being accountable to work with and not on behalf of communities, with particular attention to power, space, and race (balking

elite philanthropy models that fund most conservation land trusts). They acquired and managed their spaces in ways that built relationships, sharing growing practices with neighbors and immigrant farmers around their core sites and distribution networks. This shared development of space as well as business and advocacy networks eventually led to turnover of sites to neighbors and Twin Cities people's movement organizations, such as Tamales y Bicicletas. This trajectory of land stewardship provides a significant contrast with the much more available transient land access often proffered by developers seeking to mollify neighbors of construction sites in waiting.

The Urban Farm and Garden Alliance is a volunteer network of backyard gardeners and community gardens working to connect largely African American church and neighborhood service-provider spaces. The Alliance was established to cultivate community and neighborly relationships based on social justice and reconciliation, as well as community and leadership development, gardening, nutrition, and environmental education, and to organize backyard gardeners, in particular, to get people to know each other across different cultures and to learn to work together. They work with the state Department of Health, extension nutrition and gardening programs, health insurance providers, and community clinics to support the growth of gardening programs as spaces of repair—not only for food-access-related health issues but also for repair of stress and trauma and for racial reconciliation. Used in this way by the Urban Farm and Garden Alliance, the concept of repair becomes a boundary concept, organizing support for linked regenerative agricultural and social repair across domains that would not usually share justifications for such work. They are able to funnel devolved health care funding into community organizing, using (and creating legibility and legitimacy for) deliberately different metrics and framings; for example, antiracism training, stress and trauma amelioration, building of community health and wealth outside capital circuits, educational efforts explicitly adversarial to accumulation strategies, and the reframing of institutional contexts for building sociocultural capital as under the guidance of the neighborhood. They are known for leadership in community responsiveness training with county extension programs; for example, asking Master Gardeners to acknowledge their problematic nomenclature and legitimacy claims (based on mastery) and to simultaneously work as community and environment regenerating "land connectors" in facilitated collaboration with tenant advocates, police, and the press. In this context, their emphasis on recognition of how often-marginalized communities have already regenerated themselves in challenging conditions uses repair as a tool to redirect and modify funding streams, modes of governance, state surveillance and policing, and press coverage. In venues such as the regular Reconciliation Lunch, they ask do-gooders to reconsider their assumptions about race, societal improvement, and reform (Slocum 2007), while also attending to the harm continuing to be enacted by dominant systems.

Food Regimes of Repair? Regeneration in Relations of Value through Populist Agrifood Movements

The efforts of the Stone's Throw network to transform urban agriculture environmental governance in St. Paul—along with networked collaborations including the Urban Farm and Garden Alliance to effect recognition that prior displacements of well-established black farmers, gardeners, and orchardists to build affordable housing could have been avoided by communities' negotiating multifunctional landscapes—coincided with efforts in Minneapolis to promote the restoration of wild rice lakes on a chronically flooding publicly owned golf course as a food forest. Along with pressure by Black Lives Matter leaders, Parks and People, and other organizations and candidates to politicize the larger question of the ownership of 16 percent of the city's land by the elected Minneapolis Parks and Recreation Board, this attention to reparation ecologies brought attention to investment and disinvestment in racially differentiated neighborhoods and influenced subsequent significant turnover in park commissioner seats in the 2017 elections after repeated social media reports revealed the reluctance of the existing board to allow public participation or comment in its meetings. These efforts—often organized around public access to public land for food production—show the emergence of strong support for popularly designed and negotiated

reparative responses to environmental governance challenges.

We have described how highly networked groups of farmers, gardeners, and academic-activist organizers working in the Twin Cities have facilitated the emergence of reparative agroecologies, repairing relations with land and across communities. These efforts have built community action and resistance on the margins of capitalist development and state governance. Simultaneously, they have made demands on state, finance, and nonprofit actors for redistributive programs and reparations-based land and financial access. This dual form of organizing is seemingly paradoxical. Efforts to build local forms of power and resource governance that explicitly shun inclusion into formal political processes and turn away from seeking recognition from and making claims on the state are seemingly anathema to participation in mechanics of city governance, state funding applications, and meetings with housing development projects. Yet reparative agroecological efforts in the Twin Cities have had success in this two-pronged approach of representation and resistance. Individuals and groups make claims for repair through land and wealth redistribution, through the implementation of agroecological methods that are closely attuned to neighborhood dynamics, and through linking agrienvironmental sustainability, agrifood labor conditions, and food distribution with other social movements.

In contrast to what has been seen in many other regions, this pluralistic reparative agrarian populist political formation that we describe here has largely thwarted an alignment of urban food land efforts with singular, charismatic, popular leaders. Instead, the focus is on reparation of dispossession often involving nonhegemonic models of community land relationships, what Larsen and Johnson (2017) described in terms of a pluriversal way of making place. This is not "a populism of THE people" or recourse to legitimization by hierarchies or absolute authorities. Instead, it involves acknowledgment of the need for negotiative collaboration, mutual recognition, and consent (Whyte 2013), and it contests the shallow claims, made by what we call *extractive populism*, "that American agriculturists are rural, Christian, white, and hard working" (Martin 2014). Drawing on the agrarian populist legacy of the region evident in the Farmers Holiday case, the movement has remained critical toward the incursions of capital and land control (including via capital switching into secondary and tertiary circuits of capital [Ekers and Prudham 2017], although gentrification pressure is strengthening [Value Walk 2018]) and retained ongoing focus on antiracist politics that has linked the "food movement" to broader social and political concerns in the region (White [2011]; Sbicca [2014]; Reynolds and Cohen [2016]; and McClintock [2018] detail the dominant themes of ecogentrification and lack of movement support while also pointing to contrasts). Contesting the appropriation of the value of urban agriculture that McClintock (2018) and others describe as contributing to "racialized processes of uneven development," Twin Cities urban agriculture communities refuse and disrupt political formations framed around claims of sustainability or cultural capital, increasing property taxes or investment potential, or "frontiers," "pioneering," or scarcity, as the preceding examples demonstrate.

This is not to say that Twin Cities agriculture does not share with other U.S. cities the conjoined problems of gentrification, agrarian idealism, and nonprofit and corporate cooptation of radical agrifood politics. Twin Cities agriculture, however, also shares agrarian populist characteristics such as skepticism of expert knowledge, extralegal land tenure arrangements that challenge zoning specifications and insurability, and reclamation of personal and popular autonomy in the spheres of land, labor, and governance. Especially as a center of global agrifood finance and industry (e.g., the metro region hosts headquarters of Cargill, ADM (Archer Daniels Midland), General Mills, C. H. Robinson, Dairy Queen, Land O'Lakes, and CHS), the Twin Cities agrifood movement scene is unusual in its confluence of community-focused production and food organizing with critique and activism surrounding structural food system issues (e.g., with the Institute for Agriculture and Trade Policy headquartered here, focused on the volatility of grain and food markets and how movement actors can change enabling regulatory frameworks).

A core theme of the reparative agrarian populist efforts we have briefly surveyed is the need for public subsidy of land rents for reparative agrarian ecologies—or reparations in the form of land back to the people, particularly recognizing the disproportionate effects of dispossession on communities vulnerable to

discrimination. Seeing these acknowledgments reflected in popular agrifood movements and state responses to them has led Midwestern agrifood movement actors to explore how these aspects of populist environmental governance can be amplified elsewhere—particularly within the domain of conventional agriculture, where farmers are subject to considerable stresses and perhaps fewer entry points to critical populism than their historical or urban counterparts. As reparative efforts are threatened by the ongoing appropriation of agrarian populism by foodies, agribusiness, and increasingly financialized agrifood and agricultural land sectors, the concept and practice of regimes of repair can be useful for highlighting the difference between movement efforts that result in repair and those that use populism as a public relations strategy. Literacy about a reparative food regime, for example, might help contest Bayer's appropriation of the populist aesthetic of Farm Aid concerts for its "Here's to the Farmer" down-home country music farm tour. This campaign appears to be a public relations effort to address their fear that "'consumers remain emotionally skeptical about trusting science and research' in the field of agriculture," consequently not only trying to break the association between agrarian populism and suspicion of big businesses and banking but more ambitiously attempting "to recuperate that banker [about whom prior country musicians have sung reliably antibanker sentiments], not just including them among the neighbors but singling them out for praise and gratitude" (Murray 2018). This is a classic example of extractive populism's "moral monopoly of representation" (Müller 2016), equating "feeding the world" with Bayer-supported commodity agriculture, a false equivalence that ignores the antagonism "between the desire for autonomy or self-sufficiency and growth of capitalism, which requires people to submit to the market" (Murray 2018).

Continued populist agroecology will be necessary to repair broken socioecological relations in food and farming. Agrarian populism is a way with which people involved in agrifood movements often understand what is broken in their world. Although populism can veer right or left and therefore rightly makes many wary and requires continued critical engagement, populism has worked as an ideological vehicle for agroecological repair. As scholars and practitioners continue to build analyses supportive to reparative praxis and regenerative relations of agrifood value, a central question for farm and food systems is not whether there will be an agrarian populist understanding of the world but rather how to repair and regenerate such an understanding and practice in an ongoing way, to resist extractive appropriations, and to continue to make agrifood repair logics legible, legitimate, and actionable.

Acknowledgments

Our work on ecologies of repair and healing has been supported by our coordinating author working at an institution (Hamline University, particularly through the Food and Society Workshop) and in a city (Saint Paul) that are both engaged in the conciliative process of Truth, Racial Healing, and Transformation projects and by the network of youth, elders, friends, colleagues, ancestors, and family that make up the FreshLo-funded Art of Food in Frogtown and Rondo project. We are grateful to the many students who have worked with us on these projects, particularly Monica Saralampi and Matt Gunther; to the generous and thoughtful suggestions of Professor McCarthy and the three reviewers who clarified our observations; and to the Twin Cities Community Agricultural Land Trust, Urban Farm and Garden Alliance, and Farmers' Legal Action Group and their networks for cultivating reparation ecologies.

Funding

The work from which this article emerged was supported by four programs at the University of Minnesota: Global Programs and Strategy Alliance's Global Spotlight Program, Center for Austrian Studies, Institute for Advanced Studies AgriFood Collaborative, and the Institute on the Environment Resident Fellows program.

Disclosure

Valentine Cadieux, Alex Liebman, and Stephen Carpenter have served as board members of the Twin Cities Community Agricultural Land Trust (http://tcalt.org/), which has been consistently involved in the networks described in the second case (where Liebman also participated as a farmer and partner at Stone's Throw Urban Farm and Cadieux has served on the Urban Farm and Garden Alliance Research Team), and where they have been mentored extensively by community organizers and food and farm advocates immersed in many reparative traditions.

ORCID

Kirsten Valentine Cadieux ⓘD http://orcid.org/0000-0003-0959-8152
Bhaskar Upadhyay ⓘD http://orcid.org/0000-0001-5141-3778

References

Alkon, A., and J. Guthman. 2017. *The new food activism: Opposition, cooperation, and collective action.* Oakland: University of California Press.

Altieri, M. A., and V. M. Toledo. 2011. The agroecological revolution in Latin America: Rescuing nature, ensuring food sovereignty and empowering peasants. *Journal of Peasant Studies* 38 (3):587–612.

American Farm Bureau Foundation for Agriculture. 2015. Growing agricultural literacy. Accessed October 30, 2018. https://www.agfoundation.org/sustainability/the-necessity-of-food.

Anderson, M., J. Fisk, M. Rozyne, G. Feenstra, and S. Daniels. 2009. *Charting growth to good food: Developing indicators and measures of good food.* Arlington, VA: Wallace Center, Winrock International.

Aubrun, A., A. Brown, and J. E. Grady. 2005. *Not while I'm eating: How and why Americans don't think about food systems. Perceptions of the U.S. Food system: What and how Americans think about their food.* Battle Creek, MI: W. K. Kellogg Foundation.

Bawden, R. J., R. D. Macadam, R. J. Packham, and I. Valentine. 1984. Systems thinking and practices in the education of agriculturalists. *Agricultural Systems* 13 (4):205–25.

Beck, S. A., and Y. Bodur. 2015. Migrants, farming, and immigration: Beginning a dialogue in agricultural education. *Journal of Southern Agriculture Education Research* 65 (1):19–37.

Berry, W. 1992. The pleasures of eating. In *Cooking, eating, thinking: Transformative philosophies of food*, ed. D. W. Curtin and L. M. Heldke, 374–79. Bloomington: Indiana University Press.

Bezner Kerr, R. 2008. Gender and agrarian inequality at the local scale. In *Agricultural systems: Agroecology and rural innovation*, ed. S. S. Snapp and B. Pound, 279–306. San Diego, CA: Elsevier.

Bezner Kerr, R., C. Hickey, L. Dakishoni, and E. Lupafya. (forthcoming). Repairing rifts or reproducing inequalities? Agroecology, food sovereignty, and gender justice in Malawi.

Black Land and Liberation Initiative. 2017. Reparations for Black Land and Liberation Manifesto. May 24.

Accessed October 30, 2018. http://blacklandandliber-ation.org/uncategorized/manifesto/

Blumberg, R. 2018. Alternative food networks and farmer livelihoods: A spatializing livelihoods perspective. *Geoforum* 88:161–73.

Blumberg, R., R. Huitzitzilin, C. Urdanivia, and B. C. Lorio. 2018. *Raíces del sur*: Cultivating ecofeminist visions in urban New Jersey. *Capitalism Nature Socialism* 29 (1):58–68.

Bosworth, K. 2019. The people know best: Situating the counter-expertise of populist pipeline opposition movements. *Annals of the American Association of Geographers.* doi:10.1080/24694452.2018.1494538.

Brienen, M. 2016. A populism of indignities: Bolivian populism under Evo Morales. *Brown Journal of World Affairs* 23 (1):77–92.

Burga, F. 2016. Projects in the site planning for food justice class. *Humprey School News*, April 18. Accessed October 30, 2018. https://www.hhh.umn.edu/news/projects-site-planning-food-justice-class.

Cadieux, K. V. 2014. Reparation ecologies in a community agricultural land trust: Re-examining urban–rural edge negotiations in the context of efforts to repair social and ecological landscape relationships. Paper presented at the Annual Meeting of the Association of American Geographers, Tampa, FL, April 10.

Cadieux, K. V., and R. Blumberg. 2014. Food security in systemic context. In *Encyclopedia of food and agricultural ethics*, ed. P. Thompson. New York: Springer. https://doi.org/10.1007/978-94-007-6167-4_11-1

Cadieux, K. V., A. Liebman, M. Gunther, and M. Saralampi. 2016. Re-valuing yield: Negotiating race, values, and the agrarian question in urban agriculture. Paper presented at the Annual Meeting of the American Association of Geographers, San Francisco, March 29.

Cairns, J. 2003. Reparations for environmental degradation and species extinction: A moral and ethical imperative for human society. *Ethics in Science and Environmental Politics* 3:25–32.

Campbell, H. 2009. Breaking new ground in food regime theory: Corporate environmentalism, ecological feedbacks and the "food from somewhere" regime? *Agriculture and Human Values* 26 (4):309.

Caney, S. 2006. Environmental degradation, reparations, and the moral significance of history. *Journal of Social Philosophy* 37 (3):464–82.

Canty, J. M. 2017. *Ecological and social healing: Multicultural women's voices.* London and New York: Routledge.

Carpenter, S. 2012. The USDA discrimination cases: Pigford, in re Black Farmers, Keepseagle, Garcia, and Love. *Drake Journal of Agricultural Law* 17:1.

———. 2017. Family farm advocacy and rebellious lawyering. *Clinical Law Review* 24:79.

Coalition of Immokalee Workers. 2017. Birth of an American human rights movement? Accessed December 16, 2017. http://www.ciw-online.org/blog/2017/01/womens-march/.

Daftary-Steel, S. 2015. *Growing young leaders in East New York: Lessons from the East New York farms! youth internship program.* Brooklyn: East New York Farms.

Davis, J. E. 2010. *The community land trust reader.* Cambridge, MA: Lincoln Institute of Land Policy.

de Molina, M. G. 2013. Agroecology and politics. How to get sustainability? About the necessity for a political agroecology. *Agroecology and Sustainable Food Systems* 37 (1):45–59.

De Schutter, O. 2011. The right of everyone to enjoy the benefits of scientific progress and the right to food: From conflict to complementarity. *Human Rights Quarterly* 33 (2):304–50.

Dreher, R. 2012. Porky populism: Class war comes to dinner, and conservatives are on the wrong side. *The American Conservative*, August 6. Accessed October 30, 2018. https://www.theamericanconservative.com/articles/porky-populism/

DuPuis, E. M., and D. Goodman. 2005. Should we go "home" to eat? Toward a reflexive politics of localism. *Journal of Rural Studies* 21 (3):359–71.

Dyson, L. K. 1968. Farm Holiday movement. PhD dissertation, Columbia University, New York.

Ekers, M., and S. Prudham. 2017. The metabolism of socioecological fixes: Capital switching, spatial fixes, and the production of nature. *Annals of the American Association of Geographers* 107 (6):1370–88.

Friedmann, H. 1978. Simple commodity production and wage labour in the American plains. *Journal of Peasant Studies* 6 (1):71–100.

———. 1987. International regimes of food and agriculture since 1870. In *Peasants and peasant societies*, ed. T. Shanin, 258–76. Oxford, UK: Basil Blackwell.

Garvey, M. 2016. Novel ecosystems, familiar injustices: the promise of justice-oriented ecological restoration. *Darkmatter Journal: In the Ruins of Imperial Culture* 13:1–16.

Goetz, E. G., T. Damiano, and J. Hicks. 2017. *Racially concentrated areas of affluence: A preliminary investigation.* Accessed October 30, 2018. http://www.cura.umn.edu/sites/cura.advantagelabs.com/files/publications/DRAFT-Racially-Concentrated-Areas-of-Affluence-A-Preliminary-Investigation.pdf.

Goodwyn, L. 1978. *The populist moment: A short history of the agrarian revolt in America.* Oxford, UK: Oxford University Press.

Grant, S. 2017. Organizing alternative food futures in the peripheries of the industrial food system. *Journal of Sustainability Education* 14. http://www.susted.com/wordpress/content/organizing-alternative-food-futures-in-the-peripheries-of-the-industrial-food-system_2017_05/

Grattan, L. 2016. *Populism's power: Radical grassroots democracy in America.* Oxford, UK: Oxford University Press.

Gussow, J. D. 1991. *Chicken Little, tomato sauce, and agriculture: Who will produce tomorrow's food?* New York: Bootstrap.

Guthman, J. 2007. Can't stomach it: How Michael Pollan et al. made me want to eat Cheetos. *Gastronomica* 7 (3):75–79.

Hale, B., A. Lee, and A. Hermans. 2014. Clowning around with conservation: Adaptation, reparation and the new substitution problem. *Environmental Values* 23 (2):181–98.

Hollomon, Z., E. Bell, C. Sheehy, V. Cadieux, and the Art of Food in Frogtown and Rondo team. 2017. *The art of food in Frogtown and Rondo: A community food system planning process report.* St. Paul, MN: AEDA.

Hollomon, Z., and Hope Community Report Production Team. 2015. Feed the roots: A Hope Community listening project report. Accessed October 29, 2018. http://hope-community.org/wp-content/uploads/2017/02/HOPE-Feed_The_Roots_Finaldigital.pdf

Holmes, S. 2013. *Fresh fruit, broken bodies: Migrant farm workers in the United States.* Berkeley: University of California Press.

Holt-Gimenéz, E., and M. Altieri. 2016. *Agroecology "lite": cooptation and resistance in the Global North.* Oakland, CA: Food First Institute for Food and Development Policy.

Homegrown Minneapolis. 2018. Statement of vision. Accessed October 30, 2018. http://www.minneapolismn.gov/sustainability/homegrown/WCMS1P-130114.

Institute for Agriculture and Trade Policy. 2014. *Beyond the farm bill.* Minneapolis, MN: Institute for Agriculture and Trade Policy.

Institute for Agriculture and Trade Policy, and participants. 2012. Food + justice = democracy. Minneapolis, MN: Institute for Agriculture and Trade Policy. Accessed October 30, 2018. https://www.iatp.org/event/food-justice-democracy-0.

Jacquemet, M. 2016. Phat beets, new gardens: A community response to food gentrification. Accessed October 30, 2018. https://foodfirst.org/phat-beets-new-gardens-a-community-response-to-food-gentrification/.

Johnston, J., and S. Baumann. 2014. *Foodies: Democracy and distinction in the gourmet foodscape.* London and New York: Routledge.

Judis, J. B. 2016. *The populist explosion: How the great recession transformed American and European politics.* New York: Columbia Global Reports.

Katz, C. 2001. On the grounds of globalization: A topography for feminist political engagement. *Signs: Journal of Women in Culture and Society* 26 (4):1213–34. https://doi.org/10.1086/495653.

Kramer, D. 1956. *The wild jackasses: The American farmer in revolt.* New York: Hastings House.

Laclau, E. 2005. *On populist reason.* London: Verso.

LaDuke, W., F. Brown, N. Kennedy, T. Reed, L. Warner, and A. Keller. 2010. *Sustainable tribal economies: A guide to restoring energy and food to native America.* Minneapolis, MN: Honor the Earth.

Larsen, S. C., and J. T. Johnson. 2017. *Being together in place: Indigenous coexistence in a more than human world.* Minneapolis: University of Minnesota Press.

Le Heron, R., and N. Lewis. 2009. Discussion: Theorising food regimes. Intervention as politics. *Agriculture and Human Values* 26 (4):345.

Lindeke, B. 2015. The Seward Friendship Store sparks return of the co-op war. *Twin Cities Daily Planet,* July 10. Accessed October 30, 2018. https://www.tcdailyplanet.net/the-seward-friendship-store-sparks-return-of-the-co-op-war/.

Luoma, E. 1967. *The farmer takes a holiday.* New York: Exposition Press.

Marquis, S. L. 2017. *I am not a tractor!: How Florida farmworkers took on the fast food giants and won.* Ithaca, NY: Cornell University Press.

Martin, M. 2014. Exploring agricultural values: A workshop on different agricultural values for college students who are conducting agricultural literacy activities. *NACTA Journal* 58 (3). Accessed October 30, 2018. https://www.nactateachers.org/images/Sep14_3_Exploring_Agricultural_Values.pdf

Massey, D. B. 1994. *Space, place, and gender.* Minneapolis: University of Minnesota Press.

———. 2005. *For space.* London: Sage.

McClintock, N. 2018. Cultivating (a) sustainability capital: Urban agriculture, eco-gentrification, and the uneven valorization of social reproduction. *Annals of the American Association of Geographers* 108 (2):579–90.

McMath, R. C. 1990. *American populism: A social history 1877–1898.* New York: Hill and Wang.

———. 1995. Populism in two countries: Agrarian protest in the Great Plains and Prairie provinces. *Agricultural History* 69 (4):517–46.

McMichael, P. 2009. A food regime genealogy. *The Journal of Peasant Studies* 36 (1):139–69.

Meek, D., and R. Tarlau. 2015. Critical food systems education and the question of race. *Journal of Agriculture, Food Systems, and Community Development* 5 (4):1–135.

Méndez, V. E., C. M. Bacon, and R. Cohen. 2013. Agroecology as a transdisciplinary, participatory, and action-oriented approach. *Agroecology and Sustainable Food Systems* 37 (1):3–18.

Merkle, B. G. 2013. Whole measures only partially measures up. *Journal of Agriculture, Food Systems, and Community Development* 3 (3):187.

Minnesota Department of Agriculture. 2018. AGRI Urban Agriculture Grant RFP. Accessed October 30, 2018. https://www.mda.state.mn.us/grants/grants/urbanaggrant.

Moffitt, B. 2015. How to perform crisis: A model for understanding the key role of crisis in contemporary populism. *Government and Opposition* 50 (2):189–217.

Montenegro de Wit, M., and A. Iles. 2016. Toward thick legitimacy: Creating a web of legitimacy for agroecology. *Elementa: Science of the Anthropocene* 4: 000115. doi: 10.12952/journal.elementa.000115.

Mounk, Y. 2017. European disunion: What the rise of populist movements means for democracy. *New Republic,* July 19. Accessed October 30, 2018. https://newrepublic.com/article/143604/european-disunion-rise-populist-movements-means-democracy.

Müller, J. 2016. *What is populism?* Philadelphia, PA: University of Pennsylvania Press.

Murray, N. 2018. Agriculture wars. *Viewpoint Magazine,* March 12. Accessed October 30, 2018. https://www.viewpointmag.com/2018/03/12/agriculture-wars/.

Nagar, R., and F. Ali. 2003. Collaboration across borders: Moving beyond positionality. *Singapore Journal of Tropical Geography* 24 (3):356–72. doi:10.1111/1467-9493.00164.

Nass, D. L. 1984. *Holiday: Minnesotans remember the Farmers' Holiday Association*. Marshall, MN: Plains Press.

Orfield, M., and W. Stancil. 2017. Why are the Twin Cities so segregated. *Mitchell Hamline Law Review* 43:1.

Patel, R., and J. W. Moore. 2017. *A history of the world in seven cheap things: A guide to capitalism, nature, and the future of the planet*. Berkeley: University of California Press.

Phat Beets Produce. 2012. Neighbors outing blatant exploitation. Accessed October 30, 2018. https://youtu.be/JneAYdmDGCE.

Phillips, K. P. 1982. *Post-conservative America: People, politics, ideology*. New York: Random House.

Postel, C. 2007. *The populist vision*. Oxford, UK: Oxford University Press.

Pratt, G., and B. Yeoh. 2003. Transnational (counter) topographies. *Gender, Place & Culture* 10 (2):159–66. doi:10.1080/0966369032000079541.

Project Sweetie Pie. 2015. *Urban agriculture & legislation: The future of urban farming is in our hands*. Council on Black Minnesotans' Statewide Coalition 2015 Legislative Agenda Discussion. Accessed October 30, 2018. http://projectsweetiepie.org/event/urban-agriculture-legislation-future-urban-farming-hands/.

Racial Equity Tools. 2016. 2015–2016 Minnesota legislative report card on racial equity. Voices for Racial Justice. Accessed October 30, 2018. http://voicesforracialjustice.org/wp-content/uploads/2016/07/VRJ_ReportCard_2016_final_4web.pdf.

Reynolds, K., and N. Cohen. 2016. *Beyond the kale: Urban agriculture and social justice activism in New York City*. Athens: University of Georgia Press.

Robinson, A., and S. Tormey. 2009. Resisting "global justice": Disrupting the colonial "emancipatory" logic of the West. *Third World Quarterly* 30 (8):1395–1409.

Rochester, A. 1940. *Why farmers are poor: The agricultural crisis in the United States*. New York: International.

Rosset, P. M. 2014. Agroecology and social movements. In *The global food system: Issues and solutions*, ed. W. D. Schanbacher, 191–210. Santa Barbara, CA: Praeger.

Rosset, P. M., and M. E. Martínez-Torres. 2012. Rural social movements and agroecology: Context, theory, and process. *Ecology and Society* 17 (3):17.

Saloutos, T., and J. D. Hicks. 1951. *Twentieth-century populism: Agricultural discontent in the Middle West 1900–1939*. Lincoln: University of Nebraska Press.

Sbicca, J. 2014. The need to feed: Urban metabolic struggles of actually existing radical projects. *Critical Sociology* 40 (6):817–34.

Schneider, M., and P. McMichael. 2010. Deepening, and repairing, the metabolic rift. *The Journal of Peasant Studies* 37 (3):461–84.

Shea, P. 2013. *Small-scale food initiatives in southwest Minnesota*. University of Minnesota Institute for Advanced Study. Accessed October 30, 2018. https://umedia.lib.umn.edu/taxonomy/term/845

Shover, J. L. 1965. *Cornbelt rebellion: The Farmers' Holiday Association*. Urbana: University of Illinois Press.

Slocum, R. 2007. Whiteness, space and alternative food practice. *Geoforum* 38 (3):520–33.

Slocum, R., K. V. Cadieux, and R. Blumberg. 2016. Solidarity, space, and race: Toward geographies of agrifood justice. *Spatial Justice* 9. http://www.jssj.org/article/solidarite-espace-et-race-vers-des-geographies-de-la-justice-alimentaire/

Slocum, R., J. Shannon, K. V. Cadieux, and M. Beckman. 2011. Properly, with love, from scratch: Jamie Oliver's food revolution. *Radical History Review* 2011 (110):178–91.

Snipstal, B. 2013. *Food sovereignty: A critical dialogue*. Accessed October 30, 2018. https://www.youtube.com/watch?v=6ErTCGbggdk.

Thompson, C. D., and M. F. Wiggins. 2009. *The human cost of food: Farm workers' lives, labor, and advocacy*. Austin: University of Texas Press.

Timmermann, C., and G. F. Félix. 2015. Agroecology as a vehicle for contributive justice. *Agriculture and Human Values* 32 (3):523–38.

Ullekh, N. P. 2013. Food populism: Raman Singh offers "nutritional security" to beat Congress' "food security." *The Economic Times*, November 5. Accessed October 30, 2018. https://economictimes.indiatimes.com/articleshow/25278725.cms.

Union of Concerned Scientists, Center for Science and Democracy. 2014. *Science, democracy, and a healthy food policy: How citizens, scientists, and public health advocates can partner to forge a better future*. Minneapolis, MN: Union of Concerned Scientists, Center for Science and Democracy.

———. 2015. *Food equity and justice: Scientist-community partnerships meeting*. Minneapolis MN: Union of Concerned Scientists, Center for Science and Democracy.

Upadhyay, B., B. T. Chaudhary, D. Gautam, and B. Tharu. 2019. Maghi: A case study of indigenous Tharu cultural heritage for democratic practice and STEM education in Nepal. In *Indigenous perspectives on sacred natural sites: Culture, governance and conservation*, ed. J. Liljeblad and B. Verschuuren. London and New York: Routledge.

Upadhyay, B., G. Maruyama, and N. Albrecht. 2017. Taking an active stance: How urban elementary students connect sociocultural experiences in learning science. *International Journal of Science Education* 39:2528–47.

Value Walk. 2018. These are the 10 property investment opportunities you need to know in 2019 (Minneapolis is #6), Accessed October 30, 2018. https://www.valuewalk.com/2018/09/property-investment-opportunities-2019/.

Van Sant, L., and K. Bosworth. 2017. Race, rurality, and radical geography in the US. *Antipode Interventions*, September 14. Accessed October 30, 2018. https://antipodefoundation.org/2017/09/14/race-rurality-and-radical-geography/.

Varghese, S., and K. Hansen-Kuh. 2013. *Scaling up agroecology: Toward the realization of the right to food*. Minneapolis, MN: Institute for Agriculture and Trade Policy.

Vía Campesina. 2001. Our world is not for sale. Priority to people's food sovereignty. Accessed October 30, 2018. https://viacampesina.org/en/peoples-food-sovereignty-wto-out-of-agriculture/.

Wark, M. 2015. Inventing the future (review). *Public Seminar*, October 27. Accessed October 30, 2018. http://www.publicseminar.org/2015/10/inventing-the-future/.

Wezel, A., S. Bellon, T. Doré, C. Francis, D. Vallod, and C. David. 2011. Agroecology as a science, a movement and a practice. In *Sustainable agriculture*, ed. E. Lichtfouse, M. Hamelin, M. Navarrete, and P. Debaeke. Vol. 2, 27–43. Amsterdam, The Netherlands: Springer.

White, M. M. 2011. Sisters of the soil: Urban gardening as resistance in Detroit. *Race/Ethnicity: Multidisciplinary Global Contexts* 5 (1):13–28.

Whyte, K. P. 2013. On the role of traditional ecological knowledge as a collaborative concept: A philosophical study. *Ecological Processes* 2 (7):1–12.

Williams, J., and E. Holt-Gimenez. 2017. *Land justice: Reimagining land, food, and the commons*. Oakland, CA: Food First Books.

Wittman, H. 2009. Reworking the metabolic rift: La Vía Campesina, agrarian citizenship and food sovereignty. *The Journal of Peasant Studies* 36 (4):805–26.

KIRSTEN VALENTINE CADIEUX is Director of Environmental Studies and Assistant Professor in the Anthropology Department at Hamline University, St. Paul, MN 55104. E-mail: kvcad@post.harvard.edu. Using art and science approaches to society–environment relations and specifically the political ecology and moral economy of agrifood systems, she builds publicly engaged participatory research and exploration processes for people to learn together about differing ways of understanding and valuing environments and food systems in collaborative ways.

STEPHEN CARPENTER is Deputy Director and Senior Staff Attorney of Farmers' Legal Action Group, Inc. (FLAG), St. Paul, MN 55102. E-mail: jstephencarpenter@gmail.com. His work at FLAG has centered on discrimination in agricultural lending, debtor–creditor issues, disaster assistance, federal farm programs, sustainable agriculture and direct marketing, and the problems of farmers contracting for livestock production. He served as Senior Counsel in the Office of the Monitor in the *Pigford* case and is at present the court-appointed ombudsman for the *In re Black Farmers Discrimination* case. He has been an adjunct Assistant Professor at the University of Minnesota Law School.

ALEX LIEBMAN is a researcher of plant–soil dynamics for Lurralde, a Chilean team of activists and scientists supporting Atacameño and Ayamaran groups in their struggle for territorial sovereignty and water rights against multinational mining companies in the Atacama Desert, Seattle, WA 98104. E-mail: alexliebman@gmail.com. He holds an MSc in agronomy from the University of Minnesota and a BA from Macalester College.

RENATA BLUMBERG is an Assistant Professor in the Department of Nutrition and Food Studies at Montclair State University, Montclair, NJ 07043. E-mail: blumbergr@montclair.edu. She conducts research on alternative food networks in the United States and Eastern Europe, embodied geographies of food, critical management studies on institutional food systems, and feminist agroecologies and pedagogies.

BHASKAR UPADHYAY is Associate Professor of STEM Curriculum and Instruction at the University of Minnesota, Minneapolis, MN 55414. E-mail: bhaskar@umn.edu. His areas of interest include science as an agent for change in communities; empowerment and social justice for urban minority youth; the intersection of science and sociocultural habits in urban school settings; issues of science learning for students from immigrant families and roles of immigrant parents in their children's science learning; and social justice and equity in the way that science, food, and the environment are taught.

Development and Sustainable Ethics in Fanjingshan National Nature Reserve, China

Stuart C. Aitken, Li An, and Shuang Yang

In March 2013, several thousand delegates at China's National People's Congress voted to approve the environmentally sensitive and authoritarian Xi Jinping as president. This portended dramatic changes in environmental policies, not least of which was an offsetting of top-down development-at-all-costs dogma with a new official orthodoxy focused on a sustainable and circular economy, with inclusive and more rounded growth. This article is part of a long-term project (2008–2018) in Fanjingshan National Nature Reserve in Guizhou Province that took place as the political scene in Beijing shifted. The larger project is about human–environment dynamics and complexities focusing on the preservation of snub-nosed golden monkey habitat and the implementation of top-down grain-to-green and national forest conservation programs. This article is about the contexts of two development projects, one in the reserve and one just outside of it, with very different outcomes. Drawing on the work of Arturo Escobar, Rosi Braidotti, and Xiaobo Su, we argue for development in a time and place of rapid change as if marginalized farmers and their families mattered and the possibility of sustainable ethics with a locatable politics. The article elaborates the potency of this kind of sustainability through the stories of families living on Fanjingshan Reserve in the midst of (1) authoritarian environmental policy proclamations from Beijing and (2) boisterous local development.

2013年三月举行的中国全国人民代表大会中，数千名代表投票赞成对环境敏感的威权主义者习近平作为国家领导人。此一在环境政策上预兆式的戏剧性转变，本身不仅是以聚焦可持续的循环经济之崭新官方正统，抵销由上而下不惜一切代价的发展主义信念，同时具有包容且更为全面的成长。本文是在北京的政治场景变迁下进行的贵州省梵淨山国家自然保护区的长期计画（2008至2018年）的一部分。更大规模的计画，则关乎人类—自然动态和聚焦白鼻金丝猴栖地保育的复杂性，以及执行由上而下的退耕还林与国家森林保育计画。本文关乎结果互异的两大发展计画脉络，一个在保育区中，一个则正好落在保育区之外。运用艾斯柯巴、布莱多蒂，以及苏晓波的研究，我们支持受到边缘化的农民及其家庭在巨变的时地中好似重要的发展，以及可定位其政治的可持续性伦理之潜能。本文通过在（1）北京的威权环境政策宣言和（2）勐烈的地方发展期间，居住于梵淨山保护区的家庭故事，阐述此般可持续性的效力。关键词：中国，发展，可持续性。

En marzo de 2013, varios miles de delegados al Congreso Nacional del Pueblo de China votaron para consagrar como presidente al ambientalmente sensible y autoritario Xi Jinping. Esta decisión presagiaba cambios dramáticos en las políticas ambientales, el no menor de los cuales era la compensación del dogma de desarrollo impuesto a toda costa desde arriba como una nueva ortodoxia oficial centrada en economía circular sustentable, con un crecimiento incluyente y más robusto. Este artículo hace parte de un proyecto a largo término (2008–2018) en la Reserva Natural Nacional de Fanjingshan, Provincia de Guizhou, proyecto que se desarrolló a medida que cambiaba la escena política en Beijing. El proyecto de mayor amplitud es acerca de las dinámicas y complejidades humano–ambientales enfocadas hacia la preservación del hábitat del mono dorado de nariz chata y la implementación desde lo alto de programas nacionales de paso del grano a lo verde y la conservación de bosques. El artículo es acerca de los contextos de dos proyectos de desarrollo, uno en la reserva y otro justo afuera de ésta, con resultados muy diferentes. A partir de los trabajos de Arturo Escobar, Rosi Braidotti y Xiaobo Su, discutimos el desarrollo en un tiempo y lugar de rápido cambio, como si los agricultores marginados y sus familias importaran algo, y la posibilidad de una ética sustentable con una política localizable. El artículo elabora la potencia de este tipo de sustentabilidad por medio de historias de familias que viven en la Reserva Fanjingshan en medio de (1) la proclamación de políticas ambientales autoritarias de Beijing y (2) el estrepitoso desarrollo local. *Palabras clave: China, desarrollo, sustentabilidad.*

Figure 1. "I don't know what I should draw, and so I drew the house of my childhood. There were some small trees before. There was a river in front of the house. In the river there were small fishes. When I was a child, I liked catching fish and going swimming" (Yangyang, fourteen years old, focus group discussion, May 2010). (Color figure available online.)

We begin this article with sketches and quotes from a fourteen-year-old male student (Figure 1) and a fifteen-year-old female student (Figure 2) from Jiangkou High School, near the administrative headquarters of Fanjingshan National Nature Reserve (FNNR). Jiangkou is in the northwest corner of Guizhou Province, an area that has experienced restructuring and growth in connection to China's economic renewal (Figure 3). The young people's representations reflect well the rapid environmental changes in relatively poor peripheral areas wrought by top-down authoritarian development policies focused on a rhetoric of "grow first, clean up later" (Rock and Angel 2007). Years of this rhetoric spurred rapid and often vacuous development (cf. Shepard 2015) and ambitious rural conservation programs. Beijing launched a payment for ecosystem services (PES) project, the National Forest Conservation Program (NFCP), in 1998, seeking to reduce logging and promote afforestation through incentives paid to forest enterprises and users. One year later, another large-scale, top-down PES project, the Grain-to-Green Program (GTGP),

provided farmers with grain and cash subsidies to convert cropland on steep slopes to forestland or grassland. As two of the largest PES programs in China and in the world, the NFCP and GTGP are now implemented in twenty provinces, autonomous regions, and municipalities, generating ecological (e.g., soil erosion, droughts, floods) and socioeconomic (e.g., poverty alleviation, social development) benefits at the national and international scales, but this also comes with considerable local tensions (Liu et al. 2008; State Forestry Administration of China 2017). The students whose words and drawings begin this article were part of a larger multiyear project (2008–2018) that sought to understand the implications for families living in or close to the FNNR of environmental and policy changes emanating from larger scales of Chinese government (Aitken and An 2012; Aitken et al. 2014; Aitken et al. 2016). The project primarily involved interviews with members of farming families but also FNNR officers and administrators and communist party officials. In addition, we organized focus groups for students from local schools. Some of the emphases of the project changed as it progressed and we learned about

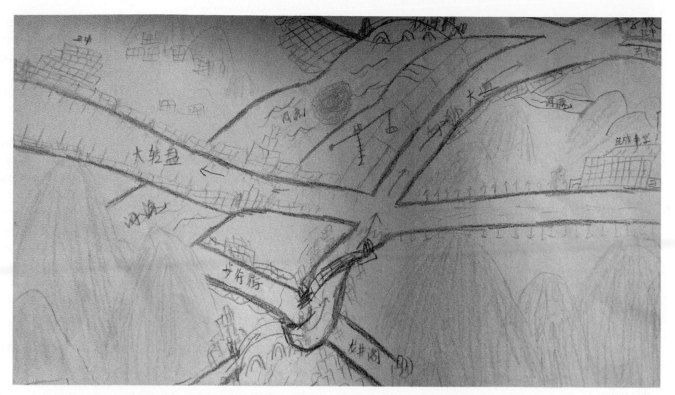

Figure 2. "I painted a road. There were a lot of trees before. Now we cannot see those dense trees, because they were all moved for the roads. The houses become higher than before" (Xiuxiu, fifteen years old, focus group discussion, March 2015). (Color figure available online.)

what was going on in FNNR, but our enduring concern was regarding local attitudes and resistances to environmental and economic change and policy reforms.

The project from which this article derives began with concerns about FNNR's endangered snub-nosed monkey species, *Rhinopithecus brelichi*. The golden monkey inhabited the higher reaches of the reserve and we were at first curious about the impact of indigenous farming practices on the elusive creature and how protection policies influenced farmers and their families. It was clear that implementation of the NFCP and GTGP was widespread in the area, so our interest broadened from indigenous farming to the impact of these national policies. This article speaks to the complexities of local and national development and environmental programs inside and on the margins of FNNR using the NFCP and GTGP as a springboard to facilitate discussion of top-down infrastructural changes and local tourist development initiatives. It is evident that the PES programs are important for county administrators and local farmers and that growth in the area spurs other kinds of development. Two specific events that were unanticipated when we started the project

guide the article's empirical discussion. In 2008, we visited the village of Zhangjiaba, situated in a remote northwestern part of the reserve, to discuss with local farmers the impact of the NFCP and GTGP on farming practices. In 2011, a ¥630 million hydroelectric dam project, funded mostly by the Yinjian County Government, was approved and building began in 2015. At completion, the dam displaced 432 people. We visited Zhangjiaba several times to talk with villagers and officials in charge of the relocations. The second event occurred just outside of the FNNR and involved the building of an internationally financed golf course. The development required the forced relocation of a village housing about 100 people. The international investment firm agreed to build a new village for the displaced people. Our discussion with farmers and officials shed light on how these developments clashed and reconciled with local lived worlds.

The question of how well national and international development goals serve local people in and around FNNR is complex in terms of human–environment relations, equity, and geography. To help with these issues, we first engage theoretically with Ecobar's (2001, 2008) ideas about how local spaces

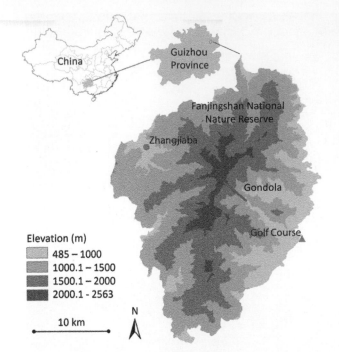

Figure 3. Fangjinshan National Nature Reserve, Guizhou Province. *Source:* Cindy (Yu Hsin) Tsai. (Color figure available online.)

are influenced by changing national and global conditions, particularly his "figured worlds" in which local practices, cultures, and identities are deployed effectively to create visibility (spontaneous, emotional, and corporeal) and what he called a defense of place. Second, we engage Su's (2012a, 2012b, 2013, 2015; see also Su and Teo 2009) theoretically sophisticated and empirically detailed work on tourist-driven development in peripheral China, which uses Escobar (2001), at least in part, to elaborate local and national reconceptualization and reterritorializations. In particular, Su (2012a) showed how people in heretofore marginalized rural areas "ground their culture in everyday life, even though they participate in trans-local networks" (32). Although Su articulated several downsides to this, we argue that this is precisely the promise of Escobar's figured worlds, which leads to a third theoretical engagement with the idea of capacity building in the sense that Braidotti (2006, 2013) meant when she argued for sustainable ethics. Her form of sustainability is not about curbing growth economics (Rees 2001), "too many people using up too much stuff" (Maskit 2009, 129), meeting the needs of the present without compromising the future, or sacrificing one area's potential for the sake of another's (Brundtland Report 1987); rather, it is about ethics that espouse

the virtue of living to a fuller potential right here and right now through politics that are locatable. In particular—and related to Su's (2012a) caution that "[t]oo much emphasis on resistance can create a frame of mind that brings endless turmoil and relentlessness to everyday life" and his admonition "that the real world cannot be reduced to the ramifications of external forces 'out there'" (33)—we argue that sustainable ethics must be derived from a complex process of negotiation and reconciliation among forces "there" and "here." Drawing from Su's (2012a, 2012b, 2013, 2015) understanding of scale, and while recognizing his concerns about strained relations between local people and county officials (see also Liu et al. 2008; Liu and Yang 2013; Guo and Liang 2017), we see some hope in the day-to-day and here-and-now for FNNR farmers.

Figuring Peripheral Chinese Development and a National Nature Reserve

The FNNR was established as a national nature reserve in 1978 to mediate environmental policies and manage the resource use relations between local people and the snub-nosed golden monkey. The year coincides with the beginning of dramatic post-Maoist changes when Deng Xiaoping launched a grand series of economic reforms focused primarily on the creation of coastal economic zones (Su 2012a). Over the next several decades, domestic demand increased and a consumer culture evolved with attendant increased mobility. The household residential registration system (*hukou*) was relaxed, enabling millions of peripheral farmers to move legally to the coastal economic zones for work.

Su (2013) pointed out that although initial development focused on the coast, there were, in time, interesting and complex repercussions for peripheral areas. Focusing on the "Ancient Town" of Lijiang, which was designated as one of Yunnan Province's main cultural heritage sites, he noted that peripheral places in "comfortable natural surroundings" became attractive to people from coastal areas who felt the loss of tradition and older ties to the land (Su 2013, 2015). As a national nature reserve, situated in one of China's primary Buddhist heritage sites, FNNR is attractive for reasons similar to those elaborated by Su but perhaps on a grander scale, one that incorporates the reserve itself, a host of peripheral tourist areas, ancient Buddhist cultural sites, and a good climate, as well as a comfortable natural rural setting. These attractions (and attendant education programs) suggest, to tourists and locals alike, the importance of the preservation of nature, protection of endangered species, and the creation of a harmonious rural idyll. The irony of the creation of an imagined rural idyll is that state-mandated programs such as NFCP and GTGP are contrary to indigenous agricultural practices and, further, that tourism creates market forces that reconstruct natural landscapes to maximize profit rather than create comfort. Nonetheless, rhetoric from Beijing increasingly spoke to local autonomy, environmental concerns, and conservation of nature.

Starting in 2011, China's 12th Five-Year Plan marked a moment that challenged top-down national orthodoxy as part of official rhetoric. Provinces were gaining more autonomy through what Ong (2006) called graduated sovereignty and Su (2012b) called rescaling with the implementation of a less centralized political economy. At the National People's Congress in March 2013, several thousand delegates voted to approve as president the environmentally sensitive but nonetheless authoritarian Xi Jinping, who now promotes the fulfillment of the Chinese dream (中国梦) through commodification and profit making as well as environmental sensitivity (Su 2015). Consequently, a new official orthodoxy was adopted, one of "sustainability and circular economy, of inclusion and more rounded growth" (Hilton 2013, 12). Changes in attitudes toward development at the highest level of Chinese government portended fundamental changes at the local level, especially in fringe areas with potential for rapid change. Accompanying the 12th Five-Year Plan was a commitment to more autonomy for local areas in the use of development and infrastructural monies and tightened assurances about environmental protections in nature reserves.

Although the 12th Five-Year Plan shifted the rhetoric in favor of careful environmental management, several influential, externally funded tourist and infrastructure development projects were already underway in and around the FNNR. The irony of the proclamations emanating from Beijing was not lost on us as we talked to family members whose ancestry and intimate connections to nature through subsistence farming in this area dated back hundreds of years. Our larger project's goal was to understand the reciprocal interactions between national development and environmental programs and the associated coupled human and natural systems over space and time. This article, drawing on the results from the household surveys of the larger project but focused mostly on interviews, speaks to the complexities of local and national development and environmental programs inside and on the margins of the FNNR.[1] By using the NFCP and GTGP as catalysts to talk to farmers about complexity, we focused discussion on top-down infrastructural changes and local tourist development initiatives. Nearly 16,000 people live within or close to FNNR. Our surveys suggested that strong values of education, self-sufficiency, and family are present in the area. Family members share views on protecting species; village elders pass onto their children what they think are important environment values. These views and principles create relationships within the region that suggest a recognition of habitat loss, dangers from flooding, and needs for preservation and ways those might relate to the practicalities of resource use but also development.

Just before we arrived to begin our 2008 pilot work, a Chinese company, using Austrian cable car technology and financing, completed an extensive aerial gondola system to transport tourists to the top of Golden Peak, one of the highest mountains in the reserve that boasts ancient temples atop craggy limestone peaks (Figure 3). Rapid development has since occurred adjacent to the visitor center at the base of the gondola. By 2013, the number of visitors to Golden Peak increased to 100,000 people annually, and by 2016 this had risen to 570,000 people annually. How well development projects such as this serve local people in and around the FNNR is complex from the perspective of spatial justice and social equity. To take another example, the amount of GTGP and NFCP compensation actually received by local farmers in the FNNR varies from village to village and is dependent on what other kind of development is ongoing, such as the gondola. Farmers in some villages received a portion of the total possible compensation (even none in extreme situations), and local leaders were able to divert the money to other purposes such as road construction (Aitken et al. 2016).

Road construction and infrastructure development enable increased access to remote areas. Part of a nationally funded multi-million-dollar superhighway system, which stretches from Shanghai to the Myanmar border, is under construction through the main north–south valley in the area just to the east

Figure 4. Zhangjiaba Valley before flooding (with an elevated access road linking to the superhighway being built in the background).
Source: Photo by Stuart C. Aitken.

of the reserve, with ancillary roads built from PES monies. The superhighway makes the region substantially more accessible to megacities elsewhere in China, but the expense of tolls on the road makes it inaccessible to most locals.[2] A palpable tension in the area, then, brings together FNNR's focus on nature preservation and long-term sustainability, the kind of fast-paced development afforded by the superhighway, and what, precisely, benefits local farmers. These tensions highlight the old "grow first, clean up later" mantra, and it is not entirely clear how the new environmental sensitivity of 12th Five-Year Plan trickles down to the local level through these effects. Clearly, there are significant issues of scale at work here that result in tensions between local well-being and specific developments (cf. Su 2012b), as evidenced by the dam and golf course projects.

The Dam

We first visited Zhangjiaba during our 2008 pilot work and then again in 2013, 2015, and 2016. The villagers we talked to in 2008 and 2013 told us that they had never seen a golden monkey in the wild but testified to the nuisance of wild boar that they

could no longer shoot because of the FNNR's ban on hunting. Some had given up their peripheral land to work areas closer to the village, which were easier to protect. Now they faced displacement by the dam and reservoir. In 2015, there were 106 households in the village, and 96 of them owned adjacent farmland. A total of 432 people were scheduled for relocation, of whom 386 were farmers. In 2014, the working group (a county negotiation team living in the village with the job of requisitioning the lands and persuading villagers to sign a relocation contract) had signed 55 percent of the people to a relocation contract. When we met with one of the working group members in 2015, he was unequivocal about the positive well-being of the villagers moving forward (Figure 4):

> The County government carries out an incentive policy [with] compensation. ... If a new house foundation is done by August 31, 2015 the house owner will get a cash award of ¥10,000 (~US$1,500). ... In addition, if a peasant moves out from the reservoir, he or she will get compensation of ¥600 per year, lasting twenty years.

Later, we sat with some villagers and had tea. The ones we spoke to said that many villagers were holding out for ¥ 1,200 per year in compensation. One man had moved to the town ten years previously, from up in the mountains, to be closer to his farmland. He told us that he had seen the golden monkey only once when he lived up in the mountains. For the last three years, all he had planted was tea trees on the slopes of the valley closer to the village. When we asked him whether this was part of the NFCP, he said that he used to be part of that but the incentive for tea comes from the local government and it is better. He gets ¥ 150 per Mu (a basic unit of land in China that varies by location but is usually around 0.16 acres) per year from the local government as opposed to ¥ 200 to ¥ 400 from the PES for his 10 Mu of land but, unlike the evergreen trees he planted for the NFCP, he can sell tea to the local tea factory at Tuanlong and make more money. He is nonetheless concerned that when the town is flooded and he relocates, it will be more difficult to get to his land. He is further concerned that the resettlement houses are selling at ¥ 2,000/m^2, whereas he is only getting ¥ 792/m^2 compensation for his current house. We asked how he might get more:

> We hold off … my house is brick and on concrete, it costs more … my family has lived here for generations … we are emotionally tied to this land, from our hearts, we do not want to move.

A large part of the figuring in Escobar's worldview relates to indigenous connections to place. Many of the families in Zhangjiaba had occupied the area for more than 300 years. Indigenous family ties are rooted vertically in place through time but they also grow horizontally in what Escobar called *redes*—a network of ties. These ties expand horizontally as an increasing number of family members migrate, for the most part temporarily, to work in coastal cities. Another important part of *redes* is hierarchical connections to other groups (e.g., nongovernmental organizations, local Communist Party groups, FNNR officials, as well as the working group). Our tea group told us that the villagers negotiated prices on an ad hoc basis with the working group, which is an ongoing negotiation:

> There were some meetings last year, but the prices were not changed after every meeting. So we wait. At the beginning, the government solicited the opinions about where we were going; they wanted us to move to Yongyi, but no one wants to live there.

Redes connections are fluid—growing, evolving, breaking, and recoupling elsewhere—depending on local and external pressures, and some are more powerful and better resourced than others. We asked about the process of relocation, and another man told us that the village was not too well organized:

> No meeting this year, it is a bit of a problem. The way they negotiate is the Working Group talking to every family one by one, asked them to sign and move. … Every three to five days they come to us.

Members of the tea group disagreed about the merits of moving to the working group's favored site of Yongyi, but they agreed that the compensation for their current houses was far too low. They said that it was not easy to "do business" with the working group and that they needed to organize and not accept compensation on a piecemeal basis. Escobar (2008) pointed out that figured worlds have growing pains and many find form through resistance, but nothing but resistance might ultimately enervate local practices and so it is important to understand the positive implications of place-based complexities. Moving to Yongyi is a difficult proposition for our tea group, but the prospect of going there raises interesting place-based complexities. We asked our tea group whether Yongyi was still in the reserve, and they said yes, so we went on to ask them about how the reserve treats them:

> It is okay. Their policies are better [than those of Beijing]. For example, the GTGP and NFCP are good as is the tea subsidy but there is no compensation if we live outside the reserve, so probably we'll stay.

The issue of connectivity continues in a related topic of conversation. They all agreed that the current biggest problem in the FNNR is poor access due to bad roads. Said a woman in the tea group:

> The tourists are much fewer in the recent two years because the roads are in bad condition. Our inn had no vacancies in summers before the roads changed, but now very few visitors come here. The roads are in such a bad shape, who would come?

The issue of local road access in the FNNR is part of a larger context of accessibility (see Aitken et al. 2016) for tourism and temporary migration. The Chinese government's past long-standing policy of restricting migration from rural to urban areas contributed to a large urban–rural income gap, which provided a tremendous incentive to migrate to urban areas (Zhao 1999). This incentive lures more and more capable and skilled rural laborers to big cities where there are higher salaried jobs. Many people in the FNNR sought better paid work and living conditions in the cities after the reduction of *hukou* residential registration restrictions.[3] Migration out of the area has quadrupled since 2000. All of the villagers in our tea group had migrated for factory work in Guangzhou, which is a fourteen-hour train ride from the FNNR, at some time or other in the past. This *hukou* policy reform is important but, as Guo and Liang (2017) demonstrated, local effects are unpredictable, so it is not easy to gauge the importance of the policy change. They noted that different degrees of citizenship for rural migrants depend on where migrants go and whether the city establishes a point system for attaining services. Guo and Liang went on to point out that the impact of *hukou* reforms seems limited, favoring more highly qualified and better educated migrants. Migrants from the FNNR go to all of the major coastal cities, so it is difficult to assess their differential ties to those places, but it is nonetheless clear that, as noted by Guo and Liang (2017), "education opportunities for their children was one of top priorities" (777). Further, it is clear that education is challenging past roles in Fanjingshan (Aitken et al. 2014). The remittances generated by migrant labor help with children's education, but the extra income might not offset the costs arising from the lost labor on farms.

Members of our tea party had stories of working in Guangzhou. Two of the women worked in a factory producing cameras to make money for their children's education, and their husbands worked in construction. To earn money for their house building in Zhangjiaba, they looked after and fed 150 pigs that were located near the factory. The women laughed:

> At the beginning, we only fed ten, and later the number increased gradually, up to seventy to eighty pigs; their fodder was the hogwash of kitchens in the factory.

The women who were part of the tea party felt that focusing on children's education not only improves their lives but also empowers families by equalizing gender and generational standings. By staying alternately with grandparents and boarding schools for much of the year and reuniting with parents at family holidays, children we talked to in the village told us that they learned the importance of extended family ties. Migration and boarding schools expanded and equalized women's and children's roles

through broadening horizons and redefining duties (Aitken et al. 2014). The broadening of local figured worlds through migration and education suggests a limitation of Escobar's defense of place, which focuses mostly on indigenous autonomy and bottom-up politics. Braidotti (2013) argued for a locatable politics that relates to Escobar's *redes* but takes it further with potentials and capacities that require an understanding of relations that are not all tied to propinquity. The tea party group all returned to the FNNR because they had no *hukou* in Guangzhou, which did not extend to them the potential for getting services through a point system, but the local government was exemplary, said one woman, in helping them remain in the FNNR:

> Without the *hukou*, it is still difficult staying there [in Guangzhou]. … The [county] government had some compensation for us [to return here] for the decoration of houses in order to create tourist spots; for example, outer wall, door and window and paintings.

When we last visited Zhangjiaba Valley in spring 2017, the dam was completed but the reservoir was not filled because more than one third of the villagers were still holding out for more compensation. We heard that the reservoir started to fill in January 2018 and that most of the villagers have moved to Yongyi and other villages in the reserve, after receiving better compensation. At the time of writing, part of the village was not yet flooded, and there were still homes standing with villagers living in them, apparently holding out for more compensation. The local roads are improved, because of the dam, increasing opportunities for tourism. The Zhangjiaba example suggests that local farmers work well with FNNR (and sometimes county) officials and gain tenacity through perceived negotiating power and hopes engendered by temporary migration and education.

The golf course example is about a development that occurred outside of the reserve using foreign investment dollars, and the outcome suggests limited negotiating power for the villagers involved.

The Golf Course

During our 2013 fieldwork, we witnessed the destruction of dozens of homes to make way for the golf course adjacent to the reserve boundary. We were told that the golf course was developed

Figure 5. New houses for villagers displaced by the golf course. *Source:* Photo by Li An.

by the Jingjia Group, a Chinese company backed by international investment dollars. A development of this kind could not happen on the reserve because of environmental planning restrictions. The involvement of international investment made it much more difficult to get information on this project, and no government official was willing to go on record about it. We do know that more than 100 people were displaced, but as the golf course neared completion, advocates for the Central Government's 12th Five-Year Plan raised issues about the land-use and aesthetic compatibility of a golf course so close to the FNNR and county permission for the development was withdrawn in 2014. The international investment firm then withdrew its financing. Some of the displaced residents moved into the half-completed new village, with no idea of when or how the project would be finished (Figure 5). The pace of this development when compared to the dam (the golf clubhouse was completed in less than a year) and the lack of liaison with villagers suggest an important distinction when local control is forfeited for

external investment monies. Escobar (2008) talked about distinctions between negotiations tied to local concerns, involving local people, and those that occur at a different scale before landing pell-mell on locals. We talked with some of the displaced villagers to get a sense of how they felt about the golf course development and their displacement. Said one woman:

> They took our land, but they paid us back the same square meters that we occupied. [My old house] is in that place in the golf field over there. … [It is gone. Our farmland] was right next to the old house.

There is a sense of loss as she described her lost land and old lifestyle. We asked them how their lives were before the houses were torn down and how things had changed:

> We planted rice and vegetables and so on. Now all we do is work on some construction sites.

The demolished village had houses built from local wood in the traditional two-story style with bedrooms above and an open central courtyard and kitchen on the bottom level. Not only were villagers displaced from this, but other aspects of their heritage were also disrupted. For example, many ancient family tombs were moved to other places, sometimes up into the mountains more than three kilometers away. The villagers were disparaging about their new homes:

> The foundation of the house is quite shallow, it is impossible to add a floor on the top of the house.

We toured a new cinder-block home with a woman and her disgust at the shoddy workmanship was quite evident. She explained that all of the villagers got together to build their old homes but outsiders built these ones. She told us that they do not know when the county, which now owned the new village, was planning to finish the job. When their old houses were bulldozed, they were told that they would have a new house within a year, but at the time of our discussion four years had passed.

We asked how the establishment of the FNNR had influenced them, but most were ambivalent:

> Hard to tell. We just do our jobs, we don't know more.

We asked whether tourism near the FNNR had some influence on their lives:

> In some places, it paid off, but in other places, it has not paid off yet.

Their reticence to discuss outcomes suggests a level of despair that we did not encounter in Zhangjiaba. They agreed that the golf course was a failure in the sense that they had not seen adequate compensation and were now in limbo.

> We cannot make a good living here. My son and his wife [migrate] out for working. They've gone to work outside every September for the last two years. They return home at the Spring Festival. The tourists are visiting here, but they are few.

By April 2016, the last time we were at the golf course resettlement (the new village was as yet unnamed), there was no noticeable improvement; there was still no water and electricity, half of the buildings were still under construction, and less than 10 percent were occupied by residents. It was clear that, unlike in Zhangjiaba, these villagers felt that they had lost opportunities for a better life and were less protected from dramatic changes. In Braidotti's (2013) terms, their diminished capacities (and politics) in one place were not compensated elsewhere.

With lost connection to their ancestral village, there came a loss of identity and place but not necessarily a loss of hope. It is important to note that some were optimistic about the tourist development in the area. When quizzed on that, one villager said: "No downside, why should there be a downside for tourism?" None of these people had ever seen a golden monkey in the wild but they all agreed that if the monkey brought in tourists then they were all for the FNNR's preservation policies.

Toward Sustainable Ethics

Escobar's (2008) figured world is about an understanding of, and emotional comfort with, the complexities of local ecosystems including human and nonhuman actors and policies put in place to manage them. The context of changing identities and autonomy is part of the creation of a figured world. Autonomy is about farmers in and around FNNR doing something (affecting) and feeling in control of their world while still under the auspices of larger environmental policies. For Braidotti (2013), affect is understood as a complex biological drive, a pragmatic effect of the relations between bodies (golden monkeys, farmers, tourists, wild boar, bamboo, grain, dams, golf courses, working groups, national policies), suggesting the potential for affecting or being affected with an impetus to increase capacities. Local feelings of belonging are not necessarily antithetical to burgeoning touristic modernity or the protection of the ethereal golden monkey, nor do they resist development (indeed, precisely the opposite) if capacities are increased. As Su (2012a) pointed out, too much emphasis on resistance and dissent misses the ways in which people "reconcile their everyday life with the social and spatial transformations in their society" (33). This idea is central to understanding how farmers build capacities to offset and mitigate the effects of change. Escobar's idea of place-based practices helps elaborate the ways in which farmers negotiate translocal change and mediate anxieties over displacements caused by tourism and infrastructure development (e.g., by focusing on earning more money and supporting their children's education). The idea of increasing local capacities aligns with Braidotti's (2013) idea of sustainable ethics, which includes recognizing possibilities and potentials.

Braidotti's sustainable ethics connects well with Escobar's figured worlds by embracing the multiplicity of relations among place, mobilities, and life in a move toward health, happiness, and the good life. Braidotti's work suggests that a locatable politics is possible if human and nonhuman agents affect outcomes positively through a multiplicity of fluid relations. She argued that sustainable ethics of this kind liberate marginalized people (and things) that would otherwise remain impotent at the center of a world that is not of their making. Zhangjiaba is witness to the potency of these ethics, when the dam bolsters villagers' resolve to make their lives better and hold out for appropriate compensation. Local government support enables road improvement projects (funded in part by PES monies), which elevates the possibility of realizing more tourist dollars. Despite the hardship of displacement, local FNNR officials seem to work well with villagers on an ultimately legible and sustainable ethical base. Clarity was much less in evidence with the golf course project outside of the reserve. A change of heart about what constituted an appropriate land use killed the golf course and left the villagers in limbo and the local government with the bill for their resettlement. Ninety percent of the displaced villagers were still without houses when we last met them and, despite the promise of new houses, there was little compensation for their predicament as market-driven forces squeezed out ethical considerations. Although tourist dollars were still within reach for them, it was clear that these villagers felt dislocated in the midst of an unclear process.

When a local world is sustained ethically, its legibility is rendered in such a way that life and political projects (from monkeys to roads to dams) are readable and can be readily translated at local scales, even when emanating in seeming authoritative ways from distant places. The question of the sufficiency of this clarity and legibility returns us in closing to the student sketches and quotes that start the article. The first sketch suggests the loss of a rural idyll with "small trees before [and] … in the river there were small fishes," and the second highlights the replacement of "dense trees" for roads and "houses becoming higher than before." Although most of those we talked to were encouraged by tourism and an influx of development monies to the FNNR, the students remarked on a cost to local life projects that is not necessarily translated back to distant places.

Notes

1. We conducted open-ended household interviews after administering a questionnaire, informed consent scripts, and a name and address coding mechanism. We ensured confidentiality throughout the study—thus, farmers' and family members' identities and personal information, as well as their answers to sensitive questions, were not revealed to individuals outside of the research project, nor were they revealed to FNNR staff or local government personnel. Further, we explained the survey purpose and reassured local people that there was no obligation to participate. If there was any sign of hesitation or discomfort during the interview, we dropped the conversation immediately. With those measures, we are confident that local peoples' participation was voluntary, our dialogues with them were equitable and trustful, and the data we collected in the survey reflected thoughts, emotions, and behaviors with regard to national development and environmental programs.

2. Tolls are approximately US$0.30 for each kilometer usage of the highway, and the average annual income of local farmers is barely US$1,000.

3. As in other rural areas of China, the residence registration (*hukou*) system demands that rural labor-oriented migrants (called a floating population; Liang 2001) move only "temporarily" (at the scale from weeks and sometimes years) to their migration destinations (often cities). Such temporary migrants keep their *hukou* and belongings (e.g., their farmland and houses) at their original villages and often come back to celebrate spring festivals at the Lunar New Year. Based on surveys in spring 2010, 2013, and 2015, we determined that an average of one third of family members in each household has done or is doing temporary work outside of the FNNR, where they mostly live in cities. It is not entirely clear whether *hukou* reform is responsible for these migrations.

References

Aitken, S. C., and L. An. 2012. Figured worlds: Environmental complexity and affective ecologies in Fanjingshan, China. *Ecological Modeling: An International Journal on Ecological Modeling and Systems Ecology* 229:5–16.

Aitken, S. C., L. An, S. Allison, and S. Yang. 2016. Nature's legacy: Children, development and urban access in Fanjingshan, China. In *Children, nature, and cities*, ed. A. M. F. Murnaghan and L. J. Shillington, 95–114. Aldershot, UK: Ashgate.

Aitken, S. C., L. An, S. Wandersee, and Y. Yang. 2014. Renegotiating local values: The case of Fanjingshan Reserve, China. In *Unravelling marginalisation, voicing change: Alternative geographies of development*, ed. C. Brun, P. Blakie, and M. Jones, 171–90. Farnham, UK: Ashgate.

Braidotti, R. 2006 *Transpositions: On nomadic ethics.* Cambridge, UK: Polity Press.

———. 2013. *The posthuman.* Cambridge, UK: Polity.

Brundtland Report. 1987. *Our common future*. Florence, Italy: United Nations World Commission on Environment and Development.

Escobar, A. 2001. Culture sits in places: Reflections on globalism and subaltern strategies of localization. *Political Geography* 20:130–74.

———. 2008. *Territories of difference: Place, movements, life, redes*. Durham, NC: Duke University Press.

Guo, Z., and T. Liang. 2017. Differentiating citizenship in urban China: A case of Dongguan city. *Citizenship Studies* 21 (7):773–91.

Hilton, I. 2013. The return of Chinese civil society. In *China and the environment: The green revolution*, ed. S. Geall, 1–14. London: Zed.

Liang, Z. 2001. The age of migration in China. *Population and Development Review* 27 (3):499–524.

Liu, J., S. Li, Z. Ouyang, C. Tam, and X. Chen. 2008. Ecological and socioeconomic effects of China's policies for ecosystem services. *Proceedings of the National Academy of Sciences* 105 (28):9477–82. doi:10.1073/pnas.0706436105.

Liu, J., and W. Yang. 2013. Integrated assessments of payments for ecosystem services programs. *Proceedings of National Academy of Sciences* 110 (41):16297–98.

Maskit, J. 2009. Subjectivity, desire and the problem of consumption. In *Deleuze/Guattari and ecology*, ed. B. Herzogenrath, 129–144. New York: Palgrave Macmillan.

Ong, A. 2006. *Neoliberalism as exception: Mutations in citizenship and sovereignty*. Durham, NC: Duke University Press.

Rees, W. E. 2001. Achieving sustainability: Reform or transformation? In *The Earthscan reader in sustainable cities*, ed. D. Satterthwaite, 22–52. London: Earthscan.

Rock, M., and D. Angel. 2007. Grow first, clean up later? Industrial transformation in East Asia. *Environment: Science and Policy for Sustained Development*. Accessed February 8, 2018. http://www.environmentmagazine.org/Archives/Back%20Issues/May%202007/Rock-abstract.html.

Shepard, W. 2015. *Ghost cities of China: The story of cities without people in the world's most populated country*. London: Zed.

State Forestry Administration of China. 2017. 退耕还林工程简报第3期（总第203期）[Grain to Grain Program Newsletter 3 (2017)]. Accessed January 23, 2018. http://www.forestry.gov.cn/main/436/content-1011771.html.

Su, X. 2012a. It is my home. I will die here: Tourism development and the politics of place in Lijiang, China. *Geografiska Annaler: Series B, Human Geography* 94 (1):31–45.

———. 2012b. Transnational regionalization and the rescaling of the Chinese state. *Environment and Planning A* 44:1327–47.

———. 2013. Moving to peripheral China: Home, play, and the politics of built heritage. *The China Journal* 70:148–62.

———. 2015. Urban entrepreneurialism and the commodification of heritage in China. *Urban Studies* 52 (15):2874–89.

Su, X., and P. Teo. 2009. *The politics of heritage tourism in China: A view from Lijiang*. London and New York: Routledge.

Zhao, Y. 1999. Leaving the countryside: Rural-to-urban migration decisions in China. *The American Economic Review* 89 (2):281–86.

STUART C. AITKEN is Distinguished Professor of Geography and June Burnett Chair at San Diego State University, San Diego, CA 92182. E-mail: saitken@sdsu.edu. His research interests include critical social theory, development, young people and families, masculinities, and film.

LI AN is Professor of Geography at San Diego State University, San Diego, CA 92182. E-mail: lan@sdsu.edu. His research focuses on complex human–environment systems, geographic information science, landscape ecology, and complex systems theory and modeling.

SHUANG YANG is an Assistant Professor of College of Harbor, Waterway and Coastal Engineering at Chongqing JiaoTong University, China. E-mail: s_yang@umail.ucsb.edu. His research focuses on complex human–environment systems, geographic information science, and complex systems theory and modeling.

A Manifesto for a Progressive Land-Grant Mission in an Authoritarian Populist Era

Jenny E. Goldstein, Kasia Paprocki, and Tracey Osborne

In this article, we offer a manifesto for a progressive twenty-first century land-grant mission in an era of rising authoritarian populism in the United States. We explore the historical context of this mode of political engagement, argue that scholars based at land-grant universities are uniquely positioned to address this political moment, and offer examples of land-grant scholars who have embraced this political obligation directly. In the midst of the U.S. Civil War, the federal government provided grants of land to one college in every state to establish universities especially with extension-oriented missions committed to agricultural research and training; today, there are seventy-six land-grant universities. Just as the constitution of these universities at a significant moment in the country's history served a political purpose, the current political climate demands a robust political response from contemporary land-grant scholars. Given the mandate for land-grant universities to serve their communities, how can a critical land-grant mission respond to the current political moment of emergent authoritarian populism in the United States and internationally? What responsibilities are entailed in the land-grant mission? We consider some strategies that land-grant scholars are employing to engage with communities grappling most directly with economic stagnation, climate change, and agrarian dispossession. We also suggest that, amid the dramatically shifting political climate in the United States, all scholars regardless of land-grant affiliation should be concerned with land-grant institutions' capacities to engage with the country's most disenfranchised populations as a means to pushing back against authoritarian populism.

我们于本文中，在美国兴起威权民粹主义的年代中，提供激进的赠地任务宣言。我们探讨此一政治参与模式的历史脉络，主张以赠地大学为基地的学者，特殊地置于应对此一政治时刻的位置，并提供直接拥抱此一政治任务的赠地学者之案例。在美国内战期间，联邦政府在每州赠地给学院来建立大学，特别是有关农业研究与训练的伸展导向任务；目前共有七十六所赠地大学。如同这些大学的组成是在国家历史上的显着时刻提供政治目的一般，当前的政治环境亦要求当代赠地学者的强烈政治回应。有鉴于赠地大学必须服务其社区，批判性的赠地任务如何能够回应美国与国际浮现中的威权民粹主义之当前政治时刻？赠地任务继承了什麼样的责任？我们考量赠地学者用来涉入最直接应对经济停滞、气候变迁与农业流离失所的社区的若干策略。我们同时主张，在美国剧烈变动的政治环境中，所有的学者，无论是否关乎赠地，皆必须考量赠地机构与该国公民权最受到剥夺的人口交涉之能力，作为反制威权民粹主义的工具。
关键词: 威权民粹主义, 高等教育, 赠地机构, 公共地理学, 美国。

En este artículo presentamos un manifiesto por una misión progresiva de concesión de tierras (land-grant) del siglo XXI en una era de creciente populismo autoritario en los Estados Unidos. Exploramos el contexto histórico de este modo de compromiso político, sostenemos que los académicos y eruditos basados en universidades del tipo favorecido por la concesión de tierras están posicionados singularmente para abocar este momento político, y ofrecemos ejemplos de eruditos de tal tipo que han abrazado directamente esta obligación política. En medio de la Guerra Civil de los Estados Unidos, el gobierno federal otorgó concesiones de tierras a un instituto universitario de cada estado para establecer universidades especialmente aquellas con misiones orientadas a la extensión comprometida con la investigación y el entrenamiento agrícola; en el momento actual, existen setenta y seis universidades del tipo land-grant. Justamente como la constitución de estas universidades en un momento significativo en la historia del país sirvió un propósito político, el actual clima político demanda también una respuesta política robusta de los eruditos contemporáneos del tipo land-grant. Considerando el mandato que se dio a las universidades land-grant de servir a sus comunidades, ¿cómo puede una misión land-grant crítica responder al momento político actual de emergente populismo autoritario en Estados Unidos e internacionalmente? ¿Qué responsabilidades van implícitas en la misión land-grant? Consideramos algunas estrategias que están empleando los eruditos land-grant para involucrarse con comunidades que luchan más directamente con el estancamiento económico, el cambio climático y la desposesión agraria. Sugerimos también que, en medio del dramáticamente cambiante clima político de los Estados Unidos, todos los académicos, independientemente de la afiliación land-grant, deben preocuparse con la capacidad de las instituciones land-grant para involucrarse con las poblaciones de mayor privación en el país, como un medio de devolver golpes al populismo autoritario. *Palabras clave: educación superior, Estados Unidos, geografías públicas, instituciones de concesión de tierras, populismo autoritario.*

America's land-grant universities were founded with the goal of serving the economic and political needs of the communities in the states in which they are based and ensuring the relevance of scholarly research to addressing practical social concerns (Bonnen 1998). Throughout their history, these institutions have been subject to repeated calls to renew this mandate and how they pursue this engagement in light of new political, economic, and demographic demands (Cochrane 1979; Campbell 1995; National Research Council; Board on Agriculture; Committee on the Future of the Colleges of Agriculture in the Land Grant University System 1997; Kellog Commission 1999; Peters 2006). Concerns about the relevance of land-grant scholarship are embedded in broader debates about expertise and democracy, the purpose of academia, and its obligations to society (McDowell 2003; Peters et al. 2008). In this article, we situate these land-grant institutions historically to explain why the current political moment demands a renewed commitment to this mandate. We demonstrate how the political challenges presented by authoritarian populism are inextricably linked with the mandate of land-grant institutions to engage more deeply and meaningfully with their communities.

Amid the Civil War and in the wake of the Industrial Revolution—a moment of social, political, and economic upheaval in the United States—Abraham Lincoln signed the Morrill Act of 1862 into law, establishing the country's first land-grant universities. With one land grant in each state, these institutions were created to support communities in the states in which they were based through research, teaching, and extension work. At their founding, land-grant universities were one of the clearest elaborations of democratic ideals in U.S. higher education: an egalitarian opportunity for all Americans, not only elites, to find pathways for

university study, particularly through engagement with the agricultural sciences. Yet, from the beginning there was ambiguity and conflict over what was entailed in this vision and how it could be achieved. Gelber (2013) wrote that at their founding in the late nineteenth century, rival visions he characterized as elitist and populist struggled over the character and content of land-grant research and education. The land-grant mission has thus always been tied to broader political currents, through continued financial support from state and federal governments, and served an important role in social and economic development locally, nationally, and even globally. Simultaneously, the land-grant mission, with its emphasis on community participation, provides opportunities for more progressive research, education, and public engagement. In this article, we argue that scholars based at land-grant universities have an opportunity to directly address rising strains of authoritarian populism in contemporary U.S. politics through their positions in these unique institutions. This means both confronting the systems of power that have shaped the current political moment and grappling with the role of knowledge generated by land-grant university scholars in that process. Even as we do so, we acknowledge that the challenges that authoritarian populism poses to our universities and communities are deeply structural and cannot be addressed by individual scholars alone. Nevertheless, we find that the land-grant mandate offers important openings for constructing spaces of resistance to authoritarian populism. We identify opportunities to forge new alliances, collectivities, and platforms for developing and pursuing grounded alternatives.

Whereas the fundamental land-grant mission has remained largely static since its inception, the political and economic context has changed dramatically, placing new demands on how land-grant scholars carry out their work. Land-grant institutions were established with a mandate to foster nineteenth- and twentieth-century fossil fuel–dependent development in the industrial and agricultural sectors, which has had serious social and environmental implications on multiple scales, including climate change. Relatedly, the communities that land-grant institutions were mandated to serve have also shifted, from being once exclusively white and agrarian to now encompassing communities of color, indigenous groups, and the rural and urban poor. Such marginalized communities have been made further vulnerable by the fallout from the imposition of international trade policy, agricultural industrialization, new energy regimes, and associated demographic and labor transformations. Such systemic economic transformation has often preceded changes in the political climate within such communities. Nationally, the authoritarian populist assault on science via the promotion of "alternative facts" has found a sympathetic audience among voters whose trust in scientific authority has dwindled, along with their trust in political elites (Canovan 1999; Brown 2014). The Trump administration's promotion of a "post-truth" alt-reality is thus part of a longer trajectory of weakening confidence in science along with rising distrust of institutions more generally (Gauchat 2012; Putsche et al. 2017). Understanding this hegemonic project will require not only interrogating the Trump administration and dominant political and economic elites who have promoted this assault but also seeking to understand the social conditions under which consent for this "post-truth" era has been authorized. We must also be reflexive in our examination of the ways in which this distrust has risen out of our own insistence on the power of experts over public participation in policymaking (Forsyth 2011), on objectivity over multiple and situated knowledges (Steinmetz 2005), and on technocracy over democracy (Guidotti 2017).

Following Scoones et al. (2018), we understand the current political conjuncture in line with the dynamics of authoritarian populism described by Hall at the height of the Thatcherite movement in the 1980s (Hall 1980, 1985). For Hall, authoritarian populism is a form of hegemonic class politics, and he therefore focused his attention on its political-ideological dimensions (cf. Jessop et al. 1984; Brubaker 2017). Hall in particular called out the "educative role" of the state in constructing this populist consensus (Hall 1980, 180; 1985, 116). We find Hall's formulation useful in that it directs us to the specifically political dimensions of knowledge production and the assembly of "common sense"[1] (Gramsci 1971). Hall's analysis of the authoritarian populism of the Thatcherite movement resonates deeply with Trump-era authoritarian populism in the sense of the crisis to which it responds and that it generates and in the racism and xenophobia that authorize it (Scoones et al. 2018). As Scoones et al. (2018) explained, authoritarian populism "typically depicts politics as a struggle between 'the people' and some combination of malevolent, racialized and/or unfairly advantaged 'Others,' at home or abroad or both" (2). In considering the role of land grants in this unique historical moment, we derive inspiration from these theorists who trace the dialogic relationship between "the knowledge of the intellectuals and popular opinion" (Crehan 2016, xii) and who thus address themselves to the responsibilities of intellectuals to political engagement. Although this populism has been manifested in a variety of national contexts across Europe and the United States (Brubaker 2017; Edwards et al. 2017; Ulrich-Schad and Duncan 2018), we focus here on the United States to highlight the unique obligations of U.S. land-grant universities.

In this context, what is the role of land-grant institutions and the land-grant mission in creating and disseminating environmental knowledge vis-à-vis rising tides of authoritarian populism in the United States? How can land-grant scholars ensure that their work serves the communities in which they are based and their most marginalized members in particular? What is the role of land-grant scholars within increasingly neoliberal universities that, given reduced public support, seek funding from corporations, which have themselves contributed to the insecurity of poor and vulnerable communities in both rural and urban environments? In this article, we argue that we need to rethink the role of land-grant institutions vis-à-vis the current political conjuncture. To that end, it is necessary to reckon with the history of the land-grant mission, the institutions tasked with carrying it out, and their role in shaping the present political climate in the United States. We then propose several ways in which scholars based at land-grant universities might confront the rising tide of authoritarian populism and its assault on scientific expertise in civic discourse.

Land-Grant Universities in U.S. Agrarian History

Although the land-grant system was not developed exclusively to serve the needs of rural communities (Bonnen 1998), this has been an important objective for many of the universities.[2] Initial interest in a university system that would educate the rural U.S. population began in the 1840s, a time at which "the so-called farmer's vote in America was becoming increasingly self-consciously political" (Brubacher and Rudy 1997, 62). This moment saw an upsurge of agrarian populism from both the left and right[3] and demands from some farm organizations that university education be made available and relevant to the agrarian class, rather than only elites pursuing liberal arts. In July 1862, as escalating Civil War battles were raging in Virginia, President Abraham Lincoln signed into law the Morrill Land-Grant College Act. The Act granted each state federal land in proportion to their state's congressional representation. The capital states earned from this land through investment or sale was to support the endowment and maintenance of at least one university in each state that upheld the objective to teach, without excluding scientific and classical research, agriculture, mechanic arts, mining, and military tactics. Such pursuits were established to "promote the liberal and practical education of the industrial classes in the several pursuits and professions in life" (2 July 1862, ch. 130, §4, 12 Stat. 504). Today there are seventy-six such institutions across the United States and its territories.

The role of these universities in serving agrarian communities was never entirely coherent, however. The focus on agricultural development in rural areas has been largely the purview of the Cooperative Extension Service, formalized in 1914 through the Smith–Lever Act to ensure the access of U.S. farmers to the insights of the agricultural sciences (McDowell 2003). The extension work of these programs is one way that the land-grant mission has been pursued; however, it is not the only platform through which the land-grant universities have been mandated to serve broader publics. University accessibility, for instance, has been an issue since the land-grants were founded, as the majority of students admitted in the first decades were white and male. The U.S. Congress sought to address this inequality by signing into law the second Morrill Act of 1890, which designated several historically black colleges as land-grant schools tasked with the same mandate as the original institutions.[4] A century later, Congress provided funding under the 1994 Improving America's Schools Act for thirty-six land-grant colleges affiliated with Native American tribes (Mack and Stolarick 2014; Halvorson 2016). Although land-grant-funded historically black and tribal colleges and universities have played an important role in serving rural populations and the wider communities within which they operate (Williams and Williamson 1988), and thus in meeting the demands we lay out in this piece, they are not without critiques. As Harper et al. (2009) discussed, designating black colleges as land-grant schools further entrenched segregation, as other land-grant universities could justify denying admission to black students. Most of the historically black land-grant colleges also fell consistently behind their peer institutions in terms of public funding rates and quality of education (Harper et al. 2009). This speaks to continued tensions since their founding over which public land-grant universities should be serving and how they might continue to do so today, as demographics in many states have shifted toward urban, non-white populations.

Although land-grant university extension work arose in response to demands of farm advocacy groups, suggesting that their role was to serve farming communities, land-grant universities were enrolled in deleterious agrarian transitions in the rural United States.[5] As agricultural economist Cochrane (1979) wrote, "The colleges of agriculture never became the training institutions for future farmers that their founders had envisaged ... [yet] they have served for at least seventy-five years as a wonderfully efficient channel for helping young men and women transfer out of agriculture and into productive nonfarm pursuits" (107).[6] In this way, Cochrane placed the land-grant colleges at the center of his narrative of rural demographic transition, alongside other systemic policy interventions that transformed the agrarian political economy of the United States (Friedman and McMichael 1989). Yet, in contrast with Cochrane's cheerful assessment of rural–urban transition, the rural economic decline that has led to this demographic shift has not always met with such celebration among rural communities. Throughout the mid-twentieth century, agricultural mechanization and a variety of forms of technological advancement (particularly in agricultural chemicals and biotechnology), largely based on fossil fuels, led to extreme pressures on the small farm, displacing both smallholders and agricultural laborers, leading to widespread dispossession of agrarian land (Buttel and Busch 1988; Williams and Williamson 1988). Researchers at land-grant institutions have been implicated in this transition directly, with several scholars and commentators noting the close relationships between land-grant institutions and agribusiness companies (Hightower 1972; Buttel and Busch 1988; Williams and Williamson 1988). A 1990 study by two Cornell University social scientists found that land-grant researchers were even more likely to have closer relationships with the biotechnology industry than their counterparts at other research universities (Curry and Kenney 1990).

Land-Grant Universities under Neoliberalism

These close linkages continue to shape research conducted at land-grant institutions in powerful ways.[7] As U.S. agriculture has shifted from small family-centered farms to agribusiness-dominated production dependent on migrant labor and global commodity chains, so, too, have the beneficiaries of land-grant extension work. Land-grant universities now depend on external funding from some of the world's largest corporations for agricultural extension, such as Monsanto-funded research on genetically modified organisms, with the resulting intellectual knowledge becoming property of the private sector rather than a public good (Glenna 2017). Through this imbrication with private investment, land-grant universities have come to play a critical role in the work of "rendering land investable" (Goldstein and Yates 2017, 209; Kenney-Lazar and Kay 2017). The effects of such privatization of agricultural knowledge, increased capital accumulation by agribusiness, and rural land consolidation have clearly expressed themselves in electoral politics, although not always in uniform or obvious ways (Lewis-Beck 1977). A growing rural–urban divide manifests itself in what Cramer (2016) called "the politics of resentment." The discontent of those affected by these processes of agrarian dispossession can exist within rural communities but also in migrant-receiving communities, in particular small towns classified as "nonmetropolitan areas," and the industrial Rust Belt (Lichter and Ziliak 2017). Rural economic distress has played a major role in the rise

of authoritarian populism today; this pattern is even clearer if we attend to these geographies produced by political economic transformation. Although media reports often erroneously flatten an analysis of voting trends in the 2016 election to suggest that rural, white, working-class voters are responsible for Trump's victory (Butler 2017; Gusterson 2017), some demographers have offered a more nuanced picture of the political economic transformations that it represents. Monnat and Brown (2017) demonstrated that the electoral shift that gave rise to Trump's 2016 victory grew out of "landscapes of despair" produced by a dramatic decline in jobs in manufacturing and natural resource industries since the 1970s. These communities within the rural–urban continuum should be understood not only as postindustrial but also, when analyzed historically, as postagrarian. The role of land-grant institutions described earlier in transforming the agrarian political economy of these communities as well as their mandate to serve their needs must be reflected on by land-grant scholars today.

In addition to the role of land-grant institutions in the transformation of agrarian political economies, this history highlights the contentious politics of knowledge production that fuels the rise of twenty-first-century authoritarian populism as well. These transformations affect public institutions in particular, but they are not limited to land-grant universities. The political economy of knowledge production within the land-grant university has developed within a broader context of privatization and neoliberalization of U.S. science and higher education over the past three decades (Mirowski 2011; Lave 2012; Newfield 2016; Busch 2017). Yet, as Prudence Carter, Dean of the Graduate School of Education at the University of California, Berkeley (a land-grant university) pointed out, even as privatization has threatened the values of education as a public good in the United States, many public universities were already largely inaccessible to many of the most marginalized constituencies they purport to serve (Carter 2018). Thus, threats to the public mission in research and education of both land-grant universities and public institutions more broadly are both acute and secular. We must understand the threats and challenges to the twenty-first-century land-grant universities within this wider historical context. Glenna (2017) suggested that deliberation on these transformations and the purpose of research within public universities is an important first step in resisting the privatization of university science.

Elements and Examples of a Progressive Land Grant Mission for the Twenty-first Century

Against this history, we offer a type of manifesto for the twenty-first-century land-grant school. We argue that the work of land-grant scholars in this moment of authoritarian populism must integrate research, teaching, and service and focus on three main points: (1) Provide inclusive education that is accessible and affordable; (2) serve the needs of the regions in which they are situated, including both rural and urban residents, in ways that support more self-sustaining, thriving communities; and (3) orient around sustainability and social justice. These are some of the characteristics of what Crow and Dabars (2015) called "The New American University," a new vision for higher education in the twenty-first century currently central to Arizona State University's charter (Arizona State University, New American University 2015). Elements of this vision are also linked with early and long-standing populist visions for the land-grant mission,[8] particularly those of accessibility (including late nineteenth-century calls for free or inexpensive higher education), and those concerned with addressing class inequality (Gelber 2013). Thus, they offer possibilities for reconciling the challenges of authoritarian populism with a more progressive vision of land-grant education and scholarship.

Accessibility

First and foremost, we must ensure that higher education is accessible to all students, particularly the rural and urban poor. Public education is under attack at all levels in the United States, and rising costs of higher education jeopardize our mission to equal access to education for all. Furthermore, the proposal to end the Deferred Action for Childhood Arrivals program threatens undocumented young people, who are among our most vulnerable students. Some colleges and universities have implemented programs to better ensure that students are not denied higher education due to financial constraints. These include New York's Excelsior Scholarship,

which provides free tuition for students and families in the state who earn less than $125,000, and Arizona State University's President Barack Obama Scholars Program, which provides free tuition and additional support for students of families that makes $42,400 or less so that they graduate debt free. These types of programs improve education accessibility for income-poor students and could be further developed within land-grant institutions.

Carter (2018) suggested that administrators at public institutions pursue this expanded accessibility by working to adapt the metrics by which the work of public institutions are measured. Ranking institutions based on the test scores of their successful applicant pool undermines the mission of democratizing accessibility. "Frankly," Carter (2018) wrote, "the strong positive correlation among test scores, socioeconomic status, and school quality raises legitimate questions about the objectivity and fairness of required admissions tests" (495). She explained that alternative metrics would help these institutions to better serve their unique mandate, which suggests different opportunities and responsibilities in education from those of private institutions. She suggested a first step could be withdrawing from the *U.S. News and World Report* national rankings, which force public institutions to compete with private institutions on unequal terms (Carter 2018).

Engaging Rural and Urban Publics

Land-grant scholars today can also work to integrate community engagement into their work at all levels, including the research methods, subjects, and questions they choose to pursue. Land-grant institutions were founded to support regional development and have largely emphasized agricultural extension. Today, however, the global population is increasingly urban with different sets of needs than those existing at the establishment of land-grant colleges and universities (McDowell 2003). In recent decades, colleges and universities have embraced engaged scholarship, expanding opportunities for faculty and students to pursue community-focused research in partnership with public entities (Boyer 1990; Barker 2004). Beyond the extension model often seen as unidirectional with regard to the flow of information and resources from the university, engaged scholarship values the coproduction of knowledge and seeks to bridge the divide between academia, government, the private sector, and community groups to improve environmental conditions and human well-being. Engaged scholarship actualizes networks with the potential to harness the untapped resources at universities to affect positive social–environmental change. Although universities have emphasized relationships with government and industry, we argue that the new mission of land-grant universities must emphasize public scholarship that engages communities in science that matters to members of the communities themselves.

Researchers at Michigan State University (MSU) fully embodied this community-oriented mission when, in 2015 they identified, exposed, and responded to the drinking water crisis in Flint, Michigan. Problems with elevated lead levels in children's blood in Flint were first identified by Dr. Mona Hanna-Attisha, a professor in the MSU College of Human Medicine who also practices as a pediatrician in Flint. Despite state officials accusing Hanna-Attisha of creating "hysteria" (Goodnough, Davey, and Smith 2016), along with two Flint public health officials and an MSU urban geographer, her published findings (Hanna-Attisha et al. 2016) ultimately forced state and federal officials to accept and address the problem, leading to the declaration of a federal state of emergency (Carravallah et al. 2017). The water crisis in Flint was itself fundamentally bound up in contemporary modes of racialized urban dispossession (Ranganathan 2016). The ideological foundations of these modes of dispossession disregard their structural underpinnings. Thus, a deeper response to the crisis currently facing Flint involves not only immediate action to repair the infrastructure responsible for this mass poisoning but also addressing the ideological hegemony sanctioning its structural drivers. For example, linking the water crisis with historically racist housing and urban development practices (Sadler and Highsmith 2016), researchers have responded with a variety of extension projects in food systems, political advocacy, and public health (MSU 2017).

Sustainable Development and Social Justice

Pursuing this progressive vision of the twenty-first-century land-grant mandate will also require centering sustainable development and social justice at the heart of this engaged research and pedagogy. Human impact on the planet and unprecedented social and environmental changes can be traced to unsustainable industrial and agricultural development, largely powered by fossil fuels. Land-grant institutions have been central to the rollout of the industrialization process now pegged as the start of the Anthropocene, the era in which humans have significantly affected Earth systems function, most powerfully evidenced by the global climate change crisis. Therefore, land-grant institutions responsive to current political and ecological dynamics must have sustainability as a central pillar of research and teaching. Environmental issues such as climate change are social and political problems; as such, they cannot be addressed in disciplinary silos but require an interdisciplinary approach. Furthermore, these problems cannot be solved by technological fixes alone; they require approaches that address the root political economic drivers of environmental change and lay the groundwork for a just transition to a more sustainable future (Paprocki 2018).

Many geographers and land-grant scholars have pursued this mission through community-oriented pedagogies (Trudeau et al. 2018). Galt et al. (2013, 130) described one model for teaching agriculture and food systems at a land-grant university (University of California, Davis) that brings a "critically reflexive research perspective to teaching" in ways that enhance student learning outcomes and raise awareness of the social justice aspects of food systems. Through development of a food systems course based around a student-centered, nonhierarchical structure, the instructors advocate for transformational learning experience in the classroom as a means to enabling students to become "active knowledge producers, engaged citizens, and democratic members of our global community—to ultimately change the food system and the world" (Galt et al. 2013, 140). Although this model could be brought to any university, it has particular salience for engaging students at land-grant institutions as a means for them to take responsibility to enact change in local communities as well as in communities at a distance. Furthermore, although upholding the original land-grant mission of bringing university knowledge to agrarian communities, the emphasis on nonhierarchical, critically reflexive learning is a more progressive way to work with nonacademic communities to advance social and environmental justice.

An example of engaged scholarship with a strong research and service focus is the Climate Alliance Mapping Project (CAMP), a collaborative effort between academics at the land-grant institution the

University of Arizona, environmental organizations, and Indigenous groups (Osborne 2017). Following the articulated values of nonacademic partners, CAMP identifies, maps, and shares information about the fundamental drivers of climate change; priority areas for keeping fossil fuels underground; and the location of pipelines and pipeline spills, especially those that cross important waterways. CAMP makes features such as pipelines more transparent and visible on the landscape to support activist campaigns, build broad-based alliances, increase public awareness, and influence climate policy to better support climate justice efforts. Inspired by decolonizing methodologies, CAMP uses a more horizontal approach to research that incorporates the interests and questions of activist partners in Arizona and abroad. In this way, the research is more meaningful to not only the researchers' nonacademic partners but the broader publics they serve.

Conclusion

To some geographers at land-grant institutions, the entreaties to maintain and broaden public education accessibility, serve the needs of urban and rural publics through a coproduction of knowledge, and emphasize sustainable development and social justice will be preaching to the choir. For others who have not grappled with the implications and responsibilities shared among scholars at such institutions, and even for those who enact the land-grant mission in practice but have not considered the mission's particular relevance to today's political climate, we call for a rethinking of the land-grant's mission within its broad historical context.

The problem of authoritarian populism, and the wider context of neoliberalization confronting our universities, is deeply structural. We have no intention of suggesting otherwise. Land-grant faculty are not individually equipped to resolve these structural concerns. Glenna (2017) rightly pointed to the folly of piling "expectations on university scientists to heroically resist science commercialization in the face of political, economic, and university pressures" (1029). Neither is this our intention. Yet we find it politically necessary to pursue strategies for confronting these challenges, and we find spaces of hope and the possibility of resistance in the promise of the land-grant mandate.

Resistance will need to be exercised on several fronts. Individual scholars will find that they are institutionally and personally equipped to pursue some of these strategies better than others.[9] Scoones et al. (2018) outlined a variety of ways in which scholars can pursue "emancipatory research" that is "open, inclusive and collaborative" (12). Their directives suggest concrete strategies through which land-grant scholars can pursue research that is politically and empirically embedded in the communities they serve (without compromising in theoretical or empirical rigor). "No single approach will do," they wrote, "each must engage in conversation with others, and respond to contextually defined questions" (Scoones et al. 2018, 12). As for land-grant scholars, the demands and capacity of individuals to pursue research that engages this political mandate will be shaped by the unique conditions of their institutions and communities. Others highlight the need for a larger process of rethinking public funding for institutions of higher education (McDowell 2003). This is work that can be done by faculty, staff, and administrators of land-grant universities both as uniquely positioned public employees and as residents of the areas in which they are based. Brady (2018) suggested that we demand a return and an expansion of public funding for public institutions due to their role in providing opportunities for more diverse student bodies; this mandate is all the more important for land-grant schools.

This work of recognizing the critical political work of scholarship embodied in the land-grant mission can be supported by all academics, regardless of their professional and institutional position. A more progressive land-grant mission aligns with the pursuit of the justice-oriented scholarship that our discipline increasingly demands in the face of the neoliberalization of universities (Lave 2015; Heynen et al. 2018). Geography as a discipline is well positioned to engage in this civic work, particularly given the discipline's commitment to field-based learning (Barcus and Trudeau 2018). For geographers based at land-grant universities, though, this commitment comes with a responsibility to grapple with the legacies of the land-grant mission, the ways in which land-grant-based work has been detrimental to many of the communities the mission was originally intended to benefit, and possibilities for aligning the land-grant mission with a more progressive politics in the face of rising authoritarian populism. The manifesto laid out in this article is not directed at land-grant institutions' cooperative extension programs exclusively; it is concerned rather with infusing this principle of engagement with the unique demands of our current political and economic moment into the work of land-grant institutions more broadly. In this sense, the concerns outlined here should be of relevance not only to extension agents, not only to land-grant scholars, but to all scholars invested in this vision of politically engaged scholarship.

Acknowledgments

The authors initiated the conversations that resulted in this piece while each of us was either a faculty member or a PhD student at a U.S. land-grant university. The authors thank James McCarthy for his helpful editorial comments as well as three anonymous reviewers for their suggestions, which greatly improved this article.

Notes

1. By *common sense*, Gramsci (and, by extension, Hall, who drew on Gramsci) referred to the diverse beliefs and seemingly self-evident truths that both derive from and entrench class-based cultural, political, and economic hegemony (see also Crehan 2016).
2. The 1887 Hatch Act established colleges of agriculture within the land-grant system and focused on serving rural communities through agricultural extension. McDowell (2003), however, importantly pointed out that this agricultural focus has never been the sole objective of the land-grant mission, writing that "for many inside and outside Land-Grant universities, the Land-Grant principle, whatever it means, is explicitly agricultural. That misunderstanding of a principle central to the Land-Grant universities continues to mislead and confound the understanding of an insight significant to the future of the academy and higher education" (33).
3. The populist politics of the Confederacy itself are perhaps the greatest example of this (Isenberg 2016); however the xenophobic "Know Nothing Party" (Formisano 2008)—the politics of which have been compared with those of Donald Trump (Reston 2015)—represented an early strand of authoritarian populism. Abraham Lincoln's own strand of populist politics resulted in the Homestead Act of 1862, under which the federal government gave small grants of agricultural lands west of the Mississippi to non-Confederate white Americans (Oliver and Shapiro 1997).
4. There are currently seventeen historically black colleges designated as land-grant institutions.
5. Indeed, voter dissatisfaction with land-grant institutions was expressed dramatically when the Connecticut General Assembly revoked the land-grant charter from Yale, which had been one of the first land-grant institutions. The large agricultural voting bloc objected to what it perceived to be elitist admissions standards and a curriculum that did not serve Connecticut farmers (Schiff 2009; Bomford 2017).
6. Compare global analysis of the agrarian question (Akram-Lodhi and Kay 2009; Edelman et al. 2014), within which the transnational role of the United States has largely been paid greater attention than rural–urban transitions within the United States itself.
7. Cornell University's Alliance for Science is one example, a program funded by the Bill and Melinda Gates Foundation that supports research on and advocacy for genetically modified crops and foods (Schnurr 2015; Antoniou and Robinson 2017).
8. Here we direct attention to the alternative populist visions that Hall (1980) referred to as "popular-democratic."
9. The capacity of individual scholars to negotiate these possibilities is deeply inequitably distributed based on professional demands and hierarchies, including requirements and incentives for hiring and promotion and the rise of contingent academic labor contracts.

References

Akram-Lodhi, A. H., and C. Kay, eds. 2009. *Peasants and globalization: Political economy, rural transformation and the agrarian question*. London and New York: Routledge.

Antoniou, M. N., and C. J. Robinson. 2017. Cornell Alliance for Science evaluation of consensus on genetically modified food safety: Weaknesses in study design. *Frontiers in Public Health* 5:1–5.

Arizona State University, New American University. 2015. Charter and goals. Accessed December 5, 2018. https://newamericanuniversity.asu.edu/about/asu-charter-mission-and-goals

Barcus, H. R., and D. Trudeau. 2018. Introduction to focus section: Out in the world: Geography's complex relationship with civic engagement. *The Professional Geographer* 70 (2):270–76.

Barker, D. 2004. The scholarship of engagement: A taxonomy of five emerging practices. *Journal of Higher Education Outreach and Engagement* 9 (2):123–37.

Bomford, M. 2017. Yale's land grant history. Paper presented at the AAG Annual Meeting, Boston, April 5.

Bonnen, J. T. 1998. The land-grant idea and the evolving outreach university. In *University-community collaborations for the twenty-first century*, ed. R. M. Lerner and L. A. K. Simon, 25–70. New York: Garland.

Boyer, E. L. 1990. *Scholarship reconsidered: The priorities of the professoriate*. Princeton, NJ: Carnegie Foundation for the Advancement of Teaching.

Brady, H. E. 2018. The argument is wrong and the message is dangerous. *British Journal of Sociology* 69 (2):498–505.

Brown, M. B. 2014. Climate science, populism, and the democracy of rejection. In *Culture, politics and climate change: How information shapes our common future*, ed. D. A. Crow and M. T. Boykoff, 129–45. London and New York: Routledge.

Brubacher, J., and W. Rudy. 1997. *Higher education in transition: A history of American colleges and universities*. London and New York: Routledge.

Brubaker, R. 2017. Why populism? *Theory and Society* 46 (5):357–85.

Busch, L. 2017. *Knowledge for sale: The neoliberal takeover of higher education*. Cambridge, MA: The MIT Press.

Butler, J. 2017. Reflections on Trump. Accessed January 3, 2018. https://culanth.org/fieldsights/1032-reflections-on-trump.

Buttel, F. H., and L. Busch. 1988. The public agricultural research system at the crossroads. *Agricultural History* 62 (2):303–24.

Campbell, J. R. 1995. *Reclaiming a lost heritage: Land-grant and other higher education initiatives for the twenty-first century*. Ames: Iowa State University Press.

Canovan, M. 1999. Trust the people! Populism and the two faces of democracy. *Political Studies* 47 (1):2–16.

Carravallah, L. A., L. A. Reynolds, and S. J. Woolford. 2017. Lessons for physicians from Flint's water crisis. *AMA Journal of Ethics* 19 (10):1001–10.

Carter, P. 2018. Self-interests, corporatization and rising educational inequality in public higher education: A review of the great mistake. *The British Journal of Sociology* 69 (2):493–98.

Cochrane, W. W. 1979. *The development of American agriculture: A historical analysis*. Minneapolis: University of Minnesota Press.

Cramer, K. J. 2016. *The politics of resentment: Rural consciousness in Wisconsin and the rise of Scott walker*. Chicago: University of Chicago Press.

Crehan, K. 2016. *Gramsci's common sense: Inequality and its narratives*. Durham, NC: Duke University Press.

Crow, M. M., and W. B. Dabars. 2015. *Designing the new American university*. Baltimore: Johns Hopkins University Press.

Curry, J., and M. Kenney. 1990. Land-grant university–industry relationships in biotechnology: A comparison with the non-land-grant research universities. *Rural Sociology* 55 (1):44–57.

Edelman, M., T. Weis, A. Baviskar, S. M. Borras, Jr., E. Holt-Giménez, D. Kandiyoti, and W. Wolford. 2014. Introduction: Critical perspectives on food sovereignty. *Journal of Peasant Studies* 41 (6):911–31.

Edwards, J., A. Haugerud, and S. Parikh. 2017. The 2016 Brexit referendum and Trump election. *American Ethnologist* 44 (2):195–200.

Formisano, R. P. 2008. *For the people: American populist movements from the revolution to the 1850s*. Chapel Hill: University of North Carolina Press.

Forsyth, T. 2011. Expertise needs transparency not blind trust: A deliberative approach to integrating science and social participation. *Critical Policy Studies* 5 (3):317–22.

Friedman, H., and P. McMichael. 1989. Agriculture and the state system: The rise and decline of national agricultures, 1870 to the present. *Sociologia Ruralis* 29 (2):93–117.

Galt, R. E., D. Parr, J. Van Soelen Kim, J. Beckett, M. Lickter, and H. Ballard. 2013. Transformative food systems education in a land-grant college of agriculture: The importance of learner-centered inquiries. *Agriculture and Human Values* 30 (1):129–42.

Gauchat, G. 2012. Politicization of science in the public sphere: A study of public trust in the United States, 1974–2010. *American Sociological Review* 77 (2):167–87.

Gelber, S. 2013. The populist vision for land-grant universities, 1880–1900. In *The land-Grant colleges and the reshaping of American higher education*, ed. R. L. Geiger and N. M. Sorber, 165–94. London and New York: Routledge.

Glenna, L. L. 2017. The purpose-driven university: The role of university research in the era of science commercialization. *Agriculture and Human Values* 34 (4):1021–31.

Goldstein, J. E., and J. Yates. 2017. Introduction: Rending land investible. *Geoforum* 82:209–11.

Goodnough, A., M. Davey, and M. Smith. 2016. Fouled water and failed politics. *The New York Times*, January 24.

Gramsci, A. 1971. *Selections from the prison notebooks*. Moscow: International Publishers.

Guidotti, T. L. 2017. Between distrust of science and scientism. *Archives of Environmental & Occupational Health* 72 (5):247–48.

Gusterson, H. 2017. From Brexit to Trump: Anthropology and the rise of nationalist populism. *American Ethnologist* 44 (2):209–14.

Hall, S. 1980. Popular-democratic vs. authoritarian populism: Two ways of "taking democracy seriously." In *Marxism and Democracy*, ed. A. Hunt, 157–85. London: Lawrence and Wishart.

———. 1985. Authoritarian populism: A reply to Jessop et al. *New Left Review* 151:115–24.

Halvorson, G. A. 2016. The role of a 1994 land grant college. *Rangelands* 38 (1):14–15.

Hanna-Attisha, M., J. LaChance, R. C. Sadler, and A. Champney Schnepp. 2016. Elevated blood levels in children associated with the Flint drinking water crisis: A spatial analysis of risk and public health response. *American Journal of Public Health* 106 (2):283–90.

Harper, S. R., L. D. Patton, and O. S. Wooden. 2009. Access and equity for African American students in higher education: A critical race historical analysis of policy efforts. *Journal of Higher Education* 80 (4):389–414.

Heynen, N., D. Aiello, C. Keegan, and N. Luke. 2018. The enduring struggle for social justice and the city. *Annals of the American Association of Geographers* 108 (2):301–16.

Hightower, J. 1972. Hard tomatoes, hard time: Failure of the land grant college complex. *Society* 10 (1):10–22.

Isenberg, N. 2016. *White trash: The 400-year untold history of class in America*. New York: Viking.

Jessop, B., K. Bonnett, S. Bromley, and T. Ling. 1984. Authoritarian populism, two nations, and Thatcherism. *New Left Review I* 147:32–60.

Kellog Commission. 1999. *Returning to our roots: The engaged institution. Third report of the Kellog Commission.* Washington, DC: National Association of State Universities and Land-Grant Colleges.

Kenney-Lazar, M., and K. Kay. 2017. Value in capitalist natures. *Capitalism Nature Socialism* 28 (1):33–38.

Lave, R. 2012. Neoliberalism and the production of environmental knowledge. *Environment and Society: Advances in Research* 3 (1):19–38.

———. 2015. The future of environmental expertise. *Annals of the Association of American Geographers* 105 (2):244–52.

Lewis-Beck, M. S. 1977. Agrarian political behavior in the United States. *American Journal of Political Science* 21 (3):543–65.

Lichter, D. T., and J. P. Ziliak. 2017. The rural–urban interface: New patterns of spatial interdependence and inequality in America. *Annals of the American Academy of Political and Social Science* 672 (1):6–25.

Mack, E. A., and K. Stolarick. 2014. The gift that keeps on giving: Land-grant universities and regional prosperity. *Environment and Planning C: Government and Policy* 32 (3):384–404.

McDowell, G. R. 2003. Engaged universities: Lessons from the land-grant universities and extension. *The Annals of the American Academy of Political and Social Science* 585 (1):31–50.

Michigan State University. 2017. MSU and Flint: Partnering for a healthier future. Accessed December 30, 2017. https://mispartanimpact.msu.edu/regional-stories-data/flint.html.

Mirowski, P. 2011. *Science mart: Privatizing American science.* Cambridge, MA: Harvard University Press.

Monnat, S. M., and D. L. Brown. 2017. More than a rural revolt: Landscapes of despair and the 2016 presidential election. *Journal of Rural Studies* 55:227–36.

National Research Council; Board on Agriculture; Committee on the Future of the Colleges of Agriculture in the Land Grant University System. 1997. Colleges of agriculture at the land grant universities: Public service and public policy. *Proceedings of the National Academy of Sciences* 94:1610–11.

Newfield, C. 2016. *The great mistake: How we wrecked public universities and how we can fix them.* Baltimore, MD: Johns Hopkins University Press.

Oliver, M. L., and T. M. Shapiro. 1997. *Black wealth/white wealth: A new perspective on racial inequality.* London and New York: Routledge.

Osborne, T. 2017. Public political ecology: A community of praxis for earth stewardship. *Journal of Political Ecology* 24 (1):843–60.

Paprocki, K. 2018. All that is solid melts into the bay: Anticipatory ruination and climate change adaptation. *Antipode.* doi:10.1111/anti.1242.

Peters, S. J. 2006. Every farmer should be awakened: Liberty Hyde Bailey's vision of agricultural extension work. *Agricultural History* 80 (2):190–219.

Peters, S. J., T. R. Alter, and N. Schwartzbach. 2008. Unsettling a settled discourse: Faculty views of the meaning and significance of the land-grant mission. *Journal of Higher Education Outreach and Engagement* 12 (2):33–65.

Public Political Ecology Lab. 2015. Home page. Accessed July 30, 2018. http://ppel.arizona.edu/.

Putsche, L., L. Hormel, J. Mihelich, and D. Storrs. 2017. "You end up feeling like the rest of the world is kind of picking on you": Perceptions of regulatory science's threats to economic livelihoods and Idahoans' collective identity. *Science Communication* 39 (6):687–712.

Ranganathan, M. 2016. Thinking with Flint: Racial liberalism and the roots of an American water tragedy. *Capitalism Nature Socialism* 27 (3):17–33.

Reston, L. 2015. Donald Trump isn't the first Know Nothing to capture American hearts. *The New Republic,* July 30.

Sadler, R. C., and A. R. Highsmith. 2016. Rethinking Tiebouth: The contribution of political fragmentation and racial/economic segregation to the Flint water crisis. *Environmental Justice* 9 (5):143–51.

Schiff, J. 2009. When Yale was a farming school. *Yale Alumni Magazine* 72:4. Accessed December 5, 2018. https://yalealumnimagazine.com/articles/2405-when-yale-was-a-farming-school

Schnurr, M. A. 2015. GMO 2.0: Genetically modified crops and the push for Africa's green revolution. *Canadian Food Studies/La Revue Canadienne Des Études Sur L'alimentation* 2 (2):201–8.

Scoones, I., M. Edelman, S. M. Borras, Jr., R. Hall, W. Wolford, and B. White. 2018. Emancipatory rural politics: Confronting authoritarian populism. *Journal of Peasant Studies* 45 (1):1–20.

Steinmetz, G. 2005. *The politics of method in the human science: Positivism and its epistemological others.* Durham, NC: Duke University Press.

Trudeau, D., L. Smith, and H. R. Barcus. 2018. Coda: Making geography relevant. *The Professional Geographer* 70 (2):333–37. https://doi.org/10.1080/00330124.2017.1366790.

Ulrich-Schad, J. D., and C. M. Duncan. 2018. People and places left behind: Work, culture and politics in the rural United States. *Journal of Peasant Studies* 45 (1):59–79.

Williams, T. T., and H. Williamson, Jr. 1988. Teaching, research, and extension programs at historically black (1890) land-grant institutions. *Agricultural History* 62 (2):244–57.

JENNY E. GOLDSTEIN is an Assistant Professor in the Department of Development Sociology, Cornell University, Ithaca, NY 14853. E-mail: goldstein@cornell.edu. Her research interests include land use politics, political ecology of climate change, and intersections between digital technology and development.

KASIA PAPROCKI is an Assistant Professor in the Department of Geography and Environment in the London School of Economics and Political Science, London WC2A 2AE, UK. E-mail: k.paprocki@lse.ac.uk. Her research interests include the political economy of development, agrarian studies, and the political ecology of climate change.

TRACEY OSBORNE is an Associate Professor in the School of Geography and Development, University of Arizona, Tucson, AZ 85721. E-mail: tosborne@email.arizona.edu. Her research interests include the political ecology of climate change mitigation in forests, climate justice, and engaged scholarship.

Index

Note: **Bold** page numbers refer to tables and *italic* page numbers refer to figures.

Printed and bound by CPI Group (UK) Ltd, Croydon, CR0 4YY

17/10/2024

01775698-0013